T0325793

The Physics of Glaciers

Fourth Edition

K. M. Cuffey

W. S. B. Paterson

AMSTERDAM • BOSTON • HEIDELBERG • LONDON
NEW YORK • OXFORD • PARIS • SAN DIEGO
SAN FRANCISCO • SINGAPORE • SYDNEY • TOKYO

Butterworth-Heinemann is an imprint of Elsevier

Butterworth-Heinemann is an imprint of Elsevier
30 Corporate Drive, Suite 400, Burlington, MA 01803, USA
The Boulevard, Langford Lane, Kidlington, Oxford, OX5 1GB, UK

Copyright © 2010 Elsevier, Inc. All rights reserved.

No part of this publication may be reproduced or transmitted in any form or by any means, electronic or mechanical, including
photocopying, recording, or any information storage and retrieval system, without permission in writing from the publisher.
Details on how to seek permission, further information about the Publisher's permissions policies and our arrangements with
organizations such as the Copyright Clearance Center and the Copyright Licensing Agency, can be found at our website:
www.elsevier.com/permissions.

This book and the individual contributions contained in it are protected under copyright by the Publisher
(other than as may be noted herein).

Notices
Knowledge and best practice in this field are constantly changing. As new research and experience broaden our understanding,
changes in research methods, professional practices, or medical treatment may become necessary.

Practitioners and researchers must always rely on their own experience and knowledge in evaluating and using any information,
methods, compounds, or experiments described herein. In using such information or methods they should be mindful of their own
safety and the safety of others, including parties for whom they have a professional responsibility.

To the fullest extent of the law, neither the Publisher nor the authors, contributors, or editors, assume any liability for any injury
and/or damage to persons or property as a matter of products liability, negligence or otherwise, or from any use or operation of
any methods, products, instructions, or ideas contained in the material herein.

Library of Congress Cataloging-in-Publication Data
Cuffey, Kurt.
   The physics of glaciers / Kurt Cuffey. – 4th ed.
      p. cm.
   Rev. ed. of : The physics of glaciers / W.S.B. Paterson. 3rd ed. 2010
   Includes bibliographical references and index.
   ISBN 978-0-12-369461-4 (alk. paper)
1. Glaciers. I. Paterson, W. S. B., Physics of glaciers. II. Title.
   GB2403.2.C84 2010
   551.31'2–dc22

2009050362

British Library Cataloguing-in-Publication Data
A catalogue record for this book is available from the British Library.

ISBN: 978-0-12-369461-4

For information on all Butterworth-Heinemann publications
visit our Web site at *www.elsevierdirect.com*

Typeset by: diacriTech, India

Printed in the United States of America
09 10 11 12 13   10 9 8 7 6 5 4 3 2 1

Working together to grow
libraries in developing countries

www.elsevier.com | www.bookaid.org | www.sabre.org

ELSEVIER   BOOK AID
              International   Sabre Foundation

Contents

Contents

References and other supplemental materials can be found on
The Physics of Glaciers companion website at:

http://www.elsevierdirect.com/companion.jsp?ISBN=9780123694614

Preface to Fourth Edition

Current concerns about global warming have produced widespread scientific interest in the behavior of glaciers in general and the polar ice sheets in particular. This increased interest, coming at a time of unprecedented advances in observational capabilities, has fueled a major expansion of the literature since the third edition went to press. A new edition to update the content and assess the current state of research was therefore overdue.

Reflecting the increased engagement of glacier studies with broad themes in environmental geophysics, the updated edition features new chapters on "Ice Sheets and the Earth System" and "Ice, Sea Level, and Contemporary Climate Change." The chapter on ice core studies is significantly expanded from the previous version and much of it is new material. The content and arrangement of chapters on glaciological fundamentals broadly follow the outline of the third edition, although many discussions have been revised extensively. All the material about flow of mountain glaciers, ice sheets, ice streams, and ice shelves has been amalgamated into a single lengthy chapter entitled "Flow of Ice Masses." Material about iceberg calving and basal melt now find their place in a chapter that reviews together all of the mass balance processes. In general the level of treatment remains unchanged, but several key topics are illuminated at a higher level of detail than in previous editions.

Many acknowledgments are due. We first must thank Shawn Marshall for conducting a first round of research and synthesis of topics presented in Chapters 4, 5, and 6. We gratefully acknowledge the scientists who reviewed individual chapters: Richard Alley, Bob Bindschadler, Jason Box, Roland Burgmann, Garry Clarke, Tim Creyts, Paul Duval, Andrew Fountain, Inez Fung, Hilmar Gudmundsson, Michael Hambrey, Will Harrison, Neal Iverson, Jo Jacka, Georg Kaser, Thomas Mölg, Tavi Murray, Tad Pfeffer, Eric Rignot, Jeff Severinghaus, Throstur Thorsteinsson, Françoise Vimeux, Ed Waddington, Joe Walder, Ian Willis, and Eric Wolff. Charlie Raymond deserves special thanks for commenting on the whole manuscript. Jeff Kavanaugh contributed helpful suggestions and graciously provided the cover photograph. Yosuke Adachi proofread the final manuscript. Mark Carey, glacier historian, suggested several of the chapter-head quotes. All of the reviewers offered excellent suggestions, some of which could not be accommodated for lack of space. We, of course, take full responsibility for the content and for the tough choices about what material to include.

Completion of the project would not have been possible without assistance from Delores Dillard and Darin Jensen of U.C. Berkeley's Department of Geography. Delores worked on digitization and manuscript acquisition while Darin took on the nearly unthinkable task of drafting more than 200 figures. KC gives additional thanks to Jean Lave and Michael Johns for their wise counsel, and to the Division of Geological and Planetary Sciences at the California Institute of Technology, and especially Jess Adkins and John Eiler, who hosted a sabbatical visit at the start of this project. Finally, we express our deepest gratitude to Lyn Paterson and Pete Lombard for their many years of support and encouragement.

Kurt M. Cuffey   W. S. B. Paterson
Berkeley, California   Quadra Island, British Columbia
February, 2010

Preface to First Edition

The aim of this book is to explain the physical principles underlying the behaviour of glaciers and ice sheets, as far as these are understood at the present time.

Glaciers have been studied scientifically for more than a century. During this period, interest in glaciers has, like the glaciers themselves, waxed and waned. Periods of activity and advance have alternated with periods of stagnation and even of retrogression when erroneous ideas have become part of conventional wisdom. The past 20 years, however, have seen a major advance in our knowledge. Theories have been developed which have explained many facts previously obscure; improved observational techniques have enabled these theories to be tested and have produced new results still to be explained.

This seems an appropriate time to review these recent developments. At present there is, to my knowledge, no book in English which does this. The present book is a modest attempt to fill the gap. To cover the whole field in a short book is impossible. I have tried to select those topics which I feel to be of most significance, but there is undoubtedly some bias towards my own particular interests.

While this book is intended primarily for those starting research in the subject, I hope that established workers in glacier studies, and in related fields, will find it useful. The treatment is at about the graduate student level. The standard varies, however, and most chapters should be intelligible to senior undergraduates.

I am much indebted to Dr. J. F. Nye for reading the whole manuscript and making many helpful suggestions. I am grateful to Drs. S. J. Jones, G. de Q. Robin and J. Weertman for reviewing individual chapters. I should also like to thank Drs. J. A. Jacobs and J. Tuzo Wilson for general comments and encouragement. The responsibility for the final form and contents of the book of course remains my own.

W. S. B. Paterson
Ottawa, Canada
March, 1968

CHAPTER 1

Introduction

"A man who keeps company with glaciers comes to feel tolerably insignificant by and by."

A Tramp Abroad, **Mark Twain**

1.1 Introduction

Glacier ice covers some 10% of the Earth's land surface at the present time, and covered about three times as much during the ice ages. Currently, all but about 1% of the ice is in areas remote from most human activities, in the great ice sheets of Greenland and Antarctica. Thus it is not surprising that the comparatively small glaciers on mountain areas were the first to attract attention, in both science and literature. Mountain glaciers have long served as natural laboratories for studying glacier processes. They are also important elements of many landscapes; they release water, scour bedrock, cool the weather in summer, and advance down valleys or retreat into high basins. Their spectacular crystalline beauty and conspicuous dynamism captivate observers. Spectacular too are the continental ice sheets, bodies of ice kilometers thick and hundreds of kilometers wide. In contrast to other topographic features of comparable size, ice sheets can change substantially over mere centuries and millennia. They command attention, in part, for their past and future influence on global climate and sea level. They also preserve an extraordinary record of past changes in Earth's climate and atmospheric composition. And the repeated waxing and waning of ice sheets and mountain glaciers throughout the last 2.6 Myr, the Quaternary, has marked vast regions of the Earth's surface with hills and valleys, lakes and river networks, soils and rocky debris – all testimony to the action of flowing ice.

The overall goal of this book is to explain the physical principles underlying the behavior and characteristics of glaciers. *Glaciers*, broadly defined, refers to all ice bodies originating as accumulations of snowfall: mountain glaciers and icefields, small ice caps, continental ice sheets, and floating ice shelves. Study of glaciers is part of glaciology, the study of ice in all its forms. Like other branches of geophysics, glaciology is an interdisciplinary subject involving physicists, geologists, atmospheric scientists, crystallographers, mathematicians, and others. Investigators trained in all these disciplines have contributed in fundamental ways to our understanding of glaciers.

Copyright © 2010, Elsevier Inc. All rights reserved.
DOI: 10.1016/B978-0-12-369461-4.00001-6

1.2 History and Perspective

Modern glacier studies began in the mid-twentieth century but were founded on a wealth of observations, measurements, and conceptual insights that accumulated since the Enlightenment. The problem of how large, apparently solid masses of ice can flow was studied and debated by many eminent scientists. Altmann, in 1751, correctly recognized that gravity was the cause of glacier motion, but he thought that movement consisted entirely of the ice sliding over its bed. Many glaciers do slide in this way but, in addition, the ice itself can flow like a very viscous fluid, as Bordier suggested in the late eighteenth century. In 1849 Thomson demonstrated ice flow in the laboratory, although the interpretation of his experiment later caused some confusion. Forbes asserted that glacier movement was viscous flow, but Tyndall opposed this view. He thought that motion resulted from the formation of numerous small fractures that subsequently healed by pressure melting and refreezing. Forbes' view prevailed, although only after much heated controversy.

Forbes also observed increases in velocity after heavy rain, which showed that water helps a glacier to slide. The two mechanisms that enable ice to slide past bedrock bumps were identified by Deeley and Parr in 1914. Studies of the flow and storage of water in glaciers date back to dye-tracing experiments by Forel around 1900.

Systematic measurements of glacier flow were begun about 1830 in the Alps. The aim of most early work was to find out how movement varied from place to place on a glacier. Agassiz showed that the velocity is greatest in the central part and decreases progressively toward each side. He also found that a glacier moves more slowly near its head and terminus than elsewhere. Reid, in 1897, showed that the velocity vectors do not parallel the glacier surface; relative to the surface, they are inclined slightly downward in the higher parts of the glacier, where snow accumulates, and slightly upward in the lower reaches to compensate for ice lost by melting. Figure 1.1 illustrates this pattern.

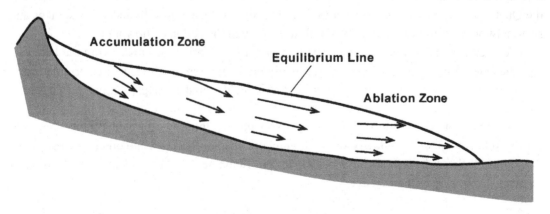

Figure 1.1: Schematic view of velocity vectors in a mountain glacier.

Ice movement at depth was long the subject of debate. *Extrusion flow* – the hypothesis that glaciers flow more rapidly at depth than at the surface – had its proponents, even in the 1930s. In the early 1900s, Blümcke and Hess used a thermal drill in a glacier in the Tyrol and attained bedrock in eleven holes, one of them more than 200 m deep. Rods left in the holes gradually tilted downhill, suggesting that ice moves more rapidly at the surface than at depth. Subsequent borehole measurements using more sophisticated methods have confirmed this pattern in general, although slow across-glacier extrusion flow sometimes occurs in narrow valleys.

Other developments about the turn of the century were the observation by Vallot of a bulge moving down the Mer de Glace and the development of mathematical models of glacier flow by Finsterwalder and others. Reid, for example, analyzed the time lag between the advance of the terminus and the increase in snowfall that produced it. Finsterwalder also pioneered photogrammetric methods of mapping glaciers.

Ahlmann, between 1920 and 1940, carried out classic investigations on the advance and retreat of glaciers in response to changes of climate; he investigated glaciers in Scandinavia, Spitsbergen, Iceland, and Greenland. Complementary studies of how a glacier surface receives heat during the melt season were begun by Sverdrup in 1934.

Work on the ice sheets of Greenland and Antarctica developed more slowly than work on mountain glaciers, but the emphasis of glacier studies has now shifted to the former. Noteworthy early work in Greenland is Koch and Wegener's study of snow stratigraphy during their crossing of the ice sheet in 1913. They also measured temperatures in the ice, in one instance down to a depth of 24 m. Wegener's Greenland Expedition of 1930–1931, which spent the winter in the central part of the ice sheet, studied the transformation of snow to ice. They also made seismic measurements of ice thickness, a method first tried a few years earlier in the Alps. The ice sheet can be regarded as a broad dome, elongated north to south, with a thickness in the center of about 3 km. Close to the edges, much of the ice flows in narrow and fast-moving outlet glaciers along bedrock troughs, a system that was not well delineated until the 1960s. Roughly half of the mass loss occurs by iceberg calving from the fronts of these outlets; the other half, by surface melt around the periphery of the whole ice sheet.

Detailed study of the Antarctic Ice Sheet began with the Norwegian-British-Swedish Antarctic Expedition of 1949–1952. The continent can be divided into three parts (Figure 1.2); the East and West Antarctic Ice Sheets, separated by the Transantarctic Mountains, and the Antarctic Peninsula, home to relatively small ice caps and valley glaciers. Near the coast in both East and West Antarctica, most of the ice flows in outlet glaciers that cut through mountains or in fast-moving ice streams set within comparatively stagnant ice. The outlet glaciers and ice streams feed floating ice shelves that surround much of the continent. Calving of icebergs accounts for most of the mass loss, but melt at the bottom of the shelves contributes significantly. Surface ablation occurs at a few places where mountains block the outflow of ice. Here, winds blow away fresh snow while ice at the surface sublimates, forming areas of "blue ice." Because calving controls most of the mass loss from Antarctica, sea level and ocean temperatures control long-timescale variations in the extent of the ice sheet.

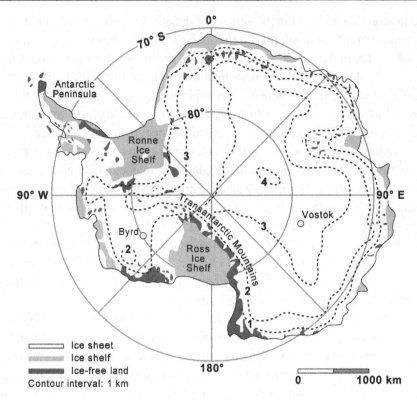

Figure 1.2: Map of Antarctica.

The modern era of glacier studies originated in the 1950s through 1970s with seminal contributions by a handful of mathematical physicists. By integrating physical principles with experimental evidence, their analyses explained many aspects of glacier form, temperature, flow, and evolution (the "glacier dynamics"), and some features of glacial structures and water systems. This body of work continues to underpin nearly all glacier studies. The most important forefront of the science shifted, however, in the 1980s and 1990s to – for lack of a proper term – the geological and climatological interfaces: the complex mix of water, sediment, and irregular bedrock constituting the glacier bed; the equally complex but more readily observed factors controlling melt and accumulation at the glacier surface; and the long and detailed histories of environmental change archived in the ice sheets and studied with ice coring.

This is not to imply that work on fundamentals of glacier dynamics has stagnated since its mid-century surge. On the contrary, glacier dynamics cannot be understood without a proper accounting for processes at the glacier surface and bed. For ice sheets, the long-term history of climate forcings must also be known. Moreover, observational capabilities and computational power have both increased dramatically, allowing analyses to revisit old questions and explore new territory. The application of numerical modelling to glacier problems began in the

1960s and has proliferated with the digital age. A model reduces a complex real situation to a simple closed system that represents the essential features and to which the laws of physics and parameterizations for complex processes can be applied. Modelling can serve three purposes: experimentation, explanation, and prediction. Experimentation, discovering the effect of changing the values of controlling variables, is generally the most useful; it can never be done in the real world. A model explanation may sometimes be illusory; the fact that a model with adjustable parameters gives plausible numerical values does not prove the validity of the underlying assumptions. Most models can be used for prediction, but first they must be tested against data. Unambiguous testing is difficult and use of all the data to "tune" the model by adjusting parameters precludes a proper assessment of its abilities. Events in the past decade gave a harsh reminder of the need to maintain skepticism about models; ice sheet models did not predict the dramatic recent increase of ice outflow from both Greenland and Antarctica. Although they are sophisticated and useful for a variety of studies, these models do not yet simulate nature adequately.

Many types of glacier studies continue to yield important results, but, above all else, the past decade has been a period of unmatched progress for observational analyses of the polar ice sheets. With data from remote sensing, the entire outflow of ice from Antarctica has been measured within a single year. So too for Greenland. Changes of surface elevation have been mapped throughout broad regions. Measurements of subtle variations in the strength of gravity, using satellites, now provide estimates of year-to-year and even seasonal variations in total mass of both ice sheets. Nearly complete maps of the velocity and thickness of ice streams permit a rigorous, if uncertain, assessment of all the forces acting on them. Arrays of shallow ice cores now show in unprecedented detail the geographical patterns of snow accumulation. With this information, atmospheric models have been calibrated and then used to map the year-to-year variations of mass gain or loss at the surface. Until recently, none of these analyses was feasible.

These capabilities, among others, will play an essential role in answering the major glaciological question of our time: how will ice on Earth respond to widespread climatic warming? At the very least, glacier changes can be monitored as they happen. At best, predictions of future changes will rigorously delimit the range of plausible behaviors and elaborate the most likely ones. Precise predictions seem impossible; even if glaciers were completely understood and characterized, the uncertainty of climate forecasts would still allow for a wide range of scenarios.

The evolution of the world's glaciers, past and future, depends on geological and climatological contingencies such as the distribution of subglacial sediments, the shapes of bedrock troughs beneath ice streams, and the interaction of mountain ranges with atmospheric winds and precipitation. Appreciation for the powerful insights in the seminal works of glaciology sometimes leads to a fetishizing of "general principles" as the goal of the science. Though invaluable, general principles by themselves do not answer the questions that matter. The fate of the ice sheets in a warming climate, for example, depends on the outflow along a few dozen outlet glaciers and ice streams in Greenland and Antarctica. These features are surprisingly diverse.

Each one needs to be understood and, ultimately, represented in models of each ice sheet. Again, the potential for melt and break-up of each major ice shelf – and how such changes relate to conditions in the nearby ocean – depends on particularities of the subshelf waters and ocean floor, the shape of the enclosing embayments, and other factors.

One particularity of great significance concerns the difference between the two ice sheets in mainland Antarctica (Figure 1.2). Although the greatest ice thickness in East Antarctica is nearly 4.8 km, most of the bedrock is above sea level. Much of the base of the West Antarctic Ice Sheet, in contrast, sits well below sea level and would remain so after isostatic rebound following removal of the ice. This suggests a possible instability; thinning of the ice past a certain level could decouple parts of the ice sheet from the bed, initiate rapid outflow, and soon set the remaining ice afloat.

Just as it is essential to match theoretical work with a realistic view of such contingencies and complexities in nature, it is essential that experimental work be coordinated with theory. Experiments should be designed to solve specific problems or characterize important components of a system. The mere acquisition of data is seldom a useful contribution in itself.

Understanding of glaciers has improved dramatically over the past half-century or so. Modern techniques of measurement would surely astonish Forbes and Agassiz, as would the abundance of information now available to evaluate theoretical models. Yet the deep inadequacies of the science should be kept in mind. Observation of the glacier bed remains a very difficult task. Interpretations of such data are usually tenuous, and theoretical views of glacier bed processes do a poor job of anticipating important new observations. Drilling boreholes in ice sheets is expensive and time-consuming; it is rarely done for dynamics studies. Of the dozens of fast-flowing polar ice streams, only one group in one sector of Antarctica has ever been studied by boreholes reaching the bed or even penetrating the deep layers of ice. Keeping track of year-to-year variations in melt and snowfall on "representative" mountain glaciers requires a consistent commitment of resources that faces vexing difficulties even in the wealthiest nations, let alone nations experiencing political upheavals. Geological reconstructions of past ice incursions provide the best hope for validating numerical models – but such studies must overcome imprecision and ambiguity. In short, much remains to be learned about glaciers, and even maintaining the current level of observation will take concerted effort.

1.3 Organization of the Book

The various aspects of glacier studies intersect one another in numerous ways, and any choice of organizational scheme is therefore somewhat arbitrary. The following summary clarifies our intentions.

Glacier ice forms by compaction and metamorphism of snow, sometimes accompanied by melting and refreezing. Chapter 2 discusses this process and the basic materials – snow, firn, and ice – that compose a glacier. One motivation for studying the transformation of snow to ice is to understand what determines the variation of density with depth near the surface of polar ice sheets. Changes in the density profile over time must be accounted for to interpret measurements of surface elevation in terms of ice mass. An understanding of the density profile also unlocks several methods for learning past climatic conditions from ice cores.

Chapter 3 continues discussion of the physical properties of ice, with emphasis on the development of grain-scale structures and on "viscous" deformation. The deformation of ice is analogous to the deformation of other crystalline solids such as metals, at temperatures near their melting points. The effective viscosity of ice depends on the grain-scale structures and on variables such as temperature and water content.

Because ice deforms viscously, the pull of gravity on a glacier causes it to flow. Slip over the bed adds another component of motion. Figure 1.3 depicts the immediate consequences of flow, for a mountain glacier and for an ice sheet with an ice shelf. The transfer of ice outward and downward thins the upper part of the glaciers but extends their termini. If flow acted alone, the surface profiles would change from the solid curves shown in the figure to the dotted ones. But glaciers also exchange mass with their surroundings. Snowfall adds material while melt and iceberg calving remove it. These mass exchange processes – introduced in Chapter 4 – tend to fill the space created by subsidence of the surface on the upper glacier (the area between the solid and dotted lines), and to remove the extended or uplifted surface on the lower glacier. The net effect of the mass exchanges, summed over the entire glacier, corresponds to the glacier-wide change in mass, or *mass balance*. The mass exchange processes depend directly on the climate and, for iceberg production, on the ice flow.

The discussion continues in Chapter 5 with a focus on the determinants of surface melt and sublimation. In temperate to subpolar settings, an increase of melt is an important driver of glacier retreat when the climate warms. The water produced by melt flows into and under a

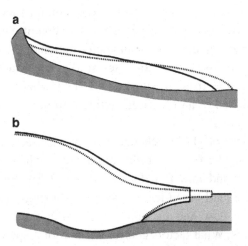

Figure 1.3: Effect of flow on (a) a mountain glacier and (b) an ice sheet ending in a floating ice shelf. Acting alone, flow would move the ice surface from the solid curves to the dotted ones. But, in addition, snowfall, melt, iceberg break-off, and related processes add and remove ice.

glacier and emerges at the front to feed rivers, an important source of water to communities downstream. Sometimes the glacier releases water in catastrophic floods. Chapter 6 reviews these and other topics related to water in glaciers.

At the bed, water also helps to govern how fast a glacier moves by basal slip. Slip motion arises from two mechanisms: deformation of subglacial sediments and sliding along the interface between the ice and its substrate. Most fast glacier flow is a consequence of rapid slip. Thus, slip is a critical process in the dynamics of ice sheets and many mountain glaciers. The ability to predict rates of slip remains poor; nonetheless, Chapter 7 spends a few dozen pages elaborating the topic.

Chapter 8 then provides a wide-ranging discussion of the general features of glacier flow: variations of flow over depth, across-glacier, and along-glacier; the different flow regimes of tidewater glaciers, ice sheet divides, ice streams, and ice shelves; and the factors controlling the overall form of ice sheets. Flow carries ice from regions of net accumulation to regions of mass loss. Acting in concert with mass exchange processes, flow controls the shape and size of glaciers and their capacity to change over time. Typical flow rates of mountain glaciers and ice sheets are tens to a few hundreds of meters per year, but some large tidewater glaciers and polar ice streams move kilometers in a year. Glacier flow thus represents an intermediate behavior in the spectrum of geophysical fluids; the Earth's lithospheric plates move only a small fraction of one meter each year, whereas rivers and winds move at meters per second.

Flow strongly influences two more characteristics of glaciers: the distribution of temperatures (Chapter 9) and the formation of large-scale structures such as crevasses, foliation, and ogives (Chapter 10). For polar glaciers, temperatures and ice flow must be analyzed together, because temperatures determine viscosity while flowing ice carries heat. Following a change of climate, the temperatures within an ice sheet and at its bed adjust slowly. Attaining a new equilibrium requires tens of millennia, but climate never remains constant for such a long period of time; in practice, an ice sheet always retains a fading memory of past conditions. Structures are worthy of attention, not only because they manifest flow and deformation of the ice, but also because they play a role in a variety of processes of central interest to glacier studies. Fracture propagation opens pathways for water to flow from the surface to the bed, where it assists basal slip. Folding disrupts the layers near the bed of an ice sheet, setting a limit on how far into the past an ice core reveals the history of climate.

The flow, extent, and shape of glaciers change over time in reaction to variations of climate, sea level, and other factors. The front of a glacier advances when annual snowfall increases or melt decreases. How far and how fast does it advance? Erosion of an ice shelf reduces the restraining forces on the land-based ice that feeds it, causing thinning and increased flow. Feedbacks between thinning, flow, and restraining forces sometimes lead to rapid and dramatic retreat of marine ice margins. What factors control this sort of response? Chapter 11 addresses these questions and others related to the mechanisms of glacier reaction. A different sort of variation over time is surging, an internal instability of glacier flow (Chapter 12). Some glaciers spend most of their time in quiescent periods, with little flow but a steady build-up of mass. But when the mass surpasses a critical level, the glacier surges forward for a few months or years. In some surges, ice velocities attain kilometers per year.

The final three chapters shift emphasis to review the role of glacier physics in several broad topics of compelling interest. Chapter 13 considers the relationship between ice sheets and global climate. Ice sheets not only react to forcings imposed on them but also influence the climate worldwide. Among other important connections, ice sheets reflect sunlight, divert atmospheric currents, and release fresh water to the ocean. The Quaternary ice ages offer the preeminent example of a global-scale coupled evolution of ice and climate. Only 20 kyr ago, continental ice covered the present sites of Stockholm, Boston, and Seattle. Aside from the obvious – they grew bigger – what role did ice sheets play in such remarkable changes?

Chapter 14 turns attention to the role of ice in contemporary and future global warming. Sea-level rise from melting ice is one of the major consequences of climate warming. By how much will it rise? Predicting the pace of future sea-level rise remains a task clouded in uncertainty – in a thick cumulus, unfortunately, rather than a thin fog.

Finally, Chapter 15 discusses ice core studies of past climate and other environmental parameters, with emphasis on the processes by which ice creates its archives. As of this writing, one ice core record from Antarctica offers a continuous 800-kyr history of temperatures, greenhouse gas concentrations, and other variables. This record spans about 30% of the entire Quaternary. Glacial ice is Earth's primary atmospheric sediment. No other geological records can match the richness of information and the continuous high-resolution chronology provided by cores from polar ice sheets. And cores from low-latitude, high-altitude sites significantly expand the geographical domain of glacial archives. Ice core studies have commanded wide attention – for demonstrating that climate has sometimes changed rapidly over broad regions of the globe, and for revealing close connections between climate and atmospheric chemistry.

Further Reading

A. Post and E.R. Lachapelle. 2000. Glacier Ice, revised edition. Univ. Washington Press, 160 pp.
   A collection of superb photographs of glaciers and glacial features, with explanatory text.

W.T. Pfeffer. 2007. The Opening of a New Landscape: Columbia Glacier at Mid-Retreat. Amer. Geophys. Union, Special Publications Series, vol. 59, 108 pp.
   Photographic documentation and accessible scientific discussion of a major Alaskan tidewater glacier undergoing rapid retreat.

R.B. Alley. 2002. The Two-Mile Time Machine: Ice Cores, Abrupt Climate Change, and Our Future. Princeton University Press, 240 pp.
   A discussion, for a general audience, of Greenland ice core analyses and implications for climate change.

R.P. Sharp. 1960. Glaciers. Univ. Oregon Press, 78 pp.
   An accessible introduction to general features of glaciers and processes of flow, from a mid-twentieth century perspective.

M. Hambrey and J. Alean. 2004. Glaciers, second edition. Cambridge University Press, 376 pp.
  D.I. Benn and D.J.A. Evans. 1998. Glaciers and Glaciation. Hodder Arnold, 760 pp.
  Two broad-ranging general introductions to all aspects of glaciers.

M.R. Bennett and N.F. Glasser. 1996. Glacial Geology: Ice Sheets and Landforms. John Wiley and Sons.
  D.E. Sugden and B.S. John. 1976. Glaciers and Landscape. A Geomorphological Approach. Hodder Arnold, 376 pp.
  Two books – one older, one newer – that review the geological and geomorphological aspects of glacier studies.

P.G. Knight (ed.) 2006. Glacier Science and Environmental Change. Blackwell, 512 pp.
  A collection of short articles that cover many areas of current glacier research.

V.F. Petrenko and R.W. Whitworth. 1999. Physics of Ice. Oxford University Press, 390 pp.
  A review of current understanding of the molecular and crystalline structure of ice and its consequences for physical properties.

E.M. Schulson and P. Duval. 2009. Creep and Fracture of Ice. Cambridge University Press, 416 pp.
  A review of current understanding of ice deformation and related topics.

G. Kaser and H. Osmaston. 2002. Tropical Glaciers. Cambridge Univ. Press, 207 pp.
  A review of the scientific understanding of glaciers in the Tropics.

J. Oerlemans. 2001. Glaciers and Climate Change. A.A. Balkema, 148 pp.
  A monograph reviewing the processes that link glacier surface balances with climate, and how one can analyze the reaction of mountain glaciers to climate changes.

R. LeB. Hooke 2005. Principles of Glacier Mechanics, second edition. Cambridge University Press, 448 pp.
  A review of many topics in glaciology, especially good for discussions of glacier bed processes.

B. Hubbard and N. Glasser. 2005. Field Techniques in Glaciology and Glacial Geomorphology. John Wiley and Sons, 400 pp.
  An introduction to many techniques used to study ice and glaciers – an important subject largely neglected in the present work.

CHAPTER 2

Transformation of Snow to Ice

"This huge ice is, in my opinion, nothing but snow, which ... is only a little dissolved to moisture, whereby it becomes more compact.... "

The Voyages of William Baffin, **R. Fotherby (17th century)**

2.1 Introduction

A fall of snow on a glacier is the first step in the formation of glacier ice, a process that is often long and complex. How snow changes into ice, and the time the transformation takes, depends on the temperature. Snow develops into ice much more rapidly on glaciers in temperate regions, where periods of melting alternate with periods when wet snow refreezes, than in central Antarctica, where the temperature remains well below the freezing point throughout the year. Thus we are dealing not with a single transformation mechanism but with different mechanisms in different areas. We have to subdivide glaciers, and even different parts of the same glacier, into different categories according to the amount of melting that takes place.

We first describe the basic materials of a glacier, and the different zones into which a glacier may be divided. The zones differ one from another in the temperature and physical characteristics of the material near the surface. Next we review the ways in which snow can be transformed to glacier ice. Finally, we discuss field observations of the rate at which density increases with time and depth, and how this process (referred to as *densification*) depends on temperature and other parameters. Crystal size also increases with time and depth as snow transforms to ice. This process of *grain growth* continues to be significant in fully formed ice. We discuss it, along with other processes active in fully formed ice, in Chapter 3.

2.2 Snow, Firn, and Ice

The term *snow* is usually restricted to material that has not changed much since it fell. We shall refer to material in the intermediate stages of transformation as *firn*. This follows common usage and fills a definite need. The original strict meaning of firn is wetted snow that has survived one summer without being transformed to ice. This narrow definition is no longer accepted, as firn also refers to altered snow on polar glaciers where no melting occurs. The broad definition, however, suffers from the lack of a clear division between snow and firn, an ambiguity that we accept. We may sometimes use "snow" when "firn" would be more appropriate. The absence

Copyright © 2010, Elsevier Inc. All rights reserved.
DOI: 10.1016/B978-0-12-369461-4.00002-8

Table 2.1: Typical densities (kg m^{-3}).

New snow (immediately after falling in calm conditions)	50–70
Damp new snow	100–200
Settled snow	200–300
Depth hoar	100–300
Wind-packed snow	350–400
Firn	400–830
Very wet snow and firn	700–800
Glacier ice	830–923

of a clear division between these terms reflects the continuous nature of snow transformation; there are no abrupt changes in physical properties, common to all glacial environments, that could serve as a basis for demarcation.

The difference between firn and ice is clear; firn becomes glacier ice when the interconnecting air- or water-filled passageways between the grains are sealed off, a process known as *pore close-off*. (A *grain* may be a single crystal or an aggregate of several.) This occurs at a density of about 830 kg m^{-3}. In glacier ice, air is present as bubbles. Compression of the bubbles largely accounts for further increases of density.

Table 2.1, taken mainly from Seligman (1936, p. 144), lists the densities of the different materials. The term *depth hoar* will be explained later.

2.2.1 Density of Ice

If bubbles account for a fraction v of the total volume, then the density of glacier ice is $\rho = v\rho_b + [1 - v]\rho_i$, where ρ_b denotes the density of fluid in the bubbles (air or water), and ρ_i the density of pure glacier ice. The latter is usually taken as 917 kg m^{-3}, a value that strictly applies only at temperatures near 0 °C and at the low confining pressures characteristic of small mountain glaciers and the upper layers of ice sheets. At mid-range depths in polar ice sheets, temperatures are −20 to −40 °C and the ice is free of bubbles. The linear thermal expansion coefficient for ice, which gives the fractional increase in distance between two points in a block of ice if it warms by one degree, is approximately $a = 5 \times 10^{-5}$ °C^{-1} at −20 °C (Petrenko and Whitworth 1999, p. 41). A temperature change of ΔT increases the volume by a fraction $[1 + a \cdot \Delta T]^3$. Thus, from the temperature effect alone, densities in ice sheets can reach values of about 922 kg m^{-3}.

Confining pressure (P) also increases the density of ice. For solid ice, the compressibility

$$\gamma = \frac{1}{\rho_i} \frac{\partial \rho_i}{\partial P} \tag{2.1}$$

is approximately 1.2×10^{-10} Pa^{-1} for isothermal compression (median of values from eleven studies; Feistel and Wagner 2006, p. 1029). Beneath 4 km of ice, a typical thickness for the center of the East Antarctic Ice Sheet, the pressure should thus increase the density from 917 to about $921 \, kg \, m^{-3}$. Because ice at these depths is within a few degrees of melting point, there is little temperature effect to increase the density further. Glacier ice should therefore attain its greatest *in situ* densities – about $923 \, kg \, m^{-3}$ – at mid-range depths in the ice sheets, where both low temperatures and moderately high pressures prevail. Gow (1970a) and Gow et al. (1997) presented density profiles based on measurements of deep ice core samples from the Antarctic and Greenland ice sheets. These were corrected for *in situ* temperatures but not for pressures.

2.3 Zones in a Glacier

Ahlmann (1935) proposed a "geophysical" classification of glaciers according to ice temperature and amount of surface melting. His categories were *temperate, sub-polar*, and *high-polar*. (A temperate glacier is at melting point throughout. On a high-polar glacier, the surface never melts.) More recent authors have subdivided some of Ahlmann's classes. However, conditions vary from place to place on a glacier and very few glaciers can be fitted into a single category. Thus, to speak about different zones in a glacier is better than trying to classify entire glaciers. The idea of zones was developed by Benson (1961) and Müller (1962).

We now describe the characteristics of the zones, starting from the highest elevations (the head of a glacier or center of an ice sheet). Very few glaciers show the entire sequence. Moreover, on any glacier the zone boundaries vary from year to year according to weather conditions. Figure 2.1 shows the features of the different zones and the underlying material in a typical year.

1. *Dry-snow zone*: No melting occurs here, even in summer. The *dry-snow line* marks the boundary between this zone and the next one.

2. *Percolation zone*: Some surface melting occurs in this zone. Water can percolate a certain distance into snow at temperatures below $0 \, °C$ before refreezing. If the water encounters an impermeable layer it may spread out laterally. When it refreezes an *ice layer* or an *ice lens* forms. The vertical water channels also refreeze, when their water supply is cut off, to form pipe-like structures called *ice glands*. Because the freezing of one gram of water releases enough latent heat to raise the temperature of $160 \, g$ of snow by one degree, refreezing of meltwater is the most important factor in warming the snow. (See Chapter 9.) As summer advances, successively deeper layers of snow warm to the melting point. The amount of meltwater produced during a summer normally increases with decrease of elevation. Thus, as we go down-glacier, we eventually reach a point where, by the end of summer, all the snow deposited since the end of the previous summer has been raised to the melting temperature. This point, the *wet-snow line*, defines the boundary with the next zone.

3. *Wet-snow zone*: In this zone, by the end of summer, all the snow deposited since the end of the previous summer has warmed to $0 \, °C$. Some meltwater also percolates into the deeper

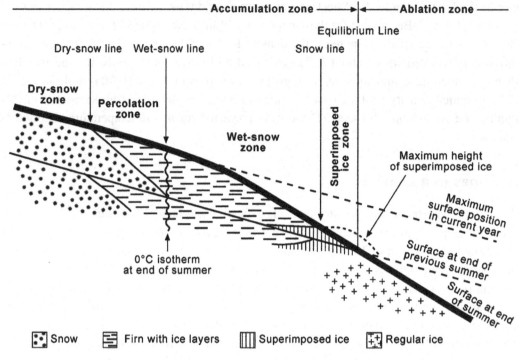

Figure 2.1: Zones of a glacier. Based on Benson (1961) and Müller (1962).

layers that were deposited in previous years, though not necessarily in sufficient quantity to raise their temperature to 0 °C. Percolation into these layers may also occur in the lower part of the percolation zone. It is important to find out where this happens because, when it does, mass-balance measurements cannot be restricted to the current year's layer. (See Chapter 4.)

4. *Superimposed-ice zone*. In the percolation and wet-snow zones, the material consists of ice layers, lenses, and glands, separated by layers and patches of snow and firn. At lower elevations, however, so much meltwater is produced that the ice layers merge to a continuous mass, called *superimposed ice*. We restrict the term *superimposed-ice zone* to the region with an annual increment of superimposed ice exposed at the surface. Superimposed ice also forms in the lower part of the wet-snow zone, but is there buried beneath firn. The *snow line* refers to the boundary between the wet-snow and superimposed-ice zones. (It has also been called the *firn line, firn edge*, and *annual snow line*.) Its location can easily be determined by observations at the end of the melt season; it is the boundary between firn and ice on the glacier surface. The lower boundary of the superimposed-ice zone is the *equilibrium line*, an important feature in mass-balance and glacier dynamics studies. Above it, the glacier has a net gain of mass over the year; below it, net loss. Some superimposed ice forms below the equilibrium line, but melts away by the end of summer.

5. *Ablation zone*: the area below the equilibrium line. Here the glacier surface loses mass by the end of the year. In typical years, the surface is ice. In years with larger-than-average melt, however, the ablation zone extends up-glacier into firn.

The preceding terms are Paterson's revision of those suggested by Benson (1961). Müller (1962), whose "percolation zone A" corresponds to our "percolation zone," divided the wet-snow zone into two parts: "percolation zone B" and the *slush zone*. These are separated by the *slush limit* or *run-off line*, the highest point from which mass escapes the glacier as flowing water.

2.3.1 Distribution of Zones

Dry-snow zones occur only in the interiors of Greenland and Antarctica and near the summits of the very highest mountains elsewhere. From observations made during extensive ground traverses, Benson (1961) found that the dry-snow zone in Greenland roughly coincides with the region where the mean annual air temperature does not exceed −25 °C. A map of the dry-snow zone on Greenland can now be constructed for each year using satellite observations. Snow containing meltwater emits and reflects significantly more microwave radiation than does dry snow at the same temperature; with water present, the snow acts more like a perfect radiator. The strength of microwave radiation (the *brightness temperature*), measured by sensors on satellites, thus indicates whether melt is present, although very small quantities may escape detection (Abdalati and Steffen 2001; Steffen et al. 2004). Effects of temperature differences can be removed by comparing emissions at two different wavelengths. Further, the polarization of the radiation provides an additional check on the presence of water. These measurements are made daily, so over a summer season the maximum extent of melt can be plotted. Figure 2.2a shows the extent of the dry-snow zone on Greenland in two successive years. The area of the dry-snow zone is smaller in years with warmer summers (Figure 2.2b).

The whole sequence of zones, from dry-snow to ablation, may be found in parts of Greenland and Antarctica. On the other hand, the major Antarctic ice shelves have only dry-snow and percolation zones; the entire mass loss results from calving of icebergs and melting at the base. Most of the sequence, with only the dry-snow zone absent, occurs on some large glaciers in northern Ellesmere and Axel Heiberg islands. In cold summers there may be dry-snow zones on the highest ice fields in these areas. The Barnes Ice Cap in Baffin Island, on the other hand, appears to consist only of superimposed-ice and ablation zones in most years. All these are "cold" glaciers; temperatures within them are below melting point.

In a "temperate" glacier the ice is at melting point throughout (see Chapter 9), except for a surface layer, some 10 m thick, in which the temperature falls below 0 °C for part of the year. Temperate glaciers cannot have percolation zones because in that zone, by definition, the temperature of part of the current year's snowpack, and thus the temperature of deeper layers, never reaches 0 °C. Again, superimposed ice forms only with firn temperatures below 0 °C. On a temperate glacier the extent of any superimposed-ice zone is insignificant, and the

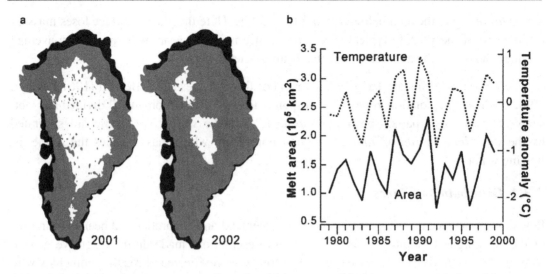

Figure 2.2: (a) The dry-snow zone in Greenland (shown as white) in two years, inferred from microwave reflections. (b) Variations of melt area, compared to variations of summer temperature averaged for coastal stations. The melt area is the area of ice sheet outside the dry-snow zone (the gray regions in panel (a)). Melt area in panel (b) was inferred from microwave emissions, and probably underestimates the true extent. Panel (a) adapted from Steffen et al. (2004). Panel (b) adapted from Abdalati and Steffen (2001) and used with permission from the American Geophysical Union, *Journal of Geophysical Research*.

equilibrium line and snow line coincide on average. A temperate glacier thus has only wet-snow and ablation zones.

2.4 Variation of Density with Depth in Firn

The progress of the transformation of snow to ice at a given place can be shown by a graph of density versus depth. Three such curves, smoothed to some extent, are shown in Figure 2.3. Byrd and Vostok are in the dry-snow zone in central Antarctica; the other location is in the wet-snow zone of a temperate glacier (Upper Seward) in the St. Elias Mountains near the Alaska-Yukon border. The curve for a percolation zone would lie between the one from the Yukon and those from Antarctica. The transformation occurs much more rapidly in the wet-snow zone than in the dry-snow zones. Firn becomes ice (density $830 \, \text{kg m}^{-3}$) at a depth of about 13 m on Upper Seward Glacier but not until a depth of 64 m at Byrd and 95 m at Vostok. The difference is even more striking if expressed in terms of time by using the rate of snow accumulation in each area. Snow transforms to ice in 3 to 5 years on Upper Seward Glacier, whereas about 280 years are needed at Byrd, and 2500 years at Vostok.

The transformation at Upper Seward Glacier appears to be unusually rapid even for temperate glaciers. In the Vallée Blanche in the Alps, Vallon et al. (1976) found the firn-ice transition at

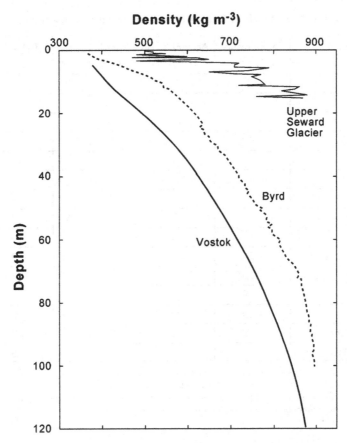

Figure 2.3: Variation of density with depth in a temperate glacier in the Yukon (Upper Seward), in West Antarctica (Byrd), and in East Antarctica (Vostok). Upper Seward data are from Sharp (1951). Byrd data are recalculated from Maeno and Ebinuma (1983). Vostok data are those of T. Sowers (Spencer et al. 2001, and pers. comm.). Some smoothing has been applied to the data; examples of point measurements can be seen in Figures 2.4 and 2.5.

a depth of 32 m, corresponding to an age of 13 years. They observed that the spring, summer, and autumn layers in the snowpack contained ice layers that made them less permeable than the winter layers. As a result these nonwinter layers retained more water, settled more quickly, and reached the density of ice at a depth of about 28 m. The winter layers had a density of only 650 kg m^{-3} at this depth; they reached the critical density at 32 m.

Table 2.2 lists the depth of the firn-ice transition, and the age of the ice there, for locations in polar regions. The table is arranged according to the mean annual temperature at each site. GISP2, Site A, and Dye 3 are sites in Greenland, Agassiz and Devon are in Arctic Canada, and the rest are in Antarctica. Vostok, Dome C, South Pole, GISP2, Byrd, Siple Dome, and Siple Station are in dry-snow zones whereas the other stations normally lie in the upper parts of percolation zones.

Table 2.2: Depth of firn-ice transition and age of ice there.

Location	Temperature (°C)	Accumulation (cm yr^{-1})	Depth (m)	Age (yr)	Reference
Vostok	−57	2.2	95	2500	(1)
Dome C	−54	3.6	100	1700	(2)
South Pole	−51	7.0	115	1020	(3)
GISP2	−32	22	77	230	(4)
Site A	−30	26	75−80	185	(5)
Byrd	−28	14	64	280	(6)
Agassiz Ice Cap	−25	16	53	235	(7)
Siple Dome	−25	11	55−60	350	(8)
Little America	−24	22	51	150	(6)
Siple Station	−24	50	70	95	(9)
Devon Ice Cap	−23	22	62	210	(10)
Dye 3	−19	49	65−70	100	(11)
S2	−19	13	38	220	(12)
Maudheim	−17	37	67	125	(13)
Roi Baudouin	−15	38	46	80	(6)

References: (1) Barnola et al. 1987; (2) Raynaud et al. 1979; (3) Kuivinen et al. 1982; (4) Gow et al. 1997; (5) Alley and Koci 1988; (6) Gow 1968a; (7) Koerner, unpublished; (8) Hawley et al. 2004; (9) Schwander and Stauffer 1984; (10) Paterson, unpublished; (11) Langway, unpublished; (12) Hollin and Cameron 1961; (13) Schytt 1958.

The transition typically occurs at depths of 50 to 80 m, and ages of 100 to 300 years. Low temperatures slow down the transformation, so both the ages and depths tend to be greater at cold sites than warm ones. Temperatures at Vostok, located in the coldest region on Earth, are about 30 °C lower than at Byrd; this explains the slower increase of density at Vostok than at Byrd, seen in Figure 2.3. The accumulation rate also matters. At a site with rapid accumulation, the depth of a firn layer, and hence the load on it, increases rapidly with age. In contrast, slow accumulation implies slow burial and hence a small load for a given age. At a given temperature, the transition therefore occurs more quickly where accumulation is more rapid (compare the values for Dye 3 and S2 in the table). The transition *depth*, on the other hand, tends to increase with accumulation rate because of the more rapid burial. On the polar ice sheets, the high accumulation sites also tend to be warm; the temperature and accumulation therefore have counteracting effects on the transition depth. Consequently, the transition depth varies only slightly over much of Greenland. The temperature effect predominates globally, however, and the transition depth is greatest at the extreme cold sites of interior East Antarctica (Vostok, Dome C, and South Pole). Moreover, the combination of low temperature and slow accumulation makes the transition ages at these sites much greater than elsewhere.

Table 2.3: Values of z_ρ in depth-density relation.

Location	z_ρ(m)	Reference
South Pole	68	Cuffey, unpublished
Dome C	61	Alley et al. 1982
GISP2	44	Cuffey, unpublished
Site A	43	Paterson, unpublished
Byrd	36	Paterson, unpublished
Little America	32	Paterson, unpublished
Maudheim	39	Schytt 1958

An empirical density-depth relation (Schytt 1958) is often useful:

$$\rho(z) = \rho_i - [\rho_i - \rho_s] \exp(-z/z_\rho). \tag{2.2}$$

Here ρ denotes the density at depth z, ρ_i the density of ice ($917\,\mathrm{kg\,m^{-3}}$), and ρ_s the density of surface snow. The parameter z_ρ is a constant for each site, and corresponds to a characteristic depth of the firn. Table 2.3 gives some typical values of z_ρ, obtained by least-squares fitting to the data. Because surface densities usually take a value between 300 and $400\,\mathrm{kg\,m^{-3}}$, a first estimate of z_ρ is given by $z_t/1.9$, for a depth to the firn-ice transition of z_t. The simple relation, Eq. 2.2, deviates most strongly from measured densities in the uppermost 20 m. Authors often report z_ρ as its inverse, $c = 1/z_\rho$, which has units of $\mathrm{m^{-1}}$.

Many discussions of densification implicitly assume that, at a given site and depth, the density does not change with time. This is sometimes called *Sorge's Law*, although "steady-state density profile" seems preferable. This assumption is plausible as long as the accumulation rate, temperature, and amount of melting, if any, remain constant.

2.5 Snow to Ice Transformation in a Dry-snow Zone

2.5.1 Processes

The transformation is analogous to the process of *pressure-sintering* or *hot pressing* in ceramics: when an aggregate of particles is heated under pressure, bonds form between them and the particles grow larger. The air space between them is reduced and the density of the aggregate increases. Indeed, polar ice sheets are a good place to study sintering because samples of the material at each stage can be obtained by coring to different depths. Below a depth of about 10 m, the process takes place at constant temperature. A graph of density against load pressure, as in Figure 2.4, best illustrates the different stages. This subject has been discussed by Anderson and Benson (1963), Shumskiy (1964, pp. 257–276), Gow (1975), Maeno and Ebinuma (1983), Arnaud et al. (2000), and Spencer et al. (2001), among others.

The transformation results from the mutual displacement of crystals, changes in their size and shape, and their internal deformation. The relative importance of these processes changes

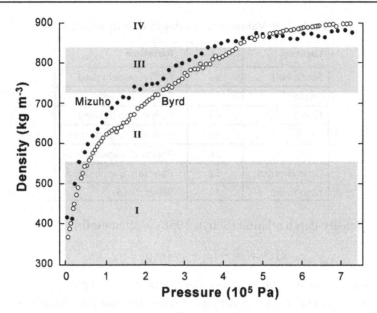

Figure 2.4: Increase of density with pressure of the overlying firn at two Antarctic stations. Adapted from Maeno and Ebinuma (1983).

as the density increases. Changes in crystal size and shape occur readily because, unlike other solids, ice in nature is usually near its melting point. Molecules are thus relatively free to move, both within the ice lattice (*volume diffusion*) and over the crystal surface (*surface diffusion*). In addition, sublimation occurs readily. (The term *sublimation* can be used in two senses. It can be restricted to the change from solid to vapor phase or used to denote the whole sequence of change from solid to vapor, movement of vapor, and change from vapor back to solid. Here we use the word in the second sense.)

The net direction of movement of molecules reflects the thermodynamic principle that the free energy of the system tends to a minimum. A reduction in surface area reduces the free energy, as does smoothing of convexities on the surface. Thus the molecules tend to be redistributed in a way that reduces the total surface area of the crystals and makes them smoother. Fresh snowflakes, with their complex shapes, are transformed to rounded particles. Breaking of the snowflakes as they strike the surface, or if they are blown along afterward, also helps to bring this about. The larger crystals tend to grow at the expense of the smaller ones because this further reduces the surface area, for a given mass.

However, the most important factor in the initial stages of transformation is settling, that is, the displacement of individual particles relative to their neighbors. The rounding of particles facilitates this. The particles slide past one another along their boundaries (Alley 1987). The increase in density brought about by settling can be estimated by considering a group of spheres, all the same size. In the closest possible packing of spheres (the *rhombohedral arrangement*), the porosity (ratio of space between spheres to total volume) equals 26%. But packing experiments with spheres show that, in practice, we can never reduce the porosity below about 40%. For spheres of ice of density $910 \, \mathrm{kg \, m^{-3}}$ a porosity of 40% corresponds to a density of $550 \, \mathrm{kg \, m^{-3}}$. Other mechanisms must be responsible for any further increase in density and so we might expect a decrease in the densification rate at this point. Figure 2.4 does show such a decrease. Note that the load, the variable used in the figure, is also a proxy for time; load pressure equals the product of accumulation rate with time.

A packed arrangement of spherical particles is not the end result. The total surface area can be further reduced by transfer of material to the points of contact between particles, to form bonds. Laboratory experiments show that sublimation, rather than diffusion, dominates in the initial stages (Hobbs and Mason 1964). This is not surprising because ice has a high vapor pressure. Vapor diffuses from the surfaces of spherical grains (which have convex surface curvature) to the necks (where the curvature is concave), because vapor pressures increase with the convexity of the adjacent surface. Bond formation typically begins well before the density reaches $550 \, \mathrm{kg \, m^{-3}}$; the uppermost layers of firn, with densities of 350 to $400 \, \mathrm{kg \, m^{-3}}$, can usually be cut into coherent blocks with a saw. (Such blocks form the structure of an igloo or snow fort.) The bonds are visible in photographs of thin sections (Gow 1975).

As the density increases and the firn becomes less porous, sublimation is greatly reduced. At the same time, the load and the area of contact between grains increase. Recrystallization and deformation become the dominant processes: molecular diffusion changes the shape and size of crystals in such a way as to reduce the stresses on them and, in addition, individual crystals deform by displacement along internal glide planes. Deformations associated with glacier flow, such as stretching or compression along the glacier, facilitate the deformation of crystals in response to the load pressure. Thus, densification of firn proceeds more rapidly if the glacier is stretching or compressing (Alley and Bentley 1988).

Densification rate decreases again at a density of about $730 \, \mathrm{kg \, m^{-3}}$ (Figure 2.4) although this transition is hard to detect in some profiles. At this point all the remaining air resides in thin channels along the intersections of grain boundaries (Maeno and Ebinuma 1983). Ice deformation ("creep") accounts for most of the densification beyond this point. Creep of ice is discussed in Chapter 3.

When the density reaches about $830 \, \mathrm{kg \, m^{-3}}$, the air spaces between grains close off. Much of the air has escaped to the surface: the remainder, about 10% by volume, is now present only as bubbles. The firn has become glacier ice. A further slow increase in density results from compression of the air bubbles by creep of the surrounding ice; Salamatin et al. (1997) discussed a model for this process.

Because air bubbles preserve samples of the atmosphere at the time of their formation, they make it possible to study such processes as the build-up of trace gases in the atmosphere as a result of human activities, and the variation of atmospheric carbon dioxide concentration with temperature during the ice ages. Because the air throughout the firn mixes with the atmosphere, however, the air in the bubbles is younger than the surrounding ice. It is important to know by how much. The ages in Table 2.2 give a first estimate for the various locations. This is an oversimplification because not all of the bubbles form at the same depth. At a given depth below the firn-ice transition, the ages of bubbles range from a few decades to a few centuries. These topics are discussed in Chapter 15.

Most of these processes are sensitive to temperature. Thus the rate of transformation varies from place to place, as described in Section 2.4. Differences in accumulation rate contribute to these variations by changing the rate at which the load on a given particle increases with time.

The stresses between neighboring crystals change continually during the transformation process. At low densities, the vertical compressive stress exceeds the horizontal components. But as the density of the firn approaches that of ice, the overall stress pattern becomes approximately hydrostatic. Thus the crystals should be no more likely to grow in one direction than in another. Examination of cores from dry-snow zones confirms that the orientations of crystals are often distributed uniformly.

2.5.2 Models of Density Profiles in Dry Firn

A model that can predict density profiles in a wide range of climatic conditions, and in the decades following abrupt climate changes, is needed for analyses of ice-core paleoclimate records. Likewise, a model of the densification process must be used to interpret glacier surface elevation changes, measured over periods of years, in terms of glacier mass changes (e.g., Arthern and Wingham 1998; Zwally and Li 2002; Helsen et al. 2008).

A first model of the densification of dry firn may be constructed from the most essential features of the process:

1. The driving force for densification is the weight of overburden (the load). Per unit horizontal area, at depth z below the surface, the load amounts to $P = g \int \rho(z)dz$. Only the ice grains, not the voids, support the load; the grain-load stress, P_*, equals $\rho_i P/\rho$. The rate of densification should increase with P_*.

2. The effective driving force depends not just on the applied load but on the difference between the vertical and horizontal stresses. This is related to the geometry of the ice skeleton. Because of voids, the ice grains are not fully supported on their sides. The stress difference decreases as the void space disappears and the firn density approaches that of solid ice. Thus the effective driving force not only increases with P_* but also decreases to zero as $\rho \to \rho_i$. The simplest such relation has the form $[\rho_i/\rho - 1] P_*$.

3. Densification rate increases with temperature because the relevant processes, including sublimation and creep of ice, are thermally activated; that is, they depend on the random agitations of molecules. The temperature sensitivity of such processes typically obeys the Arrhenius relation, $f(T) \propto \exp(-Q/RT)$, with T the temperature in Kelvin, Q the *activation energy*, and R the gas constant.

According to these concepts alone, the densification rate (or, more precisely, the rate of deformation due to densification) may be approximated as

$$\frac{1}{\rho}\frac{d\rho}{dt} = f_o \exp\left(\frac{-Q}{RT}\right)\left[\left[\frac{\rho_i}{\rho} - 1\right]P_*\right]^n,$$

(2.3)

where f_o denotes a constant coefficient. The effective driving force is raised to a power $n = 3$ because ice deforms as roughly the third power of the applied stress (see Chapter 3).

Equation 2.3 implies a relation for the variation of density with depth, assuming a steady state. In a steady state, the densification rate $d\rho/dt$ at a given depth must equal the increase of density in unit time as a parcel of firn moves a unit distance due to the vertical velocity w: thus $d\rho/dt = w \cdot d\rho/dz$ (Section 8.5.5.6). Furthermore, in a steady state $w = \dot{b}\rho_i/\rho$, for an accumulation rate $\dot{b}$ (meters of ice added to the surface per year). Combining these relations and arranging gives

$$\frac{d\rho}{dz} = \frac{f_o \rho^2}{\rho_i}\frac{1}{\dot{b}}\exp\left(\frac{-Q}{RT}\right)\left[\left[\frac{\rho_i}{\rho} - 1\right]P_*\right]^n.$$

(2.4)

This relation can be solved for $\rho(z)$ by numerical integration. Figure 2.5 illustrates the result, compared to the measured density profile at Siple Dome, Antarctica. The coefficient f_o has been adjusted to match the measurements at the bottom of the profile. The model does not show the kinks in the profile at densities of about 550 and 830 kg m^{-3} because no attempt was made to distinguish different processes. The most significant problem with the model, however, is too-slow densification near the surface. This illustrates the importance of grain settling in the near-surface layers, not simply driven by the load pressure but also facilitated by sublimation. The figure also shows the effects of increased accumulation rate and decreased temperature on the model profile; both increase the depth of the firn.

More sophisticated models of firn densification have been proposed, and calibrated against measured density-depth profiles (Barnola et al. 1991; Spencer et al. 2001; Goujon et al. 2003; Salamatin et al. 2006). These suggest that the best value for the effective activation energy Q is approximately 40 kJ mol^{-1} (Herron and Langway 1980; Spencer et al. 2001). None of these models provides satisfactory general descriptions of densification, but comparisons of modelled and measured pore close-off depths suggest the models are sometimes adequate for ice-core paleoclimate applications (e.g., Goujon et al. 2003). To calculate densification rates, the models use a generalized form of Eq. 2.3,

$$\frac{1}{\rho}\frac{d\rho}{dt} = f_1(T)f_2(\rho)P_*^c$$

(2.5)

where c is a constant and P_* again denotes the load stress. (At depths greater than pore close-off, the pressure in the bubbles must be subtracted from it.) The function $f_1(T)$ accounts for the increase of densification rate at higher temperatures and is an Arrhenius relation. The function $f_2(\rho)$ accounts for various effects of density, including the geometric requirement that

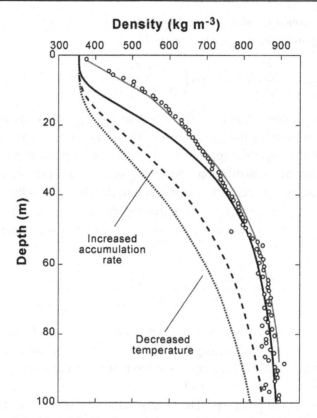

Density (kg m⁻³)

Figure 2.5: Increase of density with depth at Siple Dome, Antarctica (the circles: Hawley et al. 2004), compared to simple model (the heavy solid curve: Equation 2.4). Dashed and dotted lines show how the modelled density changes if accumulation rate is increased by a multiple of 5, or temperature is decreased by 20 °C. Thin solid curve (close to data points) shows the Herron-Langway relation for $T = -25.4\,°C$, $\dot{b} = 0.12\,m\,yr^{-1}$ (w. eq.), and initial $\rho = 350\,kg\,m^{-3}$; curve courtesy of J. Severinghaus.

densification slows as density increases, and that stress concentrates on contacts between grains. The true functional form of f_2 remains uncertain. Equation 2.5 needs to be calibrated separately for the different stages of densification. Arthern and Wingham (1998) instead used a densification model based on specific physical relations for each densification process (Maeno and Ebinuma 1983; Alley 1987); the result did not match the measured profiles very well.

Herron and Langway (1980) found a useful empirical description of steady-state firn density profiles in terms of the temperature and accumulation rate, rather than the overburden pressure. They showed that the increase of density with depth in the upper zone of the firn ($\rho < 550\,kg\,m^{-3}$) is approximately independent of the accumulation rate. Their relations for densification rate in each zone, calibrated against measured depth-density profiles, are of the form $d\rho/dt = f(T)\dot{b}^m$ $[\rho_i - \rho]$. Here $f(T)$ again represents an Arrhenius function of temperature, and $\dot{b}$ the accumulation rate. The parameter $m = 1$ if $\rho < 550\,kg\,m^{-3}$. The overburden pressure on a layer of given age increases in proportion to $\dot{b}$, so the relations imply that, in each zone, densification rate increases with pressure and decreases with age. There is no physical reason the densification should depend explicitly on age; it presumably appears as a proxy for the geometric effects of increased density. Relations for density-depth profiles were given (Herron and Langway 1980, pp. 377–378). Figure 2.5 shows the close match of this empirical relation to the measurements at Siple Dome.

2.5.3 Reduction of Gas Mobility

As the density of dry firn increases, the air-filled voids become smaller and less well connected. This inhibits the flow of air and the diffusion of gas molecules through the firn, a process of great importance to interpretations of ice core climate records. By diffusion, the vertical flux q of a gas species, from a depth (z) with a high concentration (C) to a depth with a lower concentration, is proportional to a *diffusivity*, D, such that $q = -DdC/dz$. (More generally, diffusion occurs in response not only to concentration gradients but also to temperature gradients and to gravitational settling (see Section 15.3). The flux is proportional to D in these cases, too.) Units of D are $m^2\ s^{-1}$.

How the diffusivity varies with density can be determined, approximately, from measurements of the vertical profiles of gas concentrations in firn. The atmospheric concentrations of CO_2, CH_4, and SF_6 increased dramatically in the twentieth century. The current high concentrations are now diffusing downward into the firn on polar ice sheets. The greater the diffusivity, the faster the propagation. The propagation can be modelled by using the known history of atmospheric concentrations as a boundary condition at the surface. By comparing measured concentration profiles to the modelled ones, the relation of diffusivity to density can be found. Figure 2.6 shows the relation for several sites, based on analyses of methane profiles (Fabre et al. 2000). For densities greater than about $550\,kg\,m^{-3}$, the diffusivity can be approximated as a linear function of density. Severinghaus et al. (2001) suggested the following relation. Relative to its diffusivity (D_0) in free air at standard pressure and temperature, the diffusivity of a gas in firn is approximately

$$\frac{D}{D_0} = \frac{P_0}{P_a}\left[\frac{T}{T_0}\right]^{1.85}\left[2.00\left[1-\frac{\rho}{\rho_i}\right]-0.167\right] \tag{2.6}$$

with standard values of $P_0 = 1.01325 \times 10^5$ Pa and $T_0 = 273.15$ K. (Here T and P_a are temperature and atmospheric pressure at the site.) However, Fabre et al. (2000) showed that small-scale heterogeneities, such as density differences between winter and summer strata, can significantly decrease the effective diffusivity for a given bulk density. Thus a general formulation should include not only bulk density but also firn microstructure (Albert et al. 2000).

Near the surface, pressure gradients drive air flow through firn. How much flow occurs for a given pressure gradient depends proportionately on the *permeability* of the firn. Permeability can be measured by forcing air flow through a firn sample and measuring the pressure drop across it (Albert et al. 2000). The permeability generally decreases as density increases, but no

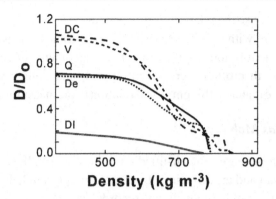

Figure 2.6: Inferred variation of effective gas diffusivity in polar firn as a function of density, at four locations in East Antarctica (DC = Dome C, V = Vostok, Q = Queen Maud Land, De = site DE08) and one site in arctic Canada (DI = Devon Island). Dome C and Vostok are the coldest sites, and Devon Island the warmest. Diffusivities are low at Devon Island because of ice layers, formed by refreezing of melt. The effective diffusivity, shown here, includes the effects of convective mixing in addition to molecular diffusion; this explains the high surface values at Dome C and Vostok. Diffusivities are normalized to the standard free-air value D_0. Adapted from Fabre et al. (2000).

clear relation exists within the range of densities characteristic of near-surface firn. Instead, the permeability is controlled by the presence of small-scale structural features such as crusts of wind-blown snow, loosely compacted snowfall layers, and depth hoar.

2.6 Hoar Layers

In certain circumstances, the transformation of recently fallen snow proceeds in a way quite different from that described so far. Instead of small, rounded forms, grains develop into large faceted crystals shaped as prisms, pyramids, or hollow hexagonal cups. Such *hoar layers* are the most coarse-grained type of firn that can form in the absence of meltwater. The average grain size is in the range of 2 to 5 mm, but some crystals can be much larger. A layer of hoar is highly porous and its density low (100 to 300 kg m^{-3}). Some hoar layers develop within the snowpack ("depth hoar"), while others grow on the surface ("surface hoar" or "hoar frost"). Rapid burial of surface hoar by a later snowfall preserves a low-density layer that may be indistinguishable from hoar formed at depth.

Hoar forms by a combined process of sublimation and vapor deposition in unconsolidated snow (Colbeck 1989; Akitaya 1974). A large vertical temperature gradient is essential; it produces a gradient of vapor pressure that drives movement of molecules from warmer layers to colder layers and out to cold air above the surface. Vapor moving through the snow condenses on ice-grain surfaces, growing faceted crystals. At the same time, sublimation removes ice from the opposite sides of the grains. To produce low-density layers within the snow, more ice must be lost by sublimation than is gained by crystal growth. Convection of air in the snowpack speeds up the formation of depth hoar. Growth of depth hoar crystals represents an increase in the internal energy of the system, supplied by the temperature gradient.

Several conditions produce the large temperature gradients needed for hoar formation (LaChapelle 1970; Fukuzawa and Akitaya 1993; Birkeland et al. 1998). A rapid onset of cold weather in autumn chills the surface layers while underlying layers are still relatively warm. Near the surface, the daily cycle of temperature variations produces large-amplitude fluctuations of the gradient, which also changes direction (Section 9.3). Solar radiation warms the snow just beneath the surface, driving vapor upward out of the snowpack and, to a lesser degree, into deeper layers. By this mechanism, hoar layers form in summer in central Greenland (Alley et al. 1990).

Depth hoar occurs widely. Layers in polar glaciers, usually only a few centimeters thick, are easily recognized in the walls of a pit and provide markers for measuring the annual snow accumulation. On the other hand, at some low-accumulation sites on the ice sheets hoar forms thick layers spanning several years' firn (Grootes and Steig 1992). In snowfields, depth hoar forms at the ground-snow interface in winter and spring. In central Alaska, Trabant and Benson (1972) found that almost the entire snowpack, 0.5 to 0.7 m thick, develops into depth hoar. Temperatures and temperature gradients in this area are higher than in polar regions and conditions suitable for depth hoar formation persist throughout the winter. Depth hoar layers are of particular concern in mountain areas because they act as fracture and sliding planes for avalanches (LaChapelle 1970).

2.7 Transformation When Meltwater Is Present

How does the transformation of snow to ice in the percolation and wet-snow zones differ from that in the dry-snow zone? There is of course a difference only during summer and down to the maximum depth to which meltwater penetrates. Elsewhere, and at other times of year, the mechanisms are the same as in a dry-snow zone. Shumskiy (1964, pp. 276–303) discussed this topic in detail.

Packing of grains is still the most important factor in the initial stages. Melting increases the rate at which grains become rounded, because the grains melt first at their extremities. The average grain size increases because the smaller grains tend to melt before the larger ones. Even without net melting, molecules move from smaller grains to larger ones because of differences in surface curvature – a process known as *Ostwald ripening* (Raymond and Tusima 1979). In addition, grains can join together in clusters by refreezing after melting. Grains join most rapidly in the surface layers because they undergo a daily cycle of freezing and thawing. Meltwater accelerates packing by lubricating the grains and permits very close packing because the surface tension of a water film tends to pull the grains together. Thus packing can achieve a higher maximum density in a meltwater area than in a dry-snow zone.

Refreezing of meltwater also speeds up the later stages of transformation. Air spaces are filled in this way. Refreezing of large quantities of meltwater to form ice layers and lenses represents a rapid transition from snow to ice. The time needed to complete the transformation varies widely between different areas according to the amount of meltwater. A superimposed-ice zone represents the extreme case in which snow transforms to ice in a single summer. Formation of ice layers accounts for the large variability of densities with depth in the firn on temperate glaciers; the site on Upper Seward Glacier, depicted in Figure 2.3, gives an example. Because melting and refreezing increase the average density of the firn, they also reduce the elevation of the surface by a small amount. This is an important process in the percolation zone of the Greenland Ice Sheet (Reeh 2008). The amount of melting varies from year to year as a function of summer temperature, so the surface elevation fluctuates even without variations in accumulation rate and mass of ice.

Further Reading

A paper by Alley (1988a) discusses the appearance and genesis of strata in dry firn. Koerner (1970) made a detailed study of how superimposed ice is formed. Baker et al. (2007) show what polar firn looks like at an extraordinarily close range, using an electron microscope.

CHAPTER 3

Grain-Scale Structures and Deformation of Ice

"All these facts, attested by long and invariable experience, prove that the ice of the glaciers is insensibly and continually moulding itself under the influence of external circumstances."

Letter to Professor Jameson,[1] **J.D. Forbes (1842)**

3.1 Introduction

Ice behaves as a thick viscous fluid; it slowly and continuously deforms under an applied stress, a process known as *creep*. Creep not only contributes to the overall motion of a glacier, but also accounts for phenomena like the closure of tunnels in ice and the flow of basal layers around bumps on the bed. The *creep relation* for ice is a formula connecting the rate and orientation of creep deformation to the applied stress. Glaciologists often call it the *flow law* or the *flow rule*, but we will use the more precise term *creep relation*[2] and refer to the parameters appearing in it as *creep parameters*. Interpretations and models of many glaciologic processes rely on this relation. It is, in particular, essential for quantitative analyses of glacier flow and evolution.

The creep relation is one form of a *constitutive relation* linking deformation and stress. A general constitutive relation for ice would describe elastic deformation and fracturing as well as creep. Fracturing sometimes occurs under a persistent tensile stress or when a large stress is applied rapidly. Elastic deformation transmits high-frequency sound waves; this is the basis for seismic soundings of ice thickness. In elastic deformation, stress and deformation are proportional. In creep, by contrast, stress and deformation *rate* are related; if proportional, the material is called "Newtonian viscous" or "linear viscous." *Perfect plasticity* defines a limiting case of creep, in which a gradually increasing stress produces no deformation until a critical value, the yield stress, is reached and deformation begins. The stress can never exceed the yield stress, but the deformation can continue at any rate. The creep of glacial ice is intermediate between Newtonian viscous and perfectly plastic behaviors.

The creep relation is a material property and has to be found empirically, by combining laboratory experiments with analyses of field measurements. Laboratory experiments reveal

---

[1] Reprinted in Forbes' *Occasional papers on the theory of glaciers* (1859).

[2] Though traditional, the term "flow law" conflates flow and deformation, two key concepts that ought to remain distinct.

Copyright © 2010, Elsevier Inc. All rights reserved.
DOI: 10.1016/B978-0-12-369461-4.00003-X

systematic variations under controlled conditions, but only provide general guidance about *in situ* behavior. Field measurements show how ice behaves *in situ*, but are generally not systematic. Moreover, ambiguities arise in their interpretation because some of the relevant variables cannot be measured. Establishing a general creep relation for ice remains a distant goal. Nonetheless, enough has been learned about the properties of creeping ice to explain many of the large-scale aspects of glacier systems. And, rather than mourning the absence of a universal relation, we recognize that glacier ice occurs as different types, subject to a variety of deformation states. The most appropriate creep relation depends on the case.

Glacier ice is polycrystalline, an aggregate of many individual crystals or grains. Interactions between grains work together with single-crystal properties to determine the bulk creep behavior of the ice. Grain-scale structural properties, including fabric and texture, influence the creep relation – often in ways that remain poorly understood. By "grain-scale" we mean dimensions of millimeters to decimeters, the scale intermediate between the *micro-structural* scale pertaining to the arrangement of molecules in crystal lattices and grain boundaries, and the *macro-structural* scale of features like crevasses and foliation (Chapter 10). Grain-scale structures evolve and develop spatial patterns. Such structures can be measured and so provide valuable information about what happens over long timescales in a glacier or ice sheet.

Grain-scale structural processes, such as the growth of crystals and their alignment along preferred orientations, interact strongly with deformation. It is therefore advantageous to consider all of these processes together, even though many of the connections between them have yet to be fully drawn. This chapter thus covers both creep deformation and grain-scale structural processes.

3.2 Properties of a Single Ice Crystal

3.2.1 Structure

The structure of ice arises from properties of the H_2O molecule (Pauling 1935). The three nuclei of this molecule can be pictured as forming an isosceles triangle with the oxygen nucleus at the apex and the hydrogen nuclei (protons) at the other two corners. This triangular configuration induces an electronic polarity, with excess positive charge at the protonic corners and excess negative charge at the apex. Because of this polarity, adjacent water molecules attach by hydrogen bonding. Specifically, each molecule attaches to four neighbors; the oxygen attaches to protons of two neighbors, while the protons each attach with a neighboring oxygen. Geometrically, these four bonds form a tetrahedron, with oxygens at each corner and one oxygen inside the volume (but close to one of the faces), as shown in Figure 3.1. Thus each H_2O molecule is surrounded by four other molecules in a regular tetrahedral arrangement. Molecules are spaced $\approx 2.76 \times 10^{-10}$ m apart.

A substance in which every atom has four neighbors arranged as a tetrahedron can crystallize hexagonally or cubically. The structure of terrestrial glacier ice, well established from X-ray diffraction measurements, is all of one type, with the oxygen atoms arranged in layers

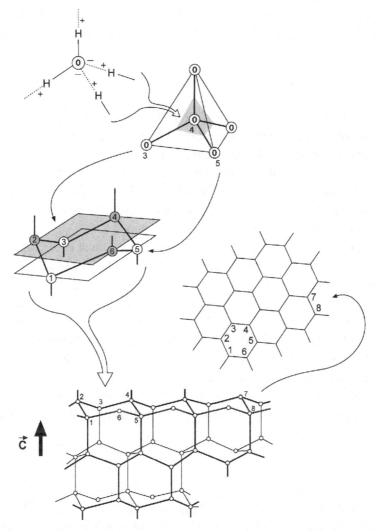

Figure 3.1: The crystalline structure of ice (adapted from drawings in Hobbs 1974). In the top-most diagram, dashed lines represent hydrogen bonds, the solid lines covalent bonds. Circles indicate oxygen atoms. Molecules arranged as tetrahedra combine to form hexagonal rings in two closely spaced planes. Atoms numbered 1 through 6 form one such ring. Viewed along the c-axis, the rings make a hexagonal mesh. Viewed obliquely, the ring layers are stacked as shown in the bottom sketch.

of hexagonal rings. The atoms in the ring layers lie in two closely spaced planes (one corresponding to the basal faces of the tetrahedra and the other to the interior oxygens). Moving around the ring, alternate oxygens lie in the upper and lower of these planes. The ice crystal structure consists of stacked ring layers, with each successive layer being a mirror image of the last, as shown in Figure 3.1. This structure resembles that of hexagonal metals such as zinc, magnesium, or cadmium. The plane of a layer of hexagonal rings is called the *basal plane* of the crystal. The direction at right angles to the basal plane defines the *c-axis*, sometimes called the *optic axis*.

3.2.2 Deformation of a Single Crystal

The susceptibility of an ice crystal to deformation depends very strongly on the orientation of the applied stress with respect to the basal planes. A shear stress on the basal planes readily causes shear deformation; the basal planes slip past one another. Deformations in other planes are much more difficult. A crystal thus behaves like a deck of cards, whose faces correspond to basal planes (McConnel 1891). This *mechanical anisotropy* of monocrystals – with easy slip on basal planes – has profound consequences for the creep behavior of polycrystalline glacier ice. Zinc, cadmium, and magnesium deform in a similar fashion.

Laboratory experiments on single crystals (reviewed by Duval et al. 1983) show that the shear strain rate $\dot{\epsilon}$ depends on the shear stress τ acting on basal planes (the *resolved shear stress*) according to a power law: $\dot{\epsilon} = k(T)\tau^m$, with power $m \approx 2$. The temperature dependence corresponds to an activation energy of about $63 \, \text{kJ} \, \text{mol}^{-1}$. Along a basal plane, slip rate does not depend on the orientation (Glen 1975). Slip on basal planes occurs by the movement of linear defects in the crystal lattice called *dislocations* (Weertman and Weertman 1964). Such defects allow planes of atoms to move past each other much more readily than in a perfect crystal. The motion of dislocations along crystallographic planes is called *dislocation glide* or, in this particular case, *basal glide* or *easy glide*. A shear strain rate, $\dot{\epsilon}_c$, arising from dislocation glide depends on the product of dislocation density ρ_d (length of linear defect per volume) and dislocation velocity v_d:

$$\dot{\epsilon}_c = b^* \rho_d \, v_d, \tag{3.1}$$

where coefficient b^* measures the lattice offset associated with a single dislocation. Equation 3.1 is known as the *Orowan relation*. Both ρ_d and v_d can vary with stress according to power laws; defining r and q as constants, $\rho_d \propto \tau^r$ and $v_d \propto \tau^q$, implying that

$$\dot{\epsilon}_c \propto \tau^{r+q}. \tag{3.2}$$

In theory, the velocity v_d of an isolated dislocation depends linearly on τ, so $q = 1$ is often assumed.

The glide rate does not always increase with the dislocation density, however. Though essential for allowing deformation at typical glaciological stresses, dislocations can also hinder deformation by interfering with one another. Such *dislocation tangles*, or pile-ups, sometimes block the movement of dislocations to the crystal boundaries, thereby reducing or preventing further strain. In this case v_d in Eq. 3.1 decreases with ρ_d.

To deform ice by shear along nonbasal planes is harder than along basal planes by at least three or four orders of magnitude (Duval et al. 1983, Figure 3.2). Such "hard deformations"

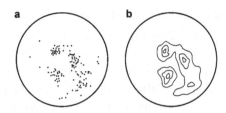

Figure 3.2: Schmidt diagram. Each point in (a) represents the intersection of the c-axis of one crystal with the surface of a hemisphere when each c-axis passes through the center of the sphere. The contours in (b) represent 1, 3, and 5% of points within 1% of area.

occur by *dislocation glide* along the prismatic and pyramidal crystallographic planes (which cut through the basal planes at high angles), or by *dislocation climb* normal to basal planes. Deformations at these hard orientations have never been measured well; even in the laboratory, it is difficult to orient samples so that zero stress acts on the much softer basal planes.

3.3 Polycrystalline Ice: Grain-scale Forms and Processes

A cubic meter of glacier ice typically consists of 10^6 to 10^9 separate grains. In fully formed ice, each grain is a single crystal. The bulk properties of glacier ice arise from the properties of single crystals, discussed earlier, combined with interactions between the grains.

3.3.1 Orientation Fabrics: Brief Description

Within a mass of polycrystalline ice, the c-axis of each crystal points in a direction given by a unit vector $\vec{c}$. The ensemble of such orientations constitutes the *c-axis fabric* of the ice, sometimes referred to as *orientation fabric*, or simply as *fabric*. Crystal orientations are plotted on a *Schmidt diagram* or *fabric diagram* (Figure 3.2); this represents the surface of a hemisphere on an equal-area projection. Each dot represents the intersection of the c-axis of one crystal with the surface of the lower hemisphere, if each c-axis passes through the center of the sphere (Figure 3.2a). Contours of the density of points per area can be drawn (Figure 3.2b).

A uniform distribution of c-axis orientations, in three dimensions, corresponds to an *isotropic* fabric; otherwise there is an *anisotropic* fabric or *preferred orientation* (equivalent to the term *lattice preferred orientation* in structural geology and seismology). Preferred orientations develop in deforming glacier ice as a consequence of strain and related processes. The patterns usually reflect the stress regime. The formation of fabrics, and their consequences for bulk creep properties, are discussed later in this chapter (Sections 3.3.4 and 3.4.7).

3.3.2 Impurities and Bubbles

Glacier ice contains soluble, particulate, and gas-phase impurities. For a solid-earth material, glacier ice tends to be exceptionally pure; the processes leading to snowfall on a glacier – evaporation followed by atmospheric transport, condensation, and precipitation – act as a distillation system. But the small quantities of impurities do affect the physical properties of the ice and provide an archive of atmospheric conditions. Furthermore, particles can concentrate on the surface of an ablating glacier, significantly affecting the absorption of solar radiation and hence the rate of melt (Chapter 5). The dirtiest surfaces are found on small glaciers in mountain settings. Factors controlling the concentration and composition of impurities in polar ice, and their association with past atmospheres, are discussed in greater detail in Chapter 15. Here we introduce the topic only to support our discussion of grain-scale processes and deformation.

3.3.2.1 Soluble Impurities

Soluble impurities in ice originate by deposition from the atmosphere, by dissolution of particle surfaces, and by incorporation of soil material into a glacier's basal layers. In ice at its melting point (*temperate ice*), percolating water removes much of the impurity content. Polar ice, on the other hand, retains most of the impurities even hundreds of millennia after deposition.

A few soluble impurities like HF and HCl can substitute for molecules in the H_2O lattice (*substitutional impurities*). Others reside in the intermolecular void (*interstitial impurities*). All impurity species are partitioned between grain interiors and grain boundaries. The concentration on grain boundaries can be many orders of magnitude larger than for interiors (Paren and Walker 1971; Wolff et al. 1988; Fukuzawa et al. 1998). The most abundant cations are H^+, Na^+, K^+, Ca^{2+}, Mg^{2+}, Al^{2+}, and NH_4^+; the anions, OH^-, Cl^-, SO_4^{2-}, and NO_3^{2-}. Typical total ionic concentrations in polar ice are $1-10\,\mu mol\,L^{-1}$ in ice deposited during Pleistocene glacial climates, and an order of magnitude less in ice from interglacial climates. Some basal layers contain very much greater amounts; concentrations as high as $800\,\mu mol\,L^{-1}$ are found in the basal layers of glaciers in the Antarctic Dry Valleys (Cuffey et al. 2000a).

3.3.2.2 Particulate Impurities

Also derived from transport through the atmosphere are microparticles, typically of size $0.1-5\,\mu m$ in polar ice. Many of these particles are fragments of the most common silicate and carbonate crustal rocks, but some are soot or organic structures. An immense diversity of particulate materials is found in the atmosphere. A similar diversity surely exists in glacial ice, but no attempt has been made to measure most types. As with soluble impurities, microparticles preferentially gather at grain boundaries (Barnes et al. 2002; Durand et al. 2006). Typical mass fractions of rock microparticles in polar ice are 10^{-8} to 10^{-6}.

Glaciers contain rock material from coarse clay to boulders. This material is either picked up from the bed or falls from the surrounding hillsides. It is found throughout mountain glaciers but only in the basal layers of ice sheets. Rocks affect the forces of contact between the base of the ice and the substrate and so influence glacier flow (Chapter 7).

3.3.2.3 Bubbles

Glacier ice contains abundant bubbles, except in the high-pressure environment of the deep ice sheets. Bubbles are generally spherical in form, but deformation of the surrounding ice sometimes stretches them to make elongated shapes (Chapter 10). Bubbles may contain liquid water, if the temperature is at melting point, but otherwise contain atmospheric gases. For unreactive gases such as diatomic nitrogen and oxygen, and the noble gases, the bubble composition in cold polar ice largely reflects the atmospheric concentrations when the ice formed (Chapter 15). Concentrations of gases such as H_2O vapor, on the other hand, are unrelated to the original air. Contact with liquid water alters gas compositions as a function of the solubilities of the different molecules. Biogenic activity may produce methane or other compounds. In polar environments, the bubbles occupy about 10% of the volume when firn turns to ice (Section 2.2). The corresponding mass fraction is about 10^{-4}. Where through-flow of melt occurs, dissolution in the water generally reduces the initial gas content.

3.3.2.4 Disappearance of Bubbles

At depth in the polar ice sheets, bubbles shrink as the pressure of the overlying ice increases. The gas pressure within the bubbles increases at the same time. At sufficient depth, the gas attains a *dissociation pressure* – a function of temperature and composition – at which the bubbles begin to disappear as the gas molecules form *clathrate hydrates* (Miller 1969). A clathrate compound is one in which a crystal lattice contains void "cages" that hold molecules of other substances. In clathrate hydrates, the lattice consists of water molecules. For the typical hydrate of air, the lattice consists of 46 water molecules in an arrangement forming eight cages containing a molecule of oxygen or nitrogen.

The disappearance of bubbles does not happen immediately but rather occurs over millennia, as gas pressures increase above the dissociation pressure. Higher pressures increase the likelihood of nucleation of new hydrates (Shoji and Langway 1987; Salamatin et al. 2001) and facilitate transfer of molecules from the bubbles into the lattice by diffusion (Ohno et al. 2004; Pauer et al. 1999). This slow process results in a broad depth range over which bubbles disappear (Figure 3.3): approximately 900 to 1600 m in central Greenland and 500 to 1200 m in central East Antarctica (Kipfstuhl et al. 2001). Within these transition zones, clathrates and bubbles coexist. Salamatin et al. (1997; 2001; 2004) constructed a plausible model for this process, albeit one that probably oversimplifies the situation (Shimada and Hondoh 2004; Kipfstuhl et al. 2001).

3.3.3 Texture and Recrystallization

The distribution of grain sizes and shapes in polycrystalline ice constitute the *texture*. Grain size is quantified by an effective diameter (D), radius, or cross-sectional area. A cross-section of ice shows an apparent average grain size that is smaller than the sizes of complete grains; the plane of a cross-section does not pass through the centers of most grains. Grain diameters reported in

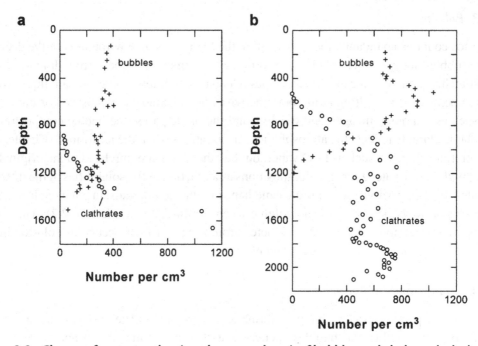

Figure 3.3: Change of concentration (number per volume) of bubbles and clathrate inclusions at depth in (a) the Greenland Ice Sheet at North-GRIP (data from Kipfstuhl et al. 2001) and (b) the East Antarctic Ice Sheet at Dome Fuji (data from Ohno et al. 2004).

the literature usually indicate apparent values, smaller than complete-grain values by a factor of 1.2 to 1.4. The distribution of grain sizes is typically log-normal.

Grain shapes are commonly close to equant. But flattening in one direction is sometimes observed, with the shortened axis parallel to the direction of finite compression in the surrounding ice (Lipenkov et al. 1989; Herron et al. 1985; Thorsteinsson et al. 1997). Sometimes, grains form complex interlocking shapes. Grain boundary geometries span the spectrum from planar to deeply interfingering.

In actively deforming ice, typical mean grain sizes range from $\overline{D} \approx 2\,\mathrm{mm}$ to $2\,\mathrm{cm}$ (Figure 3.4). The most fine-grained ice layers (excluding firn on polar glaciers) have $\overline{D} \approx 1\,\mathrm{mm}$ and occur in cold but rapidly deforming horizons, including deep layers of the Greenland Ice Sheet and basal layers of Antarctic mountain glaciers (Herron et al. 1985; Cuffey et al. 2000a). In the latter, some horizons contain mean diameters as small as $0.8\,\mathrm{mm}$. In the ice sheets, ice deposited in the Pleistocene glacial climates is consistently finer than adjacent ice from interglacial periods (Herron and Langway 1982; Durand et al. 2006). In contrast, very coarse grains are found in temperate ice that is not deforming rapidly.

Variations in grain size, like those seen in Figure 3.4, reflect a suite of *recrystallization* processes – processes that systematically change the alignment of atoms and the positions of associated grain boundaries. Such processes, acting on large numbers of grains, can alter not only grain sizes but also c-axis fabrics. We use the term *recrystallization* to refer to all such

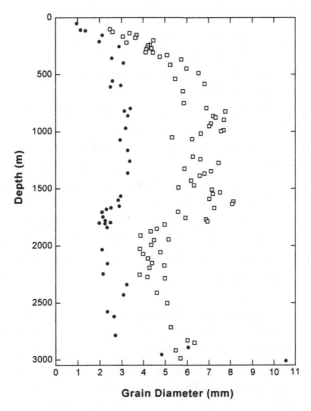

Figure 3.4: Variation of grain diameter with depth in the central Greenland Ice Sheet (GISP-2 core). Filled circles show the apparent mean grain diameter seen in the plane of thin sections (data from Alley and Woods 1996). Open squares show the mean diameter of the largest 50 grains seen in thin sections (data from Gow et al. 1997). Grain sizes increase through the upper few hundred meters. The transition from coarser to finer grains between 1500 and 1700 m corresponds to increased impurity concentrations between Holocene ice and underlying ice from the Last Glacial Maximum. Very large grains (up to 60 mm) were found in the basal layers, but are not shown here.

processes (following, for example, Placidi and Hutter 2006). The term has sometimes been used only for a subset of processes that involve nucleation of new grains.

In recrystallization, atoms escape from the crystal lattice on one side of a grain boundary and reattach to the lattice on the other side, in a position of lower free energy. By this process, the grain boundary moves through the ice. Such *grain boundary migration* is driven by differences of free energy between adjacent lattices, which arise from curvature of the boundary or from contrasts in lattice strain energy.

3.3.3.1 Normal Grain Growth

In the absence of mechanisms to form new grains, the mean grain size of polycrystalline ice increases over time by *grain growth*. This process is favored thermodynamically because of the energy penalty associated with grain boundaries. Boundaries constitute severe defects in a crystalline lattice, with a grain boundary energy per area of surface (γ_{GB}) analogous to surface tension. For boundaries that intersect basal planes at angles greater than about $20°$, the value is $\gamma_{GB} \approx 0.065 \, J \, m^{-2}$. By geometric constraints, an increase of average grain size represents a reduction in the area of grain boundary per unit volume of ice, and hence a reduction of the free energy of the system. The total grain-boundary energy, per unit volume, decreases as grains grow; the energy is approximately $3 \gamma_{GB} D^{-1}$ (units of $J \, m^{-3}$).

Grains grow by migration of their boundaries. In slowly deforming ice, migration is driven by curvature of the boundaries, a process called *normal grain growth*. Atoms on a recessed or flat grain surface are more completely enclosed by neighbors, and thus more firmly bound, than atoms on a surface that bulges out. This difference results in a net flux of atoms across a grain boundary, away from a grain that protrudes into its neighbor. Compared to large grains, small-diameter grains tend to have fewer boundaries, with smaller radii of curvature. The smaller grain thus usually occupies the protruding side of the boundary, and the boundary will migrate into it. By this process, large grains consume small grains over time, and the mean grain diameter of the polycrystal increases.

Models for this process specify the grain-boundary migration rate as proportional to the product of an intrinsic mobility, M_i, and a driving force (Cole et al. 1954; Hillert 1965). The latter depends on the grain boundary's radius of curvature, R_b, and equals γ_{GB}/R_b. In combination with geometrical constraints on the population distribution of grain sizes, this expression for driving force predicts a linear increase of mean cross-sectional area with age. Specifically, the mean area D^2 at age t is

$$D^2 = D_o^2 + k_g \, t, \tag{3.3}$$

for an initial mean area D_o^2. This relation matches measurements from the upper horizons of polar ice sheets (Figure 3.5), where grain growth begins in the firn and continues in the ice below. The growth rate k_g is proportional to the diffusivity for H_2O across grain boundaries, and varies with Kelvin temperature (T) according to

$$k_g = k_o \exp\left(-\frac{Q_b}{RT}\right), \tag{3.4}$$

where k_o indicates a constant coefficient. Q_b defines the apparent activation energy for grain-boundary self-diffusion. According to Eqs. 3.3 and 3.4, crystal size increases at a constant rate if the temperature remains constant. This is the case for the data shown in Figure 3.5, except for an initial period (a few decades long) of enhanced growth related to large temperature variations and vapor fluxes in the uppermost firn. Constant growth rates have also been observed at most locations in Antarctica and Greenland (Gow 1969; 1971; Alley and Koci 1988; Gow et al. 2004).

Table 3.1 lists temperatures and crystal growth rates in polar firn, based on measurements to a depth of at least 45 m.

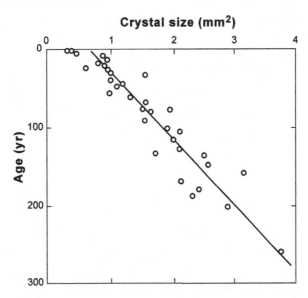

Figure 3.5: Variation, with age, of mean cross-sectional area of crystals in firn at Byrd Station, Antarctica. Redrawn from Gow (1971).

Table 3.1: Growth rates of ice crystals in polar firn.

Location	Temp. (°C)	Growth rate ($10^{-4}\,mm^2\,yr^{-1}$)	Reference
Plateau	−57	7	Gow 1971
Vostok	−57	8	Barkov and Lipenkov 1984
Dome C	−54	4	Duval and Lorius 1980
South Pole	−51	6	Gow 1969
Southice	−31	56	Stephenson 1967
Inge Lehmann	−30	70	Gow 1971
Site A	−30	78	Alley and Koci 1988
Byrd	−28	120	Gow 1971
Ridge BC	−27	85	Alley and Bentley 1988
Site 2	−25	99	Fuchs 1959
Camp Century	−24	160	Gow 1971
S2	−19	137	Hollin et al. 1961
Maudheim	−17	186	Schytt 1958

The net growth in the uppermost 100 m of polar ice sheets amounts to about a factor of four increase of D^2. Because this happens under nearly isothermal conditions, such data provide a test of Eq. 3.4. Growth rates are plotted on a logarithmic scale against the reciprocal of T, as in Figure 3.6. The points lie close to a straight line whose slope, determined by regression analysis,

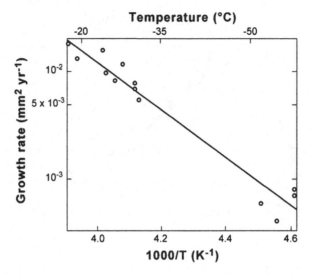

Figure 3.6: Plot of crystal growth rate against reciprocal of the absolute temperature for the data in Table 3.1.

gives an activation energy of $Q_b = 42.4\,\text{kJ mol}^{-1}$. This nearly matches the value $40.6\,\text{kJ mol}^{-1}$ obtained in a laboratory experiment (Jumawan 1972). Both are about 0.7 times the activation energy for volume self-diffusion ($61\,\text{kJ mol}^{-1}$), the process of movement of H_2O molecules through an ice lattice. The same ratio of the activation energies for grain-boundary and volume self-diffusions has been measured in metals (Cole et al. 1954).

The activation energy inferred from Figure 3.6 applies to cold polar conditions, with temperatures in the range -20 to $-50\,°C$. At higher temperatures, approaching the melting point, the apparent activation energy increases dramatically (Jacka and Li 1994). This reflects a substantial increase in the disorder of grain boundaries and the ease with which molecules diffuse along them.

The temperature effect is large but not the only control on grain growth; impurities at a grain boundary can reduce or stop its migration. Alley et al. (1986a,b) used metallurgical concepts to address this problem for ice. They concluded that soluble impurities, microparticles, and bubbles can all reduce grain growth rates, but they act in different ways. Soluble impurities reduce the grain boundary's intrinsic mobility. Microparticles and bubbles, on the other hand, reduce the effective driving force by locally pinning the boundary. Where the boundary is pinned, it migrates more slowly than in the surrounding area; the boundary thus develops a curvature that opposes further migration. In the presence of impurities, the mean grain diameter changes over time according to

$$\frac{dD}{dt} = c\,\gamma_{\text{GB}}\,M^* \left[\frac{2}{D} - \frac{1}{R_m^{(p)}} - \frac{1}{R_m^{(b)}} \right]. \tag{3.5}$$

Here the effective mobility M^* depends on the intrinsic mobility M_i and soluble impurity concentration C_ℓ according to

$$M^* = \frac{1}{M_i^{-1} + \alpha_c C_\ell}, \tag{3.6}$$

and the numbers $R_m^{(p)}$ and $R_m^{(b)}$ denote effective radii at which pinning by particles and bubbles, respectively, reduces driving force to zero. (Parameters c and α_c are constants.)

Alley and Woods (1996) analyzed correlations of grain size with impurity content in a deep Greenland ice core and concluded that soluble impurities restrict grain growth in some layers but not others. Durand et al. (2006) conducted the most complete analysis of grain sizes in an ice core from the cold central regions of Antarctica. They concluded that the prevalence of finer grains in ice-age ice compared to interglacial ice (as seen for Greenland in Figure 3.4) reflects pinning by microparticles. This mechanism restricts the growth of large grains more than small grains (Eq. 3.5). It therefore reduces the range of grain sizes (measured as a multiple of the mean grain diameter) in layers with abundant particles. The Antarctic core showed this pattern. In contrast, soluble impurities are not expected to affect the size range in this fashion. Durand et al. also concluded that, in all layers, pinning by both bubbles and clathrates further limits grain sizes.

Laboratory experiments show that soluble impurities become increasingly concentrated on grain boundaries as they migrate (Iliescu and Baker 2008). This implies that the effects of particle and soluble impurities on grain growth can interact over time.

Normal grain growth contributes to the formation of the very coarse crystals found in stagnant ice. It also accounts for the persistent enlargement of grains in the uppermost layers of the polar ice sheets (Figures 3.4 and 3.5). With increasing depth, however, grain growth is increasingly driven by differences between crystals that arise from ice deformation – the growth process is no longer "normal" but instead occurs as part of a suite of processes referred to as dynamic recrystallization, discussed next. Furthermore, the attainment of an equilibrium grain size in the middle of the ice sheet (Figure 3.4) must reflect a balance between grain growth and the formation of new grains. In ice, two processes account for formation of new grains: polygonization and nucleation.

3.3.3.2 Dynamic Recrystallization

Dynamic recrystallization refers to recrystallization that occurs during deformation. Polygonization, migration recrystallization, and nucleation are all processes of dynamic recrystallization.

Polygonization Grains sometimes experience bending stresses. For example, a grain within a region of pure shear (see Figure 3.10) tends to flatten, but also bends if neighboring grains impinge on it in different ways. Because of bending, dislocations can align and group together in a planar "wall," which forms a new boundary and subdivides the grain. (Twisting of a grain can also cause alignment of dislocations.) This is the process of *polygonization*, sometimes called *rotation recrystallization*. Polygonization has little effect on fabrics; it produces sets of neighboring grains with only slightly different c-axis orientations. Such a pattern has been observed in polar ice undergoing pure shear (Alley et al. 1995a; Castelnau et al. 1996a).

Migration Recrystallization Deformation of a grain generates new dislocations and other crystal lattice defects that distort the lattice and so increase the stored energy. This additional component of internal energy constitutes the lattice strain energy. (In general, the mechanical work done by straining produces not only defects but also heat; the term *lattice strain energy* is used here to indicate the lattice-defect component.) A difference of lattice strain energy density between two neighboring grains creates a net transfer of molecules from the higher-energy grain to the lower-energy one. Thus the grain boundary migrates into the higher-energy grain; the less-strained grain grows at the expense of the other and eventually consumes it. This is *migration recrystallization*. The different driving force – a contrast in lattice strain energy rather than a boundary curvature – distinguishes it from the normal grain growth described previously. Compared to polygonization, migration recrystallization is favored when grain boundaries move easily, the situation in warm ice (Duval and Castelnau 1995). In migration recrystallization, impurities should still inhibit the movement of grain boundaries as described for normal grain growth.

Grain Nucleation Incipient new grains form in areas of high lattice distortion and are free of lattice strain energy. If older neighboring grains contain a large density of lattice strain energy, the new grain can consume its neighbor(s) by migration recrystallization.

Large strains and large deviatoric stresses favor dynamic recrystallization in general, while high temperatures favor migration recrystallization in particular. In ice sheets, these processes are therefore most active in the deepest layers, which are warm and shear rapidly. However, dynamic recrystallization sometimes also occurs in the colder layers near the surface (DiPrinzio et al. 2005; Jacka 2009), where it coexists with normal grain growth. Although the near-surface layers deform slowly compared to ones beneath, they are relatively cold and thus stiffer, which favors the build-up of stresses.

In contrast to polygonization, migration recrystallization and nucleation affect fabrics significantly. New grains tend to form with an orientation favorable to basal glide; in other words, an orientation that aligns the basal plane with the maximum shear stress implied by the macroscopic state of stress. Thus, in some deformation regimes, the c-axes of new grains point in directions unlike those of older grains nearby. Furthermore, highly strained grains are likely to be old (and hence significantly rotated) or oriented unfavorably for shear; such grains are preferentially eliminated by migration recrystallization.

The replacement of old grains by expansion of new grains leads to grain boundaries that are often irregular and sometimes interfingering. The size of the grains depends on the rates of nucleation and boundary migration; the more rapid the nucleation relative to the migration, the smaller the average grain. This explains the small grains observed in ice deforming at high stresses; rapid deformation induces frequent nucleation, leading to an inverse relationship between grain size and stress, $D \propto \tau^{-\nu}$ (Jacka and Li 1994). (These authors hypothesized that $\nu = 1.5$ for a balance between normal grain growth and dynamic recrystallization but a value $\nu < 1$ fit their data better.) Laboratory experiments by Jacka (1984b) showed that ice deforming under a high but constant stress achieves a steady grain size. This size varies inversely with stress and is independent of the initial grain size. Experiments conducted at high stresses (400 kPa and larger) also show that nucleation occurs more frequently if the ice contains fine particulate impurities (Song et al. 2008).

If glacier ice is removed from a stressed environment, nucleation ceases but grain growth continues, a process sometimes referred to as *static recrystallization*. This leads to the development of large crystals (Steinemann 1958; Rigsby 1960). Such growth occurs, for example, in ice along glacier margins that have stagnated during retreat, and in basal ice protected by bedrock hollows. Static recrystallization is not the only way that large crystals can form, however. In warm ice, especially in temperate glaciers, grain boundaries can migrate rapidly and thus produce large crystals by migration recrystallization.

3.3.3.3 Controls on Grain Size Profiles Observed in Ice Sheets

The D profiles from Antarctica and Greenland, such as the one pictured in Figure 3.4, show an upper zone of grain growth. Initially, this occurs as normal growth; how rapidly D increases downward depends on the temperature and the age of the ice. But as depth increases, so does accumulated strain. This allows further growth by migration recrystallization, if the ice is warm enough for grain boundaries to be mobile. Accumulated strain also drives polygonization, the more likely process at low temperatures. Polygonization reduces the rate of grain growth and, if active enough, prevents the mean grain size from increasing further. This is one explanation for the transition to relatively uniform grain size. Nucleation can also contribute, and becomes more active as depth and shearing increase. Within these zones of no average growth, D varies in correlation with the climate at time of deposition; glacial-age ice is fine-grained compared to interglacial ice. Such variations probably reflect restricted grain growth due to high concentrations of particulate and soluble impurities in the glacial-age ice. At yet greater depths, close to the bed, coarse grains can develop because high temperatures increase grain growth rates and – in some locations, such as bedrock hollows and ice divides – deformation occurs slowly. At other locations, ice near the bed deforms rapidly. In these settings, grain sizes span a large range, with fine grains usually found in layers with high impurity contents and enhanced shearing.

3.3.4 Formation of C-axis Orientation Fabrics

Preferred-orientation fabrics are a consequence of recrystallization and single-crystal deformation processes interacting with bulk deformation of the polycrystalline aggregate (Kamb 1959, 1972; Budd 1972; Azuma and Higashi 1985; Alley 1988b; van der Veen and Whillans 1994; Azuma 1994). Such fabrics, which are common, profoundly influence the creep relation for glacier ice (Section 3.4.7). The common fabric types, depicted in Figure 3.7, include

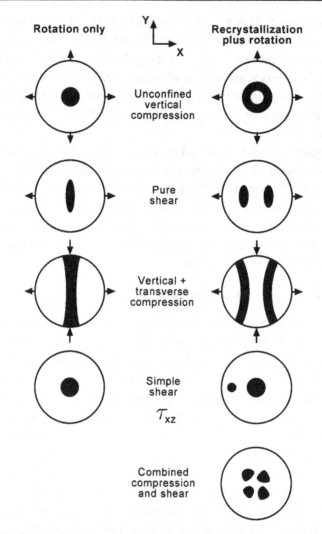

Figure 3.7: C-axis fabrics produced in simple stress systems by (a) rotation, by itself or accompanied by polygonization, and (b) rotation accompanied by migration recrystallization and nucleation. These two classes of fabrics are sometimes referred to as *deformation fabrics* and *recrystallization fabrics*, respectively. Each diagram is the projection on a horizontal plane.

single-maximum fabrics, in which the c-axes tend to point in one preferred direction; *multiple-maximum fabrics*, in which orientations cluster about several directions; *girdle fabrics*, in which c-axes preferentially point along a conical surface; and *band fabrics*, in which c-axes occupy a plane.

Individual ice crystals deform easily by basal glide (Section 3.2). Consequently, in poly-crystalline ice a crystal will deform in shear as long as a shear stress acts on its basal planes. But neighboring crystals with different orientations interfere, forcing the basal planes

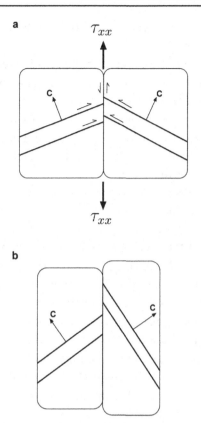

Figure 3.8: Sketch showing grain rotation from deformation of polycrystalline ice. In (a), tension τ_{xx} causes resolved shear stresses on the basal planes of two grains, which deform by basal glide. Mutual interference leads to rotation of c-axes away from the axis of tension, as shown in (b). Sliding along the grain boundary must also occur, as shown. Adapted from Alley (1992a).

of the deforming crystal to rotate (Figure 3.8). The c-axis rotates away from the direction of instantaneous extension and toward one of compression. Such rotations occur continuously as deformation proceeds. With accumulation of bulk strain, this process generates a preferred orientation, such that c-axes point along the direction of finite compression.

Figure 3.7 shows the result for simple deformations. First consider cases with rotation only. Unconfined uniaxial compression in the z-direction, balanced by extension in the x- and y-directions, produces a fabric with a single maximum centered along z. If extension in the y-direction is blocked – because the ice is confined on two sides – a single maximum oriented along z again forms, but with an elongation in the y-direction. Uniaxial extension in the x-direction, balanced by compression in both y- and z-directions, causes c-axes to rotate away from the x direction. They must therefore cluster in the yz-plane and, by symmetry, are uniformly distributed in this plane. These three cases correspond to three common flow fields in ice sheets (all with the dominant down-glacier flow following x): horizontal spreading, parallel flow, and transverse compression.

Polygonization does not change these fabrics, but migration recrystallization and nucleation alter them significantly. Old grains with orientations unfavorable for basal glide are preferentially eliminated. New grains nucleate at orientations favorable for basal glide, but then rotate as strain accumulates. Figure 3.7 shows the resulting fabrics, some of which are quite distinct from their counterparts in the absence of migration recrystallization and nucleation.

The first class of fabrics – those arising only from rotation and polygonization – are sometimes referred to as *deformation fabrics* the second class of fabrics, as *recrystallization fabrics*. To form a deformation fabric requires a bulk finite strain of at least a few tens of percent, whereas recrystallization fabrics can form more quickly if stress and/or temperature is high.

Observations of ice-core thin sections show that single-maximum fabrics commonly develop in the middle to upper layers of the polar ice sheets, a zone dominated by vertically oriented uniaxial compression (Figure 3.9; Herron and Langway 1982; Gow and Williamson 1976). This is the fabric expected from rotation and polygonization. If migration recrystallization begins in this zone a girdle fabric starts to form; DiPrinzio et al. (2005) reported such a fabric at 200 m depth in Siple Dome, and Jacka (2009) has emphasized that girdle fabrics appear in some horizons in the upper parts of several ice cores. Well-developed girdle fabrics are also characteristic of ice shelves, where rapid vertical compression and longitudinal extension occur and recrystallization is active (Gow 1970b; Budd 1972). Such fabrics have also been produced in the laboratory by unconfined uniaxial compression, as expected (Jacka and Maccagnan 1984).

In the upper layers of the ice sheet at Vostok, in central Antarctica, band fabrics have formed by uniaxial extension in one horizontal direction (Alley 1988b; Lipenkov et al. 1989).

In simple shear (see Figure 3.10) – the dominant regime in the bottom halves of mountain glaciers and ice sheets, and throughout shear margins of ice streams and shelves – the axis of instantaneous compression trends at 45° to the shear planes. Rotation of a c-axis toward this direction is, however, opposed by the bulk rotation associated with vorticity of the shearing. The axis of finite compression trends perpendicular to the shear planes. Consequently, c-axes form a single-maximum fabric, oriented perpendicular to the shear planes. This important fact means that rapidly deforming basal ice tends to have a single-maximum fabric oriented perpendicular to the glacier bed, roughly in the vertical. The effective viscosities of such layers differ greatly from those of isotropic ice studied in most laboratory experiments (Section 3.4.7). Examples of such fabrics are found at Meserve Glacier, Antarctica (Cuffey et al. 2000a), the Barnes Ice Cap (Hooke 1973; Hudleston 1977), and at Dye 3, Camp Century, and Summit on the Greenland Ice Sheet (Herron and Langway 1982; Herron et al. 1985; Gow et al. 1997; Thorsteinsson et al. 1997).

Simple shear is not confined to the bottom half of a glacier; it also occurs in the upper layers, though at a much lower rate. Such shearing can contribute to the formation of the vertically

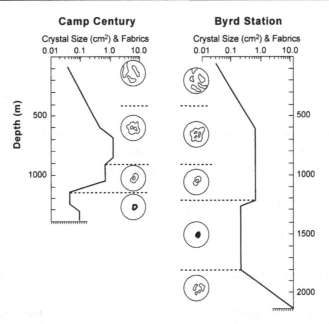

Figure 3.9: Variation with depth of fabric and average crystal size in the Greenland Ice Sheet (Camp Century) and West Antarctic Ice Sheet (Byrd Station). Adapted from Herron and Langway (1982).

oriented single-maximum fabrics observed in ice cores from this zone (Jacka 2009). Whether it is important or not must depend on the rate of shearing relative to the rate of vertical compression, which varies from place to place and with depth.

In simple shear, new grains formed by dynamic recrystallization again take orientations with high resolved stress. Some have c-axes perpendicular to the shear plane; others have c-axes parallel to the shear plane and pointing in the direction of flow. (Both orientations maximize the stress on basal planes, because of the symmetry of shear stresses.) The first type of grain reinforces the strong single-maximum fabric. The second type produces another maximum, lying on the shear plane and pointing in the direction of flow. Rotations then weaken this second maximum, perhaps to the point of vanishing.

Multiple-maximum fabrics are often observed in temperate glaciers (Kamb 1959; Rigsby 1951, 1960; Vallon et al. 1976), and sometimes in polar glaciers (Hooke and Hudleston 1980), and sometimes in basal layers of ice sheets (Gow and Williamson 1976). The maxima sometimes appear as a diamond shape in the fabric diagrams. Multiple-maximum fabrics are not satisfactorily explained, but probably arise from complex stress fields like combined shear and compression perpendicular to the shear plane, together with active migration recrystallization and nucleation. Laboratory experiments show that such a combined stress field can produce a girdle fabric or one with several maxima (Kamb 1972; Duval 1981). Multiple-maximum fabrics have also been produced in the laboratory by repeated compression of the same sample, with periods of annealing in between (Huang et al. 1985).

The preceding discussion assumes that basal glide is the dominant deformation mechanism in polycrystalline ice. The analysis based on this assumption successfully explains major features of the observed c-axis fabrics and their association with deformation regimes. This provides strong evidence for the dominant role of basal glide in deforming ice; this argument has been made quantitatively using simulations of fabric development (De La Chapelle et al. 1998; Thorsteinsson 2002; Duval and Montagnat 2002). Basal glide cannot act alone, however. We next consider the broader array of deformation mechanisms that may operate in polycrystalline ice.

3.3.5 Mechanisms of Polycrystalline Deformation

To deform polycrystalline ice, several mechanisms must operate simultaneously to accomplish deformations consistent with the bulk pattern, and to accommodate "incompatibilities": the subgrain-scale geometric mismatches between neighboring crystals with different deformation rates, orientations, and shapes. For convenience these two classes of mechanisms are sometimes referred to as *deformation mechanisms* and *accommodation mechanisms*, respectively. But they are all deformation mechanisms in the broad sense; they strain a region of the ice body, either by changing the shapes of individual grains or the relative positions of neighboring grains. Furthermore, all mechanisms that are active under prevailing conditions will likely contribute, to some extent, to both deformation and accommodation.

For a given combination of stress, temperature, texture, and fabric, a *rate-limiting mechanism* fixes the bulk deformation rate of the ice. This mechanism may operate primarily as either accommodation or bulk deformation. The rate-limiting mechanism can switch as stress, temperature, and other variables change. In the transition regime, more than one rate-limiting mechanism matters. An elementary understanding of rate-limiting mechanisms has yet to be established across the entire range of stress, temperature, fabric, grain size, and impurity contents relevant to glacier ice, although several reviews have been made from available information (Goodman et al. 1981; Duval et al. 1983; Goldsby and Kohlstedt 2001). Many experiments have resolved specific questions about ice deformation, but only at stresses higher than typical for glaciers and without regard for the range of grain sizes and fabrics.

3.3.5.1 Principal Deformation by Basal Glide

As in metals, various types of dislocation motion – introduced in Section 3.2 – account for much of the deformation of polycrystalline ice (Weertman and Weertman 1964). Basal glide is the principal deformation mechanism in conditions relevant to glacier flow, and it accounts for the development of c-axis fabrics. Basal glide is not the rate-limiting process, however, except perhaps at very low stresses. Moreover, basal glide must act in concert with other mechanisms; we next list the different scenarios that might apply. In each case, the bulk strain rate $\dot{\epsilon}$ and shear stress τ supposedly obey a power-law relationship: $\dot{\epsilon} \propto \tau^n$, with n a positive constant.

1. Basal glide is supplemented and accommodated by glide motion of dislocations in hard directions and by dislocation climb (Fukuda et al. 1987). The action of all these mechanisms together constitutes *dislocation creep*, a style of deformation favored by high stress (>200 kPa), high strain, low temperature, and coarse grains. The creep exponent has a value $n \approx 3$ to 4. In a review supplemented by new experiments, Goldsby and Kohlstedt (2001) concluded that $n = 4$ in this regime, close to the value originally suggested by Glen (1955). On the other hand, such a value contradicts several clear experiments showing a distinct $n \approx 3$ regime (Barnes et al. 1971; Russell-Head and Budd 1979; Montagnat and Duval 2004; Song et al. 2005). Dislocation creep mechanisms are governed by the Orowan relation, Eq. 3.1, which implies that n depends on $r + q$, as defined in Eq. 3.2. In a steady deformation at high stress, dislocation density supposedly increases with τ such that $r = 2$, giving $n = r + q = 3$ (Weertman 1983; Alley 1992a). Given, however, that recrystallization processes prevent deformation from being steady at the grain scale, this argument may not apply. Regardless, models of polycrystal deformation that account for dislocation motion of all types suggest that basal glide accomplishes most of the deformation, even when the stress shifts onto crystals oriented unfavorably for it (Duval and Castelnau 1995). Dislocation creep is therefore consistent with observations of fabric development, which reflect a predominance of basal glide (Section 3.3.4).

2. At low stress and strain, dislocation mechanisms account for deformation but the dislocation density maintains a constant value ($r = 0$), giving $n = r + q = 1$. This linear behavior is known as *Harper-Dorn creep* (Lliboutry and Duval 1985). Some laboratory experiments show such a regime at low stresses (Song et al. 2005). More experiments suggest $1 < n < 2$, perhaps a variant of Harper-Dorn creep with $0 < r < 1$ (Pimienta and Duval 1987). Pimienta and Duval (1987) and Montagnat and Duval (2000) suggested that grain boundary migration controls the dislocation density in this regime; the lattice that grows on the trailing side of a moving boundary should contain few dislocations. Such a process would make the effective viscosity sensitive to grain size.

3. Basal glide deformation is supplemented, accommodated, and perhaps rate-controlled by grain boundary sliding. Goldsby and Kohlstedt (1997, 2001) identified this regime using experiments on ice with extremely fine grains. They found $n = 1.8$. Low stress, high temperature, and fine grains favor this regime. The validity of extrapolating Goldsby and Kohlstedt's experimental results to realistic grain sizes for glaciers remains unclear. Regardless, grain boundary sliding seems likely if the boundaries are planar or uniformly curved. In this regime, effective viscosity should depend on grain size. Grain boundary sliding provides the simplest way for neighboring grains to rotate independently (Ignat and Frost 1987). A regime with combined basal glide and boundary sliding therefore favors the formation of c-axis fabrics (Zhang et al. 1994); Figure 3.8, for example, shows grain-boundary sliding accommodating grain rotation. If, however, grain-boundary sliding replaces basal glide as the dominant deformation mechanism, fabric development should be suppressed.

4. Basal glide deformation is accommodated by a combination of grain-boundary processes, including grain-boundary sliding, diffusion along grain boundaries, and grain-boundary migration. Which of these controls the overall rate depends on conditions, in ways not yet elaborated.

Regardless of the exact mechanisms, experimental evidence indicates that grain-boundary processes facilitate deformation at temperatures greater than about $-10\,°\mathrm{C}$ (Barnes et al. 1971; De La Chapelle et al. 1999). As temperatures approach the melting point in laboratory tests, ice softens to a much greater degree than expected from extrapolation of measurements at low temperatures. In other words, the apparent activation energy for ice creep increases at temperatures above $-10\,°\mathrm{C}$. As temperature increases, grain boundaries become wider and contain more liquid; these changes facilitate grain-boundary sliding, diffusion along grain boundaries, and grain-boundary migration.

Because the mechanisms dependent on grain size operate most readily with small grains, it is reasonable to expect that the viscosity of a fine-grained ice ($D \sim 1\,\mathrm{mm}$) depends on the grain size.

In deforming ice grains, the dislocation density evolves over time as a function of cumulative strain, the migration of grain boundaries, and polygonization (Montagnat and Duval 2000). All of these processes may therefore influence the creep properties of ice, by increasing the density of dislocations contributing to deformation or, conversely, by removing dislocation tangles.

3.3.5.2 Deformation at Low Stresses

Whereas basal glide is generally thought to dominate deformation at stresses greater than about $50\,\mathrm{kPa}$, little information has been acquired, and no consensus established, about the principal deformation mechanism at lower stresses. *Diffusional creep* operates more readily than basal glide in conditions of low stress and temperature – molecules diffuse away from grain boundaries with high compressive normal stress and toward boundaries with lower compressive stress. The molecules can move along grain boundaries or through the crystal volume. Calculations based on theoretical treatments suggest that, for fine-grained ice ($D = 10^{-3}$ m), diffusional creep is comparable to dislocation creep at $\tau = 20\,\mathrm{kPa}$, and still nontrivial at $\tau = 50\,\mathrm{kPa}$ (Duval et al. 1983; Goodman et al. 1981). (Such stresses probably occur in the middle and upper layers of the slow-flowing portions of ice sheets.) Goldsby and Kohlstedt (2001) concluded, instead, that grain-boundary sliding, accommodated by basal glide, accounts for deformation at low stresses. In the low-stress end member of this case, individual grains do not deform at all, but a polycrystal deforms as grain centers move with respect to one another.

In some conditions, however, the change to a low-stress deformation regime might occur at stresses too small to be relevant to glacier dynamics at all. Some laboratory experiments on ice specimens deformed to large strains show a consistent $n = 3$ behavior through the entire range of glaciological stresses – from about 20 to $300\,\mathrm{kPa}$ (Russell-Head and Budd 1979).

This indicated that the same mechanisms operating at high stresses also operated at low stresses. In these experiments, D was 5 mm and T ranged from -20 to $-2\,°C$.

3.4 Bulk Creep Properties of Polycrystalline Ice

Effects of the deformation mechanisms discussed in the previous section sum over a very large number of grains to give the bulk creep behavior of polycrystalline ice. Here we will briefly summarize the basic behaviors observed in laboratory deformation experiments and then review more extensively what is known or argued about isotropic and anisotropic creep relations relevant to glacier flow. We emphasize that "a universal constitutive law for ice does not exist ..." (Paterson 1983).

The creep relation for ice and the experiments used to investigate it connect strain rates with deviatoric stresses – the deviations of stress from the mean pressure acting in all directions. Many experiments have confirmed that strain rates are independent of pressure to a very good approximation.

3.4.1 Strain Rate and Incompressibility

Consider a body of ice deforming continuously as a fluid. In three dimensions, the velocity of flow, $\vec{u}$, has components $[u_x, u_y, u_z]$, usually written $[u, v, w]$. The velocity's spatial gradients determine the rate of deformation, described by the strain rate tensor $\dot{\epsilon}$. It has nine components (six independent), given by

$$\dot{\epsilon}_{jk} = \frac{1}{2}\left[\frac{\partial u_j}{\partial x_k} + \frac{\partial u_k}{\partial x_j}\right] \tag{3.7}$$

where indices j and k stand for x, y, or z; and $x_x = x$, $x_y = y$, and $x_z = z$. Thus, for example,

$$\dot{\epsilon}_{xz} = \frac{1}{2}\left[\frac{\partial u}{\partial z} + \frac{\partial w}{\partial x}\right] \tag{3.8}$$

$$\dot{\epsilon}_{xx} = \frac{\partial u}{\partial x}. \tag{3.9}$$

The diagonal components of the strain rate, $\dot{\epsilon}_{xx}, \dot{\epsilon}_{yy}$, and $\dot{\epsilon}_{zz}$, describe the stretching or compression parallel to coordinate axes (positive for stretching, negative for compression). The sum $\dot{\epsilon}_{xx} + \dot{\epsilon}_{yy} + \dot{\epsilon}_{zz}$ gives the fractional rate of volume change of the deforming ice. This number, called the *first invariant* of $\dot{\epsilon}$ and symbolized $\dot{\epsilon}_{\mathrm{I}}$, is independent of the orientation of coordinate axes. Because ice in glaciers is nearly incompressible (Section 2.2.1),

$$\dot{\epsilon}_{\mathrm{I}} = \frac{\partial u}{\partial x} + \frac{\partial v}{\partial y} + \frac{\partial w}{\partial z} = 0. \tag{3.10}$$

A more general relation, accounting for compressibility of firn and mass loss by melt, is discussed in Chapter 8.

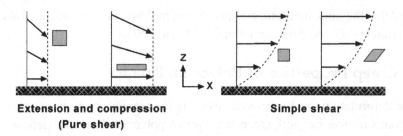

**Extension and compression
(Pure shear)**   **Simple shear**

Figure 3.10: How a block of ice (shaded gray) deforms in pure shear and simple shear as it moves in a glacier. Vectors in both cases show the pattern of ice flow in a glacier constrained by a rigid boundary, such as the bed. In these examples, the ice slides along the boundary in the pure shear case but not in the simple shear case.

Figure 3.10 illustrates how a block of ice, initially square, deforms as it moves in a glacier beside a rigid boundary. Two important situations are shown in the xz-plane. The first involves only compression and extension and is referred to as *pure shear*; it has $\dot{\epsilon}_{xx} = -\dot{\epsilon}_{zz}$ and $\dot{\epsilon}_{xz} = \dot{\epsilon}_{zx} = 0$. The second is simple shear, for which $\dot{\epsilon}_{xz} = \dot{\epsilon}_{zx} \neq 0$ but $\dot{\epsilon}_{xx} = \dot{\epsilon}_{zz} = 0$.

3.4.2 Deviatoric Stress

For an incompressible material such as ice, deformations depend not on the full stress, σ, but on the *deviations* of the stress from an isotropic state. The magnitude of the isotropic stress corresponds to the mean normal stress, sometimes called *spherical stress* and here symbolized σ_M:

$$\sigma_M = \frac{1}{3}\left[\sigma_{xx} + \sigma_{yy} + \sigma_{zz}\right]. \tag{3.11}$$

Normal stresses are defined as positive in tension. (The term *pressure* (denoted P) usually indicates the value of σ_M defined as positive for compression; hence $P = -\sigma_M$.) The deviations are called the *deviatoric stresses* and symbolized τ. Components are:

$$\begin{pmatrix} \tau_{xx} & \tau_{xy} & \tau_{xz} \\ \tau_{xy} & \tau_{yy} & \tau_{yz} \\ \tau_{xz} & \tau_{yz} & \tau_{zz} \end{pmatrix} = \begin{pmatrix} \sigma_{xx} - \sigma_M & \sigma_{xy} & \sigma_{xz} \\ \sigma_{xy} & \sigma_{yy} - \sigma_M & \sigma_{yz} \\ \sigma_{xz} & \sigma_{yz} & \sigma_{zz} - \sigma_M \end{pmatrix} \tag{3.12}$$

From this definition, the normal components of the deviatoric stresses sum to zero:

$$\tau_{xx} + \tau_{yy} + \tau_{zz} = 0. \tag{3.13}$$

Positive values of τ_{xx}, τ_{yy}, and τ_{zz} indicate deviatoric tension.

3.4.3 Bench-top Experiments: The Three Phases of Creep

Figure 3.11 shows a typical creep curve – a graph of strain rate against time or strain – obtained when a constant stress is applied to a sample of polycrystalline ice. The sample has an isotropic

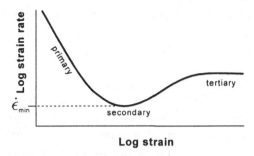

Figure 3.11: Typical shape of creep curves for polycrystalline ice. The minimum strain rate, which usually occurs at a strain of about 0.01, is commonly used as the reference for comparisons between different experiments and samples. Adapted from Budd and Jacka (1989).

c-axis fabric at the beginning of the experiment. An initial elastic deformation (not shown) is followed by a period of *transient* or *primary creep*, in which the strain rate decreases continuously until reaching a minimum value, the *secondary creep* rate. The strain rate then increases (*accelerating tertiary creep*) and, if the test continues for a sufficient time, eventually reaches a steady value (*steady tertiary creep*).

Some important points arising from analyses of many such curves are the following:

1. Given an initially isotropic fabric, the creep rate attains its minimum value at a total strain of about 1%, irrespective of the stress and temperature (Jacka 1984a). This initial stiffening originates from formation of dislocation tangles and from the shifting of stress onto crystals oriented unfavorably for deformation (Duval et al. 1983). At such a small strain, the stiffening cannot reflect a change of the fabric.

2. The minimum creep rate of the polycrystalline sample is smaller, by a factor of 100 or more, than the steady creep rate of a single crystal oriented for basal glide, if compared at the same stress and temperature (Butkovich and Landauer 1958; Duval et al. 1983). This shows that the deformation of the polycrystal is rate-limited by deformations in hard directions or by accommodation mechanisms. Thus preferred orientation fabrics can strongly influence the bulk viscosity and make it anisotropic (Lile 1978; Lliboutry and Duval 1985).

3. Serious misinterpretations can arise from experiments carried out only to small strains if a clear secondary creep minimum is not attained. The strain achieved in a given time varies systematically with the magnitude of the applied stress. But, in the initial transient stage, strain rate varies systematically with the strain. This introduces a systematic bias in the apparent relationship between strain rate and stress; the bias reduces the apparent value of n, the exponent on stress.

4. Softening mechanisms gain importance as strain accumulates. At secondary creep – when strain rate is minimized – softening mechanisms balance the stiffening mechanisms

noted in item 1. Softening results from formation of new grains by recrystallization; new grains are free of dislocation tangles and are oriented favorably for basal glide. Recrystallization begins at small total strains in ice due to intense loading on crystals with hard orientations (Duval et al. 1983).

5. The minimum creep rate is the most commonly used reference strain rate for comparisons between experiments. But glacier flow produces large strains, so tertiary creep is the more relevant phase in the shearing layers that generally control glacier flow (though not in the upper layers of ice sheets). Most experiments are not completed to such large strains, however. Values for parameters such as the exponent on stress in the creep relation may differ for secondary and tertiary creep, but experimental data for tertiary creep are too sparse for a rigorous test of this possibility. Moreover, in tertiary creep the fabric is no longer isotropic, a fact that complicates comparisons.

6. In tertiary creep, the fabric evolves by rotation and recrystallization, as described in Section 3.3.4.

7. Steady tertiary creep may be attained after total compressive strains of 10% to 15% (Gao and Jacka 1987; Jacka and Maccagnan 1984). At this stage, the creep rate significantly exceeds the minimum of secondary creep. In compression, the tertiary rate is three to four times the minimum; in shear, eight to ten times (Budd and Jacka 1989; Li Jun et al. 1996). The development of preferred orientation fabrics accounts for most of the softening.

8. In high-stress experiments, grain sizes evolve toward an equilibrium value that depends on the level of stress (Jacka and Li 1994).

9. Laboratory experiments do not achieve the high strain values expected in the dynamically important zones of glaciers. The properties of ice in glaciers, deforming at large strains, reflect an interplay of fabric, recrystallization, grain size, grain boundaries, and dislocation density and mobility. It is not clear that laboratory experiments ever reproduce these conditions. Thus, results of laboratory experiments should not be applied to problems in nature without field evaluation and calibration. Yet laboratory experiments provide the only way to systematically examine creep properties as a function of controlling variables. Comparisons of secondary creep rates across a range of values for stress and temperature give the basic power-law creep relationships discussed henceforth.

3.4.4 Isotropic Creep Behavior

3.4.4.1 Behavior in Simple Stress States: Glen's Law

Numerous laboratory experiments have examined the behavior of ice at secondary creep. They show that, at stresses important in normal glacier flow (50 to 150 kPa), the relation between a dominant shear stress τ and the corresponding shear strain rate $\dot{\epsilon}$ follows a power law:

$$\dot{\epsilon} = A \tau^{n}. \tag{3.14}$$

The *creep exponent*, n, is approximately constant, but the *creep parameter*, A – also called the *flow parameter* or the *prefactor* – depends strongly on temperature and fabric. In some cases A also varies with grain size and impurity content. We will discuss the controls on A in Section 3.4.5.

Equation 3.14 is usually called *Glen's Law*, to acknowledge the pioneering experiments by Glen (1955). Though well established, and consistent with dislocation theory, the power-law form of the relation is an empirical fit to laboratory and field data for conditions typical of glaciers. Referring to Eq. 3.14 as a "law" masks the large range of values for A and n obtained in different experiments; at a given stress and temperature, measured strain rates differ by about a factor of ten (Weertman 1973b, Figure 4). Values of n range from 1.5 to 4.2 (Weertman 1973b, Table 3.2; Weertman 1983, Table 3.1), with a mean of about 3. Because a value of 3 is also most consistent with field data, analyses of glacier dynamics usually assume that $n = 3$. Such a high value for n means that glacier flow differs markedly from that of a Newtonian viscous fluid.

3.4.4.2 Evidence Supporting $n = 3$ for Glacier Dynamics Problems

The large-scale flow of ice bodies depends primarily on deformation in their most stressed regions – usually the basal layers of glaciers, ice sheets, and ice streams; the side margins of glaciers and ice streams; and the sides and upper halves of ice shelves. In these regions, stresses typically range from 50 to 150 kPa, temperatures from -20 to $0\ ^{\circ}$C, and grain sizes from 1 to 20 mm. The effective value of n in these conditions can be discerned from analyses of field data, though the results of any one study are often ambiguous. In all such studies, stress is not measured but inferred from the balance of forces.

A particularly useful source of information is the spreading of ice shelves (Section 8.9.3), because the basal drag is known to be zero except at isolated spots where the shelf runs aground. Figure 3.12 shows the covariation of strain rate and stress found in two analyses of Antarctic ice shelves (Thomas 1973b; Jezek et al. 1985). Our analyses of the numbers indicate an n value in the range 2 to 3. The original authors argued that $n \approx 3$ is the best interpretation, but incorporating the lowest stress values from Thomas into the larger data set of Jezek et al. shifts the value close to 2.

In grounded glaciers flowing by simple shear deformation on subhorizontal planes – the usual situation in an ice sheet or along the center line of a valley glacier – the strain rate varies over depth with a pattern that depends on n (Section 8.3.1). The tilt of boreholes can be used to examine this problem. Temperate glaciers, being isothermal, are particularly useful since the dependence of A on temperature does not complicate the pattern. These studies consistently find a strong concentration of shearing in deep layers, much more than expected for a linear viscous fluid. A value $n = 3$ to 4 best describes the data (Raymond 1980).

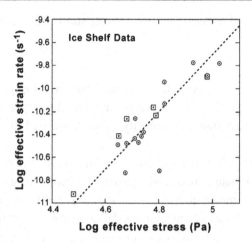

Figure 3.12: Covariation of effective stress and strain rate for locations on spreading ice shelves. Circles: Ross Ice Shelf data from Jezek et al. (1985). Squares: data for five ice shelves, reported by Thomas (1973b). Broken line corresponds to creep exponent $n = 2.5$.

As we will discuss in Section 3.4.4.5, an exponent $n > 1$ implies that every stress component acts to reduce the viscosity of the ice, and so increases strain rates of all orientations. Thus a borehole tilts more rapidly in a complex stress regime than in a simple one. Nye (1953) demonstrated this behavior using measurements of a tilting borehole in the Jungfraufirn. Raymond (1973) accounted for this effect in an analysis of the spatial variations of deformation within Athabasca Glacier, a temperate glacier in the Canadian Rockies. He found a best value of $n \approx 4$ (Raymond 1973).

In cold polar ice, the confounding effects of temperature variations can be minimized by examining small ice caps. As a further advantage, these features have frozen beds and basal slip can be assumed negligible. Martin and Sanderson (1980) analyzed flow and morphology of small Antarctic ice caps and concluded that $n \approx 3$ was needed to fit the data. Thomas et al. (1980) did the same for Roosevelt Island, Antarctica, and found n in the range 3 to 4. There was no evidence in either study for a change of n at low stresses.

Though potentially confounded by temperature variations, analyses of large ice sheet morphologies are worthwhile – especially because the overall flow depends only on the temperatures near the bed. It has long been recognized that the value of n governs the gross shape of the ice sheet surface profile, regardless of the accumulation rate distribution (Vialov 1958; Haefeli 1961; Section 8.10). In geometrically simple regions of the ice sheets, surface profiles between the margins and the central divides are consistent with n values in the range 3 to 6 (Cuffey 2006b). Reeh et al. (1985) combined flow and morphology data to analyze the flow-line leading to Dye 3 in Greenland. They found $n = 2.5$, which, given uncertainties, was consistent with $n = 3$. Likewise, Reeh and Paterson (1988) analyzed a flow-line on Devon Island ice cap and found that $n \approx 3$.

For an ice sheet's central divide, Raymond (1983) pointed out that, if $n > 1$, the deep layers should drape over a region of stiff ice near the bed, forming a bump. Such features have been observed (Section 8.9.1).

The most difficult approach for learning n is the analysis of tilt of deep ice sheet boreholes. All the variables affecting the viscosity – the temperature and fabric, all the stress components, and perhaps the grain sizes and impurities – need to be determined *and* their effects on viscosity understood. Paterson (1983) attempted such an analysis for the Byrd Station and Camp Century boreholes, and found the data to be consistent with $n = 3$. The particular interest of that study was the very low stress magnitudes addressed (20 to 45 kPa).

A number of published analyses of borehole tilt in polar settings are not credible because many of these factors could not be constrained. The general problem faced by all field assessments of n is the covariation of stress with other factors influencing the viscosity.

Field measurements of the closure of boreholes and tunnels in ice provide another constraint on n (Nye 1953; Paterson 1977). Closure occurs because of the pressure difference between a void and the surrounding ice which supports the weight of overburden. Nye (1953) developed the framework for such analyses (Eq. 6.15). (Note that the deformation in these situations is not steady, and the measured closure rates reflect a combination of primary, secondary, and tertiary creep in the surrounding ice.) These analyses have consistently supported the use of $n = 3$. Most borehole data, however, refer to situations with stresses significantly greater than 100 kPa.

In summary, our review of the available literature indicates a best value of $n = 3$ for ice dynamics problems, but a plausible range of 2 to 4. This conclusion applies to the rapidly deforming layers that control glacier motion. There are, however, many glaciological problems for which the much slower deformations in low-stress regions – like the top halves of ice sheets – need to be known. Interpretation of ice core paleoclimate records is one example. So far, field measurements addressing this problem are too few to draw any conclusions. To consider this problem further we now turn to a more detailed look at possible variations of n itself.

3.4.4.3 Behavior in Simple Stress States: Stress-dependent n

A more complex view of the creep relation would be welcome if it helps to explain the large range of A and n values observed in experiments. Polycrystalline ice deforms by dislocation creep at high stresses. A reduction in stress allows other mechanisms to gain importance, and, at a sufficiently low stress, one of these will dominate. Possibilities include grain-boundary sliding coupled with basal glide, Harper-Dorn creep, and diffusional creep (Section 3.3.5). Thus for problems encompassing both high and low stress conditions it is probably necessary to replace Eq. 3.14 with

$$\dot{\epsilon} = A_1 \tau^{n_1} + A_2 \tau^{n_2}. \tag{3.15}$$

A low-stress mechanism with $n_1 < 3$ would allow softer ice and more deformation at low stresses than predicted by Glen's Law with $n = 3$. Furthermore, for reasons given in Section 3.3.5, the low-stress mechanism is likely to depend on grain size.

The relative importance of the two terms in Eq. 3.15 would vary significantly with stress, temperature, and grain sizes over ranges relevant to glaciers. Thus, experiments focused on a subset of these ranges would yield an apparent value of n and an apparent grain-size dependence (or lack thereof) that have no general significance. Goldsby and Kohlstedt (2001) suggested that Glen's Law with $n = 3$ represents an approximation to Eq. 3.15 in a transition zone where two mechanisms contribute, and $n_1 = 1.8$ and $n_2 = 4$. They may be right, although, as discussed previously, the evidence for a distinct $n = 3$ regime is strong in some cases.

Some experiments, in addition to those of Goldsby and Kohlstedt (2001), support a relationship like Eq. 3.15. However, the data currently available lack coherence. Montagnat and Duval (2004) summarized laboratory data on polar ice that show a transition from $n = 3$ to $n \approx 2$ when stress drops below approximately 200 kPa (Pimienta and Duval 1987; Jacka 1984a; Jacka and Li Jun 2000). They explained this softening as a consequence of grain boundary migration removing dislocation tangles (a form of Harper-Dorn creep; Section 3.3.5). The experiments of Jacka (1984a), in particular, show the transition and show $n \approx 2$ in the range 50 to 200 kPa. Grains in all these experiments were fine, with $D \approx 2$ mm. In experiments on manufactured ice, with $D \approx 5$ mm, Song et al. (2005) found a clear transition from $n = 3$ at high stress to a Harper-Dorn creep ($n \approx 1.1$) at low stress. Colbeck and Evans (1973) deformed samples of fine-grained ($D \approx 2$ mm) temperate ice, which they extracted from beneath a glacier and analyzed on-site. They found $n \approx 1$ to 2 for the stress range 10 to 100 kPa. Holdsworth and Bull (1970) also analyzed fine-grained but cold ice, at Meserve Glacier, Antarctica, and concluded $n \approx 2$ at stresses below 50 kPa. In contrast to the previous results, this was based on field measurements of closure and tilt.

Such evidence for regime transitions and low n values stands in direct contradiction to some of the most careful and comprehensive laboratory experiments, which show no evidence for reduced n at low stresses. Russell-Head and Budd (1979) completed experiments to large strains and found $n \approx 3$ across the whole stress range of 20 to 100 kPa, and across a wide range of temperatures. In this ice, $D \approx 5$ mm, perhaps too coarse for grain-size-sensitive mechanisms to matter. The previously cited field measurements that encompass stresses less than 50 kPa show no reduction of n, although $n \approx 2$ appears to be consistent with the ice shelf spreading data (Section 3.4.4.2; Thomas 1973b; Jezek et al. 1985; Paterson 1983).

As this review makes clear, the factors influencing n in field and laboratory experiments remain poorly understood. At present it seems best, for ice dynamics problems, to assume a single regime (Eq. 3.14) with $n = 3$ if isotropic ice and a simple stress state can be assumed. At stresses less than about 100 kPa, a transition to some other behavior with $1 < n < 3$ should be regarded as possible, or even likely (cf. Alley 1992a). Analyses requiring a detailed creep relation for low stresses might explore equations like Eq. 3.15. Regardless, analyses at all stress regimes have to confront the very large effects of anisotropy, discussed later in this chapter.

3.4.4.4 General Stress Regimes: Nye-Glen Isotropic Relation

The state of stress in glaciers can be complex, with combined shear and normal stresses acting in all three dimensions. The creep relation given by Eq. 3.14 or 3.15 applies to simple cases with only one component of applied stress, and needs to be generalized for application to glaciers. Nye (1957) discussed how to do this, for ice that is incompressible and isotropic.

Each strain rate component is assumed to be proportional to its corresponding deviatoric stress component:

$$\dot{\epsilon}_{jk} = \lambda\, \tau_{jk}. \tag{3.16}$$

In isotropic ice, the effective viscosity does not depend on the orientation of the deformation. Thus the proportionality λ has the same value for all components, although it varies from place to place depending on factors like stress and temperature. It follows from Eq. 3.16 and the definition of τ (Eq. 3.13) that $\dot{\epsilon}_{xx} + \dot{\epsilon}_{yy} + \dot{\epsilon}_{zz} = 0$. Thus, Eq. 3.16 automatically contains the criterion of incompressibility (Eq. 3.10).

The creep relation is a physical property of the deforming material and cannot be affected by the choice of coordinate axes. The proportionality λ must therefore be a function of "invariants." For any tensor like τ or $\dot{\epsilon}$, a rotation of the coordinate axes changes the values of the individual components, but in a particular interdependent way. Just as the length of a vector does not vary with the choice of coordinate system, certain scalar measures of the magnitude of a tensor do not vary. A second-rank tensor like τ has three such invariants. The first invariant is the sum of the diagonal components – zero for both τ and $\dot{\epsilon}$ (Eqs. 3.10 and 3.13). The second invariant is an additive measure of total magnitude, analogous to the length of a vector. Glaciologists usually call the second invariants of τ and $\dot{\epsilon}$ the *effective stress* τ_E and *effective strain rate* $\dot{\epsilon}_E$. By definition:

$$\dot{\epsilon}_E^2 \equiv \frac{1}{2}\left[\dot{\epsilon}_{xx}^2 + \dot{\epsilon}_{yy}^2 + \dot{\epsilon}_{zz}^2\right] + \dot{\epsilon}_{xz}^2 + \dot{\epsilon}_{xy}^2 + \dot{\epsilon}_{yz}^2 \tag{3.17}$$

$$\tau_E^2 \equiv \frac{1}{2}\left[\tau_{xx}^2 + \tau_{yy}^2 + \tau_{zz}^2\right] + \tau_{xz}^2 + \tau_{xy}^2 + \tau_{yz}^2. \tag{3.18}$$

(Some experimental results are reported in terms of the quantity $\sqrt{2/3}\tau_E$, called the *octahedral shear stress*.)

A creep relation for complex stress systems must connect quantities that describe the overall state of stress and strain rate. Nye, following Odqvist (1934; 1966, p. 21), proposed that the effective stress and strain rate obey the observed power-law behavior for ice (Eq. 3.14), so that:

$$\dot{\epsilon}_E = A\, \tau_E^n. \tag{3.19}$$

This relation reduces to Eq. 3.14 when only one component pair, $\dot{\epsilon}_{jk}$ and τ_{jk}, is nonzero. The generalized relation is therefore consistent with the simple one. From Eqs. 3.16, 3.18, and 3.17, it follows that

$$\dot\epsilon_E = \lambda \tau_E \tag{3.20}$$

and, by Eq. 3.19,

$$\lambda = A \tau_E^{n-1}. \tag{3.21}$$

Thus the strain rates depend on deviatoric stresses according to

$$\dot\epsilon_{jk} = A \tau_E^{n-1} \tau_{jk}. \tag{3.22}$$

This is the *generalized Glen's Law* or the *Nye-Glen Isotropic Law*, the most commonly used creep relation for glacier ice. The derivation may be repeated to likewise generalize Eq. 3.15, giving $\dot\epsilon_{jk} = [A_1 \tau_E^{n_1-1} + A_2 \tau_E^{n_2-1}]\tau_{jk}$.

3.4.4.5 Implications of Non Linearity for Creep Relation

As Eq. 3.22 with $n = 3$ shows, each strain-rate component is proportional not only to its corresponding deviatoric stress but also to the square of the effective stress. The effective stress, in turn, increases with each deviatoric stress component. Thus a given deviatoric stress produces a larger strain rate in the presence of other stresses than if acting by itself. (This holds for any $n > 1$.) For example, the rate of shearing along horizontal planes (rate $\dot\epsilon_{xz}$) increases if the ice simultaneously compresses or stretches; in this case, τ_{xz}, τ_{xx}, and τ_{zz} all contribute to τ_E. As a second example, consider that a tunnel in a glacier closes under the pressure of the overlying ice. Large longitudinal compressive stresses develop at the foot of an icefall. Thus a tunnel in such a place should close up more rapidly than it would at the same depth elsewhere in a glacier, where prevailing stresses are small. This has been confirmed on Austerdalsbre in Norway (Glen 1958). Again, the flow of ice under its own weight tends to eliminate crevasses, waves, or large hummocks on the glacier surface. These processes, too, should proceed more rapidly at the foot of an icefall than elsewhere.

Moreover, the presence of additional stress components effectively changes the form of the relation between a strain-rate component and its corresponding deviatoric stress. For example, if τ_{xz} is the only nonzero deviatoric stress, then $\tau_E = \tau_{xz}$, and, from Eq. 3.22, $\dot\epsilon_{xz}$ varies as τ_{xz}^3. On the other hand, with a longitudinal deviatoric component τ_{xx} that is large compared to τ_{xz}, then $\tau_E \approx \tau_{xx}$ and $\dot\epsilon_{xz}$ varies as $\tau_{xx}^2 \tau_{xz}$; now $\dot\epsilon_{xz}$ varies linearly with τ_{xz}. This illustrates the complicated effects of a non-linear creep law. It also demonstrates the necessity of taking into account all stress components when trying to determine the value of n from measurements of deformation in a glacier. Several published analyses have failed to do this.

The creep relation is sometimes written in the inverse form

$$\tau_{jk} = 2\eta \dot\epsilon_{jk}, \tag{3.23}$$

where η defines an effective viscosity. In contrast to a Newtonian fluid, for which $n = 1$ and viscosity does not vary with stress, η for ice varies as

$$\eta = \frac{1}{2} \left[A \tau_E^{n-1} \right]^{-1} \tag{3.24}$$

In other words, ice effectively softens as the magnitude of deviatoric stress increases. Equivalent to Eq. 3.23 is

$$\tau_{jk} = A^{-[1/n]} \dot{\epsilon}_E^{[1-n]/n} \dot{\epsilon}_{jk}. \tag{3.25}$$

This formula is widely applied because field measurements determine strain rates, not stresses.

3.4.4.6 Plasticity Approximation

Together with experimentally determined values for A, the non-linearity $n = 3$ implies that deformation rates in a glacier increase strongly as the effective stress rises to about $100\,\text{kPa}$. With this level of stress, glaciers flow easily enough to redistribute mass and prevent gravitational stresses from rising much farther. It is therefore sometimes useful to idealize ice as a *perfectly plastic* material, in which a steadily increasing stress causes no deformation until τ_E attains a *yield stress*, τ_o. At this point, ice deforms readily, the stress cannot rise farther, and the strain rate takes a value determined by other factors like kinematic constraints on flow. Mathematically, this behavior represents the limit $n \to \infty$. Figure 3.13 compares this approximation to other creep relations. The yield stress τ_o for ice is of order $100\,\text{kPa}$, but the best value depends on variables affecting the viscosity. In particular, τ_o should be larger for colder ice.

3.4.4.7 Special Cases of Equation 3.22 with $n = 3$

1. Simple shear:

$$\sigma_{xx} = \sigma_{yy} = \sigma_{zz} = \sigma_o; \quad \tau_{xy} = 0$$
$$\dot{\epsilon}_{xz} = A \left[\tau_{xz}^2 + \tau_{yz}^2 \right] \tau_{xz}; \quad \dot{\epsilon}_{yz} = A \left[\tau_{xz}^2 + \tau_{yz}^2 \right] \tau_{yz}$$
$$\text{and if} \quad \tau_{yz} = 0, \quad \text{then} \quad \dot{\epsilon}_{xz} = A \tau_{xz}^3 \tag{3.26}$$

This stress system applies in the bottom layers of glaciers and ice sheets that are not slipping rapidly on their beds or negotiating complex topography.

2. Unconfined uniaxial compression in the vertical (z) direction:

$$\sigma_{xx} = \sigma_{yy} = \sigma_o; \quad \sigma_{zz} = \sigma_o + \hat{\sigma}_z$$
$$\tau_{xx} = \tau_{yy} = -\frac{1}{3} \hat{\sigma}_z; \quad \tau_{zz} = \frac{2}{3} \hat{\sigma}_z$$
$$\dot{\epsilon}_{zz} = \frac{2}{9} A \hat{\sigma}_z^3; \quad \dot{\epsilon}_{xx} = \dot{\epsilon}_{yy} = -\frac{1}{2} \dot{\epsilon}_{zz} \tag{3.27}$$

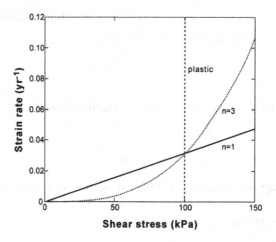

Figure 3.13: Comparison of different types of creep relations. The power-law relation with $n = 3$, appropriate for ice, is intermediate between linear viscous ($n = 1$) and perfectly plastic behaviors. Here, plastic yield stress is 100 kPa (1 bar) and viscosity for the linear relationship is about 10^{14} Pa s.

This stress system applies in most laboratory experiments, and in the near-surface layers of an ice sheet with spreading flow. Comparing Eq. 3.27 with Eq. 3.26 shows that the strain rate produced by a given compression (or tension) is only two-ninths of that produced by an equal shear stress. Laboratory measurements confirm this difference (Li Jun et al. 1996). In isotropic ice, as considered here, the difference arises because the stress, although only applied vertically, partitions into deviatoric components in both horizontal directions. The nonlinearity of Eq. 3.22 reduces the numerical values further. An even larger contrast between shear and uniaxial deformations can occur in anisotropic ice.

3. Uniaxial compression, confined in the y-direction:

$$\sigma_{xx} = \sigma_o; \quad \sigma_{zz} = \sigma_o + \hat{\sigma}_z; \quad \dot{\epsilon}_{yy} = 0$$

$$\dot{\epsilon}_{xx} = -\dot{\epsilon}_{zz}; \quad \sigma_{yy} = \sigma_o + \frac{1}{2}\hat{\sigma}_z$$

$$\tau_{zz} = -\tau_{xx} = \frac{1}{2}\hat{\sigma}_z$$

$$\dot{\epsilon}_{zz} = \frac{1}{8}A\hat{\sigma}_z^3$$

An σ_{yy} value intermediate between σ_{xx} and σ_{zz} must exist to prevent extension in the y-direction. This stress system applies in an ice shelf occupying a bay; an ice stream with a weak bed; the upper layers of valley glaciers; and the upper layers in flank regions of ice sheets shaped like ridges.

4. Shear combined with uniaxial compression, confined in the y-direction:

$$\sigma_{xx} = \sigma_o; \quad \sigma_{zz} = \sigma_o + \hat{\sigma}; \quad \dot{\epsilon}_{yy} = 0; \quad \tau_{xz} = \hat{\tau}; \quad \tau_{xy} = \tau_{yz} = 0$$

$$\dot{\epsilon}_{xx} = -\dot{\epsilon}_{zz}; \quad \tau_{xx} = -\tau_{zz} = -\frac{1}{2}\hat{\sigma}$$

$$\tau_E^2 = \frac{1}{4}\hat{\sigma}^2 + \hat{\tau}^2$$

$$-\dot{\epsilon}_{xx} = \dot{\epsilon}_{zz} = \frac{1}{2}A\tau_E^2\hat{\sigma} = \frac{1}{8}A\left[\hat{\sigma}^3 + 4\hat{\tau}^2\hat{\sigma}\right]$$

$$\dot{\epsilon}_{xz} = A\tau_E^2\hat{\tau} = A\left[\hat{\tau}^3 + \frac{1}{4}\hat{\sigma}^2\hat{\tau}\right]$$

This important stress configuration approximates the regime at many places in both valley glaciers and grounded ice sheets.

3.4.4.8 Role of Third Invariant?

Some incompressible, isotropic creeping materials exhibit more complex behavior than allowed by Eq. 3.22. When the first invariants of τ and $\dot{\epsilon}$ are zero, as implied by incompressibility, the most general creep relationship involves both the second and third invariants of τ (call them $\tau_{II} = \tau_E$ and τ_{III}) and has the form (Glen 1958)

$$\dot{\epsilon}_{jk} = C_1(\tau_{II}, \tau_{III})\,\tau_{jk} + C_2(\tau_{II}, \tau_{III})\left[\tau_{jl}\tau_{lk} - \frac{2}{3}\tau_{II}\delta_{jk}\right]. \tag{3.28}$$

C_1 and C_2 are functions of the invariants, as indicated, and $\delta_{jk} = 1$ if $j = k$ but equals zero otherwise. Nye's assumptions reduced this to $\dot{\epsilon}_{jk} = C_1(\tau_{II})\tau_{jk}$. With a nonzero C_2, the term $\tau_{jl}\tau_{lk}$ allows for compression or stretching normal and parallel to a shear plane, arising from the shear stress itself. Some fluids show this behavior, but there is no evidence of it in ice. Specifying $C_2 = 0$ is justified unless data to the contrary emerge.

The third invariant, τ_{III}, is another measure of the magnitude of τ; in contrast to τ_{II}, however, τ_{III} gauges the magnitude in a multiplicative rather than additive fashion. τ_{III} equals the determinant of τ. For example, in the absence of shear stresses, $\tau_{III} = \tau_{xx}\tau_{yy}\tau_{zz}$; there must be deviatoric normal stresses in all three directions for this quantity to be nonzero. If the flow field is two-dimensional ("plane strain"), a very good approximation in many problems of glacier mechanics, $\tau_{III} = 0$. For a more general case of three-dimensional flow, it is worth considering whether this third invariant affects deformations and should be included. A combination of shear and uniaxial compression makes τ_{III} nonzero. In a laboratory experiment, Li Jun et al. (1996) measured shear deformation while varying the magnitude of uniaxial compression, at an effective stress of $\tau_E = 200\,\text{kPa}$. The results showed no dependence on τ_{III}, consistent with most earlier experiments (e.g., Duval 1981). There are, in addition, theoretical reasons for assuming that C_1 is independent of τ_{III} if $C_2 = 0$.

Anisotropic effects might induce an apparent dependence of strain rates on τ_{III}, for some combinations of stress state and crystal fabric. An anisotropic creep relation would be needed in this case.

3.4.5 Controls on Creep Parameter A

The coefficient A in the creep relation, Eq. 3.22, ranges across several orders of magnitude in terrestrial glaciers. The higher its value, the faster the deformation at a given stress. A general understanding of how physical and chemical properties of ice affect A has not been achieved, either empirically or conceptually. Temperature, the most important variable, is best understood.

3.4.5.1 Effect of Temperature

The dependence of A on temperature is known from many laboratory experiments (e.g., Barnes et al. 1971; Weertman 1983). Ice softens as its temperature increases; best estimates from experiments are that A increases by about a factor of ten between $-30\,°C$ and $-10\,°C$, and then by another factor of five or ten to the melting point. Over the whole range of temperatures in terrestrial ice, A nominally varies by a factor of about 10^3. Such a large sensitivity means that temperatures need to be known well to predict deformation rates – especially in polar ice sheets, where temperatures range from melting point, at the bed, to as low as $-55\,°C$ at the surface. Ice softens as it warms because the mobility of dislocations increases. In addition, grain boundaries become wider and more disordered, and the amount of liquid water they contain probably increases.

At temperatures below $-10\,°C$, a simple Arrhenius relationship describes the temperature dependence:

$$A = A_o \exp\left(-\frac{Q^-}{RT_h}\right). \tag{3.29}$$

Here T_h denotes Kelvin temperature adjusted for melting point depression; if the melting point is less than the standard value 273.15 by an amount δT_m, then $T_h = T + \delta T_m$. The coefficient A_o is the *prefactor*, and Q^- the low-temperature activation energy for creep – distinct from the effective activation energy above $-10\,°C$, which we will call Q^+. Field measurements of glacier ice give a value of $Q^- \approx 60\,kJ\,mol^{-1}$ (Paterson 1977), close to the mean value for laboratory experiments (mean of 15 values is $59\,kJ\,mol^{-1}$; Weertman 1983, Table 3.1). This appears to equal the activation energy for volume self-diffusion (Weertman 1973b, Table 3), the process by which individual H_2O molecules move through the ice lattice. On the other hand, comparisons between some experimental data (Barnes et al. 1971; Goldsby and Kohlstedt 2001) suggest that the activation energy is larger for deformation at high stresses than at low stresses. The field value may therefore reflect an intermediate stress regime in which high-stress and low-stress deformation mechanisms both contribute.

Above $-10\,°C$, ice softens more than predicted by Eq. 3.29. The creep activation energy for single crystals does not increase near the melting point (Jones and Brunet 1978), so the extra softening must arise from the polycrystalline nature of glacier ice. Barnes et al. (1971) attributed it to grain-boundary sliding and the presence of liquid water on the boundaries.

Over the range -10 to $0\,°C$, the laboratory data can be approximated by increasing the apparent activation energy to a value $Q^+ = 152\,\text{kJ}\,\text{mol}^{-1}$ (mean of 5 values; Weertman 1983, Table 1). In detail, however, the apparent Q^+ increases yet more strongly near the melting point (Mellor and Testa 1969, Figure 3.3; Budd and Jacka 1989; Jacka and Li Jun 1994). A tabulated relationship might be best to use in this range, but most analyses of glacier dynamics simply use an Arrhenius relationship with a switch of activation energy from lower to higher value at $-10\,°C$. This is a reasonable approach because, at present, no field evidence supports a more complicated relationship. Using a value of Q^+ in the range 80 to 150 $\text{kJ}\,\text{mol}^{-1}$ gives a range of A values at melting point that includes most field measurements. As we will see in Section 3.4.6, however, the laboratory value $Q^+ = 152\,\text{kJ}\,\text{mol}^{-1}$ is larger than implied by most field measurements.

Ideally the prefactor A_o would be a constant. However, field data show a large variability of A not accounted for by temperature.

3.4.5.2 Effect of Hydrostatic Pressure

Hydrostatic pressure depresses the melting point of ice. For pressure P (positive in compression), the melting point decreases by BP, where $B = 7 \times 10^{-8}\,\text{K}\,\text{Pa}^{-1}$ (Section 9.4.1). Thus, Eq. 3.29 can be written in terms of the actual temperature T and the pressure P as

$$A = A_o \exp\left(-\frac{Q}{R}\frac{1}{[T + BP]}\right). \tag{3.30}$$

The temperature shift BP amounts to about $2\,°C$ beneath 3 km of ice, so high pressure can cause a small but nonnegligible softening of ice – for a given temperature – deep in the ice sheets.

Taking a more general view, the total effect of pressure can be written as

$$A = A_o \exp\left(-\frac{Q + PV}{RT}\right), \tag{3.31}$$

where V defines the *activation volume for creep*. Laboratory experiments give estimates for V of $-1.74 \times 10^{-5}\,\text{m}^3\,\text{mol}^{-1}$ (average from Weertman 1973b, Table 4) and $-1.3 \times 10^{-5}\,\text{m}^3\,\text{mol}^{-1}$ (Durham et al. 1997). To within uncertainties, these values imply the same dependence of A on pressure in Eq. 3.31 as in Eq. 3.30. Thus we can assume, as Rigsby (1958) suggested, that hydrostatic pressure does not affect the creep relation, except through its influence on the melting point.

3.4.5.3 Effect of Water Content

Water softens polycrystalline ice by facilitating adjustments between neighboring grains with different orientations – through processes like grain-boundary sliding and melting and refreezing

(Barnes et al. 1971). Within temperate glaciers the water content varies because of differences in porosity, melt rate, and drainage. Duval (1977) and Lliboutry and Duval (1985) conducted laboratory experiments on samples from a temperate glacier. They examined, in particular, how the percentage water content, W, affects the deformation rate in tertiary creep. The data were fit by

$$A = [3.2 + 5.8W] \times 10^{-24} \tag{3.32}$$

(units of $Pa^{-3}\,s^{-1}$). The average of W was 0.33% in samples from layers near the glacier's bed, the most important layers for flow. The value for A at $W = 0$ implied by Eq. 3.32 lies within the range of field-based estimates for A of temperate glaciers. Most such estimates differ one from another by less than a factor of three (Section 3.4.6). Duval's relation implies that a factor-of-three increase of A corresponds to a change of water content from zero to 1.1%. Local water contents as high as about 3% were measured in Glacier d'Argentiere (Vallon et al. 1976). These results all indicate that water content can influence significantly the viscosity of temperate glaciers.

3.4.5.4 Effect of Density

The density ρ of glacier ice and firn varies with porosity (Chapter 2). In ice with densities greater than the value for pore close-off ($\rho \approx 830\,kg\,m^{-3}$) the porosity has little effect on viscosity. At lower densities, on the other hand, the extra porosity leads to significant softening of the ice; in experiments, strain rates increase by a factor of about ten for a $150\,kg\,m^{-3}$ decrease of density (Figure 3.14; Mellor 1975; Hooke et al. 1988). In part, such softening reflects the focusing of forces onto the smaller ice skeleton, but it mostly reflects greater freedom of movement of the skeleton in the presence of larger pores. This density effect may be important for the dynamics of polar ice masses with thick firn layers, such as some ice shelves and small ice caps. Unlike ice, firn is compressible; the creep relation for firn must be more complex than indicated by Eq. 3.22.

3.4.5.5 Effect of Grain Size

Given the deformation mechanisms that operate in polycrystalline ice, it is reasonable to propose that – for fine-grained ice – the viscosity depends on the grain size, especially at low stresses and high temperatures (Section 3.3.5). In particular, the increased softening of ice at temperatures higher than $-10\,°C$ reflects grain-boundary processes (Barnes et al. 1971), suggesting that viscosity depends on grain size in this regime. Little information is available about the role of D (mean grain diameter) in the creep relation. The high sensitivity of ice viscosity to temperature and fabric obscures grain size effects in most field studies. Most laboratory experiments have been conducted at high stresses, for which no grain-size sensitivity is expected or observed (e.g., Duval and LeGac 1980). As another complication, the grain size changes considerably

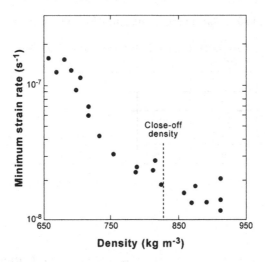

Figure 3.14: Variation of minimum octahedral shear strain rate with initial density. The tests used unconfined, uniaxial compression with an octahedral shear stress of 200 kPa, at −3 °C. Adapted from Hooke et al. (1988); data from experiments reported by Jacka (1994).

over the course of experiments carried out to tertiary creep (Jacka 1984b); the grain size cannot be controlled independently of the stress.

Goldsby and Kohlstedt (1997, 2001) proposed a grain-size-sensitive creep relation. They manufactured extremely fine-grained ice ($D < 0.2$ mm) for use in laboratory tests. The tests revealed a grain-size effect. Extrapolating the results to realistic grain sizes gave viscosities consistent with earlier laboratory experiments. At stresses $\tau > 10$ kPa their relationship reduces to a form of Eq. 3.15, specifically

$$\dot{\epsilon}_{jk} = A_1 \, D^{-p} \tau_{jk}^m + A_2 \, \tau_{jk}^n, \tag{3.33}$$

with $p = 1.4$, $m = 1.8$, and $n = 4$. The A coefficients depend strongly on temperature (see Appendix 3.1). Equation 3.33 was supported well by the experiments with very fine ice. The question is whether it applies to the much coarser ice of glaciers. The finding that $m \approx 2$ is consistent with some, but not all, of the previous experimental results (Section 3.4.4.3). Equation 3.33 implies that ice softens as grains become finer, as generally expected if grain boundary processes limit the deformation rate.

Is Eq. 3.33 supported by available field measurements in glaciers? Field studies reveal viscosity contrasts between adjacent layers in polar glaciers and ice sheets (e.g., Fisher and Koerner 1986; Dahl-Jensen and Gundestrup 1987; Cuffey et al. 2000a). The softer ice invariably has finer grains and higher impurity contents than the stiffer layers. The softer ones also usually have fabrics with strong preferred orientations, and this is a dominating influence (Section 3.4.7; Section 8.3.2; Paterson 1991). Nonetheless, fabric variations do not explain all the observed

Table 3.2: Grain-size contrast effect on strain rate.

Location	Coarse D (mm)	Fine D (mm)	Observed ξ	ξ from Eq. 3.33	Apparent p
Dye 3, W-H transition[†]	5	1	6.1	7.5	0.5
Agassiz, W-H transition	3	1	0.7	3.2	−0.15
Meserve ($\tau_E = 1.0$ bar)	1.8	0.8	0.64	1.9	0.6
($\tau_E = 1.5$ bar)[‡]	1.8	0.8	0.64	1.5	0.6
Byrd, W-H transition	8	4.5	0.0−0.25	1.1	0.39
Camp Century	7	2	−0.38	3.8	−0.38

[†] W-H = Wisconsin-Holocene transition.
[‡] Stress value is poorly constrained.

viscosity contrasts in the ice sheets (Thorsteinsson et al. 1999; Cuffey et al. 2000b) or basal layers of mountain glaciers (Cuffey et al. 2000a).

Table 3.2 provides a quick check on the usefulness of Eq. 3.33 using field measurements from polar sites. The table shows the measured contrast between shear strain rates in adjacent ice layers with different mean grain sizes. This is compared to calculations of the strain rate contrast using Eq. 3.33. For strain rates $\dot{\epsilon}_1$ and $\dot{\epsilon}_2$, the contrast is given as $\xi = \dot{\epsilon}_1/\dot{\epsilon}_2 - 1$. By comparing adjacent layers, we eliminate confounding effects of differences in stress and temperature (but not fabric). Very few data are available for this analysis, but it is apparent that Eq. 3.33 substantially overpredicts the sensitivity to grain size in glaciological settings. This conclusion is strengthened by taking fabric contrasts into account, because the fine-grained ice tends to have fabrics more favorable to shearing. (Only the last column in Table 3.2, the "apparent p" value, attempts a correction for fabrics).

The clearest field evidence for a grain-size dependence of A comes from analysis of the basal layers of Meserve Glacier, Antarctica (Cuffey et al. 2000a). This cold-based mountain glacier picks up rocks and salty frozen soil from its bed, forming basal layers with complex layering and high impurity concentrations (Holdsworth 1974; Cuffey et al. 2000c). In this situation, the grain sizes, impurities, and fabrics correlate to a lesser degree than in the ice sheets; thus the influence of each factor can be discerned better than in the ice sheet studies. The strain rate variations measured in Meserve Glacier were fit to relationships that included particulate and soluble impurity concentrations, grain sizes, and effects of c-axis fabrics. (The latter were calculated from thin-section data using anisotropic creep rules of the sort discussed in Section 3.4.7; see Cuffey et al. 2000a for a description.)

Assuming that the grain-size dependence follows the simplest power relation, $\dot{\epsilon} \propto D^{-p}$, with p a constant, the Meserve Glacier data were consistent with p in the range 0.5 to 0.75, with a best value of $p = 0.60$. This implied that grain size variations accounted for a factor of 1.7 variation of strain rates at the site – twice as large as the variation related to soluble impurity content

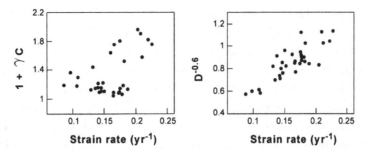

Figure 3.15: Covariation of measured strain rates with ice properties in the basal layers of Meserve Glacier. Left panel: variation with soluble impurity concentration C. Right panel: variation with grain diameter raised to the power -0.6. The coefficient γ and the exponent of D have both been optimized to give the best possible relationships. Adapted from Cuffey et al. (2000a and 2001) and used with permission from the American Geophysical Union, *Journal of Geophysical Research*.

(Figure 3.15). In this analysis, all the known possible controls on viscosity were measured and used in the relations. Only those relations including a direct dependence of viscosity on D could match the measured strain rates.

The result $p = 0.6$ is not supposed to be general, but, for illustration, consider that grain sizes of importance to ice dynamics problems range from about 1 to 10 mm. With $p = 0.6$, this corresponds to a four-fold range of ice viscosities, smaller than the potential influences of temperature and fabric, but nonetheless significant. For the sites given in Table 3.2 an effective p value can be calculated using the measured constrasts of grain size and strain rate. Specifically, for two adjacent layers, the log of the ratio of $\dot{\epsilon} \propto FD^{-p}$ provides a p value, listed in the final column in the table. Here F denotes the fabric effect, estimated from an anisotropic creep relation (Section 3.4.7). Because of limited data, F could only be calculated for Dye 3, Agassiz, and Meserve; we used $F = 1$ at the other sites.

3.4.5.6 Effect of Impurities

For a summary of the types of impurities in glacier ice, see Section 3.3.2.

Soluble impurities sometimes soften ice. Experiments on single ice crystals show that adding a few parts per million of the substitutional impurities HF and HCl increases deformation rates by at least a factor of 10 (Jones and Glen 1969). This happens because substitutional impurities create additional defects in the crystal lattice, facilitating the movement of dislocations. Interstitial impurities, on the other hand, probably do not affect deformation rates.

In polycrystalline ice, segregation of impurities to the grain boundaries contributes to formation of liquid layers, at temperatures above the eutectic point (e.g., Mulvaney et al. 1988). Liquid at boundaries softens ice by facilitating adjustments between grains and mass transport along grain boundaries; this is the effect discussed previously for "pure" ice at temperatures greater than about $-10\,^{\circ}$C (Barnes et al. 1971). By increasing the liquid content of grain boundaries, impurities could have the same effect as an increase of temperature. This idea was originally supported by Raraty and Tabor (1958), who experimented with ice containing 1% "Teepol" (sodium dodecyl sulfate). Below $-25\,^{\circ}$C the samples behaved like pure ice, whereas at higher temperatures the creep rates increased by a factor of 5. The authors suggested that $-25\,^{\circ}$C was the eutectic temperature of the system. Impurities that may influence creep rates in polar ice sheets include sodium chloride (eutectic temperature $-21\,^{\circ}$C), sulfuric acid ($-73\,^{\circ}$C), and hydrochloric acid ($-75\,^{\circ}$C).

Do soluble impurities influence creep rates in polar glaciers and ice sheets? For the ice sheets, no convincing evidence suggests a quantitatively significant direct effect of soluble impurities. The subject clearly needs more investigation. Ices with high impurity content do tend to be soft; in both the Arctic and Antarctic, impure ice-age ice is softer than adjacent Holocene ice, and impure basal layers are softer than overlying clean ice (Section 8.3.2). However, the impure layers also have strong fabrics, small grain sizes, and abundant particulates; the correlation does not indicate a direct relationship. Moreover, in the Arctic, the soft ice-age ice is alkaline because of reactions with particle surfaces, whereas the stiff Holocene ice tends to be acidic. Thus acids cannot account for softening in this case.

Soluble impurities account for some of the softening of the impure basal layers in Meserve Glacier (Holdsworth 1974; Cuffey et al. 2000a). Concentrations at this site are unusually high, with a mean total ionic content of $200\,\mu$mol$\,$L^{-1}. Moreover, they vary by a factor of approximately 20. In contrast, the strain rate varies by only a factor of about 2, and impurities appear to account for less than half of that. This indicates a low sensitivity of strain rate to the impurity content. A relationship calibrated against the data, and accounting for grain-size effects, gave the increase of creep rate as approximately $A_o = A'_o \left[1 + \gamma C_j \right]$, where C_j denotes the micromolar concentration of impurity j. The parameter γ took values $\gamma = 0.019$ if C_j refers to sulfate, and $\gamma = 0.014$ if calcium. Solute concentrations in Arctic ice caps are much lower, only a few micromolar at maximum. Thus, applying the values found at Meserve Glacier suggests that soluble impurities have almost no effect on creep rates in the ice caps. (Temperature and effective stress at the Meserve Glacier site were approximately $-17\,^{\circ}$C and 150$\,$kPa. Extrapolations to other sites are meaningful only if similar conditions apply.)

How particulate impurities influence creep rates is even less clear. In principle, fine particles enhance the generation of dislocations and so soften the ice (Song et al. 2005), but they also reduce grain boundary migration and sliding. Laboratory experiments for small concentrations of particles have produced conflicting results, with some experiments indicating softening and others stiffening (Nayar et al. 1971; Song et al. 2005). The clearest experiments that show softening were conducted at stresses much higher than those prevailing in glaciers (Song et al. 2008). At large concentrations, particles mechanically inhibit deformation. Stiffening of ice was observed for concentrations greater than about 10% in one set of experiments (Hooke et al. 1972).

As with soluble impurities, glacier layers containing abundant microparticles tend to be soft. These layers, however, are invariably fine-grained and usually have distinct, strong fabrics. In the Meserve Glacier basal layers, where particle concentrations range from 0 to 2%, there is no indication that particles affect ice viscosity directly (Cuffey et al. 2000a).

Both soluble and microparticle impurities probably have *indirect* effects on creep properties that are far more important than their direct effects. As discussed in Section 3.3.3, impurities inhibit grain boundary migration and so influence grain size, dislocation density, and dynamic recrystallization. These influences, in turn, may affect ice viscosity via fabrics, grain size, or perhaps dislocation density (Section 8.3.2).

More effort needs to be devoted to understanding possible impurity effects on ice deformation, both direct and indirect. Effects are likely to vary considerably across the range of stresses, temperatures, impurity contents, and grain sizes relevant to glaciers.

3.4.5.7 Effect of Preferred-orientation Fabric

Preferred-orientation fabrics soften or stiffen ice considerably. In contrast to temperature and other variables, however, this effect cannot be regarded, in general, as a change in the value of A; the effect depends entirely on the orientation of the deformation relative to the fabric. Fabrics simultaneously soften ice for some deformations and stiffen it for others. The form of the creep relation itself must change to accommodate this property. On the other hand, if only one component of deformation is of interest, and the stress regime simple, the fabric effect can be accounted for using a coefficient that multiplies A. For simple shear of ice with a single-maximum fabric oriented normal to the shear plane, this multiple ranges from about 0.9 to about 9. Section 3.4.7 discusses this problem.

3.4.5.8 Enhancement

The term *enhancement* provides a convenient way to talk about variations of strain rate that are not explained by stress and temperature in the Nye-Glen creep relation. For a measured strain rate $\dot{\epsilon}_m$, define the enhancement E as

$$E \equiv \frac{\dot{\epsilon}_m}{\dot{\epsilon}_o}, \tag{3.34}$$

where $\dot{\epsilon}_o$ denotes the strain rate predicted by Eq. 3.22 with A dependent on temperature alone. E is not supposed to be a physical variable, but rather a measure of our inability to predict strain rate in a particular situation. At present, effects of grain size, impurities, fabrics, and possibly other variables all lead to deviations from $E = 1$. In the case of fabrics, the value for E depends on the orientation of the deformation; thus E should not be regarded as an inherent property of the ice (see Section 3.4.7).

The layers of ice-age ice deep in the Arctic ice caps shear more rapidly than expected; measurements reveal E values averaging approximately 2.5, relative to the overlying Holocene ice (Paterson 1991; see Section 8.3.2). E values as high as 12 have been inferred for the shear margins of West Antarctic ice streams (Echelmeyer et al. 1994). Impure basal layers have E ranging from 4, at Dye 3 in Greenland, to 120 beneath a mountain glacier in China (Dahl-Jensen and Gundestrup 1987; Echelmeyer and Wang 1987). Such E values obtained in field studies may be strongly influenced by errors in estimates of stress and temperature. Some, but not all, of these E values can be explained by fabric effects (Section 3.4.7).

3.4.6 Recommended Isotropic Creep Relation and Values for A

A simple reference relationship, a calibrated form of the Nye-Glen rule, needs to be identified for use in ice dynamics analyses and for evaluating how the creep of ice depends on various factors. The relationship is supposed to apply, at minimum, to conditions relevant for global ice dynamics problems (τ between 50 and 150 kPa; T between -20 and 0 °C). It should be no more complex than demanded by the data, but must describe known features of ice creep: (1) a nonlinear dependence of strain rate on stress, with n in the range 2 to 4, and $n = 3$ an acceptable preference; (2) softening as a function of scalar stress magnitude τ_E; (3) softening as temperature increases, described effectively by an Arrhenius relationship with a switch of activation energy at a transition temperature $T_* \approx -10$ °C; (4) additional softening in specific situations such as temperate ice with a high water content and polar ice undergoing simple shear.

The relationship is, for T in Kelvin, Q in J mol^{-1}, P in Pa, and $R = 8.314$ J mol^{-1}K^{-1}:

$$\dot{\epsilon}_{jk} = A E_* \tau_E^{n-1} \tau_{jk} \tag{3.35}$$

$$A = A_* \exp\left(-\frac{Q_c}{R}\left[\frac{1}{T_h} - \frac{1}{T_*}\right]\right)$$

$$n = 3; \quad T_* = 263 + 7 \times 10^{-8} P; \quad T_h = T + 7 \times 10^{-8} P$$

$$Q_c = Q^- = 6 \times 10^4 \text{ if } T_h < T_*; \quad Q_c = Q^+ \text{ if } T_h > T_*.$$

The constant prefactor A_* (the value of A at -10 °C) must be calibrated, and judicious recommendations made for the enhancement factor E_*. For Q^+, the apparent activation energy in warm ice, there is a choice: either it too can be calibrated, or its value assumed from laboratory data. We adopt the former approach because, as we show next, the laboratory-derived values for A of temperate ice appear to be too large.

Calibration should use field data from actively flowing ice masses. Data are available for temperate glaciers, and for polar ice in the temperature range of -23 to -6 °C. The polar data can be translated to one reference temperature using activation energies. Thus the field data can be summarized as two values of A: one for ice at melting point and one for -10 °C. Table 3.3

Table 3.3: Measured and inferred values of creep parameter A at different temperatures, for $n = 3$.

T ($^\circ$C)	A (10^{-25} s^{-1} Pa^{-3})	Method	Reference
0	24	Mean of 5 calibrated models	See below[†]
	38	Mean of 5 borehole tilt values[°]	Raymond 1980
	55	Closure of tunnels	Nye 1953
0	24	**Recommended base value**	
	93	Various lab tests	Budd and Jacka 1989
−2	27	Various lab tests	Budd and Jacka 1989
−10	3.9	Ice shelf spreading	Jezek et al. 1985
	5.3	Ice shelf spreading	Thomas 1973b
	2.5–4.3	Ross Ice Shelf flow[‡]	MacAyeal et al. 1996
	1.8–3.2	Filchner-Ronne Ice Shelf flow[‡]	MacAyeal et al. 1998
−10	7.6	Borehole tilting	Fisher and Koerner 1986
	6.7	Flow-line with borehole	Reeh and Paterson 1988
	8.7	Borehole tilting	Dahl-Jensen and Gundestrup 1987
−10	3.8	Mean of ice shelf values	
	7.7	Mean of simple shear values	
−10	3.5	Various lab tests	Budd and Jacka 1989
	3.5	**Recommended base value**	

[†] Hubbard et al. 1998; Gudmundsson 1999; Adalgeirsdottir et al. 2000; Albrecht et al. 2000; Truffer et al. 2001.

[°] We have calculated A for $n = 3$ using $AR^3 = A_o \tau_o^n$, where A_o, τ_o, and n are values given by original authors, and R is Raymond's corrected stress value. Stresses are for the greatest depth of reported measurements.

[‡] Calibrated model for flow of entire ice shelf (Ross) or part of ice shelf (Filchner-Ronne). Low and high values are for effective temperatures of -15 and -20 $^\circ$C, respectively.

compiles results from the least ambiguous field analyses and, for comparison, summary values from the extensive laboratory experiments of Budd and Jacka (1989).

For $-10\,^\circ$C, we have separated the numbers for ice shelves and for grounded glaciers undergoing simple shear. The MacAyeal numbers correspond to the viscosity parameter found by optimizing finite element models against measured flow for the entire Ross Ice Shelf and one region of the Filchner-Ronne Shelf. The Thomas (1973b) and Jezek et al. (1985) numbers were derived from local spreading rates at sites on, respectively, five ice shelves and the Ross Ice Shelf. Whereas Thomas' values were derived only from measurements of surface strain rates, Jezek et al. used the height of crevasses in the underside of the shelf to deduce the stress resulting from drag of grounded parts, which tends to reduce spreading rate. These heights were measured by radar.

The Dahl-Jensen and Gundestrup (1987) analysis is of tilt of the Dye 3 borehole in Greenland, and refers to Holocene ice only. The Fisher and Koerner (1986) analysis is of tilt of a borehole in the Agassiz ice cap. It refers only to the Holocene ice immediately overlying the ice-age ice.

(We calculated the equivalent A value for the mean tilt at heights 9–12 m in their figure 1j.) The Reeh and Paterson (1988) value gave an optimal match between modelled and measured velocities along a flow-line, with measurements at depth in one borehole. Excepting a thin basal layer, the ice is all Holocene. Thus all three tabulated values refer to Holocene ice.

The A value for simple shear flow is about twice as large as that for ice shelves. This probably reflects the softening effect of fabrics in the simple shear flows; the ice shelf values thus give a better indication of isotropic ice. Indeed, the ice shelf numbers fall close to the average obtained in laboratory experiments; at this level of precision, they are indistinguishable. We thus select the laboratory number as the reference value A_*. In the laboratory, the temperature has a single, known value; in the ice shelves, it varies with position and has only been measured in a few profiles (Chapter 9).

For temperate glaciers, the five "calibrated models" represent one glacier in Iceland, two in the Alps, one in Scandinavia, and one in Alaska. These numerical models were calibrated against measured surface velocities or topographic changes. Because all of these models attempted to account for the entire stress regime, we consider their values more reliable than those from earlier studies. The dependence of ice creep on τ_E implies that neglecting any component of stress would lead to an overestimate for A. The Raymond (1980) value is the mean from four studies of tilting boreholes in mountain glaciers. We have adjusted the A numbers given by Raymond so that the measured strain rates are matched if $n = 3$, assuming that Raymond's adjusted stress values apply. The borehole tilt measurements seem to indicate softer ice than do the models of general flow; this difference may reflect softening of basal layers by fabric development or water or may reflect a too-low value for τ_E in the analyses of tilt. The same factors may explain the higher value of A found by Nye (1953) in his analysis of the closure rates of tunnels.

Regardless, the most meaningful results for glacier dynamics problems are the calibrations of full-stress models against large-scale flow; thus we take A for temperate ice as 24×10^{-25} $Pa^{-3}\,s^{-1}$. For comparison with polar ice, we note that the fabrics in temperate glaciers generally take the form of multiple maxima and broad single maxima, with strong variations from layer to layer in foliated regions (Rigsby 1960; Kamb 1959). These complicated patterns suggest that the temperate glaciers should not be compared to polar ice in simple shear, which develops strong single-maximum fabrics, but rather to ice shelves. Comparing temperate glacier with ice shelf values then gives the effective activation energy for ice warmer than $-10\,°C$.

For calibrated values we therefore recommend:

$$A_* = 3.5 \times 10^{-25}\,Pa^{-3}\,s^{-1} \qquad Q^+ = 115\,kJ\,mol^{-1} \tag{3.36}$$

$$E_* \geq 2 \quad \textbf{polar ice undergoing simple shear}.$$

The A_* value is within a factor of two of earlier estimates that included laboratory data and a broader range of less well-constrained field data (Hooke and Hanson 1986; Weertman 1983;

Table 3.4: Recommended base values of creep parameter A at different temperatures and $n = 3$.

$T(^{\circ}\text{C})$	$A(\text{s}^{-1}\text{Pa}^{-3})$
0	2.4×10^{-24}
-2	1.7×10^{-24}
-5	9.3×10^{-25}
-10	3.5
-15	2.1
-20	1.2×10^{-25}
-25	6.8×10^{-26}
-30	3.7
-35	2.0
-40	1.0×10^{-26}
-45	5.2×10^{-27}
-50	2.6

Paterson 1994). The Q^+ is smaller than the mean laboratory-derived value of 152 kJ mol^{-1} (Weertman 1983), and smaller than expected from the laboratory values of Budd and Jacka (1989, Table 3.4). In other words, temperate glaciers are stiffer than expected from laboratory measurements on temperate ice. Our value of A for temperate ice is about three times smaller than the one recommended by Paterson (1994). Of the eleven separate numbers available from field studies (five given by model calibrations, five summarized by Raymond, and one from Nye's closure analysis), none rises to or exceeds this earlier recommended value, which included results from laboratory experiments. That being said, *some* temperate ice, such as layers with high water content or strong single-maximum fabrics, is no doubt considerably softer; our number represents an effective glacier-wide value most applicable to overall flow.

Table 3.4 lists our recommended values for A as a function of temperature, according to this calibration. For polar ice undergoing simple shear, these numbers should be multiplied by a factor of 2 or more.

How well does the reference relation, Eq. 3.35, predict point measurements of deformation rates in ice bodies, if $E_* = 1$? Few unambiguous data exist for answering this question, but some of the data sets whose averages contributed to the calibration can be used, along with a few additional measurements from simple shear in ice-age ice. Figure 3.16 compares measured strain rates with predicted strain rates from Eq. 3.35, for seven ice shelf locations and nine locations in simple shear regimes near the beds of grounded polar glaciers and ice caps. We have selected these from the literature as the most numerically reliable and to represent the observed diversity.

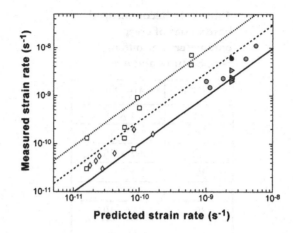

Figure 3.16: Local *in situ* measured values of strain rate compared to strain rates predicted from the recommended "base" creep relation. Solid line indicates a perfect match. Broken lines correspond to increases of measured strain rates by 3 and 9 times. Open symbols refer to polar ice; filled symbols to temperate ice. Open squares: simple shear in polar ice sheets and glaciers. Open diamonds: spreading of polar ice shelves. Filled triangles: strain rate values of temperate glaciers at a stress of 100 kPa. Filled circles: simple shear in temperate glaciers. Dark filled circle: tunnel closure in a temperate glacier. Data given in Appendix 3.2.

Also shown are several values for temperate glaciers. To assist interpretation, lines for $E_* = 3$ and $E_* = 9$ have been drawn. Some of the scatter in Figure 3.16 reflects uncertainties in estimates for temperature and stress, a difficult problem to quantify.

With a perfect relationship, $E_* = 1$. For the ice shelves – including locations on five separate shelves, analyzed by Thomas (1973b) – the mean $\overline{E}_* = 2.3$ and two standard deviations equal 2.6. It is not clear why the values obtained by Thomas are systematically larger than the mean from all ice shelf studies. For the simple shears of polar ice, $\overline{E}_* = 6.2$ and two standard deviations equal 7.6. Thus, for specific locations, strain rates deviate widely from the reference relation.

The evidence suggests that much of this deviation relates to fabrics. Measurements of thin sections show that a strong single-maximum fabric exists in the ice layers undergoing simple shear, oriented perpendicular to the shear plane. In general, the stronger the fabric, the greater the value of E_*. Likewise, girdle fabrics form in ice shelves, although their occurrence varies with depth. On average, for the numbers shown in the figure, the simple shears are a factor of 2.7 faster than the ice shelf spreadings, all else being equal. Such a difference is within the range expected for anisotropic effects related to fabrics (Section 3.4.7). Furthermore, the maximum softening by fabrics is, in theory, a factor of nine for simple shear of ice with a single-maximum fabric and a factor of three for compression and extension of ice with a girdle fabric. These limits approximately bound the data.

Concerning simple shears, the Holocene ice in Arctic ice caps typically shears faster than expected from the reference relation, by a multiple of about two. The adjacent ice-age ice and dirty basal layers shear even faster, by another multiple of 1.5 to 4 (Paterson 1991; Cuffey et al. 2000a). Present information justifies using a nominal value $E_* \approx 5$ for such ice in the Arctic, given that the typical contrast between ice-age ice and overlying Holocene ice is a factor of 2.5 (Paterson 1991). Both clean and dirty layers in Antarctic ice (at Law Dome and Meserve Glacier) show E_* in the range 5 to 12 if a strong single-maximum fabric exists. More observations are needed. We summarize ranges of observed E_* values in Table 3.5 and make a few recommendations for use in analyses when calibration against data is not possible or appropriate.

For temperate ice, factors that might reduce the viscosity include increased water content, strong fabrics, and perhaps impurities. Some observations of basal layers suggest extreme softening of temperate ice, as indicated in the table (Cohen 2000). It is surprising that the flow of temperate glaciers, overall, indicates considerably stiffer ice than that found by laboratory experiments.

As should be clear from this discussion, A must be calibrated for application to a given glacier, ice sheet, or ice shelf. In models of the Greenland Ice Sheet, for example, an E_* value of about 4 to 6 must be used to simulate the topography realistically in most models.

Insofar as the E_* values mostly reflect fabric effects, as seems likely, the reference relation cannot apply in general; it assumes isotropy. For ice dynamics calculations, the implications of this problem depend on the case. Assigning a value $E_* > 2$ is sometimes essential for accurately simulating known velocities of ice sheets and glaciers, but the same value should give the wrong answer for longitudinal stress or deformations. In large regions of mountain glaciers and ice sheets – including the bottom layers (but above the basal topography) and lateral shear margins – one component of stress and deformation sometimes dominates all others, so interactions between components matter little. In such cases, the isotropic relation can be applied without compromising simulations of the overall flow.

Table 3.5: Enhancements relative to our base relation.

Type of ice	Observed E_*	Recommended E_*
Polar ice: shelves[‡]	0.6°–2.4	1
Polar ice: simple shear		
(Holocene, Arctic)	0.9–2.5	2
(ice age, Arctic)	2–6	5
(strong single-max. fabric,		
Antarctic)	7–10	
(dirty basal layers)	4–12	
Temperate ice		
(basal layers)[†]	4–50+	

[‡] Excluding locations with $\tau_E < 20$ kPa.
° −15 °C whole-shelf value for Ross Ice Shelf.
[†] See Cohen (2000).

To discuss the applicability of the reference relation further it is necessary to first construct a model for anisotropic deformation, the subject of the next section. In Section 3.4.7.4, we return to the topic of errors introduced by using the isotropic relation.

3.4.7 Anisotropic Creep of Ice

A single ice crystal shears easily only along its basal planes (Section 3.2.2). Thus the viscosity of polycrystalline ice depends strongly on the c-axis fabric, which in glaciers often displays some degree of preferred orientation (Section 3.3.1) – a fact ignored in nearly all theoretical analyses of glacier flow. In contrast to viscosity variations related to temperature, the effects of fabric are directional; softening or stiffening depends on the orientation of the deformation with respect to the fabric. A given fabric therefore influences each component of the deformation in a different way. This means that Eq. 3.16 is incorrect and so, too, is the Nye-Glen relation (Eq. 3.22). No general creep relation for anisotropic deformation of ice has been established, but the basic properties are understood (Azuma 1994; Thorsteinsson 2002). Moreover, simple theoretical arguments lead to useful semiempirical relations, which can be calibrated against laboratory measurements.

3.4.7.1 Data and Constraints

The deformation of polycrystalline ice with a tight single-maximum fabric should resemble the deformation of a single ice crystal. Shear should occur most easily on planes perpendicular to the "fabric axis," the mean direction of c-axes. Rotating the plane of shear away from this position should make the deformation harder, until reaching a stiffest position when shear planes tilt 45° from the fabric axis. Shoji and Langway (1985, 1988) demonstrated this behavior in laboratory tests on ice from the Dye 3 core. They applied uniaxial compression to samples with a strong single-maximum fabric and varied the direction of the compression. From soft to stiff orientations, measured creep rates changed by more than a factor of 100 (Figure 3.17). In addition, compression perpendicular to the fabric axis was about four times faster than compression along the axis. Isotropic ice from the same core did not exhibit these behaviors. Experiments with ice from Dome C, Antarctica, revealed similar properties, although with different magnitudes (Duval and LeGac 1982).

The softening of ice between secondary and steady-tertiary creep, observed in laboratory experiments (Section 3.4.3), primarily reflects fabric changes resulting from grain rotation and recrystallization (Figure 3.18). In compression, initially isotropic ice evolves a girdle fabric centered on the compression axis, and the strain rate along this axis increases by a factor of about 3 (Jacka and Maccagnan 1984; Budd and Jacka 1989; Li et al. 1996). In simple shear, initially isotropic ice evolves a strong single-maximum fabric normal to the shear plane. The shear rate increases by a factor of 8 to 10 (Budd and Jacka 1989; Li et al. 1996). This degree of softening appears to be consistent with measured variations of borehole tilt across layers with different fabrics (Russell-Head and Budd 1979; Thorsteinsson et al. 1999; see Section 8.3.2).

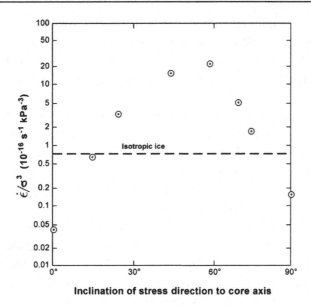

Figure 3.17: Measured variation of $\dot{\epsilon}/\sigma^3$ (deformation rate divided by the cube of the stress) for ice with a strong single-maximum fabric deformed in uniaxial compression applied at different angles to the axis of the sample. The sample axis was within a few degrees of the fabric axis. Data from Shoji and Langway (1988).

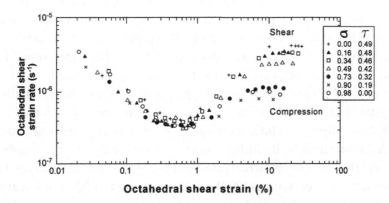

Figure 3.18: Measured creep curves for ice deformation experiments, showing softening by development of preferred orientation fabrics, in various combinations of compression and shear (magnitudes shown at right, in units of MPa). The increase of strain rate from the secondary minimum to the tertiary value is greater if shear dominates than if compression dominates. The combined octahedral stress for all tests was 0.4 MPa. Adapted from Li et al. (1996).

Experiments on ice samples taken from the Vostok core show how a band fabric influences deformation (Pimienta et al. 1987). Applying compression in the plane of the band produces a deformation rate equivalent to that for isotropic ice. Applying compression or extension perpendicular to the band, in contrast, produces deformation at a rate of only 0.06 times the isotropic one.

In deep layers of grounded glaciers, the dominant shear deformations occur on horizontal (or bed-parallel) planes. A single-maximum fabric, if oriented vertically, enhances the deformation. This explains much of the rapid shearing found at mid-range depths on the flank of Law Dome (Russell-Head and Budd 1979), and explains a large portion of the rapid shearing of ice-age ice deep in Arctic ice caps (Paterson 1991; Thorsteinsson et al. 1999). Very strong fabrics also contribute to the rapid shearing of basal ice in Meserve Glacier, a cold-based mountain glacier in Antarctica (Cuffey et al. 2000a). In that study, shear enhancements were calculated using c-axis measurements from thin sections, and an anisotropic creep relation like the one discussed here.

Another field observation explained by anisotropic deformation is the contrast, sometimes observed, between rates of borehole tilt and borehole closure (Thwaites et al. 1984). A strong, vertical single-maximum fabric enhances shear on horizontal planes, increasing borehole tilt. The closure of a borehole, on the other hand, is accomplished by shears at different orientations for which the fabric makes the ice stiffer. Thus a borehole closes more slowly in layers with such a fabric, even while tilting faster.

3.4.7.2 Model for Anisotropic Creep Behavior

How ice with a given fabric deforms anisotropically may depend on which recrystallization processes and deformation mechanisms are active. Nonetheless, important aspects of anisotropic deformation, including the experimental constraints cited in the previous section, can be explained using the assumption that bulk deformation reflects the ensemble average of properties of many individual crystals (Azuma 1994; Thorsteinsson 2001, 2002). Within polycrystalline ice, individual crystals deform primarily by basal glide. The ensemble of crystals, all shearing on basal planes, then controls the bulk deformation through geometric effects – the orientation of basal planes with respect to applied stress – and through redistributions of stress from crystal to crystal (Lile 1978). Stress focuses onto crystals oriented unfavorably – relative to neighbors – for basal glide. A crystal oriented favorably for basal glide, in contrast, is partly shielded from stress by its stiffer neighbors. Here we discuss the simplest ensemble-average model and show that its predictions are consistent with known constraints. Our treatment closely follows the theories of Azuma (1994) and Thorsteinsson (2002). Castelnau et al. (1996b) developed a more complicated theory featuring a more rigorous approach to calculating stress redistribution between neighbor crystals.

For a single crystal with orientation $\vec{c}$, a stress vector $\vec{s} = \sigma \cdot \vec{c}$ acts on the basal planes. Its components are $s_j = \sigma_{jk} c_k$. (Repeated indices always indicate summation here.) The component of $\vec{s}$ acting parallel to the basal plane defines the *resolved shear stress* *vector,* $\vec{\tau}^*$. It can be calculated by differencing $\vec{s}$ and the component of $\vec{s}$ normal to the plane:

$$\vec{\tau}^* = \vec{s} - [\vec{s} \cdot \vec{c}] \vec{c}. \tag{3.37}$$

On the right-hand side, the contribution of isotropic pressure to each term cancels, so $\vec{\tau}^*$ depends only on the deviatoric stress, which can be substituted for the full stress. The resolved shear stress therefore is:

$$\vec{\tau}^* = \boldsymbol{\tau} \cdot \vec{c} - [\vec{c} \cdot \boldsymbol{\tau} \cdot \vec{c}] \vec{c} \tag{3.38}$$

(with components $\tau_j^* = \tau_{jk} c_k - [c_p \tau_{pq} c_q] c_j$). (Alternatively, the components of this vector can be found directly from the stress tensor by a rotation of the coordinate axes into alignment with the basal planes and the c-axis.)

It is useful to have a compact way of describing the crystal orientation relative to the applied stress. This is usually done by combining, as an outer product, the *unit* vector in the direction of $\vec{\tau}^*$ with the unit-normal vector $\vec{c}$. The "outer product" of two vectors, $\vec{a}$ and $\vec{b}$, means the matrix of products $a_j b_k$ (i.e., a 3×3 matrix in three dimensions), usually symbolized $\vec{a} \otimes \vec{b}$. For the ice crystal, the product

$$S = \frac{\vec{\tau}^*}{|\vec{\tau}^*|} \otimes \vec{c} \tag{3.39}$$

defines the *Schmidt tensor*. Its components all lie in the range $[-1, 1]$. Consider, for example, a case of simple shear with $\tau_{xz} = \tau_o = \tau_{zx}$, and all other deviatoric stresses equal to zero. The x-axis is horizontal and the z-axis vertical. Then for two crystals, the first with a vertical c-axis (parallel to z), the second with a c-axis in the xz-plane but tilted at $45°$, the components of S are

$$\begin{pmatrix} 0 & 0 & 1 \\ 0 & 0 & 0 \\ 1 & 0 & 0 \end{pmatrix} \quad \text{and} \quad \begin{pmatrix} \frac{1}{2} & 0 & \frac{1}{2} \\ 0 & 0 & 0 \\ -\frac{1}{2} & 0 & -\frac{1}{2} \end{pmatrix}, \tag{3.40}$$

respectively. A block of polycrystalline ice, subjected to a given applied stress, can be viewed as an ensemble of S tensors, one for each crystal.

The Schmidt tensor has convenient properties. For a single crystal, the magnitude of the resolved shear stress on its basal planes is simply the inner product of the Schmidt tensor and the deviatoric stress tensor

$$|\vec{\tau}^*| = S : \boldsymbol{\tau} \tag{3.41}$$

or, in terms of components, the sum $S_{jk} \tau_{jk}$. For the case of simple shear considered previously, with $\tau_{xz} = \tau_o = \tau_{zx}$, the two crystals in Eq. 3.40 have $|\vec{\tau}^*| = \tau_o$ and $|\vec{\tau}^*| = 0$, respectively.

The second convenient property of the Schmidt tensor is that its components automatically give the relative sizes of the velocity gradients $(\partial u_j / \partial x_k)$ arising from shear on the basal planes of the crystal. Strain rate is the symmetrical part of the velocity gradient (Eq. 3.7); thus the components of a single crystal's strain rate are proportional to $[S_{jk} + S_{kj}]/2$. Consider the two crystals used as examples previously: in the first case, this matrix is identical to S; in the second case, the xz- and zx-terms are zero.

To proceed, the rate of shear along basal planes is assumed proportional to a power n of the resolved shear stress. Thus each crystal undergoes a strain rate

$$\dot{\epsilon} \propto \frac{1}{2} |\vec{\tau}^{\star}|^{n} [S + S^{T}], \tag{3.42}$$

where S^{T} denotes the transpose of S.

The magnitude of $\vec{\tau}^{\star}$ depends on both the bulk applied stress and the orientations of, and interactions between, crystals. The effective stress τ_{E} sets the scale for the bulk stress. To distinguish between effects of changing the magnitude of bulk stress and effects of changing the orientation and structure of the ice, we define the parameter f_{s} as the fraction of the effective stress that gets resolved onto the basal planes of a crystal:

$$|\vec{\tau}^{\star}| = f_{s}\,\tau_{E}. \tag{3.43}$$

On average, f_{s} will fall in the range $[0, 1]$, although strong focusing of stress might make it larger than 1 for a subset of crystals.

Consider further the relationship between the resolved shear stress and the bulk applied stress. We previously implied that τ in Eq. 3.38 refers to the bulk applied stress. It is likely, however, that stresses concentrate on crystals with unfavorable orientations. (The establishment of such stress inhomogeneity explains much of the stiffening of ice during primary creep; Duval et al. 1983.) Interactions between neighboring crystals reduce the $\vec{\tau}^{\star}$ on soft crystals, but increase it on hard crystals. The simplest way to account for such an effect is through an *interaction parameter*, following Thorsteinsson (2002). Define $\langle f_{s} \rangle$ as the average, over all the crystals in a volume of ice, of f_{s} from Eq. 3.43. The stress on basal planes of an individual crystal lies somewhere between this average and the value given by Eq. 3.43 itself (if in both cases, we use the bulk stress to calculate τ). The effective value for f_{s} becomes an $\hat{f}_{s}$:

$$\hat{f}_{s} = [1 - i]\,f_{s} + i\,\langle f_{s} \rangle, \tag{3.44}$$

where i, a positive number, defines the interaction strength.

Finally, we suppose that the strain rate of the polycrystal equals an ensemble average of Eq. 3.42, with $\hat{f}_{s}$ defined by Eqs. 3.43 and 3.44. Using angle brackets $\langle \rangle$ to denote the volume-weighted average over the crystals in a region of ice, the creep relation analogous to the Nye-Glen isotropic relation is

$$\dot{\epsilon}_{jk} = \frac{1}{2} A_{a}\,\tau_{E}^{n} \cdot \left\langle \hat{f}_{s}^{n} [S_{jk} + S_{kj}] \right\rangle, \tag{3.45}$$

with A_a a coefficient analogous to A in Eq. 3.22. The equivalent expression with S broken into its components gives the form most directly comparable to Eq. 3.22:

$$\dot{\epsilon}_{jk} = \frac{1}{2} A_a \, \tau_E^{n-1} \cdot \left\langle \hat{f}_s^{n-1} \left[\tau_j^\star c_k + \tau_k^\star c_j \right] \right\rangle. \tag{3.46}$$

The interpretation of Eqs. 3.45 and 3.46 is somewhat more complicated than for the isotropic rule, but nonetheless straightforward. Strain rate components increase nonlinearly with the overall magnitude of stress, τ_E. But the deformation arises only from the fraction of stress $\hat{f}_s$ that acts, on average, on the basal planes of crystals (a first effect of the crystal fabric). The geometry of the resulting deformation is then determined by S, the orientation of the crystals with respect to the applied deviatoric stress (a second effect of the crystal fabric).

3.4.7.3 Comparison with Constraints

Equation 3.45 predicts strain rates for a given applied stress acting on a given population of crystals with known orientations. One parameter, i, must be specified. There is no reason to believe, a priori, that the model should yield good results; its treatment of grain interactions is primitive – in particular by assuming that the orientation of stress acting on a crystal does not vary – and it takes no account of crystal rotations.

To compare with the data constraints summarized in Section 3.4.7.1, we apply Eq. 3.45 to simulated fabrics. These are constructed with a random number generator and appropriate relations of spherical geometry to make isotropic, single-maximum, girdle, and band fabrics. Table 3.6 shows predicted enhancements (E) of deformation rate relative to isotropic ice, calculated with a large number of model crystals (8000) and with a few different values for the interaction parameter i. The overall comparison shows an excellent qualitative match, demonstrating the essential role of the ensemble average for determining bulk polycrystal behavior. Quantitatively, the value $i = 1$ optimizes the match, and it seems reasonable to use this value at present. It should not be viewed as concrete, however, because the experimental values are still debatable.

Here, the simple shear calculation is for a single-maximum fabric, perpendicular to shear planes, with all c-axes pointing within $10°$ of the fabric axis. (The *cone angle* is said to

Table 3.6: Measured and modelled enhancements due to anisotropy.

Fabric	Deformation	Measured E	Model E $i = 0$	Model E $i = 0.6$	Model E $i = 1.0$
Single-maximum	Simple shear	8–10	4.1	6.1	8.2
Girdle	Uniaxial compression	3	1.8	2.5	3.0
Band	Uniaxial compression				
	(normal to band)	0.06	0.091	0.059	0.048
	(parallel to band)	1	0.98	0.96	0.97

be 10°.) The uniaxial compression calculation uses a girdle fabric with c-axis zenith angles uniformly distributed between 20° and 50° from the compression axis (cf. Jacka and Maccagnan 1984, Fig. 3.3, p. 279). The band fabric has a width of ±20° (cf. Pimienta et al. 1987, Fig. 3.1).

Rotation of the modelled single-maximum fabric with respect to a uniaxial compression axis gives a simulation of the Shoji and Langway (1988) experiments (Figure 3.17). Using $i = 1$, Eq. 3.45 predicts that creep rate in the softest orientation exceeds that in the hardest orientation by about 60 times, for a cone angle of 15°. Tightening the cone angle to 10° increases the multiple to about 250. This experiment thus depends sensitively on the degree of clustering of the single-maximum fabric and is not good for calibration.

The parameter value $i = 1$ means that interactions between neighboring crystals are strong enough to equalize the magnitude of resolved shear stress from crystal to crystal, regardless of the orientation of basal planes. This implies that an unfavorably oriented crystal deforms because the surrounding crystals impinge on it strongly, a result consistent with models for polycrystal deformation that resolve spatial variations of the stress field (Lebensohn et al. 2009). The anisotropic creep relation, in the case of $i = 1$, can be written

$$\dot{\epsilon}_{jk} = \frac{1}{2} A_a \, \tau_E^n \cdot \langle f_s \rangle^n \left(S_{jk} + S_{kj} \right).$$ (3.47)

Except for the use of parameter f_s, which is merely a convenience, Eq. 3.47 is identical to the one proposed by Azuma (1994); it can be called the *Azuma relation*. Thorsteinsson (2002) modelled anisotropic strain and crystal fabric development using a similar strategy to the one illustrated here, extended to calculate crystal rotations over time and to account for recrystallization processes. He found that fabrics observed in ice sheets are best explained if neighboring crystals interact strongly, as indicated here.

By comparing strain rates for isotropic ice predicted by Eq. 3.47 and the Nye-Glen relation, Azuma showed that the coefficient $A_a = 18A$. For an isotropic fabric, the anisotropic relation gives the same nonlinear dependence on stress components as the Nye-Glen relation. Consider the case of combined shear and uniaxial compression, confined in the y-direction: $\sigma_{xx} = \sigma_o$, $\sigma_{yy} = \sigma_o/2$, $\sigma_{zz} = 0$, and $\sigma_{xz} = \sigma_{zx} = \tau_o$, with other components equal to zero. The Nye-Glen relation shows that the shear strain rate $\dot{\epsilon}_{xz}$ varies as $r^3 + r/4$, for a ratio r defined as $r = \tau_o/\sigma_o$. The anisotropic relation gives the same behavior, for an isotropic fabric, as Figure 3.19 shows. This agreement shows that Nye's critical assumption, a power-law relationship between $\dot{\epsilon}_E$ and τ_E, can be justified as a consequence of the average behavior of a large number of crystals. Regardless of the fabric, the anisotropic relation – as with the isotropic one – implies that a component of deformation rate increases linearly with its corresponding stress, not as the power n, if the stress component is small compared to the dominant one.

3.4.7.4 Implications of Anisotropic Relation

Rapidly flowing ice shelves and grounded glaciers develop different fabrics (Section 3.3.4): girdle fabrics in the pure-shear regime (simultaneous extension and compression) of ice shelves

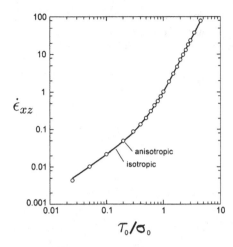

Figure 3.19: For ice with an isotropic fabric, the variation of shear strain rate due to a shear stress (τ_o) combined with a uniaxial compressive stress (σ_o) perpendicular to the shear plane. Solid line shows the prediction of the Nye-Glen relation; circles show calculations with the ensemble-average relation. In both cases, the increase is linear at low values for τ_o/σ_o but becomes cubic at high values.

and vertical single-maximum fabrics in the simple-shear regime deep in grounded glaciers (or multimaximum fabrics in a complicated regime). With strongly developed fabrics, the enhancements in pure shear and simple shear are about 3 and 9; thus ice shelf ice should appear to be harder by a factor of a few. This is consistent with inferences from field measurements (Figure 3.16). Furthermore, the ice should tend to be softer for the dominant shears than for subordinate deformations.

The anisotropic relations predict a nonlinear change of softening as a fabric strengthens; Figure 3.20 plots the functions, calculated from Eq. 3.47. Deformation rates often differ between ice layers with different degrees of fabric development. But to learn anything useful from comparisons between such layers, the shapes of the curves shown in the figure must be taken into account. There is no single "sensitivity" of deformation to fabric strength.

Without dynamic recrystallization, uniaxial compression will not produce a girdle, but rather a single-maximum fabric oriented along the compression axis. Thus the ice stiffens, not softens. This describes the situation in the upper layers of ice sheets near their central divides; due to vertical compression, fabrics evolve toward a single-maximum, oriented vertically. But as strain accumulates and the stiffening increases, compression becomes difficult, and the build-up of stress must eventually allow dynamic recrystallization. In some cases, this leads to polygonization; in other cases to the formation of girdle fabrics (Section 3.3.4). In yet other cases, recrystallization focuses in narrow zones that become structures such as shear bands with differently oriented crystals. This apparently occurs in central Greenland (Figure 10.6).

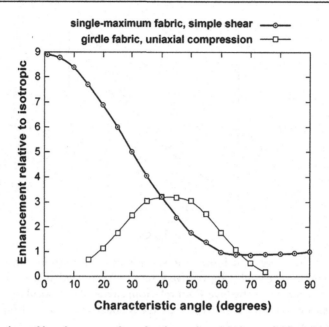

Figure 3.20: Softening of ice due to preferred orientation fabrics, calculated with ensemble-average anisotropic creep relation. Circles show the theoretical change in shear strain rate for simple shear of ice with a single-maximum fabric – with "characteristic angle" referring to the cone-angle (90° is isotropic ice). Also shown is theoretical change in compressive strain rate in uniaxial compression of ice with a girdle fabric – with "characteristic angle" being halfway between the inner and outer angles. Inner and outer angles are 30° apart.

The deformation regime throughout the mid-depth layers of grounded glaciers, and probably in many basal layers too, combines simple shear on subhorizontal planes with longitudinal and vertical extension or compression. In this case, a strong vertical single-maximum fabric (as observed in many ice cores) implies stiffer ice with respect to the extension and compression than to the simple shear. For given rates of extension and compression – which are typically fixed by constraints on the large-scale pattern of glacier flow – the magnitudes of the longitudinal stresses (τ_{xx} and τ_{zz}) must be larger than expected from the Nye-Glen isotropic relation. By how much? Consider again the case of combined shear and uniaxial extension, such that $\sigma_{xx} = \sigma_o$, $\sigma_{yy} = \sigma_o/2$, $\sigma_{zz} = 0$, and $\sigma_{xz} = \sigma_{zx} = \tau_o$, with other components set to zero. Suppose that the fabric forms a single-maximum, pointed along the z-axis, with a cone angle describing the spread about the axis. How large must τ_{xx} be for the ice to extend in the x-direction at the same rate as for an isotropic fabric? Table 3.7 lists the necessary τ_{xx} values calculated from Eq. 3.47 using randomly generated c-axis distributions. The τ_{xx} have been normalized to the value in the isotropic case.

These calculations indicate that a pronounced increase of extensional stress occurs only with a very strong fabric. Note that at cone angles greater than about 45° the fabric slightly softens the ice for extension, rather than stiffening it.

Table 3.7: Deviatoric stress
needed for extension at a
given rate.

Cone angle (deg.)	τ_{xx} (normalized)
90 (isotropic)	1.0
70	0.85
50	0.91
40	1.1
30	1.4
20	2.3
10	5.5

The nonlinearity of the Nye-Glen isotropic creep relation implies that the presence of any stress component enhances deformation in any direction (Section 3.4.4.5). As Figure 3.19 demonstrates for isotropic ice, this behavior also occurs with the anisotropic relation (Eq. 3.47). The magnitude of the effect, however, changes as preferred orientations develop, a behavior discussed by Thorsteinsson (2001). Consider again the case of combined shear and uniaxial compression, as defined previously. We illustrate, for example, the case with equal compression and shear: $\sigma_o = \tau_o$. By how much does the presence of the compression increase the rate of shear? Table 3.8 shows the fractional change of shear rate $\dot{\epsilon}_{xz}$ expected from the anisotropic relation, for different single-maximum fabrics oriented along the axis of compression, perpendicular to the plane of shear. (The fractional change means, specifically, $\dot{\epsilon}_{xz}$ calculated with $\sigma_o = \tau_o$, divided by $\dot{\epsilon}_{xz}$ calculated with the same τ_o but $\sigma_o = 0$. In all cases, the calculation uses the anisotropic relation, not the isotropic one.)

Tables 3.7 and 3.8 both suggest that – unless the fabric develops to an extreme degree and everywhere orients to the same direction – anisotropic ice retains the interaction of components implied by the Nye-Glen relation. This can be taken as justification for applying the simple isotropic relation (Eq. 3.35) to glaciers, so long as the "enhancement" multiplier E_* is assigned a value appropriate for the fabric and the dominant component of deformation (and assuming that subordinate stresses do not need to be known well). A different value could be used to calculate subordinate stresses, by themselves. On the other hand, even with moderate fabric development, some aspects of the isotropic relation are qualitatively incorrect. For example, in the case of combined shear and uniaxial compression considered before, with $\sigma_{yy} = \sigma_o/2$, the isotropic relation indicates that $\dot{\epsilon}_{yy} = 0$ because $\tau_{yy} = 0$. The anisotropic relation, on the other hand, gives a small but nonzero $\dot{\epsilon}_{yy}$ for most fabrics, even if $\tau_{yy} = 0$.

The anisotropic relation introduces new dependencies on the creep exponent n. The enhancement of deformation rates for a given fabric depends explicitly on n. The enhancement of simple shear for a tight single-maximum fabric ($E \approx 8$ if $n = 3$) ranges from $E \approx 3$ to 14 as n ranges from 1 to 4. The enhancement of pure shear for a girdle fabric ($E \approx 3$ if $n = 3$) ranges from $E \approx 2$

Table 3.8: Enhancement of shear
by simultaneous compression of
equal magnitude.

Cone angle (deg.)	Fractional increase
90 (isotropic)	1.25
60	1.45
30	1.30
20	1.13
10	1.04

to 4 as n ranges from 1 to 4. Thus Eq. 3.45 implies that the enhancement magnitudes depend on the deformation mechanism. There are no data available to test this idea; this illustrates the inadequacy of currently available experimental and field data.

3.5 Elastic Deformation of Polycrystalline Ice

If glacier ice is stressed for short periods of time or with a loading that fluctuates at a high frequency, the resulting creep deformation will be negligible compared to elastic deformations. This is the case, for example, with propagation of seismic waves, and with the response of ice shelves to tidal forcings (Anandakrishnan and Alley 1997).

Elastic deformations occur by stretching and bending of bonds in the crystalline lattice. Considering isotropic polycrystalline ice, an applied normal stress in only one direction, σ_{xx}, generates an elastic strain of magnitude $\epsilon_{xx} = E_Y^{-1}\sigma_{xx}$, where E_Y denotes *Young's modulus* (about 9×10^9 Pa). Strain also occurs in the perpendicular directions, with magnitudes ϵ_{yy} and ϵ_{zz}. These strains are a multiple of ϵ_{xx}; the multiple is known as *Poisson's ratio*, with a value approximately 0.3 for polycrystalline ice. The speed of propagation of elastic waves, both compressional and shear, depends on combinations of Young's modulus and Poisson's ratio. For glacier ice, these speeds are of order 10^3 m s^{-1}.

Elastic wave propagation through a single ice crystal is anisotropic, with fastest propagation along the c-axis. For this reason, seismic studies can identify large regions of the ice sheets where c-axis fabrics have developed preferred alignments (Blankenship and Bentley 1987).

Appendix 3.1

At shear stresses $\tau > 0.1$ bar, the Goldsby-Kohlstedt relationship is

$$\dot{\epsilon} = A_1 D^{-1.4} \sigma^{n_1} + A_2 \sigma^{n_2} \tag{3.48}$$

in which the stress σ is the "differential stress," equal to $\sqrt{3}\tau$. Each of the functions A has a prefactor A_o and an Arrhenius dependence on temperature, such that $A = A_o \exp(-\frac{Q}{RT})$ and

Q is the activation energy. The values to use, including corrections of typographical errors in Goldsby and Kohlstedt (2001), are shown in the table.

Appendix 3.1: Parameters for Goldsby-Kohlstedt relation (for T in Kelvin).

n			Prefactor		Activation energy (kJ mol^{-1})	
			$T < 259$	$T > 259$	$T < 259$	$T > 259$
$n_1 = 1.8$	A_1:		3.9×10^{-3}	3×10^{26}	49	192
			$(MPa^{-1.8} m^{1.4} s^{-1})$			
$n_2 = 4$	A_2:		4×10^{4}	6×10^{28}	60	180
			$(MPa^{-4} s^{-1})$			

Appendix 3.2: Data for Figure 3.16

Appendix 3.2: Values used for Figure 3.16. Measurements are approximate.

Location	τ_E (10^4 Pa)	T (°C)	D (mm)	Measured $\dot{\epsilon}$ (s^{-1})	Reference	Calculated $\dot{\epsilon}_{xz}$ (s^{-1})
Ice shelves:						
Ross Ice Shelf						
(low stress locations)	5	−15		3×10^{-11}	Jezek et al. 1985	2.6×10^{-11}
(high stress locations)	9	−15		1.6×10^{-10}	Jezek et al. 1985	1.5×10^{-10}
Brunt Ice Shelf						
(location R7)	3	−6		4.4×10^{-11}	Thomas 1973b	2.1×10^{-11}
(main shelf)	6	−16		6.3×10^{-11}	Thomas 1973b	4.0×10^{-11}
(main shelf)	4.5	−16		3.5×10^{-11}	Thomas 1973b	1.7×10^{-11}
Maudheim Ice Shelf	4.8	−15		5.5×10^{-11}	Thomas 1973b	2.3×10^{-11}
Amery Ice Shelf	9.5	−11		2×10^{-10}	Thomas 1973b	6.7×10^{-12}
Ward Hunt Ice Shelf	1.2	−10		3.3×10^{-12}	Thomas 1973b	6.0×10^{-13}
Grounded ice:						
Law Dome (SGF)	4.75	−5.5	6	9.2×10^{-10}	RHB	8.6×10^{-11}
Dye 3						
(Holocene)	7.2	−15	5	7.9×10^{-11}	DJG	7.6×10^{-11}
(ice age)	7.6	−14.5	1	5.6×10^{-10}	DJG	9.2×10^{-11}
Agassiz						
(Holocene)	7.5	−17	3	1.3×10^{-10}	FK	5.5×10^{-11}
(ice age)	7.5	−17	1	2.2×10^{-10}	FK	5.5×10^{-11}
Camp Century						
(basal 300 m)	4	−14	2.5	1.3×10^{-10}	FK and Paterson 1983	1.4×10^{-11}

RHB = Russell-Head and Budd 1979.
DJG = Dahl-Jensen and Gundestrup 1987.
FK = Fisher and Koerner 1986.

CHAPTER 4

Mass Balance Processes: 1. Overview and Regimes

"How much longer this little glacier will last depends, of course, on the amount of snow it receives from year to year, as compared with melting waste...."

The Mountains of California, **John Muir (1894)**

4.1 Introduction

Each year, a glacier gains ice from snowfall but also loses ice by melt and other processes. If the gains and losses are not equal, the size of the glacier – its dimensions and the mass of ice it contains – will change over time. Mass balance studies assess such changes and seek an understanding of the processes involved.

Consider the image of a glacier sketched in Figure 4.1. The mass of ice in the entire glacier can change only by processes that transfer mass between the glacier and its surroundings (the atmosphere, the land surface, the groundwater), or between ice and liquid water within the glacier. We think of these processes collectively as *mass exchanges*; they are the sources and sinks for glacier ice. The common mass exchanges are snowfall, avalanching, melt, sublimation, and the separation of ice blocks from the glacier edge, a process known as *calving*. (Calving in lakes or the ocean produces icebergs, while calving on mountain slopes generates ice avalanches.) The rates of these processes, summed over the glacier and over time, determine the *glacier mass balance*, the change in total mass of ice. In some discussions, the "total mass" also includes the mass of liquid water within the glacier and on its surface.

Ice flow redistributes the mass in a glacier. Thus the mass balance of a zone (take Zone A or B in Figure 4.1 as an example) depends not only on the mass exchanges, summed over the zone, but also on the transfer of mass by flow. Zone A – a "drainage basin" – loses mass by flow down-glacier, whereas Zone B gains mass by flow from above.

It is less obvious that the mass balance of the entire glacier depends on ice flow, but it can. The quantity of ice lost by calving usually depends on the flow rate near the glacier front. On most glaciers, an indirect relation plays an important role; mass exchanges vary with altitude, while the altitude of the glacier surface adjusts with the flow. Consider, for example, a mountain

Copyright © 2010, Elsevier Inc. All rights reserved.
DOI: 10.1016/B978-0-12-369461-4.00004-1

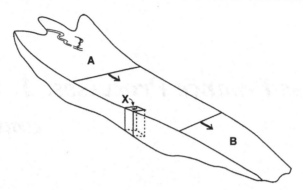

Figure 4.1: A mountain glacier, sliced lengthwise. Arrows indicate direction of ice flow, "A" and "B" are zones, and "X" locates a vertical column of unit horizontal area. "Human" figure is most likely about 100 m tall.

glacier whose flow increases abruptly. The terminus advances, expanding the area of ice surface at low altitudes where melting is strongest. Consequently, the glacier loses mass as the advance progresses.

Now focus on a single location (X in Figure 4.1), and consider a vertical column, of unit horizontal area, slicing through the glacier. The mass exchanges operating at the surface, within the ice, and at the base increase or decrease the mass in the column. The mass change per unit area defines the *specific balance* at X.[1] The specific balance constitutes the "source term" for a glacier, just as rainfall (less evaporation) constitutes the source term for water in a river basin.

Changes of glacier mass may be conceptualized in terms of a budget. Processes of gain and loss – known to glaciologists as *accumulation* and *ablation* – correspond to income and expenditure. The cumulative difference between income and expenditure gives the account balance – in this case, either the specific balance or the balance of the whole glacier. The budget may run a surplus or a deficit. On a glacier, a zone of mass surplus is where accumulation exceeds ablation; the opposite applies to a zone of mass deficit. On many glaciers the budget fluctuates over an annual cycle. The glaciers gain mass in winter and lose mass in summer. The net change over this annual cycle determines whether a glacier grows or shrinks from one year to the next. Understanding the cycle is therefore an essential component of many mass balance studies.

Mass balance is a critical factor in the relation between glaciers and climate, a topic of widespread interest and great complexity. Figure 4.2 is one way to summarize the relation. The "regional climate" in panel (a) refers to conditions averaged over a large area, such as a mountain

---

[1] Note that the specific balance equals the total mass change in the column if the column edges are attached to the ice and so move and stretch with the flow. If, instead, the column is fixed in space, ice flow passes through it. The change of mass in the column then equals the specific balance plus the net inflow of ice; Section 8.5.5 gives a formal statement of mass conservation for a column fixed in space.

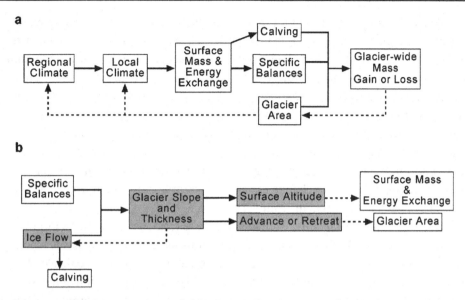

Figure 4.2: **Generalized factors determining the relation between climate, ice flow, and glacier mass balance. Feedbacks are indicated by dashed lines. The chart has been separated into two parts for clarity (panels (a) and (b)). Elements in (b) that are not shaded gray appear also in (a).**

range. The "local climate" signifies conditions on the glacier and in its immediate surroundings. Local climate determines the supply and removal of mass and heat at the glacier surface, which in turn control the specific balances. Together with the area of the glacier surface and the rate of calving, the specific balances determine the glacier-wide gain or loss of mass. The dashed lines indicate important feedbacks; as the glacier grows or shrinks, its area changes. The change in area not only affects the subsequent gain or loss of mass directly but can also modify the local climate. In landscapes with widespread glacial cover, this feedback may even affect the regional climate, and, in the case of large ice sheets, the global climate itself.

Figure 4.2b outlines another part of the glacier-climate relation (separated from panel (a) only for clarity), involving the fundamental interrelationship between mass balance and glacier dynamics. Specific balances largely determine whether a glacier expands or decays and how vigorously it flows. Glaciers form because of a mass surplus in some locality. Once a glacier is established, an increase of specific balances will, over time, thicken and steepen the glacier. This change depends on the combined action of the specific balances and the ice flow. Thickening and steepening, in turn, increase the gravitational forces acting in the ice and thus increase the flow: another important feedback. Changes of thickness and slope ultimately cause a glacier to advance or retreat, expanding or contracting the surface area. At the same time, the glacier surface rises or falls, which might change the surface climate and hence the melt. Finally, the rate of ice flow often controls the rate of mass loss by calving.

Mass balance studies are important for many reasons. Measurements of glacier mass balance reveal whether the glacial cover on a landscape is increasing or decreasing and, if there are global trends, the net effect on sea level. Measuring the mass balance of a zone sometimes reveals major changes related to ice flow; this technique has been especially important in recent studies of the polar ice sheets. Measurements of mass balance at monthly to yearly timescales have practical applications. In some regions, glacier-fed streams supply much of the water for agriculture, domestic consumption, and hydroelectric plants. Such streams have a distinctive pattern of run-off; a glacier acts as a natural reservoir, storing water in winter and releasing it in summer. Especially large quantities are released in warm summers when water from other sources is in short supply. Mass exchange processes determine how much water can be released in this way and what variations occur from year to year.

In this chapter we discuss the mass exchange processes and their general relation to altitude, climate, and other factors. We then summarize the various methods used to assess whole-glacier and zonal mass balances. Finally, we review particular features of the mass balance of mountain glaciers, the Greenland Ice Sheet, and the Antarctic Ice Sheet. The discussion of mass exchanges continues in the following chapter, which examines in greater detail the factors controlling glacier surface ablation. The interactions between mass balances, glacier evolution, and ice dynamics are the subject of Chapter 11.

4.1.1 Notes on Terminology

4.1.1.1 Specific Balance Rates

Specific balance represents a change of mass per unit area $(\mathrm{kg\,m^{-2}})$. In many applications pertaining to timescales longer than one year, the useful quantity is not the cumulative change, but the rate of mass addition per unit time. This *specific balance rate*, denoted $\dot{b}$, usually has units of $\mathrm{kg\,m^{-2}\,yr^{-1}}$. The terms "specific balance rate" and "specific balance" are in fact often used interchangeably, because measurements of specific balance are typically registered to a period of one year (Section 4.2.3 gives a precise description).

Just as precipitation rates are usually given as a water depth per unit time, the specific balance rate is often given as a thickness per unit time, using the equivalent thickness of either ice or water. Ice dynamics studies usually refer to the ice-equivalent rate (here denoted $\dot{b}_i$), the quantity most directly connected to ice thickness changes and vertical velocities. Mass balance studies, on the other hand, usually report water-equivalent rates (here denoted $\dot{b}_w$). There are therefore three common ways to express specific balance rates:

$$\dot{b}\ \ (\mathrm{kg\,m^{-2}\,yr^{-1}}); \qquad \dot{b}_i = \frac{\dot{b}}{\rho_i}\ \ (\mathrm{m\,yr^{-1}});$$

$$\dot{b}_w = \frac{\dot{b}}{\rho_w} = \frac{\rho_i}{\rho_w}\,\dot{b}_i\ \ (\mathrm{m\,w.eq.\,yr^{-1}}), \tag{4.1}$$

where ρ_i and ρ_w signify densities of ice and water, respectively. We introduce the notations $\dot{b}_i$ and $\dot{b}_w$ in these definitions for clarity. In the glaciological literature, the symbol $\dot{b}$ is commonly used for all three. Because we will want to use a variety of subscripts to denote different specific balance terms, the symbol $\dot{b}_i$ cannot be used consistently throughout this book. We will state explicitly whether $\dot{b}$ represents a thickness or a mass, but $\dot{b}_i$ will always represent an ice-equivalent thickness.

4.1.1.2 Surface, Englacial, and Basal Balances

The specific balance can be subdivided according to where along the vertical column (at X in Figure 4.1) the mass exchange occurs. Specifically, $\dot{b} = \dot{b}_s + \dot{b}_e + \dot{b}_b$, where $\dot{b}_s$ stands for the surface balance, $\dot{b}_b$ the basal balance, and $\dot{b}_e$ the *englacial* balance, the contribution from exchanges within the glacier. This subdivision is useful because different factors control each term; we discuss them separately in this chapter in Sections 4.2, 4.4, and 4.5. In most cases, the surface balance dominates the other terms, and indeed the approximation is often made that $\dot{b} = \dot{b}_s$. However, the basal balance can be important in some environments, such as floating ice shelves and glaciers on volcanoes.

If we call the total ice mass of a glacier M, and its plan-view area $\mathcal{A}$, then M changes at a rate

$$\frac{dM}{dt} = \int_{\mathcal{A}} \left[\dot{b}_s + \dot{b}_e + \dot{b}_b\right] d\mathcal{A} - \dot{B}_c,$$

(4.2)

where $\dot{B}_c$ denotes the mass per unit time lost by calving. Integrating Eq. 4.2 over an interval of time then gives the glacier mass balance, ΔM (often denoted B).

4.1.1.3 Accumulation and Ablation

Accumulation includes all processes by which snow and ice are added to a glacier. *Ablation* refers to all processes of removal. Thus, accumulation indicates a positive contribution to $\dot{b}$, and ablation indicates a negative one. "Net accumulation" means accumulation minus ablation.

4.1.1.4 Inconsistent Terminology

In the glaciological literature, the use of terms related to mass balance lacks consistency. The specific balance and the glacier balance are both frequently referred to simply as *mass balance*. The word *mass* can often be omitted, as we have done with specific balance and glacier balance." Many discussions of ice dynamics substitute the term *accumulation rate* for specific balance, in which case "accumulation rate" really means *net accumulation rate*. As noted previously specific balance and specific balance rate are often used interchangeably. The term *point mass balance* is equivalent to specific balance. Additional inconsistencies appear in discussions of the annual cycle (Section 4.2.3). Despite inconsistencies, the meaning of terms is usually made clear by the context.

4.2 Surface Mass Balance

Mass exchanges at the surface dominate the budget of most glaciers, although calving plays a comparably important role for the ice sheets and for some glaciers that reach the sea. Contributions from several processes determine the surface balance rate at a point:

$$\dot{b}_s = \dot{a}_s + \dot{a}_a - \dot{m}_s + \dot{a}_r - \dot{s} + \dot{a}_w, \tag{4.3}$$

representing snowfall ($\dot{a}_s$), avalanche deposition ($\dot{a}_a$), melt ($\dot{m}_s$), refreezing of water ($\dot{a}_r$), sublimation ($\dot{s}$), and wind deposition ($\dot{a}_w$). Sublimation can be either positive or negative. We write it as a loss term because glacier-wide losses by sublimation generally exceed the gains from vapor deposition. Wind deposition can also be either positive (deposition) or negative (scour). Its average over a glacier will usually be close to zero. Refreezing of water refers mostly to melt-water, but rain and runoff from adjacent hillslopes can also freeze. At many places on glaciers, snowfall and melt dominate the surface balance and $\dot{b}_s = \dot{a}_s - \dot{m}_s$ makes a good approximation to Eq. 4.3. On polar glaciers, however, refreezing cannot be neglected. (To calculate the surface balance for the sum of ice and water, rather than ice alone, the *net runoff* – the inflow of water minus the outflow – would replace $\dot{a}_r - \dot{m}_s$.)

4.2.1 Surface Accumulation Processes

Annual snowfall varies widely, from several meters per year in mid-latitude and subpolar maritime regions (such as Iceland, New Zealand, and Norway) to as little as a few centimeters per year high on the East Antarctic plateau. Throughout this range, snowfall essentially controls accumulation.

Three primary factors control snowfall rates:

1. Snowfall tends to increase with the water vapor content of the overlying atmosphere. This generally increases with air temperature in accordance with the Clausius–Clapeyron formula, which relates saturation vapor pressure (e_s, in Pa) to temperature of the air (T_a, in K). To a good approximation, within 1% of true values, the vapor pressure at saturation increases exponentially with T_a according to

$$\text{over liquid water:} \quad e_s = 2.38 \times 10^{11} \exp(-5400/T_a) \tag{4.4}$$

$$\text{over ice:} \quad e_s = 3.69 \times 10^{12} \exp(-6150/T_a). \tag{4.5}$$

(See Flatau et al. (1992) for precise formulae.) Roughly, a 10 °C warming doubles the vapor content of saturated air. Once air saturates, further cooling drives condensation and the formation of cloud droplets or ice particles, which precipitate if they grow large enough. By this mechanism, a cooling of given magnitude produces more precipitation from a warm air mass than from a cold one, because of the exponential form of Eqs. 4.4 and 4.6. This explains the general tendency for low accumulation rates at cold polar sites and high accumulation rates at warm temperate sites. To understand geographic patterns in more detail, however, variations of humidity must also be considered; dry air reduces the likelihood that a cooling process brings the air to saturation and forms clouds. For clouds to form, the saturation vapor pressure (e_s) must drop enough to match the actual vapor pressure (e).

2. For precipitation to reach the ground as snow rather than rain, the air in the lower atmosphere needs to be subfreezing. The relative abundance of rain and snow is referred to as the *rain-snow partitioning*.

3. High snowfall rates occur where air masses rapidly cool. Cooling results primarily from lifting of air masses, whether in frontal systems, in air flow up mountain slopes (*orographic lifting*), or in convective storms driven by ground-surface heating.

By these mechanisms, temperatures strongly influence the amount of snowfall. Whereas at low temperatures vapor content limits snowfall, at high temperatures the conversion of precipitation to rain limits snowfall. High snowfall rates thus occur where temperature conditions are intermediate; in "warm" maritime air masses at temperatures low enough for precipitation to reach the ground as snow. Where such conditions coincide with persistent atmospheric uplift, glaciers receive snow in great abundance. This is the case for windward slopes of coastal mountain ranges at mid-to-high latitudes, and along parts of the flanks of the southern Greenland Ice Sheet. Lee slopes are invariably drier than windward slopes, and continental interiors tend to be drier than coastal regions.

The west-to-east winds of the mid-latitudes cause large snow accumulations in mountain ranges on the west coasts of North and South America, northern Europe, and New Zealand. That is why southern Alaska and Patagonia are the most extensively glaciated landscapes outside the polar regions. In tropical South America, in contrast, easterly trade winds feed large snow accumulations on the eastern side of the Andes. Likewise, the Asian monsoon brings moisture to the eastern and southern sides of the Tibetan Plateau. From year to year the shifting of storm tracks along the westerly wind belts changes the annual snowfall in correlation with the quasi-periodic meteorologic phenomena known as El Niño–Southern Oscillation, the Pacific Decadal Oscillation, and the North Atlantic Oscillation. These phenomena correlate with annual variations of glacier mass balances in western North America (e.g., Bitz and Battisti 1999; Hodge et al. 1999), Europe (Nesje et al. 2000; Six et al. 2001), and Greenland (Box 2005). A shift of storm tracks simultaneously brings more snow to some regions while reducing precipitation in others; thus, for example, glacier mass balances vary inversely, from year to year, between Alaska and southern Canada (Rasmussen and Conway 2004).

Snowfall Gradient with Altitude and Elevation Deserts   In most mountain regions, precipitation rates increase with altitude in the lower few kilometers of the atmosphere. The air at sea level is usually not saturated with water; that is, $e < e_s$. Furthermore, it takes time for the water droplets and ice crystals in clouds to grow large enough to fall despite updrafts (Roe and Baker 2006). During this time, the droplets and crystals move down-wind and hence up in altitude. Moreover, at higher altitudes precipitation is more likely to fall as snow rather than rain. For these reasons, in conjunction with orographic lifting, snowfall often increases with altitude.

This increase cannot continue indefinitely, however, because the air ultimately loses most of its water to precipitation as the saturation water content declines. Precipitation and accumulation are both reduced at high altitudes. On the polar ice sheets, *continentality* – the great distance from oceanic sources of evaporation – magnifies this effect. Only a few centimeters (w.eq.) of snow falls each year in central East Antarctica, for example. The pronounced decrease of accumulation rates in high and extremely cold environments is known as the *elevation-desert effect*. It implies a self-limiting tendency for ice sheets; as an ice sheet grows, its center eventually attains a high altitude and becomes remote from moisture sources. Low snowfall rates then restrict further growth.

Redistribution by Wind and Avalanching   Accumulation may differ from snowfall at a site because winds carry snow along the surface. Interactions between topography and the winds create regions subject to wind scour or deposition. Crevasses and hollows can fill with wind-transported snow, for example. In parts of Antarctica, winds scour the surfaces adjacent to nunataks or on long regional slopes, forming zones of exposed "blue ice." Here, because of sublimation, $\dot{b}_s < 0$ even though no melting takes place.

Avalanching of snow from steep valley slopes and cirque headwalls is an important source of accumulation for some mountain glaciers. In such places, the avalanche debris builds sloping aprons along the sides and head of the glacier. These are zones of unusually large accumulation for the altitude.

Deposition   Where it freezes on the surface, rain adds a minor contribution to the accumulation. Direct deposition of atmospheric water vapor and supercooled droplets produces frost and rime, respectively. These are common but usually negligible accumulation processes. Deposition of about 0.01 m yr^{-1} occurs throughout the interior of the ice sheet in northeastern Greenland (Box et al. 2004).

Refreezing of Meltwater   Refreezing of melt forms superimposed ice on the surface or ice layers in the firn (see Chapter 2). The fraction of melt that refreezes on or within the glacier makes no contribution to the glacier mass balance. But observations and models usually give the gross surface melt at a point. The fraction that refreezes – often elsewhere on the glacier – must therefore be quantified separately and subtracted from the gross melt to assess the net loss.

Negligible refreezing occurs in ablation zones, because the meltwater drains easily. In contrast, refreezing can be significant in accumulation zones in cold regions. Jóhannesson et al. (1995) estimated that about 7% of meltwater refreezes on valley glaciers in Iceland and Norway. On polar ice fields, refreezing in the accumulation zone is estimated to average about 60% (Janssens and Huybrechts 2000). Model studies by Box et al. (2006), calibrated against available field data, suggest that about 30% of all the available meltwater on the Greenland Ice Sheet refreezes, significantly reducing the total runoff. All of these numbers include refreezing not only on the surface but also at depth in the firn, a process typically regarded as englacial or "internal" accumulation (Section 4.4).

4.2.2 Surface Ablation Processes

A glacier surface ablates mostly by melt and sublimation – processes discussed in detail in Chapter 5 – although wind scour can be locally significant. Except in the polar regions and on high mountains, melt followed by runoff accounts for most surface mass loss; summer melt rates in excess of $10 \, \mathrm{m \, yr^{-1}}$ are not unusual outside polar regions.

The net flux of energy from the atmosphere to the surface drives melt and sublimation. The important processes of energy transfer are radiation and turbulent mixing of heat and vapor in the air adjacent to the surface. If the temperature of the snow/ice surface is at melting point, the rate of melt increases in proportion to the net energy flux. Melt rates tend to increase strongly with warming of the overlying atmosphere. Melting can occur even when subfreezing air overlies the surface, provided that turbulent mixing is minimal.

Sublimation occurs at all temperatures and is the dominant ablation mechanism in very cold environments where surface temperatures seldom reach melting point even in summer (Bintanja 1999). A warm surface, dry air, and strong winds all increase sublimation. Mass loss rates, $\dot{s}$, are of order centimeters per year (e.g., Box and Steffen 2001), but sometimes amount to decimeters per year; rates of $0.2 \, \mathrm{m \, yr^{-1}}$ are typical in the Antarctic Dry Valleys, for example (Fountain et al. 2006; Kavanaugh et al. 2009a). Blowing snow also sublimates, but few direct measurements are available (Bintanja 2001a).

Sublimation and melt sometimes interact in an important way. Sublimation converts radiant and sensible energy fluxes to latent heat, thereby reducing the energy available for melting. Strong sublimation can suppress melting of the surface even if overlying air temperatures are several degrees above the freezing point (Kuhn 1987).

Another important interaction involves ablation processes and accumulation. If only a thin layer of snow accumulates in winter, firn and ice are exposed early in the ablation season. Because firn and ice absorb more solar energy than snow, this increases the total summer melt. Conversely, a thick snow cover reduces total melt. Snowfall during the summer can be particularly effective in reducing the total melt over a year.

4.2.3 Annual (Net) Balance and the Seasonal Cycle

The surface balance rate varies over hours and weeks, as individual storms and diurnal cycles of temperature and radiation affect snowfall and melt. The variability is much smaller for longer periods such as months or seasons. In some cold polar environments, such as the interior of the Greenland Ice Sheet, the mean surface balance rate even varies little from season to season (Shuman et al. 1995). In warmer settings, however, including the margins of the Greenland Ice Sheet and most mountain glaciers outside of the Tropics, the surface balance varies considerably over the seasons, dominated by accumulation in winter and ablation in summer. The seasonal cycle must be understood because its magnitude often greatly exceeds interannual variations and long-term trends.

Figure 4.3 depicts an idealized seasonal cycle. The curves are drawn smooth for clarity, but in reality the mass increases in steps corresponding to snowfall events and decreases at an

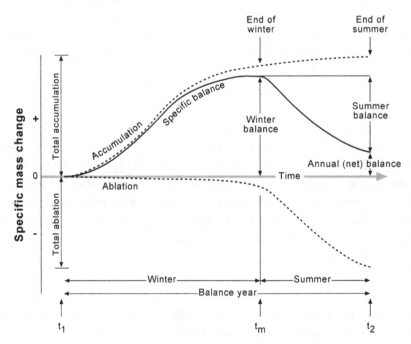

Figure 4.3: Definitions of terms in the seasonal progression of specific surface balance. "Annual" or "net" balance is, simply, the "specific balance" for one year.

irregular rate as changing weather conditions modulate the ablation. The mass added to the glacier surface generally increases over the winter season, attaining a maximum in late spring or early summer (defining the "end of winter"). Summer melting then removes mass, giving a minimum in late summer. The surface formed at the end of the ablation season is subsequently buried by renewed accumulation but can often be identified later – for example, as a dirty layer on temperate glaciers. This is the *summer surface*. If t_1 and t_2 denote the times of two successive minima, and t_m the time of the intervening maximum, the interval t_1 to t_2 defines the *balance year* (Anonymous 1969). Its length varies from year to year but on average approximates 365 days. The balance year can be divided into a *winter season* (t_1 to t_m) and a *summer season* (t_m to t_2). Note that the timing of seasons may differ between different zones on a glacier. At the end of the summer, snow accumulates on the higher parts of a glacier even while ablation continues near the terminus. Thus the length of the balance year and its two seasons varies from place to place on the glacier, mainly with elevation but also with other factors such as shading from adjacent mountains.

Integrating the accumulation and ablation rates from t_1 to t_2 gives the total annual accumulation and ablation. The *specific balance b* at any time of this year is accumulation minus ablation from the start of the balance year; in other words, the change in mass per unit area relative to the previous summer surface. The specific balance at the end of the balance year defines the *annual balance* or *net balance* for the year, b_n. Although *net balance* is the original term, usage has become confused. *Annual balance* seems clearer, and *specific balance* should suffice for periods of one year and longer. Over a period of N years:

$$\sum_{i=1}^{N} b_{ni} = \int_0^N \dot{b}_s \, dt. \tag{4.6}$$

Measurements and time series of annual b_n are widely referred to as "mass balance" data in the literature. The annual balance can be subdivided into a positive *winter balance* and a negative *summer balance*, as the figure indicates.

The preceding definitions can be applied to any location by, when necessary, allowing some terms to equal zero and arbitrarily fixing the boundaries of the budget year t_1 and t_2 to the calendar. The Antarctic Ice Sheet, much of the central Greenland Ice Sheet, and large areas of other polar glaciers lie above the run-off line, the highest elevation at which any melt leaves the glacier. In these regions no summer season exists as defined; the glacier gains mass continuously. At low elevations in the Arctic and parts of the Antarctic, summer usually consists of short periods of ablation interrupted by snowfalls. At the other extreme, ablation continues throughout the year on the lower parts of many glaciers in Iceland, New Zealand, and the tropics; snow accumulates only intermittently during storms. Ablation also continues year-round on the ablation zones of glaciers in the Antarctic Dry Valleys, where most ablation is by sublimation (Fountain et al. 2006). Glaciers near the equator can have two balance years in each calendar year; accumulation occurs in two wet seasons associated with migration of the Intertropical Convergence Zone (the thermal equator).

4.2.4 Annual Glacier Balance and Average Specific Balances

Integrating the annual specific balance over the total area of the glacier, $\mathcal{A}$, gives the balance of the whole glacier for one year, B_n:

$$B_n = \int_{\mathcal{A}} b_n \, d\mathcal{A} \quad \text{and} \quad \overline{b}_n = B_n/\mathcal{A}. \tag{4.7}$$

The quantity $\overline{b}_n$ defines the *average specific balance*, also called the *average net balance*. This quantifies the mass per unit area – or equivalent thickness of ice or water – added to or removed from the entire glacier over the year. If the glacier calves or the basal balance is not negligible, terms for these processes need to be added to Eq. 4.7.

4.2.5 Variation of Surface Balance with Altitude

The specific balance varies substantially from place to place on a glacier, with a broad upper region of mass surplus ($b_n > 0$), the *accumulation zone*, and a broad lower region of mass deficit ($b_n < 0$), the *ablation zone* (Figure 4.4 and Section 2.3). At the end of the ablation season, the surface is mostly bare ice in the ablation zone but snow in the accumulation zone. On some polar glaciers, however, superimposed ice covers the lower part of the accumulation zone (Chapter 2). In unusually warm years, the ablation zone can rise well into the area covered by old firn. The *equilibrium line* is the boundary between the two zones, where accumulation equals ablation for the year; its altitude defines the *equilibrium line altitude*, or ELA. On temperate glaciers (which have no superimposed ice) the ELA is often close to the end-of-summer snowline, but an accurate determination of the equilibrium line requires detailed field studies (Kaser et al. 2003). Often, the equilibrium line is not a distinct "line" but a transition zone where the glacier surface grades from snow, to snow patches, to bare ice.

The ratio of the accumulation-zone area to the total glacier area is known as the *accumulation area ratio*, or AAR. Haeberli et al. (2003) compiled information from 63 valley glaciers in the Alps and found that AAR values, when $B_n = 0$, range from 0.22 to 0.72 with a mean of 0.55. In general, the AAR for $B_n = 0$ reflects the area-altitude distribution of the glacier (Section 4.2.7) and how the specific balance varies with altitude (Gross et al. 1977). The AAR of tropical glaciers, for example, tends to be high because ablation increases rapidly down-glacier near the termini (Kaser and Osmaston 2002). Year-to-year variations of the ELA and AAR depend on annual climate; in warm or low-precipitation years, $B_n < 0$ and the ELA is high. Monitoring ELAs and AARs provides a convenient but imprecise way to assess mass balance changes of many glaciers in a region, though the errors are large in some years (Kuhn et al. 1999). (We give one example later in this chapter; see Figure 4.10.)

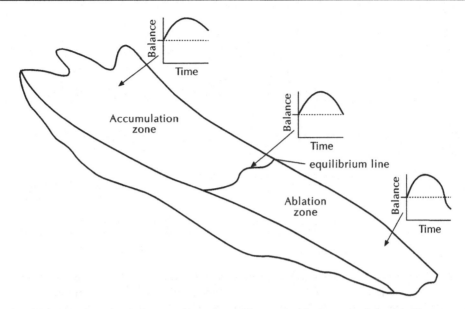

Figure 4.4: How the seasonal progression of specific surface balance varies with distance along a typical glacier. In reality, the plots of balance with time are typically stepped and rough; the smooth curves used here are idealizations.

Not all glaciers have an equilibrium line. In most drainage basins of the Antarctic Ice Sheet, the accumulation zone extends to the ice margin, where mass is lost by calving. Elsewhere, complex patterns of accumulation and ablation occur in some cases. Avalanche deposition and marine fog and cloud cover can create local patches of accumulation at unusually low altitudes.

Nonetheless, the specific balance on most glaciers correlates strongly with elevation. Figure 4.5a illustrates two examples of the variation, one from Nigardsbreen, in southern Norway (latitude 62 °N), and one from White Glacier, on Axel Heiberg Island in Arctic Canada (latitude 79 °N). Both glaciers extend from upper accumulation zones at nearly 2 km altitude to termini near sea level. Figure 4.5b shows the separate contributions of winter and summer balances to the annual total for Nigardsbreen. The magnitudes of all three balances are much larger on Nigardsbreen than on White Glacier. The climate at Nigardsbreen is some 15 °C warmer than at White Glacier, at the same elevation, and Nigardsbreen receives abundant moisture from westerly winds. Each year, snowfall and melt at Nigardsbreen far exceed those at White Glacier. Consequently, the *mass balance gradient*, the rate of change of b_n with altitude, is much larger at Nigardsbreen too.

The pattern of increasing surface balance with increasing altitude, shown in Figure 4.5a, is typical of most mountain glaciers and polar ice caps, and of most zones on the margins of the Greenland Ice Sheet. It reflects the systematic vertical variation of meteorological factors, and in particular the decrease of temperature with increasing altitude, referred to as the *lapse rate*. Values range seasonally from -4 to -10 °C km^{-1}, and typical means are -5 to -7 °C km^{-1}.

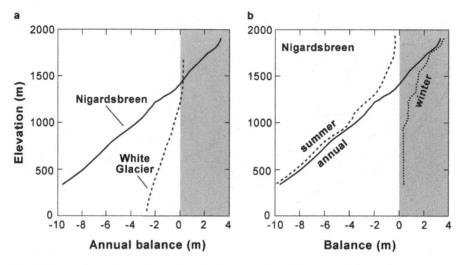

Figure 4.5: (a) Variation of specific surface balances with altitude on Nigardsbreen, Norway (1963–1964), and White Glacier, Arctic Canada (1994–1995). Data from Pytte and Østrem (1965) and *Glacier Mass Balance Bulletin 4*, respectively. (See Cogley et al. (1996) for discussion of White Glacier's mass balance). (b) The variation of seasonal balances on Nigardsbreen.

4.2.6 Generalized Relation of Surface Balance to Temperature and Precipitation

Temperature is the overwhelming control on the altitudinal gradients of annual surface balances, discussed in the previous section, for several reasons:

1. Precipitation reaching the ground switches from snow to rain as temperatures increase.

2. Melt rates increase strongly as air temperatures rise past the melting point (see Chapter 5).

3. The ice surface spends more time at the melting point if the mean temperature rises.

In very cold climates, however, with mean annual temperatures less than roughly $-15\,°C$, temperature affects the surface balance in a different way. Almost no melt occurs. Instead, the surface balance depends primarily on the amount of snowfall. An increase of temperature tends to increase the water content of the air and hence the rate of snowfall (Section 4.2.1).

All of these factors likewise influence how typical surface balance rates $\dot{b}_s$ vary with mean temperatures from place to place around the world. Figure 4.6a gives a rough conceptual view of the situation. The diagram can be divided into two parts; at low temperatures, the surface balance rate increases gradually with temperature, whereas at high temperatures the surface balance rate decreases sharply. High precipitation rates shift the $\dot{b}_s$ curve to the right; low rates shift it to the left. Figure 4.6b plots measured annual surface balance rates against estimated mean annual temperatures on a variety of glaciers. (Some of the values for temperature are not measured ones but instead are estimated from nearby weather stations and corrected

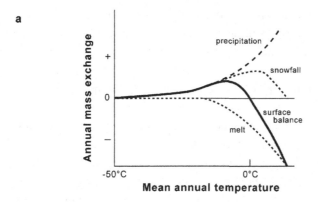

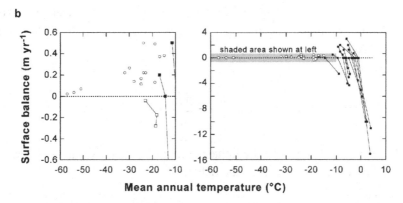

Figure 4.6: (a) Schematic relation of specific surface balance and its governing processes to climatic temperature. (b) Measured specific balances for 15 sites on the polar ice sheets (open circles), for 13 mountain glaciers (filled squares), and 3 sites in the ablation area of Taylor Glacier, Antarctica (open squares). Ablation at the Taylor Glacier sites is by sublimation, rather than melt. Sites on the same mountain glaciers are connected by thin lines. For these, temperature is known from only one location at each glacier, and estimated for the other locations using a lapse rate of $-5.5\,°C\,km^{-1}$. Data sources are Table 2.2 (ice sheets); Kavanaugh et al. (2009a) (Taylor Gl.); various issues of the *Glacier Mass Balance Bulletin* (mountain glaciers). Note the change of vertical scale between the two panels of (b).

for the elevation difference.) The plot covers a wide range of conditions, from the extreme cold of interior East Antarctica to the temperate climates at low altitudes in Norway and New Zealand. The precipitous decline of $\dot{b}_s$ at the highest temperatures, roughly a change of $-1.5\,m\,yr^{-1}\,°C^{-1}$, explains why glaciers in mid- and low latitudes do not extend far from the high mountains on which they originate and why these glaciers retreat when the climate warms.

The surface balance data do not form a single relationship. In part, this reflects how the melt rate varies not simply with mean annual temperature but with other factors such as the length of the melt season, the mean summer temperature, and the occurrence of summer snowfalls (see Chapter 5). The most important source of variability from glacier to glacier, however, is differences of annual precipitation (the "shifts" cited in the previous paragraph). Figure 4.7a, also a rough conceptual diagram, shows how the shape of the $\dot{b}_s(T)$ relation should vary; in particular, the sensitivity $\gamma_T = \partial \dot{b}_s / \partial T$, indicated on the figure, becomes increasingly negative as precipitation increases (Oerlemans 2001, pp. 50–51). Why this happens can be seen by adjusting upward the precipitation curve in Figure 4.6a. With more snowfall, glaciers extend into warmer regions. As temperature increases, the melt rate increases in a nonlinear fashion, and the precipitation changes from snow to rain. Both factors increase γ_T.

Now consider, as a special case, how the surface balance changes with time as climate fluctuates from year to year or shifts over decades. Compared to the entire range of climates represented in Figure 4.6 such changes are typically small, so a linear approximation can be used. The change of annual specific balance (Δb_n) due to changes of temperature (ΔT) and precipitation (ΔP) is

$$\Delta b_n = \underbrace{\frac{\partial b_n}{\partial T}}_{\gamma_T} \Delta T + \underbrace{\frac{\partial b_n}{\partial P}}_{\gamma_P} \Delta P + \cdots. \tag{4.8}$$

We include the "$\cdots$" in Eq. 4.8 to emphasize that additional variables (e.g., cloud cover, wind speed, and humidity) can influence the balance too.

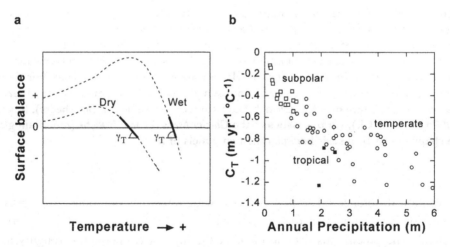

Figure 4.7: (a) Schematic diagram showing how the curve of specific surface balance against temperature varies depending on the precipitation rate. (b) Estimates for the sensitivity of glacier-wide surface mass balance to a unit temperature change, for various mountain glaciers (adapted from Braithwaite 2006). The three tropical examples are not typical of glaciers at the highest altitudes, which should be less sensitive to temperature because sublimation reduces melt.

Theoretical glacier-wide average values for the coefficients γ_T and γ_P can be defined by assuming that the climate variations ΔT and ΔP do not vary from place to place on a glacier (an appropriate assumption for small mountain glaciers but not for ice caps and ice sheets). Averaging over a glacier, of plan-view area $\mathcal{A}$,

$$\Delta \overline{b}_n = \frac{1}{\mathcal{A}} \int_{\mathcal{A}} \Delta b_n \, d\mathcal{A} = C_T \cdot \Delta T + C_P \cdot \Delta P + \cdots \tag{4.9}$$

with

$$C_T = \frac{1}{\mathcal{A}} \int_{\mathcal{A}} \frac{\partial b_n}{\partial T} \, d\mathcal{A} \quad \text{and} \quad C_P = \frac{1}{\mathcal{A}} \int_{\mathcal{A}} \frac{\partial b_n}{\partial P} \, d\mathcal{A}. \tag{4.10}$$

Parameters C_T and C_P define the *mass balance sensitivity*. As Oerlemans (2001, p. 50) has emphasized, these parameters "should be considered as basic quantities that characterise a glacier." Nonetheless, several difficulties need to be kept in mind: values of C_T and C_P are only known approximately; their values will change over time as a glacier's geometry evolves; precipitation and temperature do not necessarily change uniformly and are not the only factors influencing surface balance; and different measures of temperature can be used to define ΔT and hence C_T. (The mean temperature in the melt season is probably best.)

Values for the mass balance sensitivities can be estimated two ways: (1) from regression analysis of observations; (2) using models of the annual cycle of accumulation and melt, calibrated against observations (Oerlemans and Fortuin 1992; Oerlemans 2001, pp. 41–55; Braithwaite 2006). In principle, the second approach is the better one because it explicitly accounts for different processes. Figure 4.7b shows examples of C_T values for various mountain glaciers, estimated with this approach. As expected from the simple concepts outlined in Figures 4.6 and 4.7a, the largest sensitivities apply to glaciers in wet climates. A characteristic value for glaciers in wet temperate regions is $C_T \approx -1 \, \mathrm{m \, yr^{-1} \, °C^{-1}}$. With regard to precipitation, $C_P = 1$ gives a first estimate. This would be exact if all precipitation fell as snow and if snowfall did not affect melt rate. In fact, snowfalls reduce melt rates (snow reflects more solar radiation than does ice, and also thermally insulates underlying ice). Thus we expect $C_P > 1$ for most glaciers. Model calculations by Oerlemans (2001, p. 50) suggest a typical range of $C_P \approx 1$ to 1.5. For glaciers in the warmest and wettest conditions (mean annual precipitations greater than about $3 \, \mathrm{m}$), however, enough of the precipitation falls as rain that $C_P < 1$. Note that our definition of C_P, chosen for consistency with C_T, differs from that shown in Figure 4.13 of Oerlemans (2001).

This approach reduces the complex relationship of surface balance and climate to two parameters. Though useful for general calculations, especially when large numbers of glaciers must be considered together, this approach takes no account of variables like humidity and cloud cover that strongly influence the ablation on some glaciers. Chapter 5 provides a more complete discussion of ablation mechanisms.

4.2.7 Relation of Glacier-wide Balance to the Area-Altitude Distribution

On most mountain glaciers, the surface balance varies systematically with altitude (Section 4.2.5). Its glacier-wide average thus depends on two distinct geographical variables: the variation of specific balance with altitude and the distribution of glacier surface area as a function of altitude. Divide the range of elevations of the glacier surface into N bands. For band j, call the central elevation S_j and the surface area ΔA_j. For one year, the glacier balance is then

$$B_n = \sum_{j=1}^{N} b_n(S_j) \cdot \Delta A_j. \tag{4.11}$$

If the glacier calves, the corresponding rate of mass loss needs to be subtracted from the right-hand side.

Figure 4.8 depicts again the variation of specific balance b_n with altitude for one year on the Norwegian glacier Nigardsbreen (Figure 4.5), but now shown together with the area-altitude

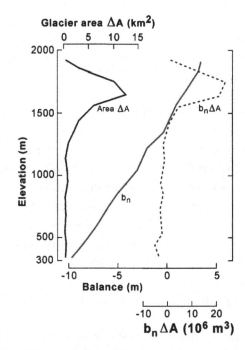

Figure 4.8: Area-altitude distribution and specific mass balance: Nigardsbreen, Norway, 1963–1964. Data from Pytte and Østrem (1965).

distribution. Here, a broad and gently sloping upper basin drains into a narrow outlet that descends steeply to a terminus near sea level (300 m elevation). About 85% of the glacier area lies between 1450 and 1950 m. In the example shown, the equilibrium line was around 1450 m, b_n was positive for 85% of the glacier surface, and $\overline{b}_n$ was +0.9 m. The glacier gained mass despite the rapid ablation of nearly $10\,\text{m}\,\text{yr}^{-1}$ at the terminus. A neighboring glacier with most of its area between 1300 and 1500 m would, in contrast, lose mass in the same year. Furthermore, if the equilibrium line were to rise to 1800 m the average balance of Nigardsbreen would be strongly negative too. A glacier with a large portion of its area close to the mean ELA is particularly likely to fluctuate between positive and negative average balances as accumulation and ablation vary from year to year. In general, if the specific balance varies nonuniformly over the glacier, the area-altitude distribution selects which specific balances contribute most to the glacier average. With a uniform variation, however, the area-altitude distribution has no effect on year-to-year changes in the average balance.

Because the area and thickness of a glacier change over time in reaction to variations of climate and flow, the area-altitude distribution ΔA_j changes too. A time series of annual glacier balances B_n thus reflects both evolution of the glacier's form and variations of the altitude-dependent specific balances $b_n(z_j)$. To discern the effects of the latter – which most closely reflect climate – a new time series can be defined as $B'_n = \sum b_n(S_j) \cdot \Delta A'_j$. Here A'_j is a *reference surface*, a chosen area-altitude distribution that does not vary over time (Elsberg et al. 2001).

4.3 Mass Balance Variations of Mountain Glaciers

The following discussion concerns mountain glaciers that terminate on land; because they do not calve, the glacier balance essentially equals the surface balance integrated over the glacier, the situation discussed so far in Sections 4.2.4 through 4.2.7. Changes in calving rate profoundly affect the mass balance of mountain glaciers, such as those in southeastern Alaska, that flow into the sea or large lakes (Section 11.5.2).

4.3.1 Interannual Fluctuations of Balance

The surface balance fluctuates from year to year with variations of temperature ΔT and precipitation ΔP (Eq. 4.9) such that $\Delta \overline{b}_n = C_T \cdot \Delta T + C_P \cdot \Delta P + \cdots$, with C_T and C_P the sensitivity parameters defined in Section 4.2.6. How much a glacier's balance varies from year to year thus depends not only on the variations of climate factors but also on values of C_T and C_P, which are functions of the mass exchange processes. The best choices for T and P in mid-latitude and subpolar settings are probably mean summer temperature and total winter precipitation, respectively. The fluctuations $C_T \cdot \Delta T$ and $C_P \cdot \Delta P$ correspond approximately to changes of the summer and winter balances defined in Figure 4.3.

Table 4.1: Factors controlling year-to-year variations of
mass balance of three glaciers in Canada (Letréguilly 1988).

Glacier	km from coast	Winter precip.	Summer temp.
		(percent of total variance)	
Sentinel	30	60	17
Place	160	28	39
Peyto	550	≈ 0	67

Summer temperatures have a significant effect on year-to-year variations of surface balance on most mountain glaciers (Oerlemans 2005). The importance of summer temperatures arises from several compounding effects of warm conditions (see Section 4.2.6 and Chapter 5), the same processes that govern the increase of specific balances with altitude.

Accumulation variations are significant in some settings. The fluctuations of climate in coastal regions differ from those far inland. Near the coast, precipitation varies more (larger absolute ΔP) while temperatures vary less. Yet because precipitation rates are comparatively high in coastal regions, we expect comparatively large values for the sensitivity C_T of coastal glaciers (Section 4.2.6). Thus it is not clear whether variations of temperature or precipitation should most strongly influence mass balance fluctuations in coastal settings. Table 4.1 shows the result of a correlation analysis for glaciers in a temperate region; accumulation dominates variability near the coasts but has no influence far inland. Note that this conclusion applies only to year-to-year variations, not to long-term climate changes.

Year-to-year variations of glacier-average balance have been studied since the mid-twentieth century on a few dozen glaciers worldwide. Figure 4.9 shows the longest record, from Storglaciären, Sweden (latitude 68°N), where observations began in 1946 (Holmlund et al. 2005). The annual balance here correlates more strongly with the summer balance ($r = 0.73$) than with the winter one ($r = 0.59$), although both contribute to interannual variability. The balance was negative for most years in the first quarter-century of observation, but since then neither positive nor negative years have predominated. Storglaciären's area has changed very little in the 60-year record, but its volume decreased. The average of $\overline{b}_n$ for the whole period was -0.26 m (w.eq.), which means the glacier thinned by about 17 m ice-equivalent, averaged over the glacier.

Figure 4.10 shows a second example, the period 1953–2005 at Hintereisferner, Austria (latitude 47°N). This 7.1-km-long valley glacier ranges in altitude from 2426 to 3710 m. The average of $\overline{b}_n$ was more strongly negative (-0.49 m) than at Storglaciären, and there was a statistically significant trend over the period toward more negative values. This glacier, like most in the Alps, is not in equilibrium with its current climate. Disequilibrium has driven a 26% reduction in the glacier area over the 53-year period and a 27 m average thinning, concentrated in the ablation zone. At Hintereisferner, fluctuations of the ELA and AAR correlate with the annual balance (Figure 4.10; $r = -0.90$ for the correlation of $\overline{b}_n$ with ELA).

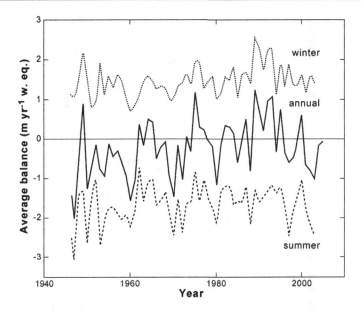

Figure 4.9: Glacier-wide average of summer, winter, and annual specific balances of Storglaciären, Swedish Lapland. (Adapted from Holmlund et al. 2005).

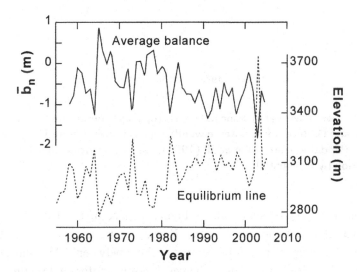

Figure 4.10: Glacier-wide average of specific balance, compared to equilibrium-line altitude, Hintereisferner. (Data from the *Glacier Mass Balance Bulletins*, compiled by S. Marshall; see Kuhn et al. 1999 for analysis of this glacier.)

4.3.2 Cumulative Balance and Delayed Adjustments

Annual glacier balances such as those shown in Figure 4.10 can be summed to make a graph of cumulative balance over time (Figure 4.11). This is a good way to find out how well the size

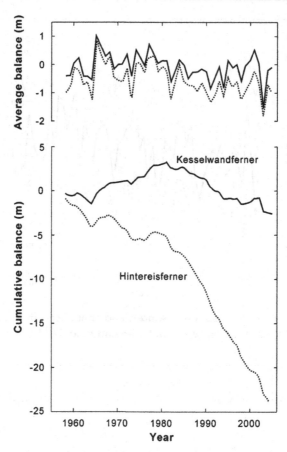

Figure 4.11: Glacier-wide average specific balances (top) and cumulative balances (bottom) on Kesselwandferner and Hintereisferner. Kesselwandferner is better adapted to present climate than is Hintereisferner. (Updated from Kuhn et al. (1985), using data from the *Glacier Mass Balance Bulletins*; compilation by S. Marshall.)

of a glacier is adapted to the present climate. Figure 4.11 compares the cumulative record for Hintereisferner with that of an adjacent glacier, Kesselwandferner. Because of their proximity, these two glaciers are subject to similar climates. Yet, between 1958 and 2005, Hintereisferner thinned by 27 m (ice-equivalent), whereas Kesselwandferner lost only 3.3 m. Kesselwandferner is the smaller glacier and its mean elevation is higher (3200 versus 3050 m). Specifically, 83% of Kesselwandferner lies above 3050 m, compared with only 49% of Hintereisferner. The long low-lying tongue of Hintereisferner, which ablates rapidly, is a relic of the Little Ice Age of the nineteenth century; the size of the glacier has not yet fully adjusted to the present climate. Persistent positive or negative values of glacier-average annual balance indicate disequilibrium between a glacier and its climate, although many years of observation may be required to discern the disequilibrium given year-to-year variations.

In addition to illustrating the importance of the area-altitude distribution for a glacier's mass balance (Section 4.2.7), Hintereisferner also shows how climate variability on timescales longer than the observations can control the glacier-wide balance. In this case, retreat of the glacier did not keep pace with climate warming, and the mass balance will continue to be negative for years to come even if the climate were to now stay constant. (In fact, continued warming has driven glaciers in the Alps even farther from equilibrium in recent decades.)

Thus, the response of a glacier to a change in specific balances may continue for many years. At Hintereisferner this lag depends, in part, on the pace of mass exchange near the terminus; with a specific balance of roughly $-3\,\mathrm{m\,yr^{-1}}$, as observed, it takes more than half a century to remove 200 m of ice. But the lag also depends on ice flow, which each year partially replenishes the 3 meters of ablation with ice from higher on the glacier. Glacier adjustments always reflect a combination of these factors – Chapter 11 analyzes the problem. Because of flow, adjustments can continue even through periods with a zero average balance; a glacier's shape can change even though the total mass remains constant. Likewise, because of changes in glacier extent and shape, the average balance can change even in periods of constant climate that follow climate shifts or anomalous years. For example, a few years of heavy snowfall will form a bulge or thickening on a glacier. This changes the ice flow and, over several years, causes the glacier terminus to advance. As the glacier extends to lower altitudes, its balance trends increasingly negative until the terminus reaches its most extended position. The balance then increases back toward zero during the subsequent retreat.

If the balance of a glacier remains zero for many years, its dimensions might eventually remain constant. The larger the glacier, the longer the period will usually have to be. The glacier is then said to be in a *steady state*. This is a useful theoretical concept – the configuration of a glacier often resembles the steady form – but an exact steady state never occurs in reality.

4.3.3 Regional Variations of Mass Balance

If mass-balance variations are caused by regional climate changes, then the annual balances of all the glaciers in a region should correlate one with another. Letréguilly and Reynaud (1990) applied a statistical model (based on Lliboutry 1974) to examine the variations of annual balances $\overline{b}_n$ between glaciers in a region and over time. They assumed that for a particular glacier j the net balance in year t can be written as

$$\overline{b}_{jt} = \alpha_j + \beta_t + \epsilon_{jt}. \tag{4.12}$$

Here α_j is a variable that depends only on the glacier whereas β_t varies from year to year but takes the same value for all glaciers. The quantity ϵ_{jt} denotes a random variable with a normal distribution and zero mean. Letréguilly and Reynaud made such an analysis for six glaciers in the Alps, seven in Norway, four in central Asia, and five in western North America. Figure 4.12

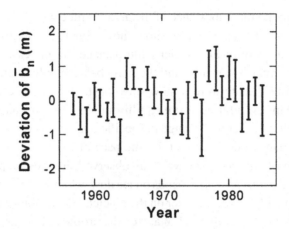

Figure 4.12: Deviation of average specific balance from its mean, for six glaciers in the Alps, for the period 1957–1985. Adapted from Letréguilly and Reynaud (1990). Each line extends from the highest to the lowest of the six values, in each year. Year-to-year variations exceed the differences between individual glaciers.

shows their results for the Alps. Although large differences between glaciers occurred in a few years such as 1974 and 1976, the year-to-year variations (β_t) were large compared to the variations between the glaciers (α_j). Results for the other areas were similar. Year-to-year variations accounted for between 73% and 85% of the total variation of annual balances within each region. Again, McCabe et al. (2000) analyzed winter balance records from 22 glaciers in the northern hemisphere. They identified broad regions of correlated interannual variations, centered on the Alps, western North America, and northeastern Asia.

These are important results for the use of glaciers as climate indicators. It suggests that measurements of year-to-year variations in mass balance on one glacier are likely to provide a record that is typical of the region. "Region," however, should be defined as an area not more than several hundred kilometers in diameter. Letréguilly and Reynaud, for example, found no correlation between mass balance variations in the Alps and Norway. Moreover, the correlation was higher for glaciers in the same valley than for those on opposite sides of a mountain range. Cogley and Adams (1998) compared records from many different glaciers worldwide and found that correlations drop exponentially with distance, such that little similarity remains for glaciers separated by more than about 600 km. On the other hand, long-term climate changes affect much larger regions than do year-to-year variations. Correlations between regions should therefore be stronger for decades and centuries than for single years.

Note that these conclusions refer only to variations of a glacier's balance about its long-term mean. Mean balances may differ substantially even on adjacent glaciers, as the example in Section 4.3.2 shows. Causes of such differences include different area-altitude distributions, different local climates, and different surface properties (e.g., cover or absence of rock debris that affects the absorption of sunlight).

4.4 Englacial Mass Balance

4.4.1 Internal Accumulation

The redistribution of mass from the surface to the glacier interior occurs by (1) freezing of meltwater or rain that percolates into cold snow or firn, (2) freezing of surface water that flows into crevasses and moulins, and (3) freezing of water that is injected into fractures in the bottom of the glacier.

The mass of meltwater refrozen in the seasonal snowpack can be measured, but there is considerable heterogeneity at scales of centimeters to meters due to the complexity of percolation and drainage processes in the snow (Pfeffer et al. 1991; Pfeffer and Humphrey 1996). This remains an important source of uncertainty in mass balance measurements and models for polar and high-altitude glaciers.

4.4.2 Internal Ablation

Melting occurs within a glacier if the temperature is at melting point and there is a source of energy. Sources include flowing water and deforming ice and, near the surface, solar radiation.

Flowing water may be warmer than the adjacent ice and so cause melting by direct heat transfer. Moreover, the water flowing through a glacier loses potential energy, which dissipates as heat. This melts ice along the boundaries of the passageways conveying the water. These processes are discussed in Chapter 6.

Ice deformation accomplishes mechanical work and the equivalent energy also dissipates as heat, a process referred to as *strain heating*. The energy ultimately derives from the loss of potential energy due to flow of the glacier, but the heat dissipation concentrates where the ice deforms. For a stress (force per unit area) in the ice of σ, and a strain rate $\dot{\epsilon}$, the strain heating rate $\dot{S}_E$ equals the deformational work per unit volume, or $\dot{S}_E = \sigma \cdot \dot{\epsilon}$. (See Section 9.7.1.1 for the precise definition of this product.) The total meltwater produced in a temperate ice layer of thickness Z_t represents an ice-mass loss (mass per unit time per unit horizontal area) of

$$-\dot{b}_e = \frac{\dot{S}_E Z_t}{L_f}, \tag{4.13}$$

for latent heat of fusion L_f. Such melting occurs throughout temperate glaciers (Section 9.4) but should be most significant in basal layers, where stresses and rates of deformation are highest.

Internal melt also occurs in temperate layers that form near the beds of some polar ice streams. Jakobshavn Ice Stream in west Greenland provides the best example (Section 8.9.2.1). This fast-flowing glacier overlies a bedrock trench. The ice thickness is 1.5 to 2.5 km along the trench's centreline, and the bottom 200 to 400 m is thought to be rapidly deforming temperate ice. Shear stresses in this layer are large for a glacier, approximately 200 to 250 kPa (Clarke and Echelmeyer 1996). These stress values, together with strain rates calculated from the creep relation for ice (Chapter 3), imply that the layer should produce a net melt of $-\dot{b}_e \approx 0.2$ to $0.6 \, \text{m} \, \text{yr}^{-1}$. Significant rates of internal melt should occur wherever ice is temperate, thick, and deforming rapidly.

4.5 Basal Mass Balance

Mass balance at the glacier base, $\dot{b}_b$, depends on the prevailing temperature, the availability of water, and the net supply rate of energy. We define the latter as, in $\text{W} \, \text{m}^{-2}$,

$$E_b = G + u_b \, \tau_b + k_T \frac{\partial T}{\partial z}. \tag{4.14}$$

Here G denotes the heat flux from the substrate. Slip of the ice over its bed (at rate u_b) adds heat by frictional dissipation, at a rate equal to the product of u_b and the shear stress on the bed τ_b. Heat conduction into the overlying ice, the final term, can either supply or remove heat, depending on the sign of the temperature gradient ($\partial T / \partial z$, with z positive upward). Parameter k_T is the thermal conductivity of ice.

Several factors influence the value of G. For floating ice G depends on temperature and circulation of the water. For grounded glaciers, G is typically similar to the crustal geothermal flux, but several processes can make the two unequal. Water flowing along the bed carries heat. Subglacial temperature gradients – and hence heat conduction – depend on the local pattern of groundwater flow, and also vary over time as climate and ice dynamics alter temperatures in the overlying ice. An example of the latter occurs when a thick glacier rapidly advances over cold terrain; the glacier insulates the substrate and warms it up.

If the temperature at the glacier base equals the melting point for the prevailing pressure, a net energy supply E_b causes melt or freeze-on, which we discuss in the next two sections. Controls on the basal temperature are discussed in Chapter 9.

4.5.1 Basal Accumulation

Basal accumulation contributes significantly to the mass balance in some regions of ice shelves and at special locations along the beds of glaciers. If $E_b < 0$ the freeze-on rate is

$$\dot{b}_b = -\frac{E_b}{\rho_i L_f} \quad (\text{m} \, \text{yr}^{-1} \text{ of ice}). \tag{4.15}$$

The condition that $E_b < 0$ typically occurs when heat conduction upward into the ice exceeds the geothermal and frictional sources. This is the case, for example, beneath polar ice streams with large vertical temperature gradients or slippery beds. (The latter reduce the frictional heating.) Freeze-on also occurs if the temperature of basal water is below the *in situ* melting point determined by the pressure (the water is said to be *supercooled*). In this case, the inflow of supercooled water maintains a negative energy flux that reduces the value of G in Eq. 4.14.

Ice accreted by freeze-on has different chemical and physical characteristics than the overlying glacier ice. Freeze-on of water flowing along the bed of a grounded glacier traps rock debris (Lawson et al. 1998). The bottom 210 m of the ice core from Vostok, Antarctica, was accreted from a subglacial lake and has a distinct isotopic composition and grain-scale structure (Section 6.4.2).

Freeze-on of supercooled water occurs where subglacial water at high hydrostatic pressure flows into a region of lower pressure. Because the melting point rises as pressure drops, such water needs to warm to maintain the same temperature as the overlying ice. If it does not warm quickly enough, the water will be colder than the ice and supercooled. Supercooling can occur in subglacial lakes, beneath ice shelves, and beneath glaciers with overdeepened beds (Röthlisberger 1972; Alley et al. 1998; Cook et al. 2006).

4.5.1.1 Accretion in Overdeepenings

Overdeepened means that the bed slopes upward in the direction of ice flow, a common situation beneath glaciers and ice sheets. The ice thickness decreases down-glacier, reducing the pressure and increasing the melting point. At the same time, the increase of bed elevation reduces the conversion of potential energy to frictional heat in water flowing down-glacier along the bed. The combination of reduced ice thickness and increased bed elevation can lead to supercooling beneath a warm-based glacier. The following paragraph explains why this happens in more detail.

The temperature of basal ice everywhere equals the melting point for the pressure. The glacier base receives heat from geothermal and ice-friction sources (call their combined energy flux E_b^*) and from frictional dissipation in the flowing water (energy flux E_w). The heat could be used to melt ice at a rate $-\dot{b}_b = [E_b^* + E_w]/\rho_i L_f$ (conduction into the overlying ice being negligible for temperate glaciers). However, a downstream decrease of pressure implies a downstream increase of ice temperature. Warmer ice, in turn, reduces the transfer of heat from water to ice; thus a portion of the available heat warms the water instead of the ice and so the water temperature T_w increases downstream too. Call the flux of water along the glacier bed q_w (volume per unit time per unit length across-glacier). For a steady state,

$$E_b^* + E_w - c_w q_w \frac{\partial T_w}{\partial x} = -\rho_i L_f \dot{b}_b, \tag{4.16}$$

where c_w denotes the volumetric heat capacity of the water, and x the direction of flow. Freeze-on, rather than melt, occurs if the inflow of cold water (the term with $\partial T_w/\partial x$) extracts more energy than is supplied by the source $E_b^* + E_w$. A downstream decrease of glacier thickness favors energy extraction, whereas an upward-sloping bed reduces the term E_w in the heat source. Both factors, acting together, are necessary for freeze-on; a precise definition of the condition for freeze-on is given in Section 6.3.2.3.

Using Eq. 4.16, Alley et al. (1998) showed that, for realistic dimensions of overdeepenings, maximum freeze-on rates of order $\dot{b}_b \sim 0.1\,\mathrm{m\,yr^{-1}}$ should occur beneath glaciers with a lot of surface melt and hence large q_w. This magnitude is consistent with observations near the front of Matanuska Glacier, Alaska. Here, several meters of ice has accreted to the base over several decades, as shown by the presence of atomic-bomb tritium in the ice.

4.5.1.2 Accretion Beneath Ice Shelves

A significant amount of water freezes to the bottoms of ice shelves at some locations, despite the occurrence of exceptionally rapid melt at other locations. Supercooling can occur where plumes of buoyant water rise along a shelf's sloping bottom. Furthermore, the base of an ice shelf is not planar but instead is marked by longitudinal grooves and ridges. Water circulating into the grooves can also supercool. In either case, the freshening of ocean water by mixing with melt contributes to the tendency for freeze-on (Jenkins and Doake 1991). Supercooling may not be necessary for basal freeze-on, however. Circulation of cold water beneath an ice shelf can shield the ice from geothermal heat and from warm waters in the nearby ocean. Freeze-on is then driven by heat conduction upward into the shelf, because the bottom layers of shelves are generally warmer than the layers above (Section 9.10).

One example of a major ice shelf with basal accumulation is the Amery Ice Shelf, East Antarctica (Morgan 1972; Fricker et al. 2001). Seen in plane view, water circulates beneath the shelf in a clockwise pattern. Saline water enters the subshelf cavity on the east side, mixes with melt and freshens. It then returns to the open ocean by rising along the base of the shelf on the west side, where it supercools and freezes. The thickness of accreted ice on the western part of the shelf is typically 50 to 150 m and reaches a maximum of about 190 m. (The accreted thickness was mapped by comparing the total ice thickness, determined from surface elevations and the requirements of flotation, to the depth of the boundary between fresh and salty ice, detected by radar surveys.) Basal accumulation is also significant beneath the Ronne Ice Shelf, one of the largest shelves in Antarctica (Thyssen et al. 1993).

Accreted ice often remelts before it reaches the shelf front. Where it survives, however, it sometimes produces icebergs with an unusual green color due to organic material derived from the original seawater (Warren et al. 1993).

4.5.2 Basal Ablation

Where $E_b > 0$, basal melting occurs at the rate given by Eq. 4.15. Typical basal melt rates of grounded glaciers are several millimeters per year. Beneath temperate ice that is not sliding, a typical geothermal heat flux of $G = 0.05\,\mathrm{W\,m^{-2}}$ gives $\dot{b}_b \approx -5\,\mathrm{mm}$ per year. For a sliding rate of $10\,\mathrm{m\,yr^{-1}}$ frictional dissipation causes an additional 3 mm per year of melt, assuming a stress of $10^5\,\mathrm{Pa}$. Some fast-flowing glaciers slide at ten times this rate, or more, in which case frictional dissipation accounts for most of the basal ablation.

Much higher rates of basal melt occur where geothermal fluxes concentrate. Such concentrations can be episodic, as with subglacial volcanic eruptions, or persistent. The latter occurs in a region of north-central Greenland. Layers in the ice sheet are warped downward, a signature of enhanced melt at the base beneath. Calculations based on this observation indicate that G exceeds $1 \, \mathrm{W \, m^{-2}}$ over an area of order $10^4 \, \mathrm{km^2}$ (Fahnestock et al. 2001). Subglacial geothermal activity in Iceland is even more intense. Heat fluxes of up to $50 \, \mathrm{W \, m^{-2}}$ in the Skaftá and Grımsvötn volcanic cauldrons generate about $5 \, \mathrm{m \, yr^{-1}}$ of basal melt in parts of the Vatnajökull Ice Cap (Björnsson 1988). Glaciers mantle many active volcanoes in the Andes, the Cascades, Alaska, Antarctica, and elsewhere; basal melt must affect the hydrology of these features appreciably.

Exceptionally high basal melt rates also occur beneath floating ice shelves, where circulation of water between warm oceans and cold subshelf cavities acts as a heat pump. Recent measurements have shown that this process is much more important for ice sheet mass balance than previously thought. Basal melt rates of 2 to $4 \, \mathrm{m \, yr^{-1}}$ appear to be typical of the large Antarctic ice shelves, and melt rates an order of magnitude higher prevail in some regions (Jacobs et al. 1996; Rignot and Jacobs 2002). Average rates of 10 to $20 \, \mathrm{m \, yr^{-1}}$ are reported on Petermann Glacier in northwest Greenland (Rignot 1996) and on floating ice tongues in the Amundsen Sea, West Antarctica (Rignot and Jacobs 2002).

Melt rates beneath ice shelves may be estimated from precise measurements of ice-thickness changes using radar (Corr et al. 2002). Usually, however, melt rates are inferred indirectly. One method is a glaciological zonal mass balance, which requires measurements of ice inflow and outflow, surface balance, and changes of shelf geometry (Section 4.7). Another method uses oceanographic measurements of water flux and salinity. Jacobs et al. (1996) and Jenkins et al. (1997) combined such oceanographic and glaciologic techniques to deduce basal melt rates of Pine Island Glacier, one of the major ice streams in the Amundsen Sea region. For the glaciological deduction, ice input was estimated from measurements of ice velocity and thickness at the grounding line (to measure the contribution flowing from the inland ice), and measurements of specific balance on its 80-km-long surface. Iceberg calving from the floating tongue took away only 50% of this mass input, and no major changes in the geometry of the floating tongue were observed over the period of study (1992–1994). The mass budget could only be closed through an area-averaged basal melt rate of $12\pm3 \, \mathrm{m \, yr^{-1}}$. This agreed with the independent oceanographic deduction (Jacobs et al. 1996), in which measurements of freshening and cooling of shelf waters that cycle through Pine Island Bay were used to estimate basal melt of 10 to $12 \, \mathrm{m \, yr^{-1}}$. Subsequent measurements showed thinning of the floating tongue. Including this information in a regional mass balance for only the inner part of the shelf revealed even higher basal melt rates (40 to $50 \, \mathrm{m \, yr^{-1}}$) near the grounding line of Pine Island Glacier (Rignot 1998; Rignot and Jacobs 2002).

The energy for such extreme melt rates is supplied by vigorous exchanges of subshelf waters with warmer open-ocean waters. From Eq. 4.15, a $\dot{b}_b$ of $-1 \, \mathrm{m \, yr^{-1}}$ requires a net heat flux of about $10 \, \mathrm{W \, m^{-2}}$. This is a factor of 200 times a typical geothermal heat flux, but ocean heat fluxes of tens of $\mathrm{W \, m^{-2}}$ are possible if the water beneath an ice shelf circulates continuously while drawing from an open-ocean source. In the Amundsen Sea, waters below 800 m are $+1\,°C$ or warmer, and hence 3 to $4\,°C$ above the local melting point (Jacobs et al. 1996). Rignot and Jacobs (2002) compared basal melt rates to "excess" water temperatures near the grounding lines at 23 sites in Antarctica. Excess water temperature refers to the difference between the nearest measured open-ocean temperature and the freezing point expected at the grounding line, given the water depth. There is a positive correlation with melt rates (Figure 4.13). These data suggest that melting of about $1\,\mathrm{m\,yr^{-1}}$ can be expected for every $0.1\,°C$ that subshelf waters exceed the *in situ* melting temperature. Further observations in the Amundsen Sea suggest this relationship might be applicable to changes over time; basal melt of the large Amundsen Sea shelves – Pine Island, Thwaites, and Smith – increased about $5\,\mathrm{m\,yr^{-1}}$ when regional ocean temperatures warmed by about $0.5\,°C$ in the 1990s (Gille 2002; Shepherd et al. 2004).

For subshelf melting, a mechanism is needed to mix and evacuate the water, and in particular to draw intermediate or deep waters into the subshelf cavity. Buoyancy can drive convection. Freeze-on produces cold, saline waters that sink, while fresh meltwater rises along the sloping base of a shelf. Important variables include the geometry of the basin beneath a shelf, and the characteristics of water masses in the surrounding region. The warmest water masses around Antarctica are at depth. They originate when sea-ice formation on the continental shelf concentrates salts, producing water with a greater density than for colder surface waters near

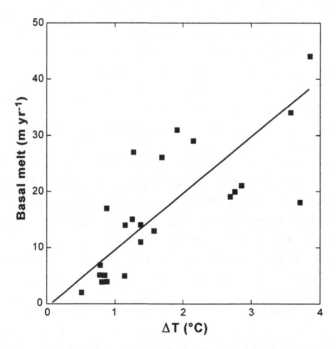

Figure 4.13: Correlation of ice shelf basal melt rate with ΔT, the excess of water temperature in nearby open oceans relative to freezing point at the grounding lines. Adapted from Rignot and Jacobs (2002).

the coast. In Antarctica, temperatures of intermediate and deep shelf waters range from -2 to $+2\,°C$, depending on their origin (e.g., Jenkins and Holland 2002). Thick ice shelves are more exposed to these intermediate and deep waters than thin ones, and therefore typically melt more rapidly.

4.6 Mass Loss by Calving

Calving is the separation of ice blocks from a glacier's margin. Most calving occurs at margins that stand or float in water; the calved blocks become icebergs. Calving at marine margins accounts for much of the mass loss from the ice sheets – more than 90% of the ablation from Antarctica and about half of the ablation from Greenland. The process is also significant for mountain glaciers that terminate in deep lakes or the ocean.

A glacier whose terminus rests on the bed but stands in ocean water is called a *tidewater glacier*. If such a glacier advances into deeper water, it may start to float, forming a *floating tongue*. A large, broad floating tongue is an *ice shelf*. Flotation occurs if ice thickness at the front margin, H_M, is smaller than the critical buoyant thickness; that is, for water of depth H_w,

$$H_M < \frac{\rho_w}{\rho_i} H_w \approx 1.1 H_w \qquad (4.17)$$

for densities ρ_w and ρ_i of water and ice. Tidewater glaciers in Svalbard, southern Greenland, and Alaska are grounded, whereas most tidewater glaciers in Antarctica and northern Greenland end in floating tongues. Floating glaciers and shelves rarely extend beyond the point where their sides are protected by the walls of a bay or fjord. Two notable exceptions, however, are the Erebus and Drygalski ice tongues in the Ross Sea sector of Antarctica, visible from planes flying to McMurdo Station.

Calving results from the initiation of fractures and their propagation through the ice thickness. Two processes are therefore essential for calving: fracturing and thinning of the ice as it flows toward the calving margin. Thinning results from ablation and from longitudinal stretching of the ice. The latter occurs in the flow regime known as longitudinally extending flow (Section 8.4). The fracturing process is not well understood; for more information about fracturing, see our discussion of crevasse formation in Section 10.9.

Several variables probably control the calving rate, but no predictive relation incorporating them has been established. Undercutting of the terminus collapses the frontal cliff, so the rate of ablation below the water line might be important (Robin 1979). (Heat transported by currents and the mechanical action of waves both contribute to rapid melt of ice at the front.) However, many fast-flowing and floating glaciers produce large icebergs that must originate with fracturing well inland of the cliff. Longitudinal and transverse stretching of the ice favor fracturing; they cause tensile failure. Such stretching usually increases as a glacier flows faster. The relation between tensile fracturing and stretching rate depends on the stiffness of the ice, a function of temperature; cold ice is stiffer and so fractures at a smaller stretching rate. Ice that is thick inhibits complete fracturing, but water from surface melting can fill fractures and help them propagate to depth. Stresses that are only important close to the ice front – related to cliff shape and buoyant forces – can also influence fracturing. In principle, all of these factors can be related to broader controls such as the force balance and flow regime of the ice, sea level and ocean bathymetry, tides, storm swell, sea ice, and temperatures of water and air.

Calving rates are specified in several ways. The total volume of icebergs generated per unit time defines the volumetric calving rate $\dot{C}$ (m^3 yr^{-1} of ice). Multiplying by the ice density ρ_i then gives the mass loss rate, $\dot{B}_c = \rho_i \dot{C}$ (kg yr^{-1}). Alternatively, the volumetric rate can be converted to an average *calving velocity*, $\dot{c}$, by normalizing to the cross-sectional area A_f of the ice front: $\dot{c} = \dot{C}/A_f$, with units of m yr^{-1}. If the ice were not flowing, the front would retreat at the rate $\dot{c}$. In practice, $\dot{C}$, $\dot{B}_c$, and $\dot{c}$ include ablation due to melting of the frontal cliff, which is difficult to separate from true calving.

4.6.1 The Calving Spectrum

The length (L) of a glacier ending at a calving front changes over time (t) if the calving velocity does not match the forward velocity of ice at the front margin (u_M):

$$\frac{dL}{dt} = u_M - \dot{c}. \tag{4.18}$$

The relative importance of terms in this relation varies from place to place and over time. This variation defines a spectrum of situations, with both end members well represented in nature. In the first case, the position of the terminus is essentially fixed, so that $dL/dt = 0$ gives a good approximation over some timescale – usually several years to decades and longer. For example, consider that many ice shelves cannot extend beyond the limits of their topographic embayments and that many tidewater glaciers cannot advance past a point where water depth increases abruptly. In these cases, the average calving rate matches the rate of ice flow; $\dot{c} = u_M$ and $\dot{c}$ varies because climatic and ice-dynamical factors modulate u_M. In the second end member, the calving rate is strongly influenced by external variables unrelated to u_M, leading to rapid changes of the terminus position (a large $|dL/dt|$). For example, increased surface melting can rapidly disintegrate an entire ice shelf (Scambos et al. 2004). Another example is a tidewater glacier whose terminus initially rests on a shoal. Retreat of the terminus increases calving as the front moves into deeper water; by this process, a glacier front can retreat a great distance while rapidly producing icebergs (Meier and Post 1987). Most calving margins are probably somewhere between these end members. Their position on the spectrum depends on the timescale considered. There is no consensus on whether calving rates are best predicted as a function of inland ice dynamics or variables like water depth and temperature at the terminus. Indeed, the observations indicate that different approaches should be applied in different situations.

4.6.2 Calving from Tidewater Glaciers

Tidewater glaciers are abundant in the mountains of southeastern Alaska, where high snowfall rates and deeply incised valleys favor thick, active glaciers. They are also widely distributed in the Arctic and the Antarctic Peninsula and nearby islands. In the southern Andes, glaciers terminate in calving fronts in large lakes; these calving fronts behave much like those in marine settings. In all these regions, the glaciers mostly terminate in fjords in shallow water, tens to a few hundred meters deep. Warm ocean water (several °C) and abundant surface melt favor calving and prevent the glaciers from developing floating tongues.

Using data from twelve Alaskan glaciers, Brown et al. (1982) proposed the empirical relationship

$$\dot{c} = k_c H_w, \tag{4.19}$$

where H_w denotes water depth at the terminus, averaged over the width of the ice front. They found a value for the coefficient of $k_c = 17 \, \text{yr}^{-1}$. Pelto and Warren (1991) found a similar relationship for a data set that included additional glaciers from Alaska, western Greenland, and Svalbard. Calving in freshwater lakes also appears to correlate with water depth, but with rates – for a given depth – an order of magnitude less than those in tidewater (Warren et al. 1995; Warren and Aniya 1999). Although there appears to be a relationship, there is no reason why calving rates should be proportional to water depth alone.

The dramatic retreat of Columbia Glacier, Alaska, commencing in the early 1980s, provided data on how calving rates change with time. These observations are broadly consistent with Eq. 4.19 (Meier et al. 1994), but at subannual timescales the calving rates are coupled to glacier velocity and thinning of the ice front (van der Veen 1996, 2002). Ice flow and calving both accelerate when the ice front thins to near a "minimum critical thickness" related to the *excess buoyancy*:

$$H_b = H_M - \frac{\rho_w}{\rho_i} H_w. \tag{4.20}$$

H_b is the thickness of the ice front in excess of that which would be floating. At Columbia Glacier, calving rates increase when H_b drops to less than about 50 m, irrespective of water depth. At LeConte Glacier, Alaska, daily to seasonal variations in calving likewise appear to vary with buoyancy; increased calving occurs when the ice nears flotation, a result of high water levels in spring tides, and of ice thinning related to periods of accelerated glacier flow (O'Neel et al. 2001).

The mechanisms connecting buoyancy with calving remain uncertain. A small H_b often leads to increased ice flow and longitudinal stretching, for reasons explained in Chapter 11. Rapid stretching, in turn, promotes tensile failure and fracture propagation, and hence calving. On the other hand, calving might increase simply because faster glacier flow delivers more ice to a location where wave action and warm water efficiently erode the ice front. The former explanation seems more likely.

This is, however, only one part of a complex problem. Rapid calving itself influences the configuration of the terminus, which in turn changes the stresses causing stretching flow (Hughes 2003). Further, the increased calving may originate externally – as a change in ocean conditions or surface melt, or from stochastic events at the calving front. For example, increased calving from the floating tongue of Greenland's Jakobshavn Ice Stream in the late 1990s was probably triggered by increases in water temperatures (Holland et al. 2008). Increased calving subsequently caused thinning, acceleration, and collapse of the floating tongue, a dynamical response that far exceeded the direct effect of terminus melt (Joughin et al. 2004a; Thomas 2004).

Moreover, the idea of a critical sensitivity to excess buoyancy is inconsistent with the existence of floating ice. Cases where calving rates are dictated by the outward ice flow and a critical thickness are therefore restricted to a subset of tidewater glaciers with H_M close to flotation.

4.6.3 Calving from Ice Shelves

Floating ice shelves occur only in the polar regions, where ocean surface waters are near-freezing ($-2\,°C$ for saltwater) and where most of the ice flowing off the land is subfreezing. Ice shelves occupy embayments all around West and East Antarctica. Small ice shelves and floating tongues also exist in Greenland and the high Arctic islands. Many ice shelves thickened and grounded during the last ice age and sometimes advanced to the edge of the continental shelf. They do not appear to have extended into the deep water and open ocean conditions beyond the continental shelf break.

The calving velocity from a steady ice shelf equals the ice flow rate at the front, u_M in Eq. 4.18. In turn, u_M is governed by longitudinal extension along the shelf – the stretching flow – and by shear on the shelf sides. (See Section 8.9.3 for an analysis of ice shelf flow.) But longitudinal stretching also drives fracturing and hence calving itself. A shelf that terminates beyond its embayment or where its embayment widens also stretches transversely (across the shelf), which contributes to fracturing. Fractures are also generated by shear along the sides of a shelf, as it flows past embayment walls, islands, or grounded spots. On the other hand, such shear zones might impede the propagation of extensional fractures from the shelf center to side margins.

Except near the side margins, the front of a broad ice shelf deforms primarily by longitudinal stretching. It is thus reasonable to propose (Alley et al. 2008) that fracturing and hence calving increase with the rate of stretching (the "longitudinal strain rate" $\dot{\epsilon}_{xx} = \partial u / \partial x$). Stretching produces tensional stresses (τ_{xx}) oriented perpendicular to the shelf front, and opens fractures that parallel the shelf front. Such fractures must develop for a large iceberg to separate from a shelf. These facts suggest a simple hypothesis – that calving rate increases in proportion to the rate of stretching, or $\dot{c} \propto \dot{\epsilon}_{xx}$. Alternatively, the calving rate might increase as $\dot{c} \propto \dot{\epsilon}_{xx} \tau_{xx}$, the rate of work per unit volume associated with the stretching. The second hypothesis is plausible because the creation of new fractures is one way for the energy of work to dissipate.

To test these hypotheses, Alley et al. obtained velocity, ice thickness, and stretching rate data from different places and times on ten ice shelves (nine in Antarctica, one in Greenland). The positions of the shelf fronts moved only slowly compared to the flow of ice along the shelves; thus, the calving rates were assumed to equal the measured velocities ($\dot{c} = u_M$). Tensile stresses were assumed to be proportional to the measured ice thicknesses at the shelf fronts ($\tau_{xx} \propto H_M$); this is a property of floating tabular ice bodies (Section 8.9.3). Over the wide range of shelves examined, inferred calving rates increased with both $\dot{\epsilon}_{xx}$ and $\dot{\epsilon}_{xx} H_M$ (Figure 4.14a). A better relationship was found by allowing the calving rate to increase also with the width of the shelf, Y (Figure 4.14b). As one example, the relation

$$\dot{c} \propto \dot{\epsilon}_{xx} H_M \frac{Y}{Y + Y_o},$$

(4.21)

with Y_o a constant equal to 200 km, explained 94% of the variance of inferred calving rates. The dependence on width presumably reflects how shear margins inhibit fracture propagation on a narrow shelf.

Although these simple relations might capture the dominant controls on calving from large ice shelves in cold settings, it is also clear that other variables affect fracturing. In particular, meltwater enhances fracture growth and the peculiar shape of an ice shelf front produces fractures. A shelf front consists of an ice cliff above the water line, as well as a much larger submerged "cliff." The upper, subaerial, cliff is not supported by pressure from the water whereas

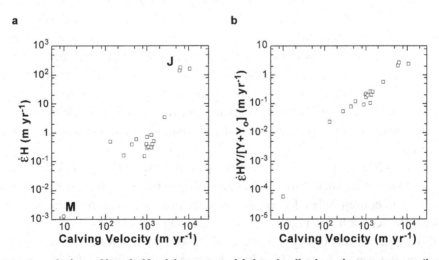

Figure 4.14: Correlation of ice shelf calving rates with longitudinal strain parameters (based on calculations of Alley et al. 2008). The end-member locations are labelled "M" for the McMurdo Ice Shelf, Antarctica, and "J" for Jakobshavn Ice Stream, Greenland.

the lower cliff is; this creates bending forces in the shelf. Melting of the cliff sometimes makes its shape irregular; in some cases, the subaqueous part erodes more than the upper cliff, while in other cases the opposite occurs. These configurations also induce stresses in the ice. So does tidal flexure. Reeh (1968) analyzed the stresses associated with bending in the presence of a straight cliff face, assuming the ice could be treated as a linear viscous fluid. He found that, in theory, the induced stresses reach a maximum at a distance of about one ice thickness from the front, with a magnitude probably great enough to cause fracture, depending on the ice thickness and temperature. The one-ice-thickness scale is consistent with observations of large blocky icebergs produced by Greenland's floating ice tongues, although small ice blocks also fall off the frontal cliffs. This style of calving seems to be typical for comparatively warm settings – perhaps because warmer ice is mechanically weaker than cold ice, or perhaps because meltwater helps fractures to penetrate downward from the surface, as described in Section 10.9.2. Such a meltwater effect might explain why marine outlet glaciers in Greenland advance in winter but retreat in summer (Sohn et al. 1998).

A different style of calving occurs on cold Antarctic shelves, where large, tabular icebergs form infrequently. They originate in fracture propagation through the shelf at a much greater distance back from the front (e.g., Fricker et al. 2002). Some of these icebergs are enormous, with areas as large as $10^4 \, km^2$. The fact that the fronts of these shelves do not produce the smaller blocky icebergs may be due to the absence of surface melt (reflecting cold air), reduced melting below the water line (cold water), or mechanically stronger ice (cold ice).

The Antarctic Peninsula is an interesting transitional setting, both climatically and glaciologically. A number of small ice shelves on the peninsula disintegrated in the last three decades of the twentieth century. Ice shelf area declined by more than $13,500 \, km^2$ in this period (Scambos et al. 2004). The loss of shelves was probably a consequence of atmospheric warming in the region, estimated to be about $3 \, °C$ over the second half of the twentieth century. Vaughan and Doake (1996) suggested that ice shelves cannot survive if mean annual air temperature rises above $-5 \, °C$. Previously, Mercer (1978) had suggested that ice shelves cannot survive if the mean temperature of the warmest month exceeds $0 \, °C$, a conclusion based on the spatial distribution of shelves on the Peninsula. Above this temperature, meltwater production helps fractures to propagate through the shelf. High air temperatures did precede the collapse of the Larsen A and Larsen B Ice Shelves in 1995 and 2002, and abundant meltwater was observed on the shelf surfaces.

Scambos et al. (2000) suggested that warming acts on the structurally weak *mélange* of water, slush, and ice that fills ice shelf "rifts," the large and open fractures that sometimes cut through a shelf from top to bottom. When frozen, the mélange provides structural integrity, but melting of it weakens the ice shelf and promotes fracture propagation. Ice shelf modelling by Larour et al. (2004) supported this theory. Fricker et al. (2005) documented a seasonal cycle in rift propagation on the Amery Ice Shelf, perhaps because summertime melt softens the mélange filling the rifts.

Warming of ocean waters may also be important. The Weddell Sea warmed in the late twentieth century, but how this contributed to ice shelf disintegrations on the Antarctic Peninsula is not known. Increased ocean temperatures should increase basal melt rates and cause an ice shelf to thin or its grounding line to retreat. Both processes can increase extensional flow (Section 11.5), possibly increasing fracturing and calving.

The presence of sea ice and icebergs packed tightly in fiords may suppress calving by damping tidal forcings and hence flexure of the floating ice. Reeh et al. (1999a) discussed the differences between tidewater outlets in southern and northern Greenland. In southern Greenland, the tidewater glaciers are grounded and calve rapidly. In northern Greenland, the outlets form extended, floating tongues of ice and ablation occurs largely by basal melting. Permanent sea ice in northeast Greenland may be one of the factors that enable the survival of floating ice tongues. Its effect is difficult to separate from the influences of colder air and ocean water.

4.6.4 Calving Relations for Ice Sheet Models

Simulations of calving fluxes in glacier and ice sheet models remain unconvincing. Absence of established relationships is the major problem. The relations identified by Alley et al. are the only ones so far proposed that might be used in models to calculate calving rate from ice dynamical variables. These relations are not well tested, however, and only apply to floating ice shelves in steady climates. The concept of a "threshold of viability" for floating ice shelves, related to climatic temperature, might be used in models. Specifically, a shelf cannot survive if summer temperatures rise beyond a threshold that triggers abundant surface melt. Originally based on Mercer's interpretation of the low-latitude limit for shelves along the Antarctic Peninsula, this idea now appears to be confirmed by recent shelf collapse events.

For lack of a better criterion, most models either let ice shelves advance to the edges of the model grid, or assume that ice shelves terminate at a prescribed water depth ($H_w = 400\,\text{m}$ is a typical limit, roughly corresponding to the edge of a continental shelf). For grounded ice, most models either assume that calving increases with water depth, as per Eq. 4.19, or constrain the ice front thickness H_M instead of the calving.

Although each of these approaches is amenable to model studies, none is satisfactory or universally applicable. This makes it difficult for ice sheet models to tackle some of the most important problems in glacier physics, from ice shelf breakup to tidewater glacier dynamics.

4.7 Methods for Determining Glacier Mass Balance

There are several strategies for determining the mass changes of a glacier system or sectors within it. Consider a region of area $\mathcal{A}$. The mass at any coordinate equals the product of vertically averaged density ($\overline{\rho}$) and ice thickness $H = S - B$, where S and B indicate elevations of the surface and bed, respectively. For a balance of the whole glacier, the change of ice mass ΔM in a time interval Δt can be written three ways:

$$\underbrace{\Delta M}_{A} = \underbrace{\Delta \cdot \int_{\mathcal{A}} \overline{\rho}\,[S - B]\,d\mathcal{A}}_{B} = \underbrace{\Delta t \cdot \left[\int_{\mathcal{A}} [\dot{b}_s + \dot{b}_e + \dot{b}_b]\,d\mathcal{A} - \dot{B}_c \right]}_{C} \quad (4.22)$$

where Δ indicates a net change over the interval. To determine the balance of a zone or sector, ice flow through the zone's boundaries must be accounted for. The following quantity then replaces term C:

$$\underbrace{\Delta t \cdot \left[\int_{\mathcal{A}} [\dot{b}_s + \dot{b}_e + \dot{b}_b]\,dS + Q_{in} - \{\dot{B}_c \text{ or } Q_{out}\} \right]}_{C'}. \quad (4.23)$$

Here Q_{in} denotes the flux into the zone – the mass per unit time entering the region by ice flow – and Q_{out} is the flux departing. If the region includes the ice margin, then the calving rate $\dot{B}_c$ replaces Q_{out}; these are not necessarily the same because the margin can migrate.

All the methods for determining mass changes from year to year give uncertain results. The best approach is to use more than one method.

The Glaciological Method   The classic glaciological method is to acquire field measurements of annual accumulated snow mass and surface wastage at a network of points along the glacier surface. Spatial interpolation over the entire glacier area then gives term C of Eq. 4.22, though any calving losses must be evaluated separately. Internal and basal balances are generally assumed to be negligible, except for refreezing within near-surface firn. The point data are usually assumed to represent characteristic values for altitude bands (Eq. 4.11, Section 4.2.7). Østrem and Brugman (1991) and Fountain et al. (1999; and papers therein) have discussed the techniques for surface balance measurements in detail.

First, consider assessments of the balance for a single year. In the accumulation zone, both the thickness and density of the annual layer must be measured. The annual layer consists of snow and ice in varying proportions, so densities vary widely. Where meltwater percolates through the current year's layer and refreezes beneath, this mass must also be measured. To locate the bottom of the annual layer, visual inspection alone often identifies the previous summer surface on temperate glaciers. The burial of stakes set in the snow can be measured from year to year. Without such surveys it is sometimes necessary to identify the annual layer using measurements of isotopic or chemical composition.

In the superimposed ice zone and the ablation zone, stakes set in holes drilled in the ice are used as references. The distance between the top of the stake and the ice surface is measured at the beginning and end of the balance year. The difference gives the thickness added or removed; multiplying by the density of ice converts it to mass. This method also works for assessments over multiple years if stakes are long compared to the annual ablation.

To estimate accumulation rates over multiple years, more constraints on the age of the firn must be acquired. On temperate glaciers, dirt layers corresponding to summer surfaces can again be identified in cores or pits. Annual cycles of isotopic or chemical layering can be measured, as can unique horizons such as volcanic ashes or, more commonly, the cesium and tritium layers associated with the final years of large-scale nuclear bomb testing in the early 1960s. If ages are well known from core studies at one site, layers can be traced to other sites using radio reflection surveys, and accumulation determined from the layer depth; this technique works well on the ice sheets.

The accuracy of mass balance measurements with these methods can be difficult to assess. The main source of inaccuracy lies in the problems of sampling. Measurements can be made at only a limited number of points and many glaciers have large crevassed areas where no measurements are possible. Layer thicknesses change over time, complicating measurements. For a full discussion, see Fountain et al. (1999).

The mass balance time series from Storglaciären and Hintereisferner, presented earlier in this chapter (Figure 4.9 and 4.10), were obtained using the glaciological method.

The Process-Model Method   This technique uses meteorological models for explicit calculations of the surface mass exchanges in Eq. 4.3; in combination, these give an estimate for term C in Eq. 4.22. The models are the same ones used for weather forecasting. Calculated snowfalls depend on temperature and precipitation, a function of the modelled wind fields and atmospheric water vapor content. Melt and sublimation are calculated from the surface energy balance and temperature. Alternatively, melt is calculated from model temperatures by assuming it increases with the number of degree-days above freezing (Section 5.4.2.1). This alternative moves the calculation one step away from a pure process model but seems to work best in practice.

For quantitatively accurate results, meteorological and mass balance variables in the model always need to be calibrated against data from the glacier. Basal and englacial terms are usually assumed to be negligible, except for refreezing in the snowpack. There is no reliable process model for calving.

The Greenland mass balance simulations of Box et al. (2006) provide a good example of the process-model method (Section 4.8.1), and the analysis of Hanna et al. (2005) exemplifies the use of degree-day calculations of melt.

The Hydrological Method   This method calculates mass balance from the measured fluxes of water to and from the glacier, independent of the glacier surface balance (Tangborn et al. 1975). Precipitation is estimated using meteorological data and models, and runoff is measured at gauging stations. To determine the glacier balance over some time period, inflows of water (precipitation, and sometimes runoff from hillslopes) are summed, and the outflow – the runoff in the glacier foreland – is subtracted. This method implicitly includes surface, englacial, and basal terms, which is an advantage. However, because many glaciers store water early in the summer and release it later (see Chapter 6), the method is not a reliable way to measure a glacier's annual mass balance.

Altimetric Methods   These methods synthesize measurements of changes in glacier surface altitude (S) to calculate term B in Eq. 4.22. Techniques for measuring altitude changes include repeat photogrammetry, ground surveys, airborne or satellite laser altimetry, and satellite interferometric radar. The latter two methods have recently become important sources of mass balance information, as they can examine large regions of remote areas including the ice sheets (e.g., Thomas et al. 2001; Davis et al. 2005; Helsen et al. 2008).

Considering timescales of years to decades, complications arise when term B is estimated from measurements of surface altitude alone – especially for polar ice sheets. A correction, usually small, needs to be made for changing bed elevation due to ongoing long-term isostatic adjustments or tectonic activity. Two other influences are more significant. In the case of ice shelves, basal elevation varies with sea level and tides. In the accumulation zone, a correction is needed for fluctuations in firn density that accompany variations of temperature and snowfall rate (Section 2.5.2). Density variations do not, of course, change the mass of the ice sheet. They do cause a vertical motion of its surface (Section 8.5.5.4 and 5). Corrections for this effect are sometimes made using a model for the rate of firn densification (e.g., Helsen et al. 2008; Reeh 2008).

Krabill et al. (2000, 2004) used airborne laser altimetry to monitor changes of the Greenland Ice Sheet; Zwally et al. (2005) did the same with a satellite laser. Arendt et al. (2002) demonstrated the usefulness of airborne laser altimetry for mapping large-scale changes of valley glaciers and icefields in Alaska. Rignot et al. (2003) used radar measurements from the Space Shuttle for a similar analysis but with better spatial coverage. They examined glaciers in Patagonia.

Geological Methods   Although not useful for constructing detailed time series, geological methods provide the only good long-term estimates of mass changes. The former extent and past surface elevations of a glacier are estimated from their geomorphological signature of ice-edge deposits and scours: moraines, outwash surfaces, kame terraces, and trimlines. These features can often be dated using historical records, lichenometry, or geochemical techniques like radiocarbon and cosmogenic nuclide dating. When a glacier advances or thickens it may override the markers of previous advances, a basic limitation of the method. Geological reconstructions have provided estimates of the net loss in ice volume of valley glaciers since the Little Ice Age (e.g., Hopkinson and Young 1998). The method has also been valuable for quantifying twentieth-century changes in Greenland outlet glaciers (Weidick 1996), including periods of rapid thinning of Jakobshavn Icestream in the early twentieth century (Csatho et al. 2008).

Flux-gate Method   The flux-gate method is used for studies of zones, and sometimes for whole-glacier balances if calving margins do not shift. In the latter case, the calving loss is assumed to equal the ice flux measured at some location inland of the margin, minus a correction for the surface balance between the inland location and the margin.

Ice flux refers to the mass or volume per unit time flowing through a given cross-section of the glacier (Section 8.1.1). The fluxes into and out of a chosen zone of the glacier are calculated from measurements of surface velocity and ice thickness, and an assumption about how the velocity varies with depth. The cross-sections used in flux calculations are usually called *flux gates*.

To find the mass balance, measurements of specific balance $\dot{b}$ from the glaciological method or the process-model method are combined with the mass-flux measurements at flux gates to find term C′ in Eq. 4.23. Alternatively, the mass-flux estimates are used with altimetry data to equate terms B and C′ in order to infer $\dot{b}$ rather than ΔM. If the surface part of $\dot{b}$ is known from the glaciological method, all of these measurements can be combined to estimate the basal balance. This method was used to infer melt rates beneath Antarctic ice shelves (Section 4.5.2).

Rignot and Kanagaratnam (2006) used the flux-gate method to quantify multi-year changes in ice sheet drainage basins of southern Greenland (Chapter 14). These changes resulted from accelerated flow in outlet glaciers, so there was no way to assess the mass balance from $\dot{b}$ alone. Another important recent application of the flux-gate method was to assess the balance of the Amundsen Sea sector of the West Antarctic Ice Sheet (Thomas et al. 2004; Section 11.5.3.2).

Gravitational Method This method infers term A in Eq. 4.22 from small changes in the gravity vector, measured with tandem satellites. At the time of writing, the Gravity Recovery and Climate Experiment (GRACE), launched in 2002, is the only satellite mission with this capability. To extract the ice sheet mass balance, effects of isostatic and tectonic adjustments of the underlying lithosphere must be taken into account; these correlate with changes of bed elevation. GRACE has provided mass balance data for Greenland, Antarctica, and Southeast Alaska (see Chapter 14; Velicogna and Wahr 2006a, 2006b; Luthcke et al. 2008). The technique resolves the mass changes averaged over distances of several hundred kilometers. How accurately it works at higher spatial resolutions is controversial.

4.8 Mass Balance Regimes of the Ice Sheets

4.8.1 Greenland Ice Sheet

Figure 4.15 plots the spatial pattern of annual specific surface balances on the ice sheet, averaged for the years 1988 to 2005. The values – which are approximate – were obtained by calibrating a meteorological process model against data from shallow cores and ablation stakes.

Precipitation occurs throughout the year, a consequence of persistent cyclonic storm activity and the abundant supply of moisture from the North Atlantic. Precipitation generally decreases from south to north, as temperatures decrease and distance from the primary oceanic moisture source increases. Most precipitation on the Greenland Ice Sheet is snow. The ice sheet is ringed by an ablation zone, which in most regions extends up to altitudes of about 1 to 1.5 km (less in northernmost Greenland). The equilibrium line is lower in the southeast than the southwest, because of high snowfall in the southeast. This, plus steep mountain topography, makes the ablation zone in the southeast narrow (too narrow, in fact, to see at the scale of Figure 4.15). Near the margin, melt rates in the south can exceed $5\,\mathrm{m\,yr^{-1}}$. Less melting occurs in the north

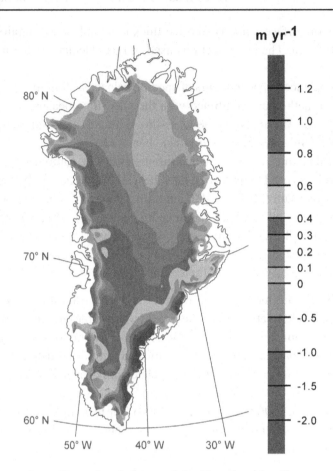

Figure 4.15: Estimated specific surface balance of the Greenland Ice Sheet, averaged for years 1988–2005 (data courtesy of J. Box; map courtesy of N. Schlegel). The balances are a revision of those given by Box et al. (2004 and 2006) using a degree-day method to calculate melt. (Refer to the insert for a color version of this figure)

(about $1 \, \text{m} \, \text{yr}^{-1}$ at sea level), and none occurs in the ice sheet interior (see Figure 2.2a). Runoff of surface meltwater removes approximately half of the yearly accumulation. Sublimation losses are about one-fourth as large as runoff. Calving and melt at marine margins account for about 50% to 60% of total ablation (total ablation is a larger number than yearly accumulation because the ice sheet has been losing mass). Approximately half of the total ablation in Greenland is associated with the discharge of about two dozen fast-flowing outlet glaciers, each of which conveys ice from a large interior basin outward to calving margins and low-elevation zones of melt (Bauer 1961; Rignot and Kanagaratnam 2006; Section 8.9.2).

Table 4.2 gives a rigorous estimate for the different surface mass exchanges, averaged over the entire ice sheet. Antarctic values are shown for comparison. The Greenland numbers are from Box et al.'s (2006) study using the process-model method (Section 4.7). They calculated

Table 4.2: Surface mass exchanges averaged for the Greenland and Antarctic Ice Sheets, for the years 1988–2004 (Box et al. 2006; Krinner et al. 2006). Units for specific rates are $m \, yr^{-1}$ (w.eq.). Units for total mass fluxes are $(10^{12} \, kg \, yr^{-1})$, based on ice sheet areas of $1.691 \times 10^6 \, km^2$ and $13.828 \times 10^6 \, km^2$ for Greenland and Antarctica, respectively. Net is the sum of snowfall and refreeze, minus melt and sublimation.

	Snowfall[†]	Sublimation (direct)	Sublimation (blowing snow)	Melt	Refreeze	Net
Specific rates:						
Greenland	0.365	0.038	0.020	0.292	0.086	0.101
Antarctica	0.175	0.013		0.002	0.002	0.162
Total exchanges:						
Greenland	617	64	34	494	145	170
Antarctica	2420	180		28	28	2240

[†] The total annual snowfall, before any sublimation.

daily weather over the ice sheet using a regional weather model with a resolution of 24 km – sufficient detail to capture the main topographic patterns of the ice sheet but not to resolve well the ablation zone in the southeast. At the boundaries of this high-resolution model, meteorological parameters were specified from *reanalyzed climatology*. Reanalyzed climatology – which provides the best available estimate for atmospheric conditions over time – incorporates all available surface and upper-air observations to guide and constrain model-based daily meteorologic reconstructions (Kalnay et al. 1996). With this forcing at the boundaries, the regional model estimates precipitation, snow-rain partitioning, and the surface energy balance everywhere on the ice sheet; the energy balance, in turn, yields estimates of melting and sublimation (Chapter 5). Results were calibrated against observations of both surface balances and meteorological conditions from a network of stations on the ice sheet. The regional weather model can thus be viewed as a physics-based interpolation scheme.

Snowfall and melt dominate the surface balance. About 34% of the melt refreezes and does not contribute to runoff from the ice sheet. Direct sublimation removes approximately 10% of the snowfall. As a rough estimate, sublimation of blowing snow is about half as large as the direct sublimation.

Table 4.3 assembles some recent estimates of Greenland's surface balance and compares it to estimated losses by calving and marine basal melt. The period of averaging is different in the different studies. The net value shows whether the whole ice sheet grew or diminished. A net value can be given only if calving-margin losses have been estimated. Recent observations show that the total calving flux from Greenland can change significantly within a single decade (Rignot and Kanagaratnam 2006), so calving losses from one period cannot be applied to others. The numbers suggest that the ice sheet was slowly losing mass in the mid- to late twentieth century. (Since year 2000, losses have increased; see Chapter 14.) Rignot et al. (2008b) suggested that the net balance was about zero in the mid-1970s, a period of reduced temperatures. The tabulated

Table 4.3: Estimated mass exchanges for the Greenland Ice Sheet, given as annual totals (10^{12} kg yr^{-1}). Net accumulation is snowfall minus sublimation. Runoff is surface melting minus refreezing. "Net" is sum of surface, calving, and basal components. We have standardized to an ice sheet area of 1.691×10^6 km^2.

Net accum.	Runoff	Surface	Calving	Basal	Net	Period	Reference
547	276	271	−239	−32	0	mid to late 20th century	Reeh et al. (1999a)
562	273	289				1961–1990	Hanna et al. (2005)
		299	−413°		−106 ± 73	1964	Rignot et al. (2008b)
		300	−404°		−97 ± 47	1996	Rignot et al. (2008b)
539	278	261				1988–1999	Mote (2003)
543	373	170				1988–2003	Box et al. (2006)
600	359	242				1993–2003	Hanna et al. (2005)

° Includes basal-melt component implicitly.

calving rate from Reeh et al. (1999a) was based on a synthesis of earlier compilations of iceberg flux observations and measured ice velocities near calving fronts. The calving rates from Rignot et al. (2008b) were calculated from observed ice flow in outlet glaciers.

The differences between these estimates arise from the different methodologies and the different time periods studied. The Box et al. (2006) study was discussed earlier. The same modelling technique gave the surface balance components used by Rignot et al. (2008b). Hanna et al. (2005) used reanalyzed climatology but, in contrast to the surface energy balance method used by Box, calculated melt from a positive-degree-day melt model. Mote (2003) used microwave emissions to map the area of ablation and combined the result with positive-degree-day melt modelling.

4.8.2 Antarctic Ice Sheet

Figure 4.16 maps the annual specific surface balance on Antarctica, estimated from a compilation of field measurements (Vaughan et al. 1999b). Values were interpolated through regions of sparse data using measurements of microwave emissions, which correlate with specific balances (Zwally and Giovinetto 1995).

Glaciologists divide Antarctica into three regions based on geography, topography, and glacier dynamics: the Antarctic Peninsula, West Antarctica, and East Antarctica. This is also a useful division for discussing its mass balance. The Antarctic Peninsula reaches lower latitudes than the rest of the continent and is strongly influenced by the surrounding ocean. Its warm, wet climate resembles that of the mountains of Patagonia rather than the Antarctic plateau. Precipitation rates exceed 1 m yr^{-1}. This, plus the rugged dissected topography and narrow form of the peninsula, lead to an alpine style of glaciation. Outlet glaciers transfer ice from high-elevation ice fields to termini along the coast or into ice shelves that fringe the peninsula. This is the

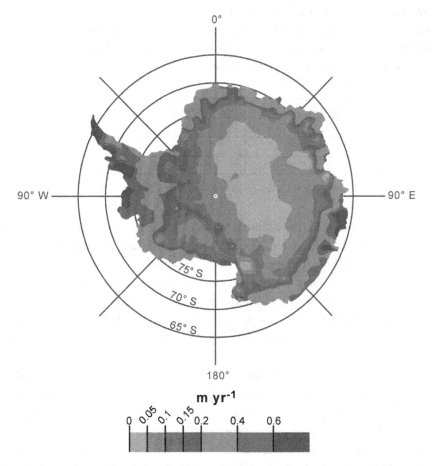

Figure 4.16: Estimated specific surface balance of Antarctica in the late twentieth century (adapted from Vaughan et al. 1999) (Refer to the insert for a color version of this figure)

only part of Antarctica with significant surface melt and the only part where recent atmospheric warming directly caused loss of ice (Vaughan and Doake 1996).

Elevations in interior East Antarctica are 3 to 4 km, and the climate is exceptionally cold and dry, with snowfalls of 0.02 to 0.1 m yr^{-1}, and recorded temperatures as low as $-89\,°C$. Compared to East Antarctica, the surface elevation of the West is much lower; elevations of 1 to 1.5 km are typical of the interior region. Consequently, the West is warmer and receives more snowfall than the East, typically 0.1 to 0.4 m yr^{-1} – similar to central Greenland.

Calving and basal melting account for most of the ablation in both East and West Antarctica, and these processes operate primarily on the ice shelves. Snowfall rates are higher on the ice shelves than on the inland ice. According to the Vaughan et al. (1999b) analysis, although ice shelves make up 12.5% of the total area of the Antarctic Ice Sheet, their annual surface mass budget amounts to 21% of the total for the ice sheet; averages were estimated as $\dot{b}_s = 0.27$ m yr^{-1} (w.eq.) on the ice shelves and $\dot{b}_s = 0.15$ m yr^{-1} (w.eq.) for the grounded ice sheet.

Table 4.4: Estimated annual mass exchanges for the Antarctic Ice Sheet, given as total (area-integrated) rates (10^{12} kg yr^{-1}). The symbols indicate snowfall ($\dot{A}_s$), sublimation ($\dot{S}_s$), net accumulation ($\dot{A}_s^{net} = \dot{A}_s - \dot{S}_s$), surface balance ($\dot{B}_s$), calving ($\dot{B}_c$), and basal balance ($\dot{B}_b$). Net balance is $\dot{B} = \dot{B}_s + \dot{B}_b - \dot{B}_c$. We used an ice sheet area of 13.828×10^6 km^2.

$\dot{A}_s$	$\dot{S}_s$	$\dot{A}_s^{net}$	Runoff	$\dot{B}_s$	$\dot{B}_c$	$\dot{B}_b$	$\dot{B}$ [†]	Period	Reference
		2144	53	2091	2016	−544	−469	20th century	Jacobs et al. (1992)
2586	290	2295	0	2295			−265	20th century	Wild et al. (2003)
2420	180	2240	0	2240			−320	1981–2000	Krinner et al. (2006)

[†] Based on these authors' estimate of $\dot{B}_s$, with values of $\dot{B}_c$ and $\dot{B}_b$ from Jacobs et al. (1992).

A climate-model estimate for the different surface exchanges, averaged for all of Antarctica, is shown in Table 4.2 (Krinner et al. 2006). Though drier than Greenland, the large area of Antarctica makes the total snowfall much greater – enough so that random interannual fluctuations might significantly influence the yearly rate of global sea-level change (Oerlemans 1981a). About 7% of the snowfall is lost to sublimation. Krinner et al. (2006) concluded that about half of the accumulation in central East Antarctica occurs through direct deposition of "diamond dust" (clear-sky ice crystal precipitation), while the other half is associated with a small number of frontal incursions.

In Table 4.4, three estimates for the net surface balance for all Antarctica, based on climate models, are compared to estimated calving and marine basal melt. Net accumulation rates roughly match the empirical estimate of Vaughan et al. (1999b). The uncertainties on the net balance are large enough that the present state of mass balance of Antarctica cannot be determined this way, except to recognize that the ice sheet is very roughly in a state of balance. We return to the topic of Antarctica's current mass balance in Chapter 14.

Further Reading

Geografiska Annaler, volume 81A (1999), features many papers about mass balance studies. Two manuals, one by Østrem and Brugman (1991), the other by Kaser et al. (2003), summarize field techniques for mass balance studies and give concise discussions of the science behind the strategies. A book compiled by Bamber and Payne (2004) discusses numerous aspects of mass balance studies, focused on ice sheets. The book by Oerlemans (2001) thoroughly discusses the mass balance of mountain glaciers, including mass exchange processes and applications to specific glaciers.

The World Glacier Monitoring Service (WGMS) was initiated in 1894 in Zurich, Switzerland, with a mandate to coordinate systematic, international observations on glacier variations. The WGMS publishes data on glacier extent and mass balance in the Fluctuations of Glaciers series, with semidecadal data compilations dating to 1959. Since 1991 the WGMS has also produced the *Glacier Mass Balance Bulletin* annually or biannually.

CHAPTER 5

Mass Balance Processes: 2. Surface Ablation and Energy Budget

"It was as hot a day as we had ever had. The temperature was 36° Fahr. in the shade. . . . This had an awful effect on the surface, covering it with pools and making it very treacherous to walk upon"

South: The Endurance Expedition, **Ernest Shackleton**

5.1 Introduction

In the complex relationship between glaciers and climate, one of the key processees is melt of snow and ice at the glacier surface. Melting followed by runoff accounts for most of the ablation on many glaciers. The runoff itself is an important water resource downstream, for agriculture, domestic uses, and power generation.

The total melt over the course of a year generally increases in warmer climates. In particular melt increases with summer temperatures. For this reason, the surface mass balance turns strongly negative when glaciers flow to low altitudes (as seen in Figure 4.6) or when ice sheets expand toward low latitudes. And increased melt drives glacier retreat when climate warms. But temperatures alone do not determine melt. Wind speed, humidity, clouds, snowfalls, surface characteristics, and other variables all play a role. Moreover, their effects are partly interdependent. As one consequence, all of these variables not only influence melt directly but also influence the sensitivity of melt to warming.

In the previous chapter we reviewed all the factors that determine glacier mass balance, but neglected the details of melt and its relation to climate – the subject of the present chapter. In very cold environments, sublimation replaces melt as the primary mechanism of surface ablation. We discuss sublimation here as well. To understand the causes of ablation by melting and sublimation, we need to examine the ways in which the glacier surface gains and loses heat and be able to measure or calculate the gains and losses under different weather conditions. The theoretical treatment of heat exchange at the surface makes drastic simplifying assumptions about how to represent rapidly fluctuating processes involving many variables. For the long-timescale problems most relevant to glacier evolution, the use of simple empirical relations is a valuable alternative. Here we outline the full theory and discuss the factors controlling

Copyright © 2010, Elsevier Inc. All rights reserved.
DOI: 10.1016/B978-0-12-369461-4.00005-3

each term in the energy budget, using field examples. We then discuss the relation between surface ablation and climate: the simplified approaches used in glacier models, the importance of variable weather conditions, and the contrasts between energy budgets of glaciers in different regions. We begin by reviewing a few fundamental concepts.

5.1.1 Radiation

All matter radiates electromagnetic energy to its surroundings. A material that emits the maximum possible amount of radiation for its temperature is known as a *perfect radiator*. Because such a material also absorbs all the radiation that falls on it, a perfect radiator is also known as a *black body*, and the radiation from it is referred to as black-body radiation. Snow behaves as a nearly perfect radiator in the infrared part of the spectrum. The intensity of the radiation emitted by a perfect radiator depends only on its temperature. The Stefan-Boltzman Law gives the radiated energy flux E (energy per unit area per unit time, $W\,m^{-2}$) as

$$E = \sigma T^4, \tag{5.1}$$

for absolute temperature T. *Stefan's constant*, the parameter σ, has a value of $5.67 \times 10^{-8}\,W\,m^{-2}\,K^{-4}$. The radiation is distributed over a band of wavelengths. The wavelength of maximum flux λ_m, the peak of the spectrum, is inversely proportional to T. For nonperfect radiators, a factor ϵ, the *emissivity*, appears as an additional coefficient multiplying the right-hand side of Eq. 5.1. The emissivity is the ratio of the actual flux to that emitted by a perfect radiator at the same temperature. Although snow reflects visible light abundantly, it nonetheless has an emissivity close to one because it emits mostly at infrared wavelengths.

To a good approximation, the sun and the Earth's surface behave as perfect radiators. The sun has an effective temperature of about 6000 K and λ_m of about 0.5 μm; this lies in the visible part of the spectrum. The Earth's effective temperature is slightly less than 300 K and λ_m falls in the infrared spectrum at about 10 μm. The spectra of solar and terrestrial radiation overlap very little; 99% of solar energy lies between wavelengths of 0.15 and 4 μm, largely in the visible range, whereas 99% of terrestrial radiation is infrared, between 4 and 120 μm. The two bands are referred to as *shortwave* and *longwave* radiation. The instruments used in field studies to measure radiation, known as *radiometers*, give values integrated over a band of wavelengths, typically 0.3 to 2.8 μm for shortwave and 5 to 50 μm for longwave.

5.1.2 Energy Budget of Earth's Atmosphere and Surface

The sun is the ultimate source of almost all heat received at the Earth's surface (geothermal sources being negligible, except locally). The flux of energy toward Earth – above the atmosphere and at normal incidence – defines the *solar constant*, with a value of approximately $1367\,W\,m^{-2}$. The Earth's surface usually receives less than this, even when the sun is directly overhead, because some radiation scatters back to space and some is absorbed by clouds, water vapor, particles, and ozone in the atmosphere. In addition, some of the radiation reaching the surface reflects back; how much depends on the nature of the surface. At the same time, the Earth's surface emits longwave radiation. Some of it escapes to space and the atmosphere absorbs the remainder, mainly in clouds, water vapor, and carbon dioxide. These clouds and gases also emit longwave radiation. Although some escapes to space and some is reabsorbed in the atmosphere, a substantial amount of reradiated longwave energy returns to the Earth's surface (the *greenhouse effect*). Because the Earth acts as a perfect radiator in the infrared, nearly all of this returned radiation is absorbed at the surface and so contributes to warming or melting of ice. The natural greenhouse effect increases the Earth's surface temperature by about 30 °C; it is a major factor in the climate.

Several additional processes transfer heat between the Earth's surface and atmosphere. Turbulent eddies bring heat to the surface if the air is warmer than the surface or remove heat if the air is colder (processes known as *sensible heat transfer*). In addition, the surface receives latent heat when atmospheric water vapor condenses on it and loses heat when moisture evaporates or sublimates from it. The amount of heat transferred by these processes increases with the vigor of turbulence in the air flowing along the surface. Underground conduction also transfers sensible heat to or from the surface, and precipitation brings additional sensible heat.

Figure 5.1 diagrams the flow of energy through the atmosphere and its exchanges with the surface; values are averaged for the planet and an annual cycle. The left-hand side of the diagram shows shortwave radiation, the right-hand side shows longwave. The longwave "back radiation" is the greenhouse effect. At some times and places, the longwave back-radiation exceeds the surface radiation, giving a net longwave flux toward the surface – opposite from the diagram. Likewise, at some times and places the fluxes of sensible heat ("thermals") and latent heat ("evapotranspiration") are directed toward the surface rather than away from it. Note the size of the greenhouse effect; on average only a fraction 4/35 (about 11%) of the longwave radiation from the surface escapes directly to space. Note too that where the surface is snow or ice, it reflects a larger fraction of total incoming solar radiation than the 9% average shown.

At the surface, the absorbed solar radiation approximately balances the upward fluxes of longwave radiation and sensible and latent heat. An exact balance, however, would occur only if the radiation, the weather, and the surface conditions did not change. Usually, in fact, the surface gains or loses energy – the surface receives a *net energy flux* that warms or cools the ground, melts snow and ice, or causes freezing. Particularly large net energy fluxes occur with the daily and seasonal cycles of atmospheric conditions.

If the surface is snow or ice, and its temperature at melting point, a positive net energy flux produces further melt. Melt on glaciers is thus favored by the following factors, in particular:

1. Strong sunlight – as at mid-day under clear skies – increases absorbed solar radiation.

2. Low reflectivity – for example, where the surface is wet ice rather than dry snow – also increases absorbed solar radiation.

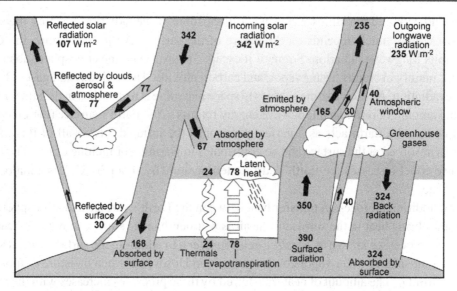

**Figure 5.1: The energy flow through Earth's atmosphere and exchange with the surface –
averaged for the globe and a year. All units are W m^{-2}. The numbers are constrained by satellite
observations at the top of the atmosphere. Partitioning within the atmosphere is estimated from
radiation models. Redrawn from Kiehl and Trenberth (1997). © American Meteorological
Society.**

3. A warm atmosphere – as during warm summer weather – increases both the greenhouse
    radiation and the sensible heat transfer to the surface.

4. A moist atmosphere increases greenhouse radiation and reduces latent heat losses from
    the surface. (On the other hand, increased moisture might enhance cloud cover and block
    sunlight.)

    Persistence of all these factors favors melt ablation, the integral of melt rate over time.

5.2 Statement of the Surface Energy Budget

Define energy fluxes as positive when they represent a flow of energy to the snow/firn/ice surface.
We are concerned with fluxes to and from a unit area of the surface – properly, each flux is a "flux
density" – so units are energy per unit time per unit area, or W m^{-2}. Assuming no horizontal
transfers of heat on the ground, the net energy flux into the surface, E_N, sums the following
components:

$$E_N = \underbrace{E_S^{\downarrow} + E_S^{\uparrow} + E_L^{\downarrow} + E_L^{\uparrow}}_{E_R} + E_G + E_H + E_E + E_P. \tag{5.2}$$

Here $E_S^{\downarrow}$ denotes the downward shortwave radiation, $E_S^{\uparrow}$ the reflected shortwave radiation,
and $E_L^{\downarrow}$ and $E_L^{\uparrow}$ the downward and emitted longwave radiations; the sum of all the radiations

gives the *net radiation*, E_R. E_G is the subsurface energy flux. E_H and E_E represent the sensible and latent heat fluxes due to turbulent mixing of the air adjacent to the surface. E_P represents the heat flux from precipitation, especially that which freezes. E_P is usually small and we will not discuss it further.[1]

In glacier studies, by convention, the latent heat flux E_E refers only to latent heat associated with vapor: that is, the heat consumed by evaporation and sublimation and released by condensation and deposition. The latent heat associated with melt and refreeze is part of E_N.

Equation 5.2 is usually referred to as the energy budget for "the surface," but practically it must apply to a layer of snow or ice with a finite thickness Δz. The term E_G measures the net energy flux through the bottom of this layer (by conduction and radiation) and its value depends on the choice of Δz. By choosing a Δz large enough so that seasonal temperature variations do not reach the bottom of the layer, we can approximate $E_G \approx 0$. Note that Eq. 5.2 does not include any terms for the energy content of meltwater flowing out of the layer; E_N already includes the energy for production of melt.

5.2.1 Driving and Responding Factors in the Energy Budget

The complexity of factors contributing to the energy budget sometimes threatens to obscure the underlying simplicity of the problem. The following summary – although not always strictly true – can be kept in mind.

Two factors primarily contribute energy to a melting glacier: sunlight and the atmosphere's heat content. The former corresponds to the downward shortwave radiation, the latter determines both the downward longwave radiation and the sensible heat transfer. Because of these energy flows to the glacier, its surface is kept "warm" – at a temperature roughly comparable to the overlying atmosphere. At the same time, the glacier continuously returns energy to its surroundings; the glacier would cool catastrophically if the inflows of energy were shut off. The biggest losses depend on the properties of the surface itself. The temperature determines the emitted longwave radiation, and the reflectivity determines the upward shortwave radiation. The atmosphere also robs some heat from the glacier by evaporation and sublimation into dry air and, at times, by sensible heat transfer into cold air. Finally, the temperature of the surface continuously tries to equilibrate with the temperature of the underlying ice, so some heat flows by conduction to or from the surface.

5.2.2 Melt and Warming Driven by Net Energy Flux

If the surface has warmed to the melting point, a positive E_N drives melt and a negative E_N drives freezing (if water is available). The total melt rate for the layer, $\dot{m}_s$, given as water-equivalent

---

[1] In temperate climates, devastating winter floods sometimes occur when rain falls on snow. It is a common misconception that such floods result from melt of snow by the rain. In fact, the principal reason for the floods is water saturation of the snowpack; rain quickly flows into the river channels because it cannot infiltrate the saturated snow and underlying soil. This is a type of *saturation overland flow*.

Table 5.1: Some thermodynamic parameters relevant to the surface energy budget.

Parameter	Symbol	Value	Units
Pure Ice (0°C)			
density	ρ_i	917	$kg\,m^{-3}$
specific heat capacity	c_i	2050	$J\,kg^{-1}\,K^{-1}$
latent heat of fusion	L_f	3.34×10^5	$J\,kg^{-1}$
latent heat of sublimation	L_s	28.3×10^5	$J\,kg^{-1}$
Water (0°C)			
density	ρ_w	1000	$kg\,m^{-3}$
specific heat capacity	c_w	4218	$J\,kg^{-1}\,K^{-1}$
latent heat of evaporation	L_v	2.5×10^6	$J\,kg^{-1}$

thickness per unit time, is related to E_N by

$$\rho_w L_f \dot{m}_s \left[1 - f_r\right] + \int_0^{\Delta z} \rho c \frac{\partial T}{\partial t}\, dz = E_N. \tag{5.3}$$

$\dot{m}_s$ is positive for melt, negative for freezing. In the first term, ρ_w is the density of water, and L_f the latent heat of fusion for ice. The parameter f_r indicates the fraction of melt that refreezes within the layer; it ranges from zero to one. The second term represents the rate of gain of sensible heat in a vertical column of unit area through the layer, with ρ the density, c the specific heat capacity, T the temperature, and t time. Values for some of the thermodynamic parameters are listed in Table 5.1. (See Chapter 2 for a discussion of densities in general.)

If temperatures have reached the melting point throughout the layer, then the second term equals zero, f_r equals zero, and the ablation rate is simply $\dot{m}_s = E_N / \rho_w L_f$: such conditions often occur in summer in the ablation zones of temperate glaciers. If the entire layer is colder than melting point, the first term equals zero and the layer's temperature changes in proportion to E_N. If the surface is at melting point, but the layer underneath subfreezing, melt produced at the surface can percolate downward and partly refreeze, forming ice layers. This process releases latent heat and warms the layer. The larger the fraction that refreezes (f_r), the more the net energy goes to warming the layer. The net ablation rate $\dot{m}_s \left[1 - f_r\right]$ and gross melt $\dot{m}_s$ are in this case given by all the terms in Eq. 5.3. The warming due to refreezing can significantly affect temperatures in the firn (Section 9.3).

5.3 Components of the Net Energy Flux

In this section we discuss the separate terms in the surface energy budget, with an emphasis on conditions in a summer melt season.

Cold winter snowpacks delay the start of the melt season, as snow must first warm to its melting point before melting occurs. This process introduces a delay to the onset of melt.

Consider a 1-m-deep snowpack with a mean temperature of $-20\,°C$, a density of $500\,\mathrm{kg\,m^{-3}}$, and a heat capacity of $2000\,\mathrm{J\,kg^{-1}\,K^{-1}}$. With a net energy flux of $E_N = 50\,\mathrm{W\,m^{-2}}$, it takes about 110 hours to warm the snowpack to melting point. This same amount of energy would melt about 0.06 m (w.eq.), much smaller than the few meters per year typical for glacier ablation zones. The transient warming of the winter snowpack is therefore a small factor in the annual energy budget, except in very cold regions.

Once the snow or ice surface starts to melt, the loss of energy by emitted longwave radiation $E_L^\uparrow$ cannot increase further because the surface temperature cannot rise above the melting point. Three factors then drive melt: increases of net shortwave radiation ($E_S^{\mathrm{Net}} = E_S^\downarrow - E_S^\uparrow$), increases of downward longwave radiation, and increases of sensible heat transfer. The latter two processes increase, in turn, with temperature of the overlying atmosphere. When melt removes snow to expose ice, the surface absorbs more shortwave radiation, increasing E_S^{Net}. For these reasons, all three energy sources contribute to increasing ablation when the summertime climate warms (Section 5.4). But ablation might also increase because of other factors, such as reduced mid-day cloud cover.

5.3.1 Downward Shortwave Radiation

The flux of solar energy toward Earth at the top of the atmosphere approximately equals the solar constant; it varies by a few percent over a year because of the eccentricity of Earth's elliptical orbit. As it traverses the atmosphere, solar radiation is partly scattered and absorbed by gases, water droplets, ice crystals, and particles. The total solar flux reaching the ground ($E_S^\downarrow$), often called the *insolation* or the *global radiation*, adds together three components: the direct solar beam, the diffuse light arriving from all directions in the sky due to scattering, and the reflected light from surrounding terrain.

Call the top-of-atmosphere solar flux $E_{So}^\downarrow$: about $1367\,\mathrm{W\,m^{-2}}$ during day, but zero at night. The direct solar beam at the surface ($E_{Sd}^\downarrow$) is a fraction of this, such that

$$E_{Sd}^\downarrow = E_{So}^\downarrow \cdot \cos Z \cdot \psi \quad \text{with} \quad \psi = \psi_o^{P/(P_o \cos Z)}. \tag{5.4}$$

Z is the zenith angle, the angular distance of the sun below a vertical line. Zenith angle varies with latitude (θ), time of year, and time of day (Oke 1987):

$$\cos Z = \sin\theta \sin\delta + \cos\theta \cos\delta \cos h, \tag{5.5}$$

in which δ signifies the solar declination, the angular distance between the sun and the equator. The hour angle, h, varies with the time of day such that $h = 0$ at local noon.

The parameter ψ, a number less than one, is the atmospheric transmissivity, a measure of the attenuation of the solar beam as it traverses the atmosphere (Oke 1987). Its value at sea level, ψ_o, is about 0.84 for a clear sky and 0.6 for thick haze. ψ_o decreases to zero under heavy cloud cover, in which case only diffuse and reflected radiations contribute to insolation. The direct radiation on a glacier surface increases with altitude, because the thickness of atmosphere traversed by the solar beam decreases. This is why the ratio P/P_o appears in Eq. 5.4; P gives the atmospheric pressure at the glacier surface, and P_o the pressure at sea level (101.3 kPa). Conversely, a large zenith angle – a slanted solar beam – decreases the transmission because the beam must traverse a greater thickness of atmosphere.

For clear-sky conditions, solar radiation is limited primarily by its direct dependence on the zenith angle rather than by transmissivity. For example, measurements in clear-sky conditions at a site on the Greenland Ice Sheet ("ETH Camp" at 1155 m altitude, 69 °N) found an average value of about 0.8 for the transmissivity (Konzelmann and Ohmura 1995). In comparison, the mean daily value of cos Z in mid-summer at this latitude is about 0.4.

Figure 5.2 illustrates the annual cycle of $E_{Sd}^{\downarrow}$ at local noon, calculated from Eq. 5.4 for the complete range of latitudes, assuming a clear sky ($\psi_o = 0.84$) and a typical altitude for an extratropical glacier (≈ 2500 m, or $P = 75$ kPa). The noon-time sun is, of course, much weaker at the poles than at low latitudes. In mid-summer, however, the daily total of flux (not shown) is about the same at all latitudes in the summer hemisphere because the length of day increases toward the poles.

Equation 5.4 applies to a horizontal surface. In general, corrections should be made for both slope and aspect; the zenith angle should be replaced by the angle between the solar beam and the normal to the local surface (Oke 1987). Surfaces that face the sun receive more direct radiation and ablate more rapidly than surfaces oriented obliquely to the sun's rays. On the other hand, in mountainous landscapes shadows eliminate all direct insolation at some times of the day, significantly reducing ablation – hence north-facing slopes are more heavily glaciated than south-facing slopes, in the northern hemisphere. Because of shading, small glaciers can survive in basins beneath steep mountain walls, even when glaciers have disappeared from everywhere else in the surrounding landscape.

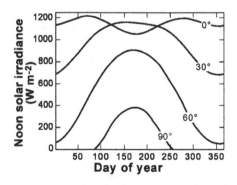

Figure 5.2: Peak daily (local noon) clear-sky direct solar radiation on a horizontal surface, as a function of latitude in the northern hemisphere. Curves were calculated from Eq. 5.4, with $P = 0.75$ bar and $\psi_o = 0.84$.

Diffuse radiation occurs in clear-sky conditions but matters most in cloudy conditions. Illumination from diffuse light explains why cloudy days are not completely dark. In clear-sky conditions at the ETH Camp in Greenland, the diffuse component was measured to be only 13–17 % of $E_S^\downarrow$ (Konzelmann and Ohmura 1995). Clouds greatly influence the summer average, however, and at this site the diffuse radiation was 40% of $E_S^\downarrow$ on average. At a cloudier, low-altitude, maritime ice cap in the South Shetland Islands, diffuse radiation accounted for about 60% of $E_S^\downarrow$ (Braun and Hock 2004). On cloudy days, the noon-time $E_S^\downarrow$ is often only 20% to 30% of its value under clear skies.

Solar radiation reflected from surrounding terrain can be significant in mountain regions, where reflective snow often mantles steep valley walls and peaks.

Summing the direct, diffuse, and reflected components gives the global radiation, $E_S^\downarrow$, the quantity of most interest to the energy budget. An effective transmissivity, ψ_*, expresses how much of the possible total solar radiation strikes the surface. Assuming a horizontal surface,

$$E_S^\downarrow = E_{So}^\downarrow \cdot \cos Z \cdot \psi_*. \qquad (5.6)$$

The time-averaged value of ψ_* depends on the factors previously mentioned, such as cloud cover, altitude, and haze. Oerlemans and Knap (1998) determined monthly-mean values from continuous meteorological data collected at a site on the lower ablation zone of Morteratschgletscher, Switzerland. For summer months, ψ_* ranged from 0.44 to 0.59. Thus, at this mid-latitude alpine site about half of all the possible solar radiation reached the ground. At ETH Camp on the Greenland Ice Sheet, monthly means of ψ_* in summer are about 0.7 (Konzelmann and Ohmura, 1995).

5.3.2 Reflected Shortwave Radiation

The backscatter of shortwave radiation from snow and ice varies with wavelength, but the total reflected energy flux ($E_S^\uparrow$) can be described as a simple fraction of the downward flux. Define the *broadband surface albedo*, α_s, according to $E_S^\uparrow = \alpha_s E_S^\downarrow$. Common albedo values for snow and ice surfaces range from 0.2 to 0.85; the albedo therefore has a very large and important influence on the total shortwave radiation absorbed by the surface, $E_S^\downarrow [1 - \alpha_s]$, and hence on ablation. Table 5.2 summarizes typical albedo values on glaciers.

Albedo at a site can be assessed by measuring the downward and upward solar radiation simultaneously, using radiometers that face up and down. Such data show the continuous variation of albedo over time. In the absence of direct measurements, albedo is often estimated from "typical" published values for snow or ice (e.g., Table 5.2; also Cutler and Munro, 1996). Most recent studies of glacier melt use parameterizations to estimate how the albedo varies through the melt season and from place to place (Oerlemans and Knap 1998; Brock et al. 2000).

Measurements reveal large variations of albedo over time on melting glaciers. Figure 5.3 illustrates how surface albedo changed through the melt season at a site near the equilibrium line

Table 5.2: Characteristic values for snow and ice albedo, from a literature review by S.J. Marshall.

Surface type	Recommended	Minimum	Maximum
Fresh dry snow	0.85	0.75	0.98
Old clean dry snow	0.80	0.70	0.85
Old clean wet snow	0.60	0.46	0.70
Old debris-rich dry snow	0.50	0.30	0.60
Old debris-rich wet snow	0.40	0.30	0.50
Clean firn	0.55	0.50	0.65
Debris-rich firn	0.30	0.15	0.40
Superimposed ice	0.65	0.63	0.66
Blue ice	0.64	0.60	0.65
Clean ice	0.35	0.30	0.46
Debris-rich ice	0.20	0.06	0.30

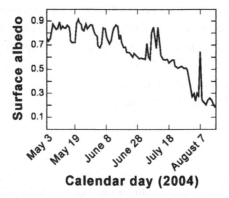

Figure 5.3: Variation of surface albedo in summer of 2004 at one location on Haig Glacier (data from Shea et al. 2005, courtesy of S.J. Marshall). Shown is the average mid-day (10:00–15:00) albedo, based on one-minute measurements by upward- and downward-looking radiometers. Rapid surface brightening (for example, on Aug. 6) occurs when new snow covers the surface. The transition from snow cover to bare glacier ice occurred July 28–30.

of a small glacier in the Canadian Rockies (Haig Glacier, 50.7°N; Shea et al. 2005). The most profound change was the reduction of albedo, hence increased energy absorption, when the seasonal snow disappeared to reveal ice. Individual snowfall events brightened the surface and temporarily reduced absorption.

Fresh snow has an albedo of 0.8 to 0.9, and such high reflectivities are typical year-round in the interiors of polar ice sheets. After a snowfall, the albedo decreases over time, particularly during melting. Several processes are responsible. Snow metamorphism increases grain size, which reduces the frequency of scattering at snow-air interfaces (Wiscombe and Warren 1980). Impurities such as rock particles and organic material concentrate on the surface, making the snow darker. Ponding of meltwater likewise increases absorption. Small quantities of water in the snow may also decrease albedo by effectively increasing the grain size (Warren 1982).

Snow albedo is very sensitive to particulate impurities; black carbon (soot) reduces albedo discernibly at concentrations of only about 0.1 ppm, and mineral dust does so at about 10 ppm (Warren and Wiscombe 1980). Natural snowpacks in tropical and mid-latitude areas often have impurity concentrations of this amount or greater (Thompson et al. 1979; Higuchi and Nagoshi 1977). Polar snowpacks are considerably more pure, with dust concentrations typically less than 0.1 ppm. On mid-latitude glacier snowpacks, the increasing concentration of impurities through the melt season reduces albedos to values as low as 0.3 by late summer.

Glacier ice typically has a lower albedo than snow, but, as Table 5.2 shows, ice albedos span a large range. Crystal size and bubble content influence albedo because they determine the amount and types of scattering surfaces in a unit volume of ice. Liquid water and impurities again increase absorption. The ice albedo varies from place to place in the ablation area; debris concentrates and water ponds in some areas; other areas are drained and flushed clean. On average, the ice exposed low in the ablation zone is the most absorptive – a consequence of high melt rates and long exposure times (Klok and Oerlemans 2002). Debris can be concentrated not just over a single melt season but cumulatively over many years.

Because albedos decrease over time in the melt season – on both snow and ice – the period of most effective energy absorption occurs later than the peak of insolation. Insolation is a maximum at the solstice (except in the tropics; Figure 5.2), but much of the winter snow still covers glaciers at this time. As the melt season progresses, darkening of the surface increases the absorbed or net shortwave radiation available for melt, $E_S^{\downarrow}[1 - \alpha_s]$, by a factor of three to four. This is one reason why, on mid-latitude glaciers, peak melt rates occur one to two months after the solstice. (Warmer air in mid-summer also contributes.)

There are many factors not discussed here, and not usually addressed in energy balance studies, that further complicate shortwave absorption. For example, albedo depends on the angle of incidence of downward radiation. This means the albedo changes with time of day, with the proportions of direct and diffuse radiation, and with surface microtopography. Thin snowpacks are translucent, so the albedo depends partly on the underlying material. Sufficient accumulation of sediment on a glacier surface (typically more than a few centimeters in thickness) insulates the underlying ice and reduces melting. Snow and ice reflectivity vary as a function of wavelength, with greater reflection of visible wavelengths than near-infrared ones; the importance of all the effects just described should, in fact, vary with wavelength.

Wavelength-dependent processes also occur in the atmosphere. Visible wavelengths constitute a greater percentage of $E_S^{\downarrow}$ when skies are cloudy or humid, because water vapor and droplets preferentially absorb the near-infrared wavelengths. Given that the visible wavelengths are reflected more by the surface, the bulk albedo should therefore rise at the onset of humid or cloudy conditions.

5.3.3 Longwave Radiation

At infrared wavelengths, snow, ice, and liquid water emit as near-perfect radiators, with typical emissivities of $\epsilon_s = 0.94 - 0.99$. The emitted longwave radiation is thus, from Eq. 5.1,

$$E_L^\uparrow = -\epsilon_s \sigma T_s^4 \approx -\sigma T_s^4 \qquad (5.7)$$

for surface temperature T_s. The negative sign indicates a loss of energy from the surface. For melting snow or ice, $T_s = 273.15$ K and $E_L^\uparrow = -315.6\,\mathrm{W\,m^{-2}}$. Within instrumental accuracy, this is the value measured by longwave radiometers in field studies (Greuell and Smeets 2001; Oerlemans and Klok 2002; Figure 5.4).

Downward longwave radiation $E_L^\downarrow$ originates as emissions from clouds and from atmospheric water vapor, carbon dioxide, ozone, methane, and other greenhouse gases. Such radiation is continually absorbed and emitted at different levels in the atmosphere. The flux at the surface depends on the amount and temperature of these constituents at different heights; longwave emissions originating throughout the bottom 1 km or so of the atmosphere reach the surface directly. The emissions are again governed by the relation $\epsilon\sigma T^4$, and so increase strongly with temperature in the lower atmosphere (Ohmura 2001). In contrast to the surface, however, the emissivity ϵ of the atmosphere deviates significantly from one, because the greenhouse gases absorb and emit only in certain wavelength bands.

The flux $E_L^\downarrow$ varies substantially and is difficult to predict without specific information about profiles of water vapor, clouds, and temperature in the lower troposphere. These data are seldom available, and $E_L^\downarrow$ must be measured or parameterized at a site. Most parameterizations define an effective atmospheric emissivity ϵ_a such that

$$E_L^\downarrow = \epsilon_a \sigma T_a^4, \qquad (5.8)$$

for a near-surface air temperature T_a. With completely cloudy skies, $\epsilon_a \approx 0.95$ (Konzelmann et al. 1994). For clear skies with dry air, ϵ_a can be less than 0.5. As we will see next (Section 5.3.4) a value $\epsilon_a \approx 0.8$ is typical for a late-summer average. Hock (2005) discussed some of the parameterizations used in glaciology. Most calculate ϵ_a in terms of humidity and air temperature measured at 2 m above the surface. Though a crude way to account for the larger-scale atmospheric conditions governing $E_L^\downarrow$, this strategy is a reasonable one. Warm, humid conditions increase $E_L^\downarrow$, and heat the surface in periods of low cloud cover. Cold, dry conditions reduce $E_L^\downarrow$, and allow the surface to cool significantly at night. The parameterizations capture these first-order effects. Moreover, more than half of the downward longwave radiation originates within 100 m of the surface, according to theoretical calculations with radiation models (Ohmura 2001).

5.3.4 Field Example, Net Radiation Budget

Figure 5.4 shows measured shortwave and longwave radiation for two intervals in the month of August 2005, at 2650 m altitude (near the equilibrium line) on Haig Glacier, Alberta, Canada. (The glacier is described by Shea et al. 2005. Figures 5.3, 5.6, and 5.7 show more data from the

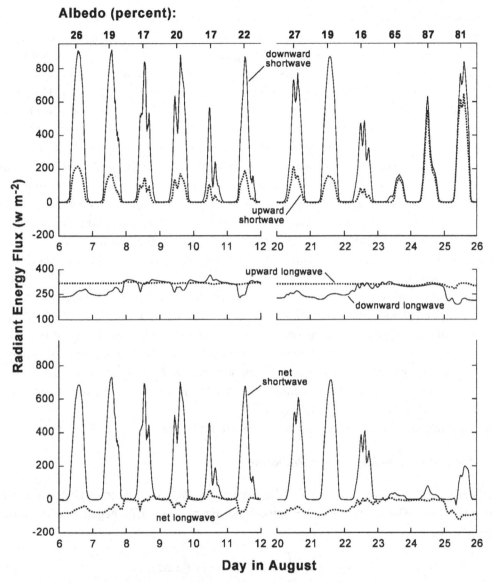

Figure 5.4: Measured radiative fluxes during two periods in August 2005, at one location on Haig Glacier. Data courtesy of S.J. Marshall. Numbers for each day above the top panel are mid-day albedos (percent).

same site.) Also shown is the net radiation E_R,

$$E_R = E_S^\downarrow [1 - \alpha_s] + E_L^\downarrow - \left| E_L^\uparrow \right|. \tag{5.9}$$

We have smoothed the curves with a two-hour center-weighted filter; the original data are 10-second measurements. Daily values of surface albedo, indicated along the top of the figure, were calculated from mean mid-day measurements (hours 10:00–15:00). The variation of albedo at this site over a longer time period was shown in Figure 5.3.

At the beginning of both periods shown in the figure, the surface was glacier ice. The weather changed significantly during both. In the first interval, initially warm and clear conditions changed to cold and partly cloudy ones (beginning early on August 8), culminating with mostly cloudy conditions and a small snowfall on August 10. In the second interval, initially warm and clear conditions ended with a cooling and major snowfall event on August 23–24. This dramatically reduced the net solar radiation because of increased albedo. The net longwave radiation increased during cloudy and partly cloudy periods, despite colder near-surface air. Net radiation was generally positive and underwent diurnal variations, especially at times of low albedo.

The top row of Table 5.3 gives values averaged for the month; the other rows give minima and maxima. (The turbulent fluxes shown in the table are discussed later, in Section 5.3.6.) As reported from other glaciers, the largest single source of energy to the surface, averaged over days, was downward longwave radiation. The net flux of longwave, however, was usually negative, because the surface emitted the $315\,\mathrm{W\,m^{-2}}$ corresponding to irradiance at the melting temperature. Thus net shortwave radiation was the main source of energy for melt (sensible heat contributed about one-third as much). Averaged over days, the net longwave E_L^{Net} varied by about $100\,\mathrm{W\,m^{-2}}$ and so significantly affected the variations of the energy budget. Low values of $E_L^{\downarrow}$ corresponded with low relative humidity and low water vapor pressure. The correlation coefficients of $E_L^{\downarrow}$ with these two factors were $r = 0.59$ and 0.63, respectively. When relative humidities were close to 100%, downward and upward longwave fluxes essentially balanced. The effective atmospheric emissivity can be calculated as $\epsilon_a = E_L^{\downarrow}/\sigma T_a^4$, which gives an average for the month of $\overline{\epsilon}_a = 0.79$. The upward longwave varied little over the period because the surface temperature remained close to the melting point.

The melt production at this site will be discussed in Section 5.4.2.

5.3.5 Subsurface Conduction and Radiation

Both heat conduction and shortwave radiation can transfer energy through the bottom of the surface layer. These two processes give a net energy flux E_G, negative when directed downward, of

$$E_G = k_{\mathrm{T}}\frac{\partial T}{\partial z} - E_S^{\downarrow}[1 - \alpha_s] \cdot f_\alpha(\Delta z). \tag{5.10}$$

Table 5.3: Mean and extreme energy budget terms ($\mathrm{W\,m^{-2}}$) for the month of August 2005, based on weather station measurements on Haig Glacier, Canada. (Data courtesy of S.J. Marshall.)

	α_s	$E_S^{\downarrow}$	E_S^{Net}	$E_L^{\downarrow}$	$E_L^{\uparrow}$	E_L^{Net}	E_R	E_H	E_E	E_N
Mean	0.42	217	121	270	−315	−45	76	37	−7	106
Minimum	0.12	0	0	187	−337	−123	−101	−27	−60	−151
Maximum	0.87	1101	892	384	−275	68	889	186	37	958

Conduction, the first term on the right, removes heat from the surface layer when it is warmer than the underlying ice. Here k_T is the thermal conductivity of snow or ice, and z is the depth below surface. With a subsurface temperature below the melting point, conduction produces daily gains and losses of heat from a shallow surface layer, decimeters in thickness. A thicker surface layer – several meters deep – gains and loses heat in the annual cycle. Negligible conductive heat transfer occurs if the ice or snow is all at its melting point. We discuss conductive heat transfer and variations of subsurface temperatures in detail in Chapter 9.

Much of the shortwave radiation penetrating snow or ice scatters from the surfaces of grain boundaries, fractures, and bubbles, and becomes the reflected flux $E_S^\uparrow$ discussed in Section 5.3.2. The remaining flux, of magnitude $E_S^\downarrow [1 - \alpha_s]$, is absorbed. Most of the absorption occurs close to the snow/ice surface, but its distribution with depth depends on the scattering and absorption of different wavelengths. The multiplier f_α, in the second term on the right side of Eq. 5.10, declines rapidly from one at the surface to near zero not far below. Its value depends on the chosen thickness Δz of the surface layer. Of the total absorbed shortwave radiant energy, a fraction $[1 - f_\alpha]$ heats or melts the surface layer directly, while the remaining fraction heats or melts the layers beneath (some of it may return to the surface layer by conduction). This distinction is important only if temperatures beneath the surface are subfreezing; to calculate the total melt in temperate ice or firn, we need to choose a large enough Δz to make $f_\alpha = 0$.

The depth to which any significant absorption occurs is apparently a few millimeters in snow, a few centimeters in ice with abundant bubbles, and as much as 20 cm in "blue ice" with fewer bubbles (Brandt and Warren 1993; Bintanja et al. 1997). Water pockets and rock particles, however, can increase absorption at depth. In cold, dry places with ice surfaces, as in parts of Antarctica, the shortwave absorption at depth is a small but nontrivial part of the energy budget (Bintanja et al. 1997; Bintanja 2000).

5.3.5.1 Why Does Glacier Ice Look Blue or White?

The focussing of absorption near the surface does not imply complete darkness at greater depths; some radiation penetrates to a few tens of meters but then mostly scatters back to the surface and contributes to the surface albedo rather than to absorption. The total down-welling radiative flux in the subsurface, $E_z^\downarrow$, declines with depth z:

$$E_z^\downarrow = E_S^\downarrow \cdot \exp(-\chi \ell(z)). \qquad (5.11)$$

The depth of penetration depends on the the extinction coefficient, χ, and the path length, ℓ, a measure of how much ice the radiation travels through. Extinction includes both scattering and absorption, and ℓ increases in proportion to both depth and density and varies with incidence angle.

Absorption within ice depends strongly on wavelength (Figure 5.5). If a beam of shortwave radiation shines into solid ice, the near-infrared wavelengths are absorbed near the surface (within

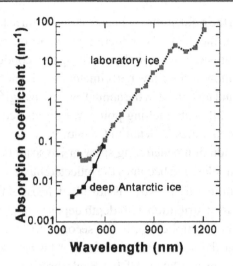

Figure 5.5: Measured absorption coefficient in pure ice, as a function of wavelength. The intensity of a beam of light shining through ice decreases with distance z, in proportion to $\exp(-\chi z)$: the extinction coefficient χ equals the absorption coefficient if there is no scattering. Laboratory measurements are those of Grenfell and Perovich (1981). The coefficient for "deep Antarctic ice" was determined by transmission of light between detectors frozen into boreholes in the bubble-free ice beneath South Pole, Antarctica (data from Askebjer et al. 1997). At the smallest wavelengths, the laboratory measurements overestimate the absorption, most likely due to scattering from microscopic fractures in the laboratory-prepared ice.

centimeters to decimeters), whereas visible wavelengths penetrate much farther. The penetration distance of visible light is inversely related to wavelength. Blue light travels several orders of magnitude farther than the near-infrared radiation (Askebjer et al. 1997). Light scattered from depths of several meters or more within a block of ice is therefore blue; this accounts for the renowned blue hue of glacial ice. In ice with abundant bubbles or other scattering surfaces, however, reflection occurs near the surface and the reflected light is still white.

5.3.6 Turbulent Fluxes

As winds flow along the surface of a glacier, turbulent eddies mix the air vertically. The temperature of the air adjacent to the glacier surface (the *skin temperature*) equals the temperature of the ice or snow. If the overlying air (the *lower boundary layer* or *surface layer*) is warmer than the surface, the mixing transfers sensible heat to the surface, at rate per unit area E_H. The moisture content of the air at the surface is determined by the saturation vapor pressure and hence the skin temperature (Eqs. 4.4 and 4.5). In the presence of drier overlying air, mixing transfers moisture away from the surface. The surface must evaporate or sublimate to maintain saturation of the air adjacent to it. This consumes latent heat, at rate per unit area E_E. In general, both of these fluxes can be either positive or negative.

5.3.6.1 Bulk Aerodynamic Approach

From the preceding description it is sensible (so to speak) to conclude that the energy fluxes E_E and E_H are roughly proportional to the contrasts of moisture and temperature between the surface and the overlying air. Call the contrasts $T_a - T_s$ and $q_a - q_s$; here T_a and T_s stand for temperatures of the lower boundary layer and the surface, and q_a and q_s stand for the corresponding absolute humidities (mass of water vapor per unit volume). Fluxes also increase with the strength of vertical turbulent exchange, an increasing function of the frequency of eddies and their distance of propagation. In bulk, the turbulent exchange of air relates to three factors; how fast the air flows (measured by the wind speed u a few meters above the surface), how rough the surface is, and whether buoyancy stabilizes the air against vertical mixing or not ("stability"). Of these, the wind speed is the most important. Thus, defining two coefficients, C_H and C_E, as the *bulk exchange parameters* for heat and moisture, a first approximation takes the form:

$$E_H = \rho_a c_a C_H \, u \, [T_a - T_s] \tag{5.12}$$

$$E_E = \rho_a L_{v/s} C_E \, u \, [q_a - q_s] \tag{5.13}$$

with ρ_a the density of air, c_a its specific heat capacity at constant pressure, and $L_{v/s}$ the latent heat of vaporization or sublimation. Using Eqs. 5.12 and 5.13 is referred to as the "bulk aerodynamic approach," because u, T_a, and q_a are regarded as representative values for the lower boundary layer, assumed to be thoroughly mixed (Garratt, 1992). In fact, all of these parameters vary with height, so the values for the exchange parameters must depend on the height at which measurements are made (Oerlemans and Klok 2002). To find values of the coefficients, for a given height, we next combine an approximate model of the turbulence with empirical constraints; this is the goal of the next four sections. Such an analysis also highlights some of the limitations of the approach.

5.3.6.2 Fluxes of Heat and Vapor: Flux-gradient Theory

Heat transfer by turbulent convection is viewed as analogous to conduction, with eddies playing the part of molecules. The vertical flux of heat E_H is therefore taken proportional to the vertical temperature gradient $\partial T / \partial z$ (with z increasing upward) and to a coefficient specifying the effectiveness of the transfer process, known as the *eddy diffusivity for heat, K_h*:

$$E_H = K_h \rho_a c_a \frac{\partial T}{\partial z}. \tag{5.14}$$

Because turbulent convection greatly exceeds molecular conduction, K_h is roughly a factor of 10^5 larger than the thermal diffusivity of air at rest. The coefficient K_h varies with height z. In periods of melting, the fixed surface temperature implies that $\partial T / \partial z$ and hence E_H increase with the air temperature in excess of $0\,^\circ\text{C}$, as in Eq. 5.12.

The vertical flux of vapor can be treated similarly. The amount of heat per unit volume, $\rho_a c_a T$, is replaced by the humidity q, the mass of water vapor per unit volume. The upward vapor flux (per unit area), Q_w, is

$$Q_w = -K_w \frac{\partial q}{\partial z}, \tag{5.15}$$

with K_w the *eddy diffusivity for water vapor*. The energy lost from the surface then amounts to $E_E = -L_{v/s} Q_w$. The most useful way forward involves rewriting this equation in terms of the water vapor pressure e rather than q, because we can calculate e as a function of temperature for the saturated air at the surface. From the gas law,

$$q = M_w \frac{e}{RT}, \tag{5.16}$$

where M_w is the molecular weight of water and R the gas constant. Further, the gas law gives an expression for RT in terms of the total pressure P,

$$\frac{P M_a}{\rho_a} = RT, \tag{5.17}$$

with M_a the molecular weight of air. It follows that

$$q = \frac{M_w}{M_a} \frac{\rho_a e}{P} = 0.622 \frac{\rho_a e}{P} \tag{5.18}$$

and so

$$E_E = \rho_a K_w L_{v/s} \frac{0.622}{P} \frac{\partial e}{\partial z}. \tag{5.19}$$

The pressure of water vapor in saturated air at the surface is $e = 611$ Pa. If the lower boundary layer has a comparatively small vapor pressure, the surface loses ice to evaporation and sublimation. The vapor pressure in the boundary layer increases with both temperature and relative humidity. In summer, air temperatures below $0\,°C$ favor evaporation and sublimation; those above favor condensation. Dry air strongly favors evaporation and sublimation, however, so relative humidity is often the most important factor. Note that the air at the surface is not always saturated, as assumed; the preceding formulation therefore systematically underestimates direct condensation (Box and Steffen 2001).

5.3.6.3 Turbulence and Flow of the Air

The turbulence of the air is measured by the *eddy viscosity*, K_m, defined by the equation

$$\tau = \rho_a K_m \frac{\partial u}{\partial z}. \tag{5.20}$$

Here τ denotes the shear stress acting on the surface and in the air above, and u indicates the wind speed at height z. This relation is equivalent to the constitutive rule of a fluid of kinematic viscosity K_m. This equation resembles those for moisture and heat fluxes; the shear stress can be regarded as a vertical flux of horizontal momentum.

Theoretical and experimental results both suggest that although K_m varies with height z the shear stress τ does not, at least in the first few meters above a horizontal surface. Equation 5.20 then implies that

$$K_m \frac{\partial u}{\partial z} = \frac{\tau}{\rho_a} = \text{constant.} \tag{5.21}$$

The quantity $[\tau/\rho_a]^{1/2}$ has the dimensions of velocity, and is known as *friction velocity*, denoted u_*. Measurements have shown that, in the first few meters above the surface, the wind speed u typically varies as the logarithm of the height and scales with u_*, so that:

$$u(z) = \frac{1}{k_o} u_* \ln\left(\frac{z}{z_0}\right). \tag{5.22}$$

k_o is a dimensionless constant, *von Kármán's constant*, with a value of 0.4; z_0 is the *surface roughness parameter*, with units of length (and often simply called "roughness"). This small quantity represents the height above the mean surface at which the wind speed is nominally zero. Its value can be determined from measurements of wind speed at several heights above the surface, using Eq. 5.22. Table 5.4 gives typical values, taken from a compilation by Brock et al. (2006).

The roughness z_0 is generally proportional to the height of bumps and undulations on the surface, but typically smaller than 5% of their amplitude. With z_0 determined, a value of u_* can be obtained using Eq. 5.22 and additional measurements of wind speed. Typical values of u_* for melting glacier surfaces are in the range 0.1 to 0.5 m s^{-1} (Kuhn 1979). Finally, Eq. 5.22 can be differentiated to show that $\partial u/\partial z = u_*/k_o z$; combining this with Eq. 5.21 gives an expression

Table 5.4: Typical ranges of surface roughness parameter in millimeters (Brock et al. 2006).

Smooth ice	0.01–0.1
New snow and polar snow	0.05–1
Snow on low-latitude glaciers	1–5
Ice in ablation zone	1–5
Coarse snow with sastrugi	11
Penitentes[†]	30
Rough glacier ice	20–80

[†] Pinnacles and blades of snow.

for the eddy viscosity,

$$K_m = u_* k_0 z. \tag{5.23}$$

Eddy viscosity thus increases with the strength of the wind (u_*); faster wind increases the vigor of turbulent mixing. The turbulent energy fluxes, E_H and E_E, should therefore also increase with u_*.

5.3.6.4 Flux-Gradient Methods for Determining Energy Fluxes

There is some evidence that, at any given height, the three eddy viscosities K_m, K_h, and K_w are approximately equal in an atmosphere with near-neutral stability. (A neutral atmosphere is one in which the temperature gradient equals the dry adiabatic lapse rate, approximately 1 K per 100 m. The assumption of equal eddy viscosities is, in any case, only plausible if wind speed, temperature, and vapor pressure all vary as the logarithm of height). By taking $K_m = K_h = K_w$, Eqs. 5.14 and 5.19 can give values for the turbulent fluxes, if combined with measurements of temperature and vapor pressure at various heights, and a value for K_m determined from wind speed measurements and Eq. 5.23. The measurements are made by instruments mounted at different levels on a mast of several meters' height.

The atmosphere above a glacier is often far from neutral. Temperature increases of 8 degrees in the first 2 meters have been measured (Holmgren 1971, Part B). Such a strong temperature inversion implies an extremely stable atmosphere and hence reduced turbulence. Grainger and Lister (1966) reviewed different laws of wind-speed variation with height and compared them with field observations. They concluded that the logarithmic law is best not only for neutral but also for extremely stable atmospheres.

How much error is introduced by assuming equality of the three eddy viscosities is unclear. Recent reviews of available data conclude that the the assumption $K_m = K_h$ is reasonable for stable boundary layers (Andreas 2002).

5.3.6.5 Calculations Using Transfer Coefficients

The flux-gradient theory can be used to formulate the bulk aerodynamic relations (Eqs. 5.12 and 5.13) into a usable form. From Eqs. 5.14 and 5.23 and the assumption $K_h = K_m$,

$$E_H = \rho_a c_a k_0 u_* z \frac{\partial T}{\partial z}. \tag{5.24}$$

Integration gives

$$E_H = \rho_a c_a k_0 u_* \frac{T - T_s}{\ln(z/z_0)} \tag{5.25}$$

for surface temperature T_s. Substitution for u_* from Eq. 5.22 gives

$$E_H = \rho_a c_a C^* u \, [T - T_s] \quad \text{with} \quad C^* = \frac{k_o^2}{\ln^2(z/z_0)}. \tag{5.26}$$

The dimensionless parameter C^* is called the *transfer coefficient*. The density of air, $\rho_a = \rho_a^\circ P/P_0$, depends on atmospheric pressure P; with ρ_a° being the density at standard pressure P_0. Substitution of numerical values $\rho_a^\circ = 1.29\,\text{kg}\,\text{m}^{-3}$, $P_0 = 1.013 \times 10^5$ Pa, and $c_a = 1010\,\text{J}\,\text{kg}^{-1}\text{K}^{-1}$ gives

$$E_H = 0.0129C^*Pu(z)\left[T(z) - T_s\right]. \tag{5.27}$$

This equation can be used with values for u and T measured at a known height z above the ground. Daily means are commonly used; in some cases they are values from weather stations nearby, adjusted for differences in elevation, rather than measurements on the glacier. This equation is normally applied to melting surfaces, in which case $T_s = 0\,°\text{C}$.

Equation 5.19, with the assumption $K_w = K_m$, leads to a similar relation for the latent heat flux

$$E_E = 0.622\rho_a^\circ L_v C^*u(z)\left[e(z) - e_s\right]/P_0 = 22.2C^*v\left[e - e_s\right] \tag{5.28}$$

involving the vapor pressures above and at the surface, e and e_s. The former is measured, the latter calculated by assuming saturation.

The value of C^* will depend on measurement height and surface roughness, but not strongly. For a measurement height z of 1 to 2 m, the values of z_0 in Table 5.4, and $k_o = 0.4$, Eq. 5.26 gives values of C^* in the range 0.002 to 0.004 for melting snow and ice surfaces. Table 5.5 lists some published values, all reduced to this dimensionless form.

The data in this table suggest that values for C^* of 0.0015 for snow and 0.002 for ice are appropriate. The value of C^* is not very sensitive to variations in roughness; increasing z_o from 1 mm to 1 cm increases the transfer coefficient by a factor of two. Using the value $C^* = 0.002$ in Eq. 5.27 suggests that, with typical conditions ($P = 80\,\text{kPa}$, $u = 5\,\text{m}\,\text{s}^{-1}$), air at $5\,°\text{C}$ provides a sensible heat flux of about $50\,\text{W}\,\text{m}^{-2}$ to a melting surface. With $e/e_s = 0.8$ and the same wind speed, the latent heat flux removes about $25\,\text{W}\,\text{m}^{-2}$ from the surface.

Table 5.5: Transfer coefficients C^* for melting snow and ice surfaces.

Surface	1000 C^*		Reference
	Heat	Vapor	
Snow	1.66	2.04	Holmgren 1971, Part D
Snow/ice	2.0	2.0	Hogg et al. 1982
Snow	1.3	1.5	Ambach and Kirchlechner 1986
Ice	1.5	1.5	Oerlemans and Klok 2002
Ice	1.9	2.2	Ambach and Kirchlechner 1986
Ice	3.9	3.9	Hay and Fitzharris 1988

5.3.6.6 Further Comments on the Turbulent Fluxes

1. The height at which air temperature equals the skin temperature is not necessarily the same as the height z_0 (Andreas 1987, 2002). In particular, microtopography of the surface should not constrain the transfer of heat as much as the transfer of momentum. Thus the lower limit of the integration leading to Eq. 5.25 should perhaps be some other number, z'_0, smaller than z_0. The definition of the transfer coefficient then would include $\ln(z/z_0)\ln(z/z'_0)$ in its denominator, rather than $\ln^2(z/z_0)$, giving a smaller transfer coefficient C^*. Hock and Holmgren (2005) argued that z'_0 for glacier surfaces is typically only $0.01 z_0$. If $z_0 = 3$ mm, this implies a 40% reduction in C^* for $z = 2$ m. Andreas (2002) reviewed evidence that z_0/z'_0 depends on the surface roughness and wind speed; the transfer coefficient apparently varies, again, within a factor of two. These arguments also apply to the transfer coefficient for latent heat.

2. Turbulent exchanges depend on the stability of the lower boundary layer. If the temperature decreases with height more rapidly than the adiabatic lapse rate (i.e., an unusually warm surface compared with the air above), buoyancy enhances vertical mixing. The usual situation on glaciers in summer is the opposite; the cold, dense air at the surface suppresses vertical mixing ("stable conditions"). In stable conditions, the wind shear $\partial u/\partial z$ concentrates near the surface more than implied by Eq. 5.22. Applying stability corrections to calculations of E_H and E_E requires a large number of theoretical assumptions, and does not appear justified for glaciers, given the sparse measurement constraints and the complexity of the surfaces (Holmgren 1971, part B; Munro 1989; Hock and Holmgren 1995; Braithwaite 1995; Oerlemans and Klok 2002). The surface roughness on most glaciers varies significantly over distances of only tens of meters; the wind profile is thus not adjusted to the underlying surface (Oerlemans 2001, p. 33). Nonetheless, it is now common practice to apply a stability correction based on the theory of Monin and Obukhov (1954) (see Stull 1988).

3. Although a logarithmic wind profile (Eq. 5.22) may be a good approximation close to the surface, many observations on glaciers show that the wind speed attains a maximum at a height of several meters above the surface (Smeets et al. 1999; Oerlemans and Grisogno 2002). This is typical of the "glacier wind," a down-slope flow of cold, dense air (Oerlemans 2001, pp. 34–38).

4. Box and Steffen (2001) measured surface vapor fluxes at a number of sites on the Greenland Ice Sheet, using several different methods. The authors concluded that a flux-gradient method using measurements from two heights more accurately gauges downward water vapor flux than does the one-height method. On the other hand, in some situations calculations with the one-height method are significantly less sensitive to measurement errors (Bintanja and van den Broeke 1995).

The formulation outlined here, using transfer coefficients, captures the essential features of the turbulent exchange processes. Quantitatively, however, the method is only accurate to within a multiple of a few, unless site-specific calibrations are applied. All methods for determining turbulent fluxes rely on gross simplifications of the fluctuating mass and heat flows near the surface; calibration is a necessity for quantitative accuracy.

5.3.6.7 Field Example of Turbulent Fluxes

Figure 5.6 shows meteorological variables and the turbulent energy fluxes for two intervals in the month of August 2005 on a mid-latitude mountain glacier: the same location and time period as shown in Figure 5.4. The onset of cool conditions on both August 8 and 23 significantly reduced

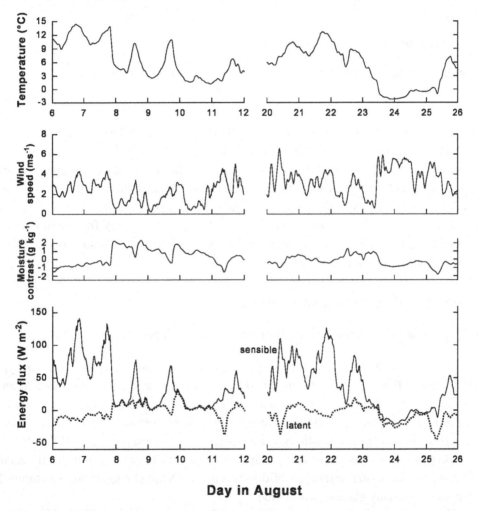

Day in August

Figure 5.6: Inferred turbulent energy fluxes during two periods in August 2005, at one location on Haig Glacier (bottom panel). Top three panels show meteorologic conditions: air temperature and wind speed at 2 m height, and depression of water vapor content at 2 m height relative to water vapor content at the surface (relative humidity was measured at 2 m; saturation was assumed at the surface). Data and calculations courtesy of S.J. Marshall.

the sensible heat flux. The latent heat fluxes were relatively small. These calculations used the transfer coefficient method with a roughness of $z_0 = 3\,\text{mm}$, and $z_0' = 0.01 z_0$ for both sensible and latent heats (see item 1 in the preceding list).

Table 5.3 gives the mean fluxes for the month. Sensible heat flux was a significant source of energy to the glacier, about half as large as net radiation, whereas sublimation and evaporation were a minor energy sink. The energy fluxes varied over time primarily because of changes in the driving variables (the temperature and humidity differences, $T_a - T_s$ and $q_a - q_s$) rather than wind speed. Applying a stability correction (see item 2 in the preceding list) using the formulation of Holtslag and de Bruin (1988), reduced the the calculated energy fluxes by about 20%.

5.3.6.8 Rate of Ablation by Sublimation

Sublimation strips material from the surface at a rate (thickness per unit time)

$$\dot{s} = -E_E/\rho L_s \tag{5.29}$$

for surface density ρ. With reference to Eq. 5.28, three processes favor sublimation: dry air (small e), a warm surface (large e_s), and strong winds (large u).

Generally, the energy consumed by evaporation and sublimation reduces the energy available for warming and melting the surface. In Greenland, the annual reduction of melt is small but not trivial. In their analysis of Greenland Ice Sheet surface balance, Box and Steffen (2001) concluded, in agreement with earlier investigators, that the energy sink from sublimation and evaporation has an important role in maintaining the ice sheet. It is also a major factor on high-altitude glaciers (Section 5.4.5.3).

5.4 Relation of Ablation to Climate

5.4.1 Calculating Melt from Energy Budget Measurements

The amount of ice melted in a given time period is controlled by the net vertical energy flux (E_N) according to Eqs. 5.2 and 5.3, integrated over time. E_N needs to be modelled to answer many questions about how glacier mass balance depends on climate and other factors. One approach uses measurements or model values for each of the separate energy fluxes that sum to E_N. In principle, this approach is the best way to rigorously connect glacier ablation to climate processes. In practice, the uncertainties in such a method can be large. To determine the accuracy requires comparisons to measurements of the melt itself. Detailed energy budget studies have now been made on many glaciers worldwide.

Figure 5.7 illustrates melt and energy fluxes for the month of August 2005 at Haig Glacier, Canada (see Figures 5.4 and 5.6 for the radiative and turbulent fluxes in two intervals within this month). The model melt rates shown here are daily values, calculated by integrating the E_N over each day, but only using times with surface temperature at $0\,°C$. For comparison, the

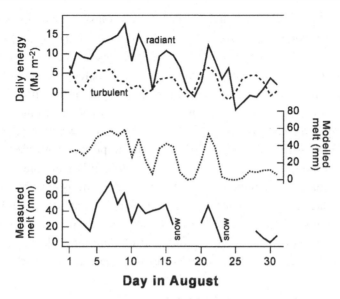

Figure 5.7: Daily-total melt parameters for the month of August 2005, at Haig Glacier. Top panel shows the net values of radiant and turbulent fluxes, integrated over each day. The sum gives a net energy flux that is converted to modelled melt (middle panel), using Eq. 5.3, when temperature is at melting point. The inferred net energy flux can be compared to actual melt, determined from surface deflation measured with a sonic ranger (bottom panel). Significant new accumulations of snow occurred at the two times labelled "snow." Subsequent compaction of the snow made it impossible to determine the melt from surface deflation; we have assumed this effect lasts for two days after snowfall ends. Data and calculations courtesy of S.J. Marshall.

measured melt rates were determined from observations of the rate of surface lowering, obtained with a sonic ranger. No values are shown during snowfalls and for two days after because snow density was not measured. The modeled and measured melt rates compare well; observed melt totalled 89 mm (w.eq.), compared to a model value of 81 mm (an integrated E_N of 284 MJ m^{-2}). Daily values correlated moderately well ($r = 0.79$). Net radiation caused more melt than did the turbulent fluxes, but both contributed to variations.

Hogg et al. (1982) made a similar comparison using data from Hodges Glacier, South Georgia. They measured downward and upward shortwave radiation and net total radiation and calculated turbulent fluxes by the transfer-coefficient method from daily mean temperatures, water vapor pressures, and wind speeds measured at one point on the glacier. The total ablation calculated from the energy fluxes was 3.45 m of water, which compared well with the 3.37 m measured at ablation stakes. Again, Braithwaite and Olesen (1990) made the comparison at two sites near the margin of the ice sheet in West Greenland. In this study, the fluxes were less constrained by measurements. Downward shortwave radiation was measured but values of albedo were assumed. Upward longwave radiation was calculated from Eq. 5.7, but downward longwave was estimated from air temperature and cloudiness using an empirical relation. Turbulent fluxes were computed using Ambach and Kirchlechner's transfer coefficients (Table 5.5). The calculated ablation rates for the season were within 1 mm day^{-1} of the observed values, although the standard deviations of the daily discrepancies at the two sites were 14 and 19 mm day^{-1}.

These results, and others, suggest that energy budgets calculated from relatively simple observations can provide reliable estimates of ablation rates, in spite of the difficulties in measuring the net radiation and the simplifying assumptions in the theories for turbulent fluxes.

It is considerably more difficult to calculate ablation for a whole glacier than for a few sites with meteorological measurements. Many of the important variables (wind, humidity, surface roughness, cloudiness, surface albedo) vary significantly from place to place and over time. Nonetheless, spatially distributed energy balance calculations have been applied to some intensively studied glaciers and appear to work well (Arnold et al. 1996; Klok and Oerlemans 2002; Hock and Holmgren 2005). So too for the Greenland Ice Sheet (Box et al. 2004, 2006). But in most studies, especially of large regions and of long-term changes, constraints from measurements are sparse, and simplified approaches must be used; we discuss these next.

5.4.2 Simple Approaches to Modelling Melt

Empirical regression equations relating measured ablation rates to air temperatures – and other variables measured by weather stations – provide a simple alternative to energy-budget calculations. The weather stations may be on the glacier or nearby. Such equations have several uses, including extending series of annual ablation measurements backward in time, estimating ablation on glaciers without measurements, and predicting the increase of ablation rate with warming. The results from these analyses need to be regarded as uncertain approximations. Ideally, uncertainties can be assessed from constraints on all the components of the energy budget. In practice, such an assessment is seldom feasible.

The most obvious variables to use in a simple melt relation are the air temperature and the radiation. The first variable is connected to the sensible heat flux, to whether the surface is warm enough to melt, and, in some complicated fashion, to the downward longwave radiation. For the second variable, either net total radiation or net solar radiation might be used.

Braithwaite (1981) used data from four energy-budget studies on two glaciers in Arctic Canada to discern correlations. He found that daily ablation correlated with air temperature but not with net radiation. This was because air temperature, in turn, correlated with the sensible heat flux, which varied more than the net radiation. Nonetheless, variations in air temperature accounted for only half the variance of ablation rate. Thus, although air temperature predicted ablation better than did net radiation, it was not a particularly good predictor at the timescale of days.

A more important question is whether simple relations can make good predictions of total ablation over a year. In this case, the mean annual temperature might be a poor indicator of melt because the ablation season lasts only a few months and the temperature during the rest of the year has little relevance. Mean summer temperature is a better choice. On climatic timescales of decades and longer, however, means of summer and annual temperature are likely to vary together.

In one analysis of year-to-year variability, Hanson (1987) obtained a correlation coefficient of -0.93 between the annual mass balance measured along a flow line on Barnes Ice Cap, Canada, and the mean summer temperature measured at a weather station 120 km away. At this ice cap, precipitation varies little from year to year and mass balance variations depend almost entirely on the summer ablation. In another case, Braithwaite and Olesen (1989) used 6 years of data from Nordbogletscher, South Greenland. They obtained a correlation with mean summer temperature of 0.84. The correlation with mean annual temperature was 0.56 and not statistically significant. These results are surprisingly good considering the simplicity of the climate metrics compared to the complexity of controls on the energy budget.

5.4.2.1 Positive Degree Days

In principle, ablation should relate yet more closely to an index that accounts not only for mean summer temperature but also for the amount of time that air temperatures exceed 0 °C. Only during such conditions will the sensible heat flux contribute directly to melt. Moreover, melt of the surface may cease entirely if air temperatures fall below 0 °C for an extended period. The most commonly used index is *positive degree days*, which we symbolize D (Krenke and Khodakov 1966; Braithwaite and Olesen 1989; Huybrechts et al. 1991; Jóhanesson et al. 1995). This index is defined, for time period Δt measured in days, as

$$D(\Delta t) = \int_{\Delta t} T_C H(T_C)\, dt \quad \text{with} \quad H(T_C) = \begin{cases} 1 & \text{if} \quad T_C > 0 \\ 0 & \text{if} \quad T_C < 0 \end{cases} \tag{5.30}$$

where T_C signifies the Celsius temperature of the air. A common approach is to calculate D using daily mean temperatures, in which case $D = \sum_j \overline{T}_{cj} H_j$, the sum for all days j in the time interval of interest. The D index is often symbolized PDD in the literature, an awkward notation for equations.

D provides a rough approximation of the integrated heat energy driving melt over the time interval of interest. For application in models, the total surface melt m_s over time Δt is parameterized simply as

$$m_s = f_m D, \tag{5.31}$$

in which f_m denotes the *degree-day melt factor*, an empirically determined coefficient relating D to observed melt. Equation 5.31 is equivalent to assuming that, during periods of melt,

$$\int_{\Delta t} E_N\, dt = \rho_w L_f f_m D(\Delta t). \tag{5.32}$$

Thus f_m should be larger for an ice surface than a snow surface, since the smaller albedo of ice results in greater net energy in otherwise similar conditions. Numerous measurements do show that ice surfaces ablate more rapidly than snow surfaces (e.g., Hoinkes and Steinacher 1975). Based on observations in Greenland (e.g., Braithwaite and Zhang 2000), values of f_m for snow and ice of, respectively, $f_{ms} = 3$ and $f_{mi} = 8\,\text{mm day}^{-1}\,^\circ\text{C}^{-1}$ are typically used in ice sheet models. For the full month of observations at Haig Glacier shown in Figure 5.7, D was 179.5 day–°C, and hence $f_m \approx 5\,\text{mm day}^{-1}\,^\circ\text{C}^{-1}$. When modelling a large region, such as the Greenland Ice Sheet, different f_m values need to be used for different regions, not only because of different surface albedos but also because the contribution of turbulent fluxes to total melt varies with such factors as wind speed (Lefebre et al. 2005). For summer totals, a common approach is to calculate degree days separately for the periods of snow cover (D_s) and exposed ice (D_i), and then the total melt as $m_s = f_{ms}\,D_s + f_{mi}\,D_i$.

In their six-year study of Nordbogletscher, Braithwaite and Olesen (1989) obtained a correlation coefficient of 0.96 between annual ice ablation and D, corrected for snow accumulation. Thus, compared to mean summer temperature, positive degree days explained about 20% more of the variance of ablation. Several other studies confirm a strong relation between ablation and D in Greenland in summer months (Figure 5.8). In general, D seems to act as a better proxy for ablation than expected from a detailed understanding of the physics; we consider the underlying reasons in the next section.

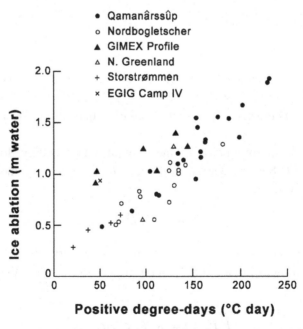

Figure 5.8: Correlation of measured ice ablation (standardized to one month) with positive degree days, in six regions of the Greenland Ice Sheet margin. Redrawn from Braithwaite (1995).

Degree-day melt models may be improved by using separate parameterizations for the influence of degree days and the influence of downward shortwave radiation (e.g., Cazorzi and Dalla Fontana 1996; Hock 1999). For example, we might write the melt in time Δt as $m_s = f_m D(\Delta t) + g_m \int_{\Delta t} E_S^{\downarrow} dt$, with g_m being a radiation-melt index that varies spatially and temporally as a function of albedo. A related approach is to use Eq. 5.31 but to make f_m an explicit function of both $E_S^{\downarrow}$ and surface albedo.

5.4.3 Increase of Ablation with Warming

The response of glaciers to climate change depends critically on how much the annual ablation increases as the climate warms. This depends on many factors, but simple analyses give useful approximations.

When Braithwaite (1981) examined the correlation of daily ablation rates with air temperatures on glaciers in Arctic Canada, he obtained an average sensitivity of $6.3\,\mathrm{mm\,day^{-1}\,{}^\circ C^{-1}}$. Earlier published values were in the range 4.5 to $7\,\mathrm{mm\,day^{-1}\,{}^\circ C^{-1}}$ (Braithwaite 1980). It is not obvious that these numbers, derived from daily variations, should be relevant to total ablation over summer months. Yet they are similar to values (f_m in Eq. 5.31) obtained by comparing cumulative melt to positive degree days. As discussed in the preceding section, f_m is typically 3 to $8\,\mathrm{mm\,day^{-1}\,{}^\circ C^{-1}}$ for the Greenland Ice Sheet; the low value is typical for snow, the high value for ice. The data shown in Figure 5.7 gave a value of $5\,\mathrm{mm\,day^{-1}\,{}^\circ C^{-1}}$ at a mid-latitude glacier, where the surface was ice but periodically covered by snow.

A rough theoretical value for the direct increase of ablation with warming can be obtained from the energy flux relations. Net energy should increase with temperature by at least the following three mechanisms:

1. Equation 5.27 suggests that the sensible heat flux should increase, for every degree of warming, by $0.0129C^*Pu$. For typical values, $C^* = 2 \times 10^{-3}$, windspeed of $5\,\mathrm{m\,s^{-1}}$, and elevation of 2000 m, the increase is $11\,\mathrm{W\,m^{-2}}$ for one degree. This flux would melt about 3 mm of ice in one day.

2. From Eq. 5.8, the increase of downward longwave radiation for a temperature change ΔT_a (small compared to T_a) should be approximately $4\epsilon_a \sigma T_a^3 \Delta T_a$. In terms of $E_L^{\downarrow}$, the total downward longwave before the change, it is $4E_L^{\downarrow}\Delta T_a / T_a$. Using values for a mid-latitude glacier (Table 5.3), this amounts to about $4\,\mathrm{W\,m^{-2}}$ for one degree of warming, enough to melt an additional 1 mm of ice per day.

3. Increased melt from the previous two factors hastens the exposure of ice as the winter snowpack disappears. Earlier exposure of ice, in turn, increases the shortwave absorption. The following calculation gives a rough idea of how important this is. For albedos of ice and snow of 0.3 and 0.6, respectively, and a daily mean downward shortwave flux of $250\,\mathrm{W\,m^{-2}}$, the ice surface would absorb about $75\,\mathrm{W\,m^{-2}}$ more than the snow surface (daily average). Mid-latitude ablation zones typically melt at 2 to 10 cm per day. The additional 4 mm per day

from sensible and longwave forcings corresponds to a 4% to 20% increase in daily melt. If it normally takes six weeks of melt for the winter snowpack to disappear, the increased melt implies two to nine extra days of ice exposure. In each of these days, the net shortwave flux is enhanced by about 75 W m^{-2}. Assume the entire melt season is 100 days. The increase of net shortwave absorption, averaged for the melt season, is thus 2 to 7 W m^{-2}. This corresponds to an additional 0.5 to 2 mm of ice per day, averaged over the melt season.

Thus, from rough theoretical considerations, the three mechanisms should increase melt by 4 to 6 mm of ice per day, given a warming of one degree. Empirical values for ice surfaces tend to be larger – about 8 mm day^{-1} °C^{-1} – probably because of additional factors that are difficult to estimate simply. Warming favors rain over snow, reducing surface albedos directly and through earlier exposure of ice. Warmer air in winter and spring results in faster warming of ice and snow and earlier onset of melting conditions.

Regardless, all of these values, theoretical and empirical, are consistent to within a factor of a few. They give a general indication of how much annual ablation should increase due to summer warming; with 100 days of melt per year, the increase is 0.4 to 0.8 m from one degree of warming. The number of days of melt will also increase with warming, however. Consider a simple description of the annual air temperature variation (Reeh 1991), such as

$$T_a = \overline{T}_a - A_T \cos(2\pi t) + G(0, \sigma) \tag{5.33}$$

where t measures time in years (zero in mid-winter), and $G(0, \sigma)$ is a random Gaussian fluctuation (mean of zero, standard deviation of σ) of daily temperature about the overall sinusoidal trend. Mean summer temperatures can increase with the mean annual temperature $\overline{T}_a$ or with the amplitude of the seasonal cycle A_T. Positive degree days can increase, in addition, with the variability of daily temperatures. The change in total melt dm_s from a change in positive degree days dD is

$$dm_s = \frac{\partial m_s}{\partial D} dD = f_m \left[\frac{\partial D}{\partial \overline{T}_a} d\overline{T}_a + \frac{\partial D}{\partial A_T} dA_T + \frac{\partial D}{\partial \sigma} d\sigma \right]. \tag{5.34}$$

Figure 5.9 shows how the positive degree days, and hence ablation, increase with mean annual temperature according to the annual cycle given by Eq. 5.33. The increase is nonlinear (over the whole range of temperatures), and the magnitude significant (considering melt-factor values) for mean annual temperatures greater than about -5 °C. This explains, succinctly, why the surface mass balance becomes strongly negative on a glacier flowing into a warm region, a fact shown clearly by the data in Figure 4.6b. Although those data also reflect differences in precipitation and other factors, they clearly show that, on average, annual melt in typical temperate-region ablation zones increases by 1 to 1.5 m per degree of warming. This observation is consistent with the simple degree-day model of Figure 5.9 for mean temperatures greater than -5 °C, assuming a surface of ice ($f_m = 8$ mm day^{-1} °C^{-1}). To see this, Table 5.6 gives the increase of annual melt per degree of warming for several environments, according to Figure 5.9. Note that,

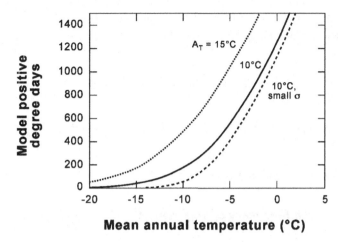

Figure 5.9: Increase of positive degree days with mean annual temperature, according to Eq. 5.33, for annual cycle amplitudes of 15 and 10 °C and $\sigma = 5$ °C. Variability was reduced to $\sigma = 2.5$ °C for the curve labelled "small σ." At temperatures greater than about −5 °C, the increase is approximately linear. Based on concepts developed by Reeh (1991).

Table 5.6: Theoretical sensitivity of melt to warming, for a surface of ice.

$\overline{T}_a$ (°C)	Melt increase (m °C^{-1})	
	$A_T = 10$ °C	$A_T = 15$ °C
−10	0.34	0.64
−5	0.86	1.1
0	1.4	

although the theoretical ablation increases nonlinearly over the whole range of temperatures, it can be well approximated by a linear function over intervals of 5 °C or so, the size of a large climate change.

An increase of ablation by 1 to 1.5 m yr^{-1} per degree of warming (and less at lower temperatures, as shown in the table) generally matches the values for the glacier-wide sensitivity of surface mass balance to temperature, the parameter C_T discussed in Section 4.2.6. In temperate and subpolar regions, that sensitivity largely manifests the increase of melt with temperature.

5.4.3.1 Total Ablation and Warming

Note that all the preceding discussion concerns the short-term effect of warming on surface ablation – specifically, the increase that could occur in a single year. The total increase of ablation due to warming can be much larger, for two reasons:

1. Warming sometimes enhances glacier flow. If sustained for years, enhanced flow can significantly reduce a glacier's surface altitude, causing additional warming and increased ablation.

2. Even without enhanced flow, a sustained increase of ablation lowers surface altitudes and so increases ablation further.

3. Warming can increase ice flow to termini in the ocean or deep lakes, significantly increasing calving.

These factors are discussed in Chapter 11.

5.4.4 Importance of the Frequency of Different Weather Conditions

The simple relations discussed in the previous section give a useful rough indication for how much the ablation increases when the climate warms. They are so generalized, however, that they obscure the diversity of processes governing the ablation – processes that vary greatly from place to place and over time. Moreover, when the climate changes, many factors change in addition to temperature. Most of the terms in the energy budget depend strongly on weather conditions. Even under similar conditions, their relative importance changes during the summer. To understand rigorously the causes of ablation, and its variations from year to year, the energy budget for the entire ablation season must be known. Thus, in principle, it is necessary to determine the relative frequency of different weather conditions, and the values of the energy budget terms associated with each at different times of the season.

Table 5.7 illustrates how the energy budget terms can vary in different weather situations at one location. These data, from Devon Island Ice Cap in Arctic Canada, are means over 7 to 12 days, at a location just above the equilibrium line (Holmgren 1971). Three common weather situations here are referred to as Types A, B, and C. Type A conditions involve cool and calm weather; little ablation occurs. Such conditions prevail when low pressure is centered east of Devon Island. In contrast, when cyclones track over Devon Island, winds are strong and warm

Table 5.7: Surface heat fluxes at Devon Ice Cap Station under different weather types. All fluxes are in $W\,m^{-2}$. Data from Holmgren (1971, Part F, p. 34).

Type	Sky	Wind	Surface	E_R	E_H	E_E	E_N
A	Clear	Light	Frozen	12	13	−5	20
A	Overcast	Light	Frozen	23	−4	−6	13
B	Overcast	Strong	Melting	45	46	10	101
C	Clear	Light	Melting	44	14	5	63
C	Clear	Light	Frozen	4	16	−6	14

Table 5.8: Energy budgets in Antarctica ($W\,m^{-2}$).

Site	Elevation	Season	E_R	E_H	E_E	Reference
Vostok	3400 m	Summer	32	−25	−2	Artemyev (1973)
		Winter	−17	15	0	
Mizuho	2230 m	Summer	20	−7	−8	Ohata et al. (1985)
		Winter	−38	37	0	
Maudheim	37 m	Summer	9	6	†	Liljequist (1957)
		Winter	−22	13	†	

† Not measured.

moist air flows over the ice cap (Type B conditions). The surface receives long-wave radiation from clouds and an increased flux of sensible heat; ablation is substantial. Type C conditions are anticyclonic, with clear skies that promote downward solar radiation. The effect of Type C conditions depends on whether the surface is melting or not. When it is, low albedos increase the absorption of solar radiation. Such conditions produce intense ablation at lower altitudes where bare ice is exposed (Alt 1978). Overall, the ice cap has a positive annual mass balance when Type A cyclonic conditions dominate the summer weather pattern. In contrast, a season in which Type C anticyclonic conditions predominate can produce enough ablation to eliminate the positive balance of five seasons of Type A.

5.4.5 Energy Budget Regimes

The example from Devon Island discussed in the previous section illustrates the energy budget near the equilibrium line on an Arctic ice cap. We now briefly discuss other environments.

5.4.5.1 Antarctica and the Coldest Sectors of Greenland

Where the climate is so cold that melt rarely occurs, net energy instead warms or cools the surface and underlying layers. Ablation occurs by sublimation, which is favored by dry air, a warm surface, and strong winds (Section 5.3.6.8). Table 5.8 shows energy budget data for three sites in East Antarctica. Vostok is in the interior, Mizuho about 375 km from the coast, and Maudheim on an ice shelf. Summer values are means for December and January, winter values for June and July.

In winter, because there is no solar radiation, the term E_R equals the net longwave radiation. At the two inland stations sensible heat fluxes almost balance the longwave radiation. Winter-time sensible heat flux is particularly high at Mizuho as a result of turbulence created by strong katabatic winds; these also produce significant evaporation in summer. Because there is little melting at Maudheim and none at the other sites, the net positive fluxes are used solely to warm the surface and the layers beneath.

Table 5.9 compares the summer energy budget, now in greater detail, for snow and blue ice sites (Bintanja 2000). Values are averages from a 37-day period. Mean elevation of snow sites was about 1500 m and for ice sites about 1200 m.

During summer, daily totals of $E_S^{\downarrow}$ are comparable to mid-latitude values because of perpetual sunlight. The surface sublimates most rapidly at this time, because its relatively high temperature tends to increase the vapor pressure gradient (the term $e_s - e$ in Eq. 5.28). Sublimation is more rapid on bare ice than on snow. Such "blue ice zones" cover about 1% of Antarctica (Winther et al. 2001). Compared to snow-covered sites, the blue ice sites absorb more solar radiation because of their lower albedos. At the blue ice sites, the summertime sublimation rate was about 0.3 m yr^{-1}, about three times greater than at the snow sites.

5.4.5.2 Mid-latitudes

On mid-latitude glaciers in summer, high values of net radiation and sensible heat flux drive rapid melt. We have already discussed the time-varying energy budget at one such site in the Canadian Rockies (Figures 5.4, 5.6, and 5.7); we now introduce another site, where we can compare energy budgets for the ablation and accumulation zones. Exceptionally thorough energy budget studies have been conducted on Pasterze Gletscher, Austria's largest glacier (latitude 47 °N). Altitudes range from 2200 to 3400 m. Table 5.10 summarizes data from this site, acquired by a network of five automatic weather stations for a 46-day period in mid-summer of 1994 (Greuell and Smeets 2001). Shown separately are averages for accumulation zone sites and ablation zone sites. At the former, snow persisted throughout the measurement period. At the latter, ice was exposed by the end of the period.

Table 5.9: Summer energy budget (W m^{-2}) for snow and blue ice areas in Dronning Maud Land, Antarctica (Bintanja 2000). Values are averages for N sites in each row.

Sites	α_s	$E_S^{\downarrow}$	E_S^{Net}	$E_L^{\downarrow}$	$E_L^{\uparrow}$	E_L^{Net}	E_R	E_H	E_E	E_G	N
Snow	0.82	334	63	185	−254	−69	−6	16	−9	−1	4
Blue ice	0.64	356	121	192	−279	−87	34	2	−26	−11	3

Table 5.10: Mean energy budget terms from a 46-day melt-season study on Pasterze Gletscher, Austria (data aggregated from Greuell and Smeets (2001) by S.J. Marshall). All fluxes are in W m^{-2}. Average altitudes of ablation and accumulation zone sites were 2312 and 3085 meters, respectively.

Sites	α_s	$E_S^{\downarrow}$	E_S^{Net}	$E_L^{\downarrow}$	$E_L^{\uparrow}$	E_L^{Net}	E_R	E_H	E_E	E_N	T_a (°C)	z_0 (mm)
Ablation	0.25	269	201	298	−315	−17	184	55	10	249	6.8	3.2
Accum.	0.60	297	120	278	−314	−36	85	22	3	109	3.4	1.7

Peak daytime values of downward solar radiation in the mid-latitudes are as high as about $1000\,W\,m^{-2}$, higher than possible in polar regions. The daily averages of $E_S^{\downarrow}$ are not very different from mid-summer in Antarctica, however. The high net radiation (E_R) at mid-latitude sites is instead, as the Pasterze Gletscher example shows, due to lower albedo and to larger downward longwave fluxes. Because longwave emissions from the atmosphere depend on both temperature and moisture, the warmer and moister atmosphere of the mid-latitudes is a strong energy source compared to the polar atmosphere. Losses of longwave from the surface, on the other hand, are constrained by the melting temperature. In addition, because of the warm atmosphere Pasterze Gletscher receives a much larger sensible heat transfer than do the Antarctic sites, although comparable to fluxes on Arctic ice caps in warm weather (Table 5.7).

On Pasterze Gletscher, net radiation decreases with increasing altitude, because of albedo; snow perpetually covers the higher parts of the glacier. Sensible and latent heat fluxes also decrease with altitude as a result of cooler air and reduced vapor pressure. The mean net energy flux in the ablation area ($249\,W\,m^{-2}$) corresponds to a melt rate of $0.07\,m\,day^{-1}$, or about $2\,m$ in one month.

5.4.5.3 Low-Latitude Glaciers

All low-latitude glaciers occur at high altitudes.

The energy budget of glaciers in the low-latitude subtropics is similar to that of mid-latitude glaciers, but with even greater magnitudes of fluxes in some regions. Consider one example from the Karakoram in Pakistan, a high-elevation region with a dry climate. At $4300\,m$ altitude on Chogo-Lungma Glacier, Untersteiner (1975) estimated daily mean net radiation of $E_R \approx 190\,W\,m^{-2}$ and a sensible heat flux of about $E_H \approx 120\,W\,m^{-2}$. Some of these high-altitude glaciers also have very large energy losses due to evaporation, a consquence of dry air and strong winds; Bai and Yu (1985) reported latent heat losses as large as $58\,W\,m^{-2}$ on firn at $4750\,m$ on Mount Muztagata in China. This is a particularly important process affecting ablation of glaciers in the tropics, discussed next.

Most tropical glaciers are in the Andes, but some exist in East Africa and New Guinea. The climate here is quite different from that of the mid-latitudes; the implications for glacier regimes have been elaborated by Kaser (2001) and Kaser and Osmaston (2002). Temperature varies little with season but precipitation and atmospheric moisture do. Seasonal migration of the intertropical convergence zone produces one wet season per year in the outer tropics but two in the inner tropics (the "inner" zone being centered on the equator). Air temperatures above melting point are rare on glaciers of the highest Andean summits and Mt. Kilimanjaro, Africa (Thompson et al. 1984, 1995; Kaser et al. 2004; Mölg et al. 2008a).

Snowfall occurs mostly in the wet seasons but, in the outer tropics, the wet season is also the primary ablation season (Wagnon et al. 1999a, 1999b, 2003). Temperatures are a little warmer in this season, but the key process appears to be reduced consumption of energy by sublimation. Sublimation occurs most rapidly in the dry season; Wagnon et al. (1999b) reported average sublimation rates of $1.1\,mm\,day^{-1}$ in the dry season in the Andes, but less than

0.3 mm day^{-1} in the wet season. Sublimation has a large energy cost in the dry season; most of the energy supplied to the surface by net radiation and sensible heat is consumed by sublimation, leaving little energy for melting. The latent heat of sublimation is 8.5 times greater than that for melt. Thus sublimation acts as a strong energy sink but an ineffective ablation mechanism. For example, an energy surplus of 50 W m^{-2} will, over one month, ablate 35 cm of ice if used for melt, but only 4.1 cm of ice if used for sublimation. Furthermore, sensible heat flux tends to be small at these sites because of persistently cold air temperatures related to the very high altitudes. The same conditions apply to the glaciers high on Kilimanjaro in the inner tropics (Mölg and Hardy 2004; Mölg et al. 2008a). On Kilimanjaro's flanks, atmospheric moisture increases down-glacier, reducing the sublimation energy sink; ablation thus increases (Mölg et al. 2008b).

On tropical glaciers with comparatively low altitudes, however, changes of atmospheric moisture act in an entirely different way. An increase of atmospheric moisture probably reduces ablation; increased cloudiness reduces the downward solar radiation, and more snowfall causes higher albedos and reduced shortwave absorption at the surface. Thus, if the climate becomes increasingly arid, not only would accumulation decrease but ablation would increase. This is one explanation for the twentieth-century retreat of glaciers in East Africa (Kruss and Hastenrath 1987; Kaser et al. 2004). On the much shorter timescale of diurnal cycles, convective cloud formation in the afternoon reduces ablation and increases precipitation on western slopes (Hastenrath 1991; Mölg et al. 2003a). This helps to explain the more extensive ice cover on western slopes in the tropics.

Table 5.11 summarizes the energy budget on Zongo Glacier in the Cordillera Real, Bolivia, at latitude 16.25 °S (Wagnon et al. 1999b). These measurements are from a site at 5150 m altitude, near the mean ELA. The net radiation and sensible heat inputs are both small compared with those on mid-latitude and subtropical glaciers in summer. The net longwave flux is particularly negative, due to the cold, dry, and low-density air at high altitude. Most notable is the large fraction of ablation energy consumed by sublimation (the variable f_E); it is about 90% in the dry season. Field measurements of ablation and runoff at this site agree well with estimates based on the energy budget.

Table 5.11: Energy balance and meteorological data from Zongo Glacier, Bolivia (Wagnon et al. 1999a). Data are for the hydrological year September 1996 to August 1997. Wet and dry seasons are November–February and May–August, respectively. All fluxes are in W m^{-2}.

Period	α_s^1	$E_S^{\downarrow}$	E_S^{Net}	E_L^{Net}	E_R	E_H	E_E	T_a (°C)	q (g/kg)	m_s (mm)	Subl. (mm)	f_E
Wet season	0.66	196	67	−54	13	4	−7	−0.3	5.8	327	27	0.41
Dry season	0.52	220	106	−95	11	9	−31	−3.8	4.4	107	117	0.90

[1] Based on the mean measured daily minimum albedo.

On tropical glaciers, many of the meteorological variables change greatly through diurnal cycles (Mölg et al. 2008a). But variations from day to day and over the seasons are much reduced compared to those on extra-tropical glaciers, and ablation is not confined to a summer season. Furthermore, as the preceding discussion should make clear, ablation at high-altitude sites only indirectly relates to temperatures, if at all. For these reasons, the positive degree-day approach to modelling ablation should not work; instead, the full energy budget needs to be calculated, at a high enough time resolution to resolve different parts of the daily cycle.

Further Reading

The book by Oerlemans (2001) is the major recent analysis of how the surface energy budget relates to glacier mass balance. It also discusses the relation between regional climate and local climate of the glacier surface, a topic not covered here. Hartmann (1994) gives a good introduction to global climatology; Wallace and Hobbs (2006) provide a readable introduction to atmospheric radiation and dynamics. The books by Stull (1988) and Garratt (1992) discuss boundary layer meteorology, a helpful resource for learning about energy budgets.

CHAPTER 6

Glacial Hydrology

"If you follow the glacier down to the river (at no slight risk of breaking your neck) you see a large torrent issuing from beneath the glacier itself to join the Chundra River...."

Journal of a Tour Through Spiti, **P.H. Egerton (1864)**[1]

6.1 Introduction

The flow of water through glaciers is a topic with extraordinary mystique. Meltwater channels form on the glacier surface and then disappear into chasms and caverns, while roaring rivers emerge from tunnels at the glacier front. Until very recently, some of the world's largest lakes remained hidden and unknown beneath the Antarctic Ice Sheet. Iceland's ice caps periodically release immense floods; the largest ones convey more water than the Mississippi and Nile combined.

Glacier hydrology is the study of water storage and transport in glaciers – and how glaciers release this water to river systems. Glacier hydrology has important practical applications:

1. Rivers fed, in part, by melt from glaciers provide much of the water supply for agriculture in regions such as the Canadian prairies and central Asia, and for hydroelectric power production in France, Switzerland, and Norway. In comparison with other types of streamflow, glacier runoff has unusual features such as large diurnal fluctuations and maximum flow during summer. Glaciers act as reservoirs that store water in solid form in cool summers and release large amounts in hot dry summers when water from other sources is in short supply. Also important is long-term storage; the mean annual flow of a glacier-fed river measured during a period of glacier retreat overestimates the amount of water available when the glaciers stabilize or start to advance. On the other hand, sustained retreat – a likely consequence of climate warming – will eventually reduce or even eliminate this source of water. To forecast the flow of glacier-fed streams requires an understanding of accumulation and ablation processes (Chapters 4 and 5), of glacier hydraulics, and of adjustments of glacier geometry.

---

[1] Quoted in *The Little Ice Age* by Jean M. Grove

Copyright © 2010, Elsevier Inc. All rights reserved.
DOI: 10.1016/B978-0-12-369461-4.00006-5

2. In the Alps and in Norway, tunnels have been drilled underneath glaciers to capture water to feed hydroelectric plants. Site selection depends on a knowledge of subglacial water flow. Is the water dispersed widely over the bed or concentrated in a few large channels?

3. The sudden drainage of glacier-dammed lakes and of water stored within glaciers – and mudflows associated with these events – has caused extensive damage in Iceland, Peru, and several other countries. Prediction of such hazards becomes increasingly important as glaciers respond to a warming climate, because many of the ice dams that currently impound lakes are thinning, and new lakes are forming.

We must also know how water flows at the glacier bed in order to understand two of the major problems in glaciology: the mechanisms by which glaciers slip over their substrates (Chapter 7) and the causes and mechanics of glacier surges (Chapter 12). Results from glacier hydraulics, which deals with the downward flow of water through its own solid, may also have application to the upward flow of melted rock through the solid rock of the Earth's mantle and crust.

In this chapter we first review the main features of the hydrological system on, within, and beneath glaciers. We then discuss the characteristics of glacier runoff, the subject most relevant to studies of water resources. This is followed by an extended discussion of glacier hydraulics, in particular the mechanics of different types of drainage systems. Finally, we briefly examine glacier floods and subglacial lakes, two of the extraordinary aspects of water in glaciers.

6.1.1 Permeability of Glacier Ice

In temperate glacier ice, the boundaries of the ice grains enclose a network of narrow water lenses and veins (Nye and Frank 1973). The diameters of these voids are controlled by salinity and *capillary effects:* changes of pressure and temperature related to curvature of the void walls (Lliboutry 1996). Tube-like voids with diameters of a few millimeters have also been observed (Raymond and Harrison 1975). Water may percolate through the ice along such features but the quantity must be very small; meltwater is often trapped in depressions on glacier surfaces to form ponds and lakes that persist for months. Furthermore, the flow of water through the ice is insufficient to flush it clean of impurities; precise measurements of temperatures show depressed values reflecting the presence of solutes (Section 9.4.1). In a study of Blue Glacier, Washington State, Raymond and Harrison estimated the flux of water draining through the ice to be about $0.1 \, \text{m}^3 \, \text{yr}^{-1}$ per square meter. This is trivial compared to the quantity of water moving through glaciers in larger voids like fractures and pipes. For most purposes the ice between such features can be regarded as impermeable. The effective permeability of a glacier thus depends on the sizes, spacing, and connectivity of fractures and other large voids. The upper horizons in the accumulation zone, however, are permeable firn.

6.1.2 Effective Pressure

Effective pressure, usually symbolized N, refers to the difference between water pressure P_w and pressure in the surrounding ice P_i:

$$N = P_i - P_w. \tag{6.1}$$

This quantity appears frequently in analyses of glacier hydraulics.

6.2 Features of the Hydrologic System

In Sections 6.2.1 through 6.2.4 our discussion follows the water from the glacier surface, through the ice, along the bed, and into the foreland. Figure 6.1 provides a schematic view of the system.

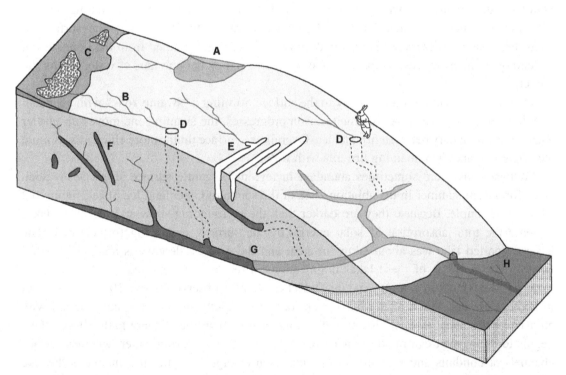

Figure 6.1: Some elements of the glacier water system: (A) Supraglacial lake. (B) Surface streams. (C) Swamp zones near the edge of the firn. (D) Moulins, draining into subglacial tunnels (for scale, white rabbit is about 10 m tall). (E) Crevasses receiving water. (F) Water-filled fractures. (G) Subglacial tunnels, which coalesce and emerge at the front. (H) Runoff in the glacier foreland, originating from tunnels and also from upwelling groundwater. Though not depicted here, water is also widely distributed on the bed in cavities, films, and sediment layers. Sediment and bedrock beneath the glacier contain groundwater. (Refer to the insert for a color version of this figure)

6.2.1 Surface (Supraglacial) Hydrology

In summer, meltwater flows on the ice surface of the ablation zone in networks of channels resembling an ordinary river system. The channels are smooth-walled and often sinuous, and convey water at speeds of a few meters per second. Some of this water flows off the sides and terminus, but most of it disappears into the glacier through cracks and vertical passageways called *moulins*. Some water returns to the surface in artesian upwellings.

Where firn mantles the surface, water percolates downward and refreezes (early in the season) or ponds on the impermeable ice or on dense firn. Runoff begins when a saturated layer forms, a process well studied in terrestrial watersheds (e.g., Colbeck 1972; Colbeck and Anderson 1982; Gray and Male 1981). In contrast to flow in the channel networks on bare ice, drainage along firn- and snow-covered regions is inefficient. Trapped water forms a saturated layer, whose upper surface defines a water table. Fountain (1989), for example, identified a water table in the firn on South Cascade Glacier (Washington State) in late summer; it was typically within about 1 m of the surface. Schommer (1976) measured summertime water levels in boreholes from 12 to 25 m below the surface of Aletschgletscher, Switzerland. He also observed that the water drained from the boreholes at the end of summer, suggesting that firn does not store significant amounts of water during the winter.

Thin firn often saturates all the way to the surface, forming a "swamp zone" with abundant puddles of standing water. As the melt season progresses, the swamp zone moves up-glacier ahead of the boundary between firn and ice. Overall, the surface drains more efficiently as more ice is exposed and the remaining firn fills with water.

Meltwater lakes are sometimes abundant in regions of gentle surface slope. Many such lakes form each summer in the ablation zone of the southern Greenland Ice Sheet; Figure 6.2 shows an example. Because they are darker than the surrounding snow and ice, these lakes increase the total absorption of solar energy by the surface, enhancing further melt. The largest reported lake was about 9 km^2 in area, and the greatest depth was about 12 m (Box and Ski 2007). Some of these lakes drain rapidly; Box and Ski reported flood volumes as great as 31×10^6 m^3 in one day, while Das et al. (2008) observed a 44×10^6 m^3 lake drain in about 17 hours. Much of the latter event occurred rapidly in one 90-minute period with an average discharge rate of 8700 m^3 s^{-1} – a larger flow than the Niagara Falls. Once a lake begins to drain, melt can rapidly enlarge the passageways conveying water, whether surface channels or conduits and fractures in the ice. Such enlargement, in turn, increases the rate of drainage. We should therefore expect some ice-bound lakes to drain rapidly, leading to floods in the glacier foreland or on the bed beneath. Whether rapid drainage occurs or not depends on how effectively the rate of melt increases when drainage begins – cold water and small lake volumes both inhibit melting and can allow stable drainage (Raymond and Nolan 2000).

Figure 6.2: A meltwater lake on the surface of the Greenland Ice Sheet, 30 km from the western margin, August 2005. The lake's diameter is 1.4 km on its long axis; the volume is about 30×10^6 m³ (Box and Ski 2007). (Refer to the insert for a color version of this figure)

6.2.2 Englacial Hydrology

Fractures produced by tension (Section 10.9), including open crevasses, provide pathways for surface water to penetrate into a glacier. Water entering crevasses may be stored in them or may drain into fracture networks or channels deeper in the glacier. Water also enters a glacier in deep vertical shafts known as *moulins*. Many surface streams flow into moulins and cascade out of sight, a visual and auditory spectacle akin to waterfalls. Melt maintains moulins and allows them to grow; as water descends through a glacier, frictional dissipation converts potential energy to heat, which melts the ice walls. Moulins initially form where surface streams intersect crevasses, and where fractures occur beneath lakes. Dye and other tracers put into moulins usually flow quickly through the glacier, especially in mid- to late summer, indicating a connection to a well-developed plumbing system (e.g., Lang et al. 1979; Nienow et al. 1998).

Inspection of boreholes using video cameras shows that numerous water pockets, fractures, and channels exist within temperate glaciers. Fountain et al. (2005) saw several dozen englacial fractures in Storglaciären, a 250-m-thick mountain glacier in Sweden. Many of these openings were elongate and appeared to be continuous across adjacent boreholes; such voids resemble subhorizontal fractures rather than conduits. Moving bubbles indicated water flow through many of them. Others showed no sign of water flow and were interpreted as blind fractures. Fountain et al. concluded that fracture networks, possibly originating as surface crevasses, could account for much of the water storage and transport within temperate glaciers. It is likely that fractures and widely spaced channels connected to moulins both carry significant flows of water.

Polythermal Glaciers Observations show that ice flow speeds up in summertime in the ablation zones of some polar glaciers, including the Greenland Ice Sheet (Battle 1951; Müller and Iken 1973; Zwally et al. 2002; Copland et al. 2003b; Joughin et al. 2008; van de Wal et al. 2008; see our discussion in Chapter 11). Except near tidewater margins – where the balance of forces fluctuates seasonally – the only plausible explanation is that surface water penetrates the glacier in moulins or fractures and then spreads out over the bed, facilitating basal slip (Iken 1972). In western Greenland, these events correspond not only to periods of rapid melt but also to drainage of large lakes on the glacier surface.

Water-filled crevasses penetrate to the bottom of a glacier if a strong supply of meltwater maintains pressures high enough to prevent closure by ice flow. In polar glaciers, the water flow and associated heat dissipation must also be strong enough to prevent closure by freezing. Observations demonstrate that surface water can indeed penetrate through a great thickness of subfreezing ice, as much as 1 km in the case of the Greenland Ice Sheet. The large lake-drainage event observed by Das et al. (2008) provides a good example from Greenland. The lake drained slowly at first, probably through tensional fractures located along one shore. Slow drainage continued for about 16 hours, during which time fractures were presumably growing downward. Then the lake level dropped dramatically, at up to $12 \, \text{m hr}^{-1}$. The entire lake drained in about 90 minutes, while the ice sheet surface simultaneously rose and shifted sideways; clearly, the fractures had reached the bed and water was rapidly escaping along the interface, some 980 meters below the surface. In the weeks following this event, new surface melt continued to flow into the glacier through the moulins formed during lake drainage.

Studies on John Evans Glacier, Ellesmere Island, have outlined a seasonal pattern of fracture penetration (Boon and Sharp 2003). This glacier is about 15 km long, and its thickness typically ranges between 100 and 200 m in the lower 4 km, the region of interest. Internal temperatures are well below freezing. Radar reflections indicate a thawed bed in the central ablation zone, but a subfreezing bed in the accumulation zone and beneath the thin ice of the margins and terminus (Copland and Sharp 2001; Copland et al. 2003b). The mean annual temperature at the equilibrium line is only about $-15\,°\text{C}$, but the surface melts extensively in summer. Surface water forms a mosaic of disconnected lakes and streams early in the melt season. No water escapes to the glacier foreland at this time. After a few weeks, lakes on the glacier expand and grow to a depth of several meters in a flat region about 4 km from the terminus. Lakes partially drain into new moulins in sporadic events, presumably when englacial passageways extend downward but do not yet connect to the bed. Some or all of the water entering the glacier refreezes, releasing latent heat and warming the surrounding ice. Eventually, through a combination of warming and accumulation of surface melt, the passageways connect through to the glacier bed, the surface lakes fully drain, and the glacier releases water to its foreland.

6.2.3 Subglacial Hydrology

Observing water flow at the glacier bed is difficult. Most information derives from bore-hole measurements of a few parameters like pressure, conductivity, and turbidity, and from measurements of water and impurities in streams emerging at the glacier front. In a few places, tunnels built for hydroelectric projects access glacier beds. The paucity of observational constraints has not hindered the development of elaborate theories, which we discuss later in this chapter. The theories yield valuable insights but should be viewed with a skeptical eye.

Several major aspects of subglacial hydrology have been learned:

1. Diverse passageways store and convey water at the glacier bed. There are ice-walled conduits; there are channels incised into rock or sediment but roofed with ice; there is water distributed widely but irregularly in thin films, small channels, and meter-scale cavities; there is water in the pore spaces of sediments and bedrock; and there are giant subglacial lakes. Channels cut upward into the ice are referred to as *Röthlisberger-* or *R-channels*. Channels incised in bedrock are called *Nye-* or *N-channels*. Visitors to glaciers commonly see R-channels where they emerge at the glacier terminus (Figure 6.3).

Figure 6.3: A tunnel (R-channel) emerging at the terminus of Pastaruri, Peru. It formed during drainage of a lake. Photo courtesy of M. Hambrey. (Refer to the insert for a color version of this figure)

2. Water supply competes with drainage; understanding the competition is key to understanding variations of the water system. Primary water sources are rain and melt from the surface, melt along the bed and within the ice, and, in mountain landscapes, runoff from valley sides. Annual melt of the surface typically amounts to a few meters, whereas annual basal melt from geothermal and frictional heat typically yields one centimeter. Surface melt provides by far the largest source for many glaciers, but is generally weak in the upper parts of accumulation zones, and is entirely absent from most of Antarctica and the center of Greenland. When drainage cannot remove all the water arriving at the bed, the volume of stored water increases and water pressures build up. The water pressure partly counteracts the weight of the glacier, and so reduces forces of contact between the ice and underlying rock.

3. Most drainage occurs along the interface between the ice and its substrate. Such drainge is fundamentally difficult, however, because of the great weight of the overlying ice. (We refer to the weight per unit area as the *overburden pressure*.) Ice tends to flow into voids and close them. For drainage to occur, basal water pressures must therefore usually rise to nearly match the overburden pressure; otherwise passageways cannot be open enough, and pressure gradients strong enough, to accommodate and drive the drainage. For example, basal water pressures measured in boreholes through West Antarctic ice streams are within 0.2 MPa of the approximately 9 MPa ice-overburden pressure – and, at many sites, within 0.05 MPa (Kamb 2001; Engelhardt et al. 1990). No surface melting occurs in this region, and basal water originates entirely from melt at the bed.

4. In one common situation, however, drainage does not require high water pressures. Where abundant surface water reaches the bed, concentrated flows melt ice, enlarging passageways and creating a plumbing system of efficient pipes. The pressure in such features is significantly less than the overburden and can even temporarily fall to the atmospheric value if the water source abruptly shuts off and conduits drain.

5. The basal water system affects a glacier's basal slip motion, and vice versa. Water on the bed lubricates the interface and facilitates slip. Lubrication arises from high values of water pressure and from separation of the ice from its substrate. Both increase sliding at the interface. High water pressures also reduce the shear strength of basal sediments, a prerequisite for their deformation. But sliding over an irregular rock bed also opens cavities that store and convey water, and bed deformation may close channels incised into sediments. Thus the connections go both ways; hydrology influences slip but slip influences hydrology. Such feedbacks remain poorly understood but play an important role in phenomena like glacier surges and seasonal variations of glacier flow.

6. In many places, the basal water system changes significantly as a function of time. Such variations partly arise from the feedbacks noted in the previous paragraph; surges, for example, are episodic phenomena involving coupled changes of basal drainage and glacier motion. The most prevalent source of variability, however, is the fluctuating water supply

from surface melt and rain. Beneath glaciers with abundant summertime surface melt, water pressures and fluxes vary greatly in daily cycles and over seasons. Thus, there must exist a system of large passageways that provides a free hydraulic connection between the glacier surface and the bed (Mathews 1964; Nienow et al. 1998). Mathews (1964) was the first to obtain a time series of basal water pressures. He measured pressures at the end of a shaft that reached the base of South Leduc Glacier, Canada, from a mine underneath the glacier. His record showed periods of stable water pressure interrupted by large and rapid rises associated with rapid snow melt or heavy rain.

Marked spatial inhomogeneity at the scale of meters also characterizes the basal water system. Water pressure at one location on the bed often differs greatly from pressure only a few tens of meters away (Murray and Clarke 1995; Fudge et al. 2008). The corresponding variations over time differ in magnitude, phase, and sign. For large differences of water pressure to prevail over short distances, parts of the glacier bed must conduct water poorly.

Figure 6.4 shows the variations in basal water pressure measured in three boreholes through Trapridge Glacier, Canada (Murray and Clarke 1995). Basal water pressure displays large variability over spatial scales of several meters and time scales of hours to days. In some boreholes, said to be *connected* to the glacier drainage system, water pressures vary greatly in direct correspondence to the daily influx of meltwater from the surface. Other boreholes are *unconnected* or isolated; these tap into a patch of the bed that responds, not directly to the diurnal influx, but to elastic deformations of the ice forced by pressure variations in the connected regions nearby. In some cases the unconnected regions are so isolated that their pressures stay nearly constant.

Measurements at other glaciers have shown that diurnal pressure fluctuations originate in conduits but also propagate laterally for tens of meters through permeable sediments (Hubbard et al. 1995).

Figure 6.4 also illustrates that water pressures frequently exceed the local overburden pressure, a situation referred to as *super-flotation conditions* or *excess water pressures*. Such conditions have long been known from glacier drilling programs; water sometimes jets out of boreholes when they connect to the basal system.

The efficiency of drainage through conduits increases greatly over the melt season. Figure 6.5 shows the changes over one month in early summer of mean velocities and transit times for water entering a single moulin and emerging at the terminus of Haut Glacier d'Arolla in Switzerland (Nienow et al. 1998). Over time, the large water fluxes from surface melt increased the size and connectivity of conduits. Moreover, the onset of such efficient drainage migrated up-glacier following the retreating snowline. Exposure of bare ice increased melt rates and eliminated water storage in snow; both factors increased the flow into moulins and, until the conduit system developed fully, increased water pressures within and beneath the glacier (Gordon et al. 1998).

Available data show that average basal water pressures are higher in winter than in summer, an observation first made by Mathews (1964). The seasonal variation is thus opposite

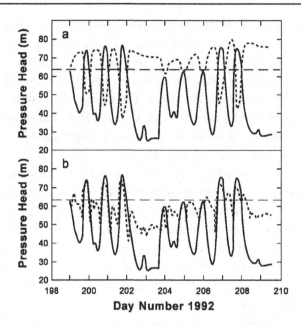

Figure 6.4: Day-to-day variation of basal water pressure measured in three boreholes, Trapridge Glacier. The solid curve in both panels shows the same borehole, repeated for comparison. Pressure variations in different boreholes may be correlated, anticorrelated, or uncorrelated. Horizontal dashed lines indicate overburden pressure. Adapted from Murray and Clarke (1995).

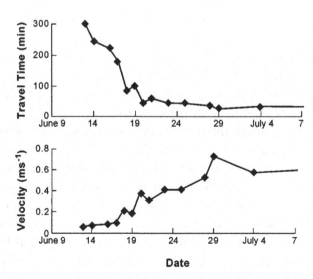

Figure 6.5: Evolution of mean velocity and travel time for water entering a single moulin and appearing at the glacier front, Haut Glacier d'Arolla. Data acquired by tracer injection. Adapted from Nienow et al. (1998).

to the diurnal one, with high pressures in periods of minimal surface melt. High pressures in winter must arise from restricted drainage.

7. Very little is known about the mechanics of water drainage along beds of sediment. A deforming sediment passively transports water in its pore spaces. Most of the water flow may concentrate in broad channels incised in the sediment, referred to as *canals*, or in a *macroporous horizon* of permeable layers at the top of the sediment. Ordinary R-channels should form if the sediment does not deform rapidly or erode easily.

The volume of water present at the glacier bed obeys a conservation relation. Define the volume per unit area as H_w, the average thickness of the water layer. H_w changes over time if the outflow of water from a given region of the bed (Q_{out}, volume per unit time) does not keep pace with the inflow (Q_{in}). For a region of bed with area $\mathcal{A}$, conservation requires that $\mathcal{A} dH_w/dt = Q_{in} - Q_{out}$. The inflow includes water from the surface, which arrives at the bed in conduits and fractures, and water draining along the bed from up-glacier. One of the major goals of glacier hydrology is to learn, in essence, how Q_{in} and Q_{out} relate to the size and shape of passageways along the bed, and to dynamic variables like the water pressure and hydraulic gradients. In this chapter, Sections 6.3.2 through 6.3.4 examine this problem. Generally, the fluxes of water increase with the slope and pressure gradients that drive flow. Water flow also increases with the efficiency of the drainage system. A system of large, connected tunnels conveys water much more readily than a system of cavities and thin layers dispersed widely on the bed. As we will see, this difference of efficiency greatly affects water conveyance along the bed.

The formal statement of water-volume conservation needed for models of the basal water system is

$$\frac{\partial H_w}{\partial t} = \dot{S}_w - \nabla \cdot \vec{q}_w, \tag{6.2}$$

where $\dot{S}_w$ denotes the source flux (water arriving at the bed, volume per unit time per unit area), and $\vec{q}_w$ the water flux along the bed (volume per unit time per unit length). Two separate mechanisms lead to accumulation of water at the bed: prolific sources of water (e.g., surface melt on warm days) and downstream decreases of drainage (e.g., a constriction in the drainage system that ponds water upstream).

6.2.4 Runoff from Glaciers

Water typically emerges at the glacier terminus in a few large streams; this indicates that the water coalesces into master channels while flowing along the bed. The pattern of flow in these streams provides information about the plumbing inside the glacier (e.g., Meier and Tangborn 1961; Mathews 1963; Elliston 1973; Nienow et al. 1998). Moreover, the stream flow – the runoff – is an important water resource for inhabitants and ecosystems downstream.

6.2.4.1 Observations

Figure 6.6 shows typical records for fine summer weather. The discharge (volume flowing in unit time) has a marked diurnal variation superimposed on a *base flow* whose volume changes more slowly. The maximum discharge may be roughly twice the minimum; less for large glaciers, more for small ones. The peak discharge comes a few hours after the peak in melt but, as summer advances, the daily rise and fall in discharge becomes more rapid and the time lag decreases. For example, at Mittivakkat Glacier in southeast Greenland the lag between peak melting and discharge decreases from 5 to 7 hours in May to 3 to 4 hours in August (Mernild et al. 2006).

In the northern hemisphere, total daily discharge usually reaches its maximum in late July or early August, when warm weather and extensive exposures of ice lead to the strongest ablation on the glacier surface (Chapter 5). When summer snowfalls shut down melting, the diurnal variation in discharge ceases and the base flow declines (Figure 6.7). When melting begins again, the base flow takes several days to reach its former level. Water flows throughout the winter in many places, but with no diurnal variation and greatly reduced base flow. Base flow generally consists of meltwater from snow and firn, water that travels slowly through the glacier, and water released from temporary storage. Groundwater and runoff from slopes above the glacier may

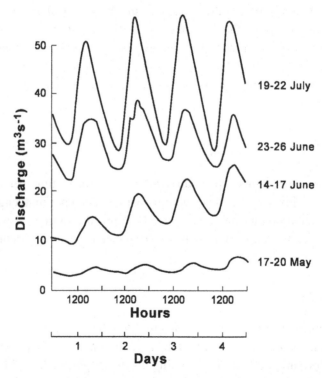

Figure 6.6: Discharge from a glacierized basin near Zermatt, Switzerland, in four periods of fine weather in summer 1959 (dates indicated beside each curve). Adapted from Elliston (1973).

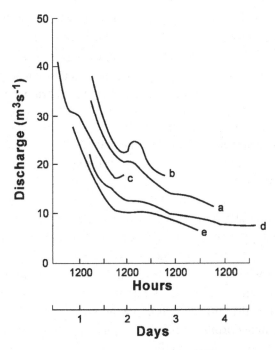

Figure 6.7: Discharge from the same basin as in Figure 6.6 after summer snowfalls. (a) 29 June–2 July 1959, (b) 30 July–1 Aug. 1959, (c) 26–28 June 1960, (d) 4–8 Sept. 1960, (e) 17–19 Sept. 1960. Redrawn from Elliston (1973). Used with permission from the International Association of Hydrological Sciences, Elliston, G.R., Water movement through the Gornergletscher, IAHS Press, International Association of Scientific Hydrology Publications, vol. 95, Fig. 2/pp. 79–84.

also contribute. The diurnal cycles largely involve meltwater originating in the ablation area and draining rapidly to the bed through moulins.

Although the diurnal cycle is strong (Figure 6.6), the runoff should not be regarded as a simple reflection of the melt production. Some water is stored within the glacier and later released. Water may take a considerable time to travel from its site of origin to the glacier terminus. From an analysis of discharge curves, Elliston (1973) inferred that at least half the meltwater spent at least one day within Gornergletscher. Lang (1973) found the lag needed to give maximum correlation between daily discharge and air temperature. He concluded that water spent, on average, two to three days in Aletschgletscher in early summer but only one day in August and September. These analyses give mean residence times for the whole drainage basin; they do not tell where most of the delay occurs. This can be determined by observing when dye, spread on the surface of the accumulation zone and thrown down moulins in the ablation zone, appears in the stream at the terminus. Ambach et al. (1974) and Behrens et al. (1976) did this on Hintereisferner. Dye from the upper part of the accumulation zone first appeared at the terminus 10 days later, with the maximum concentration after 17 days. In a similar experiment shortly before bad weather the dye was never detected. The travel time from the lower part of the accumulation zone, in contrast, was only about 20 hours, whereas times from the ablation zone varied from 0.5 to 3 hours. Thus the major delay was in the snow and firn; in the ice, water moved at rates similar to those for open channels. The reduction in the time difference between the daily peaks in discharge and melt as the season progresses (Figure 6.6) results partly from development of the subglacial and englacial drainage systems (Nienow et al. 1998) and partly from the reduction in the thickness and extent of snow cover (Willis et al. 2002).

Seasonal Variation and Storage Of the total annual runoff from mid-latitude glaciers, the vast majority occurs in the summer melt season. Østrem (1973) estimated that 85% of the runoff from Scandinavian glaciers occurs from June to August. Escher-Vetter and Reinwarth (1994) reported that 90% of the runoff from Vernagtferner, Austria, occurs from June to October. Figure 6.8 shows the runoff pattern over a melt season from a glacier in the Canadian Rockies – the same glacier whose energy budget we used as an example in Chapter 5 (Shea et al. 2005). The foreland stream flows freely by late June and shuts down in mid-September. Runoff reaches a maximum in late summer when melt production is most rapid. It diminishes when temperatures temporarily fall below freezing, and declines rapidly at the end of the melt season. One brief but intense storm precipitated 11 mm of rain in one hour on August 19 (day 232 in the figure). This generated a runoff spike during the following hour, with a maximum discharge more than twice the size of any of the other daily maxima.

Storage and release of water are important factors in the seasonal runoff pattern. Stenborg (1970) studied these processes in Mikkaglaciären. He compared the total discharge from the

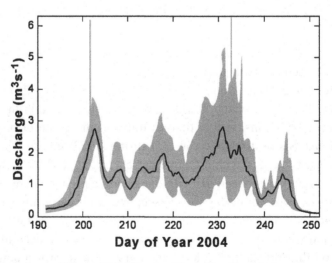

Figure 6.8: Variation of runoff from Haig Glacier (Canadian Rockies) over an entire summer season. The solid curve shows the center-weighted running mean over two days. The gray band shows the range of daily fluctuations. Two brief, exceptionally large discharge events occurred. The first peaked at 6.1 m³ s⁻¹, the second at 12.2 m³ s⁻¹. Adapted from Shea et al. (2005). Data courtesy of Shawn Marshall.

stream at the terminus with the melt estimated from meteorological parameters using regression equations. Until mid-July, melt exceeded run-off; the surplus was as much as 25% of the total summer discharge. From mid-July to early August, run-off exceeded melting, as the glacier released water stored earlier in the summer. Again, Tangborn et al. (1975) found that summer run-off from the basin of South Cascade Glacier exceeded the measured ablation by 38%. Storage occurs at different times on different glaciers, but generally corresponds to the early part of the melt season – late spring to mid-summer (Jansson et al. 2003). Although much of the water is stored in firn and slush, some is contained in cavities in the ice or at the bed. Sudden floods sometimes seen in streamflow records provide further evidence for release from cavities. Mathews (1963) identified such events in the runoff from Athabasca Glacier, Canada. The floods interrupted the normal diurnal cycle and appeared to be unrelated to weather conditions. In some cases discharge took one or two days to attain a maximum, and remained above average for up to a week. Records covering thirteen summers showed ten floods exceeding 25×10^4 m^3 and numerous smaller ones. There are no ice-dammed lakes around Athabasca Glacier so the water must have been stored within the ice. Walder and Driedger (1995) analyzed the occurrence of similar floods from South Tahoma Glacier, on Mount Rainier. They concluded that, although the water is stored within the glacier, the probability of a flood increases in periods of abundant rain and surface melt.

In the tropical Andes and the Himalayas, the seasonal cycle of runoff from glaciers differs from the mid-latitude case shown in Figure 6.8 (Willis 2005, p. 8). Alternation of wet and dry seasons, rather than warm and cool ones, dominate the seasonal cycle. High precipitation and runoff occur in the wet season. During the dry season, ablation continues to feed streamflow. The total amount is reduced but the portion originating as glacier melt increases (Mark and Seltzer 2003).

6.2.4.2 Relation to Internal Water System

In combination, observations of runoff suggest that during the melt season the glacier acts like a reservoir that is drained continuously by the streams emerging at its terminus and refilled daily by meltwater. The glacier contains a considerable amount of free water. While much of the water is in snow and firn, the ice also contains appreciable quantities in crevasses, moulins, and cavities. Some of these are isolated, at least temporarily; others are connected to the main drainage system. As long as the channels and cavities contain enough water, ice flow cannot close them. They close during the winter and are reopened by meltwater at the beginning of the following summer. This can happen in a few weeks or even days. The glacier bed seems to become flooded, fresh crevasses open, old moulins are reactivated, and new ones form. Glacier streams carry heavy loads of sediment at this time. During the initial period, some water is stored because the channels are too small to carry it all away. The channel system develops further during the summer as increasing amounts of meltwater enlarge old channels and open new ones. Thus the time that water spends in the glacier and the lag between maximum melt and maximum runoff are reduced. Initially isolated cavities connect to the main drainage system so that, for a time, runoff exceeds melting. Floods seen in streamflow records probably result from the draining of a series of previously isolated cavities; for example, when the barriers between a series of water-filled crevasses are breached successively.

6.2.4.3 Linear Reservoir Models

Picturing a glacier as a reservoir leads to the "linear reservoir model" approach for characterizing glacier runoff, a technique used in water resource analyses. Suppose that a glacier contains a volume of water V_w. For conservation of mass,

$$\frac{dV_w}{dt} = S_w - Q_w,$$ (6.3)

where the source term S_w denotes the flux into the system from melt and precipitation, and Q_w signifies the runoff (here both S_w and Q_w are volume per unit time). The key assumption of the linear reservoir approach is that the glacier releases water at a rate proportional to the stored amount:

$$Q_w = k_r \, V_w = V_w/\tau_r.$$ (6.4)

Here k_r defines the *reservoir storage constant* and its inverse τ_r is the *residence time*. Substitution of Eq. 6.4 into Eq. 6.3 gives the linear equation

$$\frac{dQ_w}{dt} + \frac{Q_w}{\tau_r} = \frac{S_w}{\tau_r}.$$ (6.5)

The solutions of this equation are broadly consistent with the observations discussed previously. A cyclic input S_w generates a cyclic variation of runoff, with peak values delayed relative to the time of maximum input. After the input shuts off, the runoff declines exponentially with time. Equation 6.4 does not allow for sudden releases of stored water and so the approach cannot account for all aspects of observed runoff.

Glaciers are often represented as a series of parallel snow, firn, and ice reservoirs, with distinct residence times τ_s, τ_f, and τ_i (Baker et al. 1982; Hock and Noetzli 1997). The runoff Q_w then equals the sum of the runoffs predicted from three separate equations like Eq. 6.5, one for each reservoir. Table 6.1 lists example residence times from studies that calibrated the parameters using hourly and daily runoff observations. In each case, runoff was measured at gauging stations within several kilometers of the glacier terminus and includes some flow from nonglacial sources.

Observations of diurnal variability, discussed in Section 6.2.4.1, demonstrate that the pathway taken by water through a glacier changes over the course of a melt season. A single set of reservoir parameters therefore should not apply at all times. For example, consider the small cirque glacier in the French Pyrenees studied by Hannah and Gurnell (2001). The runoff here was best simulated by using only two parallel reservoirs: "slow" and "fast" systems. The appropriate values for τ_r decreased over the melt season, from 45 to 18 hours for the slow system, and from

Table 6.1: Storage constants (mean residence times) adopted in applications of linear reservoir models in glacierized catchments.

Catchment	Glacierized area (km²)	τ_f (hr)	τ_s (hr)	τ_i (hr)	Reference
Vernagtferner, Austria	9	350	30	16	Baker et al. (1982)
Storglaciären, Sweden	3.1	430	30	4	Hock and Noetzli (1997)
Various catchments, Switzerland		350	120	40	Verbunt et al. (2003)
Rhône, Switzerland	20.3		125	113	Schaefli et al. (2005)
Lonza, Switzerland	28.4		96	41	Schaefli et al. (2005)
Drance, Switzerland	70.1		142	110	Schaefli et al. (2005)
Hofsjökull, Iceland	880	400	60	20	de Woul et al. (2006)

13 to 5 hours for the fast one. The relative contribution from the "slow" system declined over time and only one "fast" system was active late in the melt season.

At Black Rapids Glacier, a 20-km-long valley glacier in Alaska, 10% to 20% of the summer runoff travels through a "fast" system that responds to diurnal inputs (Raymond et al. 1995). The remainder moves through a "slow" system that varies little over weeks. Raymond et al. found that the slow-system waters contained high concentrates of solutes, indicating a long time of contact with the bed. The fast and slow systems here, then, do not reflect the difference between ice and firn, but rather the difference between surface melt from the lower ablation zone and waters moving slowly along the glacier bed.

A crude representation of the water system underlies the linear reservoir approach. Differences between the surface, englacial, and subglacial water systems are not represented. There is no accounting for exchange of water between the reservoirs. The primary assumption, Eq. 6.4, is sometimes invalid. Nonetheless, comparisons of simulated and measured runoffs show that the strategy provides a useful phenomenological description, especially for small glaciers. For water resource applications, predicting the source term – the ablation and rain – is generally a more important challenge than correcting the crude representation of drainage.

A considerably more thorough method for calculating runoff accounts for flow through different parts of a glacier, using numerical models for each part and the connections between them (Section 6.3.6.1). No runoff model accounts, however, for the occasional floods of internally stored water, which occur randomly and defy prediction (Walder and Driedger 1995).

6.2.4.4 *Water Resource Contribution to a Drainage Basin*

Glaciers can contribute significantly to the runoff from a drainage basin, but their role is often falsely characterized; enhanced runoff caused by glacier retreat must be distinguished from the "normal" runoff related to the annual hydrologic cycle. Each year, a glacier releases a total

Table 6.2: Comparison of annual precipitation with variations of mass balance.

Glacier	Precipitation $(m\,yr^{-1})$	Standard deviation of average balance $(m\,yr^{-1})$	Ratio of std. dev. to precip.
Nigardsbreen	3.8	1.06	0.3
Hintereisferner	2.3	0.54	0.2
Peyto	2.2	0.55	0.3
Storglaciären	1.5	0.75	0.5
White	0.4	0.26	0.7

Data from Oerlemans (2001), pp. 54 and 100.

runoff, Q_g, equivalent to

$$Q_g = A_g[P_g - E_g] - B_n , \tag{6.6}$$

where A_g denotes glacier area, and P_g and E_g are the mean precipitation and evaporation rates on the glacier. The annual change of ice volume, B_n, depends on the accumulation minus ablation, summed over the glacier (Section 4.2.4; B_n is the annual glacier balance). In Eq. 6.6, changes of stored water within the glacier are assumed negligible for a year, as are groundwater fluxes.

If a glacier is equilibrated to the climate, the through-flow of water, $A_g[P_g - E_g]$, normally contributes more to the runoff than do yearly volume changes. Table 6.2 lists estimated annual precipitation rates for several glaciers and compares them to typical annual volume changes per area of glacier. The latter are given by the standard deviation of the glacier-average specific mass balance, the thickness of ice removed or added to the whole glacier in a year (Chapter 4).

Thus, in a temperate region, in a typical year, the change of ice volume appears to modify the runoff by about 30% or less. Averaged over several years, the contribution would be smaller. The tabulated numbers also suggest that ice-volume changes cause a larger fraction of runoff variability in dry climates than wet ones.

If the glacier disappeared entirely, its basin would still release a runoff $A[P - E]$. Whether this represents an increase or decrease depends on how precipitation and evaporation change. Precipitation might increase with climate warming, but the establishment of vegetation can significantly increase E, which, in general, includes transpiration as well as direct evaporation.

Regardless, three factors give glaciers a critical role in runoff generation:

1. During periods of sustained retreat, the shrinking ice volume ($B_n < 0$) feeds a significant flow of water to the river system. During advance, the rivers are deprived.

2. During years of drought, melt continues to supply the rivers with water; B_n often turns strongly negative in years of low precipitation. Thus glaciers buffer the water supply against drought.

3. Over the seasonal cycle, the glacier gains mass in winter and loses mass in summer. Seasonal storage and release of surface water occur in all landscapes where snowpacks accumulate, but only in glacial landscapes does the release persist for the duration of the warm season.

In a typical alpine watershed, glaciers occupy headwater basins and high summits but much of the landscape is free of permanent ice. The annual runoff from the watershed has both glacial and nonglacial components. The fraction of runoff originating from glaciers typically decreases downstream as the fraction of total basin area covered by ice decreases. The glacial component is nonetheless significant in some populated regions, especially in the Himalayas and the Andes; for example, glaciers contribute 12% of the annual runoff in the 5000 km^2 Rio Santa watershed in the Cordillera Blanca, Peru (Mark and Seltzer 2003). But in many regions, even where glaciers are a vivid element of the landscape, the glaciers contribute little to the annual total. On the eastern slopes of the Canadian Rockies, for example, glacier melt contributes only 2% of the annual flow in the Bow River at Banff (Hopkinson and Young 1998). The situation changes drastically in periods of drought, however. Glacier inputs to the Bow River can exceed 50% in late summer of a dry year.

How might the glacial contribution to runoff change when climate warms? Consider two hypothetical climate scenarios (Figure 6.9): an abrupt warming and a progressive warming.

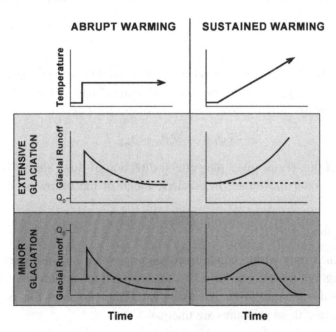

Figure 6.9: Schematic diagram showing plausible variations of glacial runoff in two climate warming scenarios, on two landscapes. "Extensive glaciation" means that glaciers cover a large fraction of the land surface and a wide range of altitudes. "Minor glaciation" means that glaciers are present only in high basins. The value Q_o indicates the relative scale of the vertical axes on the discharge plots.

Suppose that each acts on two landscapes, one with extensive glaciers covering a large area and large altitude range, the other with a small glacial cover confined to the highest basins and slopes. On the former, abrupt warming leads to a pulse of runoff production followed by reequilibration to a slightly lower value; the glacial cover has shrunk, but ablation zones have merely shifted up-slope. Under progressive warming, equilibrium lines move continuously up-slope, and the remnants of glacier tongues at low elevations rapidly ablate. On the landscape with a small glacial cover, progressive warming initially increases the glacial runoff, but runoff subsequently declines to zero as the glaciers expire. This pattern – initial increase followed by decline – is predicted for glacier runoff in the Alps over the next century as the climate warms (Braun et al. 2000).

6.3 The Water System within Temperate Glaciers

The following sections discuss theoretical analyses that provide a more detailed view of some elements of the water system introduced in Section 6.2.

6.3.1 Direction of Flow

Water flows from high to low elevations and also toward regions of reduced pressure. Thus water flows not strictly "downhill" but down the gradient of a *hydraulic potential* (Domenico and Schwartz 1998),

$$\phi_h = P_w + \rho_w \, g \, z, \tag{6.7}$$

with P_w the water pressure and z the elevation. ϕ_h equals the potential energy per unit volume of water. (Dividing ϕ_h by $\rho_w \, g$ gives a quantity with units of length, called the *hydraulic head*.) The force per unit volume driving flow, the *potential gradient*, is

$$-\nabla \phi_h = -\nabla P_w - \rho_w g \nabla z, \tag{6.8}$$

and the direction of this vector gives the general direction of flow. Shreve (1972) applied this relation to analyze water movement in glaciers; much of the following discussion uses his analysis.

6.3.1.1 Englacial Flow

Strictly, ϕ_h is defined only within conduits containing water, but for convenience we regard it as defined throughout the glacier. As a first approximation, assume that the water pressure everywhere equals the pressure of the overlying ice. (Later we show this is not true; conduits open and close because these pressures are unequal.) Thus

$$P_w = \rho_i \, g \, [S - z], \tag{6.9}$$

where ρ_i is the density of ice and S the elevation of the glacier surface. It follows that

$$\phi_h = \rho_i \, g \, S + [\rho_w - \rho_i] \, g \, z. \tag{6.10}$$

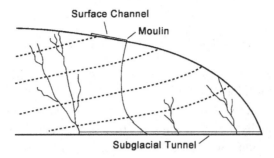

Figure 6.10: Schematic diagram of equipotential surfaces (broken lines) and related theoretical drainage pattern in a glacier. Small channels within the ice should tend to be perpendicular to the equipotential surfaces, but the inclination of a moulin is determined by that of the crevasse from which it formed. Water also flows along fractures of various orientations, in whichever direction the potential decreases. (See Figure 6.1 for a more complete depiction of features.)

Equation 6.8 shows that water flows in the direction perpendicular to the surfaces defined by constant ϕ_h. On such a surface, the gradient of ϕ_h is zero. The elevation of the equipotential surfaces z_ϕ thus varies along the glacier (direction x) as

$$\frac{dz_\phi}{dx} = \left[\frac{-\rho_i}{\rho_w - \rho_i} \right] \frac{dS}{dx} \approx -11 \frac{dS}{dx}; \tag{6.11}$$

the equipotential surfaces slope about eleven times more steeply than the surface and in the opposite direction. Figure 6.10 depicts the drainage system predicted by this analysis. However, we also expect that locally much of the flow will be diverted along conductive paths – like fractures and pipes – whose orientations do not match the theoretical potential gradient. In this case, water still flows in the direction of decreasing ϕ_h, but water pressures are perturbed in the vicinity of the conductive paths.

One type of conductive path, a moulin, forms where water flows into a crevasse. The inclinations of moulins and crevasses therefore ought to be the same when they first form. Crevasses meet the surface at a right angle, but if they penetrate to deep layers undergoing shear, they tilt; newly formed crevasses should penetrate basal layers at a down-slope angle of roughly 45° and then rotate toward the vertical due to ice deformation. The orientation of moulins may change over time as their walls melt differentially, a likely process because water runs down one side of a moulin when it is not completely filled. Holmlund (1988) found that even at a depth of 60 m moulins in Storglaciären did not trend perpendicular to the equipotential surfaces. Some moulins in the Greenland Ice Sheet remain vertical to at least 90 m depth (Peter 1996).

6.3.1.2 *Flow Along the Glacier Bed*

The following discussion applies whether the overlying ice is temperate or not. The gradient $-\nabla\phi_h$ again determines the direction of flow, but now $z = B$, the elevation of the bed. From

Eqs. 6.8 and 6.10,

$$-\nabla\phi_h = -\rho_i\, g\, [\nabla S + 0.09\, \nabla B]\,, \tag{6.12}$$

where the coefficient 0.09 is an approximate value for $\rho_w/\rho_i - 1$. The surface slope thus has a dominant role steering flow along the bed.

Basal waters also flow in the across-glacier direction. In the accumulation zone of a valley glacier, the surface is usually higher at the sides than at the center and so water flows toward the centerline. In the ablation zone, the convex surface profile impels water toward the sides. However, the steep slopes of the underlying bedrock valley walls probably permit water to drain toward the center. Thus, in either zone, drainage may concentrate along the centerline of the bed. The dependence of $\nabla\phi_h$ on bed slope implies that subglacial water tends to follow valley floors and cross-divides at the lowest point, even while moving in a general down-glacier direction given by the surface slope. *Eskers*, winding ridges of sand and gravel deposited by subglacial streams, are usually found in such places (Shreve 1985).

According to this analysis, water running along the bed must accumulate upstream of places where the bed slopes at more than 11 times the surface slope and in the opposite direction; this situation reverses the hydraulic gradient on the bed. The build-up of pressure might then force water out of passageways and into a sheet or into englacial conduits. Perhaps water passes an overdeepened section in this way. Water can also accumulate anywhere on the glacier bed if the capacity of the water system to conduct flow decreases downstream. In this case, however, the water does not stagnate.

In Eq. 6.12, the factor 0.09 indicates that the bed slope must be larger than the surface slope by a factor of 11 to have an equal influence steering the water flow. Thus water should often flow uphill at the bed of a glacier. Indeed the paths of eskers and eroded channels on formerly glaciated terrain demonstrate uphill flow. The value 0.09 only applies, however, with the assumption that water pressure equals the ice overburden; if, in effect, the ice is floating. Beneath valley glaciers with daily cycles of surface melt, this is probably a poor approximation much of the time. Lower pressures often occur in the conduits conveying the meltwater. Define the "flotation fraction" f_w as the ratio of water pressure to overburden pressure, P_w/P_i. Then Eq. 6.12 is

$$-\nabla\phi_h = -\rho_i\, g\, \left[f_w\, \nabla S + \left[\frac{\rho_w}{\rho_i} - f_w \right] \nabla B \right]. \tag{6.13}$$

If $f_w = 0.8$, the bed slope needs to be only 2.7 times greater than surface slope to have an equal influence steering the water. With $f_w = 0.56$, the flow direction responds equally to bed and surface slopes.

Equations 6.12 and 6.13 provide a useful indication of the overall direction of water flow at the bed, but not the detailed pattern. Water tends to flow in channels determined by the local bedrock topography, rather than melt out new ones in the ice. In addition, the theory applies only to conduits full of water. The pressure in partially filled channels is atmospheric, and the assumption that water pressure varies in proportion to ice pressure, already a crude approximation, is then invalid. On the other hand, conduits beneath ice sheets and long glaciers are likely to remain water-filled.

Björnsson (1982) used Eq. 6.12 to define the water drainage basins underneath three outlet glaciers of Vatnajökull, a major ice cap in Iceland. The magnitude and direction of surface slope (∇S) were measured from conventional maps, and ∇B from bedrock maps based on radar soundings. The vectors of water flow could then be plotted and the water drainage divides mapped. For comparison, the ice-drainage basins were delimited from surface contours (Section 8.2.1). For one of the glaciers, the water-drainage basin coincided with the ice-drainage basin. For the other two glaciers, the streams emerging at the termini drained water from only about half of their corresponding ice-drainage basins; the bed topography partially diverted basal waters to adjacent outlet glaciers.

6.3.2 Drainage in Conduits

Drainage in conduits was originally analyzed by Röthlisberger (1972) and Shreve (1972). Subsequent work of Nye (1976), Spring and Hutter (1981, 1982), and Clarke (1982, 2003) extended the model to describe channels with varying flows, to include some physical refinements and to analyze glacier outburst floods.

The analysis refers to flow in conduits large compared with intergranular veins and tubes. Although the conduits are initially regarded as being within the glacier, most of the analysis also applies to channels in ice at the bed.

Two opposing effects determine the size of a conduit:

1. Water flowing in a conduit enlarges it by melting ice from the walls. Viscous dissipation in the water and friction of water against the walls produce the necessary heat. In addition, water from the surface or subglacial sources may be warmer than $0\,°C$.

2. If the pressure of the overlying ice exceeds the water pressure, ice flows in and reduces the diameter of the conduit. The diameter decreases at a rate proportional to the third power of the difference between the ice and water pressures (Nye 1953). The power 3 is the exponent on stress in the creep relation for ice, discussed in Chapter 3.

By these processes, the capacity of the system adjusts, though not instantaneously, to changes in the strength of water sources such as surface melt.

The objective of the following analysis is to derive an expression for the water pressure in a conduit, assumed to be full, in terms of the discharge Q_w, the volume of water flowing in the conduit in unit time at a particular location. The water pressure is no longer assumed to be equal to the ice-overburden pressure.

6.3.2.1 Fundamental Mechanics

This analysis was largely developed by Röthlisberger (1972). For convenience, in the following discussions we use the symbol G instead of $|\nabla \phi_h|$ for the magnitude of the hydraulic gradient.

From Eq. 6.8, the force per unit volume driving water flow in a conduit is

$$G = \frac{dP_w}{ds} + \rho_w g \sin\theta. \qquad (6.14)$$

Here s denotes distance along the conduit, measured up-glacier from the exit (opposite to the water flow), and θ denotes the inclination of the conduit to the horizontal, positive when descending downstream. As before, P_w signifies water pressure.

Consider a straight, circular channel with radius R_c and cross-sectional area $\mathcal{A}_c$ (Figure 6.11). The analysis must account for three processes:

1. The tunnel contracts because the ice flows. Nye (1953) showed that a cylindrical hole, of circular cross-section, contracts under an effective pressure N at rate

$$\frac{1}{R_c}\frac{\partial R_c}{\partial t} = A\left[\frac{N}{n}\right]^n \quad \text{with} \quad N = P_i - P_w. \qquad (6.15)$$

    A and n are the creep parameters for ice (Eq. 3.35), and P_i is the ice overburden pressure, usually greater than P_w.

2. The work done by gravity and by the water pressure gradient produces heat in the water. The amount produced in a length ds of the conduit in unit time is $Q_w G$. Instantaneous heat transfer from water to ice is assumed. However, the temperature of the ice, which is always at melting point, changes along the conduit because the melting point depends on pressure. Thus some of the heat is needed to warm the water to keep it at the same temperature as the ice. In a distance ds following the water flow, the pressure in the surrounding ice drops by an amount dP_i, and the ice temperature increases by an amount $\mathcal{B}\,dP_i$. Here $\mathcal{B}$ denotes the change of melting point of ice for unit change of hydrostatic pressure (Section 9.4.1). To warm the water by this amount requires a heat $\rho_w c_w Q_w \mathcal{B}\,dP_i/ds$, for a specific heat capacity of water c_w. The walls of the conduit melt at rate $\dot{M}$ (mass per unit length of wall in unit time). With a latent heat of fusion of L_f, the heat used for melting is simply $\dot{M}L_f$.

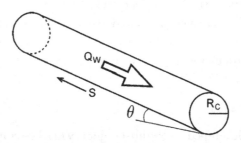

Figure 6.11: Variables for analysis of water flow in a tunnel surrounded by ice.

Heat balance thus requires that

$$\dot{M} L_f = Q_w G - \rho_w c_w Q_w B \frac{dP_i}{ds}. \tag{6.16}$$

Surface meltwater is assumed to enter the glacier at $0\,°C$; otherwise it carries heat into the ice and another term must be added.

3. The size of the tunnel, an unknown quantity, can be related to Q_w and G using empirical formulae from hydraulics. The most commonly used expression, the Manning formula, gives the mean velocity of turbulent flow in a pipe as:

$$v_w = \frac{Q_w}{\mathcal{A}_c} = \frac{1}{[\rho_w g]^{1/2} n_m} R_h^{2/3} G^{1/2}. \tag{6.17}$$

Here R_h is the hydraulic radius (cross-sectional area $\mathcal{A}_c$ divided by the perimeter) and n_m denotes the *Manning roughness coefficient*, a parameter with values in the range 10^{-2} to 10^{-1} s m$^{-1/3}$. For a circular pipe $R_h = R_c/2$ and the equation can be written

$$G = [4\pi]^{2/3} \rho_w g n_m^2 Q_w^2 \mathcal{A}_c^{-8/3}. \tag{6.18}$$

6.3.2.2 Steady-state Tunnels

At steady state, ice melted from the walls balances the inflow of ice given by Eq. 6.15. Thus

$$\dot{M} = 2\pi \rho_i R_c \frac{\partial R_c}{\partial t} = 2\rho_i \mathcal{A}_c A \left[\frac{P_i - P_w}{n} \right]^n. \tag{6.19}$$

To combine the equations, a relation is needed between the gradients of water and ice pressures, dP_w/ds and dP_i/ds. Given that water pressure is usually less than ice pressure, the value of dP_w/ds averaged over the glacier bed should also be less than the average of dP_i/ds; both pressures fall to atmospheric at the terminus. Therefore we write

$$\frac{dP_w}{ds} = f_w \frac{dP_i}{ds}, \tag{6.20}$$

where f_w, the flotation fraction, nominally ranges from zero to one. If the water pressure equals the ice overburden, $f_w = 1$, the case considered originally by Röthlisberger. With this definition, and using values $\rho_w = 10^3$ kg m^{-3}, $c_w = 4.22$ kJ kg^{-1} K^{-1}, and $B = 7.42 \times 10^{-8}$ K Pa^{-1}, Eq. 6.16 reduces to the relation

$$\dot{M} L_f = Q_w \left[\rho_w g \sin\theta + [1 - \gamma] \frac{dP_w}{ds} \right], \tag{6.21}$$

where $\gamma = 0.313/f_w$ (dimensionless). The value of B for pure water has been used; for air-saturated water, the value should be about 30% larger. In addition, we assume for now that $f_w = 1$.

Finally, elimination of the unknowns $\dot{M}$ and $\mathcal{A}_c$ from Eqs. 6.18, 6.19, and 6.21 gives a differential equation relating water pressure P_w to flux Q_w:

$$G^{11/8} - \gamma \, G^{3/8} \frac{dP_w}{ds} = \rho_i \, [\rho_w g]^{3/8} \, K_1 \, L_f \, n_m^{3/4} \, A \, [P_i - P_w]^3 \, Q_w^{-1/4}, \qquad (6.22)$$

where $K_1 = [2/27][4\pi]^{1/4} = 0.139$ and a value $n = 3$ has been used. This is the fundamental equation of steady-state tunnel theory. Note that G depends on dP_w/ds and $\sin\theta$ according to Eq. 6.14.

This equation has to be integrated numerically. An alternative approach is to simplify it by considering end members (Fowler 1987a). At the base of a glacier with ice thickness H, $P_i = \rho_i g H$. Moreover, $P_w \le P_i$, except in local patches. At a distance X from the terminus, the order of magnitude of dP_w/ds is thus $\rho_i g H / X$. With typical values for a mountain glacier's ablation zone of $H = 200\,\text{m}$, $X = 5\,\text{km}$, then $dP_w/ds \approx 400\,\text{Pa m}^{-1}$. In one end member, the bed slopes steeply down toward the terminus. For example, with slope $\theta = 10°$, then $\rho_w g \sin\theta \approx 1700\,\text{Pa m}^{-1}$. This considerably exceeds the pressure gradient, and so to a rough approximation $G \approx \rho_w g \sin\theta$. This is equivalent to assuming the water pressure stays constant along an isolated tunnel and only changes where tributaries join and contribute water. Substituting this approximation for G in Eq. 6.22 and rearranging gives the relation

$$P_i - P_w = K_2 \frac{G^{11/24} Q_w^{1/12}}{n_m^{1/4} A^{1/3}} \qquad (6.23)$$

$$\text{with} \quad K_2 = \left[[\rho_i L_f K_1]^{1/3} [\rho_w g]^{1/8} \right]^{-1}. \qquad (6.24)$$

The water pressure in the conduit is less than the pressure in the nearby ice; Eq. 6.23 gives the magnitude of the under-pressure, $P_i - P_w$, which is also the effective pressure (Section 6.1.2). In the other end member, appropriate for the lower regions of large glaciers, the bed elevation varies little. In this case $G = dP_w/ds$. Repeating the substitution gives an equation identical to Eq. 6.23 but with a different value for K_2; the relation of the effective pressure to G, Q_w, A, and n_m does not change.

According to the theory, effective pressure increases weakly with discharge and increases as roughly the square root of the driving force. Figure 6.12 shows how some of the characteristics of the tunnel flow vary as a function of discharge. The most important conclusion is that, in a tunnel system, an increase in flux *reduces* the steady-state water pressure, although the dependence is not a sensitive one. As an example, consider the values of water pressure measured by Mathews (1964) in South Leduc Glacier, Canada. Here, the mean value of $P_i - P_w$ was higher in summer than winter by a factor of 1.32. This implies a ratio of summer to winter flux of about 30. Such a large difference seems reasonable because the summer flux results from surface ablation, but the winter flux mainly from basal melting. Indeed, typical fluxes from Gornergletscher (Switzerland) are $10\,\text{m}^3\,\text{s}^{-1}$ in summer but only $0.1\,\text{m}^3\,\text{s}^{-1}$ in winter (Röthlisberger 1972).

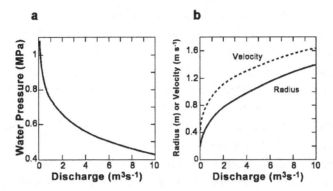

Figure 6.12: Predicted variation of water pressure, velocity, and radius for a tunnel in steady state, beneath 250 m of ice, with various discharges and one set of parameters ($\theta = 2.5°$, $A = 5 \times 10^{-24}$, $n_m = 0.1$).

The inverse relation between flux and pressure implies that in two tunnels side by side the water in the larger one has the lower pressure; it will tend to draw water away from the smaller. Thus water tends to collect into a few main arteries. Although these equations have been derived for a tunnel in the ice, they can also be applied to one at the glacier bed.

Despite these conceptual insights, the theory has limited value for calculating water pressures, because of uncertainties in the values of A and n_m, and because some of the assumptions fail. Calculations with best-guess values for the parameters often give negative values for P_w. To make it positive, the ice needs to be very soft; an A of five to ten times the generally accepted value needs to be used (Röthlisberger 1972). This is not necessarily wrong, because a water content of even 1% softens ice significantly, and basal ice layers are sometimes unusually soft (Chapter 3). On the other hand, the calculated negative pressures might reflect complexities not considered in the theory so far.

6.3.2.3 Complexities

Partial Filling of Tunnels Because the supply of meltwater fluctuates rapidly, it is possible that for much of the time water only partly fills tunnels, and the pressure is atmospheric (Lliboutry 1983a). Kohler (1995) investigated this possibility. He used tracers injected into moulins to measure average flow velocities v_w in the drainage system of Storglaciären, Sweden. He simultaneously measured water flux Q_w from the glacier. The data showed that v_w increased as a linear function of Q_w, an expected result if the cross-sectional area of the tunnels does not change with time (because $Q_w = v_w A_c$). If the tunnels were only partly filled with water, the relationship would be nonlinear. Kohler therefore concluded that water completely filled much of the tunnel system.

Channel Shape The cross-section of a tunnel at the glacier bed is probably not semicircular. The bed should retard the inward flow of the ice immediately above it. If the tunnel is not full, as is likely near the terminus, water melts ice from the sides but not from the roof. For both reasons, the tunnel is likely to be broad and low, as tunnels observed at glacier termini often are. Hooke et al. (1990) used numerical finite-element models to examine the closure of low, broad channels by ice flow. Hock and Hooke (1993) then used modelled channel geometries to interpret measurements of discharge and water transit times at Storglaciären, where they had injected tracers into moulins to measure transit times. They concluded that the subglacial drainage most likely follows an arborescent network of broad, flat channels. Cutler (1998) extended this analysis further by focusing on changes over the melt season; he found evidence that subglacial channels change from a semicircular form early in the melt season to a flatter shape as the season advances. Evidence also suggested that the channels partly fill with air during extended cold periods and at night.

Channel Roughness Low, broad channels at the bed resist the water flow more effectively than do circular channels in ice (Hooke et al. 1990; Cutler 1998). The rock material of the bed can be very rough, and the hydraulic radius of a flat channel is smaller than for a circular channel of the same cross-sectional area. Fracture systems that feed tunnels, like those observed by Fountain et al. (2005), are narrow and discontinuous and so also hydraulically rough.

The empirical relationship between flow velocity and hydraulic gradient, Eq. 6.17, depends on the parameterization for roughness. There are two common options for this (Spring and Hutter 1981; Clarke et al. 1984b), but no good observational or theoretical basis for choosing between them. Clarke (2003) found that either one works in simulations of glacier floods.

Temporal and Spatial Variability The steady-state approximation may be reasonable for conduits draining large glaciers, but extreme variations of melt and water flux occur daily in most alpine glaciers (Figure 6.6). In early morning, rapidly draining conduits are only partly filled with water. Conduits receiving the largest fluxes from the surface fill rapidly; the water backs up, increasing the pressure and its gradient. Thus water pressure and flux increase and decrease together in short period variations, not inversely as expected for steady state (Clarke 2003).

Spatial variability matters too. Constrictions of flow cause the pressure upstream to increase; it can even exceed the overburden pressure. When this happens, the ice might fracture, allowing water to leak away, or the conduit might bifurcate. Sometimes the water reaches the surface and emerges as artesian upwellings (for examples, see Stenborg 1968, and Copland et al. 2003b).

Water pressure can differ greatly between neighboring conduits. Fudge et al. (2008) measured water levels in a cluster of 16 boreholes drilled in a 60 m by 60 m region of Bench Glacier, Alaska. Mean water levels differed by as much as 100 m, equivalent to 1 MPa of pressure, even though all the boreholes showed diurnal fluctuations indicating connection to the conduit system (Figure 6.13 illustrates a period in mid-July). Moreover, as the figure shows, groups of boreholes behaved the same way, suggesting that each group was responding to conditions in a unique conduit. The most likely explanation for these observations, given the locations of the boreholes, was a system of pipes that all trend down-glacier but remain isolated from each other – and from

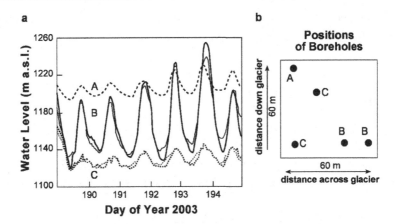

Figure 6.13: (a) Day-to-day variation of water level (a proxy for subglacial water pressure) in three sets of boreholes in Bench Glacier, Alaska. (b) A map showing the relative positions of boreholes belonging to each group. Adapted from Fudge et al. (2008).

large low-pressure conduits – for perhaps hundreds of meters. At other glaciers, dye-tracing experiments have suggested networks of parallel conduits that span much of the ablation zone (Sharp et al. 1993).

Weertman (1972) discussed a mechanism by which the water in a conduit might remain isolated from nearby conduits and from water distributed on the bed nearby. Because water pressure in a conduit is less than the ice-overburden pressure, some of the weight of the overlying glacier must be taken up by increased pressure between ice and bed on either side of the conduit. Such zones of enhanced pressures act like dams, preventing water from leaking out of the conduit or draining into it. On the other hand, a rough or porous substrate should lead to considerable spatial variability of the pressures at submeter scales, so the dams must leak. With a permeable bed, water pressure variations in a conduit should therefore propagate into surrounding regions (Hubbard et al. 1995).

Accretion by Super-cooling   According to Eq. 6.21, under some conditions water freezes to the conduit walls instead of driving melt, a process referred to as *accretion by super-cooling* (Section 4.5.1). Specifically, accretion occurs if the right-hand side of Eq. 6.21 is negative, or $\rho_w g \sin \theta < -[1 - \gamma] d P_w / ds$. (Recall that $\theta > 0$ for a conduit that descends down-glacier; $d P_w / ds > 0$ if pressure increases up-glacier; the factor $\gamma = 0.313 / f_w$; and f_w indicates the ratio of the water-pressure gradient to the ice-pressure gradient.) If the water pressure gradient is much smaller than that of ice-overburden pressure (small f_w), accretion should occur in many situations. With $f_w < 0.313$, for example, more heat is used to warm water to melting point than is generated by the work of the pressure gradient, and accretion occurs anywhere the conduit rises down-glacier ($\theta < 0$).

Even with water pressures equal to the ice overburden ($f_w = 1$), accretion occurs if the conduit rises down-glacier at a sufficiently steep slope. Consider a conduit on the glacier bed (so $\theta = \beta$, the bed slope) in a region where the bed rises down-glacier ($\beta < 0$). Such a region is said to have an "adverse slope." For accretion,

$$\rho_w g |\beta| > [1 - \gamma] \frac{dP_w}{ds}. \tag{6.25}$$

But $P_w = f_w \rho_i g H$, so Eq. 6.25 means

$$\rho_w g |\beta| > [1 - \gamma] f_w \rho_i g [\alpha + |\beta|], \tag{6.26}$$

where α denotes the surface slope, positive down-glacier. Rearranging then shows that ice accretes to the conduit walls only if the bed rises down-glacier with a slope

$$|\beta| > \left[\frac{\rho_w}{[f_w - 0.313] \rho_i} - 1 \right]^{-1} \alpha. \tag{6.27}$$

The coefficient multiplying α is 1.7 for $f_w = 1$ and 0.81 for $f_w = 0.8$.

Where accretion occurs, constriction of the conduit will inhibit water flow. Conditions favorable for rapid accretion might prevent tunnels from forming at all on certain regions of the bed.

Heat Advection and Reservoir Temperature   So far the temperature of the water has been assumed to equal the temperature of the adjacent ice. In fact, exchange of heat between water and ice is not instantaneous, and water partly retains its original temperature if a strong flow moves it quickly along a conduit. Nye (1976) and Clarke (1982) added terms to the heat balance (Eq. 6.16) to account for transport of heat energy by the flowing water ("heat advection") and exchange of heat between water and ice of different temperatures.

For a length ds of a channel, advection supplies or removes energy at rate

$$E_{ad} = -\rho_w c_w \frac{\partial}{\partial s} [Q_w T_w] \cdot ds. \tag{6.28}$$

Water draining out the fronts of glaciers carries a small amount of frictionally produced heat that would otherwise be available to melt the channel walls. Hock and Hooke (1993) found, for example, that although water entering Storglaciären as melt should be at about $0.0\,°C$, it warms to about $0.07\,°C$ as it passes through the glacier. This represents a loss of only about 1% of the heat generated by the net drop of hydraulic potential between the glacier surface and front.

More significantly, with a source water warmer than $0\,°C$ heat advection adds energy to the drainage network in the glacier. Water in lakes on and beside glaciers can warm to a few degrees above zero. Flow of such waters into the glacier supplies a significant quantity of sensible heat. Likewise, subglacial lakes on volcanic terrain supply heat. Measurements in one subglacial lake in Iceland found water temperatures of nearly $5\,°C$ (Jóhannesson et al. 2007). Clarke (1982) extended Nye's (1976) analysis of glacier floods to account for the effects of "warm" lake waters feeding conduits. The model simulates well a variety of modern and historical floods (Clarke 1982; Clarke et al. 1984b).

6.3.3 Drainage in Linked Cavities

As ice slides over a rough bed, cavities can form on the downstream sides of bumps. Examination of recently deglaciated bedrock reveals a pattern of chemical precipitates formed in subglacial water. Such patterns indicate that the cavities fill with water and link together through a network of narrow Nye channels in the bed (Walder and Hallet 1979). Water may also flow between cavities by way of conduits in the ice, or in a discontinuous water sheet or permeable sediment horizon at the interface.

Lliboutry (1969, p. 953) was the first to postulate the existence of such a linked-cavity system. Because the high normal stress in the ice on the upstream side of a bump should rapidly close conduits, the drainage system will largely avoid these regions. The result is a complex network of channels in which water flows in both across-glacier and down-glacier directions (Figure 6.14a). The following analysis of steady flow in such a system largely follows that of Walder (1986).

The main differences from tunnel flow are:

1. Cavities stay open not only because their walls melt but also because the ice slides.

2. Because cavity size depends on bedrock topography, which varies along the path, a model with uniform channel cross-section is inappropriate.

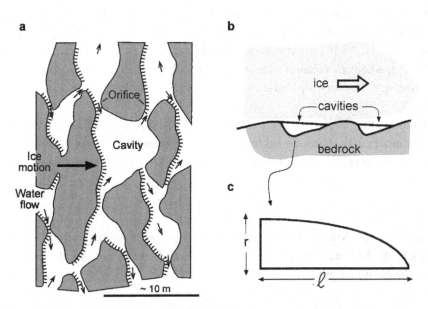

Figure 6.14: Schematic diagrams of linked cavities: (a) plan view, (b) cross-section. Panel (c) shows the idealization used in model formulation. Adapted from Fountain and Walder (1998) and Kamb (1987) and used with permission from the American Geophysical Union, *Journal of Geophysical Research*.

The following assumptions are made:

1. The bed is impermeable.

2. For simplicity, suppose that the bed is horizontal. Thus only the pressure gradient drives water flow: $G = dP_w/ds$.

3. Initially, assume a constant pressure gradient along the path.

Figure 6.14b shows the model of an idealized, water-filled ellipsoidal cavity formed in the lee of a vertical bedrock step of height r. The cavity's length ℓ is small compared with its transverse dimension, b. The glacier flows parallel to ℓ, while water flow can, in general, take any direction. In the cavity, the pressure gradient and water flow are regarded as oriented in the transverse direction (with length increment ds).

By analogy with a circular cylinder (hence Eq. 6.15) the closure rate, measured vertically, is approximately

$$w_c = A r \left[\frac{P_i - P_w}{n} \right]^n. \tag{6.29}$$

(Using ℓ instead of r in this relation ultimately gives the same qualitative conclusions.) This closure has to compensate for opening by both melting of the roof and sliding of ice over the top of the bump. Closure slows down as the water pressure rises. If the pressure approaches the overburden value, the closure rate will be too small for a stable cavity system to exist.

Let $\dot{M}$ denote the mass of ice melted from the roof in unit time per unit length in the direction of flow (kg s^{-1} m^{-1}). Melt thus removes a thickness per unit time from the cavity roof of $w_m = \dot{M}/[\rho_i P_r]$, where P_r measures the perimeter of the roof in a section perpendicular to the water flow direction. For an ellipsoidal cavity that is long and low, $P_r = k_1 \ell$, where $k_1 \approx 1.1$ (Walder 1986).

As in conduits, the energy for melt originates as viscous dissipation of potential energy. Because water moves in the transverse direction, the ice pressure and temperature can be regarded as constant and so the factor γ in Eq. 6.16 is set to zero. Thus Eq. 6.16 indicates that

$$w_m = \frac{1}{\rho_i L_f k_1 \ell} Q_w G. \tag{6.30}$$

The cavity roof moves downward with net velocity $w_c - w_m$. An ice particle in the roof will descend back to the bed in a time $r/(w_c - w_m)$. For a sliding velocity u_b, this time also equals ℓ/u_b. Hence the cavity length is

$$\ell = \frac{r u_b}{w_c - w_m}. \tag{6.31}$$

Fast sliding and rapid melt both increase the cavity size, while rapid closure by ice flow diminishes it. Assuming turbulent flow in the cavity system, the Manning formula (Eq. 6.17) again relates the water flow rate (v_w) and discharge to the pressure gradient. For cross-sectional area $\mathcal{A}_c$,

$$v_w = \frac{Q_w}{\mathcal{A}_c} = \frac{1}{[\rho_w \, g]^{1/2} \, n_m} R_h^{2/3} \, G^{1/2}. \tag{6.32}$$

For the cavity shape shown in Figure 6.14b, $\mathcal{A}_c \approx \pi r \ell / 4$ and $R_h \approx \mathcal{A}_c / 2 P_r = \pi r / 8 k_1$. Hence

$$Q_w = \frac{K_2}{[\rho_w g]^{1/2}} \frac{r^{5/3} \ell}{n_m} G^{1/2} \quad \text{with} \quad K_2 \approx \frac{\pi^{5/3}}{16 k_1^{2/3}} \approx 0.4. \tag{6.33}$$

Substitution for ℓ using Eqs. 6.29, 6.30, and 6.31, and rearranging gives an expression for the effective pressure, analogous to Eq. 6.23 for tunnels:

$$[P_i - P_w]^n = \frac{K_3}{n_m \, A} \left[r^{5/3} \frac{u_b \, G^{1/2}}{Q_w} + K_4 \, r^{2/3} \, G^{3/2} \right] \tag{6.34}$$

$$\text{with} \quad K_3 = \frac{n^n K_2}{[\rho_w g]^{1/2}} \quad \text{and} \quad K_4 = \frac{1}{k_1 \, \rho_i \, L_f}. \tag{6.35}$$

As with tunnels, the effective pressure $P_i - P_w$ increases with the pressure gradient driving the water flow. In contrast to tunnels, however, an increased discharge is transmitted with an increased water pressure, which opens the cavity further. For two cavities side-by-side, the one carrying the higher discharge will have the higher water pressure and consequently will not tend to capture water from the other. The stable configuration of a linked cavity system, in contrast to that of a tunnel system, is therefore a large number of relatively small channels distributed over a large part of the glacier bed.

Equation 6.34 can be written, alternatively, as an expression for discharge:

$$Q_w = \frac{K_3 \, r^{5/3} \, u_b \, G^{1/2}}{n_m \, A \, [P_i - P_w]^n - K_3 \, K_4 \, r^{2/3} \, G^{3/2}}. \tag{6.36}$$

The discharge that a cavity system conveys generally increases with the pressure gradient and with the sliding velocity; fast sliding opens cavities more. Note, however, that if water pressure rises to near the ice pressure, the equation breaks down; Q_w rises to infinity. This is the situation, previously mentioned, in which a stable cavity system cannot exist; the cavities expand continuously because creep closure is too slow compared with melt. For $G = 10 \, \text{Pa m}^{-1}$, $r = 0.1 \, \text{m}$, $n = 3$, and a range of plausible values for A and n_m, the critical value of $P_i - P_w$ is about 0.1–0.4 MPa. This compares with the estimate of Iken (1981, Eq. 2), based on a simple balance-of-forces argument and an idealized bed geometry, that unstable ice sliding may arise when the water pressure comes within 0.2 to 1 MPa of the overburden pressure. Iken's instability requires that the bed has a mean down-glacier slope; the quoted range of pressures applies for bed slopes between 5° and 25° and a basal shear stress of 0.1 MPa.

Because cavities are wide features that easily conduct water and because the discharge passing through a single cavity is expected to be small, the pressure gradient in cavities must usually be very small. Thus melting of the roof should typically play an insignificant role in keeping a cavity open; the closure from ice flow is balanced by opening due to sliding. Walder (1986) verified this for plausible ranges of values of G and Q_w. In general, cavities stay open largely because of sliding. The narrow passageways ("orifices") that connect them, however, are sustained by melt, because reducing the dimensions of a cavity increases the pressure gradient and the melting rate.

The analysis so far has concentrated on the cavities, although the orifices control flow through the system. The same equations should apply, although the heat used for warming water can no longer be neglected (the factor γ in Eq. 6.21 should be included in the analogous relation for melt in cavities; $1 - \gamma$ replaces 1 as the numerator in Eq. 6.30). In a steady state, the flux must be the same in both a cavity and the orifice that drains it. Because Eq. 6.33 shows that, for constant Q_w, G increases in proportion to $\ell^{-2} r^{-10/3}$, pressure gradients in the orifices can easily be 100 or even 1000 times those in the cavities. The mean pressure gradient in the network is

$$\overline{G} = \frac{G_c s_c + G_o s_o}{s_c + s_o}, \tag{6.37}$$

where s denotes length measured along the direction of water flow, and suffixes c and o refer to cavities and orifices. If $s_c/s_o \approx 10$ and $G_c/G_o \approx 0.01$, $\overline{G} \approx 0.1 G_o$. Because in a linked-cavity system water flows obliquely to the direction of ice flow, the path length may be 10 times that of a conduit in the same glacier. Thus $\overline{G} \approx 0.1 G'$ where G' is the hydraulic gradient in a tunnel. It follows that gradients in the orifices are probably comparable with those in a tunnel, whereas gradients in the cavities are very much less. Again, the equality of the fluxes and the difference in cross-sectional areas implies that water moves perhaps 100 times faster in an orifice than in a cavity. Thus the orifices control the flow but the water spends most of its time in the cavities.

In reality, the distinction between cavities and orifices must be much less sharp than assumed in this model because bedrock bumps have a wide range of sizes. Kamb (1987) argued that cavities of height $\sim$1 m and length $\sim$10 m and orifices of height 0.1 m or less are the important ones.

The preceding analysis treated the sliding velocity as an independent variable. Observations show, however, that sliding can speed up when water pressures rise. A complete model should incorporate an inverse relationship between u_b and $P_i - P_w$. Kamb (1987) proposed a model accounting for this complexity. He assumed that u_b varies inversely as $[P_i - P_w]^3$, a relation that is itself highly uncertain and inconsistent with some observations (Section 7.2.6). Kamb found that, as in the preceding analysis, flux and water pressure increase together. Processes in an orifice were considered in detail. With a small sliding velocity, a small increase in flux or water pressure might lead to rapid enlargement of the orifice. In this way, a linked-cavity system might change into a tunnel system. Conversely, a tunnel system can switch to a system of linked cavities. Because Kamb found that the pressure in a linked-cavity system is much higher than in a tunnel system carrying the same flux, such a switch might explain surges (see Chapter 12). The beds of surging glaciers are probably deformable, however, so the relevance of the proposed mechanism is unclear.

Weertman (1969a, 1972, 1986), although agreeing that surface meltwater reaching the bed in tunnels will remain in them, has argued that water formed by melting of basal ice flows in a thin sheet. A sheet of water between two flat, parallel surfaces conveys a flux proportional to powers of the thickness of the sheet and the hydraulic gradient driving flow. Values for the exponents depend on whether the flow is laminar or turbulent. Weertman originally assumed a uniform sheet thickness but later considered a sheet that pinches out on the upstream side of a bump and enlarges on the downstream side. The distinction between this and a linked-cavity system is fuzzy.

Does water flow along the bed in a sheet or in channels? Walder (1982) argued that sheet flow is unstable. If part of the sheet were to thicken slightly, more water would flow along that route, concentrating heat generation. More ice would melt from the roof and the thickness would increase further. By this feedback, a sheet would organize itself into a pattern of channels. However, rapid sliding on a rough surface or downward creep of ice (Creyts and Schoof 2009), might disrupt this process. So might a deformable bed, the next situation we consider.

6.3.4 Subglacial Drainage on a Soft Bed

If the bed consists of a layer of deformable and permeable sediments such as glacial till (a "soft" bed), some subglacial water can escape through the pores. Groundwater emerges in seeps and springs in the forelands of many glaciers. At Trapridge Glacier (Yukon), for example, fluorescent tracers injected at the glacier surface appeared years later in winter ice deposits that formed in front of the terminus (Stone and Clarke 1996). Calculations with plausible values of sediment permeability and thickness suggest that porous flow could evacuate much of the meltwater produced locally by geothermal and frictional heat (Alley et al. 1986c). However, if basal meltwater originates from a large catchment area upstream, porous flow is inadequate to remove it. Nor can flow through the pore spaces remove surface meltwater that penetrates to the bed. Furthermore, if the sediment beneath the glacier deforms, it carries some water along with it. Such "advection" accommodates the water sources only with exceptionally high sediment deformation rates and a thick deforming layer (Alley et al. 1986c). Much of the water must therefore flow along the ice-sediment interface.

Does the water at the ice-sediment interface flow as an irregular sheet or in channels carved in the ice or the sediment? Again, feedback between melt, thickening of a sheet, and concentration of flow and heating might cause a sheet to separate into some sort of channel system (Walder 1982). In addition, the flow along the thick parts of a sheet would exert a greater shear stress on the sediment than would thin parts; the thick regions would erode the sediment preferentially. These mechanisms should produce a system of channels. On the other hand, creep and displacement of the sediment itself might prevent such channels from forming, whereas rapid sliding around large rocks protruding into the ice would form cavities.

A different possibility is suggested by data from numerous borehole experiments at Trapridge Glacier. Water is rapidly redistributed along the bed, yet subglacial channels have never been

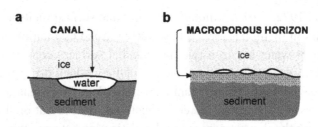

Figure 6.15: Schematic depiction of water system elements on beds of sediment.

identified at this site. Apparently, the water flows instead through a *macroporous horizon*, a thin, permeable, quasi-continuous sheet of water and sediment (Figure 6.15). This may be a combination of intergranular pore spaces in the uppermost layer of sediment, and thin films, cavities or larger gaps ("blisters") at the interface with the ice. Flowers and Clarke (2002a,b) concluded, by comparing observed water pressures to model calculations, that the water discharge (per unit width) along the bed obeys a relation analogous to Darcian flow of groundwater:

$$q_w = -\frac{k_h \, d_w}{\rho_w \, g} \frac{\partial \phi_h}{\partial s}, \tag{6.38}$$

for a hydraulic potential ϕ_h, effective hydraulic conductivity k_h, and thickness of horizon d_w. The conductivity itself varies, depending on how permeability increases as the sediment deforms and how gaps between the ice and substrate open under high water pressures.

The till beneath Trapridge Glacier is a coarse-grained sediment, typical of mountain glaciers. A macroporous horizon is less plausible for fine-grained sediments, like those beneath parts of the Pleistocene ice sheets and the modern West Antarctic ice streams. Here, and elsewhere, water might flow in channels. Because sediment is a complex material with poorly constrained deformation properties, theoretical analyses of channels in sediment are speculative. A channel cut into sediment should tend to close by creep of ice from the top and creep of sediment from the sides and bottom. Water counteracts closure by melting the ice roof and by flushing away fine-grained particles.

Walder and Fowler (1994) attempted a detailed analysis. For steady flow, these authors concluded that water could flow either in R-channels cut upward into the ice, as with a "hard" bed, or in a network of broad shallow channels, called *canals*, cut downward into the the sediment (Figure 6.15). For a canal, the approximate relationship between water flux and pressure was found to be

$$Q_w = B_c \sin^2 \alpha \, d_c^3 \, [P_i - P_w]^{-n} \tag{6.39}$$

Here α is the ice-surface slope, d_c the depth of the canal, and n the exponent in the ice creep relation. B_c depends on the thermal and mechanical properties of ice and sediment as well as the roughness of the canal's floor. The slope α is a surrogate for the hydraulic gradient in the canal.

Equation 6.39 shows that the flux in a canal increases with water pressure. Thus the larger canals do not tend to draw water away from the smaller ones; individual canals remain separate. In this respect, canals behave like linked cavities on a hard bed rather than like R-channels.

The analysis showed that low surface slopes – typical of ice sheets or ice streams on deformable beds – favor canals over R-channels. Walder and Fowler discussed how the distribution of eskers in North America supports their picture of subglacial drainage. Eskers are abundant over the Canadian Shield, where the ice sheet rested mainly on bedrock, but rare in areas mantled with thick deposits of till, such as the prairies. Eskers form in large R-channels which, according to the analysis, form under an ice sheet only if its bed is hard.

For slopes typical of valley glaciers, on the other hand, both types of features might exist. Equation 6.39 indicates that, for a given flux, water pressures in a canal system are relatively high. If both canals and R-channels exist, the channels draw water away from the canals. Thus, according to the analysis, R-channels are the preferred system in valley glaciers, even on a soft bed.

Ng (2000) proposed an alternative model for steady-state canals. He used a more detailed description of how sediment erosion and deposition depend on the water flow. Conservation of sediment mass along the channel provided an additional constraint. As in the Walder and Fowler canals, the water flux was found to increase with the pressure. Ng found, in addition, an inverse relation between pressure and the sediment flux conveyed by the canal. Sediment enters the canals because it creeps inward and erodes. The difference between ice-overburden pressure and canal water pressure drives the creep of sediment (Alley 1992b). Therefore, the lower the pressure, the greater the supply of sediment.

The available data provide little constraint on theories of canals, or other sorts of channels on soft beds of fine sediment. A meters-thick layer of water-saturated sediment underlies the Siple Coast ice streams in West Antarctica (Section 8.9.2.4). Basal water pressures measured in boreholes through three of the ice streams were within 0.2 MPa of the approximately 9 MPa ice-overburden pressure – and, at many sites, within 0.05 MPa (Kamb 2001). Such high pressures are consistent with flow in films or canals, as postulated by theory; if R-channels exist, they are widely spaced. Kamb (2001) and Engelhardt and Kamb (1997) reported on numerous experiments conducted in these boreholes. The major finding was lack of consistency from site to site. At some places, the flux of water that could be pumped out of the subglacial system or exchanged between nearby boreholes indicated the presence of a gap along the ice-sediment interface, with an average thickness of a few millimeters. At other locations, such a gap could not exist because pressures differed between nearby boreholes by as much as 0.2 MPa. Such differences would be neutralized by flow along the interface if such a gap were present. In these regions. the authors concluded, a system of canals must exist to transport the meltwater from upstream. The most likely explanation for the diverse observations is a highly complex and inhomogeneous water system, with both films and channels.

Recent observations of this region in West Antarctica reveal that large volumes of ponded subglacial water move from one place to another, on timescales of a few years. Specifically, satellite altimetry measurements show that the ice sheet surface swells and subsides throughout regions of several kilometers' dimension. The regions are located on Whillans and Mercer Ice Streams and near the grounding zone of the Ross Ice Shelf (Gray et al. 2005; Fricker et al. 2007). The observed vertical displacements amounted to several meters over the years 2003 to 2006. The most plausible explanation is partial filling and draining of subglacial lakes. According to Fricker et al.'s estimates, one subglacial lake near the Engelhardt Ice Ridge (formerly Ridge B/C) drained an estimated $2\,km^3$ of water to the Ross Sea over a period of about three years. The basin subsequently started to refill. Roughly $1.6\,km^3$ of water was redistributed between other subglacial basins over the same period. Such volumes are comparable to the total amount of water supposedly produced over a few years by subglacial melt in the entire catchment of Whillans Ice Stream ($0.53\,km^3\,yr^{-1}$). Thus, it appears that most subglacial water in this region moves through a network of subglacial lakes before draining to the Ross Sea.

By what mechanism does water flow in these events? The fluxes appear to follow gradients of hydraulic potential, a useful fact to know but not a helpful one for distinguishing between possible drainage system types. A well-developed tunnel system would accommodate faster transfers of water than those observed. Till canals, a macroporous horizon, or a free water sheet are all, in principle, consistent with the observations.

6.3.5 Summary of Water Systems at the Glacier Bed

Four systems whose characteristics have been analyzed in the preceding sections (with varying degrees of success) carry water along the beds of glaciers.

1. A tunnel system. The pressure and its gradient vary inversely with the flux over periods long enough for steady-state theory to apply. Thus, of two adjacent tunnels, the larger one has the lower pressure and so draws water from the smaller one. The system therefore develops into one, or a few, large tunnels – at least near the front of the glacier. Over short periods, like diurnal cycles, the opposite behavior applies: water accumulates in the upper parts of the tunnel system when inputs are strong, leading to simultaneous increases of pressure, pressure gradient, and flux. At either timescale, pressures may differ considerably in nearby tunnels. Although they have a tendency to join eventually, two tunnels can run parallel to each other for a distance much greater than the spacing between them.

    The tunnel system carries a large flux at low pressure in approximately the same direction as the ice flow. Dye tracing experiments in the ablation zone show that water travels from surface to terminus in a few hours at velocities of the order of $1\,m\,s^{-1}$, which is comparable with that in an open channel (Lang et al. 1979). Because it evacuates water quickly, a tunnel system weakens or nullifies the influence of variable water sources on the flow of a glacier (Chapter 7).

2. A linked-cavity system or, more precisely, a distributed system of passageways whose diameter fluctuates widely along the path. The wide parts are cavities on the downstream

sides of the larger bedrock bumps; the narrow parts occur downstream of small bumps or as channels cut in the bed. The steady-state water pressure, unlike that in a tunnel, increases with the flux and so larger channels should not grow at the expense of the smaller ones. The system is therefore a complicated meandering network of many interconnected cavities. The existence of such networks is confirmed by observations of chemical precipitates on bedrock exposed by retreating glaciers. Transit over a given distance takes much longer than in a tunnel because the path length is longer and the narrow orifices throttle the flow. Measurements of flow in linked cavities can be obtained by injecting tracers into boreholes; for example, Kamb et al. (1985) measured a mean velocity of only $0.02\,\mathrm{m\,s^{-1}}$ in a dye-tracing experiment at Variegated Glacier (Alaska) during a surge (see Section 12.3.2). Because sliding tends to keep cavities open, the sliding velocity influences their size. At the same time, the sliding velocity depends on the water pressure and volume in the cavities (see Chapter 7), so complex feedbacks link together the processes of sliding and water drainage.

3. A network of broad, shallow "canals" in the surface of a soft bed. The water pressure nearly matches the ice-overburden pressure. Canal drainage is expected only beneath ice sheets and ice streams with low surface slopes. The sediment might need to be fine-grained.

4. A permeable macroporous horizon of saturated granular material at the ice-bed interface. Water flows through the pore space, generally at high pressure, and perhaps also along films at the interface. Hydraulic conductivity of the horizon depends on sediment properties and water pressure. This is the soft-bed equivalent of a linked cavity system. Water in the macroporous horizon is distributed widely on the bed and has a large influence on flow of the glacier by basal sliding and sediment deformation (Chapter 7).

Drainage in a valley glacier in summer, except during a surge, is probably by way of distributed linked cavities and macroporous horizons between widely spaced branches of a tunnel system. Because a tunnel can carry a large flux at low pressure, the distributed systems normally receive little of the surface melt. The amount of melt varies diurnally, seasonally, and according to the weather, however, and so the system is never in a steady state.

The size of passageways within the ice and at the bed can adjust, though not instantaneously, to the volume of water. The adjustment time probably ranges from a few days to one or two weeks depending on position in the glacier and the amplitude and rate of change of flux. The system cannot adjust to diurnal variations, but for seasonal variations, the system adjusts to the average conditions. Pressures then vary inversely with the flux, and hence water pressure is higher in winter than in summer. Case studies showing how the drainage system evolves over the summer season are discussed by Hock and Hooke (1993) for Storglaciären, and Nienow et al. (1998) for Haut Glacier d'Arolla.

During winter the passageways shrink and many probably close completely, trapping water within the glacier or at its bed. When melting begins in spring, the water pressure initially increases with the flux and starts to enlarge the tunnels and cavities. Whether water initially drains through the distributed systems may depend on conditions in that particular year. In most cases, however, the tunnel system, with its characteristic low pressure, eventually takes over the bulk of the drainage. The short travel times measured by dye tracing confirm this. Again, water reaching the bed in tunnels connected directly to moulins should remain in tunnels except after sudden increases in flux, when the resulting high pressure may drive some water into the cavities and pore spaces. In late summer, the reduced water supply may be insufficient to fill the tunnels. In this case the pressure in them will drop to atmospheric, and, except beneath shallow ice, they will begin to close rapidly. The distributed systems are kept open by sliding, by high water pressures, and, where water passes through orifices, by melt.

6.3.6 System Behavior

To understand how a glacier's plumbing system works, and not just the local behavior of individual elements within it, the entire pathway taken by water through the glacier needs to be analyzed as a complete system. Fluxes, pressures, and pressure gradients all depend, in general, on conditions upstream and downstream as well as on the interplay of melt, ice flow, and sediment movement at a site.

The simplest case is a single pipe running through a glacier (Röthlisberger and Lang 1987). The pressure at the downstream exit is atmospheric or, if the glacier terminates in a lake or the sea, the static water pressure at the exit. At a steady state, the water flux at any point equals the total of upstream sources – melt of conduit walls and inputs from moulins. Upstream integration of the pressure gradient given by the conduit formulae gives the pressure variation along the pipe. Obtaining a solution requires iteration, because melt and hence flux depend on conditions in the conduit. The complexity increases greatly for time-dependent problems and systems with multiple interacting elements.

Clarke (1996) proposed a method to corral the complexity of time-dependent problems into a mathematically tractable form (though with no explicit accounting for spatial variations). He made an analogy between a hydraulic path through a glacier and an electrical circuit, with hydraulic potential taking the place of voltage and with conduits and distributed pathways represented by resistors of various types. With this approach, water pressures obey a system of linked ordinary differential equations in time. The approach allows for diverse elements to be added to the circuit, representing discharge sources, open and closed cavities for water storage, upward and downward steps in the elevation of the passageways, switches between different types of parallel flow paths, and more. The analysis showed how simple forcings – such as regular periodic inputs of water – lead to diverse and complex pressure and flux variations beneath a glacier, because of the interaction of different types of elements. Complex behaviors observed in numerous boreholes reaching the bed of Trapridge Glacier motivated the analysis.

In general, the flux of water (per unit width across-glacier) transmitted by the basal water system increases with both the size of passageways and the size of the hydraulic potential gradient. This leads to a phenomenological view of the water system's behavior as follows. (For an application of this strategy using numerical techniques, see Alley 1996.) Assume the water flux is $q_w = -c_1 H_w \, \partial \phi_h / \partial x$, with H_w the volume of water per unit area, x pointing down-flow, and c_1 taken as constant. Substitution into the relation for conservation of water volume (Eq. 6.2) gives, in one spatial dimension,

$$\frac{\partial H_w}{\partial t} = \dot{S}_w + c_1 \frac{\partial H_w}{\partial x} \frac{\partial \phi_h}{\partial x} + c_1 H_w \frac{\partial^2 \phi_h}{\partial x^2}. \tag{6.40}$$

But, on the bed, $\phi_h = P_w + \rho_w g \beta$, so that $\partial \phi_h / \partial x = \partial P_w / \partial x - \rho_w g \beta$, where β denotes the bed slope, positive for a bed descending down-glacier. Combining expressions gives a new relation with both H_w and P_w. But these two variables are also interdependent. Consider, specifically, the case of rapid fluctuations – with periods of minutes to perhaps a few days. In a distributed system, increasing the water pressure relative to the ice-overburden pressure P_i opens wider the cavities and pore spaces; thus, H_w and P_w increase together. To illustrate the implications, take the simplest possible relation: $H_w = c_2 P_w / P_i$. Substituting and rearranging gives the relation

$$\frac{\partial P_w}{\partial t} = \left[c_2^{-1} P_i \right] \dot{S}_w + \overbrace{c_1 P_w \frac{\partial^2 P_w}{\partial x^2}}^{A} +$$

$$\underbrace{c_1 \left[\frac{\partial P_w}{\partial x} - \rho_w g \beta \right] \frac{\partial P_w}{\partial x}}_{B} - \underbrace{\left[c_1 \rho_w g \frac{\partial \beta}{\partial x} \right] P_w}_{C}. \tag{6.41}$$

Thus variations of water pressure at the bed are driven by sources of water ($\dot{S}_w$), and propagate diffusively (term A). This means that zones of high water pressure tend to flatten out and spread and that fluctuations of pressure at one location propagate along the bed. Propagation can also occur as simple waves (term B if $\rho_w g \beta > \partial P_w / \partial x$) or nonlinear waves (the other case for term B). In addition, water tends to collect in bedrock hollows (term C), until the water pressure builds up to permit drainage. This discussion illustrates how the water system can respond in diverse ways to varying sources and pressure boundary conditions – even if the simplest relations between quantities are assumed. Descriptions of diffusive and wave behaviors in other contexts can be found in Chapters 9 and 11.

6.3.6.1 Integrated Modelling

Because a glacier's water system has several different components, each with a complex spatial distribution, a thorough analysis requires numerical modelling with a geographically realistic framework. Flowers and Clarke (2002a, 2002b) made an ambitious attempt to model the entire hydraulic system of a glacier. The model simulates variations over time and in both horizontal dimensions, on a domain covering the entire glacier. The supraglacial, englacial, basal, and groundwater systems were each represented by a conservation relation of the form $\partial H_w / \partial t = \dot{S} - \nabla \cdot \vec{q}_w$ (Eq. 6.2), where H_w and $\vec{q}_w$ take values unique for each system, at every coordinate. Source terms $\dot{S}$ included melt and rainfall on the surface, melt at the bed, and porosity changes in the groundwater system. In addition, all four equations included source terms that represent gains and losses due to exchanges with the other systems.

For example, surface meltwater, calculated from meteorologic data using the positive degree-day method, flows downhill on the model glacier to regions with crevasses or moulins, which carry it to the bed. A Darcian relation like Eq. 6.38 describes its subsequent movement along the bed, from which it either feeds runoff at the terminus directly or goes into an underlying aquifer. At the point of exchange from englacial to basal systems, and from basal to aquifer systems, the rate of exchange depends on a difference in hydraulic potential between them. For surface to englacial exchange, all water that flows to crevassed or moulined regions goes into the glacier, unless englacial reservoirs are full of water already. Transfers from surface to englacial (and englacial to basal) systems are not instantaneous but occur at a rate inversely proportional to time constants, which are adjustable parameters.

The method was applied first to hypothetical glaciers of simple form. Calculations showed that ice surface and bed topographies largely steer the flows of water, as we should conclude from simpler theoretical considerations. The presence of the glacier, which constitutes a source of pressurized water at the surface of the aquifer, strongly influenced the direction of groundwater flow; water was driven away from the glacier (unless the glacier sat in a deep valley) and emerged in the foreland. Diurnal cycles propagated through all the systems at a rate that depended on the time constants in the source-term exchanges. The authors (2002b) subsequently applied the model to Trapridge Glacier, Yukon, for which parameters could be calibrated by comparison to observed basal water pressures. The calibrated model then afforded a detailed look at how the summertime hydraulic pattern developed and terminated.

A related effort to calculate seasonal runoff variations from integrated modelling was made by Arnold et al. (1998) and applied to Haut Glacier d'Arolla, Switzerland. This model treated the subglacial water system using relations for conduits, rather than a Darcian relation as used by Flowers and Clarke; conduits are an important part of the water system at Haut Glacier d'Arolla. Arnold et al. connected surface melt to meteorological conditions using energy balance relations of the sort discussed in Chapter 5.

6.4 Glacial Hydrological Phenomena

6.4.1 Jökulhlaups

A *jökulhlaup* is the sudden and rapid draining of a glacier-dammed lake or of water impounded within a glacier. Such events, which cause extensive flooding and pose a great hazard to people downstream, have been reported from Iceland (where the name *jökulhlaup* originates), Norway, Switzerland, Pakistan, Greenland, Alaska, Canada, South America, and New Zealand. Glacier-dammed lakes form in various situations, but most commonly when a glacier blocks the stream draining a side-valley. A particularly dangerous case, because of the large volume of water impounded, is when a surging or other advancing glacier blocks the drainage from a major valley. A related phenomenon occurs in Peru, where lakes dammed by terminal moraines have formed in the wake of retreating glaciers; collapsing ice sometimes falls into a lake and generates waves large enough to overflow the moraine. The largest frequently occurring jökulhlaups are those released from ice-bound lakes in volcanic areas of Iceland. In particular, large lakes form beneath and beside the Vatnajökull and Myrdalsjökull ice caps. Water to fill the lakes originates as melt at the base of the ice caps, driven by very high heat fluxes from beneath. The largest floods in Iceland occur during volcanic eruptions (Björnsson 1992, 2002). Outside Iceland, well-studied examples of periodic floods include the aptly named Hazard Lake in Yukon Territory (Clarke et al. 1984b; Clarke 2003), Hidden Creek Lake on Alaska's Kennicott Glacier (Anderson et al. 2006; Walder et al. 2006), and Lago Argentino, Patagonia, a lake regularly dammed by advances of Glaciar Perito Moreno (Skvarca and Naruse 2006; Stuefer et al. 2007).

Jökulhlaups may occur once per year or only every several years. The lake starts to drain when it reaches a certain level and the flood may stop before the lake is empty. Occasionally, water from the lake flows across the glacier surface and cuts a channel in it. In the great majority of cases, however, the lake drains under the ice before the water level has reached the ice surface. The floods are sometimes extremely large. The peak flow of the 1996 outburst of Grimsvötn in Iceland was about $5 \times 10^4 \, \mathrm{m}^3 \, \mathrm{s}^{-1}$, or about one-quarter the flow of the Amazon; $3.4 \, \mathrm{km}^3$ of water drained in just 40 hours (Snorrason et al. 2002). A flood of comparable size occurred here previously, in 1934 (Thorarinsson 1953). Yet larger floods have occurred in prehistoric times. Flows greater than $10^7 \, \mathrm{m}^3 \, \mathrm{s}^{-1}$ were released from glacial Lake Missoula during the last ice age (Pardee 1942; Bretz 1969; O'Connor and Baker 1992). These floods formed the "channelled scablands," a $5000 \, \mathrm{km}^2$ area of eastern Washington State that is marked with gigantic channels scoured into the bedrock. Sediment deposits reveal that more than forty such floods occurred during the last ice age alone (Waitt 1980; Atwater 1984).

Fundamentally, glacier floods occur because of feedback between melt and the ability of drainage paths to convey water. Figure 6.16 summarizes the key process (Nye 1976; Walder and Costa 1996). Discharge of water, whether through subglacial tunnels or supraglacial streams, enlarges or deepens the channels because frictional heat production causes melt. In turn, a larger tunnel or more deeply incised stream conveys a larger discharge for a given water level in the lake. Thus the melting and discharge increase together. This positive feedback stops only after significant depletion of the volume or pressure of the source water. (The feedback may fail to develop, however, if the lake is small and cold.)

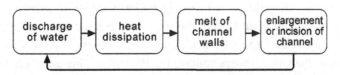

Figure 6.16: The positive feedback mechanism responsible for catastrophic drainage in glacial floods. The feedback can fail to develop, however, if the volume of the reservoir is too small or the water too cold.

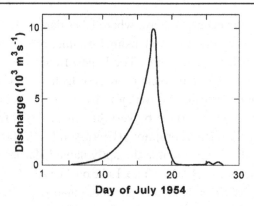

Figure 6.17: Hydrograph of the 1954 jökulhlaup from Grimsvötn, Iceland. Redrawn from Rist (1955).

Grimsvötn, a subglacial lake in the middle of the Vatnajökull Ice Cap, provides one of the best studied cases (Thorarinsson 1953; Björnsson 1974; Nye 1976). A volcanic center supplies heat flow in excess of $50\,W\,m^{-2}$, which melts ice from the base to form a lake. Melt rates typically reach $5\,m\,yr^{-1}$. Basal melt also produces a depression in the upper surface, focusing meltwater from the surrounding region into the lake. A floating "ice shelf," about 220 m thick, covers most of the lake. The lake drains catastrophically to the coastal plain through a subglacial passage 50 km long. Drainage stops when the water level has fallen by about 100 m, although the lake is much deeper than this. The water regains the critical level after five or ten years. Figure 6.17 shows the characteristic features of many of the floods: drainage at an increasing rate for about ten days followed by rapid cut-off that takes about two days. In the former phase, drainage increases because melt enlarges the passageways (Liestøl 1956). Collapse or closure of the passageways or the existence of a rock threshold can explain why drainage often stops before the lake empties. Grimsvötn starts to drain when the water is still about 20 m below the depth necessary to float the ice dam (Nye 1976); the observations suggest that drainage occurs when effective pressure in the dam falls below some threshold. As the lake fills, water pressure increases, the drainage system enlarges, and the potential gradient increases; these interdependent processes determine the onset of the jökulhlaup.

Not all floods begin like the one shown in the figure, however. In the exceptionally large jökulhlaup of 1996, the discharge increased rapidly at a nearly constant rate and reached its peak value in only 16 hours. Melt of passageways cannot explain the rise; instead, water probably advanced from the lake as a broad turbulent sheet with pressures high enough to deform the overlying ice and open further the gaps between ice and bed (Jóhannesson 2002).

During a jökulhlaup, water drains from a large reservoir that effectively fixes the water pressure at the entrance to the passageways conducting the water. This differs from the steady-state tunnel case analyzed in Section 6.3.2; that analysis assumed a constant discharge and then found the pressure needed for a balance between tunnel closure and melting. In the present case, when the reservoir starts to drain, the initial discharge is determined by the pressure gradient, which scales with the difference between pressures in the reservoir and at the exit of the tunnel. If, with this discharge, melt does not balance the inflow of the tunnel walls, the tunnel contracts. This may prevent the flood from developing. If, on the other hand, the initial discharge exceeds that required for balance, the tunnel enlarges and the discharge increases, causing yet further enlargement, as sketched in Figure 6.16. This fundamental instability explains the shape and dimensions of the rising limbs of the typical flood (Nye 1976). As the reservoir empties, water pressure decreases at a rate dependent on the shape of the reservoir, and the tunnel starts to close. The closure, and hence the termination of the flood, occurs rapidly – primarily because closure rate depends on the third power of the effective pressure in the tunnel (Eq. 6.15).

To simulate the progress of the Grimsvotn jökulhlaup, Nye (1976) used the equations for nonsteady flow in a conduit: equations for heat balance (Eq. 6.16) and turbulent flow (Eq. 6.18) and two relations for changes of conduit size. First, the conduit expands if melt exceeds inflow of ice, with a rate given by the difference of terms in Eq. 6.19:

$$\rho_i \frac{\partial \mathcal{A}_c}{\partial t} = \dot{M} - 2 \rho_i \mathcal{A}_c A \left[\frac{P_i - P_w}{n} \right]^n.$$

(6.42)

In addition, conservation of water mass within a unit length of conduit requires that

$$\frac{\partial \mathcal{A}_c}{\partial t} = \frac{\dot{M}}{\rho_w} - \frac{\partial Q_w}{\partial s}.$$

(6.43)

This states that, if the conduit is full, the change of cross-sectional area must equal the volume of water melted from the unit length of wall minus the difference between the volumes entering the upstream and leaving the downstream end of the unit section.

Nye assumed that the water was close to $0\,°C$ when it left the lake. This is not plausible for most glacier-dammed lakes. Even subglacial lake waters can be "warm" in volcanic areas (Jóhannesson et al. 2007); the lake water feeding the 1996 Grimsvötn jökulhlaup was about $4\,°C$. Clarke (1982) added one more equation to take account of such positive water temperatures and found that release of thermal energy by the water was the major factor in enlarging the tunnel. His model obtained a good reproduction of the discharge-time relation for Hazard Lake, Yukon. On the other hand, several observations of the 1996 Grimsvötn flood showed that the flood waters lost their heat to the ice much more quickly than predicted by theory (Jóhannesson 2002); for example, the water was supercooled when it appeared at the terminus. In this flood, high water pressures rather than melt drove the opening of passageways. High pressures in jökulhlaups sometimes force water into the overlying ice along fractures and conduits, which emerge at the surface (Roberts et al. 2000).

Clarke (1982) used his theory to develop formulae for the maximum discharge, based on some simplifying assumptions. However, an equally good prediction of peak discharge can be obtained from an empirical relation that Clague and Mathews (1973) fitted to jökulhlaup data from ten lakes:

$$Q_{\text{MAX}} = 75 \left[V_o/10^6 \right]^{0.67}.$$

(6.44)

Here Q_{MAX} is peak discharge ($m^3 \, s^{-1}$) and V_o the volume (m^3) of the lake just before it starts to drain. This formula, which takes no account of such quantities as tunnel shape and slope, hydraulic roughness, and ice-flow parameters, "confounds understanding but seems to give reasonable results" (Clarke 1986), a humbling thought for theoreticians. The most recent estimates for the multiplier and exponent in this relationship are 46 and 0.66 for floods draining through tunnels (Walder and Costa 1996).

A relationship of the same form applies to catastrophic drainage of lakes impounded by glaciers blocking side valleys (Walder and Costa 1996). Such floods commonly escape through a breach between the ice dam and the bedrock valley wall. The floods tend to be much larger than for subglacial ones, for a given lake volume; the best-fit multiplier and exponent in Eq. 6.44 are 1100 and 0.44, respectively. The similarity of these values to those describing floods from earthen dams indicates that, regardless of the drainage process that starts the flood, the hydraulics of flow through a rapidly forming subaerial breach controls the subsequent discharge.

Jökulhlaups transport sediment. In Iceland, where the volcanic substrate is fragmentary or highly fractured rock, the floods move great quantities of sediment which are deposited in broad outwash plains called *sandurs* at the periphery of the ice caps. A May 1721 eruption of the volcano Katla beneath the Myrdalsjökull Ice Cap produced a jökulhlaup that buried the nearby sandur in about 90 m of debris and extended the coastline southward by some 5 km. Sediment moving in subglacial conduits and erosion of conduit floors are probably significant processes affecting the size and shape of subglacial passageways and hence their ability to convey the flood. These processes were not included in Nye's theory. Fowler and Ng (1996) incorporated erosion of sediment along the conduit sides into a model for jökulhlaups, assuming a sediment bed. In their view, competition between widening of the conduit by erosion and narrowing by sediment creep determines the conduit shape, which is broad and flat. Erosion was assumed proportional to the rate of energy expenditure by the flowing water.

A further complication for models is that sometimes the flood emerges in several outlet channels and sometimes as a broad irregular sheet several kilometers wide. Clarke (2003) suggested that multiple channels develop because overpressurization of the main conduit at constrictions causes ice fracture or lateral drainage. But in the cases like the 1996 Grismvötn event – in which floods propagate as a sheet of overpressurized water – flow in the sheet and interaction between the sheet and developing pipes are both important processes (Flowers et al. 2004), as is flow upward into the glacier (Roberts et al. 2000).

6.4.2 Antarctic Subglacial Lakes

In Chapter 9 we explain how the thick ice and low accumulation rates in Antarctica allow the base of the ice sheet to warm to melting point despite the frigid surface climate. Many lakes exist beneath the Antarctic Ice Sheet. The lakes were originally identified from bright horizontal reflecting surfaces seen in radio-echo soundings (Oswald and Robin 1973). At least

150 subglacial lakes have now been identified in Antarctica (Siegert 2005; Fricker et al. 2007). Most occupy bedrock basins. Their abundance in Antarctica can perhaps be explained by three factors. First, surface slopes of the ice sheet are small, implying a small typical hydraulic gradient; this permits basins to trap water. Second, subaerial erosion has not acted on the bedrock for millions of years. Thus, numerous bedrock basins have developed by tectonic processes and subglacial erosion, but have not been breached or filled with sediment. Third, the absence of surface meltwater streams implies that no water reaches the bed as a concentrated source capable of sculpting a network of large and abundant tunnels.

Lake Vostok, beneath 3750 m of ice, is the most thoroughly investigated subglacial water body in Antarctica but has not yet been entered or directly sampled (as of 2008). With an area of 15,690 km^2 and an estimated volume of 5400 km^3, it is the world's seventh largest lake. Equally remarkable is that the seminal Russian and French ice-coring studies were conducted at Vostok Station for many years before anyone knew of the lake beneath. Only in the late 1990s did the coring penetrate deeply enough to recover bottom ice, a 210 m-thick layer clearly identifiable as lake water frozen onto the base of the ice sheet (Jouzel et al. 1999; Siegert et al. 2001). The transition from the ordinary glacier ice to the lake ice is abrupt and defined by a switch to large and undeformed crystal textures, with distinct isotopic and chemical compositions. This lake, and others like it in East Antarctica, attract much interest because they presumably have been sealed from direct contact with the atmosphere for millions of years. The meltwater passing through the lakes, however, is derived from ice "only" a few hundred thousand years old.

The ice floats freely on large lakes. Large lakes can be identified because the overlying ice sheet surface is nearly flat in the direction of ice flow. (We discuss ice flow in the vicinity of these lakes in Section 8.9.5.) Water sometimes moves between lakes as slow subglacial "floods," events identified by deflation and swelling of the surface. Such events have been identified in soft-bed regions of West Antarctica (Section 6.3.4) and in East Antarctica in a region of unknown bed type (Wingham et al. 2006). In the latter case, about 1.8 km^3 of water moved within 16 months from one subglacial lake to a group of lakes more than 290 km distant. The drainage path followed a subglacial trench. The positive feedback between melt and drainage (Figure 6.16) makes drainage events inherently unstable. Consequently, it is reasonable to expect that lakes drain episodically rather than continuously (Wingham et al. 2006; Evatt et al. 2006). A lake might steadily fill up as it accepts water from subglacial catchments. This increases the water pressure as the ice surface swells. It also increases the surface slope on the downstream side, increasing the hydraulic gradient. As with the lakes beneath Icelandic ice caps, rapid drainage begins when the hydraulic gradient directs water out of the lake and, simultaneously, effective pressures are low enough to open passageways.

Large transfers of water between subglacial lakes might influence ice sheet dynamics; such events could change the resistance of the bed to sliding motion over large regions, on timescales of years. In addition, lakes underly the upper ends of some of Antarctica's major ice streams (Bell et al. 2007). Water draining from the lakes lubricates the bed downstream, allowing rapid ice flow. Whether the presence of the lakes actually influences rapid ice flow or not remains unknown. Over many years, the amount of water drained along the bed is determined by the basal melt in the catchment; the presence of lakes is largely irrelevant. Perhaps lakes influence ice flow downstream only if they drain episodically.

Although the ice sheet surface over Lake Vostok is flat in the direction of ice flow, in the transverse direction the surface elevation decreases by about 40 m over the 200 km span of the lake (Siegert et al. 2001). By the requirements of floating equilibrium, the ice thickness and basal elevation must also vary. At the same time, hydrostatic equilibrium in the lake beneath requires uniform pressure in the lake on surfaces of constant elevation. Thus if H_1 denotes ice thickness above one side of the lake, and H_2 (smaller than H_1) the thickness above the other side, then $\rho_i H_1 = \rho_i H_2 + \rho_w \Delta B$, with ΔB the elevation difference of the ice base between the two locations. Call the surface elevation difference ΔS. Then $H_2 = H_1 - \Delta S - \Delta B$. The elevation along the roof of the lake therefore rises from one side of the lake to the other by $\Delta B = \Delta S \rho_i / [\rho_i - \rho_w] = -11 \, \Delta S$. In other words, the base of the ice slopes eleven times more than the surface and in the opposite direction; it follows an equipotential as determined in Section 6.3.1.1. But water can nonetheless flow along this surface because of density contrasts. Increased pressure lowers the melting point, so higher temperatures should occur under the thin-ice side. Water produced by melt on the thick-ice side of the lake supercools as it flows along the roof to the thin-ice side, and so freezes on. This explains the 200 m of accreted lake ice at the Vostok core site. Water can circulate in the lake because its density depends on both salinity and temperature. Within the lake, waters should be slightly warmer at the bottom than at the roof because of geothermal heat. The warmer waters should rise buoyantly. Depending on the salinity of the lake, the fresh water produced by melt of the overlying ice might also be buoyant.

Further Reading

Fountain and Walder (1998) reviewed water flow in temperate glaciers, and the formation of features like moulins. Jóhannesson et al. (2007) presented results of direct measurements in a volcanic sub-glacial lake in Iceland. Hubbard and Nienow (1997) and Tranter (2005) reviewed measurements and interpretations of sediments and solutes in proglacial streams and subglacial waters – topics of central interest to studies of erosion. Willis (2005) wrote a summary of the hydrology of watersheds containing glaciers.

CHAPTER 7

Basal Slip

"... only the weight of as much ice as is forced up above the natural level of the floating mass will press on the prominence."

On the Power of Icebergs ..., **Charles Darwin (1855)**

7.1 Introduction

Glaciers move by three mechanisms: plastic deformation of the ice, sliding of ice over the bed, and deformation of the bed itself. Sliding and bed deformation are not mutually exclusive; displacement often occurs at the interface between ice and a deforming bed. However, the mechanics of sliding on a deformable bed differ completely from the case of a rigid bed. In the former case, sliding activates deformations in a thin layer of the substrate. The ability of the bed to reshape itself then limits the forces acting between it and the sliding ice. Thus, rather than classifying glacier beds as sliding or deforming, it is more meaningful to characterize them as rigid or deformable. These cases are often referred to as "hard beds" and "soft beds." To complicate matters, basal ice sometimes contains a lot of debris, making it hard to define the ice-substrate interface.

Glaciologists often refer to the combined motion arising from sliding and bed deformation as *basal motion*. Because sliding and deformation both occur in a thin layer at the glacier sole, it is appropriate to use the term *slip* as a synonym for basal motion, a practice we follow here.[1]

Basal slip at a rate significant to glacier flow occurs only when special conditions apply. The bed deforms rapidly only if it consists of sediments saturated with water at high pressure, close to the weight of the overlying ice. Likewise, rapid sliding occurs only where basal temperatures attain the melting point.

Understanding basal slip remains a major problem in glacier physics. To analyze glacier flow we need a *slip relation* (including a "sliding law") – a relation between basal velocity, shear stress, and characteristics of the bed such as water pressure and volume, bedrock topography, and sediment properties. Correct formulation of a slip relation is essential for predicting the overall motion of a glacier and its reaction to changes of mass balance or terminus position. Deformable sediments underlie large regions of the ice-age ice sheets in Europe and

---

[1] The term "slip" avoids awkward or unclear language constructions associated with verb forms of "basal motion."

Copyright © 2010, Elsevier Inc. All rights reserved.
DOI: 10.1016/B978-0-12-369461-4.00007-7

North America, and basal slip accounts for nearly all the flow of the ice streams that drain ice from Antarctica. Understanding basal processes is thus a prerequisite for explaining such phenomena as the rapid retreat of ice sheets 10 to 15 kyr ago, and predicting whether global warming will lead to disintegration of the West Antarctic Ice Sheet. Moreover, inadequate knowledge of basal slip severely restricts our understanding of glacier surges and subglacial erosion. Whether the slip rate of most glaciers can ever be predicted or not is an open question.

A theory of slip might also have applications to other problems in geophysics involving a velocity discontinuity between a moving mass and a fixed substratum – for example, the emplacement of thrust sheets. As with faults in crustal rock, episodic slip along a glacier's bed releases seismic energy. Monitoring such events is one way to learn about conditions underneath the ice.

The difficulty of observing basal slip poses a major obstacle to learning how the process works. Observations in subglacial cavities and tunnels, though valuable, may not be typical of subglacial conditions; most direct observations have been made near the edge of the glacier or in icefalls. Boreholes through a glacier allow several methods for measuring basal slip: inserting anchors, tilt sensors, and other devices into subglacial sediment; monitoring the bed with video cameras; and repeating measurements of borehole tilting (*inclinometry*). The latter gives the component of velocity due to deformation within the ice, which must be subtracted from measured surface velocities to find the basal component.

The surface velocity of many glaciers increases early in the melt season and sometimes after heavy rain. This important observation indicates, together with borehole measurements of basal conditions, that increasing the amount or pressure of water at the bed lubricates the ice-bed interface. Slip interacts strongly with subglacial hydrology, and, for a complete analysis, the basal water system and the glacier motion must be treated as coupled problems.

In this chapter we summarize measurements of basal velocity and review the most important aspects of theoretical treatments. We discuss sliding on hard beds first. In spite of, or perhaps because of, the scarcity of observations, complex mathematical theories of sliding have been developed. They have little predictive value but contain important insights. In the second half of the chapter we turn attention to deformable beds. We summarize the important processes, review field and laboratory measurements of deformation behavior, and consider the problem of sediment conservation. Finally, we discuss simple phenomenological relations that might be used in glacier dynamics models.

7.1.1 Measurements of Basal Velocity

Table 7.1 lists rates of basal slip, arranged from fastest to slowest and by bed type. Measurements were obtained with a variety of methods. Most values are measurements over periods ranging from a few hours to a few weeks and were obtained from observations in subglacial cavities and tunnels or by down-borehole photography. The rates for Glacier d'Arolla and Black Rapids Glacier and for Variegated Glacier during its surge were determined from measured surface velocities by subtracting a component of internal deformation determined with inclinometry.

Table 7.1: Measured rates of basal slip.

Glacier	Thickness (m)	Slip rate (mm per day)	Reference
Sliding Over Bedrock:			
Argentière	100	200–1100	(22)
Grindelwald	40	250–370	(4)
Blue	65	350	(18)
Bondhusbreen	160	71–106	(12)
Engabreen	210	120	(5), (6)
Østerdalsisen	40	29–97	(23)
Mer de Glace	100	30–80	(26)
Grinnell	15–20	33	(13)
Mont Collon	65	30	(10)
Casement	20	24	(21)
Blue	26	16	(17)
Blue	65	10	(18)
Vesl-Skautbreen	55	8	(19)
Motion Over Sediment:			
Variegated[‡] (surge)	385	13150	(16)
Columbia	~500	6000–9000	(20)
Columbia	~950	3000–5000	(11), (20)
Whillans Ice Stream	1030	1100	(9)
Breidamerkur	~125	~270	(2), (3)
Variegated[‡] (no surge)	356	250	(8)
Black Rapids	600	90–130	(24), (25)
Storglacier	100	40–100	(15)
Trapridge	65	40–80	(1)
Blue	120	3–30	(7)
Arolla[†] (mixed bed)	130	16	(14)

[†] Bed is sediment, with patches of bedrock. These values are for the borehole closest to the glacier center-line, reported by Harbor et al. 1997.

[‡] Boreholes found sediment, but thickness of the layer was not determined. Retreat of the glacier exposed till-mantled bedrock near the front of the glacier (C. Raymond and W. Harrison, pers. comm.). The bedrock is easily eroded, fractured rock. Thus it is likely that the bed is a layer of sediment, but this has not been demonstrated conclusively.

References are: (1) Blake et al. 1994; (2) Boulton et al. 2001; (3) Boulton and Dobbie 1998; (4) Carol 1947; (5) Cohen et al. 2000; (6) Cohen et al. 2005; (7) Engelhardt et al. 1978; (8) Engelhardt et al. 1979; (9) Engelhardt and Kamb, 1998; (10) Haefeli 1951a; (11) Humphrey et al. 1993; (12) Hagen et al. 1986; (13) Hallet et al. 1986; (14) Harbor et al. 1997; (15) Hooke et al. 1997; (16) Kamb et al. 1985; (17) Kamb and LaChapelle 1964; (18) Kamb and LaChapelle 1968; (19) McCall 1952; (20) Meier et al. 1994; (21) McKenzie and Peterson 1975; (22) Reynaud et al. 1988; (23) Theakstone 1967; (24) Truffer et al. 2000; (25) Truffer and Harrison 2006; (26) Vivian 1977.

A similar method gave rates for Storglaciären, for Columbia and Trapridge Glaciers, and for Whillans Ice Stream, but the internal deformation was estimated from models, validated by inclinometry measurements elsewhere on the glacier or at other times. In these places, internal deformation was small compared to the total.

The slip rates vary widely in space and time. Individual values, excluding surges, differ by a factor of more than 1000. The velocity of Variegated Glacier changed by a factor of 50 over a period of two days during a surge. The velocities of 10 and 350 mm per day for Blue Glacier were measured over the same period at points only 15 m apart. The ice was separated from the bed over a wide area around the faster point, but was in contact at the other location. The two locations on Columbia Glacier were separated by 7 km along-glacier; the faster flow occurred in thinner ice, closer to the terminus. In most places, the velocity varied with time. Pronounced diurnal variations were observed at Trapridge and Columbia Glaciers and at Storglaciären.

Such variations over time demonstrate that basal slip rates depend profoundly on factors in addition to the gravitational forces acting on the glacier; gravitational forces usually remain nearly constant over periods of days to months. Furthermore, the gravitational forces on rapidly slipping glaciers range from small values (as at Whillans Ice Stream) to large ones (as at Columbia Glacier). The primary control on slip rate is therefore not the magnitude of gravitational force but the nature of the glacier bed – whether it is well lubricated or not. From Table 7.1, it appears that slip tends to be more rapid on beds of sediment than on hard rock. As we will see, however, if the water pressure is much below the pressure exerted by the ice a sediment bed acts rigidly and allows only slow sliding.

How slippery are the beds of glaciers? Basal slip generates a shear stress, acting on the glacier sole, that opposes the motion. A slippery bed implies that little shear stress arises despite rapid slip. The simplest measure of lubrication is the apparent *drag factor*, ψ, a positive number defined by

$$\tau_b = \psi u_b \qquad\qquad (7.1)$$

for a basal shear stress τ_b and rate of slip u_b (e.g., MacAyeal et al. 1995). Table 7.2 shows approximate values of ψ in a variety of settings spanning the entire range of observed behaviors for glaciers with melting beds. Values are approximate because stresses are inferred, not directly measured (Chapter 8). The most slippery beds occur where glaciers and ice streams flow into the sea. Such glaciers can be regarded as an intermediate case between typical glaciers on land and floating ice shelves.

Table 7.3 lists measured ratios of slip rate to total velocity, for glaciers with basal ice at melting point. Some of these values were obtained by inclinometry of boreholes that may have reached only to the top of dirty basal ice layers rather than to the bed (Aletsch, Athabasca, Salmon, and Tuyuksu). In these cases, the values are upper limits of the ratio. Table 7.3 shows that the ratio varies widely, even on the same glacier. On the average, however, slip accounts for roughly half the total. As we should expect, the fraction tends to be larger for fast-flowing glaciers than for slow ones. Where basal temperatures are below melting point, slip is usually but not always insignificant (Section 2.8).

7.1.2 Local vs. Global Control of Basal Velocity

As a glacier slips over its bed, at rate u_b, the displacements and deformations generate a force per unit area that opposes the motion: the basal shear stress τ_b. The slip relation $u_b(\tau_b)$ and

Table 7.2: Apparent drag factors.

Glacier	Terminus type	τ_b (kPa)	u_b (m yr^{-1})	ψ (kPa (m yr^{-1})$^{-1}$)
Various ice shelves		~0	100–1000	0
Whillans Ice Stream	marine	3	400	8×10^{-3}
MacAyeal Ice Stream	marine	14	400	0.04
Columbia[†]	marine	100–150	1000–2000	0.05–0.2
Storglaciären	land	40	30	1
Engabreen	land	100	44	2
Trapridge	land	80	30	3
Haut Gl. d'Arolla	land	120	6	20
Variegated (no surge) (surge)	land	130	90 4800	1 0.03

[†] About 10 km from the terminus.

the corresponding resistance relation $\tau_b(u_b)$ depend on mechanical properties of the interface and underlying sediment. They could be written in terms of variables like water pressure P_w, water volume per unit area H_w, and sediment strength τ_*: for example, as $u_b(\tau_b, P_w, H_w, \tau_*, \ldots)$. These relationships are *local* in the sense that they describe a property of a region of the glacier bed, valid regardless of the large-scale setting and overall dynamics of the glacier.

But to understand what factors determine the slip velocity u_b, and to make sense of observed rates, it is not sufficient to know these local relationships. Two end-member situations must be distinguished:

1. In the first end member, the magnitude of τ_b is not a function of the slip rate. Instead – in order for the glacier to maintain mechanical equilibrium – τ_b is entirely determined by the gravitational force acting on the glacier, per unit area. This situation applies when the bed provides all or most of the resistance to the glacier's flow. In this case, the local relation $u_b(\tau_b)$ determines the rate of slip. If conditions at the bed change, so will the slip rate. For example, an increase of basal water pressure makes the bed more slippery and increases u_b. We refer to this first case as *local control* of basal velocity. (It is sometimes referred to as *dynamic control*.)

2. In the second end member, the bed provides only a small fraction of the resistance to the glacier's flow (a "weak bed"). The magnitude of τ_b at a point is small compared to the gravitational force per unit area. A floating ice shelf offers the extreme example, because the viscosity of water provides essentially no resistance. Forces produced by deformation of the ice in a large surrounding region of the glacier control the flow. The most important processes are shearing of ice on the glacier's sides and flow over localized resistant patches

Table 7.3: Ratio of basal velocity to surface velocity.

Glacier	Thickness (m)	Velocity ratio	Reference
Sliding Over Bedrock:			
Vesl-Skautbreen	50	0.9	(19)
Blue	26	0.9	(17)
Blue	65	0.88	(18)
Østerdalsisen	40	0.65	(23)
Bondhusbreen	160	0.26	(12)
Engabreen	210	0.15	(5), (6)
Blue	65	0.03	(18)
Motion Over Sediment:			
Whillans Ice Stream	1030	1	(9)
Trapridge	65	0.9−1	(1)
Columbia	~950	~0.8−1	(11), (20)
Variegated (surge)	385	0.95	(16)
Storglacier	100	0.7−0.9	(15)
Arolla	130	0.6	(14)
Black Rapids	600	0.6	(25)
Variegated (no surge)	356	0.53	(8)
Blue	120	0.07	(7)
Bed Type Unknown:			
Athabasca	316	0.87	(29)
Athabasca	322	0.75	(30)
Athabasca	265	0.67	(29)
Tuyuksu	52	0.65	(31)
Aletsch	137	0.5	(27)
Salmon	495	0.45	(28)
Athabasca	209	0.1	(30)

References in addition to those cited in Table 7.1: (27) Gerrard et al. 1952; (28) Mathews 1959; (29) Raymond 1971; (30) Savage and Paterson 1963; (31) Vilesov 1961.

of the bed, such as bedrock hills that rise into the ice. This second case we refer to as *global control* of basal slip. (It is sometimes referred to as *kinematic control*.) Suppose, for example, that a reduction of ice viscosity increases shearing at a glacier's sides. If the bed is weak, the glacier also speeds up on its centerline and basal slip increases. On the other hand, a change of conditions at the bed along the centerline has little effect on u_b, but instead changes τ_b according to the local relation $\tau_b(u_b, \ P_w, \ H_w, \ \tau_*, \ \ldots)$.

Glaciers can lie anywhere on the spectrum between these end members of "local" and "global" control. How resisting forces are partitioned between a glacier's bed and elsewhere is discussed in Chapter 8.

7.2 Hard Beds

7.2.1 Weertman's Theory of Sliding

Modern study of hard-bed sliding dates from an analysis by Weertman (1957a). The problem is to explain how ice, assumed to be at the melting temperature, moves past bumps in the glacier bed. According to Weertman and also, much earlier, Deeley and Parr (1914), two mechanisms operate. This was originally a postulate but is now an observed fact.

The first mechanism is *regelation*. All the ice is assumed to be at its pressure melting point, which varies along the bed. The largest pressures occur on the upstream sides of bumps, which must provide the resistance to glacier movement. Thus the ice is colder on the upstream side than on the downstream side. This temperature contrast drives heat flow upstream through the bump. With all the ice at its melting point, the heat must cause melting on the upstream side. The meltwater then flows around the bump, in the direction of lower pressure, and refreezes on the downstream side where melting points are higher. This mechanism does not work for large bumps (of the order 1 m or more in length) because the heat conducted through them is negligible.

The second mechanism is enhanced creep. All the ice deforms viscously. A bump interferes with the flow, however, and produces an excess stress that increases the deformation – specifically, deformations occur that allow the ice to stretch and compress to move over and around the bump. How fast the ice moves over the bump depends on the product of strain rate and distance over which enhanced straining occurs. The larger the bump, the larger the region of enhanced stress upstream, and thus the greater the velocity. Consequently, this mechanism operates more effectively for large bumps than for small ones.

The glacier moves over its bed by a combination of these two processes; neither is adequate alone. Figure 7.1 depicts the setting. The basic assumptions of Weertman's equations are:

1. Clean ice rests on rough, undeformable, impermeable rock.

2. A thin film of water (about $1\,\mu$m in thickness) everywhere separates ice and rock. The interface therefore supports no shear stress, only a normal stress.

3. Ice and rock are nowhere separated by more than the water film; no cavities form.

4. The glacier bed is idealized as an inclined plane with cubical bumps on it. (This is sometimes called the *tombstone model*; each cube can be regarded as a memorial to a now-discredited assumption about the glacier bed.) Each cube has sides of length a. A distance λ, measured from the center of each cube, separates neighboring bumps. The bumps are far enough apart that processes at one do not affect processes at its neighbors.

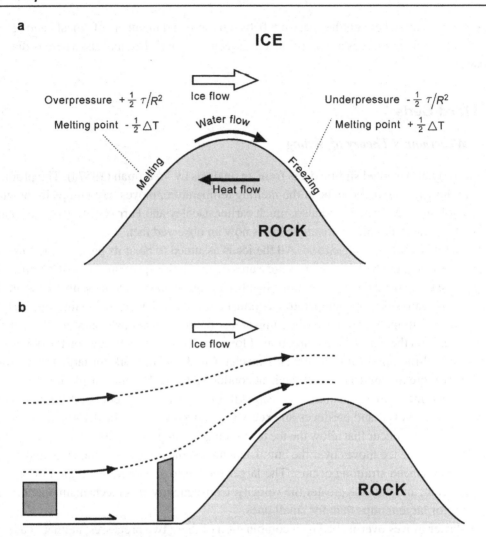

a

ICE

Overpressure $+\frac{1}{2}\,\tau/R^2$ Ice flow   Underpressure $-\frac{1}{2}\,\tau/R^2$

Melting point $-\frac{1}{2}\triangle T$ Water flow   Melting point $+\frac{1}{2}\triangle T$

Melting   Freezing

Heat flow

ROCK

b

Ice flow

ROCK

Figure 7.1: (a) The regelation mechanism of glacier sliding. Symbols are defined in the text. (b) The creep mechanism of sliding. Dashed curves represent flowlines. A parcel of ice first compresses longitudinally as it approaches the bump, as shown. In general, the parcel also shears.

We first calculate the velocity due to regelation. Let τ_b denote the average shear stress that the glacier exerts on its bed. As there is one cube in an area λ^2, each cube feels an average force $\lambda^2\tau_b$, acting along the bed. Relative to the hydrostatic pressure, which is much larger than τ_b, this force produces an elevated compressive stress on the upstream face (area a^2) and, provided the ice maintains contact with it, a reduced compressive stress on the downstream face. By symmetry, the elevated and reduced stress anomalies are equal and so each takes the value $\frac{1}{2}\tau_b\lambda^2/a^2$. Their difference amounts to $\tau_b\lambda^2/a^2$, which implies that the melting point of ice is reduced on the upstream side of a bump by an amount

$$\Delta T = \mathcal{B} \tau_b \lambda^2 / a^2 \tag{7.2}$$

relative to the downstream side. Here $\mathcal{B}$ is a constant; for a hydrostatic pressure, $\mathcal{B} \approx 7.4 \times 10^{-5}$ K $(kPa)^{-1}$ (this value can vary by a few tens of percent, depending on the concentration of dissolved air in the water). Because the ice remains at melting point throughout, the temperature difference given by Eq. 7.2 exists across the cube.

Call the velocity due to regelation u_1. A volume of ice $u_1 a^2$ melts on the upstream face in unit time. The water flows around the bump and refreezes on the downstream face. Refreezing releases latent heat of amount $\rho L u_1 a^2$ in unit time, where L denotes the specific latent heat of fusion and ρ the density of ice. This heat is conducted through the bump to the upstream face where it produces the melting. The heat conducted in unit time is $a k_r \Delta T$, with k_r the thermal conductivity of the bedrock. It follows that

$$\rho L u_1 a^2 = a k_r \Delta T \tag{7.3}$$

and, by Eq. 7.2,

$$u_1 = \frac{k_r \mathcal{B}}{\rho L a} \frac{\tau_b}{R^2} \quad \text{with} \quad R = \frac{a}{\lambda}. \tag{7.4}$$

The quantity R can be regarded as the *roughness* of the bed. Equation 7.4 shows that u_1 varies as $1/a$. In other words, regelation works most effectively for small bumps.

Now consider the second mechanism: movement of the ice around bumps by creep deformation. Call the velocity due to this mechanism u_2. We have already shown that the presence of the bump produces an additional longitudinal stress equal to $\frac{1}{2}\tau_b/R^2$. It is compressive on the upstream side of a bump and tensile on the downstream side. This stress produces a strain rate proportional to $\left[\frac{1}{2}\tau_b/R^2\right]^n$ where $n \approx 3$. (See Chapter 3; we use the simple form of the creep relation because R is usually small compared with unity and thus $\frac{1}{2}\tau_b/R^2$ greatly exceeds the mean shear stress τ_b.) To determine u_2 we need to know the distance x over which the excess longitudinal stress acts. A plausible assumption is that the volume of ice affected by the bump approximates the volume of the bump itself. So we take $x = a$. It follows that

$$u_2 = 2\,[3]^{-[n+1]/2}\, a\, A \left[\frac{\tau_b}{2R^2}\right]^n, \tag{7.5}$$

where A signifies the creep parameter for ice. The numerical factor arises because the stress is uniaxial (Eq. 3.27). This velocity increases in direct proportion to a, the dimension of each cube.

The combination of mechanisms therefore gives the sliding velocity

$$u_b = u_1 + u_2 = \frac{C_1}{a} \frac{\tau_b}{R^2} + C_2 a \left[\frac{\tau_b}{R^2}\right]^n, \tag{7.6}$$

where C_1 and C_2 are constants.

So far we have referred to bumps of only one size, but in general there must be a wide range of sizes. Regelation allows the ice to flow easily around small bumps while stress enhancement does the same for large ones. Thus bumps of some intermediate size are the main hindrance to motion and so largely determine the sliding velocity. Weertman called such a bump a *controlling obstacle*. Its size (a_c) is that for which u is minimum and hence $u_1 = u_2$. Thus, from Eqs. 7.4 and 7.5

$$a_c = 2^{[n-1]/2} 3^{[n+1]/4} \tau_b^{[1-n]/2} R^{[n-1]} \left[\frac{k_r B}{\rho L A} \right]^{1/2}. \tag{7.7}$$

Note that a_c depends on τ_b and R, unless $n = 1$. The quantity a_c is called the *controlling obstacle size* and the corresponding λ_c defines the *controlling obstacle spacing*.

We can now derive the formula for the sliding velocity. As a first approximation, assume that the controlling obstacles completely determine the sliding velocity. (We have already made this assumption implicitly by writing τ_b in Eq. 7.7. More precisely, the shear stress in Eq. 7.7 should be the fraction of τ_b supported by the controlling obstacles.) In this case, the sliding velocity u_b equals $u_1 + u_2$ and, for the controlling obstacles, $u_1 = u_2$. Therefore, from Eqs. 7.4 and 7.7, or Eqs. 7.5 and 7.7,

$$u_b = 2^{[3-n]/2} 3^{-[n+1]/4} \left[\frac{\tau_b^{1/2}}{R} \right]^{n+1} \left[\frac{k_r B A}{\rho L} \right]^{1/2} \tag{7.8}$$

or

$$u_b = \text{constant} \cdot \left[\frac{\tau_b^{1/2}}{R} \right]^{n+1}. \tag{7.9}$$

Because n is about 3, sliding velocity nominally varies as the square of the basal shear stress and inversely as the fourth power of the roughness.

With a smooth bedrock surface lacking any bumps of the controlling obstacle size and smaller – a possible consequence of abrasion – regelation no longer matters and sliding occurs only by enhanced creep. In this case, sliding velocity varies as the third power of the basal shear stress as shown by Eq. 7.5. In either case, sliding velocity depends sensitively on the value for roughness.

The value of the "constant" in Eq. 7.9 is of the order 10^{-20} (in S.I. units, for $n = 3$). Assuming a typical value for basal drag of $100 \, \text{kPa}$, a roughness of $R = 0.1$ then implies a sliding velocity of about $30 \, \text{m} \, \text{yr}^{-1}$. Increasing the roughness to $R = 0.2$ decreases the rate to $2 \, \text{m} \, \text{yr}^{-1}$. The corresponding drag factors are $\psi \approx 3$ or 50 for the smoother and rougher beds, respectively (units same as in Table 7.2). Comparison to values in Table 7.2 shows that Weertman's formulation is quantitatively consistent with observations of high-drag, slow-sliding mountain glaciers.

Weertman (1964) later refined his analysis. The main improvement was to consider the effect of bumps of different sizes, namely, ... $10^{-2}a_c$, $10^{-1}a_c$, a_c, $10a_c$, $10^2 a_c$, ... Weertman assumed that all sizes of bumps had the same roughness (*white roughness*). He also assumed that the ice moved past bumps smaller than a_c purely by regelation and past those larger than a_c purely by creep. The contributions of the two processes to the velocity could therefore be added. Weertman also considered the effect of air-filled cavities between ice and rock on the downstream sides of bumps and made other minor improvements to the theory. The new analysis merely changed the value of the constant in Eq. 7.9.

7.2.2 Observations at the Glacier Sole

7.2.2.1 Characteristics of Basal Ice

Kamb and LaChapelle (1964) confirmed that both regelation and enhanced creep were taking place at the base of Blue Glacier. Observations were made in a tunnel close to the side of the glacier near the top of an icefall. Bedrock was reached at a vertical distance of 26 m below the surface. The surface sloped at 28°, and the bed sloped at 22°, increasing to 55° a short distance farther down the icefall. Observations included:

1. The basal ice layer had smaller grains and contained fewer air bubbles but much more dirt than the ice above. Its thickness varied from almost zero on the tops of bumps to about 3 cm. Its upper boundary, which was distinct, appeared as a straight line in sections cut parallel to the direction of flow and coincided with the crest of the first bump upstream. This layer was identified as ice formed by refreezing on the downstream side of the bump and of water melted on the upstream side.

2. The ice in this *regelation layer* appeared to be undeformed by flow. However, there were foliation planes (Chapter 10) in the ice at distances of 50 cm and more above the bed. These planes, and the regelation layer as a whole, were bent downward in the downstream direction, corresponding to the steepening of the glacier bed. This demonstrated the existence of creep flow around the larger bumps. These observations also suggested that the controlling obstacle size lies between 3 and 50 cm.

3. The basal ice froze to the rock when the pressure was released by cutting out the ice immediately above. This confirmed the existence of a thin layer of water, at freezing point, at the ice-rock interface.

4. The ice was separated from bedrock over a distance of 10 m down-glacier from the tunnel. As the tunnel reached bedrock near the crest of a large rise, this separation could be described as a large cavity in the lee of the rise. Large cavities would be expected because the hydrostatic pressure, which would cause the ice to flow in and close them, was low.

5. The measured sliding velocity was 1.6 cm per day. The velocity cannot be calculated from Weertman's theory for comparison because it is not clear how to calculate the roughness of the bed.

7.2.2.2 Measurements on a Bump

Cohen et al. (2000) used tunnels excavated in bedrock to reach the bottom of the Norwegian Glacier, Engabreen, beneath 210 m of ice. They installed on the bed an artificial bump, inside a cavity melted in the glacier sole. The bump was a flat-topped cone of concrete, 0.15 m high and 0.25 m in diameter at its base. The cone was instrumented to measure stresses and temperatures. The glacier slid about 13 cm per day. After the ice made full contact with the bed, enclosing the bump, measurements were consistent with theory:

1. Compressive normal stress was 1.8 to 2.9 MPa on the upstream side of the bump but only 0.1 to 0.45 MPa on the downstream side.

2. The temperature on the downstream side was 0.1 to 0.2 °C higher than on the upstream side.

3. Regelation, estimated from the measured temperature gradient across the bump, was found to be negligible. Thus the controlling obstacle size was smaller than the bump; Cohen et al. suggested a size of about 5 cm, based on an extension of Weertman's theory due to Lliboutry.

7.2.3 Improvements to Weertman's Analysis

Nye (1969a, 1970) and Kamb (1970) developed hard-bed sliding theories with a more realistic model of the bed than Weertman's. The results are not fundamentally different; the important controlling parameters and their effects on sliding rate remain the same. The improved realism, however, makes possible a theoretical estimate of sliding rate from measured bed topography. The improved theories can also be used in models of other basal processes, such as glacial abrasion (Hallet 1979).

The Nye-Kamb theories consider an area $\mathcal{A}$ of the bed, large enough for its roughness to represent the bed as a whole, yet small enough for the sliding velocity over it to be uniform. Such an area may not exist in reality. The small-scale features of the bed are regarded as *roughness* and described statistically. A statistical description allows a calculation of sliding velocity without knowing the details of every bedrock bump and hollow.

Define as xy-plane the least-squares fit to the bed over the area $\mathcal{A}$, with x-axis in the direction of sliding. The z-coordinate points upward into the ice. Let $\tilde{z}(x, y)$ signify the deviation of the bed from the xy-plane. In the analysis, $\tilde{z}$ is decomposed into its spectral components by Fourier transform methods. The quantity $\langle \tilde{z}^2 \rangle = \mathcal{A}^{-1} \int \int_A \tilde{z}^2(x, y) \, dx \, dy$ is a measure of the roughness of the bed over the area. However, the analysis finds that sliding depends not so much on $\langle \tilde{z}^2 \rangle$ but on the distribution of roughness a/λ over the wavelength spectrum; here a denotes the amplitude and λ the wavelength of a particular Fourier component. Much of the theory assumes *white roughness*; each segment of the bed, when referred to its mean as datum plane, looks the same at all magnifications and in all directions.

Nye and Kamb assumed that the roughness is small. They also made Weertman's first three assumptions; a thin water film everywhere separates clean ice from rough, undeformable bedrock. In addition, they made the important simplifying assumption that ice deforms like a Newtonian viscous material ($n = 1$ in the creep relation).

Let $P_0(x, y)$ indicate the normal stress exerted by the bed on the base of the ice. The average value of P_0 is the stress needed to support the weight of the glacier. Because of the thin water film, the rock-ice interface supports no tangential stress along the interface. The mechanical equilibrium of the glacier parallel to the mean bed must therefore depend on bedrock bumps projecting upward into the ice. To balance the gravitational force per unit area parallel to the mean bed (the "driving stress" defined in Section 8.2.1), $P_0(x, y)$ must fluctuate about its average, with greatest compression on the upstream sides of bumps and least compression on the downstream sides. For overall equilibrium, parts of the bumps must slope uphill.

For the ice to remain in contact with the bed, as assumed, there must be a certain combination of melting, refreezing, and up-and-down and sideways motions for a specified sliding velocity. The analysis finds the distribution of $P_0(x, y)$ that produces the necessary heat flow and ice deformation. This pressure distribution $P_0(x, y)$ results in drag between the ice and the bed, a force acting parallel to the xy-plane. The average drag force per unit area of the xy-plane constitutes the basal shear stress τ_b. It equals the average of the x-component of $P_0(x, y)$ applied to the bedrock surface. For sliding at the observed velocity u_b, this shear stress must equal the overall stress implied by the glacier's driving stress (the force per unit area due to down-slope weight and pressure gradients in the ice). This problem has an exact solution under the stated assumptions.

These analyses led to conclusions similar to Weertman's. They identified a *transition wavelength* λ_c of the order 0.5 m. Ice slides past bumps of scale less than λ_c mainly by regelation and past bumps of scale larger than λ_c mainly by creep. This result is equivalent to Weertman's finding of a controlling obstacle size, because Weertman could equally well have used the controlling obstacle spacing. Calculated values of controlling obstacle spacing and transition wavelength are of the same order of magnitude.

Nye showed that the total drag equals the sum of the separate drags of each of the Fourier components of the bedrock relief. This justifies the use of a single statistical parameter as a measure of the bed roughness. He also found that, for the case of white roughness, obstacles in the wavelength range $\lambda_c/13 < \lambda < 13\lambda_c$ account for 90% of the drag. Thus Weertman's assumption that the sliding velocity depends only on obstacles of the controlling size is a reasonable approximation. Nye and Kamb both derived relations between u_b, τ_b, and statistical measures of roughness; the relations have the same form as Eq. 7.9 with $n = 1$. Kamb, and later Lliboutry (1975) using a related analysis, tried to include a nonlinear flow relation, but had to make questionable assumptions to do so.

To use a nonlinear creep relation, Fowler (1979, 1981) analyzed flow over a bed with no small-scale roughness; he ignored regelation and assumed the ice to move past bumps by deformation alone. For a sinusoidal bed of wavelength λ and amplitude a, with large λ

(to neglect regelation) and small roughness $R = a/\lambda$, the sliding velocity follows

$$u_b = c\frac{\lambda}{R^{n+1}} A \tau_b^n. \tag{7.10}$$

Here A and n are the creep parameters, τ_b the basal shear stress, and c a coefficient dependent on n and the ratio of λ to the ice thickness. Kamb (1970) and Lliboutry (1987) derived relations of the same form.

That sliding should depend on τ_b^n rather than $\tau_b^{[n+1]/2}$ is of course expected from Weertman's theory when ice moves past bumps solely by ice deformation. Equation 7.10 predicts that u_b varies as $R^{-[n+1]}$ whereas the term in Eq. 7.5 is R^{-2n}. This difference arises from the difference between the tombstone and sinusoidal models of the bed.

Gudmundsson (1997) also treated ice as a nonlinear fluid and ignored regelation but used a numerical finite-element technique to simulate sliding of ice over an undulating bed. The numerical approach circumvented the approximation of small roughness and avoided some of the questionable assumptions related to the treatment of nonlinearity. The calculations showed that, for small roughness, sliding velocities vary inversely with roughness raised to the power $n + 1$, as predicted by the analytical approaches. With a large roughness, however, the parameter c in Eq. 7.10 increases with the roughness, if $n > 1$; sliding occurs more rapidly on a rough bed than predicted by the approximate theories. For $n = 3$ and a rough bed (take $R = 0.4$), the value of c is about ten times larger than for a smooth bed. Results also showed that, for sinusoidal bed undulations, the ice can keep recirculating in the troughs and never join the main flow, if the ratio of amplitude to wavelength exceeds a critical value of about 1:3.

7.2.4 Discussion of Assumptions

Are these analyses based on realistic assumptions?

1. Cavities, filled with air or water, can form on the downstream sides of bumps. The preceding theories, based on the assumption that only a thin water film separates ice and rock, are not valid in this case. Section 7.2.6 discusses how cavities facilitate sliding.

2. The base of a glacier is often not a sharp interface between clean ice and solid rock. Glaciers would be unlikely to erode their beds if the basal ice were clean. Examination of the bases of temperate glaciers usually reveals a layer of dirty ice, up to several meters thick. Rocks in the ice cause friction on bedrock and abrade it. In a tunnel under Glacier d'Argentière, Boulton and Vivian (1973) observed that boulders in the ice moved more slowly than the ice itself. They also saw rocks being rolled along between ice and bedrock. As a result of these processes, sliding velocities calculated from the theory will be too high. Friction between debris in the basal ice and the bedrock invalidates one of the fundamental assumptions of the theory – that no tangential stress acts along the interface. Direct measurements beneath one

glacier show that tangential stresses can be quantitatively significant (Iverson et al. 2003), a topic we discuss in Section 7.2.7.

3. The white-roughness spectrum does not represent bedrock realistically. Because bumps near the controlling size provide most of the resistance to motion, abrasion should flatten them (Kamb 1970, p. 711). Indeed glaciated bedrock often lacks small bumps with wavelength scales of about half a meter or less. Perhaps hard-bed sliding is appreciable only over polished bedrock with no obstacles near the controlling size.

4. The basal ice layers usually differ from the remainder of the ice in grain size, fabric, amount of water, air bubbles, and impurities. Some of these factors change the ice's mechanical properties (Section 3.4.5).

5. Thin deposits of calcite and silica are widespread on the lee side of small bedrock bumps exposed by retreating glaciers (Hallet 1976a). They form by expulsion from the regelation water when it refreezes on the downstream side of the bump. The increased concentration of impurities lowers the melting temperature and so reduces the rate of heat conduction to the upstream side. Hallet modified Nye's theory to take account of this effect, which should reduce sliding rates.

6. Robin (1976b) suggested that patches of the basal ice may freeze to the bed, reducing the overall sliding velocity. If water produced by pressure melting within the ice were squeezed out of it, the ice loses a potential heat source and the ice might cool below melting point. Changes in water pressure might also lead to temporary freeze-on.

7.2.5 Comparison of Predictions with Observations

We have already shown (Section 7.2.1) that, given realistic values for bedrock roughness, Weertman's analysis predicts sliding rates and drag factors with the same order of magnitude as observed for the slowest sliding mountain glaciers. Predicting sliding velocity from Eq. 7.9 and related equations in any specific case, however, is a very uncertain exercise because of the high sensitivity to the value of bed roughness – a quantity not well known or always well defined. Kamb (1970) discussed nine measurements of sliding velocity, four made in tunnels and the others in boreholes. For each case he used the creep relation to calculate the basal shear stress from the strain rates measured in the basal ice. He then calculated, from his equations, the roughness needed to give the measured sliding velocities. In comparison with the measured roughness of typical bedrock, most calculated values were too low by factors of two to four. In other words, the sliding was more rapid than the theories would predict for these beds. Likewise, Hallet (1976b) made spectral analyses of bedrock profiles measured in front of retreating glaciers and calculated the roughness. For a typical basal shear stress of 100 kPa, Nye's formulation predicted sliding velocities of less than $1 \, \mathrm{m\,yr^{-1}}$, much lower than most measured sliding velocities. Although they predict fast sliding on smooth surfaces, the theories do not appear to explain the upper range of observed rates (Table 7.1) from realistic bed roughnesses.

A more important problem is that the theories predict a constant sliding rate over a given bed if the basal shear stress remains constant. The theory cannot therefore explain the numerous observations of short-period changes in surface velocity. Iken (1977b), for example, observed substantial fluctuations in velocity of several glaciers in the Alps during summer. In one instance, the velocity increased by a factor of five in two days. It also fluctuated with a diurnal cycle. Such variations cannot result from variations in ice deformation rate because ice thickness and surface slope cannot change significantly in such short time periods. They must therefore represent changes in basal velocity. Changes in the water at the glacier bed provide the most plausible explanation, a subject we take up next.

7.2.6 How Water Changes Sliding Velocity on Hard Beds

The preceding analyses suppose that all the basal water resides in a thin layer between the ice and the bedrock and originates in the regelation process or from geothermal and frictional heat. In reality, water can also concentrate in pockets called *cavities* in the lee of bedrock bumps, where pressures are lowest (Chapter 6). An irregular system of small channels links the cavities together. Furthermore, large quantities of surface meltwater penetrate to the beds of temperate and some cold glaciers. When more enters than can drain away, the volume and pressure of water on the bed increases. This enlarges the cavities and increases the area of separation between ice and rock. Stresses concentrate on the remaining areas of contact, increasing the sliding rate (Lliboutry 1968). In addition, increased water pressures could reduce the frictional drag produced as rocks in the basal ice scrape against the bed (Eq. 7.30); such a change would also increase sliding rate.

Lliboutry (1958b, 1968, 1978, 1987) emphasized the importance of water-filled cavities in sliding; that sliding velocities exceeding about 20 m yr^{-1} can be explained only by cavity formation. His views, at first highly controversial, are now accepted for the case of sliding on a hard bed. It is now recognized, however, that fast sliding also occurs because of weak deformable substrates.

7.2.6.1 Effective Pressure

The difference between the normal stress exerted by the weight of overlying ice, P_i, and the water pressure, P_w, is usually referred to as the *effective pressure* and symbolized by N. At the glacier bed, it must be close to the value

$$N = \rho g H - P_w, \tag{7.11}$$

with ρ the average density of the glacier, of thickness H. The effective pressure as defined here and commonly invoked by glaciologists is a special case of the *effective stress*, defined later in Section 7.3.2.2.

7.2.6.2 *Condition for Cavity Formation*

The compressive normal stress P_0 exerted by the ice on an inclined rough bed varies from point to point. The mean value of P_0 must equal P_i, the ice-overburden pressure. To support the component of the weight of the ice in the direction of the average bed, however, P_0 must attain relatively high values on the upstream sides of bumps and low values on the downstream sides. (Here all pressures are defined as positive in compression.) At any place with P_0 less than the water pressure P_w, water can force its way between ice and bedrock to form a water-filled cavity. The minimum value of P_0 is called the *separation pressure*, denoted P_s; ice starts to separate from the bed when $P_w = P_s$. An increase in P_w above P_s increases the area of separation and enlarges the cavities. Moreover, water in the lee of a large bump may submerge a small bump immediately downstream.

An expression for P_s can be derived as follows. Take the x-axis along the average bed slope, with x positive in the direction of ice flow, and z-axis positive upward. Write

$$P_0(x) = P_i + \tilde{p}(x), \tag{7.12}$$

where $\tilde{p}$ denotes a fluctuating term, with mean of zero, positive (compressive) on the upstream sides of bumps and negative on the downstream sides. Thus if the bed elevation varies about its mean value as

$$\mathrm{B}(x) = a \sin(2\pi x/\lambda), \tag{7.13}$$

the pressure term is 90° out of phase, namely

$$\tilde{p}(x) = \tilde{p}_m \cos(2\pi x/\lambda). \tag{7.14}$$

Here a denotes the amplitude and λ the wavelength of the bedrock undulations, and $\tilde{p}_m$ is the maximum value of $\tilde{p}$. For basal shear stress τ_b, balance of forces parallel to the mean bed requires that

$$\lambda \tau_b = \int_0^\lambda \tilde{p}\left[\frac{d\mathrm{B}}{dx}\right] dx. \tag{7.15}$$

Substitution from Eqs. 7.13 and 7.14 and integration gives $\tilde{p}_m = \lambda \tau_b / a\pi$. The minimum value of P_0, namely $P_i - \tilde{p}_m$, occurs at the inflexion points of the downstream faces of the bumps. This is the separation pressure

$$P_s = P_i - \frac{\lambda \tau_b}{a\pi}. \tag{7.16}$$

The quantity a/λ is the roughness of the bed. Separation begins when P_w reaches P_s. Thus – for a given value of basal shear stress – the rougher the bed, the higher the water pressure needed to produce cavities. Moreover, for a given bed roughness a small τ_b inhibits cavity formation. As τ_b increases, the pressure fluctuations around the bumps also increase, thereby decreasing the water pressure needed to form cavities.

7.2.6.3 Faster Sliding Related to Low Effective Pressures

In theory, the rate of sliding should increase as cavities expand. With larger cavities, the area of contact between the ice and the bedrock is reduced, effectively decreasing the roughness of the bed and the drag it exerts on the ice. To maintain large cavities, the volume of water per area of bed must remain large enough to fill the cavities, and the water pressures must be high enough to prevent closure of the cavities by ice creep. Specifically, the closure rate depends inversely on the effective pressure, N (Eq. 7.11), the difference between the ice overburden pressure and the water pressure in the cavity. When $N \to 0$, the glacier floats on its bed.

Figure 7.2a shows measured values of both horizontal velocity and borehole water level, a proxy for subglacial water pressure, on Findelengletscher in the Swiss Alps (Iken and Bindschadler 1986). Surface velocity was measured at four lines of stakes several times a day and water pressure recorded in eleven boreholes for five weeks early in the melt season. Variations of velocity and pressure correlate clearly over timescales of days to a few weeks. Figure 7.2b shows the relationship as a function of the effective pressure; velocity varies approximately as N^{-1}.

One simple theoretical explanation for this observation is as follows. Define the ratio τ_b/N as the *bed-separation index* (Bindschadler 1983). The value of N at which cavities start to form varies in proportion to τ_b (Eq. 7.16). With this criterion, combined with the expectation of slower cavity closure at low N, it is plausible that a larger τ_b/N implies more extensive separation between ice and bedrock, and hence faster sliding. Multiplying the right-hand side of Weertman's relation (Eq. 7.9) by the bed separation index yields

$$u_b = k\tau_b^p \, N^{-q}, \tag{7.17}$$

with $p = 3$ and $q = 1$, and k dependent on the thermal and mechanical properties of ice and inversely on the bed roughness. More generally, a relation of the form Eq. 7.17 – with positive values for p and q – applies in theory as long as $u_b \propto \tau_b^{p-1} H_w$ and $H_w \propto \tau_b/N$, where H_w denotes the average thickness of the basal water layer (Weertman and Birchfield 1982, 1983; Alley 1989, 1996). Equation 7.17 was tested using observations of Variegated Glacier, Alaska, as it built up during the years following a surge (Raymond and Harrison 1987). To explain the observed velocities required positive values for p and q, but no single relationship accounted for velocity variations over both time and distance, and no optimal values for p and q could be identified. (In addition, the bed of this glacier is at least partly mantled with sediment; it is not clear that hard-bed theories apply at all.)

Although the simple formula (Eq. 7.17) provides a useful improvement to Weertman-type relations, both data and theory indicate that it lacks generality. The first problem is that the basal drag cannot increase indefinitely with the sliding rate, as the formula implies. Consider again the effect of increasing water pressure on the sliding rate (continuing from Section 7.2.6.2). As P_w increases above the separation pressure (P_s) it attains a second critical value, P_c, at which sliding becomes unstable. This situation is analogous to flow of a landslide; unstable slip occurs

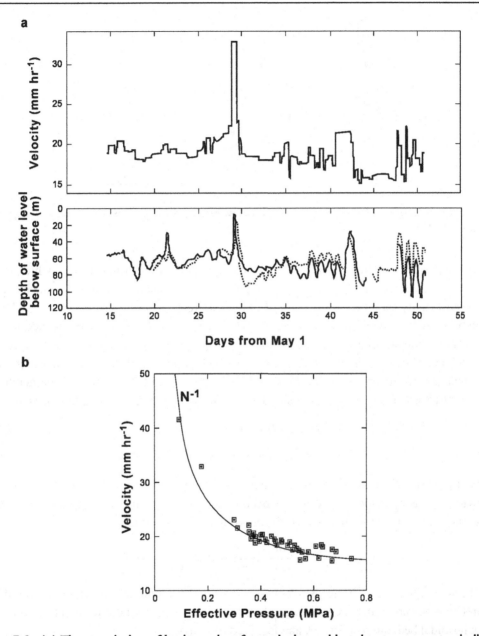

Figure 7.2: (a) The covariation of horizontal surface velocity and basal water pressure – indicated by the water level in several boreholes – on Findelengletscher. The plot shows a selection of the data given by Iken and Bindschadler (1986). (b) The corresponding relation between velocity and effective pressure, N, the ice overburden minus water pressure. The curve is the regression of velocity on N^{-1}. Adapted from Iken and Bindschadler (1986).

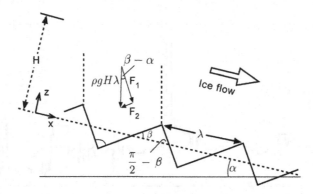

Figure 7.3: Model of bed for derivation of critical water pressure for unstable sliding. Adapted from Iken (1981).

because the basal shear stress equals the frictional strength of the sliding surface, which cannot increase no matter how fast the slip. The critical pressure for unstable sliding of a glacier can be derived most easily for a bed in the form of a tilted staircase as in Figure 7.3 (Iken 1981). The upstream face of each bump makes an angle β with the mean bed, the downstream face an angle $\frac{\pi}{2} - \beta$. Call the mean slope of the bed α and the spacing between each bump λ. An ice column of length λ and thickness H rests on each bump. The component of its weight, per unit width, perpendicular to the upstream face is $F_1 = \rho g H \lambda \cos(\beta - \alpha)$, and the component perpendicular to the downstream face is $F_2 = \rho g H \lambda \sin(\beta - \alpha)$. The average pressures on the faces are

$$P_1 = \rho g H \cos(\beta - \alpha)/\cos\beta, \tag{7.18}$$

$$P_2 = \rho g H \sin(\beta - \alpha)/\sin\beta. \tag{7.19}$$

If a water pressure $P_w > P_2$ opposes P_2, a net force moves the ice upward along the upstream faces with increasing velocity; sliding has become unstable. Because $P_i = \rho g H \cos\alpha$ and, for a wide glacier, $\tau_b \approx \rho g H \sin\alpha$, the critical pressure takes the value

$$P_c = P_i \frac{\sin(\beta - \alpha)}{\sin\beta \sin\alpha} \approx P_i - \frac{\tau_b}{\tan\beta}. \tag{7.20}$$

It can be shown that Eq. 7.20 holds for two-dimensional bed undulations of any shape, if β is interpreted as the angle between the mean bed and the steepest tangent to the upstream faces. For a sinusoidal bed, $\tan\beta = 2\pi a/\lambda$ and Eq. 7.20 becomes

$$P_c = P_i - \frac{\lambda \tau_b}{2\pi a}. \tag{7.21}$$

It follows from Eq. 7.16 that $P_c = (P_i + P_s)/2$.

Defining the critical effective pressure as $N_c = P_i - P_c$, Eq. 7.20 also states

$$\tau_b = N_c \tan\beta. \tag{7.22}$$

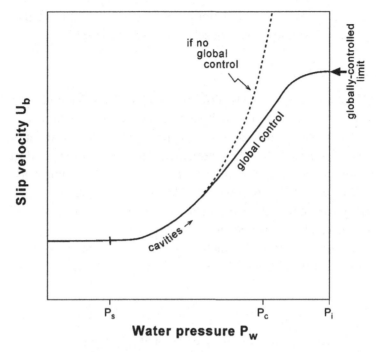

Figure 7.4: Relationship between sliding rate and water pressure: a schematic view.

Formally this is the same as the Coulomb friction criterion; hence the analogy to landsliding. However, whereas a glacier slides even at water pressures lower than P_c, with Coulomb friction no motion occurs when $\tau_b < N_c \tan \beta$.

We can now make a schematic hypothesis for the variation of sliding velocity with water pressure (Figure 7.4). First consider the "local" problem only, in which the value of τ_b is fixed by gravitational forces (see Section 7.1.2). When $P_w < P_s$, the sliding velocity has a constant value determined by the processes of regelation and enhanced ice deformation. When $P_w = P_s$, water-filled cavities start to form and sliding velocity then increases with water pressure until $P_w = P_c$. At this point the sliding velocity could increase drastically. Most observations, however, show that the velocity remains stable even as P_w approaches P_i (for example, Figure 7.2). The discrepancy may arise because of incomplete lubrication of the bed, restricted water supply to low-pressure zones, or the presence of vertical faces. But there is a more general explanation for this behavior; as the bed becomes very slippery, the velocity is increasingly controlled by deformations in the surrounding glacier – the "global" control defined in Section 7.1.2. Faster flow increases the resisting forces due to shearing on the sides of the glacier, due to stretching along the glacier, or due to flow over anomalous sticky patches on the bed. As these forces increase, the bed takes up less of the gravitational force, and basal shear stress decreases over most of the bed. Although unstable sliding can occur – the fronts of glaciers on steep mountains sometimes fall down as ice avalanches (Section 12.6) – the velocity of most glaciers should tend toward a globally controlled limit as shown schematically in Figure 7.4.

7.2.6.4 Relations for Basal Drag?

The importance of global controls on fast-flowing glaciers means that no single relationship for sliding rate as a function of drag can be written. Given a sliding rate, however, the mechanical properties of the interface should determine the magnitude of basal drag. Relations for basal drag are still poorly constrained, but some of their qualitative properties can be deduced theoretically.

Consider a crude model for sliding in the presence of cavities. (For extensive theoretical treatments, see Lliboutry 1987 and Fowler 1987a.) Low effective pressures result in larger cavities because they reduce the rate of closure by ice creep, allowing the ice of a cavity roof to move farther down-glacier before it reconnects with the bed. In the limiting case of a cavity with no through-flow of water, implying that melt of the cavity roof can be neglected, the nominal length of a cavity is (see Eqs. 6.29 and 6.31)

$$\ell = \frac{n^n u_b}{A N^n},$$ (7.23)

where A and n are the creep parameters for ice. As cavities grow larger, the shear stress ($\hat{\tau}_b$) in the remaining areas of contact between ice and bedrock must increase. A simple description of this intensification is

$$\hat{\tau}_b = \tau_b \left[1 + \frac{\ell}{\lambda_o} \right],$$ (7.24)

with λ_o a length scale corresponding to the size of cavities needed to double the stress in the regions of contact.[2] Analogy with Eq. 7.9 gives the sliding rate, assuming a bed smooth enough for regelation to be negligible, as

$$u_b = \frac{C}{R^\mu} \hat{\tau}_b^n = \frac{C}{R^\mu} \tau_b^n \left[1 + \frac{\ell}{\lambda_o} \right]^n.$$ (7.25)

Substituting from Eq. 7.23, dividing through by N^n, and rearranging give

$$\frac{\tau_b}{N} = K \frac{\hat{u}^{1/n}}{k_0 + \hat{u}} \quad \text{where} \quad \hat{u} = \frac{u_b}{N^n}$$ (7.26)

and with the coefficients given by

$$k_0 = \frac{\lambda_o A}{n^n} \qquad K = k_0 \, C^{-1/n} \, R^{\mu/n}.$$ (7.27)

---

[2] In nature, the complexity of a glacier bed's geometry prevents complete separation of ice from substrate at a single critical cavity size. Thus we do not describe the stress intensification as proportional to $\tau_b/[\lambda_o - \ell]$.

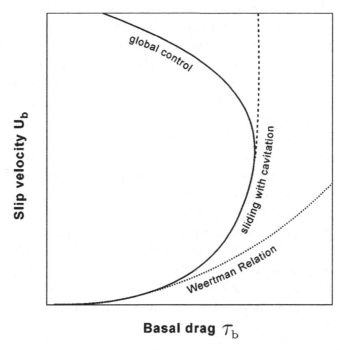

Basal drag $\mathcal{T}_b$

Figure 7.5: Covariation of basal drag and sliding rate: a schematic view. A similar relation applies for deformable beds, so the vertical axis can be regarded as the slip velocity in general.

According to Eq. 7.26, as effective pressure decreases from the ice-overburden pressure to near zero, τ_b/N first rises to a maximum but then decreases. Because of cavity formation, the basal drag can never rise above a limiting value – corresponding to the limit identified by Iken based on the balance of forces. More sophisticated analyses give similar results. Schoof (2005) used mathematical arguments to demonstrate that basal drag cannot exceed a limiting value unless the bed undulations include vertical faces. This analysis neglected regelation and treated ice as a linear viscous fluid. By using numerical finite-element models, Gagliardini et al. (2007) reached the same conclusion for a more realistic nonlinear fluid. The latter study examined an idealized bed, but Schoof proved his result for a bed of arbitrary form.

These arguments imply a relation between basal drag and sliding rate, for a given bed, as sketched in Figure 7.5. Sliding rate and drag increase together, heading toward a critical limit. As sliding increases, however, resisting forces other than basal drag become important; the sliding rate is controlled globally, at least in part, and the basal drag decreases as the glacier speeds up.

7.2.6.5 *Temporary Increase of Sliding by Transient Cavity Growth*

So far we have discussed how the existence of cavities facilitates rapid sliding, but have ignored the direct effects of cavity growth. An increase of water pressure should lead to expansion of cavities, displacing ice both downstream and upward. Observations of cyclical daily variations show that, although glaciers slide most rapidly at times of high water pressure, they slide faster when water pressures are rising than when declining. Figure 7.6 shows an example, from Lauteraargletscher, Switzerland (Sugiyama and Gudmundsson 2004). These observations most likely indicate that, when cavities grow rapidly, the corresponding displacement of ice temporarily accounts for a significant fraction of the sliding (Iken 1981). Because the effect is only temporary, transient cavity growth cannot be responsible for sustained rapid sliding.

Observations of the surface elevation of a glacier provide a measure of the size of subglacial cavities; large cavities elevate the overlying surface (Iken et al. 1983). However, longitudinal variations of glacier flow – which compress or stretch the ice – also affect surface elevations. The importance of this effect must be determined before surface elevations can be used as a proxy for cavity size. Sugiyama and Gudmundsson (2004) used borehole measurements of vertical ice deformation to do this. They found that, at some times, surface uplift was not caused by ice deformation. At such times, which mostly occurred early in the melt season, fastest sliding coincided with the greatest rate of surface uplift. This supports Iken's idea that cavity growth temporarily contributes directly to sliding. At other times, however, ice deformation caused at least part of the surface uplift; surface elevations are not, by themselves, a reliable indicator of subglacial cavity size.

7.2.6.6 Faster Sliding Not Caused by Low Effective Pressures

Although an important variable, effective pressure is not the only factor that controls separation of the ice from its bed. Both melt and the rate of sliding influence cavity size. The distribution of water between cavities, tunnels, and films probably changes in response to variations of water inputs and sliding rate. Consider a drainage system along the glacier bed that suddenly receives an increased influx of water from the glacier surface. Unable to flow through the existing passageways, the water may back up in the glacier, increasing the pressure at the bed; over time, cavities then expand, increasing the sliding rate and the drainage. But suppose instead that the extra water, rather than being trapped in existing passageways, floods across the glacier bed – perhaps by melting new passageways or by occupying new space created by increased sliding (Iken and Truffer 1997). In this second scenario, drowning of roughness on the bed, and hence increased sliding, might occur without any change of effective pressure.

A bed with the staircase geometry shown previously in Figure 7.3, and used to derive Iken's criterion for unstable sliding, provides a limiting case (Humphrey 1987). The picture now includes a water-filled cavity in the lee of a step, as shown in Figure 7.7. The cavity's bed-parallel length is X_w and the water pressure is P_w. Following Humphrey, we now assume that, in contrast to the assumption made previously to derive Eq. 7.20, the pressures on the bedrock faces are no longer fixed by the glacier geometry. The part of the down-glacier force taken up by drag on the bed (per unit area, this force equals τ_b) can adjust through changes of the ice deformation in the surrounding glacier – the "global" controls defined in Section 7.1.2. With these assumptions, a balance of forces parallel and normal to the mean bed requires that, for no

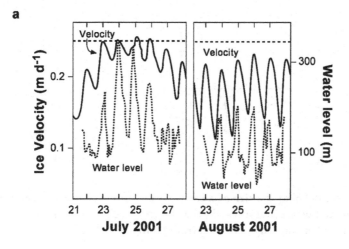

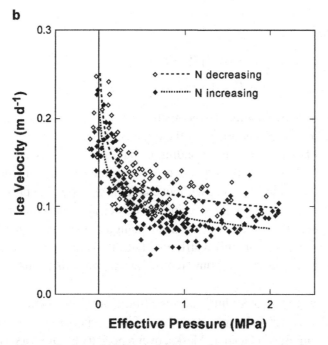

Figure 7.6: Measured covariation of sliding velocity and basal water pressure at one site on Lauteraargletscher. Adapted from Sugiyama and Gudmundsson (2004). (a) The day-to-day variations in two periods. Dashed line indicates the overburden pressure. (b) The corresponding relationship of sliding velocity with effective pressure, N, showing faster sliding at times of rising water pressure (N decreasing) compared to times of falling water pressure (N increasing).

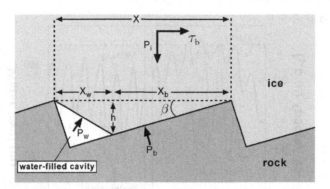

Figure 7.7: Variables defining the size of a cavity formed on a "tilted staircase" bed. P_i and P_w are ice-overburden and water pressures, P_b the compressive normal stress where ice rests on rock, and τ_b the overall basal drag. Adapted from Humphrey (1987).

acceleration of the ice:

$$X\tau_b = h[P_b - P_w], \tag{7.28}$$

$$X P_i = [X - X_w] P_b + X_w P_w \tag{7.29}$$

for variables as defined in the diagram. By geometry, $h = [X - X_w] \tan \beta$. Combining relations shows that $P_w = P_i - \tau_b / \tan \beta$, regardless of the dimensions of the cavity. (In other words, the water pressure always has a value corresponding to Iken's critical limit for the staircase bed geometry.) For this situation, an increased flow of water to the glacier bed would cause cavities to expand, increasing the sliding rate, but without changing the effective pressure. The absence of any relation between P_w and cavity size in this model is a consequence of the staircase geometry, which prevents the direction of P_b from changing. On a real glacier bed, curvature of the bedrock surface circumvents this restriction, but the model nonetheless demonstrates a behavior – minimal change in P_w as a function of cavity size – that must approximate some situations.

Observations demonstrate that sliding rate sometimes varies independently of the effective pressure (Iken and Truffer 1997; Harper et al. 2005, 2007). Harper et al. measured subglacial pressures at 51 sites along Bench Glacier, Alaska, over a period of four years. Borehole observations also established that the glacier rests on a hard bed. Figure 7.8 shows the variations of ice motion, effective pressure, and bed separation during a speed-up event early in the melt season; similar events occurred in the other years. Variables were inferred by averaging measurements at 17 boreholes clustered in the middle of the glacier. The bed separation was inferred from measurements of surface elevation change, corrected for vertical deformation within the ice. Speed-up of the glacier occurred in two episodes. In neither period were effective pressures reduced. In the first episode, bed separation increased, suggesting growth of cavities. In the second episode, bed separation neither increased nor was particularly large. However, several

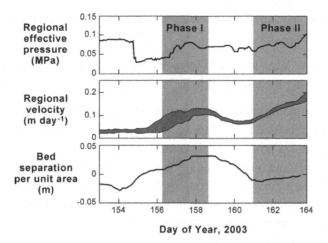

Figure 7.8: Covariation of measured basal effective pressure, surface velocity, and bed separation in a 60 m × 100 m region of Bench Glacier, Alaska. Velocity increased in two phases. The first occurred despite no reduction in effective pressure. The second occurred despite no reduction in effective pressure and no increase in the average separation of ice and bed. The band of velocity values shows the range of measurements at the borehole sites. The other curves are averages for the 17 sites. Adapted from Harper et al. (2007) and used with permission from the American Geophysical Union, *Geophysical Research Letters*.

lines of evidence showed that a connected water system covered a greater fraction of the bed than at other times of the year: basal pressures varied synchronously over regions hundreds of meters in dimension; tests using water injected into boreholes showed that the basal system transmitted water rapidly; and the episode terminated with flooding in the outlet stream, implying that the bed released a large volume of water at this time. After the flood, effective pressures on the bed rose significantly and diurnal variations began.

The data from Bench Glacier also showed that effective pressures were low at the onset of speed-up episodes, but no lower than during winter. In addition, the first response of the basal water system to springtime melt occurred within a few days of when the surface snow warmed to melting point throughout its whole thickness.

7.2.6.7 Perspective on the Role of Water in Hard-bed Sliding

Water at the glacier bed increases the sliding rate in at least four different ways, when water pressures rise to near the ice-overburden value:

1. Low effective pressures allow the formation of large cavities, which reduce the area of contact between ice and bed.

2. Even without decreasing effective pressure, water can spread out on the bed, or cavities can expand, in response to increased water delivery from the surface.

3. Reduced effective pressures could reduce the frictional drag due to rocks in the basal ice (see Section 7.2.7).

4. An increase of water pressure results in cavity expansion, temporarily displacing ice down-glacier and increasing the rate of sliding.

As a consequence of these mechanisms, the effective pressure N and the volume of water, per area of bed, need to be included as independent variables in sliding relations. For this reason, glacier sliding is a complex process. As discussed in Chapter 6, the water pressure and volume depend on water sources and on the drainage system that develops along the bed. This system can change rapidly, and moreover depends on the sliding motion itself. Cavities opened by sliding, if linked to one another, are conduits for water flow. But sliding can also disrupt the network of tunnels along the bed that drains most of the surface meltwater. Sediment on the bed, either in isolated patches or a continuous cover, further complicates the situation as discussed in Section 7.3. For these reasons, theoretical analyses of sliding with cavities have no predictive value. Furthermore, no simple relation like Eq. 7.17 can be used to infer how u_b and τ_b correlate in nature – not only because the water pressure and volume depend on u_b, but also because u_b depends on global constraints. In other words, u_b depends on particularities of the situation such as the width of the glacier and spatial variations of bed properties; how such variables affect flow fields are discussed in Chapter 8.

7.2.7 Sliding of Debris-laden Ice

Most basal ice contains rock debris in varying concentrations and sizes, a fact ignored in all the analyses we have discussed so far. These analyses assume that ice and rock are separated everywhere by at least a thin layer of water, incapable of supporting shear stresses. Thus no tangential force resists the flow of ice over the bed; the drag is only the resultant, along the average line of the bed, of the pressure distribution over the undulations. In reality, friction between the bed and debris in the ice reduces the sliding velocity. In effect, the basal shear stress τ_b in the previous analyses should be replaced by $\tau_b - \tau_f$, where τ_f denotes the frictional drag stress.

A friction law relates τ_f to properties of the basal ice and the bed. The form of the friction law depends on the debris concentration and whether the ice contacts the bed directly or transmits its weight through the particles. In one end-member case, the concentration of debris is high and melt of the ice prevents the ice from touching the substrate beneath the debris. Coulomb friction acts between the debris particles and the bed, giving a total force $\mu P_c \mathcal{A}_c$. Here μ is a coefficient of friction (typically about 0.6 for rock-rock contact), P_c is the pressure at the regions of contact, and $\mathcal{A}_c$ is the area of contact. The force per unit area $\mathcal{A}$ of bed constitutes the drag stress $\tau_f = \mu P_c \mathcal{A}_c / \mathcal{A}$. The normal stress at the bed must balance the weight of overlying ice, P_i. Thus $P_i \mathcal{A} = P_c \mathcal{A}_c + P_w \mathcal{A}_w$, where P_w denotes pressure in the water at the interface and $\mathcal{A}_w$ the wetted area. Assuming there are no areas where ice contacts the bed directly, $A = A_c + A_w$ and the drag stress can be written

$$\tau_f = \mu \, [P_i - f_w P_w], \tag{7.30}$$

with f_w being the fraction of bed surface that is wet (Schweizer and Iken 1992). If debris touches the bed only at minute contacts, this formula reduces to the classic Mohr-Coulomb relation, $\tau_f = \mu \, [P_i - P_w]$. (As before, the pressure difference $N = P_i - P_w$ defines the effective pressure.)

In general, however, ice can penetrate through a debris layer by regelation (Iverson 1993); the rock particles, surrounded by ice, no longer bear the weight of the glacier. This situation has been analyzed most completely for sparse debris concentrations – less than about 15% by volume (Hallet 1981; Shoemaker 1988; Cohen et al. 2005). Each rock particle in contact with the bed contributes to the drag by an amount proportional to the coefficient of friction μ and to the force of contact. The latter depends on the movement of the ice; flow toward the bed impels stronger contact between entrained particles and the substrate. Such flow occurs where the ice undergoes longitudinal or transverse extension and where geothermal and frictional heat melt the glacier's base. Ice pushes particles toward the bed only on the upstream side of a bump, however; regelation ice forms on the downstream side and tends to cover up debris particles (alternatively, cavities form on the downstream side). At the same time, ice flowing parallel to the bed swarms around the particles. The rate of ice flow past a rock increases with the contact force between rock and substrate. Interference of the flow causes a drag, which must equal the frictional drag at the particle-bed contact. A plausible friction rule thus takes the form (Shoemaker 1988)

$$\tau_f = k_1 \, \mu \, C \, \eta \, \overline{w}. \tag{7.31}$$

Here $\overline{w}$ is the average component of ice velocity perpendicular to the mean bed, C the debris concentration, μ the coefficient of rock-rock friction, η the effective viscosity of ice, and k_1 a parameter dependent on the bed geometry and simplifications in the other terms. Note that, in contrast to Eq. 7.30, τ_f does not depend on the weight of the overlying ice. If contact forces are large, however, the ice might flow rapidly around particles, forming cavities on their lee sides. Cavity formation will increase the contact force by an amount proportional to $N = P_i - P_w$ (Cohen et al. 2005).

Frictional stress on a glacier bed has been measured at one location, the Norwegian glacier Engabreen (Iverson et al. 2003; Cohen et al. 2005). Tunnels constructed to capture water for a hydroelectric project give access to the bottom of the glacier. Iverson and collaborators installed a granite tablet in bedrock at the glacier bed. The tablet's upper surface was even with the surface of the surrounding bedrock. Load cells connected to the tablet allowed calculation of both normal and shear stresses. Measurements were acquired over three periods, one of which lasted for about 340 days. Rock particles accounted for between 2% and 11% of the basal ice by volume (mean of 5.3%), with the higher concentrations occurring closest to the bed. Ice thickness above the site was 213 m.

The measurements showed that the shear stress τ_f on the tablet had sustained values in the range 300 to 500 kPa (Cohen et al. 2005). In comparison, typical basal shear stress values, averaged over the glacier bed, are only about 100 kPa. The measured τ_f values also greatly exceeded those expected from a simple friction rule like Eq. 7.30; with the measured effective pressures of about 200 kPa, and a rock-rock friction coefficient of $\mu = 0.6$, the equation gives $\tau_f \approx 120$ kPa. The large measured τ_f values resulted, most likely, from downward motion of the ice ($\overline{w}$ in Eq. 7.31). Melt rates, estimated from heat flux measurements, always exceeded 0.12 m yr^{-1}. The measurements also showed that τ_f varies with effective pressure; the correlation gave an apparent friction coefficient of $\mu = d\tau_f/dN = 0.04 - 0.08$. This small value for μ shows that normal friction as given by Eq. 7.30 accounts for only a small fraction of τ_f.

The large measured values of τ_f are not likely to be typical of glacier beds; they occurred because heat flux through the experimental apparatus caused abnormally rapid melt. To explain the values, and extrapolate to more realistic melt rates, Cohen et al. formulated a model of particle-bed contact forces arising from the processes discussed in the third paragraph of this section. (The model is a modification to that of Hallet 1979 and 1981.) The analysis featured numerical finite-element calculations of ice flow past a particle adjacent to a rigid boundary. Results indicated a strong enhancement of the particle-bed contact force due to ice flowing downward around the particle and pressing it against the bed. The presence of the rigid boundary greatly increases the force on the particle; the boundary effectively increases the pressure difference between the particle's top and bottom. In addition, Cohen et al. made calculations for a variety of n values – n being the exponent in the creep relation for ice – and concluded that ice deformation around the particle must be a linear-viscous process ($n = 1$). With a typical value of $n = 3$, contact forces were not nearly as high as the measured values.

Having identified model parameters that gave a good match to measured values of τ_f and $d\tau_f/dN$ for conditions at Engabreen, Cohen et al. estimated the values of τ_f for conditions typical of mountain glacier beds. Choosing 10% basal debris and typical melt rates gave τ_f of about 100 kPa. Given that total basal shear stresses are also about 100 kPa, this result implies that frictional forces can equal or exceed the drag forces from regelation and creep around undulations on the bed. If correct, this model implies that particle friction severely limits sliding rates, if the basal ice contains about 10% rock debris. The major weakness of the Cohen et al. model is the absence of a thermodynamic component; for example, the effects of freezing in the low-pressure zones near particles are not evaluated.

One main difficulty in reconciling sliding rate data with the hard-bed theories is that sliding often occurs more rapidly than predicted. The observations of large τ_f beneath Engabreen make this problem more severe, from the theoretical standpoint. Regardless, it appears that friction must be included in theoretical analyses of sliding unless the ice contains no debris.

7.2.8 Sliding at Sub-Freezing Temperatures

Laboratory experiments have shown that regelation occurs even at temperatures a few degrees below melting point (Telford and Turner 1963; Gilpin 1980). This suggests that glaciers may slide at sub-freezing temperatures. Such regelation is attributed to the existence of a liquid water layer surrounding foreign solids in ice (Dash et al. 2006). These layers, whose presence has been demonstrated using various experimental techniques, can persist even to temperatures of tens of degrees below melting point. They exist because, in essence, intermolecular forces create a repulsion across an interface between ice and rock. The presence of a thin layer of water (sometimes called *premelt* or an *interfacial film*) reduces the chemical potential of the ice-rock boundary, because the surface energy of a dry rock-ice interface exceeds the sum of the surface energies of the ice-water and rock-water interfaces (Wettlaufer and Worster 1995). Thus the water layer exists at thermodynamic equilibrium, despite temperatures well below the freezing point of bulk water.

The pressure in the water layer is less than the pressure in the adjacent ice and rock. The greater the temperature depression below melting point, the greater the pressure difference, and the thinner the layer. Because the water layer is very thin at low temperatures (less than 10 nm at $-10\,^\circ$C), processes involving transport of water along the layer must be very slow; in many respects it is still valid to view the bed as frozen and immobile.

Shreve (1984) analyzed the consequences of premelt layers for glacier sliding. He used a relationship between melting temperature T, pressure P, and layer thickness h, proposed by Gilpin (1979):

$$T - T_0 = -\mathcal{B}[P - P_0] - b\,h^{-\alpha} \tag{7.32}$$

where T_0 stands for the melting temperature at pressure P_0 for ice in contact with an infinitely thick water layer (e.g., $0\,^\circ$C at 101.3 kPa), b and α denote constants, and $\mathcal{B}$ is 7.42×10^{-5} K kPa^{-1}. The constants b and α were calibrated to match experimental regelation data. This relation, though not derived from first principles, can be regarded as an empirical approximation adequate for crude analyses of processes dependent on transport in the liquid layer.

Shreve emphasized that, because the liquid layer between ice and rock can never be infinitely thick, it must always be colder than the nominal pressure-melting temperature, even in a temperate glacier. Its temperature is controlled by its thickness and this in turn depends on whether more ice melts into the layer than refreezes. In a situation with net melting, the thickness of the layer, and therefore also its temperature, depends on the rate at which the excess water escapes to the subglacial water system. In a situation with no net melting, the temperature of the ice and rock determine the temperature of the liquid layer, whose thickness adjusts accordingly. Shreve modified the sliding theory of Nye (1969a) to take account of the layer. He calculated a sliding speed of 0.3 mm day^{-1} at $-1\,^\circ$C for a basal shear stress of 100 kPa and a plausible bed roughness. Sliding at sub-freezing temperatures is thus a negligible component of glacier motion. On the other hand, movement of this amount could, in time, produce bedrock striations.

Hallet and others (1986) measured intermittent movement of ice over a bedrock ledge at $-1\,^{\circ}C$ in a cavity under thin ice at Grinnell Glacier, Alaska Echelmeyer and Wang (1987) observed sliding of clean ice over a large boulder embedded in basal debris in a tunnel in Urumqi Glacier No. 1, China. The velocity, 0.5 mm day^{-1} at $-4.6\,^{\circ}C$, was several times that predicted by Shreve's analysis. Cuffey et al. (1999) measured slow sliding in a tunnel beneath Meserve Glacier, Antarctica. The ice slid at about 0.02 mm day^{-1} over basal boulders, about ten times faster than predicted by Shreve's analysis. Cuffey et al. attributed the discrepancy to highly concentrated solutes in the liquid layer; for a given temperature, solutes increase the layer thickness. The very small water layer thickness at such low temperatures limits the sliding velocity because it limits the flow of water from melting sites to freezing sites in the regelation process. Experimental evidence and theoretical considerations both show that the thickness of an interfacial film depends strongly on impurities in the water, on the chemical composition of the solid, and other factors (Dash et al. 2006); the film thickness, and hence sliding rate, cannot be predicted in practice.

7.2.9 Hard-bed Sliding: Summary and Outlook

Several features of the hard-bed sliding processes have been established:

1. Ice moves past bedrock bumps by a combination of regelation (dominant for small bumps) and creep flow (dominant for large bumps). The controlling obstacle size, for which both processes are equally effective, is probably in the range 10 to 100 mm.

2. When the pressure of water at the bed exceeds a certain value (the "separation pressure") that depends on the roughness of the bed, cavities form in the lee of bumps and this increases the sliding velocity. On a hard bed, rapid sliding is not normally possible without cavities. Once cavities form, variations in the pressure and volume of water at the bed influence sliding rates by several mechanisms.

3. As water pressure continues to rise, cavities become more extensive, sliding becomes faster, and the drag force provided by the bed cannot increase further. At this point sliding becomes unstable, unless it is restrained by forces acting at other locations such as the glacier sides (which usually happens; see Chapter 8).

4. If the drag force provided by the bed is small, because water pressures nearly match the ice overburden value, the sliding rate cannot generally be determined from local characteristics of the bed. Instead, the forces produced by deformation of the ice in a large surrounding region of the glacier must be considered.

5. Friction between the bed and rock particles in the basal ice reduces the sliding velocity below that expected at a sharp interface between clean ice and rock. The frictional force may be large for moderate concentrations of rock in the basal ice (order 10%).

6. Sliding occurs at sub-freezing temperatures, but at a rate negligible to glacier flow.

No existing theory provides a quantitative basal boundary condition for the problem of glacier flow on a hard bed. A comprehensive sliding theory must treat the processes of sliding, rock-rock friction, water flow at the bed, and cavity formation simultaneously. Laboratory experiments with careful control of variables such as heat flow and temperature may be useful.

7.3 Deformable Beds

Until the mid-1980s, theoretical analyses of basal motion assumed the bed to be rigid and impermeable. In fact, glacier retreat often reveals not bedrock but glacial deposits ("till" or, more generally, sediments). As a large-scale example, the Laurentide Ice Sheet rested on sediments throughout its southwestern sector, in what is now the Great Lakes basin and the prairies, and in its center over the present Hudson Bay. Geologists have long known that till deforms in shear under certain conditions (e.g., McGee 1894; MacClintock and Dreimanis 1964).

More recently, observations in Iceland and Antarctica revealed that deformable beds play a central role in the dynamics of large, active glaciers. Boulton (1979) and Boulton and Jones (1979) showed that near the margin of Breidamerkurjökull shear in a layer of water-saturated till accounted for 90% of the velocity measured at the glacier's surface. The yield stress for shear of this till is only 3 to 8 kPa at low effective pressures (Boulton and Dent 1974); such stresses would produce no significant deformation in ice. Seismic reflection surveys on Whillans Ice Stream (formerly known as Ice Stream B) in West Antarctica showed that the bed is mantled by a layer of till, several meters thick, highly porous, and saturated with water (Alley et al. 1986c; Blankenship et al. 1986). The shear strength of this layer was later determined from samples recovered through boreholes and found to be only a few kPa (Kamb 1991, 2001). Deformation of this layer, mostly as sliding along its surface, explains why the ice stream moves several hundred meters per year even with a driving stress of only about 20 kPa. A nearby ice stream, Bindschadler, behaves in the same fashion except that most of the motion arises from shearing within the sediment, not sliding on its surface (Kamb 2001). (On both ice streams, the partitioning of soft-bed motion between sliding and shearing has been observed at only a few sites examined in borehole studies, so these observations should not be generalized.) Furthermore, drilling programs have revealed deforming tills beneath mountain glaciers (Humphrey et al. 1993; Iverson et al. 1994; Truffer et al. 2000). Conditions in subglacial till also control the surging behavior of some glaciers (Clarke et al. 1984a; Porter et al. 1997).

If a glacier rests on a deformable substrate, two processes contribute to the net rate of basal slip u_b: sliding of ice along the top of the substrate and deformation at depth within the substrate.

Regardless of which process dominates, we refer to this situation as a *deforming bed* or a *soft bed*. The sliding motion, called *soft-bed sliding*, must involve deformation of a thin layer of substrate wherever ice or rock inclusions touch the bed. Thus the difference between the two processes is not fundamental. Soft-bed sliding differs fundamentally, however, from sliding over a rigid substrate (whether bedrock or unconsolidated material that cannot deform under prevailing conditions); in soft-bed sliding, the strength of the substrate limits the forces by which the bed resists movement of the ice. The more readily the substrate deforms, the smaller the resisting forces. This statement also applies if deformation occurs at depth in the substrate. With a deforming bed of either type, a weak substrate allows the ice to move rapidly even with a small basal shear stress (a "well-lubricated" bed).

Deformation at depth can take the form of discrete slip planes or distributed shear of a horizon. In addition, how the deformation is distributed with depth depends on the timescale considered. At any moment, deformation may concentrate on a single surface. But this surface can move with time, distributing the strain over a range of depths. The focus of slipping can also shift between the interior of the substrate and the interface. Furthermore, rocks partly embedded in the ice and projecting downward can plough through the substrate; the interface need not be a distinct surface.

Glaciological studies of deformable beds have several major goals. First, we must be able to predict the shear stress that a deforming bed exerts on a glacier; such stresses resist the flow of the ice and govern the dynamics of many ice streams and mountain glaciers. A related goal is to develop relations for predicting the slip rate u_b as a function of substrate properties, conditions at the bed, and the large-scale configuration of a glacier. A third major goal is to understand what controls the quantity of rock material transported by deforming beds, in order to assess how deformable beds form, persist, and are depleted. None of these goals has been achieved. In many cases the greatest difficulties arise because the bed deformation process depends sensitively on the water pressure at the glacier base; without accurate predictions of basal water pressure, there is no possibility of predicting the bed deformation itself.

7.3.1 Key Observations

Glacier beds of deformable material are widespread, and active deformation of them appears to be common but not universal. Landforms and deposits show that deformable beds underlie vast areas of the Pleistocene ice sheets on North America and Europe, especially near their margins. Boreholes and tunnels have provided direct access to the beds of only a few modern glaciers, but actively deforming beds are well represented in this group. Table 7.4 summarizes observations of deforming beds from borehole studies.

As the table shows, both soft-bed sliding and deformation within the substrate contribute to total slip rates. Deformation typically occurs to a few decimeters below the interface. On average, sliding contributes most. The West Antarctic ice streams (Whillans and Bindschadler) and the large Alaskan tidewater glacier (Columbia) move rapidly. Most of the mountain glaciers

Table 7.4: Locations known to have subglacial deformation and observed or inferred properties.

Glacier	u_b (m yr^{-1})	Character of basal motion	Deforming depth (m)	Bed stress (kPa)	N (kPa)	References
Whillans	400	>80% is sliding	~5	<6	−20–150	(1), (5), (6)
Bindschadler	360	<20% is sliding	~5	<10	20–40	(10)
Columbia	~2000		0.65	~100		(8)
Trapridge	~30	50–80% is sliding	0.3	80		(2), (7)
Black Rapids (year 1997) (year 2002)	35–45 35–45	shear deep in till mostly sliding	>2			(13) (12)
Storglaciären	15–35	sliding dominates	0.3	40	110	(9)
Breidamerkur	~100	sliding varies from 10–80%	~0.5	80	0–200	(3), (4)
Bakaninbreen (surge front)	~3	mostly deformation		60		(11)

References are: (1) Alley et al. 1986c; (2) Blake et al. 1992; (3) Boulton et al. 2001; (4) Boulton and Dobbie 1998; (5) Engelhardt et al. 1990; (6) Engelhardt and Kamb 1998; (7) Fischer and Clarke 1994; (8) Humphrey et al. 1993; (9) Hooke et al. 1997; (10) Kamb 2001; (11) Porter and Murray 2001; (12) Truffer and Harrison 2006; (13) Truffer et al. 2000.

move at moderate rates, with basal stresses a few tens of percent lower than typical for glaciers. The very fast flow of Columbia Glacier and the West Antarctic ice streams probably reflects persistently high water pressures; all of these glaciers flow into the sea. As discussed later, high water pressures (or, properly, low effective pressures) reduce the strength of a deforming bed.

Figure 7.9 shows an example of how the net horizontal displacement of material within a deforming bed varies with depth. Displacements are largest near the upper surface and decrease rapidly downward. Though limited, most evidence indicates that the displacement accumulates over time in a very irregular fashion (Figures 7.10 and 7.11). Conditions at the glacier bed change rapidly and fluctuate widely, especially if surface meltwater reaches the bed. The rate and location of deformation seem to be sensitive to such changes.

7.3.2 Till Properties and Processes

The term *till* as used by glaciologists and in the present discussion refers to deformable subglacial sediments, irrespective of their origin. To a geologist, till is a poorly sorted sediment deposited by a glacier.

7.3.2.1 Granulometry, Fabrics, and Stress Bridging

The particles in till range from clay and silt (*fines*) to pebbles, cobbles, and boulders (*coarse clasts*). Many individual till deposits contain this entire range of sizes. Soils engineers usually

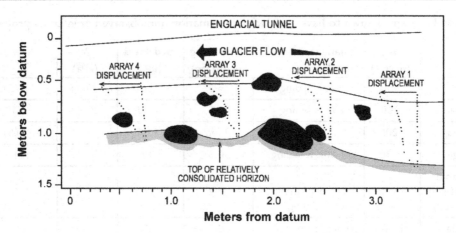

Figure 7.9: Positions of displacement markers (shown as dots) embedded in till under Breidamerkurjökull, beneath about 10 m of ice. Large boulders are shown. Markers were inserted into the till using small-diameter drill holes from a tunnel excavated in the ice. After 5.7 days, the till was dug out. Adapted from Boulton and Hindmarsh (1987) and used with permission from the American Geophysical Union, *Journal of Geophysical Research*.

divide sediment into fines and "granular material" (or clasts) based on differences in mechanical behavior. Water, contained in pores within a matrix of the fine-grained material, is an essential component of a deformable till.

The abundance of fines varies from till to till (Iverson et al. 1997). There are clay-rich tills such as the Two Rivers Till, a Laurentide Ice Sheet deposit containing 32% clay and 38% sand and gravel. In contrast, the sandy till beneath Storglaciären has only 4% clay but 75% sand and gravel. Such differences strongly affect the ability of water to move through the till; water moves more readily through a coarse matrix. Differences in grain size distribution also influence the mechanical strength of the till at high confining pressures; to shear a coarse-grained till requires a larger applied stress.

The grains in a till are not perfectly spherical and so can be oriented to form fabrics. Hooyer and Iverson (2000) and Iverson et al. (2008) demonstrated, using laboratory experiments, that shearing a till strengthens its fabric. The fabric develops quickly and attains a maximum strength at moderate strains. Because the same fabric persists as strain accumulates further, it is not a good indicator of the total strain.

In a deforming till, grains evolve toward finer sizes by abrasion and crushing. Crushing occurs when grains bear a large load – most likely when grains of similar size form chains of contact. Consequently, crushing leads to dispersion of the grain size distribution so that large grains are buffered by numerous smaller grains (Figure 7.12; Hooke and Iverson 1995). A deformed till thus develops roughly equal volume fractions of large and small grains. Many tills have this property, which can be described as a power-law or fractal distribution of grain numbers (e.g., Clarke 1987; Hooke and Iverson 1995; Tulaczyk et al. 1998). The number of grains with

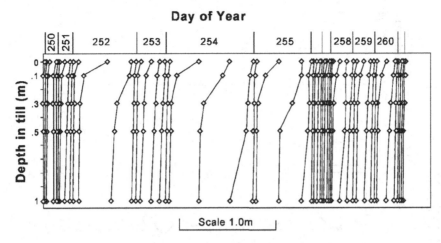

Figure 7.10: Horizontal displacements, at six-hour intervals, of strain markers embedded in till under Breidamerkurjökull, beneath about 100 m of ice. The spacing between each pattern in the figure is arbitrary, chosen for clarity. Deformation occurs episodically, concentrates near the surface, and tends to propagate downward. Markers were inserted at the bottom of a borehole, using a specialized drill head that emplaced a till core containing the markers, which were attached to drag spools. Adapted from Boulton et al. (2001).

diameter D varies in proportion to D^{-d}, with a fractal dimension d of 2.84 to 2.96 (Hooke and Iverson 1995). Compared to granular materials made purely by crushing, for which a typical $d \approx 2.6$, the tills include a slight excess of fines produced by abrasion.

As a till deforms, chains of contact between grains temporarily form and then vanish. Such *stress bridging* transmits forces through the till more effectively than does the disorganized matrix. Stress fluctuations arising from this behavior have been observed in laboratory experiments (Iverson et al. 1997).

7.3.2.2 Effective Stress

Effective stress, a central concept of soil and rock mechanics, expresses how frictional forces are reduced by the presence of pressurized water. Just as deviatoric stress is fundamental to ice deformation, effective stress is fundamental to deformation of till, soil, and rock.

The forces applied to a till are distributed between the solid matrix and the water in the pores. The effective stress, denoted N, measures the force per area provided by contacts between grains. For stresses defined as positive in compression, the components are

$$N_{jk} = \sigma_{jk} - P_{\mathrm{w}} \, \delta_{jk}, \tag{7.33}$$

where σ_{jk} denotes a component of the total stress and P_{w} the water pressure. Consider any surface in a till that might shear. Call the normal stress on the surface σ_n. The quantity $N = \sigma_n - P_{\mathrm{w}}$ defines the *effective normal stress* acting on the surface. On a horizontal shear plane,

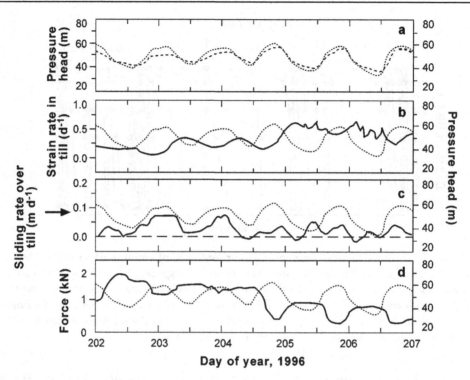

Figure 7.11: Subglacial instrument records during a five-day period in July 1996, under Trapridge Glacier. (a) Two measurements of pressure, given as equivalent water depth. (b) Shear strain rate within till (solid curve), obtained from a tilt sensor, compared to one water pressure record. (c) Sliding rate (solid curve), obtained with an anchor on a spooled wire, compared to the same water pressure record. (d) The force on a ploughmeter (solid curve) inserted 0.14 m into the till, compared to the same water pressure record. Some smoothing of the data has been applied. Adapted from Kavanaugh and Clarke (2006) and used with permission from the American Geophysical Union, *Journal of Geophysical Research*.

for example, $N = \sigma_{zz} - P_w$, if z denotes the vertical coordinate. Its value at the ice-till interface, N_o, is usually approximated by setting σ_{zz} equal to the ice-overburden pressure: $N_o \approx \rho g H - P_w$. In glacier studies, the term *effective pressure* (Eq. 7.11) is often taken as a synonym for "the effective normal stress acting on a plane parallel to the bed."

Equation 7.33 and the expression for N_o strictly apply only if the pore water is stagnant. Percolation generates seepage pressures. Downward seepage increases the effective stress on horizontal surfaces; upward seepage does the opposite.

7.3.2.3 Porosity, Consolidation, and Dilatancy

Porosity refers to the fraction of the total volume occupied by voids (or pores). Typical values are $v \approx 0.2$ to 0.4 (Boulton and Dent 1974, Table 7.2). The upper value appears to be typical

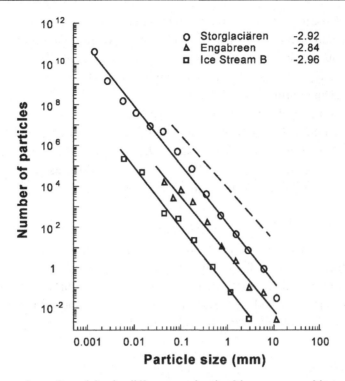

Figure 7.12: The number of particles in different grain-size bins, measured in samples of tills from three different locations. The slope of a line corresponds to the fractal dimension; the legend gives values. The dashed line shows a slope of 2.6 for comparison. Adapted from Hooke and Iverson (1995).

of a saturated till deforming subglacially, but is not diagnostic of active deformation. The *void ratio*, $v/(1-v)$, provides another measure of pore space.

Increasing the effective pressure on a till compresses it and reduces the porosity. A larger change occurs if the pore water can escape than if trapped. The combined process of porosity reduction and water expulsion is *consolidation*. Decreasing the effective pressure causes the inverse process of decompression or *swelling*, but this changes the porosity by only a small amount compared to compression (e.g., Tulaczyk et al. 2000a). Thus, low porosity in a till sometimes represents a memory of earlier periods of high load. Furthermore, a history of repeated loading generally decreases the porosity. It can only be regenerated by dilatancy accompanying shearing. If a shear deformation continues for a long period, porosity will attain a constant value dependent on the effective pressure and the particle size distribution.

Dilatancy, a property of granular materials, increases the complexity of till deformation processes (Reynolds 1885; Andrade and Fox 1949). A dilatant material expands when it deforms in shear; the grains rearrange themselves in a way that reduces their interlocking and hence their resistance to movement past one another. Dilation increases the pore space and also changes the water pressure in a shearing zone. If water cannot readily flow from elsewhere to fill the new pore space, the pressure in the shearing horizon decreases. Reduced pressure, in turn, effectively strengthens the till and might prevent further deformation and dilation. In contrast, if water flows efficiently to the dilating layer, pressure remains high, deformation continues, and the enhanced porosity might weaken the till.

Dilatancy allows a consolidated till to recover porosity and permeability when it deforms. On the other hand, a dilated till subject to an applied load collapses and stiffens when deformation stops. Boulton et al. (1974), Tulaczyk et al. (2000a), and Moore and Iverson (2002) have described dilatant behavior of tills in laboratory tests. Sediments beneath the fast-flowing Antarctic ice streams are dilated (Alley et al. 1986c; Kamb 2001).

7.3.2.4 Intergranular Water Flow

The percolation of water through intergranular pore spaces in a till is an important process controlling the effective stress and hence the resistance to shear.

Elevation differences and pressure gradients drive water flow through a porous medium (as within a glacier and along the bed; see Chapter 6). The effects of elevation z and water pressure P_w combine to give the *hydraulic potential*, ϕ_h, defined as

$$\phi_h = \rho_w\, g\, z + P_w, \tag{7.34}$$

for water of density ρ_w at pressure P_w. Normalizing the potential to the unit weight of water defines the *hydraulic head*: $\phi_h/[\rho_w g]$, with units of length. *Darcy's Law* states that the flux of fluid through a porous medium (volume per unit time passing through a unit area normal to the flow direction) varies directly with the gradient of ϕ_h and inversely with the viscosity of the fluid. Specifically, the flux of water $\vec{q}_w$ (m s^{-1}) obeys

$$\vec{q}_w = -\frac{k_h}{\eta_w}\nabla\phi_h \tag{7.35}$$

where η_w signifies the fluid viscosity (Pa s), and k_h the *hydraulic permeability*, with units of m^2. The permeability increases with porosity. Measured values of permeability for deformable tills range from 4×10^{-15} to 2×10^{-13} m^2 (Boulton et al. 1974; Brown et al. 1987; Clarke 1987). A related quantity is the *hydraulic conductivity* $\kappa = \rho_w\, g\, k_h/\eta_w$, with units of m s^{-1}. For water at $0\,°C$, $\eta_w = 1.787 \times 10^{-3}$ Pa s and so $\kappa = 5.5 \times 10^6\, k_h$.

As a specific example of Eq. 7.35, consider a parallel-sided subglacial till layer lying on an impermeable substrate of slope β. The water flux along the layer would be

$$q_w = \frac{k_h}{\eta_w}\left[\rho_w\, g\, \sin\beta + \frac{\partial P_w}{\partial s}\right], \tag{7.36}$$

where s is a distance coordinate measured opposite to the direction of flow (because a down-slope decrease of water pressure enhances the flow). In steady conditions, no flow occurs across the till layer, and the gradient of ϕ_h in that direction is zero; a difference in water pressure between the top and bottom of the till balances the elevation difference – a hydrostatic condition. However, P_w often fluctuates at the till's upper surface, as shown in Figure 7.11. Such variations induce temporary vertical gradients of ϕ_h. By Darcian flow, these P_w fluctuations propagate into the layer according to the diffusion equation (Section 9.3). For one dimension this states that

$$\frac{\partial P_w}{\partial t} = d_T \frac{\partial^2 P_w}{\partial z^2}, \tag{7.37}$$

where t denotes time, z depth, and d_T the *hydraulic diffusivity* of the till (dimensions of length squared per unit time). A larger d_T implies a faster propagation of pressure signals. Such propagation increases with increasing hydraulic conductivity κ, but is buffered by the till's capacity to increase the storage of water in a unit volume – primarily a function of the compressibility of the grain skeleton. Thus the till's compressibility, porosity, and k_h all control d_T (Walder 1983). d_T is sometimes called the *coefficient of consolidation*. Values range from 5×10^{-9} to $4 \times 10^{-4} \, \mathrm{m^2 \, s^{-1}}$ (Boulton and Dent 1974; Brown et al. 1987; Iverson et al. 1997; Porter and Murray 2001).

The propagation of changes in water pressure is analogous to the propagation of temperature changes as discussed in Chapter 9. According to Section 9.8, a sinusoidal fluctuation at the surface, of period T, diminishes to a fraction $1/e$ (or 37%) of its surface amplitude at a depth $[d_T T / \pi]^{1/2}$. For a daily pressure fluctuation, such as occurs beneath glaciers where surface meltwater reaches the bed, this depth equals 0.17 m if $d_T = 10^{-6} \, \mathrm{m^2 \, s^{-1}}$. For the annual cycle, the depth increases to 3 m.

Hydraulic diffusivities of till vary considerably with grain size. Iverson et al. (1997) measured d_T on a clay-rich and a sandy till and found values of about 6×10^{-9} and $3 \times 10^{-6} \, \mathrm{m^2 \, s^{-1}}$, respectively, a factor of about 500 difference. Consequently, rapid pressure fluctuations at the surface of a till should propagate readily into a sandy till but not into a clay-rich one.

7.3.2.5 Seepage Pressure and Piping

The flow of water through pore spaces, governed by Eq. 7.35, also exerts a force on the solid grains, called the *seepage pressure*. It equals $\nabla \phi_h$. A seepage pressure large enough to overcome the grains' weight sometimes allows flowing water to excavate large voids. This is the process of *piping*, infamous for its role in catastrophic failures of earthen dams. Piping could occur in subglacial tills if a large hydraulic gradient develops.

7.3.2.6 Ploughing, Regelation, and the Ice-till Interface

The small-scale geometry of the ice-till interface matters greatly. A smooth and planar interface should allow the glacier to move easily by soft-bed sliding. In contrast, penetration of ice into the till and interlocking with the grains should inhibit sliding. Regelation, the process of melting and refreezing discussed in Section 7.2.1, provides a mechanism for such penetration.

Iverson and Semmens (1995) demonstrated this process in tills using laboratory experiments. They also outlined a theory, which was extended considerably by Rempel (2008). The downward motion by regelation is driven by the effective stress, and so should be negligible where water pressures nearly equal the weight of overburden. Penetration of ice occurs to a depth determined by a balance between such downward motion and the basal melt rate. Penetration rates measured in the laboratory tests were about $0.1 \, \text{m yr}^{-1}$ at $N \approx 100 \, \text{kPa}$. The theory predicts penetration depths on the order of $1 \, \text{cm}$ to $1 \, \text{m}$ for the ranges of typical subglacial melt rates and N values. However, the grain size also matters; in fine-grained materials, surface tension effects should limit the penetration by ice. At N of about $50 \, \text{kPa}$, surface effects should stop the regelation process entirely in a till with abundant fine silt and clay. Thus a fine-grained till is more likely than a coarse one to form a smooth interface with the ice – and thus the ice is more likely to move by soft-bed sliding.

Large clasts at the interface may be partly embedded in ice and partly in till. Such clasts will bear more than the average shear stress along the contact. Ice is stronger than deforming till and so will push such clasts forward and plough them through the till (Brown et al. 1987; Iverson 1999; Thomason and Iverson 2008). Such *ploughing* constitutes a mechanism of motion intermediate between soft-bed sliding and distributed deformation.

7.3.3 Constitutive Behaviors

Till is a complex material and its deformation depends not only on applied shear stress but also on effective pressure, porosity, volume fraction of fines, and strain history. It is inhomogeneous and may be anisotropic if deformation has produced a preferred orientation in the grains. Its characteristics may change with time as fines are washed out and clasts are ground down. To determine how all of these variables affect the mechanical properties remains a task for the future. In the present discussion, the till is assumed to be water-saturated, ice-free, and isotropic. Despite the potential complexities, the constitutive behavior of a till subject to a large deformation of constant orientation may be quite simple: the till can behave as a perfect plastic at a critical state.

7.3.3.1 Plastic Deformation

Most laboratory experiments show that, after a brief period of initial stress adjustment and small deformation, a shearing till attains a yield stress τ_* with a value essentially independent of the deformation rate (Figure 7.13; Kamb 1991; Iverson et al. 1998; Tulaczyk et al. 2000a; Truffer et al. 2000). This is characteristic of a perfectly plastic material (Section 3.4.4.6). Deformation occurs as slip along a plane or as shear of a narrow zone called a *shear band* (Iverson et al. 1997). Continuing the deformation to a large strain results in no significant changes of stress or porosity; the till is said to be at a *critical state* (Schofield and Wroth 1968). The stress τ_* defines the *residual strength* or *ultimate strength*. Plastic behavior of this sort occurs in other saturated granular materials like soils; its occurrence in tills is not surprising (Kamb 1991).

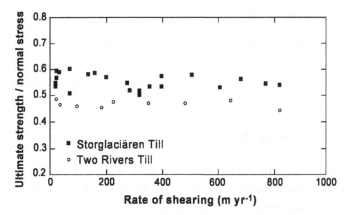

Figure 7.13: Laboratory measurements of till shear strength, τ_*, as a fraction of normal stress. Two different tills were deformed to large strains. The strength does not depend on the rate of shearing, here measured as the rate of displacement at the top of the sample. From Iverson et al. (1998).

More difficult to assess are the *in situ* properties of actively deforming till. Nonetheless, subglacial measurements also indicate that till deformation is best described as plastic. Drag forces can be measured on objects (called *fish*) inserted into the till beneath boreholes and subsequently dragged through the till by motion of the glacier. The dragging induces deformation of the surrounding material. The measurements show little variation of the drag force even as the rate of motion changes, a manifestation of plastic behavior (Hooke et al. 1997). In an experiment beneath Trapridge Glacier, Yukon, instruments installed in boreholes simultaneously recorded basal sliding, water pressure, till strain rate, and till strength (Figure 7.11; Kavanaugh and Clarke 2006). These records show that basal water pressure varies in anticorrelation with the drag force on *ploughmeter* rods dragged through the till by motion of the overlying ice. In Kavanaugh and Clarke's calculations, the observed relationship could be reproduced by assuming plastic deformation of the till, with a τ_* dependent on water pressure. With a viscous till instead of a plastic one, the drag force on the ploughmeter would increase during episodes of fast flow – corresponding to periods of high water pressure – and disrupt the anticorrelation. Furthermore, the measured shear was focused too strongly near the top surface of the till to be explained by a viscous behavior. Plastic behavior, though providing a best match to these data, did not explain the measurements perfectly either, however. The simple plastic model predicted a more strongly episodic shearing of the till than observed.

The fact that till behaves plastically, the simplest constitutive behavior, does not mean that the behavior of a glacier bed mantled with till is simple or always behaves plastically. As discussed later in this chapter, little is known about the large-scale behavior of beds of deforming till. We have discussed some of the causes of complexity in Section 7.3.2 and pursue the issue further in Section 7.3.4.

7.3.3.2 Plastic Yield Strength

The slip plane of a deforming till supports a stress no greater than τ_*, the residual strength. To initiate the slip usually requires a stress only slightly different from τ_*. Laboratory experiments establish that the residual strength of till obeys the *Mohr-Coulomb rule* (Iverson et al. 1998),

$$\tau_* = c_o + fN \approx fN, \tag{7.38}$$

$$\text{with}\quad N = \sigma_n - P_{\text{w}} \quad \text{and}\quad f = \tan\varphi. \tag{7.39}$$

Here c_o is the apparent cohesion, f the friction factor, and N the effective normal stress on the slip plane. A compressive normal stress of magnitude σ_n acts across the plane. (Here we define stress as positive in compression for convenience; both τ_* and N are taken as positive quantities.) Equation 7.38 states that compression across the shear plane makes it harder to deform a till but that the pressure of pore water counteracts such strengthening (Figure 7.14). High pore-water pressures (low N) weaken a till. For the same reason, elevated water pressures are an essential factor enabling shear beneath landslides and on thrust faults (Terzaghi 1950; Hubbert and Rubey 1959). As noted previously, N might deviate from Eq. 7.39 if water flows normal to the shear plane, inducing a seepage pressure.

The value of c_o depends on the till constituents; clay has appreciable cohesion whereas sand has very little. In deforming tills, however, the cohesion is probably small enough to be neglected in most problems. In shear at critical state, the material has no true cohesion; the c_o values measured in experiments are thus referred to as "apparent cohesion."

The friction factor f is often written, by analogy with ordinary friction, as the tangent of a friction angle φ. Measured φ values range between about 10° and 40°; hence f ranges from

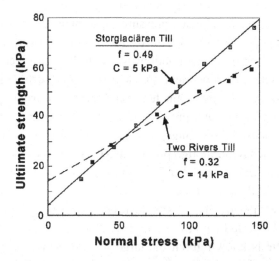

Figure 7.14: Laboratory measurements showing linear increase of till shear strength, τ_*, with the effective pressure, for two tills deformed to large strains. From Iverson et al. (1998).

0.17 to 0.84. Primarily this range reflects differences in the clay content, with $\varphi < 20°$ for a clay content $> 25\%$, as a general rule. In experiments on two tills, one clay-rich and one sandy, Iverson et al. (1998) found that τ_* of the clay-rich till was less than half that for the sandy till, with $\varphi \approx 8°$ and $26°$, respectively.

A key unanswered question about the influence of grain size distributions on τ_* concerns the role of large clasts, which are removed for laboratory experiments. Large clasts act as undeformable inclusions in the till, and should disrupt shear planes intersecting them. Such disruption presumably strengthens the till overall.

Though not shown explicitly in Eq. 7.38, porosity influences τ_*; for a given N, higher porosity implies a weaker till. In critical state deformation, however, porosity is interdependent with both τ_* and N. The values for c_o and φ implicitly include the effects of porosity already and porosity does not represent an extra degree of freedom (Tulaczyk et al. 2000a). Such interdependence reflects dilatancy (Clarke 1987). If less porous than appropriate for critical state, a deforming till will dilate. If more porous, it will compact. Before attainment of critical state – or after departure from critical state – porosity influences τ_* strongly. A till weakens dramatically if porosity rises above the critical state value.

The dependence of τ_* on P_w in Eq. 7.38 is the most important implication of the Mohr-Coulomb behavior of tills. We discuss this further next.

7.3.3.3 Values for τ_*

To measure residual strength in a laboratory, till is sheared in an apparatus that controls the confining pressure and either the stress or displacement rate. An apparatus in which the sample fills a rotating circular ring – a "ring-shear device" – permits experiments to large strains. To assess τ_* *in situ* beneath a glacier requires measuring forces on an object dragged through the till by motion of the overlying ice. The object may be a rod that sits in the ice borehole and has a tip extending down into the till (a ploughmeter) or may be a more equant object (a fish) inserted into the till and connected to the overlying ice only by a cable. To derive τ_* values from such *in situ* measurements, plastic behavior of the till must be assumed. The same data can be used to infer effective viscosities, using a different assumption about the till's constitutive behavior.

Table 7.5 shows results of both laboratory and *in situ* measurements. These values refer to residual strength except in the case of the New England tills, for which strength at initial yield was reported. For experiments giving values of φ and c_o, we report τ_* at $N = 50$ kPa as an example. Notice that the range of φ values is not great, and apparent cohesion is small. Thus the effective stress N provides the overwhelming control on τ_*.

7.3.3.4 Implications of P_w Dependence and φ Values

Much as temperature strongly controls the viscosity of ice, water pressure strongly controls the strength of till. In fact, the latter relationship is even stronger than the former: variations of

Table 7.5: Measured strength values for tills.

Glacier	τ_* (kPa)	τ_* at $N = 50$ (kPa)	φ	Apparent c_o (kPa)	% Clay	Analysis	Reference
Whillans Ice Stream	2–4				35	lab[a]	(7), (8)
Whillans Ice Stream		22	24°		35	lab[b]	(10)
Bindschadler Ice Stream	~1					lab[a]	(8)
Columbia	5.5–13					in situ[c]	(4)
Trapridge	48–57				<20	in situ[d]	(3)
Storglacier	52				4	in situ[e]	(5)
Storglacier		24	26°	5	4	lab[b]	(6)
Black Rapids		43	40°	1.3	<6	lab[b]	(9)
Breidamerkur		35	32°	3.7		in situ[f]	(2)
Laurentide: (Two-Rivers)		31	18°	14		lab[b]	(6)
(New England)		30–80	31–39°	0–40		lab[g]	(1)

[a] Measurement supposed to be indicative of *in situ* properties, as it was an undrained test on freshly acquired sample.
[b] Measured with ring-shear device.
[c] Subglacial bent drill rod.
[d] Subglacial ploughmeter.
[e] Subglacial fish.
[f] Model fit of data to measured strain rates.
[g] Cited as undrained consolidated tests.

References are: (1) Brown et al. 1987; (2) Boulton and Hindmarsh 1987; (3) Fischer and Clarke 1994; (4) Humphrey et al. 1993; (5) Hooke et al. 1997; (6) Iverson et al. 1998; (7) Kamb 1991; (8) Kamb 2001; (9) Truffer et al. 2000; (10) Tulaczyk et al. 2000a

water pressure can activate and shut down the shearing of till altogether. Due to the difficulty of water drainage along the contact between a glacier and its bed or through underlying aquifers, subglacial water pressures often rise to near the ice-overburden value (Chapter 6). Beneath Whillans Ice Stream in West Antarctica, for example, P_w matches the 9 MPa overburden to within 1% (Kamb 2001). Such low effective stresses explain why subglacial tills are often very much weaker than ice. Furthermore, the ice overburden itself greatly strengthens a subglacial till; σ_n in Eq. 7.39 reaches nearly 10 MPa beneath 1 km of ice. Thus till deformation can *only* occur with high values of P_w (Brown et al. 1987). Without low effective stresses, till would act essentially like a porous rock, given that typical shear stresses beneath glaciers are only about 100 kPa. Using a typical value of $\varphi = 30°$ from Table 7.5, a 100 kPa shear stress could cause widespread till deformation only with effective stresses less than about 170 kPa. Thus, beneath a glacier of 100 m thickness, the water pressure would need to exceed about 83% of the ice-overburden pressure, and about 98% beneath 1 km of ice.

Furthermore, τ_* varies over time and from place to place because of fluctuations in subglacial water pressures. Thus till should deform episodically, with most of the action occurring at times of high water pressure. Even with active deformation occurring elsewhere beneath a glacier, till might stagnate in regions of reduced water pressure. Water pressure fluctuations will, in addition, influence the pattern of deformation at depth in a till (Section 7.3.4.2). Taken together, the capacity for such variations implies that till deformation is strongly controlled by hydrological variables: the supply of water to the glacier bed, the subglacial drainage system, and boundary conditions such as water depth at marine termini.

The level of effective stress also modulates differences in glacier bed behavior related to the till's frictional character, the value of φ. To illustrate, suppose that $N = 170$ kPa. Then, for a high-friction till ($\varphi = 35°$), the strength of the bed is $\tau_* \approx 120$ kPa, but for a weak till ($\varphi = 18°$), it is only $\tau_* \approx 55$ kPa. Such differences vary in proportion to the effective stress; they vanish when the glacier floats on its bed ($N = 0$).

Water also mediates a potentially important interaction between till strength and dilatancy. If till behaves as a perfect plastic, faster shearing does not generate increased stresses that oppose flow. However, when a consolidated till begins to shear it dilates. Dilation can reduce the water pressure locally, increasing τ_*. Such a negative feedback makes possible a kind of stable creep ("sub-critical-state creep"). Moore and Iverson (2002) discussed this possibility and reported observations of it in laboratory experiments. They found that such stable creep occurs initially as a till deforms. But when the porosity and residual strength attain values appropriate for critical state, the deformation suddenly switches to unrestrained slip. Such a process would introduce further complexity in deforming subglacial tills.

Finally, water pressures can build up in front of rocks as they are ploughed through till. The rocks thus move forward into a zone of low effective stress and weak till. The faster the ploughing, the weaker the till, and the less resistance to motion (Thomason and Iverson 2008). This situation, a type of *velocity weakening*, also occurs on faults in crustal rocks and plays a role in earthquakes.

7.3.3.5 Apparent Viscosities

Apparent viscosity is another measure of a deforming till's strength. This can be calculated, as with τ_*, from the drag forces on objects moving through subglacial tills. It can also be calculated from basal slip rate if both the applied stress and the thickness of the deforming layer are known or assumed. Table 7.6 lists estimates of apparent viscosities, assuming till behaves as a Newtonian viscous material. There is no evidence that till behaves viscously at a local scale, but it does at larger scales in some situations (Anandakrishnan et al. 2003) but not others (Tulaczyk 2006). Of the values in this table, only the one from Bindschadler Ice Stream was derived from the large-scale behavior of a deforming till bed. For comparison, the effective viscosity of glacier ice at a stress of 100 kPa ranges from about $5 \times 10^{12} - 10^{15}$ Pa s, depending on the temperature and other variables.

7.3.4 Slip Rate u_b on a Deformable Bed

How rapidly a glacier moves over a deformable bed can perhaps be predicted from glacier geometry and ice properties in the limiting situation of "global" control (Section 7.1.2). In other

Table 7.6: Estimates of apparent viscosity of till.

η (Pa s)	Glacier	Method	Reference	Comment
$5 \times 10^{10} - 5 \times 10^{11}$	Breidamerkur	Strain profile data	(3)	
$3 \times 10^{10} - 1.5 \times 10^{11}$	Trapridge	Strain profile data	(2)	
$3 \times 10^{9} - 3 \times 10^{10}$	Trapridge	Ploughmeter	(4)	
$2 \times 10^{8} - 5 \times 10^{8}$	Columbia	Bent drill rod	(5)	
$1 \times 10^{12} - 4 \times 10^{12}$	Bakaninbreen	Strain rate data	(6)	
$7.5 \times 10^{6} - 1.5 \times 10^{8}$	BIS[†]	Fits to model of tidal response	(1)	Depth not measured (range 0.5–10 m)

[†] BIS = Bindschadler Ice Stream.
References: (1) Anandakrishnan et al. 2003; (2) Blake et al. 1992; (3) Boulton and Hindmarsh 1987; (4) Fischer and Clarke 1994; (5) Humphrey et al. 1993; (6) Porter and Murray 2001.

situations, u_b so far defies prediction. Thus, the following discussion concerns not formulae for calculating u_b but rather the factors influencing it.

Total slip motion arises from soft-bed sliding (at rate u_{ss}) and from deformations at depth in the till:

$$u_b = u_{ss} + \int_Z \frac{\partial u}{\partial z} \, dz, \tag{7.40}$$

where the integration spans the thickness of till that deforms (Z), and $\partial u / \partial z$ is twice the shear strain rate. But a plastic material tends to deform on discrete "slip planes." Observations on a variety of granular materials suggest that slip planes are really shearing horizons, but with widths of only about ten grain-diameters (Stevenson and Scott 1991; Hobbs and Ord 1989); it is reasonable to regard them as discrete planes. Thus the net basal motion should perhaps be viewed as a summation of slip events, with continuous sliding and ploughing at the ice-till contact being one term in the sum (Kamb 1991; Hindmarsh 1997; Iverson et al. 1998). One option for rewriting Eq. 7.40 takes the form

$$u_b = u_{ss} + \int_Z n_s(z) \bar{s}(z) \, dz + \int_\Omega \frac{\partial u}{\partial z} \, dz, \tag{7.41}$$

where n_s denotes the number of active slip planes per unit depth (units m^{-1}), and $\bar{s}$ the average slip rate on them (units $m\ s^{-1}$). The final term is retained because, even with active slip planes, distributed deformation might arise from processes like ploughing, disruption of slip planes by coarse clasts, and feedbacks between water pressure and dilatancy. (The integration now ranges over the horizons of till, Ω, exclusive of slip planes.)

The next three sections discuss how this summation is realized in deforming beds.

7.3.4.1 Large Variability

Observations show that till deformation varies significantly on timescales of hours to days beneath Trapridge Glacier, Storglaciären, and Breidamerkurjökull (Kavanaugh and Clarke 2006; Iverson et al. 1995; Boulton et al. 2001). Fischer and Clarke (1997) reported that motion beneath Trapridge occurs episodically. Several factors explain such variability. One is the granular nature of till; slip planes form quickly and then dissipate, and dilatant effects depend on the supply and mobility of water. A second cause of variability is geometric; coarse clasts, basal topography, and water channels all might interfere with slip planes. A third cause, well established by measurements, is that subglacial water pressures fluctuate widely and rapidly, especially where surface meltwater reaches the bed in large quantities (Blake et al. 1992; Iverson et al. 1995; Kavanaugh and Clarke 2006). Observations even show diurnal fluctuations of water pressure beneath West Antarctic Ice Streams, where no surface melt reaches the bed (Engelhardt and Kamb 1997; Kamb 2001). Such fluctuations presumably arise from tidal influences at the grounding line.

Many of the observations show that periods of rapid slip of soft-bed glaciers correspond to periods of high basal water pressure (Boulton et al. 2001; Kavanaugh and Clarke 2006). But sometimes the most rapid motion coincides with rising pressure (Blake et al. 1994). And at other times, there appears to be no relationshp at all (Kavanaugh and Clarke 2006; Porter and Murray 2001).

Beneath some glaciers, till properties vary spatially; the bed is a mosaic of weaker and stiffer regions. In such cases, the stiff regions exert larger shear stresses on the sole of the glacier than do the weak regions, but the relative contributions can shift over time. An increase of stress in the stiff regions decreases the stress in the weak regions, reducing deformation in the latter. Such a trade-off occurs beneath Storglaciären. Here Iverson et al. (1995) observed an inverse relation between till deformation and water pressure – opposite to the relation expected if τ_* alone controls the deformation rate. High water pressures lift the glacier, releasing stress from weak regions and shifting stress onto stiff parts of the bed, which might be higher-standing regions or areas of exposed bedrock.

7.3.4.2 Depth Profile of Deformation

How deformation varies with depth in a till is important; the deeper the deformation, the greater the quantity of rock material moved in unit time (the "till flux"), a key factor controlling the evolution of till layers (Section 7.3.6). Furthermore, where local conditions determine u_b, a thicker deforming horizon might imply faster glacier flow.

The strength τ_* typically increases downward in a till. The weight of the till itself increases the compressive stress σ_n across potential shear planes (Eqs. 7.38 and 7.39). At depth z_t below the ice-till interface, the effective stress approximates

$$N(z_t) = N_o + [1 - v][\rho_r - \rho_w] g \, z_t \tag{7.42}$$

in the absence of seepage pressures or pressure fluctuations. (Here ρ_r and ρ_w are the densities of rock and water, and N_o the effective pressure at the interface. The second term on the right constitutes the *buoyant weight* of the till.) In contrast, the stress applied to the till by the glacier only increases by a trivial amount over the same depth. Thus the till strengthens downward while the stress acting on it does not. This predisposes the ice-till interface and the upper part of the till to shearing (Alley 1989; Boulton and Hindmarsh 1987; Iverson and Iverson 2001). In other words, the abundance of slip planes and the rate of motion on them, the critical factors in Eq. 7.41, should tend to be largest near the top of the till. This is consistent with observations. First, deforming till layers are thin, typically a few decimeters. Second, much of the motion occurs as soft-bed sliding (Table 7.4). The magnitude of strengthening with depth, $d\tau_*/dz_t$, is $fg\,[\rho_r - \rho_w][1 - v]$ or about 1 to 5 kPa m^{-1}.

In a plastic material, deformation concentrates in shear bands at the weakest horizon. By the preceding arguments, the weakest subglacial horizon should normally be the interface. However, the depth profile of till strength varies over time. Water pressure fluctuations at the ice-till interface generate pressure waves that propagate into the till by diffusion (Section 7.3.2.4 and Eq. 7.37). Thus the location of the weakest till migrates downward with the peak pressure (assuming homogeneity of the till in other respects), and the zone of most active shearing should migrate downward too (Iverson et al. 1998; Tulaczyk et al. 2000a). Such a process helps to explain why, over time, the deforming horizon spans a thickness of decimeters or greater. Tulaczyk (1999) pointed out that negative pressure fluctuations – periods of low water pressure – drive consolidation and strengthening at the surface of the till. Thus, the weakest horizons might occur below the surface at most times. Interference of shear bands by coarse clasts provides another mechanism to distribute deformation downward, and ploughing clasts at the interface have this effect in the uppermost horizons (Brown et al. 1987). Dilation of shear zones may also distribute the deformation; dilation reduces water pressure in the shearing zone, locally strengthening the till and possibly shifting the deformation to another horizon.

Finally, there may be persistent gradients of hydraulic potential across the till layer. Upwelling water, for example, implies enhanced pressures deep in the till, a situation that would focus shearing far below the surface. One set of measurements, from Black Rapids Glacier, Alaska, suggested active shearing occurring at depths greater than 2 m (Truffer et al. 2000). Though not typical, such "deep shearing" is consistent with fault-plane-type features observed in till outcrops (Ehlers and Stephen 1979; Benn 1995).

The partitioning of basal slip between soft-bed sliding (u_{ss} in Eq. 7.41) and deformation within the till likely depends on at least three factors; the propagation of pressure fluctuations, the effective stress N_o at the interface, and the coupling of ice and till. A fine-grained till with a smooth upper surface surely favors sliding (Thorsteinsson and Raymond 2000). For example, at one site in the clay-rich till beneath Whillans Ice Stream most of the basal motion – at least 80% – results from soft-bed sliding (Engelhardt and Kamb 1998). (There is, however, no good explanation for why the nearby Bindschadler Ice Stream slides much more slowly at the site examined.) In contrast, where ice regelates into coarse-grained till, or where coarse clasts plough through the till, the rough interface should inhibit sliding.

Iverson (1999) and Iverson et al. (1995, 1999) have analyzed the importance of effective stress in this problem for the case of Storglaciären, introduced previously. When water pressure is high (N low), sliding and ploughing dominate; water floods the ice-till contact, concentrating stress on clasts bridging the interface and shifting stress to nearby stiffer sections of the bed. When water pressure decreases the ice makes stronger contact with the till, increasing deformation at depth. If water pressure decreases past a threshold, however, deformation within the till stops and slow hard-bed sliding occurs. Support for this picture was provided by indirect measurements of the stress at the ice-till contact (Iverson et al. 1999). Such stresses lead to elastic deformations of the underlying till that, although small, were measurable. Results suggested that the contact stress varies in proportion to $N^{1.7}$.

Though the patterns appear more complex, observations reveal a similar behavior beneath Breidamerkurjökull (Boulton et al. 2001). Declining water pressures correlate with a shift from sliding to shearing at depth. These observations are significant because, compared to Storglaciären, Breidamerkurjökull is a broader glacier with a more extensive cover of till on its bed.

7.3.5 Large-scale Behavior of Soft Beds

How do deformable beds behave at the scale of large sections of glaciers and ice streams? Though an essential question for studies of ice dynamics, little is known about the answer, except that:

1. Beds with a continuous till cover sometimes behave as a system at a critical state, akin to crustal faults that produce earthquakes (Bindschadler et al. 2003; Kavanaugh 2009). The ice slips episodically, and sometimes produces small earthquakes (Anandakrishnan and Alley 1997). Slip events sometimes involve large areas of the glacier, but smaller events are more frequent. This behavior is revealed indirectly by brief, high-magnitude pulses of water pressure recorded in boreholes in mountain glaciers. In addition, measurements of the velocity of some polar ice streams demonstrate episodic slip directly (see later). Episodic slip is expected if the glacier bed behaves as frictional material with a yield strength. Thus perfect plastic behavior describes not only the local properties of till but also, in some cases, large sections of the bed.

2. Beds with a continuous till cover and low effective pressures are weak, a consequence of low τ_* values of the till. A "weak bed," again, means that the magnitude of basal drag remains low despite rapid slip motion of the glacier.

3. The basal drag varies considerably from place to place on the beds of polar ice streams (see Figure 8.15). Variations include localized "sticky spots" and systematic regional trends.

Variations presumably reflect differences in τ_*, and, in regions of thin till, varying degrees of bedrock exposure.

The beds of the West Antarctic ice streams offer an extreme example of the effects of weak till (Section 8.9.2.4). Slip rates are 0.2 to 1 km yr^{-1}, despite basal shear stresses of only 2 to 20 kPa over large regions (or 2% to 20% of typical glacier values). Connected to the Ross Ice Shelf, these ice streams provide a unique opportunity to examine large-scale behavior of till beds; the whole system responds to changing forces associated with ocean tides. Two types of responses have been observed. First, the flow of the lower regions of Whillans and Bindschadler ice streams varies diurnally, tens of kilometers inland from the grounding line (Anandakrishnan et al. 2003; Bindschadler et al. 2003). Second, observations on Kamb Ice Stream, a nearly stagnant feature currently freezing to its bed, reveal fluctuations of basal seismicity – small earthquakes generated by sudden slip of the ice over patches of the bed (Anandakrishnan and Alley 1997). The ice itself responds elastically over the short timescales of tidal variations. The observed flow perturbations propagate along the ice streams at only a few meters per second on Bindschadler and Kamb Ice Streams and at about 90 m s^{-1} on Whillans Ice Stream. These rates are much slower than elastic wave transmission in ice (order 10^3 m s^{-1}), and so must reflect a delaying effect from coupling of ice to the till.

The propagation speed along Whillans Ice Stream is comparable to elastic shear wave speeds in the till. The ice stream moves in a stick-slip pattern; the ice moves rapidly (at up to 1 m per hour), but only during events lasting 10 to 30 minutes (Figure 7.15). The observations match modelled behavior of a friction-locked fault with a strength of 4 kPa, a value similar to τ_* measured on samples taken from boreholes (Kamb 2001). Thus the entire till bed here behaves plastically (Tulaczyk 2006).

Much slower propagation occurs along Kamb and Bindschadler ice streams. On Bindschadler, ice flow at the grounding line varies by a factor of three over a day. These variations propagate inland, but the amplitude decreases and there is a time delay. If the till behaved as a perfect

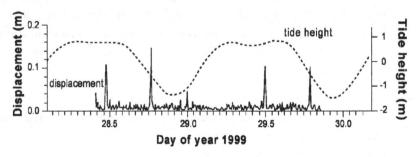

Figure 7.15: Horizontal displacement over 5-minute intervals at one site on Whillans Ice Stream, showing brief periods of active slip. The dashed line shows the modelled ocean tide for a nearby location. Adapted from Bindschadler et al. (2003).

plastic, it would stop responding as the signal decays, in contradiction to the observations. The observed behavior indicates instead a quasi-viscous fluid or a rate-dependent frictional slip. How such behavior arises from the plastic but complex till material is uncertain; some possible contributing processes were discussed earlier.

In systems with weak till beds, such as Whillans Ice Stream, what resisting forces balance the gravitational force that drives flow? Possible resisting forces include viscous drag arising from shearing of ice along the glacier margins or from stretching of ice where it enters the fast-flowing regions (Chapter 8). Alternatively, enhanced stresses on localized strong patches of the bed could account for the resistance (Kamb 1991). Alley (1993) and Stokes et al. (2007) suggested several origins for such *sticky spots*:

1. Hills on the bed that interfere with the ice flow. Compressive normal stress on the upstream side of a hill produces a "form drag" that opposes the flow.

2. Discontinuous till cover. Hard-bed sliding occurs on regions of bare rock set within an otherwise till-covered bed.

3. Efficient drainage or increased elevation lead to locally reduced water pressures and hence patches of strong till (high τ_*).

4. Regions where water freezes to the base of the glacier.

Stokes et al. (2007) summarized the evidence about sticky spots on both modern and past ice streams. The existence of sticky spots under Kamb Ice Stream was first revealed by basal seismicity (Anandakrishnan and Alley 1994). Subsequently, variations in the surface ice flow have been used, together with ice flow models, to map sticky spots beneath large regions of several ice streams. Sticky regions occur in a complex and irregular pattern (Figure 8.15). On the Antarctic ice streams, in general, u_b and τ_b correlate poorly because the rate of slip depends on global controls.

What would a large-scale deformable-bed slip relation look like? Considering the importance of global controls on slip rates, it is sometimes preferable to approach the question by asking instead what relation describes the basal drag. Such relations must depend on the situation: a relation appropriate for a vast and featureless bed with a pervasive cover of till is unlikely to apply to a bed of rough topography and discontinuous till. A simple partitioning of forces provides a starting framework. The basal drag (stress τ_b) averaged over a large region sums the resistance provided by weak regions of till (average stress $\tau_b^{(w)}$) with the resistance on sticky spots (average stress $\tau_b^{(s)}$). If we call the fraction of bed area that is weak or sticky a_w or a_s, then

$$\tau_b = a_w \tau_b^{(w)} + a_s \tau_b^{(s)}. \tag{7.43}$$

One possibility is that weak patches are too weak to support any force and that movement of the glacier over the sticky spots entirely controls the movement over the weak patches. In this case $\tau_b \approx a_s \tau_b^{(s)}$. If the patches are small compared to ice thickness, a plausible relation is $u_b \propto [\tau_b/[a_s N_o]]^m$, with the parameters determined by a process like hard-bed sliding on the sticky regions.

For a different end member – representing the lower part of Whillans Ice Stream where it merges with the floating ice shelf – consider a bed with no sticky spots ($a_w = 1$) and a weak, uniform, plastic till layer. The basal stress cannot exceed τ_*. In this case, as long as slip occurs,

$$\tau_b = \tau_b^{(w)} = \tau_*(N_o). \tag{7.44}$$

Ice deformation across the whole glacier width and the value of τ_* together determine u_b by global control. As discussed in Section 8.6.1, a larger value of τ_* implies that a smaller fraction of the gravitational force is taken up by stresses in the ice. This makes u_b vary inversely with τ_b, a relation opposite to that for hard-bed sliding.

A third possibility is that widespread weak plastic till provides a lot of the resistance to flow, but that sticky spots also contribute significantly. Thus, because of interaction of the ice with the sticky spots, $\tau_b > \tau_*$ on average, and the bed as a whole does not behave like a perfect plastic. The drag exceeds the average till strength by an amount dependent on the rate of slip:

$$\tau_b - \tau_* = [S_o u_b]^{1/m} \quad \text{for} \quad \tau_b \geq \tau_*, \tag{7.45}$$

with m a parameter that may be infinite. S_o defines an effective stiffness for the bed. With m positive and of order one, Eq. 7.45 describes an apparent viscous behavior. Such a relationship applies to the responses of Kamb and Bindschadler ice streams to tidal forcing. Anandakrishnan and Alley (1997) and Anandakrishnan et al. (2003) estimated S_o from a model fit to the observed signals of seismicity and flow rate and found $S_o = 1.5 \times 10^7$ Pa s m^{-1} for Bindschadler Ice Stream and a value ten times higher for Kamb Ice Stream (using an exponent $m = 1$ and $\tau_* = 0$). The higher value for Kamb Ice Stream was expected because, unlike Bindschadler, this ice stream is nearly stagnant and partly frozen to its bed.

It is not known for certain that such quasi-viscous behavior reflects interactions of ice flow with sticky spots, the idea favored here, rather than averaged properties of the deforming till. Numerous analyses of ice dynamics and basal processes have used relationships similar to Eq. 7.45 (or formulae for strain rate that imply such a relation) (e.g., Clarke 1987; Alley et al. 1987a; Hindmarsh 1997). Hindmarsh (1997) suggested that an apparent viscous behavior would arise from the episodic character of slip events – the frequency of such events increasing as τ_b increases and N decreases. The deformation of sea icepacks offers a possible analogue. Individual blocks of ice interact elastically or frictionally, but the motion of sea ice over large regions can be described using an interaction force dependent on the large-scale strain rate (Hibler 1979). On the other hand, the absence of apparent viscous behavior on the broad and homogeneous plain of Whillans Ice Stream suggests that without sticky spots the bed behaves plastically.

A simple relation that can accommodate the variety of behaviors likely for deforming beds is

$$\tau_b = \tau_* \left[1 + \gamma u_b^{1/m}\right], \tag{7.46}$$

with γ a function that depends on the situation, and m a parameter whose value ranges widely; it is positive for quasi-viscous behavior, infinity in the limit of perfect plasticity, and negative for velocity weakening.

7.3.6 Continuity of Till

Given the profound influence of a deformable substrate on glacier flow, it is important to understand how deformable sediments accumulate, persist, or dissipate. A subglacial till layer evolves according to a mass balance or "continuity" relation (Alley et al. 1987a; Cuffey and Alley 1996). Like the continuity relation for a glacier's ice thickness (Section 5.2), the relation for till states that differences between sources and outflows of material produce changes of layer thickness, H_B, at a coordinate:

$$\frac{\partial H_B}{\partial t} = \dot{s}_i + \dot{s}_b - \nabla \cdot \vec{q}_B. \tag{7.47}$$

Here $\dot{s}_i$ denotes the quantity (thickness per unit time) of rock material derived from melt of the basal ice and $\dot{s}_b$ the quantity from erosion of underlying material. Both of these terms can also be negative, representing loss of rock by freeze-on into the ice or deposition beneath the deforming layer. $\vec{q}_B$ is the flux of till (volume per unit time per unit width) conveyed by deformation of the layer. Unfortunately, none of these terms can be predicted. Subglacial erosion is a notoriously difficult process to constrain. The magnitude of till flux depends critically on the distribution of shearing with depth and thus must depend on variables like the variability of water pressure and the coupling between ice and till (Section 7.3.4.2).

Although shearing focuses at or near the ice-till interface, observations show that deformation typically extends to depths of decimeters and probably much farther in a few cases (Table 7.4). Thus till fluxes cannot be negligible, a conclusion supported by additional observations. For example, the outflow of sediment from Whillans Ice Stream accumulates in a deposit at the grounding line; its size suggests a till flux per unit width of about 100 $m^2 yr^{-1}$ (Anandakrishnan et al. 2007). Surges of soft-bed glaciers deposit thick sequences of muds (Boulton et al. 1996). Molded sediment covers the beds of former ice streams (Ó Cofaigh et al. 2002).

Deformable beds need not originate subglacially; glaciers often advance over sediments deposited in previous glaciations, by rivers, or in lakes and marine basins. Without replenishment, such materials will be depleted subglacially over time, by till flux and by the action of water flowing in channels.

Where the ice contacts bedrock, glaciers erode their beds by four mechanisms: quarrying, abrasion, fluvial incision, and dissolution. The first three produce sediment. Quarrying, the separation and removal of rock fragments, occurs when forces exerted by the ice focus on bedrock ledges or other prominences, increasing stresses in the substrate and driving the growth of fractures (Iverson 1991; Hallet 1996). Factors favorable to quarrying likely include extensive cavities, variable water pressures, and abundant preexisting planes of weakness in the rock.

Table 7.7: Inferred rates of erosion by modern glaciers. Data from Hallet et al. (1996), except as noted.

Region	Erosion rate (mm yr^{-1})
Norway and Svalbard	0.05–1
Swiss Alps	0.2–2
Southeast Alaska[‡]	1–15
Iceland, Asia, New Zealand	0.1–10
West Antarctica (Whillans Ice Stream)[†]	0.1
Transantarctic Mountains (Meserve Glacier)[°]	0.001

[†] Anandakrishnan et al. (2007).
[°] Cuffey et al. (2000c).
[‡] original numbers have been divided by 3.5 to account for enhanced short-term erosion during rapid glacier recession (Koppes and Hallet, 2006).

Abrasion occurs when sliding of the glacier drags rock particles across the bed. The rate of abrasion increases with the rate of sliding and the frictional force of contact (Section 7.2.7; Hallet 1979). Fast-flowing water erodes the bed by fluvial abrasion and quarrying but only along narrow channels that convey large quantities of water.

Worldwide, typical rock erosion rates ($\dot{z}_b$) averaged over drainage basins are $\dot{z}_b \sim 0.01 - 0.1 \, \text{mm yr}^{-1}$. Regions with steep slopes or great topographic relief – usually areas of active tectonic uplift – erode more rapidly, typically at $\dot{z}_b \sim 1$ to $10 \, \text{mm yr}^{-1}$. Some glaciers erode more slowly than these typical values; others, more rapidly. Table 7.7 summarizes estimates of glacial erosion rates, calculated by dividing observed sediment fluxes at the fronts of glaciers by the area of the contributing basin. The fastest glacial erosion occurs in Southeast Alaska, a tectonically active setting with very high rates of snowfall. An Antarctic mountain glacier with a frozen bed exemplifies the opposite end of the spectrum.

For a subglacial till layer to persist, the erosion upstream must replenish the material transported downstream by bed deformation. Given erosion rates as shown in the table, is a steady state plausible? Consider a temperate mountain glacier. Choose a cross-section where the glacier width is W_o. The net influx of rock material derived by erosion upstream totals $f \, W_o \, L \, \dot{z}_b$ (volume per unit time), with L the length of the drainage basin upstream, and f an order-one geometric factor related to basin shape. For comparison, the net outflux by till deformation equals $H_B \, \bar{u} \, W_o \, [1 - \nu]$, with H_B the depth of the deforming layer, and $\bar{u}$ the rate of till motion averaged over the layer thickness. Equating the expressions and rearranging gives the rate of till motion that can be sustained at steady state:

$$\bar{u} = \frac{1}{1-\nu} \frac{f \, L \, \dot{z}_b}{H_B}. \tag{7.48}$$

Taking typical values of $H_B = 0.1$ m, $L = 5$ km, $f = 3$, and $v = 0.4$ gives $\bar{u} \approx 25$ to 2500 m yr^{-1} for erosion rates of 0.1 to 10 mm yr^{-1}, respectively. Such velocities are comparable to, or higher than, typical rates of basal slip for mountain glaciers. Thus, maintaining a deforming bed appears to be feasible in a wide range of conditions, largely because of the small thickness of the deforming layer. The fact that sliding accounts for much of the basal motion reinforces the conclusion. Of course, this simple mass balance does not consider other processes like entrainment of eroded rock material into the ice or erosion by subglacial streams (Alley et al. 1997c). We have also not considered how an eroding region of bedrock beneath the upper part of a glacier grades downstream into a region mantled by sediment.

MacAyeal (1992) suggested that accumulation and redistribution of basal till might induce irregular oscillations of the West Antarctic Ice Sheet. Changes in till distribution would act in combination with variations of basal temperature to control the changing pattern of slippery-bed regions over time. MacAyeal's model ice sheet displayed periods of stability interrupted by episodes of fast flow and disintegration; the latter only occurred at times of extensive and continuous till cover. This model, though hypothetical, illustrates the potential for complex behaviors of a coupled ice-till system on geological timescales.

7.3.7 Additional Geological Information

Landforms and deposits of the Pleistocene ice sheets are a potentially rich source of information about systems with deformable beds. We here note only a few highlights from this substantial area of research.

Vast regions of till-covered landscapes, formerly the beds of ice sheets, are sculpted into streamlined drumlin forms and lineations (e.g., Boulton et al. 2001; Ó Cofaigh et al. 2002). Such morphologies reflect interactions between ice flow and the deformable substrate, but how bed deformation contributed to the shaping of landforms remains unclear. Hindmarsh (1997) argued that some of the characteristic forms reflect viscous flow of the till acting under the constraint of mass continuity requirements (Eq. 7.47).

Outcrops of till commonly reveal deformation structures like drag folds and shear plane surfaces and evidence for a variety of styles of deformation (Benn and Evans 1996). Distributed deformation to a large strain homogenizes a till, making it difficult to distinguish from some tills originating by other processes. A subglacial till undergoing internal deformation can transport a large quantity of rock material as till flux, q_B (Alley et al. 1989; Alley 1991a; Boulton 1996). Alley (1991a) argued that such transport provides the most plausible explanation for the large volume and rapid emplacement of till sheets near the margin of the Laurentide Ice Sheet. Hooke and Elverhoi (1996) likewise explained the rapid accumulation of a Pleistocene submarine fan deposit by a large q_B. The contribution of other processes to such large sediment fluxes needs further investigation, however.

A weak deformable bed – such as that beneath the modern West Antarctic ice streams – limits how thick and steep a glacier can be; the gravitational force, per area of bed, cannot greatly exceed the plastic strength of the till. A weak deformable bed explains why parts of the southern Laurentide Ice Sheet were thin and flat, yet flowing rapidly. Geological reconstructions reveal that lobate tongues of ice advanced rapidly across the gentle topography of the North American prairies. Yet gravitational stresses on these tongues were exceptionally low, of order 10 kPa (Clark 1992; Mathews 1974).

7.4 Practical Relations for Basal Slip and Drag

To establish general relations for basal slip and drag from physical principles is a desirable goal, but one that faces insurmountable obstacles posed by the large range of bed properties and the diversity and complexity of interactions between water, sediment, and ice at the glacier bed. To address problems of glacier flow, an empirical approach must be adopted and relations calibrated for application to individual glaciers. In general, values for slip rate (u_b) and basal drag (τ_b) must be determined simultaneously by using an empirical relation between them as the basal boundary condition in a flow model of a large region of a glacier with known patterns of surface velocity and thickness. The following discussion offers some ideas about what types of empirical relations might prove useful.

In the simplest relation, $\tau_b = \psi u_b$, the coefficient ψ takes values from zero to about 20, for stress in kPa and velocity in $m\,yr^{-1}$ (Table 7.2). The lowest values reflect near-flotation conditions. Thus a first modification is to write $\psi \propto N$, the effective pressure (for flotation, $N \to 0$). Allowing for the nonlinearity of ice deformation then suggests that $u_b = k\tau_b^n N^{-1}$, a form of Eq. 7.17. The implied inverse relation between u_b and N matches some but not all observations from mountain glaciers (Figures 7.2, 7.6, and 7.8). An increase of u_b with τ_b has been observed in the time evolution of Variegated Glacier between surges, the best test case so far available (Raymond and Harrison 1987). No single relationship of this form, however, is consistent with the observations of Variegated Glacier, of other mountain glaciers, or of polar ice streams (Bentley 1987). This failure is not surprising, given that (1) the form of the relation is not consistent with theoretical limits for fast sliding with cavities or for homogeneous beds of plastic sediment; (2) varying types and amounts of sediment underly glaciers; (3) N is usually measured at only a few points or not at all; and (4) inferred values for τ_b are uncertain and require accurate modelling of glacier flow (including, in the general case, a correct slip relation itself).

Hard-bed Slip   Theoretical considerations indicate that a relation $u_b \propto \tau_b^p N^{-q}$ cannot be general. At a sufficiently high value of N no cavities form and, on a hard bed, Weertman's original relation or the improved versions of it should apply. Thus, for slow sliding u_b should not depend on N at all. In the case of fast sliding on a hard bed, cavity formation increases with sliding rate. Theory then suggests that the drag τ_b attains a constant value or even decreases as sliding rate increases further (Section 7.2.6). These slow and fast end members can be accommodated with a simple phenomenological relation stating that slip occurs at a rate expected for sliding without cavities ($k_1 \tau_b^m$) plus an enhancement proportional to the nominal size of cavities. Using Eq. 7.23 to define the latter effect suggests that

$$u_b = k_1 \tau_b^m \left[1 + k_2 \frac{u_b}{N^n} \right]. \tag{7.49}$$

Here n denotes the creep exponent for ice and m, k_1, and k_2 are parameters to be determined empirically. In principle k_1 and k_2 depend on bed roughness and ice viscosity and so can be expected to vary in space and time. Setting the exponent to $m = n = 3$ reduces the number of free parameters and matches theoretical expectations when regelation provides negligible drag. (Equation 7.49 with $m = n$ is equivalent to an expression suggested by Schoof (2005, p. 626) from a mathematical analysis of sliding with cavities.) Equation 7.49 implies that – for rapid sliding – the drag cannot rise beyond a limiting value that is independent of u_b but proportional to $N^{n/m}$. Call this limiting value τ_b°.

On a homogeneous and "flat" bed (one with small-scale roughness but no larger-scale hills and valleys), the limiting value τ_b° would occur throughout the region of fast flow. More generally, however, drag concentrates on broad patches of reduced water pressure or rough bedrock and on hills that produce form drag by interfering with the glacier's flow. If the drag τ_b is averaged over an area large enough to encompass such "sticky" regions, a dependence on u_b would probably reappear. Choosing the simplest plausible expression,

$$\tau_b = \tau_b^\circ \left[1 + k_3 u_b^{1/p} \right], \tag{7.50}$$

where k_3 and p represent new parameters (but $p \approx 3$ if ice deformation controls the flow over sticky regions). Empirical calibrations of Eq. 7.50 could treat τ_b°, k_3, and p as three free parameters, or τ_b° could be replaced by values for τ_b obtained from Eq. 7.49 using calibrations of k_1 and k_2.

Soft-bed Slip For a soft bed, the strength of the substrate, τ_*, constrains the drag (Section 7.3.5). The drag can be regarded as the sum of resistances provided by the widespread weak substrate and by flow over sticky spots, as expressed previously by Eq. 7.46:

$$\tau_b = \tau_* \left[1 + \gamma u_b^{1/p} \right] \quad \text{with} \quad \tau_* = fN. \tag{7.51}$$

In one end member the second term on the right is negligible compared to the first and thus $\tau_b \approx \tau_*$, which describes parts of the West Antarctic ice streams (Figure 8.15). If sticky-spot inhomogeneities dominate the resistance, the opposite applies and

$$\tau_b \approx \gamma_o N u_b^{1/p}, \tag{7.52}$$

a form of Eq. 7.17. A candidate location where such an approximation may work is Columbia Glacier, Alaska (Section 8.8); its overall basal drag appears to be much greater than the strength of sediments measured in a borehole. Equation 7.52 is also consistent with behaviors observed at Variegated Glacier, discussed previously. Variegated Glacier is underlain by sediments, at least in part, and its slow phase (between surges) plausibly represents a "sticky" situation.

Note that Eqs. 7.50 and 7.51 are identical in form, except that different factors control the background stress (τ_* or τ_b°) in each case. In both cases, the form of the enhanced-drag term $1 + \gamma u_b^{1/p}$ is hypothetical.

Additional Considerations Calibrated forms of any of the preceding simple relations might prove useful in analyses of glacier flow but need to be evaluated with case studies. Some additional considerations follow.

1. Without extensive borehole studies, the nature of a glacier's bed – sediment, bedrock, or mixed – is usually unknown.

2. Observations of glacier flow and thickness can be used together with a flow model to analyze the balance of forces in a glacier (Section 8.5); this allows maps of both τ_b and ψ (Eq. 7.1) to be drawn. Such analyses have now been completed for large regions of ice streams in Antarctica and Greenland. Using these maps, borehole studies could be designed to measure basal conditions in different regions with different τ_b values. Data so acquired would provide valuable constraints on relations for basal slip and drag.

3. It is now clear that basal water affects sliding rates, not only through effective pressure, but also through the degree of dispersion of water across the bed, which sometimes varies independently from effective pressures (Section 7.2.6.6). None of the empirical relations take account of this.

4. All these relations need values for N. Ideally, N should be measured in boreholes connected to the subglacial water system, but such measurements are not usually available. In most cases, N must be estimated instead from theoretical considerations of the subglacial water system, discussed in Chapter 6. The water pressure, and hence N, depends on the sources of water, the drainage systems along the bed, and the boundary condition at the front of the glacier (air pressure for a glacier terminating on land, but water pressure for a glacier terminating or going afloat in the ocean or a lake). To find N from theoretical relations requires integrating the water pressure gradient up-glacier along the bed from the glacier terminus. The results of such calculations cannot be considered accurate in the absence of comparisons to measured water pressures at a few locations. Measurements show that N beneath warm-based ice sheets and large glaciers is small compared to the ice-overburden. Because water drains along the bed, usually down a pressure gradient, we should always expect subglacial water pressures to be higher than the pressure at the terminus.

For glaciers that end in the sea or a lake, subglacial water pressure is sometimes calculated from

$$P_w = \rho_w g d. \tag{7.53}$$

Here ρ_w is the density of water ($1028 \, \mathrm{kg} \, \mathrm{m}^{-3}$ for seawater), g the gravitational acceleration, and d the depth of the glacier bed below sea or lake level. This is the same as taking

$$N = \rho_i \, g H^*, \tag{7.54}$$

where H^* denotes the height of the glacier surface above buoyancy. Equation 7.54 has been used to calculate N and thus the basal velocity in flow models of the Antarctic Ice Sheet (Budd et al. 1984; Huybrechts 1990a).

The water pressure can be calculated in this way only if the subglacial water has a free connection to the ocean or lake, a plausible assumption near the grounding line or terminus but implausible at 50 or 100 km from it. Moreover, Eq. 7.53 implies a zero subglacial water pressure for a bed above water level. This is generally false. The equation also gives incorrect predictions for places where it can be checked. For example, measured effective pressures were 0.14 MPa at Byrd Station and 0.16 MPa at Camp UpB on Whillans Ice Stream (Alley et al. 1987b; Engelhardt et al. 1990). In comparison, values predicted from Eq. 7.54 are 12.7 MPa and 2.8 MPa, respectively. The water pressures greatly exceed the predicted values. For these reasons, this method of calculating water pressures should be abandoned.

One simple approach, sometimes used to specify sliding velocities in models of mountain glaciers, makes no attempt to calculate water pressures but instead considers the trade-off between motion by sliding and by internal deformation. The latter depends on the integral of the rate of shearing throughout the ice thickness, and so depends on $\tau_b^n H$, whereas sliding depends only on a power of τ_b. The stress varies little from place to place along the centerline of a typical mountain glacier, whereas the ice thickness varies considerably. Thus the fraction of motion due to sliding should be larger in thin-ice regions such as steep ice falls and near the glacier head and terminus. Papers by Budd et al. (1979) and Oerlemans (2001) give examples of sliding parameterized as an inverse function of ice thickness, supported by an assortment of justifications.

Further Reading

Clarke (2005) made a useful summary of many of the topics addressed here; some in greater detail, all expressed with clarity. Fowler (2003) discussed the plastic vs. viscous character of till deformation at the large scale most relevant to glacier dynamics and landforms. Wiens et al. (2008) analyzed the long-period seismic signals produced by stick-slip motion of Whillans Ice Stream. Bennett and Glasser (1996) wrote a fine introduction to glacial geology and geomorphology.

CHAPTER 8

The Flow of Ice Masses

"I was looking upon the counterpart of the great river systems of Arctic Asia and America . . . a plastic, moving, semi-solid mass . . . ploughing its way with irresistible march through the crust of an investing sea."

Elisha Kent Kane, on the Humboldt Glacier, Greenland (1854)[1]

8.1 Introduction

Glacier dynamics, the subject of this chapter and Chapters 11 and 12, deals with the flow and evolution of mountain glaciers, ice sheets, ice streams, and ice shelves. Typical questions are What controls the volume and surface topography of glaciers and how are they affected by accumulation and ablation, ice temperature, and the nature of the bed? How does velocity vary along a flow line, across a glacier, and with depth? How old is ice at different locations? These questions can be answered by applying the methods of continuum mechanics. Conservation statements for mass, momentum, and energy are combined with information specific to the glacier problem: the deformation properties of ice, hydrological and geological influences on basal slip, and the geographical framework of topography, sea level, and climate.

In this chapter we examine the rate and spatial distribution of flow within glaciers (the "flow field"), and what factors control the flow field at a given moment in time. We also examine, as a special case, the topography and flow fields for glaciers in a steady state. How glaciers react to external forcings, such as changes in snowfall and temperature, is discussed in Chapter 11. Chapter 12 discusses the phenomenon of surges, an internal instability.

Material in this chapter accumulates toward a reasonably complete picture of flow dynamics. We first summarize basic quantities relevant to flow: material fluxes, gravitational driving forces, and resisting forces. Second, we consider how flow varies with depth and examine the most basic properties of flows that vary along-glacier. Next we review the fundamental general relations for mass conservation and force balance. This provides a basis for a more complete examination of flow variations along and across glaciers. The variation along a tidewater glacier is considered as a special case. The unique components of ice sheet systems are then reviewed: divides, ice

---

[1] Quoted in E.B. Bolles (1999), *The Ice Finders*.

Copyright © 2010, Elsevier Inc. All rights reserved.
DOI: 10.1016/B978-0-12-369461-4.00008-9

streams, ice shelves, and subglacial lakes. Finally, we discuss the broad features of ice sheet topography, and some of the local effects of bed irregularities on the form of a glacier's surface.

8.1.1 Ice Flux

At a location on a glacier, the quantity of ice conveyed by flow, per unit time, is the flux. Define a coordinate system with x pointing along the glacier in the direction of flow, y pointing across the flow, and z pointing upward. Call the elevations of bed and surface B and S. The integrals through the ice thickness

$$Q = \int_B^S \rho(z)\, u(z)\, dz = \overline{\rho u} H \quad \text{and} \quad q = \int_B^S u(z)\, dz = \overline{u} H \tag{8.1}$$

define the fluxes of mass and volume, per unit width across-flow. Dimensions of q are area per unit time; for Q, mass per unit time per unit length. Here u is the rate of flow along-glacier, ρ the density, and $H = S - B$ the ice thickness. $\overline{u}$ and $\overline{\rho u}$ indicate depth-averaged values. The approximation $Q = \overline{\rho}\, q$ is usually very good. Furthermore, in many cases $\overline{\rho}$ closely approximates the value for fully dense glacier ice, $\rho_i \approx 917\,\mathrm{kg\,m^{-3}}$. Thus, for many glacier dynamics analyses no distinction needs to be made between Q and $\rho_i q$. This assumption may fail, however, on thin ice shelves or mountain glaciers and on small polar ice caps, where firn accounts for a significant portion of the ice thickness.

To find the total outflow of ice from a drainage basin, integrate q or Q across the basin outlet: $\int_Y Q\, dy$, for an outlet of width Y. The mass M of ice in the basin upstream changes if the total outflow differs from the mass added throughout the basin by snowfall and removed by melt and related mass exchange processes. The rate of change of M is

$$\frac{dM}{dt} = \rho_i \int_A \dot{b}_i\, dA - \int_Y Q\, dy, \tag{8.2}$$

with A the area of the basin that drains through section Y, and $\dot{b}_i$ the ice-equivalent specific mass balance (thickness per unit time added to the glacier). Equation 8.2 expresses a principal reason for the importance of glacier flow: a change of flow and hence Q can drive glacier evolution. In Chapters 4 and 5, we examined the processes governing $\dot{b}_i$. In this chapter we examine the processes governing Q.

Table 8.1 gives flux values q at representative locations on the centerlines of some important or extensively studied glaciers. Fluxes vary along a glacier; values here typify the most active regions. The tabulated fluxes vary by more than three orders of magnitude. Both thickness and velocity variations contribute to this range, but the latter are more important. Data sources are Storglaciären (Hanson and Hooke 1994); McCall (Rabus and Echelmeyer 1997); Blue (Meier et al. 1974); Worthington (Harper et al. 2001); Taylor (Kavanaugh et al. 2009b); Columbia

Table 8.1: Measured fluxes for some major and minor glaciers.

Glacier	Type	Surface velocity $(m\,yr^{-1})$	Ice thickness (m)	Flux, q $(10^4\,m^2\,yr^{-1})$	Distance from head (km)
Storglaciären	mountain	15	200	0.3	2
McCall	mountain	15	170	0.26	3
Blue	mountain	50	250	1.25	2
Worthington	mountain	75	200	1.5	4
Taylor	ice sheet outlet	17	1000	1.7	50
Columbia (year = 1977) (year = 1995)	mountain, tidewater	730 2900	950 800	70 230	52
Jakobshavn (40 km above g.l.)[†] (7 km above g.l.)[†]	ice sheet outlet	1050 3800	2500 1760	250 670	500 530
NEGIS	ice sheet outlet	300	1200	36	620
Whillans	ice sheet outlet	440	1030	45	500
Pine Island	ice sheet outlet	1500	1800	270	300
Jutulstraumen	ice sheet outlet	550	1300	70	550

[†] g.l. = grounding line.

(O'Neel et al. 2005); Jakobshavn (Clarke and Echelmeyer 1996); NEGIS (Joughin et al. 2001); Whillans Ice Stream (Engelhardt and Kamb 1998); Pine Island (Thomas et al. 2004a); Jutulstraumen (Rolstad et al. 2000).

Storglaciären, McCall, Blue, Worthington, and Columbia all occupy valleys in mountainous terrain. Columbia is much bigger than the others and ends in tidewater. Storglaciären is in northern Sweden, McCall in northern Alaska, Blue in coastal mountains of Washington State, and Worthington and Columbia in coastal southern Alaska. Taylor Glacier descends through the Transantarctic Mountains; it drains a small dome (area $\sim 500\,km^2$) on the flank of the main East Antarctic ice sheet. The rest are major outlets of the polar ice sheets: Jakobshavn and the Northeast Greenland Ice Stream (NEGIS) in Greenland, Whillans Ice Stream and Pine Island Glacier in West Antarctica, and Jutulstraumen in East Antarctica. All five of these outlets are "major" in the sense that they drain a large area extending from the central divide all the way to the ice sheet margin. Their total flux $\int_Y q\,dy$ constitutes a significant fraction of the total outflow from their ice sheets.

What determines the flux q? We can address this question in two ways. One is the approach based on mechanics. Gravitational forces cause the ice to flow, while flow modulates resisting forces within the ice and along boundaries. Flow must be fast enough that the gravitational and resisting forces balance. How much resisting force arises from a given rate of flow depends on the properties of ice, the properties of the substrate, and the geometry of the glacier. This approach is always valid and is essential for interpretations, but uncertainties often make it useless for predictions of flux – especially if the glacier slips over its bed.

The second approach to understanding flux values recognizes that a glacier tends toward a steady state, adjusted to its role in the hydrological cycle. Snowfall adds mass to the glacier, which is eventually lost as meltwater and icebergs. The fluxes convey ice from locations of net snowfall to locations of net loss. If the fluxes are too small to transport all the snowfall, the glacier will build up, as indicated by Eq. 8.2. The forces and the ice thickness then increase until transport rates match the snowfall. Thus, fluxes tend to be determined by the mass input to the system (a function of the snowfall, melt, and basin geometry) and by the geometry of the outlet. This second approach fails dramatically if the glacier is far from steady state – for example, when a glacier surges or responds rapidly to a climate change. Nonetheless, this approach explains many observed characteristics of glacier flow.

8.1.2 Balance Velocities

In many situations, even when glaciers respond to changing climate, the rate of mass gain or loss dM/dt is small compared to the other terms in Eq. 8.2, provided we consider averages over a period of years (the short-term variability of snowfall and melt must be averaged out). Thus setting $dM/dt = 0$ gives an approximation for the rate of flow, averaged over depth, based on the equality of inputs and outflow:

$$\rho_i \int_{\mathcal{A}} \dot{b}_i \, d\mathcal{A} = \rho_i \dot{b}_i \, X \, W = \overline{\rho} \, H \, U_\text{B} \, Y. \tag{8.3}$$

Here H is the thickness and Y the width of a section of glacier between two adjacent flow lines (Figure 8.1). The section drains a basin of area $\mathcal{A}$, with length X, average width W, and average specific mass balance $\dot{b}_i$ (the $\mathrm{m\,yr^{-1}}$ of ice added to the surface each year). This relation means that the annual flux of ice through a cross-section, perpendicular to the surface and to the flow

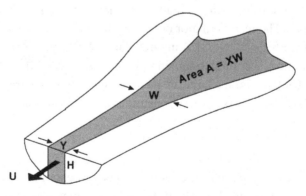

Figure 8.1: Situation sketch for derivation of balance velocity; the mass accumulated in area $\mathcal{A}$ flows through the the part of the glacier's cross-section bounded by H and Y. The glacier should be regarded as wide; in a narrow valley, the flux gate does not necessarily have straight and vertical sides.

direction, must equal the total annual accumulation upstream. This assumption defines U_B, the depth-averaged *balance velocity*, indicated in Eq. 8.3 (Budd et al. 1971). Later in this chapter we show that depth-averaged velocities and surface velocities typically differ by less than 20%. With an estimate for this difference, U_B can be converted to a *surface balance velocity* for direct comparison to measurements of surface velocity. This is a way to assess the overall state of a glacier; if measured velocities and balance velocities differ significantly, the glacier must be growing or shrinking.

Because balance velocities approximate true velocities in many glacier systems, a map of balance velocities reveals a lot about the structure of flow fields, including typical rates and dominant spatial patterns. Such a map is constructed from data on surface topography, ice thickness, and specific balance by using Eq. 8.3. To delimit the drainage basin upstream of each location, flow vectors are drawn perpendicular to surface elevation contours and pointing down-slope, a usually accurate assumption regardless of whether or not the glacier is in a steady state (Section 8.2.1).

Figures 8.2a and 8.2b show balance velocity maps for Greenland and Antarctica. U_B is small in the interior regions and increases toward the margins. Three factors drive the increase, with reference to Eq. 8.3: the area drained ($\mathcal{A} = XW$) increases, the ice thickness (H) decreases, and the specific balance ($\dot{b}_i$) is positive. Near the margins in Greenland, however, $\dot{b}_i$ decreases and turns negative. The maps show a striking feature of the ice sheets; toward the margins, the flow concentrates into narrow, fast-flowing zones. These are the ice streams or outlet glaciers, discussed in Section 8.9.2.

A typical mountain glacier is divided into an upper accumulation zone with $\dot{b}_i > 0$ and a lower ablation zone with $\dot{b}_i < 0$ (Section 4.2.5). In the accumulation zone, in a steady state, the annual flux of ice through a cross-section must equal the total annual accumulation upstream. Similarly, the flux through a cross-section in the ablation zone must equal the mass of ice lost by ablation downstream from the cross-section. Thus the flux must increase steadily from the head of the glacier to the equilibrium line and decrease from there to the terminus, where it equals zero unless the glacier calves or advances. The pattern of U_B is similar, but modified by variations in the glacier's thickness and width.

8.1.3 Actual Velocities

Figure 8.3 shows measured surface velocities along the centerline of McCall Glacier, a valley glacier in northern Alaska. Broadly, its pattern is the one expected from balance velocity. Detailed analysis shows, however, that the velocity reaches a maximum at about 2 km down-valley from the position of maximum flux, because the glacier thins and narrows. Furthermore, the measured velocities near the terminus are smaller than needed for balance in the current climate, and the glacier is retreating.

Figure 8.4 maps the measured surface velocities of the Northeast Greenland Ice Stream. Overall, the ice stream looks similar to its equivalent on the balance velocity map (Figure 8.2a).

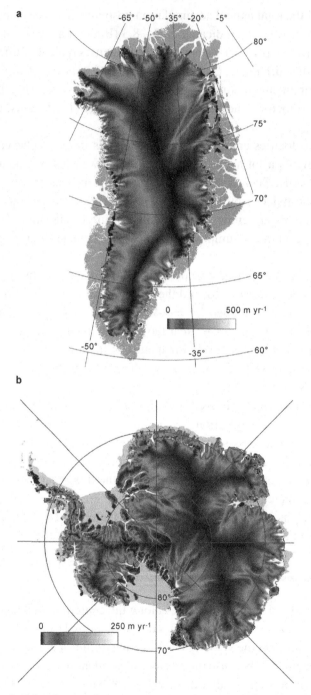

Figure 8.2: Balance velocities for (a) the Greenland Ice Sheet (Bamber et al. 2000a) and (b) the Antarctic Ice Sheet (updated from Bamber et al. 2000b by J. Bamber). Images courtesy of J. Bamber. (Refer to the insert for a color version of this figure)

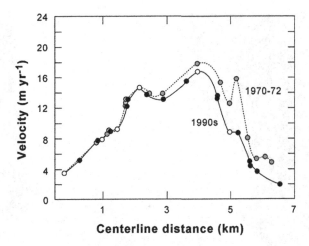

Figure 8.3: Mean annual velocities along the centerline of McCall Glacier, Alaska, in two periods, 1970–1972 and 1993–1995. Zero marks the glacier head. Velocities decreased between the two periods as the glacier thinned and retreated. Open circles on the 1990s curve are estimated from calculations; the others are measured. Adapted from Rabus and Echelmeyer (1997).

In fact, analyses using the flux-gate method discussed in Chapter 4 indicate this ice stream is close to a state of balance, except at its lower end (Joughin et al. 2001). Flow generally increases along the whole length of the ice stream – much of the mass is lost by calving at the terminus rather than by melt in the ablation zone. Across the ice stream, velocities vary rapidly in the side margins and define a broad maximum in the middle. Factors controlling the variations along-flow (the *longitudinal profile*) and across-flow (the *transverse profile*) are discussed later in this chapter (Sections 8.6 and 8.7).

Consider again the observed fluxes of different glaciers listed in Table 8.1. Much of this variation can be explained by Eq. 8.3, using HU_B as an estimate for flux q. This table also shows the distance X from the point of measurement to the head of the glacier, as a proxy for the basin area. In general, q increases with X. (Recall that we selected the point of measurement, and hence X, to characterize the active parts of each glacier; slow-flowing regions near the heads of the glaciers and near the fronts of land-terminating glaciers are not represented.) The smallest glacier, Storglaciären, has the smallest flux. The ice streams draining large basins all have large fluxes. Despite a larger basin, however, the flux of NEGIS is much smaller than that of Jakobshavn. In part, this reflects how the ice entering Jakobshavn converges into a narrow channel (large W/Y). It also reflects the climatic setting; NEGIS drains the dry (small $\dot{b}_i$) northeastern sector of Greenland, whereas Jakobshavn drains the wetter west-central region. Again, the basins of Blue Glacier and Storglaciären are nearly equal in size but Blue's flux is four times larger, primarily because of abundant snowfall in its coastal mountain setting. Worthington and Columbia also catch large quantities of snow. However, Columbia drains a much larger basin than the other mountain glaciers. In flux, Columbia matches some of the large

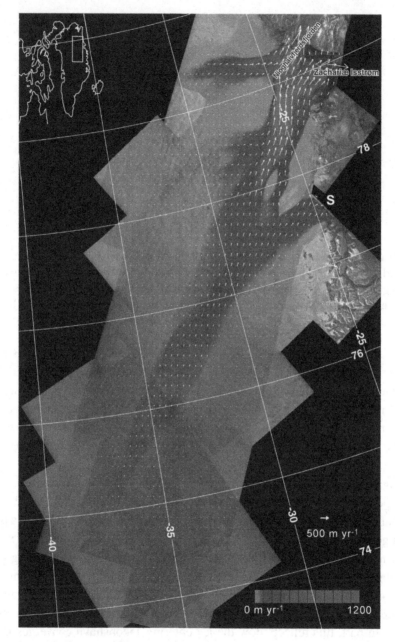

Figure 8.4: Surface velocity map of the Northeast Greenland Ice Stream and surroundings, measured by InSAR (Joughin et al. 2001). Inset panel at upper left shows the location of the figure. Image courtesy of I. Joughin and used with permission of the American Geophysical Union, *Journal of Geophysical Research*. (Refer to the insert for a color version of this figure)

polar ice streams. In contrast, at a point where its basin is similar in size to Columbia's, the polar Taylor Glacier carries a flux only a few percent as large as Columbia's. Taylor Glacier drains a dry region of Antarctica, where snow accumulates at only about 1% of the rate in Columbia's basin in coastal Alaska.

The two flux values for Columbia Glacier represent two different years but the same location. After a period of stability in the 1970s, the terminus of Columbia began to retreat rapidly in the early 1980s. None of the factors controlling balance velocities can explain the three-fold increase in flux from 1977 to 1995. This change is a noteworthy example of a transient response. So too for Pine Island Glacier. This major outlet of West Antarctica drains a large basin, so its flux is expected to be large at steady state. But, in addition, this zone of Antarctica is currently losing mass because the flux of Pine Island Glacier has increased by some 50% in recent decades (Thomas et al. 2004a).

Velocities of the glaciers in Table 8.1 range from order 10 m yr^{-1} to a few km yr^{-1}. The very rapid flows originate in different ways in different places. In some cases (Whillans, Columbia), the bed is slippery and the glacier glides rapidly along its base (a slippery bed is said to be "well lubricated"). In other cases, ice deformation generates the rapid motion; the best example is the upper location along Jakobshavn, where large stresses, thick ice, and warm ice at depth all contribute to rapid shearing.

In general, the ice flow velocity $\vec{u} = [u, v, w]$ at a point is controlled by processes in a broad region of the surrounding glacier. Stresses, ice mass geometry, ice creep properties, and bed properties all play a role. With regard to the vertical variation of flow, the part of the motion due to basal slip, $\vec{u}_b$, is distinguished from the part due to *internal deformation* of the ice:

$$\vec{u}(z) = \vec{u}_b + \int_B^z \frac{\partial \vec{u}}{\partial z} \, dz. \tag{8.4}$$

The second term on the right is called the *deformational velocity* or *shear velocity*. Note, however, that internal deformation occurs not only in association with vertical gradients of the flow but also with horizontal gradients, $\partial \vec{u}/\partial x$ and $\partial \vec{u}/\partial y$. Such deformations completely control the flow of glaciers in some situations. The surface velocity, $\vec{u}(S)$, is by far the easiest component of flow to measure, and in fact data on flow rates at depth are still rare.

8.1.4 How Surface Velocities Are Measured

The most accurate measurements are obtained by installing on the glacier a network of markers, usually metal poles, and surveying their positions. Repeating the survey at a later time gives displacements and hence velocities. Until the early 1990s, surveys used optical techniques, a process notorious for producing more cold fingers than data points – or, in bad weather, for producing no data at all. Optical surveys still sometimes have a useful role for low-cost, frequently repeated measurements. But the advent of the Global Positioning System (GPS) technique has revolutionized surveying; now, positions of survey markers can be determined by placing a receiving antenna on the marker and recording, for a period of minutes, the signals from GPS network satellites. The signal from each satellite includes information about the time of transmission and the orbit; combining information from several satellites fixes the receiver location by trilateration. Because of imprecision of the receiver's clock, signals from at least four satellites are usually required to "trilaterate" accurately. By acquiring data simultaneously at several survey markers and a base station located on bedrock near the glacier, position accuracies of better than 1 cm are typically achieved. Lower but still impressive accuracies can be achieved without a nearby base station; thus the technique works even in the central parts of the ice sheets. Glacier movement can also be measured as a continuous function of time, at a point, by leaving a GPS receiver on-site for continuous signal recording.

Such survey techniques yield surface velocity measurements with great accuracy, but only at points in an established network. A more complete map can be acquired with two techniques of remote sensing (Massonet and Feigl 1998; Bamber 2006). Glacier flow carries features such as crevasses and small surface hummocks. In *feature tracking*, images of the glacier surface are used to map their locations. Comparing locations in successive geo-located images gives displacements and hence velocities. Usable images include photographs from aircraft and ground stations and radio- and visible-wavelength scenes acquired by satellites. The second method is InSAR: *Interferometric Synthetic Aperture Radar*. A satellite transmits microwave radiation and records the pattern of reflections from the surface, along a swath. Greater return times correspond to greater distances from the satellite. Undulations of the surface modulate the phase of the returned signal; for example, reflection from a hillside facing the satellite compresses the signal. Now suppose the satellite repeats the measurement, a few days later, from exactly the same orbit. If the undulations have moved – because they are carried by a flowing glacier – the pattern of phases will have shifted. Differences between the earlier and later phase patterns give a quantitative measure of the displacements and hence velocities in the direction viewed by the satellite. (To compute the vector components of velocity, a similar comparison is made with data acquired from an intersecting orbit.) In practice, the repeat measurements are not taken from identical orbits. Thus phase differences between observation pairs also record the topography of the surface itself. Enough repeat measurements must be acquired to solve for both topography and velocity. Alternatively, topographic data acquired independently can be assimilated and the radar data used to calculate velocities alone. Because phase information is cyclic rather than unique, the absolute signal – the total velocity – can only be determined by a spatial summation of cycles from a point of known velocity, a process known as *phase unwrapping*. (Whereas a child might prefer to unwrap gifts, a glaciologist delights in unwrapping phases.)

The InSAR method will fail if melt or other processes change the surface enough to disrupt the coherent pattern of phase differences arising from flow and topography. Noisy data cause the same problem. Overall, InSAR techniques offer a remarkably complete and accurate view of flow in large, previously obscure zones of the ice sheets, especially the major ice streams. Figures 8.4 and 8.28 illustrate examples.

8.2 Driving and Resisting Stresses

Glaciers move slowly enough that accelerations cause negligible inertial forces. A glacier is effectively in static equilibrium, even while in motion, so forces can be assumed to sum to zero. A *force balance* analysis examines the terms in this sum, usually in the direction along-glacier, parallel to flow.

8.2.1 Driving Stress and Basal Shear Stress

Gravity pulls a glacier vertically downward. This vertical force causes a primarily *horizontal* flow of ice for two different reasons: because it sets up pressure gradients in the glacier and, in some cases, because the glacier rests on a sloping bed. In either case, the down-glacier force effect of gravity can be estimated from the same formula, as we now show.

As a first model of a glacier, consider a parallel-sided slab of ice, thickness H, resting on a rough plane of slope α (Figure 8.5a). The length and width of the slab are long compared with H. Consider a column of ice perpendicular to the plane and of unit cross-section. The weight of the column has a component $\rho g H \sin \alpha$ parallel to the plane, where ρ is the density and g the gravitational acceleration. This component is the *driving stress*, τ_d. Glaciologists refer to τ_d as a "stress" because it is a force per unit area of the glacier. However, the force in question acts on the whole column, not on a particular surface, so τ_d is not a true stress. For equilibrium, resisting forces must balance the driving stress. On most glaciers, the largest resisting force is the *basal drag*, the shear stress τ_b across the base of the column (also referred to as the *basal shear stress* or the *basal resistance*). Thus

$$\tau_d = \rho g H \sin \alpha \quad \text{and} \quad \tau_b = f' \tau_d, \tag{8.5}$$

where f' denotes a number usually of order one.

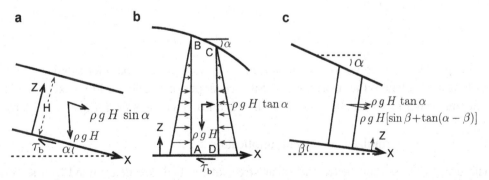

Figure 8.5: Gravitational forces composing the driving stress: (a) the down-slope component of weight, (b) the pressure gradient force, and (c) the combination.

An ice mass with surface slope α, but resting on a horizontal plane, better represents an ice sheet or the lower zone of a large mountain glacier (Figure 8.5b). Consider the equilibrium of a column ABCD of unit thickness in the direction normal to the xz-plane. Because ice behaves as a fluid, a normal pressure acts on AB and CD with a magnitude given, to a very good approximation, by the hydrostatic head. (Note that shearing in the ice does not affect this approximation provided that the planes of shear are horizontal.) The pressure increases from zero at B to $\rho g H$ at A, where $AB = H$. Integrating the pressure gives the normal force on AB, or $\frac{1}{2}\rho g H^2$. The normal force on CD therefore amounts to $\frac{1}{2}\rho g H^2 + d/dx\left[\frac{1}{2}\rho g H^2\right]\delta x$. The difference between the two is a horizontal force to the right (the driving stress), again balanced in part by the basal drag:

$$\tau_{\mathrm{d}} = -\rho g H \frac{dH}{dx} = \rho g H \tan\alpha \quad \text{and} \quad \tau_{\mathrm{b}} = f'\tau_{\mathrm{d}}, \tag{8.6}$$

where, as before, f' denotes a number usually of order one. A real glacier does not match either of these simple cases, but the precise shape has little influence on the force that drives flow. At depth, in any glacier, there is always a horizontal gradient of hydrostatic head proportional to $-dS/dx = \tan\alpha$, where S is elevation of the ice surface and α the surface slope. This means that a vertical column will always be pushed by a *horizontal* driving stress of magnitude $\tau_{\mathrm{d}} = \rho g H \tan\alpha$, regardless of the bed slope. Gravity therefore always pushes a glacier horizontally, in the direction of the downward surface slope. The horizontal component of basal drag and other resisting forces must balance the horizontal τ_{d}.

Now consider the balance of forces *parallel* to the bed on a wedge-shaped section with sides perpendicular to the glacier bed (Figure 8.5c; Nye 1952b). Assume a small bed slope, and measure H perpendicular to the bed. In the downhill direction, the component of weight $[\rho g H \sin\beta]\delta x$ adds to the hydrostatic gradient force (Eq. 8.6), $-\rho g H [dH/dx]\delta x$, to give a driving force $\tau_{\mathrm{d}}\delta x$. Again, τ_{d} is balanced in part by the uphill basal drag $\tau_{\mathrm{b}}\delta x$:

$$[\rho g H \sin\beta]\delta x - \rho g H [dH/dx]\delta x = f'\tau_{\mathrm{b}}\delta x. \tag{8.7}$$

But for small angles $dH/dx = \beta - \alpha$ and $\sin\beta = \beta$. Thus

$$\tau_{\mathrm{d}} \approx \rho g H \alpha \quad \text{and} \quad \tau_{\mathrm{b}} = f'\tau_{\mathrm{d}}. \tag{8.8}$$

Therefore, provided that the slopes are small, τ_{d} is the same as for a parallel-sided slab or a flat bed, as long as α refers to the surface slope. Surface slopes on glaciers seldom exceed $20°$, and most are much smaller. For a $20°$ slope, $\alpha = 0.35$ and Eq. 8.8 differs from Eq. 8.5 or 8.6 by less than 5%.

These important formulae have several implications:

1. The driving stress, and hence the shear stress at the bed, are determined by the surface slope. Ice therefore tends to flow in the direction of maximum surface slope even if the bed slopes in the opposite direction. This agrees with observations. Ice flowing from central

West Antarctica to the coast at Pine Island Bay crosses a basal rise with a height of nearly 1 km. Mountain glaciers erode and flow through overdeepenings, which impound lakes when the ice retreats. For example, Pleistocene glaciers scoured the bottom of Washington State's Lake Chelan to nearly 500 m below the elevation of its outlet. (In detail, however, the directions of flow and basal drag can deviate from the surface slope because different parts of the surrounding glacier can push or pull the ice in different directions. This effect only occurs on distance scales of a few ice thicknesses; see Section 8.7.2.)

2. It follows that flow lines can be determined from a contour map of the ice surface, provided that "slope" is interpreted as the average value over distances of several times the ice thickness. Small-scale features such as hummocks do not affect the flow.

3. The driving stress can be calculated from measurements of ice thickness H and surface slope α. Using $\rho = 917 \, \mathrm{kg \, m^{-3}}$, Table 8.2 gives estimates for τ_d at the same sites listed in Table 8.1. As with fluxes, τ_d varies along a glacier, but these sites are chosen to be typical of broad zones. (Note, however, that typical stresses can differ greatly from zone to zone on the same glacier.)

   The characteristic magnitude of τ_d is about 100 kPa (or 1 bar); the tabulated values average 140 kPa and represent the whole range observed for major glaciers, although driving stresses decline to zero in zones adjacent to ice shelves.

   The thick line separates the table into slow-flowing ($<100 \, \mathrm{m \, yr^{-1}}$) and fast-flowing glaciers. For the slow glaciers, driving stresses tend to exceed 100 kPa but do not deviate

Table 8.2: Measured driving stress.

Glacier	Type	Surface velocity ($\mathrm{m \, yr^{-1}}$)	Driving stress (kPa)	Distance from head (km)
Storglaciären	mountain	15	130	2
McCall	mountain	15	160	3
Blue	mountain	50	170	2
Worthington	mountain	75	80	4
Taylor	ice sheet outlet	17	140	50
Columbia	mountain, tidewater (mean of 2 years)	1830	190	52
Jakobshavn	ice sheet outlet (40 km above g.l.)[†] (7 km above g.l.)[†]	1050 3800	410 360	500 530
NEGIS	ice sheet outlet	300	23	620
Whillans	ice sheet outlet	440	15	500
Pine Island	ice sheet outlet	1500	20	300
Jutulstraumen	ice sheet outlet	550	210	550

[†] g.l. = grounding line.

from it very much. Indeed, values in the range 50 to 200 kPa have been measured on a great number of glaciers. On the ice sheets, a range 30 to 120 kPa applies to the broad "flank" regions upstream of, and between, the fast-flowing outlets (Cooper et al. 1982). For the fast-flowing glaciers, however, the driving stresses vary by a factor of about 30.

4. Measured driving stresses provide a rough estimate for the basal drag τ_b because, in most situations, $0.5 < f' < 1.5$. (The exception is floating ice shelves[2] and some of the ice streams that feed them, for which $f' \approx 0$.) Because τ_b values estimated from $\tau_b = \tau_d$ normally lie in the narrow range 50 to 200 kPa, a reasonable approximation is to regard slow glacier flow as a process of *perfectly plastic* deformation (Section 3.4.4.6), with τ_b everywhere equal to τ_o, the *yield stress* for ice, usually set to $\tau_o = 100$ kPa. However, this approximation should not be used for many polar ice streams – the tabulated values for τ_d range from 15 to 410 kPa.

5. If we assume perfect plasticity, we can write

$$H = \frac{1}{f'} \frac{\tau_o}{\rho_i\, g\alpha}. \tag{8.9}$$

If $\tau_o = 100$ kPa and $f' = 1$, then $\tau_o/\rho_i g = 11$ m. With this value, a crude estimate of ice thickness can be obtained from measurements of surface slope alone.

6. Equation 8.9 implies a roughly constant value for $H\alpha$. Thus a glacier is relatively thin where the surface is steep and thick where the surface flattens. Inflection points on the surface correspond to the tops of bedrock hills.

7. A constant value for $H\alpha$ implies certain overall shapes for glacier profiles. The surface slope α is the sum of the bed slope β and the ice thickness gradient, dH/dx. Consider a glacier resting on a bed with a total elevation range of ΔB. If the glacier is thin compared to ΔB, a common situation for mountain glaciers, then $\alpha \approx \beta$, and the glacier can be pictured crudely as a long slab of nearly uniform thickness. On the other hand, if the glacier is thick compared to ΔB, then $\alpha \approx dH/dx$ and the product $H\, dH/dx$ is nearly constant. Roughly, the glacier takes the form of a parabola: H varies as $\sqrt{x}$, with x the distance up-glacier from the terminus (see Section 8.10.1).

8. Inspection of the derivations for τ_d shows that, with a constant density, the driving stress $(\hat{\tau}_d)$ acting at a depth $H - z$ within the glacier increases linearly from zero at the surface to τ_d at the base:

$$\hat{\tau}_d(z) = \tau_d\left[1 - \frac{z}{H}\right]. \tag{8.10}$$

---

[2] The surface of an ice shelf usually slopes, so it has a driving stress as defined by τ_d. Where a shelf's surface is flat, however, $\tau_d = 0$ and f' is undefined.

The plastic approximations – that $\tau_b \approx 100\,\text{kPa}$ and $H\alpha \approx$ constant – are appropriate for many glaciers. On the other hand, the $30\times$ range of τ_d values for the large and fast-flowing glaciers shows that to achieve even a crude understanding of such features we must consider additional factors. Most important are the nature of the resisting forces and their relation to flow. Understanding these factors will also give mechanistic constraints on values for the parameter f'.

8.2.2 Additional Resisting Forces and the Force Balance

In Eqs. 8.5 to 8.8, we indicated that the equality between basal shear stress and driving stress is only approximate; other forces act on a section of glacier. To understand the situation, consider a rectangular section of glacier delimited by vertical sides as in Figure 8.6, with length ΔX, width ΔY, and axes as shown. The base is irregular, as it conforms to the subglacial topography. The glacier flows in direction x.

We now summarize the different forces acting on the section. Because the ice is fluid, the weight of overlying ice produces compressive normal stresses that push against all of the vertical sides. We will use the term *overburden stress* for this static, isotropic component and denote it Λ. It is one part of the true stress σ. Flow and deformation of the ice produce additional components of stress – shear stresses and normal deviatoric stresses – referred to as *resistive stresses*. The true state of stress is the sum of overburden and resistive components. Integrating stresses over the areas bounding the section gives forces acting on it. We are concerned here only with forces parallel to x, along-glacier.

Gravity effectively pushes the glacier in direction x. The sources of the push were explained in Section 8.2.1. One part of the push is the down-slope component of the section's weight, if the bed slopes: $Mg\sin\bar{\beta}$, for slope $\bar{\beta}$, gravitational acceleration g, mass $M = \rho g\overline{H}\cdot\Delta X\Delta Y$, and mean thickness $\overline{H}$. The other part of the push arises from the difference in overburden stresses pressing against the down-glacier face of the section (labelled B) and its opposite, up-glacier

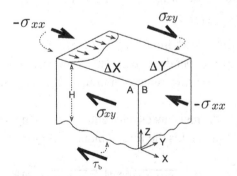

Figure 8.6: Stresses acting on a section of glacier. (Gravity acting on the interior is not shown.) Normal stresses are defined as positive in tension. Note that σ_{xx} sums an overburden component Λ_{xx} with terms related to stretching and compressing of ice. Specifically, $\sigma_{xx} = \Lambda_{xx} + \tau_{xx} - \tau_{zz} = \Lambda_{xx} + 2\tau_{xx} + \tau_{yy}$ (see Section 8.5.3).

face (call it B′). The net driving force is thus

$$\overline{\tau}_d \cdot \Delta X \, \Delta Y = \rho g \overline{H} \, \sin \overline{\beta} \cdot \Delta X \, \Delta Y + \int_{B'} [-\Lambda_{xx}] \, dy \, dz - \int_{B} [-\Lambda_{xx}] \, dy \, dz, \tag{8.11}$$

where $\overline{\tau}_d$ indicates the average of driving stress in the region, and, for consistency with other stresses, the overburden stress Λ_{xx} is defined as positive in tension (thus $\Lambda_{xx} < 0$ in the glacier). Acting up-glacier are three types of resisting forces:

1. All the resistances on the base combine to make the basal drag force of magnitude $\overline{\tau}_b \cdot \Delta X \Delta Y$.

2. Shearing of ice on the lateral walls of the section (the face labelled A and its opposite A′) generates a resistance called the *side*, *lateral*, or *wall* force. Specifically, for the axes shown in the figure, the up-glacier force is

$$F_{wall} = \int_{A} \sigma_{xy} \, dx \, dz - \int_{A'} \sigma_{xy} \, dx \, dz. \tag{8.12}$$

(The shear stress σ_{xy} is identical to the corresponding deviatoric stress component τ_{xy}.)

3. Stretching and compression of ice at the front and back of the section (again, the face labelled B and its opposite B′) generate an up-glacier *longitudinal resistive force* of magnitude

$$F_{long} = \int_{B'} [\sigma_{xx} - \Lambda_{xx}] \, dy \, dz - \int_{B} [\sigma_{xx} - \Lambda_{xx}] \, dy \, dz. \tag{8.13}$$

Notice that the force on face B pulls down-glacier if the integrand $\sigma_{xx} - \Lambda_{xx} > 0$, because positive stress indicates tension. (We will show later that the resistive stress $\sigma_{xx} - \Lambda_{xx}$ depends on how fast the ice deforms; it can be written in terms of deviatoric stresses as $\sigma_{xx} - \Lambda_{xx} = \tau_{xx} - \tau_{zz} = 2\tau_{xx} + \tau_{yy}$.)

Splitting the true stress (σ_{xx}) into overburden (Λ_{xx}) and resistive ($\sigma_{xx} - \Lambda_{xx}$) components proves useful because it separates terms dependent on gravity from terms dependent on ice deformation (van der Veen and Whillans 1989a). In reality, of course, the true stress is the one that acts in the ice.

For equilibrium, the sum of the resistances must equal the driving force. Equating and dividing through by the area $\Delta X \Delta Y$ thus yields the force-balance relation

$$\overline{\tau}_d = \overline{\tau}_b + \frac{1}{\Delta X \, \Delta Y} F_{wall} + \frac{1}{\Delta X \, \Delta Y} F_{long}. \tag{8.14}$$

All of these forces are normalized to the map-plane area and so have units of stress. We will usually refer to the three terms on the right-hand side as *drags* or *resisting stresses*. In general, however, the lateral and longitudinal terms take either positive or negative values; in some situations they enhance the driving force rather than the resistance. As a convenient notation, we rewrite Eq. 8.14 as

$$\tau_d = \tau_b + \tau_w + \tau_L, \tag{8.15}$$

where τ_w is the *side* or *wall drag* and τ_L the *longitudinal drag*. Another notational convenience, already introduced in the previous section, is to define a correction factor f' so that $\tau_b = f'\tau_d$ (as in Eq. 8.9). Thus

$$f' = 1 - \frac{\tau_w + \tau_L}{\tau_d}. \tag{8.16}$$

On most mountain glaciers and throughout broad regions of the ice sheets, f' is within the range 0.5 to 1.5. On valley glaciers, in particular, f' normally falls in the range 0.5 to 0.9 if averaged along the valley (Section 8.6.2), but fluctuates more widely over short distances (Section 8.7.2). On ice shelves and some polar ice streams, however, τ_d greatly exceeds τ_b and f' approaches zero (Sections 8.9.2 and 8.9.3).

In the limit of small ΔX and ΔY, these relations give the force balance for a vertical column of ice, a topic developed in Section 8.5.3. There we will see that one additional term appears in Eqs. 8.14 and 8.15, called the *bridging effect*; it can be neglected at length scales greater than about one ice-thickness. That analysis also leads to precise definitions of driving and overburden stresses and shows how their values depend on the choice of coordinate system.

8.2.3 Factors Controlling Resistance and Flow

What factors explain the different driving stress values for fast-flowing glaciers shown in Table 8.2? How much of the driving stress is balanced by basal drag? The relative magnitudes of the terms in the force balance, Eq. 8.15, define the *force partitioning*. Table 8.3 lists estimates of the force partitioning for a small mountain glacier (McCall) and some fast-flowing glaciers: the large tidewater mountain glacier (Columbia) and several of the ice sheet outlets. For McCall, Jakobshavn, and Columbia, the values refer to a strip along the central flow line. For the others, the values refer to averages for large rectangular regions, spanning most of the glacier width and extending tens of kilometers along-flow. The methods for obtaining these estimates are discussed in Section 8.5.3.2. The values are approximations; forces are not directly measured in any of the studies.

Ice is fully grounded at all of these sites, except two: the "grounding zone" of Jutulstraumen and the "ice plain" of Whillans Ice Stream. Both of these sites lie within transition zones between grounded ice and floating ice shelves. (The term *ice plain* refers to a region near the grounding line of an ice sheet where the surface is nearly flat but the ice remains mostly grounded.) In fast-flowing glaciers, resistance is evidently a combination of basal and side drags. At most locations, the basal drag dominates. It typically accounts for 70% of total resistance, but its contribution varies greatly. To understand this variation, and the variation in τ_d, it is necessary

Table 8.3: Averaged force balance terms. Values of widths and force ratios are approximate.

Glacier	Width (km)	τ_d (kPa)	τ_b/τ_d (%)	τ_W/τ_d (%)	τ_L/τ_d (%)	Reference
McCall	0.8	160	65	35	~0	Rabus and Echelmeyer (1997)
Jutulstraumen						Rolstad et al. (2000)
(inland)	30	214	96	5	−1	
(grounding zone)	20	52	53	38	9	
MacAyeal Ice Stream	75	21	69	31	0	Joughin et al. (2004b)
Bindschadler Ice Stream						Joughin et al. (2004b)
(tributary)	30	45	86	14	0	
(upper-main)	20	23	4	96	0	
(middle)	40	12	34	66	0	
Whillans Ice Stream						
(B2)	35	15	18	82	0	Joughin et al. (2004b)
(ice plain)	90	2.6	15	85	~0	Bindschadler et al. (1987)
Jakobshavn (40 km)	4	410	60	40	~0	Clarke and Echelmeyer (1996)
NEGIS						Joughin et al. (2001)
(middle)	50	60	75	25	0	
(ice plain)	45	23	60	40	0	
Columbia						O'Neel et al. (2005)
(average over 8 km)	5	200	70	30	0	
(average over 0.6 km, in steep section)	3	350	63	23	14	

to examine the relation of resisting stresses with flow, a task we pursue throughout this chapter. The remainder of this section provides an overview of basic concepts.

The longitudinal term has almost no effect on the large regions shown in the table. It is, however, an important force on floating ice shelves and for local-scale variations on grounded glaciers.

8.2.3.1 Mechanisms of Resistance to Flow

The driving force in a glacier determines how large the resisting forces must be, according to Eq. 8.15. How fast the glacier flows then depends on how effectively the resisting forces inhibit motion. The following discussion summarizes information from Chapters 3 and 7, which should be consulted for explanations.

One mechanism of restraint is the creep deformation of ice, by shearing near the glacier bed and on side margins, and by stretching or compression along the direction of flow. Each of these deformations gives rise to deviatoric stresses τ according to the creep relation for ice. For a given rate of deformation, deviatoric stresses increase with the effective viscosity η. The relevant relations are (recalling that $\tau_{jk} = \sigma_{jk} - \frac{1}{3}\sigma_{ii}\delta_{jk}$):

$$\tau_{jk} = 2\eta\dot{\epsilon}_{jk} \quad \text{with} \quad 2\eta = \left[A\,\tau_E^{n-1}\right]^{-1} = \left[A\,\dot{\epsilon}_E^{n-1}\right]^{-1/n}, \tag{8.17}$$

in which

$$\tau_E^2 = \frac{1}{2}\left[\tau_{xx}^2 + \tau_{yy}^2 + \tau_{zz}^2\right] + \left[\tau_{xz}^2 + \tau_{xy}^2 + \tau_{yz}^2\right], \tag{8.18}$$

$$\dot{\epsilon}_E^2 = \frac{1}{2}\left[\dot{\epsilon}_{xx}^2 + \dot{\epsilon}_{yy}^2 + \dot{\epsilon}_{zz}^2\right] + \left[\dot{\epsilon}_{xz}^2 + \dot{\epsilon}_{xy}^2 + \dot{\epsilon}_{yz}^2\right], \tag{8.19}$$

and

$$\tau_{xx} + \tau_{yy} + \tau_{zz} = 0. \tag{8.20}$$

Deformations, given as strain rates, correspond to the gradients of velocity within the ice (Section 3.4.1):

$$\dot{\epsilon}_{jk} = \frac{1}{2}\left[\frac{\partial u_j}{\partial x_k} + \frac{\partial u_k}{\partial x_j}\right]. \tag{8.21}$$

The glacier bed also restrains flow. The magnitude of basal resistance, τ_b, depends on the rate at which the glacier slips on its bed (u_b). For a given magnitude of basal resistance, slip increases the total flow, sometimes by a very large amount. No broadly applicable relationship between slip and resistance exists. A phenomenological description of the process is:

$$\tau_b = \tau_b^\circ\left[1 + \gamma_1 u_b^{1/p}\right] \tag{8.22}$$

$$\textbf{hard bed:} \quad [\tau_b^\circ]^m = \frac{\gamma_2 u_b}{1 + \gamma_3 u_b/N} \tag{8.23}$$

$$\textbf{deformable bed:} \quad \tau_b^\circ = \tau_* = C + N\tan\varphi \tag{8.24}$$

or Eq. 8.23 applies, if $\tau_b^\circ < \tau_*$, in which

$$N = P_i - P_w. \tag{8.25}$$

Here $\gamma_1, \gamma_2, \gamma_3$, and p indicate poorly known parameters that depend on context. N is the effective pressure at the bed, the excess of the ice-overburden stress P_i compared to water pressure P_w. The stress (τ_*) necessary to deform the bed, also called *yield stress*, increases with N and with two material properties, the cohesion C and friction angle φ. For rock or lithified sediment, C is large enough that glaciers do not cause deformation.

Rather than repeating these relations in subsequent formulae, we will sometimes use shorthand notations such as:

$$\tau_b = \frac{u_b}{\lambda'} \quad \text{or} \quad \tau_b^m = \frac{u_b}{c_o\,\lambda}, \tag{8.26}$$

where the lubrication parameters, λ' or λ, describe how slippery the bed is. "Slippery" means the bed provides little resistance even as the ice moves rapidly over it. In terms of the drag factor ψ for the bed (defined in Section 7.1.1), the parameter $\lambda' = \psi^{-1}$.

8.2.3.2 *Qualitative Relation of Resistance to Flow*

The action of these mechanisms on large regions of the bed and within the ice determines how the resisting forces on a section of glacier relate to its flow. Here we discuss qualitative aspects of this relation based on the force balance expressed by Eq. 8.14; the precise formulation is discussed in Section 8.5.3.

Ice is a rather stiff fluid, with an effective viscosity of about 10^{14} Pa s at a stress of 100 kPa and temperature of $-10\,°C$. (For comparison, the viscosity of a familiar substance – peanut butter at room temperature – is about 10^2 Pa s.) Thus, forces acting on one part of a glacier can be transmitted throughout a broad region. Regions of slowly flowing ice restrain nearby fast-flowing regions, while the fast regions pull or push the slow ones forward. We must account for such effects to understand the variations of flow across ice streams and valley glaciers, and along glaciers with variable bed topography or lubrication. The single most important control on ice motion, however, is often the nature of the bed.

To illustrate essential points, consider a mountain glacier or ice stream flowing at rate U down a valley or through slow-moving ice on either side. Call the width Y and ice thickness H. Flow at the glacier surface is the sum of basal slip and internal shear (Eq. 8.4), or $U = u_b + \Delta U$. The average shear rate between the glacier's surface and bed, $\Delta U/H$, is proportional to the shear stress divided by effective viscosity η. Thus, using $u_b = \lambda' \tau_b$ (Eq. 8.26), $U = \lambda' \tau_b + c_z H \tau_b/\eta$, with c_z a coefficient related to the shape of the shear profile. Substituting into Eq. 8.14 and using Eq. 8.8 gives the balance of forces as:

$$\rho g H \alpha = \underbrace{\frac{1}{\lambda' + c_z H/\eta}}_{\tau_b} U + \tau_w + \tau_L. \tag{8.27}$$

In the margins, the resistive stresses scale with the rate of across-glacier shearing ($\partial u/\partial y$) and with the ice viscosity. Thus the side-drag τ_w increases with the centerline velocity U and the ice viscosity η, but decreases with Y. There is an additional decrease of τ_w with Y because the total driving force and the total drag force on the bed increase with the glacier width, whereas the forces contributed by the margins do not; τ_w contributes a progressively smaller fraction of resistance as Y increases. On the other hand, for a given value of resistive stress the side-drag force increases with the ice thickness H: the higher the sides, the greater the total force produced by shearing along them. Thus, in combination, the side-drag scales in the present case as

$$\tau_w \sim \eta U \frac{H}{Y^2}. \tag{8.28}$$

The longitudinal term τ_L depends on variations of stretching and compressing along a flowline – and associated tensional and compressive stresses. In most cases, the variations of flow along a glacier are smaller than those across-glacier, for comparable length scales, and τ_L therefore contributes less to the force balance than do τ_b and τ_w. In any case, the longer the section of glacier considered – the larger the ΔX in Figure 8.6 – the smaller the longitudinal force compared to the total driving and basal drag forces. Just as the relative importance of τ_w decreases with glacier width, the importance of τ_L decreases with the length of the section being examined.

Together with examples drawn from Tables 8.2 and 8.3, Eqs. 8.14, 8.27, and 8.28 illustrate some major aspects of glacier dynamics:

1. The force partitioning gives some indication of which processes most strongly determine the flow. Where τ_b dominates the right-hand side of Eq. 8.27, basal slip and shearing of basal ice layers control the flow. In this case, for example, an increase of water pressure that makes the bed more slippery will lead to faster flow of the glacier. Where τ_w dominates, on the other hand, shearing of ice against valley walls or ice stream margins controls the flow. In this case, a reduction of ice viscosity in the shearing margins causes the whole glacier to speed up.

2. Because the ice thickness and surface slope set the driving stress, they ultimately set the magnitude of the total resisting force as well. Hence, in combination with material properties of ice and bed, the ice thickness and surface slope govern the flow. This describes a fundamental aspect of glacier dynamics; glaciers build up in regions of accumulation, increasing the driving stress and increasing the flow, until the flux is large enough to carry away all the accumulation (this statement constitutes a dynamical counterpart to Eq. 8.2).[3]

   The driving stress needed to establish the balance between flux and accumulation differs from place to place, as shown by the range of values in Tables 8.2 and 8.3. We next discuss several of these examples from Table 8.3.

3. McCall Glacier is a typical valley glacier. The basal drag is approximately 100 kPa. The valley is narrow enough that resistance from the sides matters, and the basal drag balances only a fraction $f' = 0.65$ of the driving stress.

4. The inland site on Jutulstraumen provides an example of a wide glacier ($Y \gg H$) with a large basal resistance ($\tau_b \sim 200$ kPa). Drag on the bed counteracts almost all of the driving stress. In general, for any region of an ice sheet that is both wide and long compared to ice thickness, the force balance reduces to $\tau_b \approx \tau_d$, corresponding to $f' = 1$ (Eqs. 8.5–8.8).

---

[3] For example, where τ_w and τ_L can be neglected, a simple model for the flux is, from Eq. 8.27,

$$q = U H = \rho g H^2 \alpha \left[\lambda' + c_z H / \eta \right]. \qquad (8.29)$$

5. Farther down Jutulstraumen, where it transitions to the Fimbul Ice Shelf, the bed resistance drops dramatically ($\tau_b \sim 30\,\text{kPa}$), probably because basal water pressures approach the ice overburden pressure (hence, $N \to 0$). The slippery bed allows the glacier to convey its flux with only about 25% of the driving stress needed at the inland site. Furthermore, the resistance has partly shifted from the bed to the sides.

6. Whillans and Bindschadler Ice Streams illustrate what happens when the bed is very slippery for a great distance inland of the grounding line. Weak sediments and high water pressures are the lubricants. Consider first the tributaries feeding these ice streams. Their beds provide most of the resistance and the magnitudes of driving stress and basal drag are moderate; the tributaries do not differ much from normal flow on the flank of an ice sheet. In contrast, the basal drag on the trunk ice streams is exceptionally low (τ_b in the range 1 to $10\,\text{kPa}$), and the driving stress needed to convey the flux falls to about 10 to $20\,\text{kPa}$. Drag against the side margins provides much of the resistance, even though the ice streams are wide. Farther downstream, where the ice begins to float, the driving stress is even lower and the bed strength drops to near zero.

7. In contrast to these weak-bed ice streams is Jakobshavn, with a basal stress of 200 to $300\,\text{kPa}$. Unlike the other strong-bed case (the inland site on Jutulstraumen) the fast flow at Jakobshavn is confined to a narrow axis overlying a bedrock trough. With this config-uration, shearing zones on the sides generate large resisting forces. The ice stream must develop a very high driving stress to convey its large flux despite the strong bed and narrow trough.

8. The other sites in the table illustrate some of the underlying complexity. Columbia Glacier, confined to a mountain valley, superficially resembles Jakobshavn; a rather high driving stress is needed to convey a large flux against significant resistances from sides and bed. At Columbia, however, almost all the motion is by basal slip, which is not significant at the inland site on Jakobshavn. Despite the fact that weak sediment partly mantles the bed of Columbia, the average basal resistance is still of order $100\,\text{kPa}$. Perhaps bedrock knobs sticking into the ice account for the basal resistance.

    MacAyeal Ice Stream and the "ice plain" site on NEGIS represent an intermediate sort of ice stream, where the bed is weak ($\tau_b \sim 10$ to $20\,\text{kPa}$) but still accounts for more resistance overall than the side drags.

9. The observed tendency for the sides to take up a larger fraction of total resistance when τ_b decreases suggests that shearing in the glacier margins may be regarded as a perfectly plastic deformation of ice. If the stress τ_{xy} in each margin equals a yield strength τ_o, the total side resistance, per length of glacier, is roughly $2H\tau_o$. The force balance for a glacier of width Y is thus roughly $\tau_d = \tau_b + 2H\tau_o/Y$. Rearranging gives an expression for the apparent plastic yield strength: $\tau_o = [\tau_d - \tau_b]Y/2H$. Table 8.4 lists some examples. All values for τ_o so determined are in the range 130 to $220\,\text{kPa}$. (In terms of the scaling in

Table 8.4: Apparent yield stress of side margins.

Glacier	Thickness H (km)	Width Y (km)	$\tau_d - \tau_b$ (kPa)	Nominal yield stress τ_o (kPa)
McCall	0.17	0.8	56	130
Jutulstraumen (grounding zone)	1.2	20	20	170
NEGIS (middle) (ice plain)	2.0 1.2	50 45	15 9	190 170
MacAyeal Ice Stream	1.0	75	6.5	130
Bindschadler Ice Stream (upper-main)	1.0	20	22	220
Jakobshavn (40 km)	2.5	4	160	130
Columbia (8-km average)	0.9	5	60	170

Eq. 8.28, perfect plasticity implies that the viscosity η decreases in proportion to an increase of U; this is, of course, an approximation for the true non-linear creep behavior of ice.)

10. Over short distances ($\Delta X \sim H$), the resistance τ_L can be significant, but its value fluctuates about zero. The second set of values for Columbia in Table 8.3 shows the force balance in a short region of the glacier (600 m long) that slopes steeply compared to the adjacent regions up- and downstream. Here τ_L contributes to resistance.[4]

Equation 8.27 lacks precision. Better formulations for the terms relating resistance to flow are needed; in this chapter, Section 8.3 discusses the relation of τ_b with shearing over the bed, and Sections 8.6 and 8.7 discuss the across-glacier and along-glacier variations of flow, respectively, which account for τ_W and τ_L. Section 8.5.3 discusses the precise formulation of Eq. 8.27 in terms of deviatoric stresses.

8.2.4 Effective Driving Force of a Vertical Cliff

Vertical cliffs are common features at ice margins, especially at the termini of tidewater glaciers and floating ice shelves and along the lower margins of land-terminating mountain glaciers in cold polar settings. At a cliff, a gravitational driving force pushes the ice mass outward. We refer to it as a "driving force" because it causes the ice to deform and flow

---

[4] The fact that τ_L can usually be neglected in the force balance of large regions does not mean that the longitudinal stress τ_{xx} is necessarily small. The term τ_L depends not on the magnitude of τ_{xx} but on its variability along the glacier (Sections 8.5.3 and 8.7).

horizontally toward the cliff. We picture the force as horizontal, given its effect on the ice, but in fact the true force acts vertically – the horizontal component is deviatoric, that is, related to deviations of stress from the mean. It is a special case of the longitudinal forces that determine τ_L in Eq. 8.15; we treat it separately because its magnitude depends explicitly on gravity.

The force originates, like τ_d, in the gradient of hydrostatic pressure, but Eq. 8.8 does not work because slope α is undefined. Define z as true vertical as in Figure 8.7. At depth $S - z$ the weight of overlying ice causes a vertical compression on horizontal surfaces, of magnitude $\Lambda_{zz} = \overline{\rho} g [S - z]$, with $\overline{\rho}$ the density averaged from z to the surface. A horizontal compression of the same magnitude would prevent the ice from deforming. But, in fact, the horizontal compression is smaller; it must be zero above the level of any water or sediment. Below the water line, it equals the pressure of water pushing against the cliff (a sediment barrier can act the same way). The difference between vertical and horizontal compressions, integrated over the ice thickness, is the driving force. (Note that the difference between vertical and horizontal compressions also equals $\tau_{xx} - \tau_{zz}$ and hence $2\tau_{xx} + \tau_{yy}$.)

For a marginal cliff of height H_M, the driving force, per unit width across-glacier, amounts to

$$\phi_M = g \int_0^{H_M} \overline{\rho} \, [S - z] \, dz - F_M, \qquad (8.30)$$

where F_M denotes the force acting back against the cliff. For a constant ice density, ρ_i, then

$$\phi_M = \frac{1}{2} \rho_i g H_M^2 - F_M. \qquad (8.31)$$

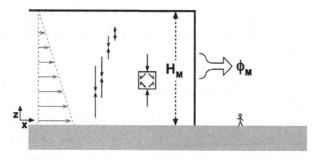

Figure 8.7: Schematic of an ice cliff at a glacier margin. The weight of ice produces a compressive stress that increases with depth as shown by the vertical arrows. The weight is not balanced by compression from the sides. Thus the ice deforms, as shown by the shear couples, and flows toward the cliff. The deviatoric force driving the flow can be visualized as a horizontal pull on the cliff of magnitude ϕ_M. The ϕ_M can also be visualized as the total force due to a hydrostatic pressure that pushes the ice horizontally, as shown by the arrows on the left.

If the terminus stands or floats in water, the back force F_M is found by integrating the water pressure (Sections 8.8 and 8.9.3).

Consider a section of glacier, length Δx and unit width across-flow, ending at a terminal cliff. In general, the slope of the glacier surface upstream of the cliff gives a driving stress $\tau_d = \rho g H(x) \alpha(x)$ at each point. The terminal cliff increases the total driving force (per unit width) on the section, from $\int \tau_d \, dx$ to $\int \tau_d \, dx + \phi_M$. For a glacier on a flat bed and with no back force, this sum simply equals $\frac{1}{2} \rho g [H(x)]^2$, the hydrostatic force at each point; it makes no difference – for the driving force – whether the ice surface declines toward the margin as a gentle slope or a terminal cliff. With a terminal cliff, however, a given thickness and driving force is attained closer to the margin than without a cliff. For equilibrium, the resisting stresses here must also be larger because of the terminal cliff; a cliff therefore increases flow and deformation near the margin.

For glaciers with strong beds and typical basal drags of order 100 kPa, the effects of a terminal cliff extend only a short distance up-glacier – less than 10 to 20 ice thicknesses, the normal "longitudinal coupling length" explained later, in Section 8.7.2. On the other hand, if resistance from both the bed and side walls is weak, the effects of a terminal cliff can be felt a great distance inland.

8.3 Vertical Profiles of Flow

The variation of velocity with depth in a glacier defines the *shear profile*, one of the factors determining the relation between ice fluxes and glacier properties. Following Nye (1952a), we can derive the basic characteristics of the shear profile by combining the relation for ice creep with a relation for stress. This analysis gives an equation connecting the flow and the basal shear stress – a proper formulation of the term with τ_b in Eq. 8.27.

8.3.1 Parallel Flow

Define axes as in Figure 8.5a. Let u be the x-component of velocity, and H the ice thickness. The simplest model assumes the glacier deforms in simple shear; that is, the only nonzero deviatoric stress component is τ_{xz}. The flow lines are therefore parallel, a situation sometimes called *laminar flow* in the older literature. It follows that the z-component of velocity equals zero, and so $\dot{\epsilon}_{xz} = \frac{1}{2} \, du/dz$. The creep relation gives

$$\frac{1}{2} \frac{du}{dz} = A \, \tau_{xz}^n. \tag{8.32}$$

Following Eq. 8.10, it is reasonable to assume a linear increase of shear stress with depth. Thus,

$$\tau_{xz} = \tau_b \left[1 - \frac{z}{H} \right], \tag{8.33}$$

where τ_b is the value of τ_{xz} at the bed (where $z = 0$). With this expression for τ_{xz}, integration of Eq. 8.32 from the bed up to z gives the velocities

$$u(z) = u_b + \frac{2A}{n+1} \tau_b^n H \left[1 - \left[1 - \frac{z}{H} \right]^{n+1} \right] \tag{8.34}$$

$$u_s = u_b + \frac{2A}{n+1} \tau_b^n H. \tag{8.35}$$

Here u_s and u_b are the velocities at the surface and base, and $u(z)$ is the velocity at arbitrary depth $H - z$. The present theory deals only with deformation within the ice and gives no information about the slip velocity u_b. Equation 8.34 shows that the velocity increases continuously with height and, because $n \approx 3$, most of the increase takes place in the layers near the bed. A second integration, now of Eq. 8.34 from bed to surface, gives the flux and the depth-averaged velocity $\bar{u}$ as:

$$q = u_b H + \frac{2A}{n+2} \tau_b^n H^2 = \bar{u} H. \tag{8.36}$$

For most glaciers, $\tau_b \propto \rho g H \alpha$ and so, with $n = 3$, the flux depends on $H^5 \alpha^3$, and the surface velocity on $H^4 \alpha^3$. Thus even a small change of glacier geometry alters the shear flow significantly.

These relations also show that surface velocity and depth-averaged velocity are similar. If $u_b = 0$, Eqs. 8.35 and 8.36 give $\bar{u}/u_s = [n+1]/[n+2] = 0.8$ for $n = 3$. If, on the other hand, motion is entirely by slip, $\bar{u} = u_s = u_b$. This sets limits on the likely value of the depth-averaged velocity.

The creep parameter A was treated as a constant in this analysis. In fact, it depends on temperature, crystal fabric, water content, and perhaps other variables (Chapter 3). Temperate glaciers are by definition nearly isothermal. In other glaciers, the highest temperatures, and thus the highest values of A, are found in the basal layers. This concentrates the shear deformation even closer to the base than in a temperate glacier. The crystal fabric in basal layers also tends to enhance the shearing. Because A multiplies stress raised to the power n (Eq. 8.32), the velocities are sensitive to the A value near the bed, but insensitive to its value in the upper half of the ice thickness.

Velocities at depth can be determined by measuring the rate at which a borehole tilts. Figure 8.8 illustrates several velocity profiles obtained in this way. Panel (a) compares theoretical curves for $n = 3$ and for $n = 1$ to a measured profile from Worthington Glacier, Alaska (Harper et al. 2001). Worthington Glacier is temperate and hence nearly isothermal. As is typical for measured profiles in glaciers, the linear viscous ($n = 1$) curve fits the observations poorly, and irregularities appear near the bed. Figure 8.8b illustrates profiles from polar

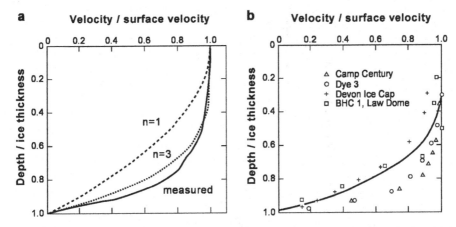

Figure 8.8: Measured and theoretical vertical shear profiles. (a) A temperate glacier. Measured velocities in Worthington Glacier taken from the deeper of the two profiles shown in Figure 7a of Harper et al. (2001). (b) Polar ice sheets and ice caps, and theoretical curve for $n = 3$ and uniform temperature. Data sources are Law Dome (Etheridge 1989), Camp Century (Gundestrup et al. 1993), Devon Ice Cap (Reeh and Paterson 1988), and Dye 3 (Dahl-Jensen and Gundestrup 1987).

ice sheets. Differences in the shapes of the profiles result from differences in temperature and other variables affecting A and deviations from parallel flow.

8.3.2 Observed Complications in Shear Profiles

Apart from the effects of temperature, several important types of deviations of measured shear profiles from the theoretical curves are observed (the following discussion assumes familiarity with the information about ice deformation given in Chapter 3):

1. *Enhanced deformation of ice-age ices.* Figure 8.9 shows how shear strain rate, normalized to stress cubed, varies as a function of depth at Dye 3 in southern Greenland (Dahl-Jensen and Gundestrup 1987). The shearing rate increases by a factor of about 2.5 at the boundary between Holocene ice and the underlying ice from the last (Wisconsin or Würm) glaciation. The boundary is identified by a sharp change in the oxygen-isotope ratio of the ice. Similar increases of deformation rate at this boundary are observed in the tilt or closure of boreholes at Camp Century in Greenland, at Agassiz and Devon Island ice caps in Arctic Canada, and Byrd Station in Antarctica. Mechanical tests on samples from the Dye 3 ice core also show this softening, provided the stress orientation is chosen to match that in the ice sheet; ice-age ices deform about 3.5 times faster than Holocene ice (Shoji and Langway 1987). This is sometimes expressed as "the enhancement factor, in shear, of ice-age ice relative to Holocene ice is 3.5."

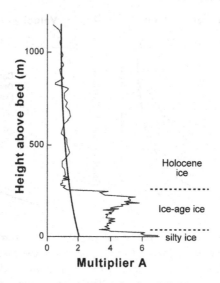

Figure 8.9: Multiplier $A(T)$ in the creep relation obtained from measured shear strain rates and calculated stresses in the lowest 1150 m of the borehole through the Greenland Ice Sheet at Dye 3. The thin, irregular curve shows the true pattern. The rapid increase at 250 m above the bed occurs at the transition from Holocene to ice-age ice. The lowest 25 m, which also has enhanced deformation, is "silty ice" containing material derived from the bed. The smooth curve is the value of A fitted to the Holocene data; it increases with depth because the temperature increases. Adapted from Dahl-Jensen and Gundestrup (1987).

The boundary at all these locations is also marked by significant changes in the ice. In comparison with the Holocene ice above, the ice-age ice has smaller crystals with a stronger single-maximum fabric (oriented vertically), and concentrations of chloride and sulfate higher by factors of two to six. The concentration of "dust" (wind-blown particles in the size range 0.1 to 2 μm) in ice-age ice is roughly 10 times that in Holocene ice.

Laboratory experiments on ice from Vostok, Antarctica, provide a useful contrast (Pimienta et al. 1988). This is a very different setting. The depth of the Holocene/Wisconsin boundary is less than 10% of the ice thickness. Instead of clustering about a vertical axis, the c-axes lie approximately in a vertical plane oriented transverse to the flow direction. For the stress system believed to prevail *in situ*, the Vostok ice deforms at rates of only about one-tenth of that for ice with randomly oriented crystals at the same stress and temperature. This contrasts with the results at Dye 3 where the ice has an enhancement factor, in shear, of about 5 relative to randomly oriented ice.

Thus ice-age ice is not consistently soft but rather softer or harder depending on the crystal fabric and the orientation of stress. The enhanced deformations of ice-age ice observed in the Arctic and at Byrd therefore arise primarily from their strong vertically oriented single maximum fabrics (Paterson 1991). For reasons discussed in Section 3.4.7, when such a fabric deforms in shear parallel to the basal planes of the crystals, it is "soft." Some authors have questioned the importance of fabric as a determinant of ice viscosity because measured variations of fabric strength and strain rate do not always correlate. As Figure 3.20 shows, however, the relation between fabric strength and deformation rate is nonlinear. Theory predicts that no correlation should be observed over a wide range of fabrics.

More recent analyses of the tilt of the Dye 3 borehole applied anisotropic creep relations to evaluate more precisely the role of fabric at this site (Thorsteinsson et al. 1999; Cuffey et al. 2000b). These studies concluded that most, about 70%, of the shear enhancement is indeed due to fabric effects. The remainder – revealed by variations of tilt that anisotropy cannot explain – most likely arises from a softening effect of small grain sizes.

Several analyses show that soluble and insoluble impurities, at concentrations typical of ice-age ice, have no measurable effect on deformation rate (Section 3.4.5.6). The impurities may, however, play an essential role for initiating the development of soft ice fabrics. Impurities impede grain-boundary migration and so reduce the rate of crystal growth (Section 3.3.3.1). This explains the small crystal sizes. Further, the high concentration of dust particles may increase the dislocation density in the ice-age ice, because particles block the movement of dislocations. This favors development of new grains by nucleation and migration recrystallization. For reasons not well understood, fabric develops more rapidly in the fine-grained ice, perhaps due to this recrystallization process or perhaps because of a deformation mechanism that depends directly on grain size. These processes need only produce a small initial difference in fabric strength because a feedback mechanism amplifies any small difference in viscosity (Pimienta 1987, pp. 124–126). Specifically, as the fabric grows stronger, deformation by basal glide within crystals increases. This causes further rotation of c-axes, further strengthening the fabric and so increasing the deformation rate again. Moreover, the deformation keeps the average grain size small because new grains nucleate and grow.

2. *Shear bands.* In some places, the strain rate $\dot{\epsilon}_{xz}$ concentrates in narrow horizons. Such shear bands probably originate with minor differences of fabric, grain size, or impurity content between the layers. The feedback process described in the preceding section then amplifies such differences.

   Figure 8.10 shows measurements from a location near the margin of Law Dome, Antarctica. Most of the shearing occurs in two broad bands (Etheridge 1989). The enhanced shearing cannot be related to any special properties of ice-age ice, but is probably related to strong fabrics (Russell-Head and Budd 1979).

   Gow and Williamson (1976) described apparent shear bands, seen in the Byrd Station core; compared to the surrounding ice, the bands had higher particle concentrations, stronger single-maximum fabrics, and smaller crystals.

3. *Apparent stress shadows.* Some observations reveal anomalously low shearing rates near the bed. One example is the bottom 100 m of ice (out of a total of 370 m) at a location

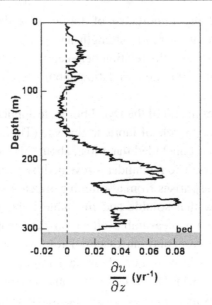

Figure 8.10: Shear strain rate measured in Law Dome, Antarctica, at borehole BHC1. Adapted from Etheridge (1989).

on the flank of Law Dome (Russell-Head and Budd 1979). Another, from Devon Island, can be seen in Figure 8.11 later in this chapter. Most likely, these horizons show that τ_{xz} decreases near the bed where topography interferes with the flow. Immediately upstream or downstream of a hill on the bed, for example, τ_{xz} and $\dot{\epsilon}_{xz}$ should be reduced. Where the bed undulates, the maximum shearing is theoretically expected to occur above the bed by a distance proportional to the spacing of the undulations, if the wavelength of undulations is considerably less than the ice thickness (Kamb 1970; Gudmundsson 2003).

4. *Enhanced deformation of basal ice.* In other places, basal layers shear more rapidly than expected. Ice layers adjacent to the beds of ice sheets sometimes contain a large quantity of rock debris, seasoned with soluble impurities. Such *silty ice* layers are found at the bottom of cores from Byrd Station, Camp Century, Dye 3, Agassiz Ice Cap, and central Greenland. Similar layers occur beneath mountain glaciers in the Antarctic Dry Valleys and elsewhere. Small crystals and strong single-maximum fabrics favorably oriented for the observed shearing also characterize these layers.

Measurements usually show that these layers deform rapidly compared to adjacent horizons of clean ice. For example, the basal silty ice at Dye 3 deforms about four times as fast as the clean ice above (Figure 8.9). Thorsteinsson (1990) compared the two types of ice. He found that the crystal size changes at this boundary from 3 mm in the clean ice to 1 mm in the silty ice, but that the single-maximum fabric did not strengthen significantly. He ascribed the enhanced deformation to unspecified impurities, but small grain sizes might be more important. Holdsworth and Bull (1970) described rapidly shearing "amber ice" layers, less than one meter thick, at the base of Meserve Glacier, Antarctica. These layers contain about 1% rock particles. Salt concentrations are much higher and grain sizes smaller than in the clean ice. Cuffey et al. (2000a) analyzed additional measurements from this site and concluded that both small grains and soluble impurities, but not particulates, soften the ice (Section 3.4.5.5). In these layers, the concentration of ions is about 100 times greater than typical concentrations for ice-age ice in the Arctic.

Echelmeyer and Wang Zhongxiang (1987) observed rapid shearing, with an enhancement of approximately 100, of frozen rock-rich ice at the bed of Urumqi Glacier in the Tien Shan. Even if thin, such a soft layer can contribute significantly to the overall flow of the glacier. Temperature was higher at this site ($-4\,^\circ$C) than at Meserve Glacier or Dye 3. Sliding was also observed, suggesting an unusual amount of liquid water for the temperature, possibly related to concentrated solutes (Cuffey et al. 1999).

8.4 Fundamental Properties of Extending and Compressing Flows

8.4.1 General Concepts

In the parallel flow model, the horizontal velocity component (u) does not vary with distance x, and the component perpendicular to the surface (w) is zero. Such conditions rarely occur in reality; instead, u varies along a glacier's length. Because of the nonlinear creep properties of ice, such variations modify the shape of the vertical shear profile discussed in the previous section. A more profound consequence is that the along-glacier gradient of u largely determines the vertical velocity w. A glacier cannot remain in a steady state unless there is a velocity component w to compensate for accumulation and ablation. The difference between w and specific mass balance at a point determines the rate at which a glacier thickens or thins at that location. Furthermore, in polar and subpolar glaciers the vertical velocity affects the temperature within the ice, and hence modulates the viscosity.

Longitudinal deformation in glaciers follows a systematic pattern overall. As discussed in Sections 8.1.2 and 8.1.3, the velocity u tends to increase along the accumulation zone, where flux increases, and then decreases through the ablation zone. In other words, the longitudinal strain rate $\dot{\epsilon}_{xx} = \partial u / \partial x$ tends to be positive in the accumulation zone and negative in the ablation zone. Because ice is nearly incompressible, extension in the x-direction must be accompanied by an equal shortening in the z-direction. This can keep the thickness constant in spite of accumulation. Similarly, in the ablation zone, compression along the glacier leads to vertical stretching. This description applies strictly to a valley glacier where the walls prevent lateral expansion. More generally, incompressibility requires that the strain rates at a point in any three mutually perpendicular directions cancel each other:

$$\frac{\partial u}{\partial x} + \frac{\partial v}{\partial y} + \frac{\partial w}{\partial z} = 0. \tag{8.37}$$

Thus the gradients of the horizontal velocity components u and v govern the vertical velocity component w. Integrating Eq. 8.37 up from the bed

$$w(z) = w_b - \int_B^z \left[\frac{\partial u}{\partial x} + \frac{\partial v}{\partial y} \right] dz, \tag{8.38}$$

where w_b is the vertical velocity at the base (often negligible, but important if the glacier slides rapidly on a sloping bed). Ice therefore tends to flow toward the bed in regions of horizontal extension and upward in regions of horizontal compression.

Flow is said to be *extending* or *compressing* according to whether the longitudinal strain rate is positive or negative. It follows from the steady-state condition $\bar{u}H \approx \dot{b}_i x$ (cf. Eq. 8.3) that $\partial u / \partial x$ can be estimated roughly from:

$$\dot{\epsilon}_{xx} = \frac{\partial u}{\partial x} \approx \frac{\dot{b}_i}{H}, \tag{8.39}$$

if H is approximately constant (Eq. 8.94 will give a more complete formulation). Typical values of $\dot{\epsilon}_{xx}$ range from $10^{-2} \, \text{yr}^{-1}$ in temperate glaciers to $10^{-5} \, \text{yr}^{-1}$ in the central parts of the ice sheets. Some fast-flowing glaciers stretch as rapidly as $\dot{\epsilon}_{xx} \approx 0.3 \, \text{yr}^{-1}$.

Although accumulation and ablation are the most consistent factors controlling the state of flow, others are important; Section 8.7.1 provides a formal discussion. Where a glacier flows down a step in its valley, extending flow occurs upstream of the step and compressing flow downstream. At one such place in the ablation zone of Blue Glacier, Meier et al. (1974) measured $\partial u / \partial x = 0.03 \, \text{yr}^{-1}$ upstream and $\partial u / \partial x = -0.08 \, \text{yr}^{-1}$ downstream. In comparison, the mean $\partial u / \partial x$ due to ablation was $-0.015 \, \text{yr}^{-1}$. Compressing flow occurs in the lower layers of a glacier upstream of a large bump in the bed. The glacier therefore thickens upstream of the bump, and then thins by extending flow as it crosses the bump; together, these effects steepen the surface above the bump and hence increase the stress, driving the ice over the obstruction. Again, valley walls converging in the downstream direction cause transverse compression in the glacier ($\partial v / \partial y < 0$), which is compensated by longitudinal extension. Conversely, longitudinal compression occurs where a valley widens, or a glacier tongue spills out of a confined valley onto an open plain.

Except near side margins, flow in grounded glaciers is primarily a combination of shearing, which predominates near the bed, and longitudinal extension and compression, which predominate near the surface. Figure 8.11 shows one of the few cases with measurements of both

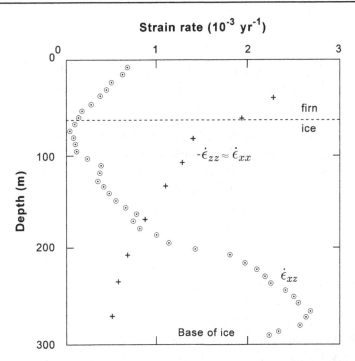

Figure 8.11: Normal ($\dot{\epsilon}_{xx}$ and $\dot{\epsilon}_{zz}$) and shear ($\dot{\epsilon}_{xz}$) strain rates measured in a borehole through Devon Island ice cap, Arctic Canada. Note the relatively high shear in the firn and the reduction in shear just above the bed, which may result from a bump immediately downstream. From Paterson (1976) and unpublished data.

components (Paterson 1976). The borehole was in the accumulation zone at a place where the flow lines along the surface were approximately parallel; the transverse strain rate was negligible. The basal temperature was $-18.5\,°C$. In the upper half of the ice thickness, the longitudinal strain rate exceeded the shearing.

8.4.2 Uniform Extension or Compression

We now illustrate the basic properties of extending and compressing flow using a simple case. The theory was developed by Nye (1957), who gave analytical solutions for special cases. $\dot{\epsilon}_{xx}$ varies with depth, but is assumed uniform with distance along-glacier. There is no transverse straining ($\dot{\epsilon}_{yy} = \tau_{yy} = 0$). The latter implies that $\tau_{xx} = -\tau_{zz}$. We further assume that the flow is unaffected by side margins. With these conditions, $\tau_{xz} = \tau_b\,[1 - z/H]$ and $\tau_b = \tau_d$ are good assumptions.

Replace Eq. 8.32 with the two relations

$$\dot{\epsilon}_{xz} = \frac{1}{2}\frac{\partial u}{\partial z} = A\,\tau_E^{n-1}\,\tau_{xz} \qquad \dot{\epsilon}_{xx} = \frac{\partial u}{\partial x} = A\,\tau_E^{n-1}\,\tau_{xx}, \qquad (8.40)$$

in which

$$\tau_E^{n-1} = \left[\frac{1}{2} \left[\tau_{zz}^2 + \tau_{xx}^2 \right] + \tau_{xz}^2 \right]^{[n-1]/2} = \left[\tau_{xx}^2 + \tau_{xz}^2 \right]^{[n-1]/2}. \tag{8.41}$$

(More generally, in Eq. 8.40, $\partial u/\partial z = 2\dot{\epsilon}_{xz} - \partial w/\partial x$, but the term $\partial w/\partial x$ is negligible if the glacier bed approximates a plane.) Notice that, for given τ_{xz}, the longitudinal stress deviator τ_{xx} increases the strain rate at all depths compared to its value for parallel flow. The difference between the velocities at surface and bed will therefore increase too. This reinforcing of one stress by another arises because the creep relation is nonlinear; it does not happen in a Newtonian viscous material ($n = 1$).

The depth-variation of $\dot{\epsilon}_{xx}$ depends on the situation. A plausible "typical" case – if $\dot{\epsilon}_{xx}$ arises from accumulation or changes of valley slope – is $\dot{\epsilon}_{xx} \propto u(z)$, implying that the upper layers not only flow faster than the layers beneath but also stretch more rapidly. We will use this assumption to solve Eq. 8.40. Defining r_o as the value of $\dot{\epsilon}_{xx}$ at the glacier surface, then $\partial u/\partial x = r_o u(z)/u_s$. This gives a relation for $u(z)$ that can be set equal to $u(z)$ expressed as the integral of the first formula in Eq. 8.40. We then solve the set of equations numerically for τ_{xx} and $u(z)$. Finally, we calculate $w(z)$ from Eq. 8.38, with $\dot{\epsilon}_{yy} = 0$. Figure 8.12 illustrates a few examples of velocity and stress components, for parameters typical of a zone of extending flow in a temperate glacier. The central panel depicts horizontal velocity profiles for three cases, labelled A, B, and C: (A) a case of no extending flow and no sliding; (B) the same situation but with extending flow of $r_o = 0.02$ yr^{-1}; and (C) the case with $r_o = 0.02$ yr^{-1} and 10 m yr^{-1} of sliding. Vertical velocities are shown, on the left, for the two cases with extending flow; the

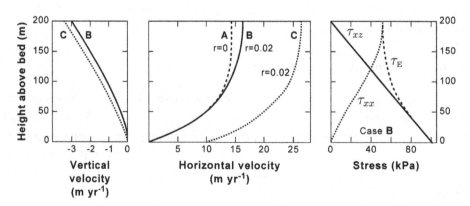

Figure 8.12: Theoretical velocity and stress profiles illustrating effects of extending flow. Middle panel shows the velocity profiles for the three cases: (A) has no extending flow, but (B) and (C) do. Basal slip also occurs in (C). Left panel: resulting vertical velocities, negative because flow is downward. Right panel: stress terms for (B).

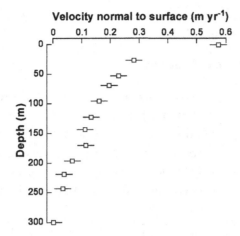

Figure 8.13: Downward velocity perpendicular to surface as a function of depth, Devon Island ice cap. The bars represent one standard error on each side of the mean. Adapted from Paterson (1976).

right panel plots stresses for case B. The increase of u due to the stress τ_{xx} is not very large, provided that τ_b approximates the typical value of 100 kPa. With a longitudinal strain rate of 0.05 yr^{-1}, a moderately large value, the deformational velocity increases by about 20%, mostly because the shearing $\partial u / \partial z$ increases in the middle of the ice column. This effect should be larger where the ratio τ_{xx}/τ_b is larger – perhaps, for example, near the head of a glacier.

The calculated vertical velocity is downward, a consequence of extending flow. The $w(z)$ profile is quasi-linear. For comparison, Figure 8.13 illustrates one of the very few measured profiles for $w(z)$ in a glacier. The good match to the theory (left panel of Fig. 8.12) is not surprising, since the integral form of Eq. 8.38 should smooth irregularities.

The calculations here and in Nye (1957) assume isotropic ice. A crystal fabric that favors vertical shear is unfavorable for vertical compression and longitudinal extension, A strong fabric likely develops in the deep, rapidly shearing layers, reinforcing the tendency for $\dot\epsilon_{xx}$ and $\dot\epsilon_{zz}$ to diminish near the bed. At the same time, the vertical compression must tend to be large enough for vertical velocity at the surface to match accumulation. This suggests that the development of a fabric increases the stresses τ_{xx} and τ_{zz}. The importance of such effects in glaciers remains largely unknown.

In Section 8.7, we will discuss how extending and compressing flow fit into the larger-scale pattern of variations along a glacier.

8.5 General Governing Relations

In this section we review the general equations for force balance and mass conservation. These relations provide a framework for most analyses of ice dynamics. The vertically integrated forms show how the evolution of ice masses and the spatial variation of their flow relates to quantities like ice thickness and specific mass balance.

8.5.1 Local Stress-equilibrium Relations

Because the acceleration term in Newton's second law can be neglected, the equations of motion reduce to equations of static equilibrium expressing the balance between the forces applied to the surface of a body and the gravitational force acting on its mass. Considering a small parcel of ice within a glacier, the surface forces are, per unit area, the stress tensor σ. Gradients in stress across the parcel give net forces that must balance gravity. As usual, we consider coordinates defined as x along-glacier, y across-glacier, and z positive upward. The equations are

$$\frac{\partial \sigma_{xx}}{\partial x} + \frac{\partial \sigma_{xy}}{\partial y} + \frac{\partial \sigma_{xz}}{\partial z} = -\rho g_x = -\rho g \sin \overline{\beta}, \tag{8.42}$$

$$\frac{\partial \sigma_{xy}}{\partial x} + \frac{\partial \sigma_{yy}}{\partial y} + \frac{\partial \sigma_{yz}}{\partial z} = -\rho g_y = 0, \tag{8.43}$$

$$\frac{\partial \sigma_{xz}}{\partial x} + \frac{\partial \sigma_{yz}}{\partial y} + \frac{\partial \sigma_{zz}}{\partial z} = -\rho g_z = \rho g \cos \overline{\beta}, \tag{8.44}$$

where g_x, g_y, and g_z are the components of gravity in directions x, y, and z. The normal stresses, $\sigma_{xx}, \sigma_{yy}, \sigma_{zz}$, are positive in tension, negative in compression. With x-axis horizontal, $g_x = g_y = 0$ and $g_z = -g$. More generally, x can be chosen to tilt at angle $\overline{\beta}$ down from horizontal, giving the terms indicated by the second equality of each expression. The choice of $\overline{\beta}$ can be arbitrary – the mathematics works regardless – but terms are most simply related to deformation and flow in the glacier by choosing $\overline{\beta} = 0$ if the bed elevation has no mean trend or if it rises down-glacier. If the bed tilts down-glacier, setting $\overline{\beta}$ to the mean slope magnitude is preferable.

8.5.1.1 Definitions of Other Stresses

As discussed in Chapter 3, and summarized by Eq. 8.17, ice deforms only if the stress deviates from an isotropic state; strain rates depend on the deviatoric stress, not the full stress σ. Call the mean of the normal stresses σ_M:

$$\sigma_M = \frac{1}{3} \left[\sigma_{xx} + \sigma_{yy} + \sigma_{zz} \right]. \tag{8.45}$$

The σ_M follows the same sign convention as the stresses; it is negative for compression. (The term *pressure*, often used in this context, refers to the same quantity but defined as positive in compression. Hence $P = -\sigma_M$.) Deviatoric stress, τ, is the deviation of stress from an isotropic value of magnitude σ_M; for example, $\tau_{xx} = \sigma_{xx} - \sigma_M$ and $\tau_{xz} = \sigma_{xz}$ (Section 3.4.2). By this definition,

$$\tau_{xx} + \tau_{yy} + \tau_{zz} = 0. \tag{8.46}$$

An arbitrary component of τ can be written with subscripts j and k, each of which can signify x, y, or z:

$$\tau_{jk} = \sigma_{jk} - \sigma_M \delta_{jk}, \tag{8.47}$$

where δ means the matrix with components $\delta_{jk} = 1$ if $j = k$, and $\delta_{jk} = 0$ otherwise.

As an alternative, consider the *resistive stress*, $\boldsymbol{R}$, with components

$$\boldsymbol{R}_{jk} = \sigma_{jk} - \Lambda \delta_{jk}. \tag{8.48}$$

Here Λ signifies the normal stress on the xy-plane needed to balance the weight of overlying ice, or $\Lambda = -\rho g[S - z] \cos \bar{\beta}$. Λ is a useful reference stress because, unlike σ_M, it does not depend on the state of deformation. In most situations, $\sigma_{zz} \approx \Lambda$; thus $R_{zz} \approx 0$, even in a flowing glacier. In contrast, τ_{zz} takes a value dependent on the flow and is seldom zero. Using resistive stresses in place of deviatoric stresses simplifies the force balance relations (van der Veen and Whillans 1989a). (Note that when z is true vertical ($\bar{\beta} = 0$), Λ is sometimes called the *lithostatic stress*. The resistive stress was originally defined by van der Veen and Whillans for this case of $\bar{\beta} = 0$, so no term $\cos \bar{\beta}$ appeared in its definition. They symbolized it L.)

Either τ or $\boldsymbol{R}$ can be substituted for σ in Eqs. 8.42 through 8.44 to write new forms of the stress-equilibrium relations. For example, the vertical equation in terms of τ is

$$\frac{\partial \tau_{zz}}{\partial z} + \frac{\partial \sigma_M}{\partial z} + \frac{\partial \tau_{xz}}{\partial x} + \frac{\partial \tau_{yz}}{\partial y} = -\rho g_z. \tag{8.49}$$

8.5.2 General Solutions for Stress and Velocity

To estimate the stress and velocity throughout an ice mass, the three local stress-equilibrium relations (Eqs. 8.42–8.44), with τ substituted for σ, can be solved together with the six equations of the creep relation that connect strain rates to stresses (Eq. 8.17). Recall that strain rates are related to velocity gradients by $\dot{\epsilon}_{jk} = [\partial u_j / \partial x_k + \partial u_k / \partial x_j]/2$. The stresses acting on the glacier surface are assumed zero. Kinematic requirements connect the vertical velocities at the surface and the bed to elevation changes and specific mass balance (Eqs. 8.65 and 8.76, discussed later). The rate of basal slip must be specified directly or through a relation to bed stress such as Eq. 8.25. With such boundary conditions, the equations are sufficient to determine the six components of stress and three components of velocity at each point of an isothermal glacier. Only temperate glaciers can be considered as isothermal, however; in other cases, temperature and the equation of heat transfer have to be included in the analysis. Note that a relation for mass conservation does not provide an independent equation: it is implied in the creep relation because the normal deviatoric stresses sum to zero, $\tau_{xx} + \tau_{yy} + \tau_{zz} = 0$. When the bedrock topography, specific mass balance, surface temperature, geothermal heat flux, and creep parameters n and $A(T)$ are given, the surface elevation corresponding to a steady state is found as part of the solution. In many modelling studies, however, particularly those concerned with the interpretation of field data, the ice thickness is prescribed at its present value, which is assumed to correspond to a steady state. In such a case the velocity averaged over the ice thickness can be prescribed by mass conservation.

To solve these partial differential equations, even by numerical methods, approximations are necessary and simplifying assumptions are often made. The *finite difference* and *finite element* methods use discretized forms of the equations, which are written in terms of values at the nodes of a grid (finite difference) or integrated within the elements of a grid (finite element). The problem is often reduced to two dimensions by assuming a state of *plane strain*, in which the transverse velocity component $v = 0$, and u and w are independent of y. Thus all strain-rate and deviatoric stress components involving y are zero. This may be a good assumption at many places in an ice sheet but, as shown later, it is not valid in a valley glacier or ice stream. A further simplification of the two-dimensional problem is to assume that the ice is deforming in *simple shear*, that is, $\dot{\epsilon}_{xz}$ and τ_{xz} are the only nonzero components. This assumption is not valid near an ice divide or the ice margin, if the bed is bumpy, or if the flow regime varies over short distances along-glacier.

An often useful simplification of the equations is the *shallow-ice approximation* (Fowler and Larsen 1978; Hutter 1981). Because the thickness of an ice sheet is small compared with its lateral extent, the x- and y-derivatives of stress, velocity, and temperature are small compared with the z-derivatives. New coordinates $X = \varepsilon x$, $Y = \varepsilon y$, and $Z = z$ are introduced, with the parameter ε defined as the ratio of the mean ice thickness to the lateral extent. Typical values are 10^{-3} for Antarctica and 5×10^{-3} for Greenland. The parameter appears in the equations and boundary conditions when they are written in terms of the new coordinates X, Y, and Z. The powers of ε show the relative importance of the different terms. The shallow-ice approximation means that $\varepsilon \ll 1$ and the force balance reduces to the equality $\tau_b = \tau_d$. Models that also include τ_w and τ_L are said to be "higher-order," because they retain terms with higher powers of ε.

8.5.3 Vertically Integrated Force Balance

Vertical integration of the equilibrium equations gives the force balance for a location on a glacier. Such an analysis shows how the driving stress is partitioned into resisting forces, as represented by Eqs. 8.14 and 8.27, and allows the basal drag to be mapped. Figure 8.14 depicts a section of a flow line with some of the important terms illustrated. Our treatment most closely follows the analysis of van der Veen and Whillans (1989a), but most of the method and interpretations were made earlier by Budd (1968, 1970) and Nye (1969b).

Start with the stress-equilibrium equations, written in terms of τ. The mean normal stress σ_M and all components of τ are functions of x, y, and z. First integrate the vertical-axis relation,

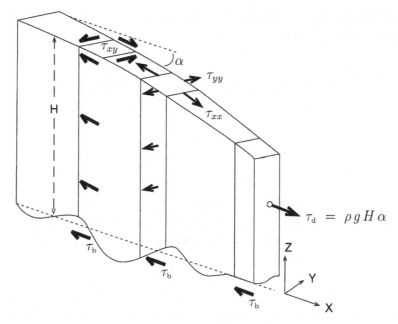

Figure 8.14: Cartoon of part of a flow line, showing important stress terms acting on columns. Note that we define τ_b as the stress acting on the base of the ice in direction x; it is sometimes defined differently, as the stress acting parallel to the local bed.

Eq. 8.49, from some arbitrary z to the surface at S, to obtain an expression for $\sigma_M(z)$. Assume that $\sigma_{zz} = 0$ at the surface. Next differentiate with respect to the horizontal coordinate x or y. For x this gives

$$\frac{\partial \sigma_M}{\partial x} = \overline{\rho} g_z \frac{\partial S}{\partial x} - \frac{\partial \tau_{zz}}{\partial x} + \frac{\partial}{\partial x} \int_z^S \Delta \sigma_{zz} \, dz \tag{8.50}$$

in which $\overline{\rho}$ is the average density between z and the surface, and, for brevity, we have written

$$\Delta \sigma_{zz} = \frac{\partial \tau_{xz}}{\partial x} + \frac{\partial \tau_{yz}}{\partial y}. \tag{8.51}$$

Then substitute Eq. 8.50 into the stress-equilibrium equation for a horizontal direction (x or y), such as

$$\frac{\partial \tau_{xx}}{\partial x} + \frac{\partial \sigma_M}{\partial x} + \frac{\partial \tau_{xz}}{\partial z} + \frac{\partial \tau_{xy}}{\partial y} = -\rho g_x. \tag{8.52}$$

Also make a substitution for τ_{zz} using Eq. 8.46. Rearranging the result and, finally, integrating from bed to surface gives, for the x-direction,

$$\int_{B}^{S} \left[\bar{\rho} g_z \frac{\partial S}{\partial x} + \rho g_x \right] dz = \tau_{xz}(B) - \tau_{xz}(S) - \int_{B}^{S} \frac{\partial}{\partial x} \left[2\tau_{xx} + \tau_{yy} \right] dz$$

$$- \int_{B}^{S} \frac{\partial \tau_{xy}}{\partial y} dz - \int_{B}^{S} \frac{\partial}{\partial x} \int_{z}^{S} \Delta\sigma_{zz} \, dz' \, dz. \tag{8.53}$$

Now make a distinction between the total basal resistance in the x-direction, τ_{bx}, and the value $\tau_{xz}(B)$. This is necessary because the bed undulates and so does not parallel the xy-plane at an arbitrary point. The total basal resistance in the x-direction, per unit area, is

$$\tau_{bx} = \tau_{xz}(B) - \tau_{xx}(B) \frac{\partial B}{\partial x} - \tau_{xy}(B) \frac{\partial B}{\partial y}. \tag{8.54}$$

A similar relation applies at the ice surface, but with a sum of zero; no resistance acts on the glacier surface. Using these relations, and moving derivatives outside of integrals, gives

$$\underbrace{\int_{B}^{S} \left[\bar{\rho} g_z \frac{\partial S}{\partial x} + \rho g_x \right] dz}_{\tau_d} = \underbrace{\tau_{bx}}_{\tau_b} - \underbrace{\frac{\partial}{\partial x} \int_{B}^{S} \left[2\tau_{xx} + \tau_{yy} \right] dz}_{\tau_L}$$

$$\underbrace{- \frac{\partial}{\partial y} \int_{B}^{S} \tau_{xy} \, dz}_{\tau_W} \underbrace{- \int_{B}^{S} \int_{z}^{S} \frac{\partial \Delta\sigma_{zz}}{\partial x} \, dz' \, dz}_{\tau_T}, \tag{8.55}$$

in the x-direction. An analogous relation applies to the y-direction. This expresses the force balance for the vertical column, with the separate resisting terms of Eq. 8.15 identified by the braces. It is equivalent to Eq. 8.14, the force balance for a rectangular block, in the limit of zero length and width (X and Y). One new term, τ_T, appears here. For large sections of a glacier, this factor is negligible; we ignored it in the earlier discussion.

The term on the left-hand side, the driving stress, is equivalent to Eqs. 8.5 through 8.7 for the appropriate geometries. In terms of the surface slope measured from true horizontal, α, the factor $\partial S/\partial x = -\tan(\alpha - \bar{\beta})$. Therefore, $\tau_d = \rho g H \left[\cos(\bar{\beta}) \tan(\alpha - \bar{\beta}) + \sin(\bar{\beta}) \right]$, for a uniform density. The product $\rho g H \alpha$, with H measured perpendicular to x, gives a close approximation to τ_d in most situations (Eq. 8.8).

In general, each of the terms in the column force balance is a vector, with Eq. 8.55 and its y-direction analogue specifying the (x, y) components. For example, given a horizontal xy-plane and a uniform density, driving stress is the vector

$$\vec{\tau}_d = -g \int_B^S \rho(z) \, dz \cdot \nabla S = -\rho g H \cdot \nabla S. \tag{8.56}$$

Van der Veen and Whillans (1989a) showed that repeating the entire derivation using R (defined by Eq. 8.48) in place of τ yields a more compact form of Eq. 8.55,

$$\tau_{dx} = \tau_{bx} \underbrace{-\frac{\partial}{\partial x} \int_B^S R_{xx} \, dz}_{\tau_L} \underbrace{-\frac{\partial}{\partial y} \int_B^S R_{xy} \, dz}_{\tau_W} + \underbrace{\frac{\partial}{\partial x} \int_B^S R_{zz} \, dz}_{\tau_T}. \tag{8.57}$$

Thus each of the "resisting" terms emerges, in a simple fashion, from a gradient of a resistive stress.

The integrals in Eqs. 8.55 and 8.57 can be written as the products of ice thickness and depth-averaged values. Thus, for the x-direction,

$$\tau_d = \tau_b - \frac{\partial}{\partial x} \left[H \left[2\overline{\tau}_{xx} + \overline{\tau}_{yy} \right] \right] - \frac{\partial}{\partial y} \left[H \overline{\tau}_{xy} \right] + \tau_T \tag{8.58}$$

In other words, the resistances τ_W and τ_L depend on gradients of ice thickness (e.g., $\partial H/\partial x$) and gradients of stress (e.g., $\partial \overline{\tau}_{xx}/\partial x$). Because τ_W and τ_L dynamically link fast- and slow-moving regions, their effects are referred to as *stress-gradient coupling* of the flow.

We can now outline how the flow field connects to the force balance. First, use the ice creep relation to substitute strain rates for deviatoric stresses in the terms τ_W and τ_L in Eq. 8.55 (and take $\tau_T \approx 0$). Second, write the basal drag τ_b in terms of velocity, using relations for basal slip and the vertical shear profile (Section 8.3). The resulting relation, between driving stress and the flow field, is the precise form of Eq. 8.27. The (relatively) simple case of a glacier moving only by basal slip illustrates the relation (MacAyeal 1989). Velocities u and v equal their averages over depth. Define the depth-averaged effective viscosity η from Eq. 8.17 such that $\overline{\tau}_{xx} = 2\eta \, \partial u/\partial x$, and $\overline{\tau}_{xy} = \eta \left[\partial u/\partial y + \partial v/\partial x \right]$. The relation for the x-direction is then (with α the slope in direction x)

$$\rho g H \alpha = \tau_b(u) - \frac{\partial}{\partial x} \left[2\eta H \left[2\frac{\partial u}{\partial x} + \frac{\partial v}{\partial y} \right] \right] - \frac{\partial}{\partial y} \left[\eta H \left[\frac{\partial u}{\partial y} + \frac{\partial v}{\partial x} \right] \right]. \tag{8.59}$$

A similar relation applies to the y-direction. If measurements of surface flow (u, v) and ice geometry (H, α) are available throughout a region of a glacier, Eq. 8.59 can be used to make a map of the basal drag τ_b. (In the general case, the calculation must resolve vertical variations of velocity and viscosity, which requires a three-dimensional flow model rather than Eq. 8.59.) Information about temperature is needed to constrain effective viscosities. Figure 8.15a is a

a   **Rutford Ice Stream**

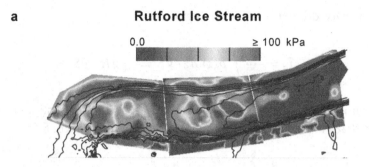

b   **Bindschadler and MacAyeal Ice Streams**

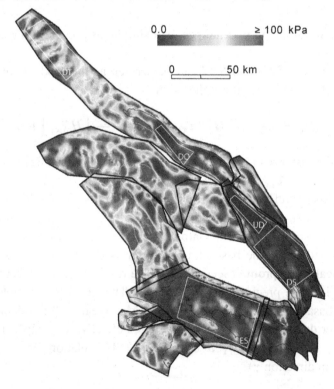

Figure 8.15: Maps of basal drag τ_b beneath three major ice streams in Antarctica. Values are only approximate, and are obtained by calibrating a flow model against measured velocities (Section 8.5.3.2). (a) Rutford Ice Stream flows from left to right in the picture. Adapted from Joughin et al. (2006). (b) Ice streams D (Bindschadler) and E (MacAyeal) flow toward the lower right. Adapted from Joughin et al. (2004b). Images courtesy of I. Joughin, and used with permission from the American Geophysical Union, *Journal of Geophysical Research*. (Refer to the insert for a color version of this figure).

map of basal drag inferred by this method for a 150-km-long section of Rutford Ice Stream, Antarctica (Joughin et al. 2006). Drag concentrates beneath the side margins of the stream and in patches of high friction. Understanding the origins and variability of such "sticky spots" is a key challenge for glacier physics; little is known about them (Section 7.3.5). Figure 8.15b gives a second example, a 400-km-long section of Bindschadler and MacAyeal Ice Streams, West Antarctica (Joughin et al. 2004b). This shows in detail the change from the inland tributary region with significant basal drag to the main ice streams with large areas of weak bed (see Sections 8.2.3 and 8.9.2.4). Sticky spots appear in this region too.

8.5.3.1 Interpretation of Resisting Forces on the Column

The basal shear stress, τ_b, is the apparent friction along the bed. At the scale of meters, the forces acting between the ice and the substrate vary considerably along the interface. Forces concentrate on the up-flow faces of boulders and bedrock bumps. Flow over and around larger-scale hills and valleys induces broad patterns in the average of the forces ("form drag"). Summing the forces' x- and y-components gives, per unit area, τ_b. (Note that an alternative definition of τ_b is the sum of forces parallel to the surfaces of the hills and valleys, rather than along the basal plane defined by the x and y coordinates.)

The second resistance term, τ_L, arises from gradients in the viscous stresses accompanying the longitudinal deformations $\dot{\epsilon}_{xx}$ and $\dot{\epsilon}_{yy}$ (Nye 1969b; Budd 1968, 1971). If the longitudinal stress τ_{xx} becomes increasingly compressive along the glacier, the column feels a push backward; the flow is restrained. On the other hand, an increasingly tensile longitudinal stress pulls the column forward and effectively adds to the driving stress. The force component τ_L, often called $2G$ in the older glaciological literature, significantly influences the dynamics of grounded glaciers in several situations: where driving stress changes rapidly along-flow (in rugged mountain valleys, for example), near an ice divide, and near a glacier terminus. τ_L is an important term in ice shelves, where τ_b is often negligible.

The third resistance term, the side drag or wall drag τ_W, arises from gradients across-flow in the viscous stresses accompanying shearing on vertical planes ($\dot{\epsilon}_{xy}$). This force component especially influences glaciers and ice shelves whose lateral margins act as rigid walls; viscous drag on the margins opposes fast flow in the center of the glacier or shelf. Likewise, slow-flowing ice adjacent to an ice stream or near the edge of a valley glacier is pulled forward by the faster ice nearby: in the slow-flowing ice, τ_W adds to τ_d rather than opposing it. The "rigid walls" may be the sides of a bedrock valley or may be zones of ice immobilized by a frozen bed.

The fourth term, τ_T, is best regarded not as a resistance but as a correction to the driving stress. These "bridging effects" are small enough to ignore in most analyses (Kamb and Echelmeyer 1986b). "Bridging" means that the weight of a column of ice is not fully supported by the normal stress σ_{zz} acting on its base. As an analogue, consider a bridge spanning a gully (van der Veen and Whillans 1989a). The weight of the central span cannot be supported by a normal stress σ_{zz} along its base, where the air exerts little force, but instead transfers laterally to the bridge abutments. Beneath the abutments, in contrast, σ_{zz} exceeds the overburden. Large shear stresses (τ_{zx}) in the transition from abutment to span accomplish the lateral transfer. Glacier ice is fluid enough that very little stress transfer of this sort occurs over distances greater than about one ice thickness. For this reason, most analyses omit the term τ_T. In special situations, however, like steep ice falls, bridging effects may play an important role. The term τ_T has been referred to as "T" in the older glaciological literature.

8.5.3.2 Methods for Estimating the Force Balance

Stress-at-boundaries Method   The force balance for a large rectangular region of the glacier, width Y and length X, was given as Eq. 8.14 (Section 8.2.2). This is equivalent to Eq. 8.58 integrated over the region, and is

$$\overline{\tau}_d = \overline{\tau}_b + \frac{1}{Y} \cdot \Delta\left(H\,\overline{\tau}_{xy}\right) + \frac{1}{X} \cdot \Delta\left(H\left[2\overline{\tau}_{xx} + \overline{\tau}_{yy}\right]\right), \tag{8.60}$$

where the notation $\Delta(F)$, for any F, indicates the difference between F on opposite sides of the block. This relation is applied by first using measurements of the surface velocity field to calculate strain rates, $\dot{\epsilon}_{xx}$, $\dot{\epsilon}_{yy}$, and $\dot{\epsilon}_{xy}$, along the boundaries of the region. The creep relation for ice gives the corresponding surface values for the stresses τ_{xx}, τ_{yy}, and τ_{xy}. These provide the upper boundary conditions for a calculation of stresses at depth, with a flow model or with simple assumptions about the depth variation. Ice thickness and surface topography must be known to calculate $\overline{\tau}_d$ for the whole region. The mean bed stress is calculated as a residual. In Table 8.3, the values for Jutulstraumen were obtained by this method.

Spatially Distributed Methods   The vertically integrated force balance is calculated for all the elements of a grid covering a region, and the results are mapped or averaged. For the calculation at grid elements, two methods can be used. In the first, measurements of surface strain rates again constrain the values for the stresses τ_{xx}, τ_{yy}, τ_{xy} at depth. From Eq. 8.58, basal drag is calculated as a residual. In the second method, a flow model accounting for all stress components calculates the surface velocity field. Such a model implicitly enforces the column force balance relation. (Eq. 8.59 can be used if most of the motion is basal.) To find the basal drag, the match of model and measured velocities is optimized, often by adjusting a coefficient ψ defined by $\tau_b = \psi u_b$ (MacAyeal 1989). These methods gave the values for Columbia, NEGIS, and Whillans, Bindschadler, and MacAyeal Ice Streams in Table 8.3. The problem of inverting surface measurements for conditions at depth in a glacier is fundamentally ill-posed because, at spatial scales comparable to the ice thickness, different patterns of basal drag and basal slip and different depth variations of velocities can lead to nearly identical patterns of velocity at the surface (Bahr et al. 1994). Simplifying assumptions must always be made and basal quantities found by applying a formal technique for optimization (e.g., Truffer 2004; Maxwell et al. 2008; Kavanaugh and Cuffey 2009).

Simple Estimate for a Centerline   If the force balance for the centerline of a glacier is sought, averaged along a segment very much longer than the ice thickness, τ_L can be assumed zero and the force balance reduces to the problem of how τ_d is partitioned between τ_b and τ_w. Furthermore, if flow is confined to a long channel, this partitioning can be estimated, roughly, from the shape of the channel alone, using results from model calculations of idealized cross-sections. (See Section 8.6.2.) In Table 8.3, the values for Jakobshavn were obtained by this method.

8.5.3.3 Uncertainties in Estimates of Vertically Integrated Force Balance

Inaccuracies in calculated force-balance terms derive from several sources. The combined uncertainty is difficult to quantify. Numbers such as those in Table 8.3, mapped in Figure 8.15, and discussed in Section 8.9.2 later in this chapter, reveal major characteristics of the glaciers but lack quantitative precision. Uncertainties arise from inadequate constraints on:

1. *Ice viscosity.* The effective viscosity of ice in a polar glacier is not known, a priori, to better than a multiple of about five (Section 3.4.6). Values for both of the parameters in the ice creep relation, A and n, are uncertain. A depends strongly on temperature, which must usually be estimated from models. It also depends on the c-axis fabric. Deformation of ice is anisotropic; depending on the fabric, the effective viscosity may differ greatly for deformations of different orientations. This is almost never accounted for, and it is not clear how to do so without measurements of fabric within the glacier, which are almost never obtained. Authors sometimes calculate different versions of the force balance using different assumed creep relations. This provides a valuable illustration of the sensitivity of the results but does not quantify the full uncertainty.

2. *Ice thickness and surface elevation* and hence the driving stress. These are typically known well if the force-balance calculation is made for a large region. The patterns seen in a map like Figure 8.15, however, are dubious except at spatial scales considerably larger than the spatial resolution of accurate measurements of thickness and slope. The importance of accurate measurements can be illustrated by comparing two versions of the inferred basal drag at MacAyeal Ice Stream. The version published in 2004 differed significantly from an earlier one, published in 1995, because by 2004 the ice stream's surface topography was better known (MacAyeal et al. 1995; Joughin et al. 2004b).

3. *Ice surface velocities and strain rates.* Surface velocities are usually constrained well by measurements. Strain rates are less certain, because they are calculated from differences of velocities. The force balance depends on gradients of strain rates (e.g., $\partial^2 u/\partial x^2$); even with good velocity measurements, the inferred values of such quantities might contain considerable noise. In glaciers that move primarily by basal slip, the velocity and strain rates are constant with depth, a situation that is optimal for force balance analyses. In other situations, errors in measured strain rates must be taken into account to avoid overfitting of

modelled flow to surface observations; such errors are amplified in estimates of velocities and forces at depth (Bahr et al. 1994). This is a major and unavoidable source of uncertainty in maps like Figure 8.15 if the spatial resolution is comparable to the ice thickness. This problem diminishes as the grid spacing increases.

Joughin et al. (2004b) used "identical twin" experiments to illustrate how well their method – an optimized flow model – resolves the spatial pattern of basal drag. The flow model was first used to calculate velocities from a known ice geometry and a specified map of basal drag. To simulate the effects of uncertain measurements, random noise was added to the resulting velocities and to surface elevations and other variables. The basal drag was then reconstructed using the model optimization and compared to the original pattern. This exercise demonstrated that the method performed well for mapping large-scale features.

8.5.4 General Mass Conservation Relation (Equation of Continuity)

We now shift the topic of discussion from the balance of forces to the conservation of mass (also referred to as "continuity"). Many important characteristics of glacier flow and evolution are manifestations of the requirement that mass be conserved.

8.5.4.1 Local Conservation Relation

Consider a small parcel of ice moving with the flow in a glacier. The parcel's boundaries are attached to the ice. The parcel's density ρ changes if ice flow closes bubbles and other voids or melt creates new voids. The rate of density change of the parcel, $D\rho/Dt$, defines the *densification rate*. The parcel's mass is $m = \rho V$, for a volume V. Differentiating and dividing by V gives $D\rho/Dt = V^{-1} Dm/Dt - \rho V^{-1} DV/Dt$. The term $V^{-1} DV/Dt$ is the *volumetric strain rate*, the rate of volume change per unit volume. It equals the sum of extensions and compressions in all three directions or the divergence of the velocity: $\nabla \cdot \vec{u} = \partial u / \partial x + \partial v / \partial y + \partial w / \partial z$. The mass of the parcel remains constant unless there is melt. Call the rate of mass loss by melt $\dot{\mu}_i$, measured as mass per unit time per unit volume. Then, combining the previous expressions shows that the densification rate must relate to volumetric strain and internal melt such that

$$\frac{D\rho}{Dt} = -\rho \left[\frac{\partial u}{\partial x} + \frac{\partial v}{\partial y} + \frac{\partial w}{\partial z} \right] - \dot{\mu}_i. \tag{8.61}$$

(Here density refers, specifically, to the mass of ice per unit volume, rather than to the mass of ice plus water. Note that a negative $\dot{\mu}_i$ would indicate formation of new ice by freezing.)

Now suppose that rather than watching a moving parcel we observe a small region fixed in space, volume V, with ice flowing through it. The mass within the region changes at rate $V \partial \rho / \partial t$, because of melting or because ice flow removes more mass than it brings in. To calculate the flow effect, suppose that the small region defines a cube with sides of lengths δx, δy, δz. In unit time, a mass of material $\rho u \, \delta y \delta z$ flows in the x-direction into one face of the cube and a mass $[\rho u + \delta x\, \partial(\rho u)/\partial x]\,\delta y\delta z$ flows out the opposite face. The difference is $[\partial(\rho u)/\partial x]\,\delta x\delta y\delta z$. Similar expressions give the differences in flux through the other pairs of faces perpendicular to y and z. Per unit time, the change of mass in the cube by flow amounts to $[\partial(\rho u)/\partial x + \partial(\rho v)/\partial y + \partial(\rho w)/\partial z]\,\delta x\delta y\delta z$. This, plus the rate of melt, must equal $-[\partial\rho/\partial t]\,\delta x\delta y\delta z$. The minus sign indicates that if the flow terms are positive, more material flows out of the cube than flows in and the density decreases. Thus

$$-\frac{\partial\rho}{\partial t} = \frac{\partial[\rho u]}{\partial x} + \frac{\partial[\rho v]}{\partial y} + \frac{\partial[\rho w]}{\partial z} + \dot\mu_i. \tag{8.62}$$

Defining $\vec{u} = [u, v, w]$, the equation can be written

$$\frac{\partial\rho}{\partial t} + \nabla\cdot[\rho\,\vec{u}] + \dot\mu_i = 0. \tag{8.63}$$

With constant density and no melt, this equation reduces to $\nabla\cdot\vec{u} = 0$, an approximation introduced previously as Eq. 8.37. Comparison of expressions relates the local rate of density change, $\partial\rho/\partial t$, to the densification rate:

$$\frac{D\rho}{Dt} = \frac{\partial\rho}{\partial t} + \vec{u}\cdot\nabla\rho = \frac{\partial\rho}{\partial t} + u\frac{\partial\rho}{\partial x} + v\frac{\partial\rho}{\partial y} + w\frac{\partial\rho}{\partial z}. \tag{8.64}$$

In a steady state, $\partial\rho/\partial t = 0$ and the densification rate determines the density gradient, $\nabla\rho$. To a good approximation, this describes the situation in firn. As new accumulation buries firn, the additional weight drives densification (Chapter 2). Rapid densification leads to a large gradient of density with depth and hence to thin firn; slow densification leads to thick firn.

8.5.4.2 Approximation of Incompressibility

A block of ice without voids is essentially incompressible over the ranges of temperature and pressure prevailing in terrestrial ice bodies (Section 2.2.1). On glaciers, the firn layer compresses readily because it contains interconnected voids. In the ice beneath, closure of bubbles and fractures permits slow densification, and internal melt permits slow volumetric strain. Quantitatively, however, these processes below the firn are trivial compared to strain rates associated with the large-scale patterns of ice velocity. Thus, in most glacier dynamics problems, the volumetric strain rate can be set to zero; $\nabla\cdot\vec{u} = 0$ replaces Eqs. 8.61 and 8.64.

8.5.5 Vertically Integrated Continuity Equations

A glacier's surface elevation, S, and thickness, $H = S - B$, directly relate to glacier mass, to forces, and to ice flux. Furthermore, accurate and spatially extensive measurements of S and H can be obtained. We now derive relations for these important quantities, by combining vertical integration of the local conservation relation with other information. Whillans (1977) and Reeh et al. (1999b) discussed some aspects of the derivations in greater detail than possible here.

8.5.5.1 Kinematic Relation for Surface Elevation Change

At any coordinate, the elevation of the glacier surface S changes with time because of surface accumulation and ablation and ice flow. The kinematic requirement, independent of the dynamical situation, states that

$$\frac{\partial S}{\partial t} = \frac{\dot{b}_s}{\rho_s} + w_s - u_s \frac{\partial S}{\partial x} - v_s \frac{\partial S}{\partial y}, \tag{8.65}$$

with u_s, v_s, and w_s the components of ice velocity at S. Here $\dot{b}_s$ is the surface specific balance rate, the mass per unit time per unit area added to the surface by accumulation or removed by ablation (see Chapter 4). Dividing by the density of the surface material ρ_s converts $\dot{b}_s$ to a thickness per unit time. The surface may be snow, firn, or ice.

If $w_s > 0$, upward flow raises the surface, unless ablation ($\dot{b}_s < 0$) removes all the upwelling material. Conversely, downward flow $w_s < 0$ lowers the surface, unless accumulation prevails. But the surface also rises or falls as the horizontal flow (u_s, v_s) carries surface topographical features along the glacier. The final two terms account for such *advection of topography*.

8.5.5.2 Mass Change of a Vertical Column

To obtain a relation for the mass change of a vertical ice column, we integrate Eq. 8.62 directly:

$$\int_B^S \frac{\partial \rho}{\partial t}\, dz = -\int_B^S \left[\frac{\partial}{\partial x}(\rho u) + \frac{\partial}{\partial y}(\rho v) \right] dz - w_s\, \rho_s + w_b \rho_b \underbrace{- \int_B^S \dot{\mu}_i\, dz}_{\dot{b}_e}. \tag{8.66}$$

The term $\dot{b}_e$, the englacial mass balance, is the total melt in the column (times -1). Next bring the derivatives outside the integrals. This introduces new terms:

$$\int_B^S \frac{\partial \rho}{\partial t}\, dz = \frac{\partial}{\partial t} \int_B^S \rho\, dz - \rho_s \frac{\partial S}{\partial t} + \rho_b \frac{\partial B}{\partial t}, \tag{8.67}$$

$$\int_B^S \frac{\partial}{\partial x}(\rho u)\, dz = \frac{\partial}{\partial x} \int_B^S \rho u\, dz - \rho_s u_s \frac{\partial S}{\partial x} + \rho_b u_b \frac{\partial B}{\partial x}, \tag{8.68}$$

and a similar relation for ρv. Combining with Eq. 8.65 and a relation like it for B (Eq. 8.76) gives the fundamental relation for the change of mass:

$$\frac{\partial}{\partial t} \int_{B}^{S} \rho dz = \underbrace{\dot{b}_s + \dot{b}_e + \dot{b}_b}_{\dot{b}} - \underbrace{\left[\frac{\partial Q_x}{\partial x} + \frac{\partial Q_y}{\partial y} \right]}_{\nabla \cdot \vec{Q}}. \tag{8.69}$$

Ice flow influences the column mass only through the divergence of the mass flux per unit width, $\nabla \cdot \vec{Q}$. (See Eq. 8.1 for definition of a component of Q.) Here $\dot{b}$ is the specific mass accumulation rate for the entire ice thickness, the sum of surface, englacial, and basal terms (Chapter 4), given in units of mass per unit time per unit area.

8.5.5.3 Change of Ice Thickness and the Ice-equivalent Thickness

If density is a constant, ρ_i, then the change of column mass is simply $\rho_i \, \partial H / \partial t$. Taking x parallel to the flow and assuming constant density, Eq. 8.69 gives the relation for ice thickness as

$$\frac{\partial H}{\partial t} = \dot{b}_i - \frac{1}{\rho_i} \frac{\partial Q}{\partial x}, \tag{8.70}$$

where $\dot{b}_i$ denotes the specific mass balance in length per unit time ($\dot{b}_i = \dot{b}/\rho_i$, for an ice density ρ_i). Figure 8.16 illustrates the processes adding or removing mass from the column.

For the general case of variable density, the *ice-equivalent thickness*, $\hat{H}$, is often used. This refers to the thickness if the glacier were compressed to a uniform density ρ_i. That is, $\rho_i \hat{H} = \bar{\rho} H$. The left-hand side of Eq. 8.69 becomes

$$\frac{\partial}{\partial t} \int_{B}^{S} \rho dz = \rho_i \frac{\partial \hat{H}}{\partial t}. \tag{8.71}$$

Most ice sheet models are formulated in terms of $\hat{H}$.

Avoiding expressions with $\bar{\rho}$ simplifies the mathematics. Thus we define a correction factor h_ρ, such that $\hat{H} = H - h_\rho$. Specifically,

$$h_\rho \equiv \frac{1}{\rho_i} \int_{B}^{S} [\rho_i - \rho] \, dz. \tag{8.72}$$

This integral calculates the missing mass in a vertical column through the glacier, compared to a solid ice column of the same height. We refer to h_ρ as the *column mass deficit*. A general relation for thickness changes is then, from Eq. 8.69,

$$\rho_i \frac{\partial H}{\partial t} = \rho_i \frac{\partial h_\rho}{\partial t} + \dot{b} - \nabla \cdot \vec{Q}. \tag{8.73}$$

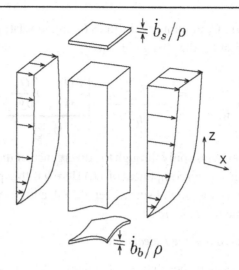

Figure 8.16: Illustration of the main processes adding or removing mass from a vertical column through the ice thickness: surface and basal specific mass balances and divergence of flux (the case here shows positive divergence and hence removal of mass). Schematic adapted from Whillans (1977).

A zero value for the term with $\partial h_\rho / \partial t$ occurs in two situations: a constant density and a density profile that moves up and down with changes in surface elevation (i.e., a density profile invariant as a function of depth $S - z$).

8.5.5.4 Profile of Vertical Velocity

The vertical velocity $w(z)$ at an arbitrary elevation within the glacier differs from w at the base of the ice mass, $w(\text{B})$, by the integral of the vertical normal strain rate:

$$w(z) = w(\text{B}) + \int_{\text{B}}^{z} \dot{\epsilon}_{zz} \, dz. \tag{8.74}$$

Into this we substitute an expression for $\dot{\epsilon}_{zz} = \partial w / \partial z$ from Eq. 8.61. This gives w at any depth $S - z$ as

$$w(z) = w(\text{B}) - \int_{\text{B}}^{z} \left[\frac{1}{\rho} \frac{D\rho}{Dt} + \frac{\dot{\mu}_i}{\rho} \right] dz - \int_{\text{B}}^{z} \left[\frac{\partial u}{\partial x} + \frac{\partial v}{\partial y} \right] dz. \tag{8.75}$$

This is the general form of Eq. 8.38; a new term provides a correction for densification and internal melt. This correction normally is negligible – except in the firn layer, where densification significantly increases the magnitude of w.

Equation 8.75 needs an expression for $w(B)$. Kinematics requires that at the lower boundary

$$w(B) = \frac{\partial B}{\partial t} + \frac{\dot{b}_b}{\rho_i} + u_b \frac{\partial B}{\partial x} + v_b \frac{\partial B}{\partial y}. \tag{8.76}$$

Terms are changes in the basal elevation, at rate $\partial B/\partial t$; thickness per unit time added or removed by basal melt or freeze-on (basal mass balance rate $\dot{b}_b$); and basal slip (u_b, v_b) up or down a sloping bed.

8.5.5.5 Full Relation for Surface Elevation Changes

Combining Eqs. 8.75 and 8.76 with Eq. 8.65, bringing derivatives outside the integrals, and using the flux $q = \bar{u}H$ (Eq. 8.1) show that

$$\frac{\partial S}{\partial t} - \frac{\partial B}{\partial t} = \underbrace{\frac{\dot{b}_s}{\rho_s} + \frac{\dot{b}_b}{\rho_i} - \int_B^S \frac{\dot{\mu}_i}{\rho} \, dz}_{\text{Accumulation Terms}} - \underbrace{\nabla \cdot \vec{q}}_{\text{Flux Divergence}} - \underbrace{\int_B^S \frac{1}{\rho} \frac{D\rho}{Dt} \, dz}_{\text{Densification}} \tag{8.77}$$

$$\text{with} \qquad \nabla \cdot \vec{q} = \frac{\partial q_x}{\partial x} + \frac{\partial q_y}{\partial y} \tag{8.78}$$

This relation is equivalent to Eq. 8.73, but now showing how specific processes govern changes in surface elevation, if the bed elevation changes, the density varies, and accumulation or loss occur within the ice and at the surface and base.

8.5.5.6 Discussion of Vertically Integrated Continuity Relations

1. In most situations, the dominant terms in Eqs. 8.73 and 8.77 are the surface balance rate $\dot{b}_s$ and the flux divergence. The former depends on climate. The latter describes how spreading of flow thins the ice. These expressions thus describe a competition between climate and ice flow. At centers of accumulation, glacier growth is opposed by outflow. In regions of ablation, inflow opposes glacier shrinkage. For ice shelves, not only climate but also oceanographic conditions matter; they control the basal mass balance $\dot{b}_b$, a sometimes large term.

2. Most analyses of the long-term evolution of an ice mass, or its steady-state properties, can use the following relations:

$$\frac{\partial H}{\partial t} = \dot{b}_i - \nabla \cdot \vec{q} \qquad \text{with} \qquad \dot{b}_i = \frac{1}{\rho_i} \left[\dot{b}_s + \dot{b}_b \right]. \tag{8.79}$$

(As before, $\dot{b}_s$ and $\dot{b}_b$ are mass per unit time per unit area while $\dot{b}_i$ is length per unit time.) This approximation assumes (1) a negligible internal mass balance compared to surface and basal ones, and (2) a uniform density with value ρ_i. The latter allows $q = \bar{u}H$ to replace $Q/\rho_i = \overline{\rho u}H/\rho_i$. How much error does this introduce? The question centers on whether $\bar{\rho}$, the average over depth, differs much from ρ_i. In an ablation zone, or a thick ice sheet, the two are essentially identical. On a small polar ice cap or a thin ice shelf, the firn thickness may be 40 m, with an average density of $650 \, \text{kg m}^{-3}$. For a total ice thickness of 500 m, the error from using $\overline{\rho}$ instead of ρ_i is about 2%. For a total ice thickness of 200 m, the error increases to about 6%. Such errors are often small compared to other uncertainties but not always negligible.

3. In the firn layer, variable density significantly influences vertical velocities. In steady state, the mass flux ρw is constant with depth and the velocity within the firn follows $w = \rho_f w_f / \rho$. Here ρ_f and w_f indicate density and velocity at the bottom of the firn ($\rho_f \approx 830 \, \text{kg m}^{-3}$). For a typical surface density of $\rho_s \approx 300 \, \text{kg m}^{-3}$, w at the surface is nearly three times w_f. The densification term in Eq. 8.77 accounts for the difference. Figure 8.17 illustrates the observed variation of vertical velocity within the firn column at Siple Dome, Antarctica (Hawley et al. 2004). The pattern is close to the one expected for steady state.

4. Over annual and yearly timescales, surface elevations fluctuate because of variable firn density (Eq. 8.77). This complicates studies of mass balance. As discussed in Chapter 4, these studies attempt to measure the mass change given by Eq. 8.69, integrated over a

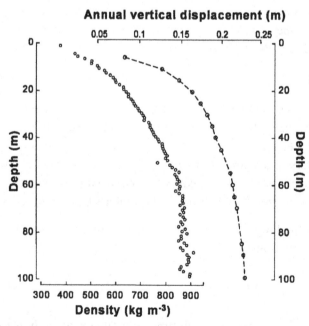

Figure 8.17: Variation of density and vertical velocity through the firn, measured in a borehole at Siple Dome, Antarctica. Data from Hawley et al. (2004). The vertical velocity was measured as displacement relative to the top of the borehole casing (approximately the surface), using a video camera to view metal bands placed on the borehole walls. Velocity is downward.

large region of an ice sheet or glacier. One method is to measure changes of surface altitude S from aircraft or satellite. As Eqs. 8.73 and 8.77 show, to convert such measurements to a mass change requires correcting for variations in the densification rate or density profile (e.g., Zwally and Li Jun 2002; Helsen et al. 2008). An expression for the densification rate (Chapter 2) as a function of variables like overburden, temperature, and ρ can be substituted into Eq. 8.75 or 8.77.

5. Do changes in the basal elevation, $\partial B/\partial t$, matter? They can be neglected in many analyses. However, if a glacier slides, B can fluctuate over hours and days as basal cavities open and collapse. The basal elevation of ice shelves is determined by ice thickness and buoyancy and varies with tidal cycles and trends in sea level. The bed elevation of ice sheets varies over decades to millennia because of isostatic adjustments of the underlying lithosphere, which must be accounted for in ice sheet models and multiyear altimetry studies. On geological timescales, erosion and tectonic uplift are major factors.

8.5.5.7 Applications of the Vertically Integrated Continuity Equations

Equation 8.69 (or 8.79) provides the fundamental relation for analyzing the response of glaciers to changes in mass balance and other forcings, the subject of Chapter 11. Integrating this equation over a region of a glacier leads to several strategies for quantifying glacier mass changes, discussed in Chapter 4. The steady-state form of the equation $\dot{b}_i = \nabla \cdot \vec{q}$ is useful for understanding the surface profiles of ice sheets (see Section 8.10). The general version gives a relationship showing the factors that control longitudinal strain rate (Section 8.7.1) and an expression for thickness changes of an ice shelf (Section 8.9.3.8).

8.5.5.8 Emergence Velocity and Altitude Changes of a Marker

The term *emergence velocity* refers to the upward or downward flow of ice relative to the glacier surface at a fixed x, y coordinate.[5] It is a "submergence velocity" in the accumulation area. It equals the sum of the flow terms in Eq. 8.65: $w_s - u_s \partial S/\partial x - v_s \partial S/\partial y$. Equation 8.75, with $z = S$, gives the expression for w_s. At a fixed coordinate, the glacier surface elevation changes according to the sum of emergence velocity and the amount of ice added to the surface.

Emergence velocity differs from the vertical motion of a marker, such as a stake, that moves with the ice. Choose x and z coordinates to be true horizontal and true vertical, as usually obtained in surveys (hence z refers to the altitude), and let x be oriented along the flow (implying $v_s = 0$). The altitude of the marker changes at rate w_s. In terms of the surface slope α, the surface gradient is $-\partial S/\partial x = \tan \alpha$. Thus, the altitude change of the marker equals the emergence velocity minus $u_s \tan \alpha$. In other words, even with zero emergence velocity, ice flow transports the marker downward along the sloping surface of the glacier, decreasing the marker's altitude.

---

[5] An emergence velocity can also be defined, however, for a coordinate that moves with the flow.

8.6 Effects of Valley Walls and Shear Margins

As valley glaciers, ice streams, and ice shelves flow, they shear against lateral margins. Such deformations restrain the flow along the centerline and support some of the driving stress of the ice mass. Analyses of measured flow indicate that such side drag accounts for typically about 30% of the resisting force acting on major ice streams but ranges from only a few percent to more than 90% (Table 8.3). The large values also apply to some ice shelves. Here we discuss the across-glacier ("transverse") variation of flow influenced by lateral margins.

Consider a valley glacier or an ice stream, with surface slope α. Define the x-axis as positive in the direction of flow, the z-axis positive upward, and the y-axis pointing across the glacier. Depending on the case, the origin is positioned either on the centerline or at one side margin. Assume that u is the only nonzero velocity component. We are interested in how u varies with y and z and so assume, in addition, no variation with x. The vertically integrated force balance (Eq. 8.58) then relates basal drag τ_b, driving stress, and a term accounting for the side drag:

$$\tau_b = \rho g \, H\alpha + \frac{\partial}{\partial y}\left[H \, \overline{\tau}_{xy} \right]. \tag{8.80}$$

A substitution for $\overline{\tau}_{xy}$ using the viscous relation $\overline{\tau}_{xy} = \overline{\eta} \, \partial u / \partial y$, with $\overline{\eta}$ the effective viscosity of ice (Eq. 8.17), shows that on the centerline

$$\tau_b \approx \rho g \, H\alpha + \overline{\eta} \, H \, \frac{\partial^2 u}{\partial y^2}. \tag{8.81}$$

(The centerline is an axis of symmetry along which $\partial H / \partial y$ and $\partial \overline{\eta} / \partial y$ can be assumed zero.) Thus the bed stress and hence the flow along the centerline are reduced in proportion to the curvature of the across-glacier velocity profile ($\partial^2 u / \partial y^2$). The curvature is negative, since u attains its highest value in the center. Furthermore, even a small curvature may correspond to a large force; with $n = 3$ in the creep relation, small stresses imply a large viscosity η.

Determination of the velocity distribution requires a viscosity appropriate for ice, $\eta = \frac{1}{2}[A\tau_E^{n-1}]^{-1}$. With u independent of x, and u the only nonzero velocity component, $\dot{\epsilon}_{xx}$, $\dot{\epsilon}_{yy}$, $\dot{\epsilon}_{zz}$, and $\dot{\epsilon}_{yz}$ are all zero. It follows that τ_{xx}, τ_{yy}, τ_{zz}, and τ_{yz} are all zero, and so

$$\tau_E^2 = \tau_{xy}^2 + \tau_{xz}^2. \tag{8.82}$$

Deformation rates obey the relations

$$\frac{\partial u}{\partial y} = 2 \, A\tau_E^{n-1} \, \tau_{xy} \tag{8.83}$$

$$\frac{\partial u}{\partial z} = 2 \, A\tau_E^{n-1} \, \tau_{xz}. \tag{8.84}$$

In reality, some longitudinal strain usually occurs at rate $\dot{\epsilon}_{xx}$. Along the sides and near the bed, however, $\dot{\epsilon}_{xx}$ is generally small compared to $\dot{\epsilon}_{xy}$ and $\dot{\epsilon}_{xz}$; it therefore does not affect the transverse velocity profile in any essential way.

8.6.1 Transverse Velocity Profile Where Basal Resistance Is Small

By "small" we mean that τ_b is both a small fraction of driving stress and much less than the plastic yield strength for ice ($\sim 100\,\text{kPa}$). This situation applies to ice streams with very slippery beds. For example, large regions of the West Antarctic Ice Streams (Whillans and Bindschadler), underlain by weak deformable sediment, have bed stresses less than $10\,\text{kPa}$ and in the range 5% to 30% of driving stress (Joughin et al. 2004b; Figure 8.15b). Side drag balances most of the driving stress, and a simple analysis suffices to illustrate the features of the across-glacier velocity profile $u(y)$ (Nye 1952a; Raymond 1996; Joughin et al. 2004b).

The ice stream margins overlie boundaries between slippery-bed regions (under the ice streams) and regions with essentially no sliding. The zone of rapid shearing at the boundary is known as a *shear margin*. Here we assume that the edge of the shear margin acts as a rigid wall, implying zero ice velocity in the no-sliding regions bordering the ice stream. This is not strictly true. (See Raymond 1996 for a more sophisticated analysis that does not require this assumption.)

For the slippery-bed region, assume that basal resistance is fixed at a value $\tau_b = \tau_*$ corresponding to the yield stress of sediment mantling the bed. A small value for τ_* then implies that τ_{xz} is everywhere small enough to neglect (1) the variation of u with z (Eq. 8.84) and (2) the contribution of τ_{xz} to τ_E. (τ_{xz} is small compared to τ_{xy} in Eq. 8.82, except near the centerline, where $\partial u / \partial y$ must be negligible anyway.) Further, picture the ice stream as a slab with a rectangular cross-section, resting on a uniform bed; τ_*, ρ, H, and α, and hence τ_d, do not vary across the glacier. As we are ignoring longitudinal stress gradients, the force balance reduces to (Eq. 8.80)

$$\rho g \alpha - \frac{\tau_*}{H} = \frac{\partial \overline{\tau}_{xy}}{\partial y}. \tag{8.85}$$

Both sides have been divided by H.

Call the ice stream width Y^*, and the half-width $Y = Y^*/2$. Set $y = 0$ at one side margin. The centerline, at $y = Y$ is an axis of symmetry, where $\tau_{xy} = 0$. Using this boundary condition, Eq. 8.85 gives $\tau_{xy} = [\rho g \alpha - \tau_*/H][Y - y]$. Integrating the creep relation, now $du/dy = 2\,A\,\tau_{xy}^n$, then gives the across-glacier velocity profile as

$$u(y) = \frac{2A}{n+1} \left[\rho g \alpha - \frac{\tau_*}{H} \right]^n Y^{n+1} \left[1 - \left[1 - \frac{y}{Y} \right]^{n+1} \right]. \tag{8.86}$$

Thus the shape of the transverse velocity profile is the same as for the theoretical depth profile discussed in Section 8.3.1. In the present case, shearing concentrates strongly near the margin, where stresses are highest, whereas the central region moves nearly uniformly (a *plug flow*). In both cases, velocities depend on $[\rho g \alpha]^n$. In the present case, velocities scale with Y^{n+1} rather than H^{n+1} as in the depth profile. Ice stream and ice shelf widths often greatly exceed even the largest ice thicknesses. Thus, their flow tends to be very rapid given the level of driving stress. Indeed, many ice shelves and ice streams flow at hundreds of meters per year or faster, despite driving stresses of only 10% to 30% of the typical value for glaciers.

In Eq. 8.86 the bed resistance appears in the term $[\rho g \alpha - \tau_*/H]^n$. Because this is raised to the power $n \approx 3$, the basal drag significantly slows the ice, even in a weak-bed ice stream. With $\tau_* = 0.2\tau_d$, basal drag reduces the speed at the centerline by about 50% compared to the case with $\tau_* = 0$. Note that the speed of the ice stream depends inversely on the magnitude of basal drag, a reversal of the usual situation implied by Eq. 8.35.

The low values of τ_* inferred for soft-bed ice streams (Table 8.3) represent averages over large regions. Within these regions are high-resistance patches, the "sticky spots." Such patches should create irregularities in the transverse velocity profiles.

Joughin et al. (2004b) compared measured velocity profiles across Whillans Ice Stream to curves of the form given by Eq. 8.86. Figure 8.18 illustrates two examples. The overall match of the measured shapes to the theoretical curve with $n = 3$ is very good. However, irregularities due to sticky spots appear in the measurements (compare to Figure 8.15b, which shows a nearby ice stream). Furthermore, in a few of the profiles examined by Joughin et al. the shear zone was narrower than predicted by the theory. This might reflect softening of the ice in the margins due to higher temperatures or favorably oriented crystal fabrics. Some of the profiles also show an inflection point near the margin; here the slow ice outside of the shear zone is being pulled forward by the faster ice nearby (Raymond 1996).

8.6.2 Combined Effects of Side and Basal Resistances

In most glaciers the basal drag, not the side drag, provides the largest resisting force. But analyses must include the side drag if the ice thickness or the slipperiness of the bed vary across the glacier. The simple analysis of the previous section does not work for this case, because neither τ_{xz} nor $\partial H/\partial y$ can be neglected. For the centerline of a glacier, the net force effect of side drags

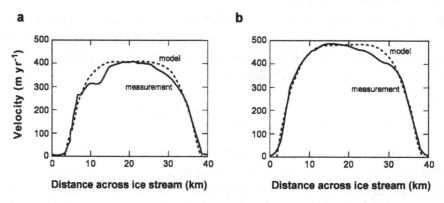

Figure 8.18: Theoretical vs. measured variation of velocity across a soft-bed ice stream, Siple Coast, Antarctica. Measured values from InSAR. (a) Ice stream "B2." (b) Ice stream "B1." Both are branches of Whillans Ice Stream. Adapted from Joughin et al. (2004b), Figures 9b and 9d.

is often summarized using the parameter f, defined by $\tau_b = f\tau_d$. So defined, f is a special case of the parameter f' defined in Section 8.2; specifically, $f = f'$ if $\tau_L = 0$ (Eq. 8.16). Because the valley walls support some of the driving stress of the glacier, f is less than 1. For a valley glacier, f depends on the valley cross-sectional form, as we discuss next; f is therefore usually called a *shape factor*.

8.6.2.1 Velocities Across a Valley Glacier: Theory

Nye (1965a) studied flow down a channel of uniform cross-section and slope; we now summarize his analysis. Both the bed and the surface tilt at an angle α. As before, assume u is the only nonzero velocity and independent of x. Set the origin on the bed and on the centerline, with the x-axis pointing down-glacier, parallel to the bed and the surface, y across-glacier, and z positive upward. Again, we wish to examine how u varies with y and z. Equations 8.82 through 8.84 give τ_E and the strain rates $\dot\epsilon_{xy}$ and $\dot\epsilon_{xz}$. It is assumed that $n = 3$. With the plausible assumption that

$$\sigma_{zz} = -\rho g\,[H - z]\cos\alpha, \tag{8.87}$$

for depth below the surface $H - z$. The stress-equilibrium equations (Eq. 8.42) reduce to the single formula

$$\frac{\partial \tau_{xy}}{\partial y} + \frac{\partial \tau_{xz}}{\partial z} = -\rho g \sin\alpha\,; \tag{8.88}$$

in other words, the down-slope weight of the glacier is supported by shearing along both horizontal and vertical planes. This relation, together with the expressions for strain rates (Eqs. 8.83 and 8.84) are sufficient to determine τ_{xy}, τ_{xz}, and u. For boundary conditions, take τ_{xz} as zero on the upper surface and u as zero on the lower surface. The solutions also apply with a sliding velocity uniform at all points.

Nye obtained an analytical solution for a semicircular channel and numerical solutions for the cases when the channel cross-section is a rectangle, half of an ellipse, and a parabola. Of these, a parabola probably best approximates a glacial valley; the other shapes imply that the valley walls are vertical at the glacier surface. The principal results are as follows:

1. On the centerline of a semicircular channel,

   $$\tau_{xz} = \frac{1}{2}\,\rho g\,[H - z]\sin\alpha \tag{8.89}$$

   at all depths; this is one-half the value for a very wide channel. For channels of other shapes but same ratio of width to depth, Eq. 8.89 applies near the surface.

2. In channels of other shapes, the increase of τ_{xz} with depth is not linear, but on the centerline it approximates the linear function

   $$\tau_{xz} = f\rho g\,[H - z]\sin\alpha. \tag{8.90}$$

Table 8.5: Shape factor f for calculation of shear stress on the centerline. (W = half-width/thickness).

W	f		
	Parabola	Semi-ellipse	Rectangle
1	0.445	0.500	0.558
2	0.646	0.709	0.789
3	0.746	0.799	0.884
4	0.806	0.849	
∞	1	1	1

Here the shape factor f is chosen to give the correct value of surface velocity when Eq. 8.84 is integrated over depth. Table 8.5 gives values of f.

Another way to derive a shape factor is to consider a spatially averaged balance of forces for a cross-section of the glacier. Call $\mathcal{A}_c$ the area of the cross-section and p the length of its perimeter, excluding the surface. Force balance then requires that

$$p\,\overline{\tau}_b = \rho g\,\mathcal{A}_c \sin\alpha, \tag{8.91}$$

where $\overline{\tau}_b$ is the shear stress averaged along the perimeter. If we assume that τ_b on the centerline equals this average, then $f = \mathcal{A}_c/Hp$ for a centerline thickness of H. The two methods give the same value of f for $W = 1$; otherwise the values differ by, on average, about 10% (Budd 1969, p. 45). Thus the value of f depends to some extent on the basal boundary condition – uniform velocity versus uniform shear stress.

3. Figure 8.19 illustrates how surface velocity varies on a line across a glacier of parabolic cross-section, for different values of $W = Y/H$. Velocity profiles for rectangular and semielliptical channels are broadly similar to those for the parabolic form. In each case the velocity changes little across the central part of the glacier and decreases rapidly toward the sides.

4. The inflection points in Figure 8.19 show that the shear strain rate component $\partial u/\partial y$ attains a maximum at a short distance from the edge, not at the edge itself. This explains why crevasses are often found near the edge of a glacier while the ice at the edge may be unbroken.

5. The drag of the valley walls reduces the velocity considerably. For $W = 2$, for example, the centerline velocity drops to only about one-quarter the velocity in a very wide channel. The reduction varies in proportion to f^n.

6. For a parabolic valley, the mean velocity over a cross-section (mean over both depth and width) matches, to within a few percent, the mean surface velocity on a profile across

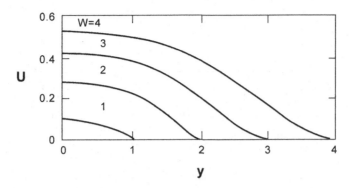

Figure 8.19: Computed variation of surface velocity across a glacier in various parabolic channels. y is the distance from the centerline and W is the half-width of the glacier, both divided by the ice thickness on the centerline (hence $W = Y/H$). U is the ratio of the velocity to the velocity – for same thickness and slope – in a very wide channel. Adapted from Nye (1965a).

the whole glacier. This theoretical result holds for values of W between 2 and 4, which includes the majority of valley glaciers. This is an important practical result. It means a good estimate of ice flux can be obtained without needing to measure velocity at depth. Instead, flux can be estimated as the product of cross-sectional area with the mean velocity obtained from measurements across the glacier surface. The cross-section needs to be known from radio-echo sounding or another method.

8.6.2.2 Velocities Across a Valley Glacier: Observations

Raymond (1971b) used five boreholes to measure velocity in a transverse section of Athabasca Glacier, Canada. Figure 8.20 compares the observed velocity distribution to the theoretical distribution for a parabolic channel with $W = 2$. There are major differences:

1. In the theoretical model, sliding velocity was assumed uniform. The measured sliding velocity, on the other hand, is high in the central part of the section and decreases to near zero at the edge. This variation of sliding velocity accounts for two more differences, listed next.

2. The observed contours of constant velocity are approximately semicircular and differ significantly from the shape of the channel and from the contours in Nye's model.

3. The shear strain rate $\partial u/\partial y$ near the sides exceeds, by a multiple of two to three, the shear strain rate $\partial u/\partial z$ at the base on the centerline. Thus friction near the sides supports the glacier more strongly than does basal friction near the centerline. This implies that the normally used value of the shape factor would be too high. This effect may be more pronounced on Athabasca Glacier than on most other glaciers because sliding contributes an unusually high proportion – about 80% – of the total motion on the centerline in the region studied.

a

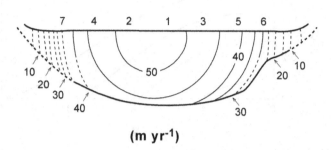

(m yr⁻¹)

b

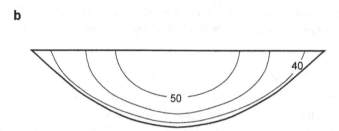

Figure 8.20: (a) Distribution of longitudinal velocity in a cross-section of Athabasca Glacier, measured by Raymond (1971). Points 1 to 5 represent positions of boreholes. Only the surface velocity was measured at points 6 and 7. (b) Velocity distribution computed by Nye (1965a) for parabolic channel with $W = 2$. A constant sliding velocity has been introduced so that the range of velocities is about the same as observed.

Reynaud (1973) showed that, with a different basal boundary condition, the theoretical model could match the velocity distribution measured by Raymond. Reynaud assumed that the drag between ice and bedrock was proportional to the effective pressure at the bed, the normal pressure of the overlying ice minus the pressure of water at the interface. (In other words, $\tau_b = \tau_* = N \tan \varphi$.) The water pressure was calculated assuming a horizontal "water table" in the glacier, at a fixed depth below the ice surface. With these assumptions, the predicted sliding rate attains a maximum in the valley center and is much reduced near the margins, as Raymond observed.

Whether the bed of Athabasca Glacier consists of bedrock or deformable sediment is unknown; "sliding" in the preceding discussion should be interpreted as basal slip of unknown type. The velocity pattern measured at Athabasca and Reynaud's theoretical explanation of it are consistent with a bed of deforming sediment. The yield strength of a sediment decreases as the effective pressure decreases (Eq. 8.25). Low effective pressures in the valley center would reduce the sediment strength and facilitate rapid slip.

One glacier known to be underlain by sediment is Black Rapids Glacier, Alaska. Truffer et al. (2001) modelled and measured its flow. They assumed a perfectly plastic sediment, so the basal shear stress equalled the yield strength of the substrate. They found a good match to the measured transverse velocity profile, provided that some slip was allowed to occur along the side margins. The model was also consistent with observed fluctuations of velocity over time, at daily and seasonal timescales, which arose from variations of basal water pressure. At times of high water pressure, the sediment strength decreased, the glacier moved faster, and more of the driving stress was supported by side shearing; $\partial u / \partial y$ increased near the side margins.

Harbor et al. (1997) conducted an experiment similar to Raymond's; they measured flow at depth along a transverse line of eight boreholes in Haut Glacier d'Arolla, Switzerland. The overall pattern of flow in the cross-section resembled that at Athabasca. At Arolla, however, basal slip accounted for only about 65% of motion on the glacier's centerline. Moreover, the fastest slip occurred not in the center but closer to the shearing side margins, at a location where water levels in boreholes fluctuated over time.

These results emphasize that the distribution of velocity in a cross-section depends on mechanical properties of the basal interface as well as on the shape of the section. How the properties of the interface vary along the cross-section can differ considerably from place to place.

In his study of Athabasca Glacier, Raymond also measured a transverse velocity component v of up to $2\,\mathrm{m\,yr^{-1}}$, even though the valley walls are parallel. This flow compensated for excess ablation near the valley walls. Higher elevations in the central part of the glacier, relative to the side, provided the surface slope to drive this flow. Moreover, the value of v was greater at depth than at the surface, a pattern also observed in Worthington Glacier, Alaska (Harper et al. 2001). Flow greater at depth than at the surface is called *extrusion flow*. Theoretical work has shown how local patches of extrusion flow can develop when ice slides over an undulating bed (Gudmundsson 1997), but its origins in the present case are rather obscure.

8.6.2.3 Flow Around a Bend

For simplicity, all the preceding analyses focused on straight channels and ignored planform curvature. However, many mountain glaciers flow around bends in valleys, and flowlines usually curve where tributaries enter a trunk glacier. Nor is such curvature unique to mountain glaciers; the flow lines of topographically controlled ice streams can curve strongly. Some tributaries of the Jakobshavn Ice Stream, Greenland, turn by nearly 90° where they enter the main ice stream.

Meier et al. (1974) discussed flow of valley glaciers around bends. For the case of uniform curvature, the theory was elaborated by Echelmeyer and Kamb (1987). Compared to the straight channel discussed in the previous section, the following apply:

1. The surface slope, in the direction of flow, is larger on the inside of the bend than the outside. The driving stress is therefore larger on the inside too. This causes an asymmetry of the

velocity profile across the glacier surface; the ice flows faster on the inside of the bend than on the outside, at the same distance from the margin.

2. Thus the shearing $\dot\epsilon_{xy}$ is most rapid at the inside margin, where it concentrates in a narrow band adjacent to the valley wall.

3. Part of the motion around the bend is a rigid-body rotation, akin to a rotating wheel. For this component, the strain rate $\dot\epsilon_{xy}$ (which equals $[\partial u/\partial y + \partial v/\partial x]/2$) is zero, but the velocity gradient $\partial u/\partial y$ is not. Superimposed on such rotation is the component of $\dot\epsilon_{xy}$ due to shearing against the valley walls. From this combination, the strain rate and τ_E on the glacier surface both attain a zero value not at the centerline, as in a straight channel, but at a location closer to the inside of the bend (called the *stress axis*). The maximum of the velocity u occurs between the stress axis and the centerline.

All of these features are consistent with measurements of flow rate, surface topography, and crevasse distributions around a bend of Blue Glacier, Washington State (Meier et al. 1974; Echelmeyer and Kamb 1987). At this glacier, however, large longitudinal deformations associated with local changes of the slope complicate the observed strain rate patterns.

8.7 Variations Along a Flow Line

Extending and compressing flows were introduced in Section 8.4, and Section 8.4.2 discussed their characteristics at a single location on a glacier. Here we discuss the broader scale variations of flow along a glacier or an ice sheet flow line.

8.7.1 Factors Controlling Longitudinal Strain Rate

Differentiation of the expression $q_x = \bar u H$, where $\bar u$ is the value of u averaged over the ice thickness H, gives

$$\frac{\partial q_x}{\partial x} = H\,\frac{\partial \bar u}{\partial x} + \bar u\,\frac{\partial H}{\partial x}. \tag{8.92}$$

Substituting this expression, and a similar one for $\partial q_y/\partial y$, into the continuity relation (Eq. 8.79) gives

$$\frac{\partial H}{\partial t} = \dot b_i - H\left[\frac{\partial \bar u}{\partial x} + \frac{\partial \bar v}{\partial y}\right] - \bar u\,\frac{\partial H}{\partial x} - \bar v\,\frac{\partial H}{\partial y}. \tag{8.93}$$

If the x-axis follows a flow line, $\bar v$, but not $\partial \bar v/\partial y$, is zero. Hence, in a steady state ($\partial H/\partial t \approx 0$),

$$\frac{\partial \bar u}{\partial x} = \frac{\dot b_i}{H} - \frac{\bar u}{H}\frac{\partial H}{\partial x} - \frac{\partial \bar v}{\partial y}. \tag{8.94}$$

This shows the factors that determine the longitudinal strain rate averaged over depth. Extending flow is favored by accumulation ($\dot{b}_i > 0$), by downstream thinning of ice ($\partial H/\partial x < 0$), and by downstream narrowing of the valley or convergence of flow lines ($\partial \bar{v}/\partial y < 0$). Because specific mass balance ($\dot{b}_i$) is positive in accumulation zones and negative in ablation zones, extending flow predominates in accumulation zones, while compressing flow predominates in ablation zones (Section 8.1.2 and Figure 8.3). The thinning term often overwhelms this pattern, however, where ice thickness varies considerably across irregular bedrock topography or across transitions between sticky and slippery regions of the bed. The transverse strain rate, usually small compared with the longitudinal one, can be estimated from

$$\frac{\partial \bar{v}}{\partial y} = \frac{\bar{u}}{Y} \frac{dY}{dx}, \tag{8.95}$$

where Y denotes valley width or, on an ice sheet, the distance between adjacent flow lines (Nye 1959).

Equation 8.94 gives the depth-averaged value of $\partial u/\partial x$. If a glacier moves entirely by basal slip, $\partial u/\partial x$ does not vary with depth. Otherwise, variation is expected. In some situations, such as extending flow driven by accumulation or by increasing valley slope, extension or compression prevail throughout the ice thickness, except in a basal layer influenced by bed roughness. A different pattern occurs in the presence of large-amplitude basal topography. For example, upstream of a prominent subglacial hill the glacier's upper layers must undergo longitudinal extension even while the lower layers compress.

8.7.2 Local-scale Variation: Longitudinal Stress-gradient Coupling

Over distances short compared to the length of a glacier, the flux per unit width $q = \bar{u}H$ is nearly constant, except where a valley abruptly narrows or widens. Thus a glacier flows fastest through thin-ice regions overlying topographic highs; increased surface slope and driving stress impel the faster flow. Conversely, the ice moves slowly through thick regions overlying topographic depressions; the surface slope and driving stress both decrease. For a mountain glacier descending a valley, long-wavelength variations of surface slope mimic variations of the valley slope. The fastest flow and thinnest ice occur in steep regions; the slowest flow, in regions of gentle slope. Differences in the slipperiness of the glacier bed can also affect the flow regime; thinner ice and faster flow occur in regions of well-lubricated bed.

The extending or compressing flow associated with transitions between such regions enhances the rate of vertical shearing, $\partial u/\partial z$, because of the nonlinear creep properties of ice (Section 8.4.2). A different effect, especially important in mountain glaciers, is the adjustment of flow and basal shear stress by longitudinal stress gradients (Orowan 1949; Budd 1968, 1971; Nye 1969b; Kamb and Echelmeyer 1986a). From Eq. 8.58, with $\tau_{yy} = 0$ and with no side drag or bridging effect, the balance of forces requires that

$$\tau_b = \tau_d + 2\frac{\partial}{\partial x}[H\bar{\tau}_{xx}] = \tau_d - \tau_L, \tag{8.96}$$

with τ_L signifying the "longitudinal drag" as previously defined. The deviatoric stress $\bar{\tau}_{xx}$ increases with the stretching rate and the effective ice viscosity $\bar{\eta}$ (defined by $\bar{\tau}_{xx} = 2\bar{\eta}\,\partial u/\partial x$). Thus

$$\tau_b \approx \rho g H \alpha + 4H\bar{\eta}\frac{\partial^2 u}{\partial x^2} + 4\frac{\partial\bar{\eta}H}{\partial x}\frac{\partial u}{\partial x}. \tag{8.97}$$

(If necessary, the side drag can be accounted for by multiplying the right-hand side of these relations by a shape factor f as in Section 8.6.2.)

Consider flow through a region with a local maximum of u and τ_d, such as an icefall (Figure 8.21). Ice moving along the glacier first extends as it enters the icefall and then compresses as it exits. In the middle of the icefall the glacier is restrained by both tension upstream and compression downstream; longitudinal drag $\tau_L > 0$ and τ_b decrease relative to τ_d. Upstream of the icefall, in contrast, tensional stresses pull the glacier forward; τ_L acts in the same direction as gravity ($\tau_L < 0$), and τ_b increases. Downstream of the icefall, compressional stresses push the glacier forward, and τ_b again increases. Thus the variations of τ_b resemble a spatial average of the variations of τ_d; as the figure illustrates, the action of τ_L smooths the pattern of τ_d to give the pattern of τ_b. Such an interaction of forces and flow up-glacier and down-glacier is referred to as *longitudinal stress-gradient coupling* (see also Section 8.5.3).

Such variations of τ_b are then reflected in variations of the glacier velocity u, since relations for the vertical shear profile and the slip rate connect u with τ_b (Section 8.3). Thus the variation of u through the icefall also follows a smoothed version of the driving stress. In other words, the faster flow induced by the icefall spreads up- and down-glacier. But, in the icefall itself, the glacier moves more slowly than expected from the driving stress. (With reference to Eq. 8.97, the smoothing derives from the term $\partial^2 u/\partial x^2$.)

An analogous smoothing of $u(x)$ occurs at a transition from slow to fast basal slip. The transition marks a zone of increased tension. Upstream of the transition, the glacier is pulled forward by the tension in the transition zone, so $\tau_L < 0$ and $\tau_b > \tau_d$. In contrast, the faster flowing ice downstream is held back by tension acting in the transition zone.

Kamb and Echelmeyer (1986a) analyzed Eq. 8.97, by substituting for τ_b a relation in terms of u. The relation they used is appropriate for flow by internal deformation or slow sliding; the analysis does not apply to rapid soft-bed sliding. Their results are approximations but illustrate the essential features of how longitudinal coupling acts as a spatial-averaging process. (More complex analyses of the problem and criticisms of the simple theory are given by McMeeking and Johnson 1985, Kamb 1986, and Gudmundsson 2003.) Call the mean value of u along the

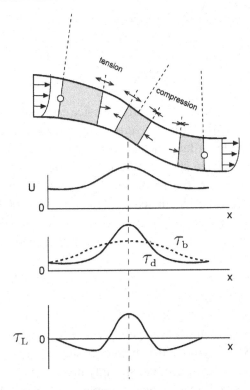

Figure 8.21: Schematic showing longitudinal variations as a glacier flows through an ice fall. Top panel, thick lines, show the surface and bed of the glacier (vertical scale highly exaggerated). Long dashed lines delimit regions of tension and compression, associated with the velocity pattern U. The three shaded regions represent sections of glacier, with arrows depicting the stresses acting at boundaries. The blocks are being pushed down-glacier (the left and right sections) or pushed up-glacier (the middle one). Bottom two panels show driving stress, basal drag, and longitudinal drag along the glacier.

glacier u_o and the variations about this value Δu. For small variations, the linearized form of the equations yields the solution

$$\Delta u(x) = \frac{u_o}{2\ell} \, F * J, \tag{8.98}$$

where F is a forcing function, J a filter function, and $*$ denotes convolution. The deviation of $\tau_d^n H$ (or, equivalently, of $\alpha^n H^{n+1}$) about its mean value defines the forcing F at any location along the glacier. This accounts for the direct contribution of driving stress variations to flow variations. (A different forcing proxy, perhaps $\alpha^m H^m$, would be used if sliding dominates.) The function J then averages these contributions over distance. J is center-weighted. To a good approximation, the distribution of weights defines a triangular shape, with a total width along its base of 4ℓ. (The term *longitudinal averaging length* usually means 4ℓ.) The theory gives the length scale ℓ as

$$\ell = 2H \sqrt{nf\,\overline{\eta}/3\,\eta_v},\tag{8.99}$$

in which n is the creep exponent, $\overline{\eta}$ the effective ice viscosity for longitudinal deformation, η_v the effective ice viscosity for the vertical profile of shear, and f the shape factor (Section 8.6.2). For temperate valley glaciers, Eq. 8.99 predicts a length scale ℓ of 1 to 3 times the ice thickness H. For ice sheets, cold ice near the surface makes the glacier stiffer to stretching than to basal shear; the predicted ℓ therefore increases to 4 to 10 times H. A slippery bed also increases the length scale (Gudmundsson 2003); it has the same effect as decreasing the viscosity of basal layers and hence decreasing η_v in Eq. 8.99. The more a glacier moves by slip instead of internal deformation, the greater the distance of stress transmission.

Consider "local" variations of stress and flow along a glacier – those over distances of about $10H$ or less ($20H$ or less for ice sheets). In mountainous topography, variations in both H and α lead to large variations of driving stress along a glacier. The velocity, however, varies much less; it varies only with the center-weighted average of H and α, as specified by Eqs. 8.98 and 8.99 and illustrated in Figure 8.21. Kamb and Echelmeyer showed that measured velocities along two valley glaciers, Blue and Variegated, are strongly modulated by this effect. This has been confirmed elsewhere. Figure 8.22 shows one example, from McCall Glacier in Alaska; only the broad pattern of variations in driving stress appears in the flow. The considerable variations of driving stress over short distances do not produce similar variations of the velocity (Rabus and Echelmeyer 1997). At this site, the ice thickness was $H \approx 150\,\text{m}$, and the velocity data were best matched if $\ell \approx 3H$. Kavanaugh and Cuffey (2009) examined longitudinal coupling along a larger and colder glacier in Antarctica. Taylor Glacier is a 0.5- to 1-km-thick outlet

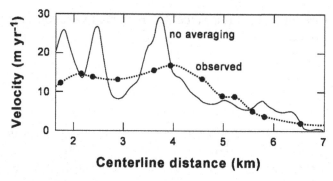

Figure 8.22: Observed surface velocity along McCall Glacier, Alaska, compared to variations of driving stress (expressed as the equivalent velocity, calculated with no spatial averaging of the driving stress). Adapted from Rabus and Echelmeyer (1997).

glacier of the East Antarctic Ice Sheet. Here, the relation between velocity and driving stress variations implied an ℓ of about $2H$ to $4H$, with larger values applying to the colder, inland part of the glacier. The analysis confirmed that ℓ varies with H; the length scale of averaging decreased where the glacier flows over bedrock steps and thins. In contrast to the linearized theory, however, the forcing function needed to match the velocity data was $\tau_d H$, not $\tau_d^3 H$. The linearized theory does not necessarily work well, in detail, for the large-amplitude variations that occur on real glaciers.

To summarize: the variations of flow along a glacier or ice sheet depend on thickness and surface slope values averaged over a distance of several ice thicknesses along-flow. The preceding analysis gives guidance as to the width and shape of the averaging function, but the analysis is semiquantitative.

The analysis further implies that flow variations over length scales greater than $10H$ to $20H$ are *not* influenced by longitudinal stress gradients (consistent with values given in Table 8.3), except in the unusual case that both basal resistance and side drags are ineffective. We next consider an example of the typical large-scale longitudinal variation.

8.7.3 Large-Scale Variation

Variations in the flow regime over large scales ($>20H$) arise from all the terms in Eq. 8.94. Mass balance drives persistent increases of flow in accumulation zones and decreases in ablation zones. Transverse strain significantly increases flow where the discharge from broad drainage basins funnels into narrow ice streams or valley glaciers. H systematically decreases toward the margins of ice sheets and decreases on the inland side of major subglacial mountain ranges; both situations favor extending flow. In contrast to the local-scale variations discussed in the previous section, however, these trends are spread over too great a distance to influence the force balance significantly.

We now consider the general variations along a flow line from the center of an ice sheet to the margin, using a theoretical analysis. The flow line is "simple" in that it does not cross any major subglacial landforms, the width does not vary, and there is no restraint from thinner or slower ice on its sides ($\tau_w = 0$). The distributions of stress and velocity versus depth vary along the flow line due to systematic variations in mass balance, flux, and ice thickness. Because the deformation rate of ice and the occurrence of basal sliding depend strongly on temperature, the analysis must consider both ice flow and heat flow.

Dahl-Jensen (1989a) made one such analysis. Assumed conditions are typical of southern Greenland: a flow line about 200 km long, maximum ice thickness around 2000 m, and surface temperature ranging from $-20\,°C$ at the ice divide to $-8\,°C$ at the margin. Accumulation and ablation rates vary with elevation. The bed is taken as a horizontal plane, but surface elevations are obtained as part of the solution, which is a steady state. Where the bed warms to melting point, ice slides according to Weertman's relation (Section 7.2.1). The analysis is two-dimensional (plane strain).

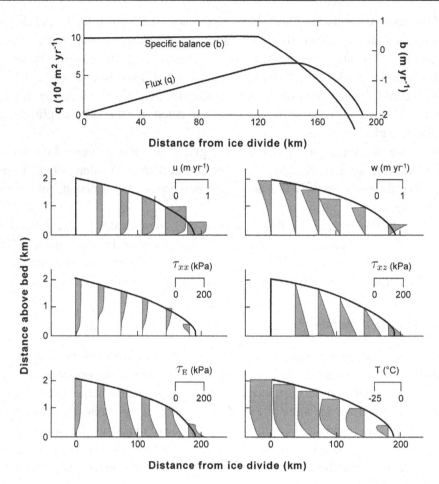

Figure 8.23: Calculated surface profile and variation, with depth, of velocities, stress components, and temperature along an ice sheet flow line (with properties similar to southern Greenland). The scale for each variable is shown on its figure. The equilibrium line is at a distance of 138 km. Adapted from Dahl-Jensen (1989a) and used with permission from the American Geophysical Union, *Journal of Geophysical Research*.

Figure 8.23 shows the result. The surface profile resembles those derived from simple analyses, discussed later in this chapter (Section 8.10). Basal temperature increases with distance from the ice divide until the melting point is reached; however, it decreases again near the margin. These temperature variations, discussed in Chapter 9, mostly derive from variations in strain heating. Because of the temperature pattern, the concentration of shear deformation against the bed becomes more pronounced toward the ice sheet margin. Vertical velocity decreases approximately linearly with depth; its surface value depends on the specific balance, horizontal velocity, and surface slope. It changes sign at the equilibrium line. The longitudinal deviatoric stress τ_{xx} is not negligible; it exceeds the shear stress in the upper third of the ice column. However, the longitudinal stress *gradient* $\partial \tau_{xx}/\partial x$ is negligible. The basal shear stress does not vary much along the flow line, which shows that the ice can be regarded as perfectly plastic, for rough calculations. Note that this model gives velocity and stress distributions at an ice divide; we discuss this setting in Section 8.9.1.

Models like this reproduce the broad-scale features of stress and velocity distributions in ice sheets. Velocities measured in boreholes, on the other hand, reflect local-scale conditions related to the form of the bed and the presence of ice layers with different properties (Figures 8.8 to 8.11). Figure 8.11, for example, suggests that shear is less concentrated near the bed than this model would suggest. Interpretation of borehole data requires detailed modelling (e.g., Hvidberg et al. 1997).

8.8 Flow at Tidewater Margins

Many glaciers in coastal Alaska, Patagonia, and the Arctic terminate in the ocean at calving fronts but remain grounded (Section 4.6.2). Far inland from the calving front, these *tidewater glaciers* undergo "normal" glacier flow. Near the terminus, their flow has unique features.

8.8.1 Theory

Meier and Post (1987) and Hughes (2003) discussed the dynamics of glaciers with marine margins; the following summary uses parts of their analyses. The front of a tidewater glacier is a vertical cliff. Call the ice thickness here H_M and the submerged thickness H_S, which equals the water depth. Near the cliff, the vertical force due to the static weight of the ice exceeds the horizontal static force due to the water (see Section 8.2.4). This difference gives the margin stress, the net force effect per unit width, as (from Eq. 8.31)

$$\phi_M = \frac{1}{2} \rho_i g H_M^2 - \frac{1}{2} \rho' g H_S^2, \tag{8.100}$$

for density of water ρ'. The quantity ϕ_M can be regarded as a force that effectively pulls seaward on the tidewater end of the glacier.

Define the x-axis as horizontal, pointing in the direction of flow. Consider a longitudinal section of the glacier, from an upstream position x to the front margin at $x = X_m$. The section is not only pulled seaward by the force ϕ_M but also is pushed in the same direction by driving stresses acting between x and X_m arising from the surface slope (Eq. 8.6). Opposing these gravitational forces are basal drag (τ_b) and drag from the valley walls (τ_W). The residual equals a net longitudinal force F_L on the section, acting down-glacier:

$$F_L(x) = \phi_M + \int_x^{X_m} [\tau_d - \tau_W - \tau_b] \, dx. \tag{8.101}$$

In turn, this force must be supported on the inland side (x) by a tensional resistance R_{xx}, integrated through the ice thickness, deriving from extending flow. Assuming negligible transverse deformation ($\tau_{yy} = 0$), then $R_{xx} = 2\bar{\tau}_{xx}$, and F_L at x takes the value (from Eq. 8.55)

$$F_L(x) = 2 H \bar{\tau}_{xx} = 4\bar{\eta} H \frac{\partial u}{\partial x}. \tag{8.102}$$

The second equality uses a viscous creep approximation to set $\bar{\tau}_{xx}$ proportional to the along-flow stretching rate, $\partial u/\partial x$, and to an effective viscosity, $\bar{\eta}$. Large and fast-flowing tidewater glaciers move primarily by basal slip (Meier and Post 1987), so the vertical variation of u can be ignored.

Although no known relations accurately predict slip rates, one established fact is that low effective pressure at the bed (a near-zero value for $N = P_i - P_w$) allows rapid slip, whether the bed is deformable or not. Thus, we insert into the force balance statement a simple phenomenological slip relation $u = \lambda_* \tau_b^m N^{-1}$, where the lubrication parameter λ_* depends on bed properties. Thus:

$$4\bar{\eta} H \frac{\partial u}{\partial x} = \underbrace{\phi_M + \int_x^{X_m} \tau_d \, dx}_{\text{Driving Force}} - \underbrace{\int_x^{X_m} \left[\overbrace{[\lambda_*^{-1} N u]^{1/m}}^{\tau_b} + \tau_W \right] dx}_{\text{Resisting Force}}. \tag{8.103}$$

Both components of the resisting force should increase as the glacier flows faster: τ_W because of the larger shear strain rates against the valley walls, and τ_b because of increased drag around bed irregularities and localized patches of high friction. On the other hand, a drop in basal effective pressure decreases τ_b and, at the limit of flotation, lets $\tau_b = 0$. If this limit is approached, the glacier can flow very rapidly without generating much resistance from the bed. Extending flow ($\partial u/\partial x > 0$) arises from the driving force exceeding the resisting force in Eq. 8.103. A large ice thickness (hence large ϕ_M and τ_d) and small N thus favor extension.

It is not obvious from Eq. 8.103 by itself, however, if a glacier stretches and flows rapidly; observations must be made. The greatest ambiguity concerns the surface slope leading to the terminus. A steep surface increases τ_d and hence the driving force, but also facilitates subglacial water flow and hence favors high N and increased resistance.

In Eq. 8.103, $4\bar{\eta} H \, \partial u/\partial x$ and ϕ_M together account for the longitudinal force, the integral of τ_L along the flow line. It is meaningful to separate them as shown because ϕ_M depends explicitly on gravity and represents the effect of forces that drive flow. (We say the "effect of forces" because the true forces – the weight of ice near the terminal cliff – act vertically. See Section 8.2.4.) The sum labelled "driving force" represents, specifically, the force that causes extending flow on the glacier (stretching rate $\dot{\epsilon}_{xx} = \partial u/\partial x$). The velocity u near the front of the glacier increases with the driving force because $u = u_o + \int \dot{\epsilon}_{xx} \, dx$. Here the integral is calculated along a flow line and u_o indicates the velocity at a point far enough inland for the effects of the terminus and its slippery-bed region to be negligible.

Note that Eq. 8.103 applies to any distance inland from the terminus and indeed to any glacier moving by slip (letting $\phi_M = 0$ in the absence of a margin force). Where the bed is strong, however, the basal drag, velocity, and driving stress vary along a glacier in an interdependent way such that the $\int \tau_d \, dx$ and $\int [\tau_b + \tau_w] \, dx$ cancel one another over distances of about 10 to 20 ice thicknesses; such interdependence arises from the style of longitudinal coupling discussed in Section 8.7.2. Thus for tidewater glaciers we must distinguish different cases: (1) If the bed is strong (τ_b of order 100 kPa) all the way to the terminus, then the effect of the margin force only extends inland about 10 to 20 ice thicknesses. Moreover, its importance rapidly diminishes inland relative to the other forces. (2) A slippery-bed region near the terminus allows for a large region of stretching and fast flow. How much ϕ_M contributes to the driving force relative to τ_d – and where the stretching occurs – depends on the geometry and bed characteristics in any specific case.

8.8.2 Observations: Columbia Glacier

Columbia is a large and well-studied tidewater glacier in southern Alaska (Tables 8.1–8.3). Icebergs released from it are a hazard to ships travelling to the port of Valdez, the outlet of the Alaska Pipeline. Before 1980, the terminus rested on a sediment shoal at a distance of about 63 km from the glacier head. The lower 20 km or so of the glacier sat in a basin with bed elevations below sea level by some 100 to 700 m. Since the early 1980s, the terminus has retreated more than 15 km into the basin. The following discussion uses information given by Meier and Post (1987), Meier et al. (1994), and O'Neel et al. (2005).

Before retreat (data from 1977), the water depth at the terminus, on the shoal, was only a few tens of meters, and the margin stress must have been small; using estimated values of 50 m for elevation near the terminus and 20 m for water depth, $\phi_M \approx 10^4$ kPa m. Along its lower 10 km, the glacier flowed at about 0.9 km yr^{-1}, with little longitudinal extension. Even before retreat, basal slip accounted for most of the glacier's velocity.

By 1985, retreat was under way. The terminus had retreated only about 1 km, but now stood in the deeper water behind the shoal (water depth of about 200 m). With a surface elevation near the terminus of about 50 m, this implies $\phi_M \approx 8 \times 10^4$ kPa m, a much larger force than before. But the glacier had also thinned near the terminus, flattening the surface and reducing driving stresses. In combination, the net driving force (Eq. 8.103) had increased near the terminus but apparently was about the same farther inland. Yet velocities had increased markedly along the entire lower glacier. At 5 km from the terminus, for example, the velocity in 1985 was about 2.2 km yr^{-1}, more than 1 km yr^{-1} faster than before retreat. Moreover, from here to the terminus the glacier stretched rapidly, at a rate $\dot{\epsilon}_{xx} \approx 0.3$ yr^{-1}.

Between these two times of observation, the surface elevation at 5 km from the terminus decreased from about 250 m above sea level to about 180 m. The bed here sits more than 300 m below sea level. Together, these observations suggest that the basal effective pressure N dropped significantly, increasing the effective lubrication of the bed. (In Eq. 8.103, the term $\lambda_*^{-1} N$ decreased.) This is the most plausible explanation for the large increase in slip rate and extension.

Closer to the terminus, the velocity and longitudinal strain rate increased even more. Increases of both lubrication and force ϕ_M probably contributed. However, observations showed that during short-period fluctuations velocity and iceberg calving both increased at times when the terminus nearly went afloat (van der Veen 1996). This indicates that lubrication by low effective pressures was the more important variable, since ϕ_M decreases when the water level rises quickly. (Water pressure on the terminal cliff pushes up-glacier.)

Borehole studies conducted on the lower glacier in 1987, in the zone 5 to 10 km from the terminus, confirmed that effective pressures at the bed were very low and sometimes fell to zero. The boreholes encountered deforming sediment of low strength, estimated as 5 to 13 kPa *in situ* (Chapter 7). The basal resistance must therefore have focused on sticky spots of some sort, perhaps bedrock knobs. At timescales of hours to days, increased water input from surface melt and rainfall caused the glacier to speed up (see Figure 11.15). This provides further evidence that effective pressures on the bed control the velocity.

The situation at Columbia Glacier is particularly interesting because of the analogy with many polar ice streams, discussed in the next section. Most of the large ice streams draining the Greenland and Antarctic ice sheets end in the ocean as tidewater glaciers or ice shelves. As at Columbia, the effective pressures at their beds must be close to zero – an inevitable consequence of deep water at termini, bed elevations well below sea level, and gentle surface slopes (in turn, a consequence of thick ice and slippery substrates). We should therefore expect the polar ice streams to flow rapidly near their grounding lines. This is observed. Furthermore, the Columbia example suggests that thinning of ice at a marine margin, from whatever initial cause, lowers effective pressures and leads to faster flow with increased longitudinal stretching. The latter contributes to further thinning, a positive feedback with dramatic consequences for the glacier. We consider this mechanism of glacier response more fully in Chapter 11.

8.9 Ice Sheets: Flow Components

The flow field of an ice sheet (e.g., Figure 8.2) can be regarded as an assemblage of components. Large regions of slow to moderate flow, varying gradually with distance, define the ordinary *flank flow* regime. Flow lines begin at *ice divides*, the boundaries between the drainage basins. Near the margins, the flow in many drainage basins concentrates into narrow and fast-flowing *ice streams* or *outlet glaciers*. Most of the large outlets terminate as tidewater glaciers or feed floating *ice shelves*.

Ordinary flank flow is characterized by moderate driving stresses, typically 30 to 120 kPa, and resistance from basal drag. The vertical shear profile should be broadly similar to the theoretical curve discussed in Section 8.3, and to the examples in Figures 8.8 to 8.11, although with significant variations related to temperatures, fabrics, and local bed topography. The stresses and velocities along a typical flow line presumably resemble those discussed in Section 8.7.3

and shown in Figure 8.23. Observations show that, in these regions, surface velocities correlate with $\tau_d^n H$, as predicted from the shear profile relations (Cooper et al. 1982). Thus, although basal slip no doubt occurs where the bed reaches melting point, the bed is typically not very well lubricated. Large *subglacial lakes*, however, do not fit this generalization at all. A few dozen large lakes exist beneath the East Antarctic Ice Sheet (see Chapter 6). Over lakes, basal resistance is absent and longitudinal and lateral viscous forces control the ice flow.

We now discuss the flow associated with each of these "atypical" components of ice sheets: ice divides, ice streams, ice shelves, and large subglacial lakes. We then turn attention, in Section 8.10, to the factors controlling an ice sheet's large-scale profiles of surface elevation and thickness.

8.9.1 Flow at a Divide

Glaciologically, ice divides are unusual places. The surface slope and hence the driving stress vanish. The direction of driving stress and flow on one side of the divide opposes that on the other. Pulled from both sides, the ice at the divide deforms in longitudinal extension ($\dot{\epsilon}_{xx} > 0$), accommodated by vertical compression ($\dot{\epsilon}_{zz} < 0$). (We define the horizontal coordinate x as perpendicular to the divide.) In contrast to most locations in grounded ice, the dominant deviatoric stresses are τ_{xx} and τ_{zz} rather than τ_{xz} or τ_{xy}. From Eq. 8.94, the depth-averaged magnitudes of compressive and extensional strain rates approximate $\dot{b}_i / H$, the ratio of annual accumulation to ice thickness; typical values are $5 \times 10^{-6}\,\mathrm{yr}^{-1}$ in central East Antarctica and $10^{-4}\,\mathrm{yr}^{-1}$ in central Greenland. Many divides follow long ridges, so the transverse deformation rate $\dot{\epsilon}_{yy}$ is small and thus $\dot{\epsilon}_{xx} \approx -\dot{\epsilon}_{zz} \approx \dot{b}_i / H$. Some divides intersect at domes, where $\dot{\epsilon}_{yy} \approx \dot{\epsilon}_{xx}$; in this case $\dot{\epsilon}_{zz} \approx -\dot{b}_i / H$, but the horizontal terms are each only about half as large.

On both flanks of the divide, the flow varies with depth as depicted, roughly, in Figure 8.8. The shape of this velocity profile influences horizontal stretching at the divide itself; stretching is most rapid near the surface and smallest near the bed. The nonlinearity of ice creep (Eq. 8.17) magnifies this contrast. Because the driving stress is zero at the divide, there is no shear stress component τ_{xz} that increases with depth. Instead, the layers deep beneath the divide reside in a zone of low stress and, with $n > 1$, the creep relation implies exceptionally stiff ice. There exists, in effect, a wedge of almost stagnant basal ice beneath the divide (Nye 1951). The stretching rate $\dot{\epsilon}_{xx}$ thus declines more rapidly with depth than does the flow rate $u(z)$ on the flanks. Model calculations suggest that beneath the divide $\dot{\epsilon}_{xx}$ (and hence $-\dot{\epsilon}_{zz}$) decreases approximately as a linear function of depth. Far from the divide, in contrast, $\dot{\epsilon}_{xx}$ and $\dot{\epsilon}_{zz}$ have nearly constant values through the upper half of the ice column. These contrasting patterns distinguish the *divide flow* and *flank flow* regimes.

Raymond (1983) and Reeh (1988) examined flow near an ice divide using numerical model calculations. Reeh and Paterson (1988) discussed flow at the ice divide of Devon Island ice cap, Canada, and Hvidberg et al. (1997) for the central Greenland divide. These models indicate that the divide flow regime transitions to the more typical flank flow regime within a distance of approximately $4H$ from the divide. Close to the divide, the enhancement of tension τ_{xx} in the stiff basal wedge partly supports the driving stress. Thus the basal stress τ_b is not only zero under the divide but also is generally less than τ_d in a zone around the divide. The models suggest that this zone extends out to a distance of only about $2H$ (Raymond 1983).

The basal zone of stiff ice changes the depth profile of vertical velocity (Section 8.4.2); compared to flank locations, the downward flow at the divide is slower at a given depth. As one consequence, a "warm spot" should develop on the bed beneath the divide, because the downward advection of cold ice is reduced (see Chapter 9). As a second consequence, the isochrones, the surfaces connecting ice of the same age, should drape over the stiff basal zone, forming an upwarp beneath the divide, known as a *Raymond bump*.

Such upwarps appear in radio-echo sounding profiles of internal layers beneath ice divides at several locations in Antarctica, including Fletcher Promontory and Roosevelt Island (Vaughan et al. 1999a; Conway et al. 1999). Figure 8.24 shows the prominent bump of isochrones within Roosevelt Island. (The reflectors are acidic layers produced by volcanic eruptions.) Observing bumps of this sort leads to two conclusions: that the divide has not moved in recent millennia (Nereson and Waddington 2002) and that the ice creep relation is nonlinear, with exponent $n > 1$. An alternative explanation for the upwarp, that it develops from locally reduced accumulation on the divide crest, has been discounted at these sites but may apply elsewhere.

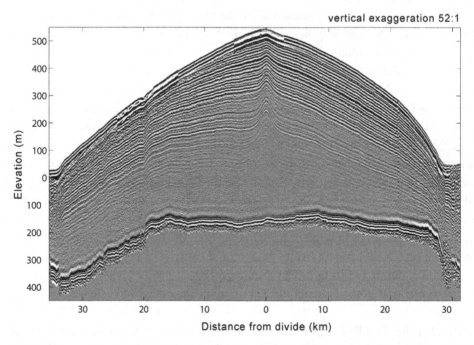

Figure 8.24: Cross-section of Roosevelt Island, Antarctica, showing internal strata and the Raymond Bump beneath the divide (Conway et al. 1999). Image courtesy of H. Conway.

No Raymond bump exists under the central Greenland divide, an observation consistent with model calculations showing divide migration during the last major climate transition (Marshall and Cuffey 2000).

Direct measurements of vertical strain within Siple Dome, West Antarctica, confirm the existence of distinct divide-flow and flank-flow regimes (Zumberge et al. 2002; Elsberg et al. 2004). Siple Dome is an ice ridge, approximately 1 km thick, that stands between Kamb and Bindschadler Ice Streams. Vertical strain was measured using changes in the lengths of cables frozen into boreholes. Lengths were monitored with both fiber optic cables and wire-resistance devices. Measurements from a borehole array at the Siple Dome divide and an array located 7 km away on the flank show the expected difference in flank and divide profiles of $\dot{\epsilon}_{zz}$; Figure 8.25 illustrates the fiber-optic measurements.

Because many ice cores are taken at divides, a detailed understanding of the divide-flow regime is important. Factors that significantly influence divide flow probably include the anisotropic character of ice creep and changes in the deformation mechanisms at low stresses (Pettit and Waddington 2003). These topics need more investigation. We discuss the calculation of depth-age relations at a divide in Chapter 15.

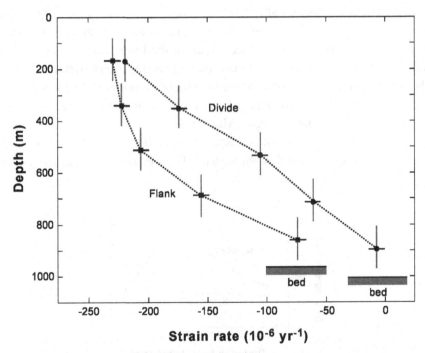

Figure 8.25: Measured vertical strain rates ($\dot{\epsilon}_{zz}$) beneath the Siple Dome ice divide and a location 7 km away on the flank. At each point, vertical bars show the depth range over which the measurement was taken; horizontal bars show the uncertainty in the strain-rate values. Adapted from Zumberge et al. (2002).

8.9.2 Ice Streams

An *ice stream* may be defined as a region of a grounded ice sheet in which the ice flows much faster than in the regions on either side. An ice stream in the strict sense has no visible rock boundaries; if it has it is called an *outlet glacier*. This is not a practical distinction, however. For example, Jakobshavn in West Greenland starts as an ice stream and ends as an outlet glacier while Rutford Ice Stream in West Antarctica is bordered by mountains on one side and ice on the other. We use "ice stream" in the broad sense that includes fast-flowing outlet glaciers. Most ice streams lie in deep channels with beds below sea level and end either as a floating glacier tongue or by joining an ice shelf.

Although outlet glaciers and ice streams comprise only 13% of the Antarctic coastline, they drain more than 90% of the accumulation from the interior (Morgan et al. 1982). Similarly, much of the discharge from Greenland is concentrated in some two dozen large outlet glaciers (Bauer 1961; Rignot and Kanagaratnam 2006). The flux from Jakobshavn, in particular, is about 7% of the annual mass loss from the ice sheet. The state of the ice sheets therefore depends profoundly on flow in ice streams.

The surface profile of an ice stream and the variation of driving stress along it differ from an "ordinary" ice sheet flow line. The profile of an ice sheet crudely resembles a parabola, and the slope and driving stress increase steadily while thickness diminishes with distance from the ice divide (e.g., Figure 8.23). The slope of an ice stream, in contrast, attains a maximum at its head and then decreases downstream; the surface profile of the lower part is concave upward. The maximum slope usually corresponds to a maximum in driving stress (Bentley 1987, Figures 8 and 10). The velocity, however continues to increase to the grounding line. Figure 8.26 illustrates schematically the contrast between the distributions of driving stress along a flow line in an ice stream and an "ordinary" ice sheet margin (Mae 1979).

Ice streams display a broad range of characteristics and behaviors. At one extreme are outlet glaciers, such as Byrd Glacier, that flow through the Transantarctic Mountains into the Ross Ice

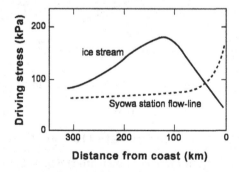

Figure 8.26: Driving stress along flow lines in an ice stream and elsewhere in an ice sheet ("Syowa station flow line"). Adapted from Mae (1979).

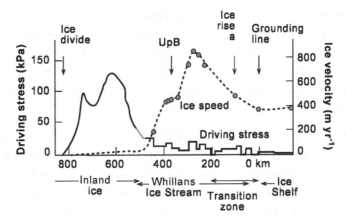

Figure 8.27: Driving stress (solid line) and surface velocity (dashed line) along a flow line following Whillans Ice Stream. Redrawn from Alley and Whillans (1991).

Shelf. At the other are the five major ice streams on the Siple Coast, on the opposite side of the shelf. Heavily crevassed shear zones separate these ice streams from ridges of slow-moving ice on either side. In comparison with other ice streams, they are wider and the bedrock channels they occupy are shallower and less well defined, with no pronounced downward step at the head. Their surface slopes are exceptionally low and a typical driving stress is only 20 kPa (Figure 8.27). In spite of this, the ice streams attain maximum velocities of close to 1 km yr^{-1}. Tables 8.1 to 8.3 outline the range of flow regimes on ice streams. Compared to the typical 50 to 150 kPa value for glaciers and ice sheet flanks, driving stresses on ice streams range from the very low (10 to 20 kPa) to the very high (300 to 400 kPa).

The different behaviors of individual ice streams suggest that there are different flow mechanisms. This is true, as we discuss in the next few sections. Several factors account for fast ice stream flow:

1. *Topographic focussing.* Most ice streams flow along troughs in the bed. Basal topography influences the flow field of an ice sheet most strongly near the margin, where the ice is thin. Ice flux strongly focuses into paths where bed elevations are low. Large driving stresses develop to impel large ice fluxes into and through topographically controlled ice streams. Such stresses are attained largely by increased surface slopes in the onset region and by large ice thicknesses in the deep troughs. If the ice stream is narrow, side drags provide resistance against the flow that must be overcome by further increases of driving stress. Large driving stresses indeed characterize the inland portions of many ice streams. However, topographic focussing cannot explain the continued and usually increased velocities farther downstream.

2. *Reduced ice viscosity.* Focussing of ice flux leads to rapid flow and large stresses and thus enhances frictional heating (Chapter 9). This heat source increases temperatures at depth

and hence softens the ice. The viscosity of ice decreases by a factor of about seven as ice warms from -10 to $0\,°C$ (Chapter 3). A layer of temperate ice may form at depth, and water contained in such a layer makes it even softer. In the lower reaches of ice streams, basal shear stresses typically decline, which reduces the heat generation. But high stresses and heat production persist in shear margins that control the overall flow. Heat produced by longitudinal straining may also be important.

3. *Basal lubrication.* Most ice streams flow rapidly because their beds are slippery. The lower part of Columbia Glacier, Alaska, discussed in Section 8.8, provides an analogue. Basal resistance is low even though slip is rapid. For this to be the case, basal water pressures P_w averaged over a large fraction of the bed must be close to the ice overburden pressure P_i. (In other words, the effective pressure $N = P_i - P_w$ must be close to zero.) All major ice streams flow into ice shelves or floating ice tongues. At the transition point, the effective pressure equals zero and the water pressure is determined hydrostatically by the water depth. Inland, basal water pressures must be even higher, since a water pressure gradient must develop to drain subglacial melt. In most cases, we expect that effective pressures remain close to zero for a great distance inland. The flatter the surface, the more accurate this expectation will be. (Except for one set of boreholes in the Siple Coast ice streams, no data have ever been obtained to test these statements.) If the bed slopes steeply downward toward the coast, the elevation gradient rather than the pressure gradient controls the water drainage, and effective pressures might be higher and the bed less slippery.

Effective lubrication also arises from anomalously large water sources. Intense surface melt may be important at lower elevations in Greenland. A high geothermal flux enhances basal melt. The Northeast Greenland Ice Stream (NEGIS) exists because of a localized region of such melt. High geothermal fluxes throughout West Antarctica, a tectonically active region, contribute to the formation of ice streams there.

Given a low basal effective pressure, the next question is whether the bed consists of deformable sediments or not. The frictional properties of sediments inherently limit the basal resistance, and sediment yield strengths are proportional to N. Sediments are known to underlie the Siple Coast ice streams. Sediments, molded into streamlined forms, cover vast areas of former ice stream beds exposed by the Holocene retreat of the Antarctic Ice Sheet. Whether sediments are present beneath other modern ice streams remains unknown. This seems likely, especially where driving stresses are low and sediment accumulations are observed at the grounding line (Alley et al. 1989). A high driving stress might indicate a bedrock substrate but is also consistent with a sediment bed if effective pressures are high or if basal resistance concentrates on localized spots of high friction.

8.9.2.1 Jakobshavn Ice Stream

The following discussion of this major Greenland ice stream uses information from Echelmeyer and Harrison (1990), Echelmeyer et al. (1991b, 1992), Iken et al. (1993),

Funk et al. (1994), Clarke and Echelmeyer (1996), Sohn et al. (1998), Lüthi et al. (2002), Thomas et al. (2003), Thomas (2004), and Joughin et al. (2004a).

The drainage basin of Jakobshavn extends to the highest part of the Greenland Ice Sheet, a distance of about 550 km from the terminus on the west coast of the island. The ice stream proper begins about 80 km from the terminus, where ice sheet flow converges on a subglacial bedrock trench. The ice stream follows this deep and narrow trench all the way to the terminus, which is in a rock-walled fiord. Heavily crevassed margins border the ice stream, which is some 6 km wide. Along most of its length, the centerline ice thickness ranges from 1.5 to 2.6 km, which is 1 to 1.5 km thicker than the adjacent ice sheet. Most of the bed along the centerline lies below sea level. Seasonal surface melt occurs along the whole length, and most of the ice stream traverses the ablation zone. Melt rates near the terminus reach a few meters per year. On the Antarctic ice streams, in contrast, the surface does not melt.

The ice stream terminates in tidewater as a floating glacier tongue – an ice shelf much longer than its width – with a rapidly calving front. Ice thickness at the front was about 550 m in 1990, which implies an outward deviatoric force at the front of $\phi_M \approx 1.4 \times 10^5$ kPa m per unit width across-glacier (Eq. 8.100). From 1960 to 1997 the floating tongue maintained a length of about 14 km. Rapid thinning and enhanced calving starting in 1997 led to rapid retreat of the terminus. Retreat continued, interrupted by one period of readvance, until by 2003 most of the floating tongue had disintegrated. Retreat has shifted the terminus farther inland than any previously recorded positions. Rapid retreat, however, is not unprecedented; between 1850 and 1960 the terminus retreated by some 30 km from a more seaward position (Sohn et al. 1998). Rapid retreats are characteristic of tidewater glaciers in fiords (see Section 8.8 and Chapter 11).

For at least two decades before 1997 the ice stream flowed at about 4 km yr^{-1} near the grounding line, increasing to 7 km yr^{-1} at the floating terminus. At the time, this was the fastest flow observed for any nonsurging glacier. From 1997 to 2006, however, velocities increased along the lower 40 km or so of the ice stream, coincident with disintegration of the floating tongue (Joughin et al. 2004a; Thomas 2004). By 2003, flow rates at the terminus attained 12.6 km yr^{-1}, and the rate near the grounding line was about 8 km yr^{-1}. The lower 40 km of the ice stream underwent rapid extension of $\dot{\epsilon}_{xx} \approx 0.3$ yr^{-1} in 2003.

The long upstream section of the ice stream moves at a less dramatic but still impressive pace of 1 to 2 km yr^{-1}. Driving stresses here are exceptionally high ($\tau_d = 300$ to 420 kPa), due to the large ice thickness above the trench combined with a surface slope resembling that of the surrounding ice sheet. Side drags balance 10% to 50% of τ_d. The remainder indicates basal stresses of $\tau_b \approx 200$ to 260 kPa (Clarke and Echelmeyer 1996) – very high values that indicate a strong bed. (The upper and lower bounds for basal and side drags were obtained by Clarke and Echelmeyer using the shape factor approximation, $\tau_b = f \tau_d$, with two values of f; see Section 8.6.2. One f value was calculated with the assumption that the shear zones in the ice overlying the trough edges act like rigid walls. The other assumed that these shear zones contribute no drag.)

The large stresses drive rapid internal deformation. Low ice viscosities near the bed also contribute. Because of strain heating, the bed is at melting point. Iken et al. (1993) suggested that temperate ice occurs in a basal layer, a few hundred meters thick. Such a layer would form because transversely convergent flow at the ice stream head and along its length drives vertical extension, and hence thickens basal layers. A high rate of strain heating because of the large stresses also contributes. No boreholes have penetrated the deep layers of the central part of the ice stream. Boreholes in the margins, however, confirm that the bed is at melting point. One such borehole found a basal temperate layer, about 30 m thick, which matched predictions for the site (Lüthi et al. 2002). This borehole also penetrated a basal layer of Pleistocene-age ice, with an enhanced rate of shearing of the sort discussed in Section 8.3.2. Moreover, radar soundings of internal strata showed that this layer becomes considerably thicker in the middle of the ice stream.

Given the soft basal ice and the large basal stresses, internal deformation of the ice, alone, explains the observed velocities along most of the ice stream (Clarke and Echelmeyer 1996). In the lower 10 km or so of the grounded ice stream, however, velocities increase significantly; basal slip becomes important (Fastook et al. 1995).

8.9.2.2 Outlet Glaciers of Southeast Greenland

Rapidly flowing outlets drain the southeastern flank of the Greenland Ice Sheet, a region of high accumulation rate. The two largest glaciers, Helheim and Kangerdlugssuaq, each carry a total ice flux (volume per unit time) similar to that of Jakobshavn. Few details of their dynamics are known, but they appear to be broadly similar to Jakobshavn; they occupy deep topographic troughs, are a few kilometers wide, and terminate in tidewater. The topography is more mountainous than in west Greenland, and bedrock valley walls rather than shearing ice define the side margins for a greater distance inland. Moreover, the bed rises inland from the terminus, making these glaciers thinner and steeper than Jakobshavn. Observed surface velocities attain values as high as $5 \, \mathrm{km \, yr^{-1}}$ on Kangerdlugssuaq and 8 to $11 \, \mathrm{km \, yr^{-1}}$ on Helheim (Thomas et al. 2000; Howat et al. 2005). Both glaciers retreated and accelerated in recent years (Luckman et al. 2006), a process we will discuss in Chapter 11.

8.9.2.3 Northeast Greenland Ice Stream (NEGIS)

The NEGIS was first identified by Fahnestock et al. (1993) from satellite images showing the texture of the ice sheet surface. The NEGIS is a remarkable demonstration of how basal lubrication affects ice sheet flow (Figure 8.4). Streaming flow occurs because water exists beneath a region of the ice sheet that is otherwise frozen to its bed. Basal melt produces water in a localized source at the head of the ice stream. Melt rates of up to $15 \, \mathrm{cm \, yr^{-1}}$ are inferred from analyses of radar stratigraphy (Fahnestock et al. 2001). Such rates require a geothermal heat flux of 10 to 30 times a typical crustal value. This suggests volcanism. Joughin et al. (2001) made a thorough analysis of the dynamics of NEGIS; we took much of the following information from that study.

The NEGIS commands our attention for its unusual characteristics. It starts very close to the central ice divide, because this is the position of the water source. No other ice stream does so, probably because strain heating and basal lubricants – water and sediment – normally do not concentrate in the upper parts of a drainage basin. The bed is weak in the upper section of NEGIS; basal resistance amounts to only about 20 to 30 kPa, or about 60% of the local driving stress. Side drags balance the rest. From its source, the ice stream follows a nearly linear path until near its end, where bedrock topography splits it into two distributaries. Although the ice stream overlies subglacial troughs in a few areas, no trough is present beneath most of the ~300-km-long linear section. This too contrasts with most ice streams. It suggests that either the ice stream is a new feature or its path varies over time; otherwise, subglacial erosion would carve a trough. The path of the ice stream is probably governed by how water drains from its upstream source. The drainage, in turn, is largely steered by the ice surface topography (Section 6.3.1.2), itself a function of climate, the locations of ice sheet margins, and dynamic feedback with the flow of the ice stream.

The flow regime varies with distance. In an extensive central section, τ_b balances most of τ_d, and fluctuations of τ_b correlate with fluctuations of τ_d – traits characteristic of a hard bed. In this section, the ice stream moves at only 50 to 100 m yr^{-1}. The ice stream then transitions downstream to a lower ice plain, with fast flow (up to 400 m yr^{-1}), low driving stress ($\tau_d \approx$ 23 kPa), significant side drags (about 40% of τ_d), a weak bed ($\tau_b \approx 14$ kPa), and basal elevations well below sea level. This ice plain region, defined by a nearly flat surface and low driving stresses, resembles parts of the Siple Coast ice streams, discussed in the next section; in particular it appears similar to MacAyeal Ice Stream and some of the tributaries feeding into Kamb and Bindschadler Ice Streams. However, in contrast to the Siple Coast ice streams, a short zone of high driving stress separates the ice plain from the grounding line downstream. This occurs where the ice stream passes through a gap in the coastal mountains.

By comparing inferred ice fluxes to estimated accumulation rates Joughin et al. (2001) concluded that the 300-km-long middle section of the ice stream is in mass balance to within ±6 cm yr^{-1}. However, Storstrømmen glacier, one of the three outlets from the ice stream, surged between 1978 and 1984 (Reeh et al. 2003; see Chapter 12). The glacier is now slowly returning to its presurge dimensions. (The satellite imagery used by Joughin et al. was acquired in 1995–1996.) Nioghalvfjerdsfjorden Glacier, another of the ice stream's outlets, ends in a floating tongue that undergoes massive calving at intervals of several decades (Reeh et al. 2001). Calving occurs when the sea ice that normally prevents it breaks up. Such changes at the outlets are likely to affect flow upstream by mechanisms described in Chapter 11.

8.9.2.4 Siple Coast (or Ross) Ice Streams

Figure 8.28 maps the measured flow of the five major ice streams draining the Ross Sea side of West Antarctica (the Siple Coast) and their tributaries. The measurements were acquired by satellite-based interferometry (Joughin et al. 1999, 2004b). Figure 8.27 showed the variation of driving stress and velocity along one of them. Boreholes have reached the bed in this region.

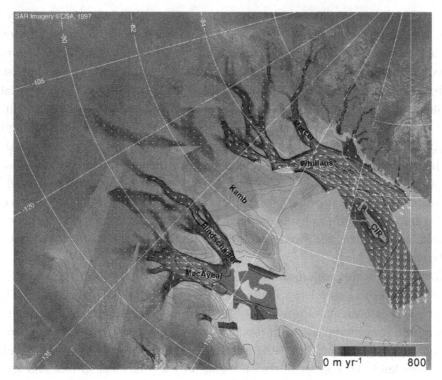

Figure 8.28: Surface velocities of the Siple Coast region, Antarctica, superimposed on radar imagery. From analysis of Joughin et al. (2004b). Image courtesy of I. Joughin. (Refer to the insert for a color version of this figure)

Table 8.6: Old and new designations for Siple Coast ice streams.

Old designation	New designation
A	Mercer
B	Whillans
C	Kamb
D	Bindschadler
E	MacAyeal

The following discussion uses information from Alley et al. (1986c), Alley and Whillans (1991), Bindschadler et al. (1987), Blankenship et al. (1986), Engelhardt et al. (1990), Kamb (1991, 2001), Shabtaie and Bentley (1987, 1988), Whillans and van der Veen (1997), Raymond et al. (2001a), and Joughin et al. (2004b). The Siple Coast ice streams were originally designated with letter codes, a system used in much of the literature. The ice streams have recently been renamed to honor glacier scientists; Table 8.6 gives the correspondence.

The Siple Coast ice streams are wide and flow by basal slip over beds of weak sediment. Based on the measurements beneath Whillans and Bindschadler Ice Streams, reviewed in Chapter 7, "slip" is anywhere from 20% to 80% soft-bed sliding, with shearing in the substrate accounting for the remainder. Driving stresses are exceptionally low ($\tau_d \approx 2$ to 20 kPa), especially considering the rapid motion (300 to 800 m yr^{-1}). Over large regions, drag from lateral shear margins provides much of the resistance to flow; dynamically, these regions resemble ice shelves in elongate embayments. Overall, however, basal resistance is not trivial. It derives largely from localized high-friction "sticky spots," set amidst broad swaths of extremely weak substrate providing resistance of only a few kPa (Figure 8.15b). Samples of the saturated sediments underlying Whillans Ice Stream were obtained from deep boreholes (Kamb 1991, 2001). Mechanical tests showed that the yield strength of these sediments is only a few kPa, confirming the low values for basal drag inferred by force balance studies. The existence of sticky spots has been affirmed by observations of seismicity at the bed of Kamb Ice Stream; stress builds up on sticky patches but is released in sudden slip events (Anandakrishnan and Alley 1994).

An extensive system of numerous tributaries feeds the upper ends of the ice streams (Figure 8.28; Joughin et al. 2004b). Tributaries coincide with topographic depressions. Where examined in detail using geophysical techniques, the onsets and side margins of the tributaries also correspond to the boundaries of sediment-filled basins (Peters et al. 2006; Bell et al. 1998; Anandakrishnan et al. 1998). This pattern suggests that streaming flow begins where the bed first becomes deformable. It further suggests that erosion in the onset regions supplies lubricating sediment to the main ice streams. In the tributaries, driving stresses are 45 to 90 kPa, of which basal drag balances about 60% to 90% (Joughin et al. 2004b). The bed alternates between weak and strong patches, a pattern that might reflect feedbacks between ice surface slope, drainage of basal water, and slipperiness of the bed.

Farther downstream, the main ice streams are a few hundred kilometers in length, 20 to 100 km in width, and bounded by heavily crevassed shear margins with widths of a few kilometers. Between the streams stand interstream ridges, where the bed is frozen and the ice moves only a few tens of meters per year. The streams rest in subtle topographic troughs, with depths only 10% to 20% of the ~1 km ice thickness. On all the ice streams, driving stresses averaged over large sections, tens of kilometers in dimension, are 10 to 20 kPa, and decrease as the streams grade into the Ross Ice Shelf along flat regions called *ice plains* (Table 8.3; labelled "transition zone" in Figure 8.27). In comparison to such driving stresses, the inferred basal drags are only 2 to 6 kPa on Whillans Ice Stream, 9 kPa on the active part of Kamb, 1 to 4 kPa on Bindschadler, and 15 kPa on MacAyeal. Basal drag thus accounts, on average, for about 20, 50, 20, and 70% of driving stress on these four ice streams, respectively. Such large differences make it clear that, even within the Siple Coast, generalizations about ice streams can be misleading.

Longitudinal resistance influences the ice streams very little, except in local patches; although longitudinal strain rate is high in some regions, it does not change rapidly with distance along-flow. On Whillans and Bindschadler Ice Streams in particular, shearing in the side margins and basal drag on sticky spots control the fast flow. We discussed the across-glacier velocity profiles for weak-bed segments of these ice streams in Section 8.6.1.

The basal slip mechanisms of the active fast-flowing ice streams were discussed in Chapter 7. To summarize: the sediment beds are deformable, with low yield strengths reflecting small effective pressures. This, in turn, permits rapid sliding along the ice-sediment interface and deformation at depth in the sediment, without generating much resisting force. The sediment behaves plastically, but the overall relation between basal drag and velocity, which remains essentially unknown, must depend on the nature of sticky spots. Furthermore, the motion is not steady. In the lower sections of Whillans and Bindschadler Ice Streams, tidal variations at the grounding line induce large variations of ice stream motion over days and weeks. On the ice plain of Whillans Ice Stream, large sections of ice periodically lurch forward in a stick-slip behavior reminiscent of the earthquake cycle on crustal faults.

The Siple Coast ice streams are not currently in a steady state; the region is gaining mass, largely because one of the ice streams stalled within the last few centuries and another is slowing down. Furthermore, several lines of evidence reveal that large changes occur over centuries. First, the lateral shear margins shift, causing large sections of ice streams to stagnate while other regions, previously stagnant, are entrained in the streaming flow. This behavior is inferred from observations of old abandoned shear margins, now buried under centuries of accumulation, and from reconstructions showing major changes in flow direction (Conway et al. 2002; Retzlaff and Bentley 1993; Gades et al. 2000; Clarke et al. 2000). Temperature profiles and the pattern of visible flow structures on the Ross Ice Shelf both preserve a legacy of these changes (Fahnestock et al. 2000; Joughin et al. 2004c; see Chapter 9 for temperature information).

Second, active ice streams stagnate; this happened to the lower section of Kamb Ice Stream approximately 200 years ago. Whillans Ice Stream is currently slowing and may stagnate within one or two centuries (Joughin et al. 2005). Stagnations are apparently caused by strengthening of the weak bed. If the yield strength of the basal sediment approaches the driving stress, the ice stream cannot continue to flow rapidly; stresses are far too low for significant horizontal-plane shearing in the ice.

Several interrelated causes for the variability of the Siple Coast ice streams have been proposed. First, velocities on an ice stream centerline depend strongly on ice stream width, as shown by Eq. 8.86. Widening, by outward migration of a margin, would lead to significantly accelerated flow. Jacobson and Raymond (1998) suggested that migration reflects an inherent instability of an ice stream margin whose position is thermally controlled – that is, fixed by a boundary between frozen and thawed regions of the bed. Heat dissipation in the shear zone should warm the ice and deliver heat to the bed, expanding the thawed region and leading to outward migration of the margin. On the other hand, the inflow of cold interstream ice opposes this process. Geological and topographical controls might stabilize some of the margins; the extent to which this happens is not clear.

Second, because the yield strength of sediment depends strongly on water pressure, processes that remove water might strengthen basal sediments and decelerate an ice stream. Two processes for water removal are upstream diversion and freeze-on (Alley et al. 1994; Tulaczyk et al. 2000b). Surface elevation differences between neighboring ice streams are small; thus minor changes of the surface topography can alter the subglacial hydraulic gradients and divert water from one ice stream to another (the *water piracy* hypothesis). Freeze-on is important because of large vertical temperature gradients in the basal layers (Chapter 9), a consequence of rapid downstream advection of cold ice combined with thinning by longitudinal extension in the onset region for streaming flow. Steep temperature gradients drive freeze-on of water to the glacier base, especially beneath Whillans and Kamb Ice Streams (Joughin et al. 2004b). Engelhardt (2001) observed a basal layer of debris-rich, frozen-on ice, more than 10 m thick, in deep boreholes through Kamb Ice Stream. Basal freeze-on might terminate slip altogether. But basal water is also replenished by subglacial melt originating upstream and draining along the bed. Given that freezing conditions prevail, the entire system of ice streams can only be maintained by this replenishing flow of subglacial meltwater (Parizek et al. 2002).

Raymond (2000) discussed the interaction of basal water with the flow of weak-bed ice streams. Basal water lubricates an ice stream bed and is necessary for fast flow. It reduces the yield strength τ_* and hence τ_b below the driving stress, shifting stress from the bed to the shear margins, and increasing basal slip u_b (see Section 8.6.1). This implies, however, that frictional heat production might also shift to the margins. Melt generation at the bed, which depends on the product $\tau_* u_b$, is reduced if τ_* declines too much relative to the increase of u_b. Raymond identified an intermediate state in which the bed retains some strength but melt generation is a maximum. This occurs, in theory, when $\tau_b \approx 0.25\tau_d$. Consider the case of MacAyeal Ice Stream and the ice stream tributaries, features for which $\tau_b > 0.25\tau_d$. According to the theory, a speed-up of the ice stream would increase melting; the faster flow would continue. Whillans Ice Stream illustrates the contrary case, because $\tau_b < 0.25\tau_d$. Here a faster flow reduces melt production, and increased flow cannot persist unless the supply of water from upstream also increases. This illustrates that the ice streams, though similar in broad characteristics, might evolve in very different ways in response to similar perturbations.

Despite the variability of the Siple Coast ice streams on timescales of centuries, geological evidence from the Ross Sea floor shows that the ice streams existed late in the last glacial period and extended all the way to the outer edge of the grounded ice sheet (Anderson and Shipp 2001). Moreover, they persisted throughout the major retreat of the grounding line over the last 10 kyr or so, with flow lines emanating from their present positions. Thus, stabilizing factors must operate in the system despite the short-term variability. Raymond discussed one possible stabilizing factor. If water drainage along the bed increases, water pressures are reduced. This would increase τ_* and disrupt fast flow. But reduced water pressures would normally lead to constriction of the drainage passageways, trapping water and stabilizing the system against such a change (Weertman 1972). Parizek et al. (2002) emphasized that the throughflow of water originating beneath the interior of the ice sheet is a persistent process on long timescales and so constitutes a reliable way to maintain basal lubrication. Another persistent factor maintaining lubrication is high water pressure at the downstream end of the basal drainage system; the effective pressure must be zero at the grounding line. Hulbe and Fahnestock (2004) considered feedbacks between ice flow and the vertical profile of temperature. Acceleration of an ice stream enhances the thinning of ice as it passes through the onset region. Thinning, in turn, increases basal temperature gradients, driving more basal freeze-on and hence reducing bed lubrication. While this mechanism may produce variability of ice stream flow on short timescales, over the long term it should help to stabilize the whole system by preventing catastrophic outflows of ice.

8.9.2.5 Amundsen Sea Sector Ice Streams

The following information is from Shepherd et al. (2001, 2002), Rignot et al. (2002), Schmeltz et al. (2002), Thomas et al. (2004a), and Payne et al. (2004).

Pine Island and Thwaites Glaciers are the fastest flowing West Antarctic ice streams, with grounding-line speeds greater than $2 \, \mathrm{km \, yr^{-1}}$. They drain the interior of West Antarctica on the East Pacific side – the opposite side of the ice sheet from the Siple Coast. In contrast to the Siple Coast streams, which end in the massive Ross Ice Shelf, the Amundsen Sea streams end in "small" shelves that are only a few tens of kilometers in dimension. Pine Island's shelf is laterally confined by ice, but Thwaites' is not. The drainage basins of both ice streams overlie deep subglacial depressions. The bed beneath Pine Island Glacier is 600 m below sea level at the grounding line, and drops steadily to about 1300 m below sea level only 25 km inland. The bed remains this low for a further 200 km inland, along a broad trough-shaped basin. At the Thwaites grounding line, the bed lies about 1 km below sea level and slopes gradually back into the deep Byrd Subglacial Basin.

These ice streams are probably the most vulnerable part of the West Antarctic Ice Sheet to climate and sea-level changes (Hughes, 1981), given (1) their deeply sunken beds extending far inland and (2) the absence of a large ice shelf to shield the grounding lines from external influences. Currently, both ice streams are thinning and their grounding lines retreating – most likely as a response to increased ocean temperatures (Chapter 11).

Figure 8.29 shows measured ice velocities and topography along Pine Island Glacier, together with estimates of driving stress and basal drag (Payne et al. 2004). In a long interior section, 70 to 250 km from the grounding line, low driving stresses prevail (20 to 40 kPa), supported by a combination of basal and side drags. Basal drags apparently are about 10 to 30 kPa; this is a weak-bed ice stream. No measurements of bed properties have been obtained here, but the weak bed suggests the presence of sediments. The bed is not so weak, however, as beneath the Siple Coast ice streams. And, in contrast to the Siple Coast, where driving stresses decline steadily toward the grounding line, Pine Island Glacier passes through a 50-km-long zone in which the surface steepens and driving stress increases, peaking at 150 kPa. Basal drag also rises in this section, suggesting increased effective pressures. Perhaps the steep surface here increases the hydraulic potential gradient, facilitating drainage of water along the bed and reducing water pressures. This strong-bed zone acts as a frictional dam that maintains the thickness of the

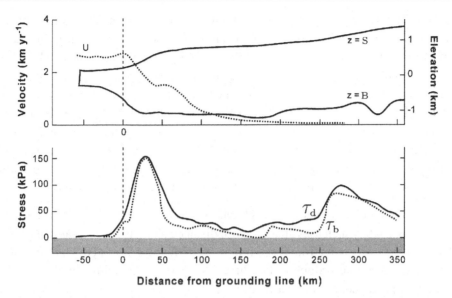

Figure 8.29: Longitudinal transect along Pine Island Glacier, West Antarctica. Distance zero denotes the grounding line. Top panel shows velocities U measured by InSAR and approximate surface and bed profiles. Bottom panel shows driving stress and basal drag estimated from a flow model. Adapted from Payne et al. (2004).

interior ice. Changes in bed properties here could radically affect the future evolution of the West Antarctic Ice Sheet, yet almost nothing is known about the state of the bed and the processes controlling it.

8.9.2.6 Other Antarctic Ice Streams

This category covers Antarctic ice streams in the broad sense other than those on the Amundsen Sea and Siple Coasts. Many occupy fiord-like channels, with beds up to 2 km below sea level (Rutford Ice Stream, for example). In some cases these topographic troughs are entirely subglacial, in other cases they rise above the ice as valley walls (for example, Byrd Glacier, which cuts across the Transantarctic Mountains). The heads of ice streams, where flow lines converge and the velocity increases substantially, often coincide with the head of the fiord (McIntyre 1985). Driving stresses span a large range, 10 to 300 kPa (Bentley 1987, Figures 8,10,12; Joughin et al. 2006; Pattyn and Derauw 2002); the 300 kPa value applies to Shirase Glacier. The position of the maximum in driving stress between the head and the grounding line (sketched in Figure 8.26) varies considerably from glacier to glacier. Basal resistance also varies considerably, though estimates of it are not available for most ice streams. Joughin et al. (2006) constructed maps of basal resistance for eight large ice streams flowing into the Filchner and Ronne Ice Shelves. Their beds appear to be markedly inhomogeneous, a mosaic of weak- and strong-bed regions. The largest area of weak bed, which spans about 300 km, is beneath the Recovery Ice Stream, which drains about 8% of the East Antarctic Ice Sheet and flows into the Filchner Ice Shelf. Several large subglacial lakes occupy the head of this ice stream (Bell et al. 2007). Water lubricating the bed passes through these lakes. Bell et al. suggested that latent heat released by refreezing of water downstream of the lakes keeps the bed at melting point and permits the ice stream to maintain continuity between its water source and its fast-flowing lower section – a process probably relevant for the NEGIS too (Section 8.9.2.3).

Ice-stream velocities are typically a few hundred meters per year but reach 2.5 km yr^{-1} on Shirase Glacier (Pattyn and Derauw 2002; Bentley 1987; Morgan et al. 1982; Rignot 2006). Although basal ice deformation is probably important in the high-stress inland parts of ice streams, fast flow is generally attributed to basal slip for the following reasons:

1. If ice deformation predominated, the velocity would increase with the driving stress. However, velocity usually increases steadily with distance along the ice stream whereas driving stress declines downstream from a maximum, as in Figures 8.26 and 8.27. Decreasing effective pressure, as is likely along a flow line leading to a grounding zone, can account for increased basal slip despite decreased driving stress.

2. Calculations in specific cases suggest that, given measured ice thicknesses and driving stresses, ice deformation could not produce a velocity of several hundred meters per year even if all the ice were at melting point (e.g., Frolich et al. 1989; Pattyn and Derauw 2002).

3. The heads of some ice streams are associated with water on the bed. The large subglacial lakes at the head of the Recovery Ice Stream are one example. As a second, McIntyre (1985) inferred ponded water at the head of Byrd Glacier from radar echoes.

An empirical slip relation frequently used for modelling flow in ice streams is

$$u_b = \frac{k\,\tau_d^n}{N}. \tag{8.104}$$

Here u_b is slip velocity, τ_d driving stress, N effective pressure, and k an adjustable parameter (see also Sections 7.3.5 and 7.4). Bentley (1987, Table 1) showed that values for k differ greatly from one ice stream to another. To apply this relation, not only does k need to be calibrated, but also N needs to be calculated using a model for the basal water system, accounting for water sources and drainage. Basal melt produced under an ice sheet will accumulate until water pressure gradients are sufficient for drainage to the grounding line. Some models have calculated N by assuming water pressure is uniform and equal to its grounding line value. This is not a physically plausible assumption, although it provides a lower bound on water pressure. Furthermore, as shown by Table 8.3 and many analyses (e.g., Joughin et al. 2006), the resistance to flow provided by side drags cannot be neglected on most ice streams that have been studied. Using a relation for u_b in terms of τ_d – but without accounting for ice stream width and the relative values of τ_b and τ_W – is not a meaningful way to simulate ice stream flow.

8.9.3 Ice Shelves

An ice shelf is a large thick sheet of ice floating on the sea but attached to a grounded glacier or ice sheet. The place where the ice starts to float is called the *grounding line* or *grounding zone*. Ice shelves surround much of Antarctica (Figure 2.2). They are nourished by flow from the ice sheet, by snow accumulation on their surfaces, and in some places by freezing of seawater to the base. Calving of icebergs and basal melting are the normal forms of ablation (Chapter 4). The ice spreads under its own weight as it moves out to sea. Most ice shelves are confined in bays, the shores of which produce a drag on the moving ice. Grounding on shoals and islands also restricts movement. Grounding over an appreciable area produces a dome-shaped *ice rise* with its own radial flow pattern. Less extensive grounding results in surface irregularities called *ice rumples* that can move with the shelf.

The Ross Ice Shelf is the world's largest, with an area of about $525 \times 10^3 \, \mathrm{km}^2$ – similar to the area of France. Its thickness varies from over $1000 \, \mathrm{m}$ at the grounding line to about $250 \, \mathrm{m}$ at the seaward margin, where an ice cliff or *barrier* rises about 30 to $50 \, \mathrm{m}$ above water level. The height of this cliff is limited by the yield strength of ice, just as the yield strength limits the depth of a crevasse (Chapter 10). The ice in a shelf typically flows at a few hundred meters per year, increasing to about $1 \, \mathrm{km \, yr}^{-1}$ at the front.

As in the case of mountain glaciers and ice sheets, the distribution of stress and velocity in an ice shelf and the shape of the steady-state profile can be understood by combining force balance and mass continuity constraints with relations describing the properties of ice. An ice shelf is a particularly simple case because boundary conditions are well defined, conditions change only slowly with horizontal distance, and horizontal velocities are independent of depth. Weertman (1957b) and Robin (1958) were the first to analyze ice-shelf spreading. Thomas (1973a), Morland and Shoemaker (1982), and van der Veen (1986) extended the theory. Sanderson (1979) calculated steady-state profiles and van der Veen time-dependent ones, in both cases by numerical methods. Recent work includes numerical simulations of the complete flow fields in the Ross, Ronne, and Larsen B ice shelves (Humbert et al. 2005; Larour et al. 2005; Vieli et al. 2006).

Except where the ice shelf locally grounds, the bottom boundary provides no resistance to flow; this is the principal dynamic difference between ice shelves and grounded glaciers. Thus the balance of forces, at most locations on the shelf, relates driving stress τ_d only to the sum of longitudinal and side drags, $\tau_L + \tau_W$. Furthermore, because τ_b contributes little drag, forces acting at the shelf edge and sides influence the flow even at locations far from these boundaries. To construct a meaningful analysis of the shelf dynamics, the force balance relation needs to be integrated over distance.

To simplify the mathematics we will, in the following discussion, assume uniform density. This assumption does not obscure any of the basic principles. It is not realistic, however, because a significant proportion of an ice shelf consists of firn. Thomas (1973b) showed that using the mean density instead of the observed variation of density with depth could result in an overestimate of the effective deviatoric stress by a factor of two. A paper by Sanderson (1979) gives the equations in the case of variable density.

8.9.3.1 Driving Force on an Ice Shelf

Figure 8.30 defines axes and variables. The outer edge of an ice shelf is a vertical cliff or *barrier*. Call the ice thickness here H_M, the thickness of the submerged portion H_S, and the vertical coordinate z. In contrast to the case of a tidewater glacier, considered previously, H_S does not equal the water depth but rather follows from the flotation condition,

$$H_S = \rho H_M / \rho', \tag{8.105}$$

with ρ and ρ' the densities of ice shelf and seawater, respectively. This assumes that buoyancy alone determines the flotation level; stresses due to bending of the ice are negligible.

From Eq. 8.31, the frontal cliff creates a deviatoric force per unit width that effectively pulls the ice outward (Section 8.2.4):

$$\phi_M = \int_0^{H_M} \rho g [H_M - z] \, dz - \int_0^{H_S} \rho' g [H_S - z] \, dz = \frac{1}{2} \rho g H_M^2 - \frac{1}{2} \rho' g H_S^2. \tag{8.106}$$

Using Eq. 8.105,

$$\phi_M = \frac{1}{2} \rho g \, H_M^2 \left[1 - \frac{\rho}{\rho'} \right] = \frac{1}{2} \rho g \, H_M h. \tag{8.107}$$

Here $h = H_M - H_S$ defines the *freeboard*, the height of the frontal cliff above the water line (also equal to the value of surface elevation S at the margin). Along a flow line leading to this cliff, any variation in ice thickness gives rise to a surface slope, since $S = H \left[1 - \rho/\rho' \right]$. Thus a local

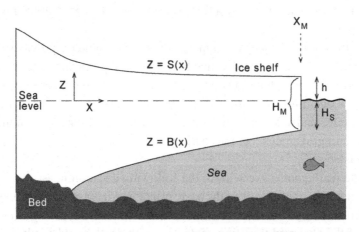

Figure 8.30: Coordinates and variables for analysis of an ice shelf (dimensions not to scale).

driving stress $-\rho g H \, dS/dx$ (Eq. 8.6) – also acts on the shelf at every location. Defining the x-coordinate as horizontal and pointing in the direction of flow, and the position of the margin as $x = X_m$, the total *nonlocal driving force* (per unit width across-flow) pulling on the ice shelf at location x amounts to

$$F_D = \phi_M + \rho g \int_x^{X_m} H \frac{-dS}{dx} \, dx. \tag{8.108}$$

But the condition of floating equilibrium requires that

$$H \frac{dS}{dx} = \left[1 - \frac{\rho}{\rho'} \right] H \frac{dH}{dx}, \tag{8.109}$$

which when integrated over x and added to ϕ_M gives the driving force as

$$F_D(x) = \frac{1}{2} \rho g \left[1 - \frac{\rho}{\rho'} \right] H^2 = \frac{1}{2} \rho g \, H(x) \, S(x). \tag{8.110}$$

Thus the driving force does not depend on how the thickness varies between a point on the shelf and the terminus.

Resistances oppose this driving force. Shear of the ice shelf against the embayment walls generates resisting stresses τ_W, and grounded spots contribute a basal drag τ_b. The net longitudinal force pulling the ice shelf outward is thus

$$F_L = F_D - \int_x^{X_m} [\tau_W + \tau_b] \, dx = \phi_M + \int_x^{X_m} \left[\rho g H \frac{-dS}{dx} - \tau_W - \tau_b \right] dx. \tag{8.111}$$

For mechanical equilibrium, a tensional resistance R_{xx}, integrated through the ice thickness at x, must support this force. The tension is associated with stretching of the shelf. From Eq. 8.58, F_L is the value at x of

$$F_L = H \left[2 \bar{\tau}_{xx} + \bar{\tau}_{yy} \right], \tag{8.112}$$

where $\bar{\tau}_{xx}$ and $\bar{\tau}_{yy}$ signify averages over the ice thickness of the horizontal deviatoric stresses. (Note also that the quantity $F_L - \phi_M$ equals $\int \tau_L \, dx$.) From the creep properties of ice, $\bar{\tau}_{xx}$ relates to the stretching rate along-flow ($\partial \bar{u}/\partial x$), and $\bar{\tau}_{yy}$ to the transverse stretching ($\partial \bar{v}/\partial y$). Writing the depth-averaged effective viscosity as $\bar{\eta}$, Eq. 8.112 becomes

$$4 \frac{\partial \bar{u}}{\partial x} = -2 \frac{\partial \bar{v}}{\partial y} + \frac{F_L}{\bar{\eta} H}. \tag{8.113}$$

The second term on the right-hand side defines a driving stress for longitudinal spreading of the shelf (Hughes 2003). With F_L given by Eq. 8.111, this summarizes the main dynamics of an ice shelf; the elevation of the ice surface above water level causes stretching, but the stretching rate is reduced by side drags and patches of basal resistance.

The primary goal of the following theoretical development is to obtain an expression for the longitudinal stretching rate $\dot{\epsilon}_{xx} = \partial \bar{u}/\partial x$. The velocity in the shelf is then, if the x-axis follows a flow line,

$$u(x) = u_o + \int_0^x \dot{\epsilon}_{xx}\, dx, \tag{8.114}$$

where u_o signifies a known velocity at one end of the shelf.

8.9.3.2 Unconfined Shelf of Uniform Thickness

Weertman (1957b) analyzed this case. The driving force is only that due to the terminal cliff, ϕ_M. The sides and bed provide no resistance. The shelf is unconfined at the sides, so there is complete symmetry of all stress and strain-rate components in the x- and y-directions. Thus $\tau_{xx} = \tau_{yy}$, and hence $\tau_{zz} = -2\tau_{xx}$. Combining Eq. 8.111 and Eq. 8.112 then gives

$$\bar{\tau}_{xx} = \frac{\phi_M}{3H}. \tag{8.115}$$

The shear stresses τ_{xz}, τ_{xy}, and τ_{yz} are all zero, so (Eq. 8.18)

$$\tau_E = \frac{1}{\sqrt{2}} \left[\tau_{xx}^2 + \tau_{yy}^2 + \tau_{zz}^2 \right]^{1/2} = 3^{1/2}\, \tau_{xx}. \tag{8.116}$$

By the creep relation,

$$\dot{\epsilon}_{xx} = A\tau_E^{n-1} \tau_{xx} = 3^{(n-1)/2} \left[\frac{\tau_{xx}}{A_{\text{inv}}} \right]^n, \tag{8.117}$$

where the multiplier A has been replaced by A_{inv}^{-n} as a notational convenience. Because $\dot{\epsilon}_{xx}$ is independent of z,

$$3^{(1-n)/2n}\, \dot{\epsilon}_{xx}^{1/n} \int_B^S A_{\text{inv}}\, dz = \int_B^S \tau_{xx}\, dz = \frac{\phi_M}{3}. \tag{8.118}$$

Using Eq. 8.107 and $n = 3$, this gives

$$\dot{\epsilon}_{xx} = \frac{\partial u}{\partial x} = \frac{1}{9} \left[\frac{\rho g h}{2 A_{\text{inv}}} \right]^3, \tag{8.119}$$

where $\overline{A_{inv}} = H^{-1} \int_B^S A_{inv}\, dz$. Equation 8.119 gives the spreading rate of an unconfined ice shelf, or an iceberg of uniform thickness, with surface at height h above the water line. Any horizontal line in the shelf stretches by the same amount. This expression is always positive. The shelf can spread to maintain constant thickness in spite of accumulation, but ablation always results in thinning. Because an average value of the creep parameter $\overline{A_{inv}}$ is used, the shelf is regarded as isothermal in this analysis.

If the shelf is confined in the y-direction so that $\dot{\epsilon}_{yy} = 0 = \tau_{yy}$, then $\tau_{xx} = -\tau_{zz}$ and $\tau_E = \tau_{xx}$. From Eq. 8.112, the depth-averaged τ_{xx} equals $\phi_M/2H$, and the spreading rate becomes

$$\dot{\epsilon}_{xx} = \left[\frac{\rho g h}{4\overline{A_{inv}}}\right]^3. \tag{8.120}$$

This is about 10% faster than for an unconfined shelf. This calculation assumes that the confining side walls exert no drag on the ice, an unusual situation.

With values typical of the Ross Ice Shelf, $h = 50\,\text{m}$, and $\overline{A_{inv}} = 5 \times 10^5\,\text{Pa yr}^{1/3}$ corresponding to a temperature of $-15\,^\circ\text{C}$, Eq. 8.120 predicts $\dot{\epsilon}_{xx} = 4 \times 10^{-2}\,\text{yr}^{-1}$, which is 10 to 30 times the measured values. This error occurs because the analysis ignores the drag of the side walls (τ_W) and the restraining effect of ice rises. A more precise analysis by Jezek et al. (1985) suggested that the net effect of these resistances on the Ross Ice Shelf is to reduce $\dot{\epsilon}_{xx}$ by a factor of about ten. To overcome these resistances, the ice shelf builds a thickness gradient and hence a surface slope, which increases the driving force (Budd 1966). We discuss this case in Section 8.9.3.4.

8.9.3.3 Unconfined Shelf of Nonuniform Thickness

A column of ice moving along the shelf thins by longitudinal spreading, which is not fully compensated by accumulation. Thus the thickness of an unconfined shelf normally decreases outward, especially near the grounding line. Equation 8.119 still gives the spreading rate if S, the surface elevation at any point, is used in place of h, the elevation at the front. (The additional assumption needs to be made that the slope of the bottom surface of the shelf is small so that the stress τ_{xz} will be negligible.) Thus an unconfined shelf should extend most rapidly on its inland side and at a smaller rate near its outer edge; spreading rate is proportional to S to the third power, and S is proportional to thickness. Observations indeed show this pattern. Figure 8.31a plots measured velocities along the Erebus Ice Tongue, an unconfined shelf flowing from Ross Island, Antarctica (Holdsworth 1982).

8.9.3.4 Confined Shelf of Nonuniform Thickness

Most ice shelves are confined in embayments. The following analysis of such features is based on that of van der Veen (1986). As before, Figure 8.30 gives the coordinate system. Surface elevation S and ice thickness H vary with x but not with the transverse coordinate y. The ice

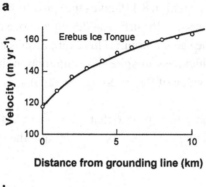

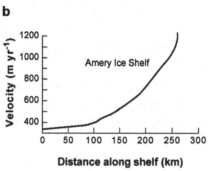

Figure 8.31: Measured velocities along central flow lines of two ice shelves. (a) An unconfined shelf. Adapted from Holdsworth (1982). (b) A confined shelf. Adapted from Budd et al. (1982). Note the difference in scales: in comparison to the Erebus Ice Tongue, the Amery Ice Shelf is very large and flows rapidly.

shelf is assumed to be confined in a bay whose walls are parallel vertical planes, distance $2Y$ apart. The flow lines therefore parallel the x-axis, and $\tau_{yy} = 0$. The walls exert a shear stress τ_{xy} on the sides of the shelf.

The assumptions are:

1. τ_{xy} at the sides, averaged over the ice thickness, has a limiting value τ_o, independent of x. Suggested values are similar to basal shear stresses in grounded ice sheets, namely 50 to 150 kPa. Our analysis in Table 8.4 indicates even higher values might be appropriate for rapidly deforming margins.

2. $\partial \tau_{xy}/\partial y$ does not depend on y so that

$$\frac{\partial \tau_{xy}}{\partial y} = \frac{\tau_o}{Y} \quad \text{and} \quad \tau_{\mathrm{w}} = H\frac{\tau_o}{Y}, \tag{8.121}$$

because τ_{xy} must be zero on the centerline. It follows that $\partial \tau_{xy}/\partial x = 0$.

3. Because there is no basal resistance, the shear stresses τ_{xz} and τ_{yz} equal zero, provided that the slope of the shelf base is small, as observed. Measurements confirm that shearing on horizontal planes in ice shelves is negligible (Sanderson and Doake 1979).

Equations 8.111 and 8.112 thus give the force balance as

$$2H\bar{\tau}_{xx} = \phi_{\mathrm{M}} + \rho g \int_x^{X_m} H \frac{-dS}{dx} dx - \int_x^{X_m} H \frac{\tau_o}{Y} dx. \tag{8.122}$$

Applying the flotation condition (Eqs. 8.109 and 8.110) then yields the following expression for the deviatoric stress of stretching:

$$\bar{\tau}_{xx} = \frac{1}{4}\left[\rho g H\left[1 - \frac{\rho}{\rho'}\right]\right] - \frac{\tau_o}{2H}\int_x^{X_m} \frac{H}{Y} dx. \tag{8.123}$$

Note that, in the general case, Y varies with x; it does not in the present analysis because we have assumed parallel embayment walls.

We have now determined $\bar{\tau}_{xx} = -\bar{\tau}_{zz}$. On the centerline $\tau_{xy} = 0$ and therefore $\tau_E = \tau_{xx}$. From the creep rule the spreading rate on the centerline is thus

$$\dot{\epsilon}_{xx} = \left[\frac{\bar{\tau}_{xx}}{A_{\mathrm{inv}}}\right]^n \quad \text{with} \quad H\,\overline{A_{\mathrm{inv}}} = \int_B^S A_{\mathrm{inv}}\,dz \tag{8.124}$$

and $A_{\mathrm{inv}}^n = 1/A$.

On the right-hand side of Eq. 8.123, the second term describes the restraining force per unit area due to shear at the side walls. The first term – the driving force – is the same as previously derived for the case of a confined shelf with side-shear disregarded, but with H a function of x rather than equal to its value at the ice front. Sanderson (1979) postulated that these two forces should increase with x at about the same rate in a stable ice shelf. He showed that, because H changes only slowly with x, the side-shear term changes with x at a rate of approximately $\tau_o/2Y$. Equating this to the rate of change of the first term gives

$$\frac{dH}{dx} = 2\tau_o\left[\rho g Y\left[1 - \frac{\rho}{\rho'}\right]\right]^{-1}. \tag{8.125}$$

This approximate relation for the thickness gradient is independent of accumulation and ablation rates, ice thickness, velocity, and flow parameters. It predicts that the thickness gradient varies inversely to the width of the shelf, a prediction confirmed by field data as Figure 8.32 shows. The value chosen for τ_o was 90 kPa, a typical plastic yield strength for ice. This suggests simply that the driving stress of an ice shelf is largely balanced by drag on the walls at the same longitudinal position. In this case the velocity along an ice shelf centerline should increase strongly with

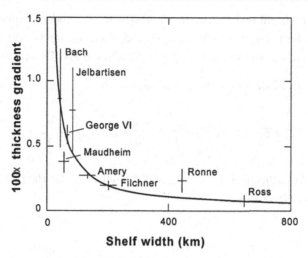

Figure 8.32: Relation between longitudinal thickness gradient and width of ice shelves. The bars show the range of values found in each. The solid line is the theoretical relation Eq. 8.125. From Sanderson (1979).

the width of the shelf, as shown by the analysis of weak-bed ice streams in Section 8.6.1. Such a relationship indeed describes the Amery Ice Shelf (Budd et al. 1982), which occupies a bay about 500 km long. Figure 8.31b shows measured velocities along this shelf. Velocities increase steadily downstream as the width increases from about 70 km to 170 km. The stretching rate increases toward the front even though the shelf thickness decreases, a pattern opposite that observed on the unconfined Erebus Ice Tongue. At Amery, ice entering from tributary glaciers fills most of the increased width; however, some transverse stretching also occurs, and flow lines diverge along the shelf. We next consider the effects of such additional deformations.

8.9.3.5 Diverging and Converging Flow

If the sides of the confining bay are not parallel, the flow lines in the ice shelf are usually not parallel either. In addition, ice rises and areas of thick ice at the mouths of ice streams influence the direction of flow. The effects of transverse strain $\dot{\epsilon}_{yy}$ and variations of the vertical-plane shears $\dot{\epsilon}_{xy}$ need to be included for a more general understanding of shelf flow. Their effects can be illustrated by introducing scaling parameters (Thomas 1973a). Define parameters a and b by $\dot{\epsilon}_{yy} = a\dot{\epsilon}_{xx}$ and $\dot{\epsilon}_{xy} = b\dot{\epsilon}_{xx}$. For analyses of field data, the a and b are empirical coefficients determined from measurements of flow. It follows that $\dot{\epsilon}_{zz} = -[1+a]\dot{\epsilon}_{xx}$ and $\tau_{\mathrm{E}}^2 = [1+a+a^2+b^2]\tau_{xx}^2$. Equation 8.124 becomes

$$\dot{\epsilon}_{xx} = [1+a+a^2+b^2]\left[\frac{\bar{\tau}_{xx}}{A_{\mathrm{inv}}}\right]^3.$$

(8.126)

The additional deformations therefore soften the ice and increase the spreading rate. In addition, $[2+a] H \bar{\tau}_{xx}$ replaces the term $2H\bar{\tau}_{xx}$ in Eq. 8.122, and Eq. 8.123 becomes

$$\bar{\tau}_{xx} = \frac{\rho g S}{2[2+a]} - \frac{\tau_o}{H[2+a]} \int\limits_{x}^{X_m} \frac{H \cos \psi}{Y} \, dx. \tag{8.127}$$

Here ψ is the angle that each side wall makes with the centerline. These equations are not restricted to the centerline. They can also be applied to the band between two flow lines, distance $2Y$ apart, rather than to the whole ice shelf.

8.9.3.6 General Flow of an Ice Shelf

Numerical methods are used to simulate flow throughout an ice shelf, with a realistic representation of its geography. If measurements of ice thickness and surface topography are available, the entire ice shelf can be modelled using Eq. 8.59, derived from the vertically integrated balance of forces (MacAyeal 1989). The key difficulty is that the effective ice viscosity depends on temperature, fabric, and possibly other variables. Thus, in practice, measurements of the flow must be used together with the model simulations to make a map of the viscosity variations, which can then be used in further modelling. Larour et al. (2005) analyzed the viscosity distribution in the Ronne Ice Shelf using this approach. They found that ice is soft along the shearing side margins due to frictional heating and stiff where large glaciers transport cold ice from the continent into the shelf. Rack et al. (2000) found that models reproduced the overall pattern of flow on the northern Larsen Ice Shelf well, but did not reproduce high observed longitudinal strain rates near the ice shelf front where rifting occurs; here the ice no longer behaved as a viscous fluid. Humbert et al. (2005) used a model of the entire Ross Ice Shelf to show how the large ice rises, Roosevelt Island and Crary Ice Rise, significantly restrain the outward flow of the shelf.

8.9.3.7 Back Force of an Ice Shelf

The *back force* is the total force transmitted upstream that opposes spreading of an ice shelf, originating from sources other than the pressure of seawater at the ice front. On a section of shelf from x to the front margin at X_m, the back force (per unit width across-shelf) is the resistive force in Eq. 8.111, or:

$$F_B = \int\limits_{x}^{X_m} [\tau_w + \tau_b] \, dx = \int\limits_{x}^{X_m} \left[\int\limits_{B}^{S} \frac{\partial \tau_{xy}}{\partial y} \, dz + \tau_b \right] dx. \tag{8.128}$$

In other words, back force arises from side drag exerted by lateral walls and by slower-flowing ice and from basal resistance on grounded spots. A special case is a flow line leading to the upstream end of a grounded ice rise. At the ice rise a compressive force pushes back on the shelf, with magnitude $F_g = H \left[2\bar{\tau}_{xx} + \bar{\tau}_{yy} \right]$. At locations upstream of the ice rise, the back force is then $F_g + \int \tau_w \, dx$. Lingle et al. (1991, Eq. 8.5) gave an empirical formula for the total back force due to an ice rise, from drag along its sides and stronger compression on its upstream end compared to the downstream end. MacAyeal (1987) extended the concept of back force to a curved channel.

The total back force at any point can be inferred from measurements of surface strain rates $\dot\epsilon_{xx}$ and $\dot\epsilon_{yy}$. Such measurements are used to substitute for the stresses $\overline\tau_{xx}$ and $\overline\tau_{yy}$ in Eq. 8.112. The substitution requires calculating the effective ice viscosity from the creep relation by using the measured strain rates and estimates of temperatures at depth. The resultant net longitudinal force F_L (Eq. 8.112) is subtracted from the net driving force (as in Eq. 8.111) to find the back force:

$$F_B = \frac{1}{2}\rho g \left[1 - \frac{\rho}{\rho'}\right] H^2 - 2\overline\eta H \left[2\dot\epsilon_{xx} + \dot\epsilon_{yy}\right], \tag{8.129}$$

where the viscosity $\overline\eta$ is the average, over depth, of $\frac{1}{2}\left[A\,\dot\epsilon_E^{n-1}\right]^{-1/n}$. Strain rates can be assumed uniform with depth.

Figure 8.33 illustrates the distribution of back force in the Ross Ice Shelf calculated from thickness and strain-rate measurements. The *back stress* (F_B/H, or force per unit width divided by thickness) ranges from about 40 kPa near the ice front to 200 or 300 kPa near the grounding line. The back stress is highest upstream of Crary Ice Rise and Roosevelt Island and where Beardmore Glacier flows into the ice shelf. The value at the mouth of Kamb Ice Stream is low because this ice stream is inactive.

Flow in the section of ice shelf west of Roosevelt Island, with back stress increasing regularly away from the ice front, resembles that predicted for an ice shelf in a channel with parallel sides. Elsewhere, however, flow is more complex than in the longitudinal models. Thickness and velocity do not vary uniformly over the ice shelf (Robin 1975). Outlet glaciers and ice streams persist for some distance into the shelf as zones of thick, fast-moving ice until drag of the surrounding ice retards them.

8.9.3.8 Ice Shelf Profiles

The specific mass balance and spreading rate determine the thickness of an ice shelf according to the requirement of mass conservation (Section 8.5.4). In an ice shelf, velocity and strain-rate components are independent of depth, and, if the x-axis follows a flow line, v, but not the transverse strain rate $\dot\epsilon_{yy}$, is zero. Specific balances at both the surface ($\dot b_s$) and the base ($\dot b_b$) are important. The vertically integrated mass conservation equation (Eq. 8.69) thus takes the form

$$\frac{\partial}{\partial t}[\overline\rho H] = \dot b_s + \dot b_b - u\frac{\partial}{\partial x}[\overline\rho H] - \overline\rho H \left[\dot\epsilon_{xx} + \dot\epsilon_{yy}\right]. \tag{8.130}$$

Increases of specific balance thicken the shelf. Advection of inland ice onto the shelf also thickens it. Increases of spreading rate thin it.

Van der Veen (1986) used this relation to compute transient profiles of ice shelves. He concluded that an ice shelf responds much more rapidly to changes in specific balance than

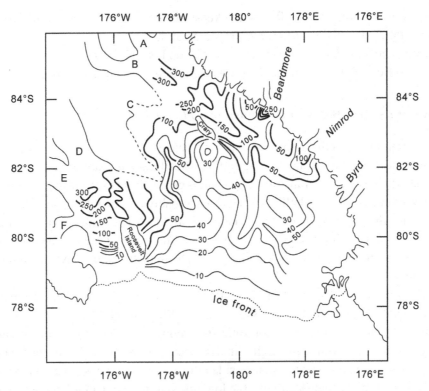

Figure 8.33: Lines of equal back force (MN per unit width of ice shelf) on the Ross Ice Shelf; back force is due to shearing past ice margins, compression upstream of ice rises, and compression and shearing at the mouths of ice streams and glaciers. Adapted from Thomas and MacAyeal (1982).

does an ice sheet. This is expected because changes of velocity propagate quickly throughout a shelf. At most places in a grounded ice sheet, the local driving stress – and hence ice thickness and surface slope around a point – determines the velocity because the basal drag provides most resistance. (Here "local" is an average over a distance of a few ice thicknesses, as discussed in Section 8.7.2.) In an ice shelf, in contrast, basal drag provides little resistance and the velocity and the longitudinal deviatoric stress of the spreading depend on the shelf geometry over a large region (Eqs. 8.111 and 8.112). Thus a change in stress in one part of an ice shelf produces almost immediate changes in stress, strain rate, and velocity throughout the shelf. A change in stress at one point in an ice sheet, on the other hand, only affects the stress and velocity at a distant point when a change in geometry has propagated to it. Chapter 11 discusses such time delays in grounded ice.

For a steady state and a uniform density Eq. 8.130 can be written

$$u\frac{\partial H}{\partial x} = \frac{1}{\rho}\left[\dot{b}_s + \dot{b}_b\right] - H\left[\dot{\epsilon}_{xx} + \dot{\epsilon}_{yy}\right]. \qquad (8.131)$$

Because ice shelves spread ($\dot{\epsilon}_{xx} > 0$), the thickness always tends to decrease toward the front margin. Snowfall on the upper surface and freeze-on at the base partly counteract this thinning trend. Relations for the spreading rate (Eqs. 8.126 and 8.127) and flow (Eq. 8.114) can be substituted into Eq. 8.131 to calculate the thickness profile $H(x)$ (e.g., Sanderson 1979).

8.9.4 Transition Zone Between Grounded and Floating Ice

This is the region in which the deformation pattern changes from the basal shear characteristic of an ice sheet to the combination of longitudinal stretching and transverse shear typical of a floating ice shelf. The extent of transition zones varies greatly from location to location. Where an ice sheet feeds directly into the shelf, the transition probably occurs within a distance equivalent to only a few ice thicknesses. On the other hand, ice streams such as those on the Siple Coast – with low surface slopes, low basal shear stresses, and little vertical shear – are essentially long transition zones between the inland ice and a shelf. The transition thus may extend hundreds of kilometers inland.

Table 8.3 gives two examples of the force balance in seaward parts of transition zones (for Jutulstraumen entering the Fimbul Ice Shelf and for Whillans Ice Stream entering the Ross). Going from the inland ice to the transition zone, the driving stress decreases and resistance shifts from the bed to the sides. Although longitudinal resistance may have little impact on the balance of forces, the variations of basal resistance, side drag, and ice thickness lead to large gradients of velocity along-flow. At some locations, the longitudinal strain rate in ice streams matches the highest strain rates measured in the large ice shelves; about 3×10^{-3} yr^{-1}, for example, along the upper part of Whillans Ice Stream. It is therefore essential to include the longitudinal deviatoric stress as well as the side drags in analyses of flow in ice streams or other transition zones. Recent numerical models of flow in the transition zone that include effects of both side and longitudinal forces include Payne et al. (2004), Dupont and Alley (2005, 2006), and Pattyn et al. (2006). How the simulated flow varies with distance through the transition depends critically on the choice of parameterizations for basal resistance and slip. Thus these models are illustrative, not predictive.

Processes in the transition zone may control the stability of a grounded ice sheet with its base below sea level (a *marine ice sheet*). Such an ice sheet is buttressed by its surrounding ice shelves, and their shrinkage or removal increases outflow from the grounded ice. One hypothesis is that the operation of feedbacks makes such a process irreversible. Hughes (1973) suggested that the West Antarctic Ice Sheet might be disintegrating now as a continuation of recession since the last ice age. Mercer (1978) suggested that it could happen as a consequence of global warming. In Chapter 11, we discuss how changes of an ice shelf perturb the adjacent grounded ice. To summarize that discussion: the back force from an ice shelf partly controls the ice thickness and flow in the transition zone. In this zone, the longitudinal strain rate depends on the excess of driving force over back force, $F_D - F_B$, calculated for a flow line extending from the transition zone to the ice shelf front (Eqs. 8.111, 8.113, and 8.128). A reduction of back force increases the stretching in the transition zone and hence thins the ice. What happens next depends on the situation. Thinning might simply reduce the driving force until a new balance with the back force is established. On the other hand, thinning might cause the grounding line to retreat or might increase slip motion in the transition zone by reducing effective pressure at the bed.

An unconfined ice shelf with no grounded regions exerts no back force on the inland ice. Changes on such a shelf should have no influence on the spreading rate and ice thickness at the grounding line.

8.9.5 Flow Over Subglacial Lakes

The large subglacial lakes of East Antarctica (Section 6.4.2) are regions in the ice sheet interior with negligible basal resistance. Over a lake, the ice velocity rises to a local maximum. Flow is analogous to that in a confined ice shelf; longitudinal strain and side shear ($\dot{\epsilon}_{xx}$ and $\dot{\epsilon}_{xy}$) dominate the deformation. However, in the case of a lake, there is not one but two transition zones: an extending zone at the upstream edge and a compressing zone at the downstream edge. The latter prevents the floating ice over the lake from spreading pervasively like an ice shelf. The extending, compressing, and lateral shearing deformations continue into the ice beyond the perimeter of the lake itself because of stress-gradient coupling (Sections 8.7.2 and 8.6.1).

To model flow in this situation requires solving for all components of stress and velocity. Figure 8.34 illustrates results of calculations by Pattyn (2003) for flow over a hypothetical lake 50 km long and 25 km wide. The extending flow upstream of the lake thins the ice, and the compressing flow downstream thickens it. These warp the surface topography as shown. For comparison, the figure also shows measured surface topography across Lake Vostok (Bell et al. 2007). The overall patterns are the same as in the model. However, the measured transitions appear more complicated than the ones calculated by Pattyn's model, which has a rather low spatial resolution. The measurements show a narrow trough at the upstream edge and a narrow ridge at the downstream edge. These may be related to the disappearance and then reformation of a vertical shear profile upstream and downstream of the lake.

8.10 Surface Profiles of Ice Sheets

Here we discuss the steady-state form of ice sheets. Changes over time are discussed in Chapter 11.

8.10.1 Profile Equations

First consider an ice sheet on a horizontal bed. Figure 8.35 represents a cross-section and shows the coordinate system. The total width is $2L$, and the thickness is H in general and H_* at the divide on the centerline. The ice sheet is pictured as a long ridge perpendicular to the plane of the diagram.

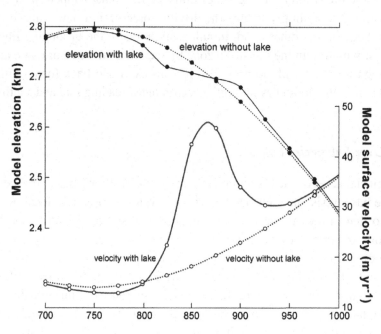

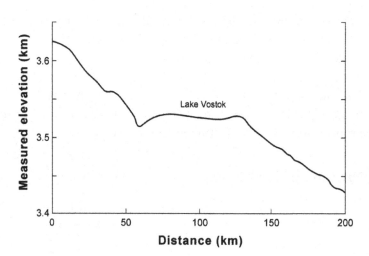

Figure 8.34: (a) Theoretical effect of a large subglacial lake on the elevation and surface velocity of an ice sheet. The model lake is 50 km long and 25 km wide. Adapted from Pattyn (2003). (b) Measured surface elevation of the Antarctic Ice Sheet on a transect crossing Lake Vostok. Adapted from Bell et al. (2007) and used with permission from the American Geophysical Union.

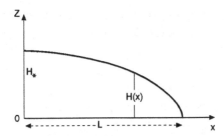

Figure 8.35: Coordinate system and variables for an ice sheet section.

Because the base (at elevation B = 0) is horizontal, the surface slope is $-dH/dx$, a positive quantity. Assume that the basal shear stress (τ_b) balances the driving stress. Thus

$$\tau_b = -\rho g H \frac{dH}{dx}, \tag{8.132}$$

with ρ the density, assumed constant, and g the gravitational acceleration.

The simplest case is to treat ice as a perfectly plastic material; the ice thickness adjusts itself so that at every point τ_b equals the yield stress τ_o. Then integrating Eq. 8.132 with $\tau_b = \tau_o$ yields the equation for the surface profile (Nye 1951),

$$H^2 = \frac{2\tau_o}{\rho g}[L-x], \tag{8.133}$$

which is a parabola. This equation also applies to a perfectly plastic ice sheet on a horizontal base with (1) a circular plan, if L is the radius and (2) a plan of irregular shape if $L-x$ is the distance from the edge measured along a flow line (Nye 1952a).

Equation 8.133 implies a thickness at the central divide of $H_* = [2\tau_o L/\rho g]^{1/2}$. As a first check, this formula, with $\tau_o = 100\,\mathrm{kPa}$, was applied to central Greenland. It gives $H_* = 3.15\,\mathrm{km}$ compared with a true value of about 3.20 km. For East Antarctica, using a typical dimension, $L = 1000\,\mathrm{km}$ gives $H_* = 4.7\,\mathrm{km}$. Although there are a few spots in Antarctica with ice this deep, it is too high as a typical value; most ice thicknesses along the central divides are in the range 3 to 4 km. A different value for τ_o could be used – perfect plasticity is, after all, a crude approximation. Although driving stresses in valley glaciers usually are 100 to 150 kPa, a mean value of 100 kPa is too high for ice sheets. Using a value $\tau_o = 75\,\mathrm{kPa}$ instead of 100 kPa in Eq. 8.133 reduces H_* by 14%. Though very rough, these calculations are still remarkable for their ability to explain a major characteristic of the ice sheets from a single property of ice, determined in laboratory experiments.

Where the ice sheet bed is slippery over a large area, τ_o should not be the plastic strength of ice but rather the effective yield strength of the bed, a smaller number if effective pressure at the base is low. A smaller τ_o implies a lower and flatter surface (Boulton and Jones 1979). For example, compared to present-day Greenland and East Antarctica, slopes were appreciably smaller on the southwestern lobes of the Laurentide Ice Sheet in North America (Mathews 1974); so too on the flanks of the modern West Antarctic Ice Sheet, where driving stresses are in the range 20 to 40 kPa.

Next we write a more general formulation; the bed elevation is allowed to vary, and we replace the assumption of perfect plasticity with the creep relation for ice and a relation for basal slip. To derive the profile in this general case we use the mass conservation constraint for steady state, and an equation for the velocity averaged over the ice thickness (Sections 8.1.2 and 8.3). Specifically, for a distance x_o from the ice divide, we write the volumetric ice flux through a section of width Y_o as

$$q^* = \bar{u} H Y_o = \left[c_o \lambda \tau_b^m + \frac{2A}{n+2} \tau_b^n H \right] H Y_o,$$ (8.134)

where the first term on the right accounts for flow by slip and the second term for internal deformation (Eqs. 8.35 and 8.26). Here the parameter c_o sets the rate of "normal" slip, and the parameter λ, with value ≥ 1, allows for fast flow in regions of well-lubricated bed. As usual, A and n are the creep parameters for ice (Chapter 3). At steady state, the ice flux also matches the total accumulation upstream:

$$q^* = \int_0^{x_o} \dot{b}_i(x) Y(x) \, dx,$$ (8.135)

where $\dot{b}_i$ represents the specific balance (m yr^{-1} of ice) and Y the width of the flow band upstream of x_o. Using the assumption that $\tau_b = \tau_d$, recognizing that

$$\tau_d = - \rho g \, [S - B] \frac{dS}{dx},$$ (8.136)

and setting Eq. 8.134 equal to Eq. 8.135 gives a relation for H or S.

The *Vialov profile* is the special case of no sliding, a flat bed, uniform accumulation, and constant flowband width (Vialov 1958). All ablation occurs by calving at the edge. The equation for $H(x)$ is, in terms of distance from the divide x,

$$\dot{b}_i x = \frac{2A}{n+2} \left[-\rho g H \frac{dH}{dx} \right]^n H^2.$$ (8.137)

Its solution defines the topographic profile

$$H^{2+2/n} = K \left[L^{1+1/n} - x^{1+1/n} \right],$$ (8.138)

with

$$K = \frac{2 \, [n+2]^{1/n}}{\rho g} \left[\frac{\dot{b}_i}{2A} \right]^{1/n}.$$ (8.139)

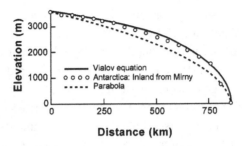

Figure 8.36: Profile of Antarctic Ice Sheet inland from Mirny compared with theoretical profiles: a parabola, and Eq. 8.140. Data from Vialov (1958).

But $H = H_*$ when $x = 0$ so we can write the profile as

$$[H/H_*]^{2+2/n} = 1 - [x/L]^{1+1/n}. \tag{8.140}$$

Perfect plasticity corresponds to $n \to \infty$, which reduces this equation to the parabola derived before (Eq. 8.133).

Figure 8.36 shows that Eq. 8.140 provides a good fit to a measured profile in East Antarctica (a profile not near a major outlet glacier). However, this applies only to the shape; the known value of H_* has been used rather than a value for K calculated from Eq. 8.139. The whole profile is scaled by the ice thickness at the divide, predicted to be

$$H_* = K_v \sqrt{L} \left[\frac{\dot{b}_i}{2A} \right]^{1/[2n+2]} \quad \text{with} \quad K_v = \left[\frac{2[n+2]^{1/n}}{\rho g} \right]^{n/[2n+2]}. \tag{8.141}$$

Some important conclusions can be drawn from this analysis. Because $n \approx 3$, Eqs. 8.141 and 8.140 show that ice thickness increases as the 1/8 power ($= 0.125$) of the accumulation rate. Thus the steady-state thickness of an ice sheet is only weakly sensitive to snowfall rate, unless snowfall also changes the ice sheet's span L. Equations 8.140 and 8.141 show further that H varies inversely as the 1/8 power of A, a function of ice temperature. For the same accumulation rate, colder ice implies a thicker ice sheet. But the dependence is not very sensitive: a decrease in temperature from -10 to $-30\,°C$ would increase H by about 35%. This applies only if basal temperature remains below melting point; a switch from a frozen to a melting base can cause a large change in the dimensions of a glacier or ice sheet. Changes in the ice sheet span strongly affect the thickness; H varies as the square root of L, regardless of the value for the creep exponent n. If an ice cap with $L = 50\,km$ grows into an ice sheet with $L = 500\,km$, the thickness more than triples. An ice sheet margin may be fixed, if it flows into the sea, or it may be free to advance across land. In the latter case, the volume and thickness of an ice sheet change much more than in the former.

The preceding equations describe the broad features of measured profiles on ice masses across a wide range of sizes, ice temperatures, and accumulation and ablation rates. (See, for example, Martin and Sanderson 1980 for small ice caps on the Antarctic Peninsula; Thomas et al. 1980 for Roosevelt Island; and Cuffey 2006b for the main ice sheets on Antarctica and Greenland.) This supports the prediction that the creep properties of ice largely determine their shapes. A single regularly shaped dome is nevertheless a very crude picture of the ice sheets. The Greenland Ice Sheet is fringed by mountains that channel the flow into large outlet glaciers. In central Antarctica, buried mountain ranges produce irregularities of the ice surface, whereas near the perimeter much of the flow is channelled into ice streams by bedrock troughs and sedimentary basins. Moreover, many of these ice streams flow by rapid slip. This was also likely the case for large sectors of the Pleistocene ice sheets.

8.10.2 Other Factors Influencing Profiles

We now examine the influence of some of these factors in more detail, by abandoning the special restrictions that lead to Vialov's profile. We will show profiles calculated by integration of Eqs. 8.134 through 8.136 along an axis inland from a margin to the divide. The thickness at the divide is supposed to be unaffected by the other side of the ice sheet; this is equivalent to assuming symmetry of the ice sheet and its basal topography. The ice deformation term used in Eq. 8.134 and the assumption that $\tau_b = \tau_d$ are valid only with small longitudinal strain rates and negligible longitudinal stress gradients. Thus we mean to address only the large-scale features of the profiles: variations over distances greater than about 20 times the ice thickness, the longitudinal coupling length (Section 8.7.2). The relation of surface and bed topographies at smaller scales is discussed later, in Section 8.10.5.

In comparison to the Vialov profile:

1. No ice sheet is ever in a steady state, as was assumed. However, the important factor is the relative size of the terms in the equation of mass conservation (Eq. 8.79):

$$\frac{\partial H}{\partial t} = \dot{b}_i - \frac{\partial q}{\partial x}. \tag{8.142}$$

If the flow term $\partial q / \partial x$ is comparable with $\dot{b}_i$, $\partial H / \partial t$ is small and the ice sheet should be near enough a steady state for the theory to apply. If, however, $\partial q / \partial x$ is small, the cumulative ablation and accumulation govern the profile. Lliboutry (1965, pp. 458–460) has drawn a similar distinction between *drainage glaciers*, in which flow is important, and *reservoir glaciers*, in which it is negligible. The Meighen Ice Cap in arctic Canada, which has an area of $80\,\mathrm{km}^2$ and a maximum thickness of about $125\,\mathrm{m}$, is an example of a reservoir glacier (Paterson 1969). A reservoir glacier may, in time, build up to such a thickness that flow becomes important.

The greatest observed imbalance for a large sector of a major ice sheet is in the basin of Pine Island Glacier in West Antarctica. Here $\partial q / \partial x$, averaged over the basin, appears to be some 60% greater than $\dot{b}_i$ (Thomas et al. 2004). However, this imbalance largely reflects

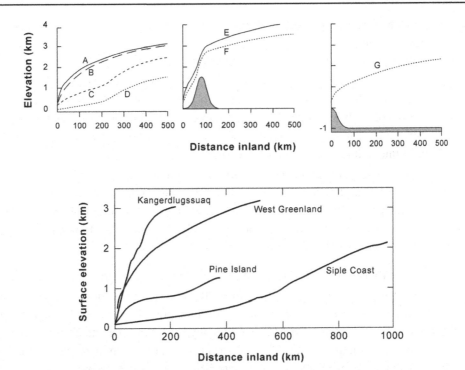

Figure 8.37: Theoretical and observed ice sheet surface profiles. Top three panels (drawn with identical vertical scales) show theoretical surface elevation profiles for the seven cases discussed in the text. In cases A, B, and C the bed is flat and at zero elevation. In case D, the bed is is flat and at 900 m below sea level. Shading indicates the bed in other cases. Bottom panel shows measured profiles following the ice surface gradient. Data for West Greenland from Reeh et al. (1999b); for others, taken from topographic maps.

recent events, so the profile may still be close to a steady-state one. The profiles discussed next all represent steady states.

2. Vialov's profile assumes an accumulation rate uniform over the whole ice sheet. Curve A in Figure 8.37 is a Vialov profile, with uniform accumulation of $0.3 \, \mathrm{m \, yr^{-1}}$. For comparison, curve B shows the effect on the steady profile of including an ablation zone, defined so that the flux decreases to zero at the margin. Ablation takes a uniform value within this zone. (Weertman 1961b and Paterson 1972a gave formulae for this case.) The ablation zone has little effect on the profile, partly because of the nonlinear relation between ice flux and thickness, and partly because the ablation zone is narrow. To match conditions in western Greenland, we chose an ablation rate of four times the accumulation rate; thus the ablation zone spans only 20% of the ice sheet.

3. Curve C in Figure 8.37 illustrates the effect of basal slip in a well-lubricated marginal zone. For this example, we specified a gradual transition of basal conditions between

220 and 120 km, from a hard-bed region in the ice sheet interior to a lubricated region near the margin, where $\lambda = 10^3$. (As used here, the lubrication parameter λ specifies the enhancement of glacier motion due to slip at the bed: $u = u_d[1 + \lambda]$, with u the total velocity, and u_d the velocity due to ice deformation alone. "Slip ratio" is another term for λ.) This value of λ was selected so that $\tau_d \approx 25$ kPa, a typical value for the Siple Coast of West Antarctica and a plausible estimate for the southwestern Laurentide Ice Sheet. Comparison to the hard-bed curves shows the large impact on the volume of the ice sheet. Curve D represents a flow line that terminates at a 1-km-thick grounding line but otherwise is identical to case C; this situation most closely resembles the Siple Coast.

A value $\lambda = 10^3$ signifies a very large enhancement of flow above "normal" for the level of stress. A physically based model for λ, free of ad-hoc assumptions, is desirable but impractical (Chapter 7). Two end-member cases give insight as to how such an analysis might proceed, if calculation of profiles is the goal:

a. One possibility is that deformable sediment, with yield strength τ_*, mantles the bed in the slippery zone. If the bed is uniform in the direction perpendicular to the plane of the profile, then $\tau_d \approx \tau_*$, a function of longitudinal position in general. Then integrating Eq. 8.136 gives the profile, without needing any information about accumulation or ice properties. (Alternatively, the effect of variations perpendicular to the profile's plane can be accounted for by using a characteristic value for f' in $\tau_d = \tau_*/f'$.) Because effective pressures at the bed most strongly control τ_*, the basal hydrologic system must be modelled accurately, a major challenge (see Chapter 6).

b. As a second end member, the bed in the slippery zone can be regarded as too weak to support any of the driving stress. The profile follows an ice stream with $\tau_* = 0$ and driving stress equal to side drags generated by shear margins. This provides a rough approximation for the Siple Coast (see Section 8.9.2.4). The velocity profile across an ice stream with a very weak bed is given by Eq. 8.86. Integrating that equation across-flow gives an expression for ice flux to replace Eq. 8.134 (Raymond et al. 2001). Equating this flux with the upstream accumulation then leads to a steady-state profile, as in the derivation of the Vialov profile. Cuffey (2006b) showed that the observed profile of West Antarctica can be matched with this method. However, the model profile is then sensitive to new variables; ice thickness depends inversely on the width of the ice stream and inversely on a geometric factor f_s raised to the power $1/[n + 1]$, where f_s represents the areal fraction of streaming flow. Additional models must specify the ice stream width, the factor f_s, and the location of the transition between ice streams and normal inland ice. These variables may be controlled by the regional geology.

4. The Vialov profile assumes a flat bed. Mountain ranges are major topographic barriers to ice sheet flow, especially in southeastern Greenland and in Victoria Land of East Antarctica. Consider a mountain range near an ice sheet margin. Above the crest of the range, the ice

surface elevation increases compared to the case with a flat bed. Inland of the range, the ice is therefore thicker and flatter, and the surface higher; the mountain range acts like a dam. Curve F in Figure 8.37 illustrates this effect. Conversely, if the ice sheet rests in a broad basin, surface elevations are reduced (curve G). But ice *thickness* again increases compared to the case of a flat bed, and so the surface remains flatter inland. The upward-sloping bed near the margin thus also acts like a dam.

Recall that these profile calculations implicitly assume symmetry of the surface and bed profiles about the ice sheet center. In reality, one side of the ice sheet will usually have a lower surface, which draws ice from the central region toward it and shifts the divide in the opposite direction.

An ice sheet growing on an initially flat surface will, over time, lead to isostatic subsidence of the underlying lithosphere (Weertman 1961a). The weight of the ice sheet induces pressure gradients in the viscous mantle beneath; in response, mantle material creeps until horizontal pressure gradients are neutralized. For local isostatic equilibrium, a thickness H of ice depresses the bedrock by $H\rho/\rho_m$, with ρ the density of ice and ρ_m the density of the upper mantle. The latter is 3300 $\mathrm{kg\,m^{-3}}$ and so $\rho_m = 3.7\rho$. For perfectly plastic ice, with yield stress τ_o, Weertman showed that the equilibrium surface profile follows (compare Eq. 8.133),

$$S^2 = [1 - \rho/\rho_m] \frac{2\,\tau_o}{\rho g} [L - x], \tag{8.143}$$

with S the elevation of the surface above the original flat bed. The ice thickness is S $\rho_m/[\rho_m - \rho] \approx 1.4\,\mathrm{S}$. Note that local isostatic equilibrium, assumed for these relations, is only expected if the ice sheet span exceeds a few hundred kilometers. For a small ice sheet, $H\rho/\rho_m$ overestimates the isostatic depression because a small ice sheet's weight is partly supported by shear stresses in a surrounding region of the underlying lithosphere.

5. Ice flowing from an ice sheet interior toward the margin is usually focused into ice streams and outlet glaciers. In Eqs. 8.134 and 8.135, this means that the width Y increases inland. Curve E in Figure 8.37 is for the same situation as curve F but now showing the effect of such flow-line convergence; Y/Y_o decreases from a value of four in the interior to a value of one where the flow line crosses the mountain range. Spatial focussing increases the flux that a flow line transmits and so, by itself, makes the ice sheet profile steeper and thicker. However, as discussed in Section 8.9, the focussing of flux also concentrates strain heating, warms the ice, and concentrates basal water and sediment, all of which make the flow easier. There are thus competing effects on the profile. The latter effect, easier flow, generally dominates near marine margins.

    Where topography forces the width Y_o to be narrow, side drags balance some of the driving stress and, as discussed previously, τ_b in Eq. 8.134 is reduced from τ_d to the smaller value $f\tau_d$. This influences the profile in the same way as an increase of the focussing (Y/Y_o).

Measured surface profiles plotted in Figure 8.37 show some of the features discussed here. The West Greenland profile follows the EGIG survey line, which overlies a rather flat bed except for a low mountain range near the coast. This profile resembles that of Vialov. The Kangerdlugssuaq profile follows a major outlet glacier from the eastern coast of Greenland up to the divide. The large coastal mountain range here is analogous to that of the middle panel in the figure. The Siple Coast profile follows Whillans Ice Stream and then proceeds up-gradient to near the West Antarctic divide. This illustrates the effect of the well-lubricated marginal zone. Pine Island Glacier rests in a broad topographic basin and, for most of its middle section, overlies a slippery bed.

Variations in the surface form of an ice sheet also reflect the interacting effects of bedrock topography and irregularities in the positions of the margins. The essential features can again be understood from a perfectly plastic model (Reeh 1982). Flow lines are assumed to be orthogonal to the surface contours. The basal stress, set to yield stress τ_o, balances driving stress. Equation 8.136 is applied along a flow line:

$$\left[\frac{dS}{d\ell}\right]^2 = \left[\frac{\partial S}{\partial x}\right]^2 + \left[\frac{\partial S}{\partial y}\right]^2 = \left[\frac{\tau_o}{\rho g\,[S-B]}\right]^2 \tag{8.144}$$

Here x and y are coordinates in a horizontal plane, and ℓ is horizontal distance from the margin, measured along the flow line. For a given bedrock topography and position of the ice margin, this equation is solved by the method of characteristics to give the projection of the flow line on the xy-plane and the elevation along it.

Solutions for idealized beds illustrate certain features of ice sheet topography and flow:

1. Flow converges toward sections of the margin that are concave (recessed). These regions, equivalent to outlet glaciers, largely control surface elevations in the central part of the ice sheet.

2. Flow diverges toward convex sections of the margin; an ice divide running inward from the margin may develop.

3. An ice divide forms over a bedrock ridge oriented perpendicular to the margin, whereas a bedrock trough channels the flow.

4. A flow line, when followed upstream from the margin, tends to follow the direction of greatest uphill slope of the bed. Because the bed has most effect where the ice is thin – to affect the flow, the bed needs to perturb the surface slope (Section 8.2.1) – flow lines usually show the greatest curvature in regions near the margin.

Reeh applied his model to central Greenland, with present ice margins, realistic bedrock topography, and a yield stress of $\tau_o = 90\,\text{kPa}$. All the existing ice divides and outlet glaciers were reproduced as were the elevations along the central ice divide.

Although the plastic model is successful at explaining these large-scale features, a more refined analysis of surface topography requires a flow model incorporating a realistic creep rule, a coupled model for temperature, and information about basal motion. Three-dimensional ice sheet models do a good job of simulating the surface topography of Greenland – it would be surprising if they didn't, given the success of plastic models. They do, however, need to be tuned to observations using an adjustable coefficient on the ice viscosity. (See our introduction to ice sheet evolution models in Chapter 13.)

8.10.3 Relation Between Ice Area and Volume

Integration of Eq. 8.133 shows that, for an ice cap of circular plan with radius L and maximum thickness H_*, the volume V is proportional to $L^2 H_*$. (This is also true for the profile of Eq. 8.140.) But $H_* \propto L^{1/2}$ by Eq. 8.133 and the area $\mathcal{A} = \pi L^2$. Thus $V \propto \mathcal{A}^{1.25}$. Figure 8.38 is a plot of volume against area for the ice sheets and four ice caps whose volumes are known well from soundings. The regression line is

$$\log V = 1.23\,[\log \mathcal{A} - 1] \qquad (8.145)$$

with V in km^3 and $\mathcal{A}$ in km^2. The regression coefficient is close to the theoretical value. The constant 1 corresponds to a yield stress of about 50 kPa. Chizhov and Kotlyakov (1982) found that the relation $\mathcal{A} = \text{constant} \times H_*^4$ provided a good fit to data from 15 ice caps with areas between 10 and $10^4\,\text{km}^2$. This relation is equivalent to a value of 1.25 for the regression coefficient of $\log V$ on $\log \mathcal{A}$.

Bahr et al. (1994) considered the relationship between volume and area for mountain glaciers; data from 144 glaciers gave $V \propto \mathcal{A}^{1.36}$. To explain the value of this exponent, the assumption of

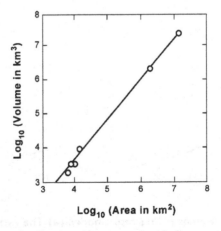

Figure 8.38: Logarithmic plot of volume against area for six ice caps and regression line (Eq. 8.145). From Paterson (1972a) with permission from the American Geophysical Union, *Reviews of Geophysics and Space Physics*.

plasticity needed to be replaced by scaling relations based on realistic flow relations. Bahr et al. showed that the exponent depends on how glacier width scales with length and how specific mass balance varies along a glacier.

8.10.4 Travel Times

For an ice sheet in steady state, with thickness profile $H(x)$, the velocity averaged over depth equals the balance velocity U_B (Section 8.1.2). Consider a flow band, of width Y in general, and Y_o at distance $x = X$ from the divide. Mass conservation implies a simple relation for U_B using Eqs. 8.134 and 8.135: $\langle Y \dot{b}_i \rangle X = U_B H Y_o$, where $\langle Y \dot{b}_i \rangle$ denotes the average of specific balance times width along the flow band upstream of X. Knowing the velocity permits a calculation of the time needed for ice to travel from one place to another. Ice deposited as snowfall at a location x_o arrives at location x after a time T_t of approximately

$$T_t = \int_{x_o}^{x} \frac{1}{U_B} \, dx = \int_{x_o}^{x} \frac{H Y_o}{\langle \dot{b}_i Y \rangle} \frac{1}{x} \, dx. \tag{8.146}$$

Thus the transit of ice is fastest if the ice sheet is thin, the accumulation rate high, and the flow lines convergent. The transit time increases roughly as the logarithm of x/x_o and the ratio $H/\dot{b}_i$ sets the scale. (The same ratio defines the characteristic value for $\dot{\epsilon}_{xx}^{-1}$ and $\dot{\epsilon}_{zz}^{-1}$ in extending and compressing flow.) Figure 8.39 depicts a few examples of transit times for ice parcels starting at two arbitrary locations, 250 km and 450 km inland from the margin

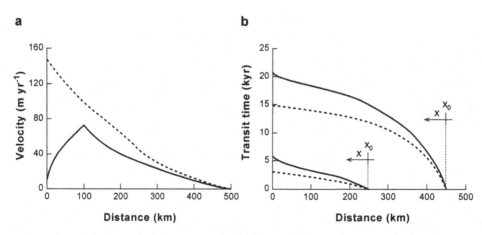

Figure 8.39: Theoretical travel times from simple model. (a) The variation of balance velocity with distance from the coast, for an ice sheet with (the solid line) and without (the broken line) an ablation zone. (b) Corresponding transit times for ice parcels originating at two locations. Flow is from right to left in these figures.

of an ice sheet with a 500 km span. Two cases are depicted for each: a Greenlandic case (solid curves), corresponding to profile B of Figure 8.37, and a West-Antarctic case (dashed curves), corresponding to profile D. The latter has no ablation zone. For accumulation rate and dimensions like those of Greenland and West Antarctica, as chosen, transit occurs in order 10^4 years. For East Antarctica, this interval would be several times longer; for Arctic ice caps, several times shorter. Note that Eq. 8.146 cannot be applied to origins x_o near the ice divide; such trajectories pass close to the bed, making U_B a poor approximation for velocity.

A three-dimensional ice sheet model, equipped with "particle tracking" algorithms, gives a detailed view of the ages of layers inside an ice sheet (Clarke et al. 2005). Instead of using the balance velocity in Eq. 8.146, the model velocity field $\vec{u}(x, y, z, t)$ is used to calculate how far, and in what direction, a parcel of ice moves in unit time. Integrated over time, this gives a particle path and the ages along it. Repeating this calculation for many paths in the ice sheet, with different points of origin, generates a three-dimensional map of the internal distribution of ages.

8.10.5 Local-scale Relation of Surface and Bed Topography

This discussion applies to both ice sheets and mountain glaciers. The ice sheet profiles discussed so far represent only the large-scale pattern of surface elevations, averaged over a distance of greater than 20 times the ice thickness or so. Over such large distances, the average bedrock slope is small, a few percent or less, even for major mountain ranges. If averaged over distances smaller than about 20 km, in contrast, basal topography can be rugged. The flow of ice over a rugged landscape produces undulations in the surfaces of ice sheets and glaciers. Gudmundsson (2003) and Raymond and Gudmundsson (2005) have summarized and extended the theoretical treatment of this problem (originating with Budd 1970 and Hutter et al. 1981). The 2005 study used numerical methods so that a realistic creep relation for ice ($n = 3$), and flow over large-amplitude undulations, could be analyzed.

Consider a bedrock topography with a characteristic amplitude and wavelength (peak-to-peak distance). Typical amplitudes are 5% to 20% of the wavelength. If the wavelength is small, less than the ice thickness, the surface topography displays essentially none of the bed's relief. Flow variations are confined to a basal layer, while the upper layers move uniformly. The viscosity of the ice is large enough in the upper layers, because stresses are low, to prevent much extending and compressing deformation at the small wavelength of the bed topography. Observations of flow using closely spaced boreholes confirm this picture (Harper et al. 2001); flow in the basal layers varies around bumps in the bed, but these fluctuations largely do not propagate to the surface. (In the small mountain glacier studied by Harper et al., however, flow at the surface is not uniform. But the observed variations relate not to bed undulations but to larger-scale variations of slope and ice thickness along and across the glacier.)

As the wavelength of bed undulations increases, longitudinal deformations become feasible in the upper layers of the glacier. Thus the velocity at the ice surface starts to vary, and undulations appear in the surface. Their amplitude increases with the wavelength and amplitude of the bed topography, but thick ice suppresses them. The amplitude of bedrock undulations does not increase indefinitely, however; very long undulations are effectively planes that do not perturb the surface. The most prominent surface undulations are therefore at some intermediate wavelength, predicted to be a few times the ice thickness. Observations are consistent with these predictions. Comparisons of bedrock and surface topographies in Dronning Maud Land, Antarctica, show that the surface most strongly expresses bedrock features with a wavelength of about three times the ice thickness (Beitzel 1970). Surface undulations of the predicted wavelength are also observed in West Greenland but die out in the central part of the ice sheet (Budd and Carter 1971).

Where the mean slope of a glacier surface is small, as for an ice sheet, the prominent surface undulations lie about 90° out of phase with the underlying bedrock undulations (assuming no basal slip). Thus the surface is steepest where the ice is thinnest, as expected from the perfect plasticity approximation (Section 8.2.1).

Whether the glacier moves primarily by slip or not has a major effect on the development of surface undulations. With significant slip, the maximum surface slope shifts downstream of the bedrock highs. At slip rates very much higher than deformational velocities, the surface and bedrock undulations coincide. Moreover, slip generally enhances the surface expression of basal topography. Choose a wavelength of $5H$ for an example. Then according to Raymond and Gudmundsson's (2005) analysis, the amplitude of surface undulations is several times larger on a slipping glacier compared to one moving only by internal deformation. On the latter glacier, the amplitude of surface undulations remains less than 10% that of bed undulations, for all wavelengths smaller than $10H$ (at slopes < 1°). In contrast, on the slipping glacier the amplitude of surface undulations exceeds 50% that of bed undulations, for all wavelengths greater than $3H$. This prediction is consistent with observations; prominent surface undulations appear on fast-flowing ice streams and surging glaciers.

The preceding discussion concerns bed topography that undulates in one direction, parallel to flow. Gudmundsson (2003) analyzed what should happen where a glacier flows over an isolated bedrock hill. The hill was modelled as a Gaussian function with radial symmetry. The ice piles up on the upstream side of the hill, forming a bulge in the surface, centered over the upstream flank of the hill. A surface depression forms on the downstream side.

Further Reading

Raymond (1980) discussed the dynamics of temperate glaciers in more detail than was possible here. A chapter by Bamber (2006) concisely introduces the satellites, sensors, and techniques used for InSAR. The monograph edited by Alley and Bindschadler (2001) contains many papers about the West Antarctic Ice Sheet. The book by Hutter (1983) provides a rigorous theoretical discussion of some of the fundamentals of glacier dynamics. The collection edited by Knight (2006) includes many papers summarizing recent work on ice dynamics.

CHAPTER 9

Temperatures in Ice Masses

"The action of heat is always present, it penetrates all bodies and spaces, it influences the processes of the arts, and occurs in all the phenomena of the universe."

The Analytical Theory of Heat, **Joseph Fourier (1822)**[1]

9.1 Introduction

The temperature distribution in glaciers and ice sheets deserves attention both for its intrinsic interest and its relation to other processes. The present variation of temperature with depth provides information about past variations of surface temperature. The deformation rate of ice depends sensitively on temperature; cooling from $-10\,°C$ to $-25\,°C$ increases the viscosity by a factor of five. A glacier, previously frozen to its bed, starts to slip when its base warms to the melting point; the terminus then advances, perhaps unstably. The interaction between heat flow and ice flow adds considerable complexity to glacier dynamics and evolution. Properties such as the velocity of seismic waves and the absorption of radio waves, on which depend methods of measuring ice thickness, also vary with temperature. In particular, by scattering radio waves, water in the ice makes radar sounding difficult.

What controls the temperature distribution? Through the energy balance, the climate determines the surface temperature, as described in Chapter 5. Geothermal heat and frictional heating from basal slip warm or melt the base. Ice deformation and refreezing of meltwater warm the interior. Heat is transferred within the glacier by conduction, ice movement (advection), and, in some cases, water flow.

Geothermal, frictional, and deformational heat sources typically concentrate at or near the base of a glacier. The glacier itself is an effective thermal insulator. Thus the ice near the bed of a glacier tends to be comparatively warm. But the different heat sources and processes of transfer vary in importance. Moreover, glaciers thrive in a diverse range of climatic conditions. Consequently, four main types of temperature distributions occur:

1. All the ice is below melting point.

2. Only the bed reaches melting point.

---

[1] Translation by A. Freeman (1878).

Copyright © 2010, Elsevier Inc. All rights reserved.
DOI: 10.1016/B978-0-12-369461-4.00009-0

3. A basal layer of finite thickness is at melting point.

4. All the ice is at melting point except for a surface layer, about 15 m thick, where temperatures fluctuate with the seasons.

Glaciers with the first two types of distributions are said to be *cold*, those of the third type are called *polythermal*, and the last category is *temperate*. Reality is more complicated than this terminology suggests, however, for different types of temperature distributions may be found in different parts of the same glacier.

In this chapter we discuss the general problem of heat transfer in glaciers. But we first discuss the thermal properties of snow and ice, the factors controlling temperatures near the surface, and the characteristics of temperate glaciers.

9.2 Thermal Parameters of Ice and Snow

For a geological material, glacier ice is unusually pure. Consequently its thermal properties at subfreezing temperatures are well known from measurements on pure ice. Table 9.1 gives values obtained from a comprehensive data review by Yen (1981).

The following empirical formulae – expressed in S.I. units – relate the specific heat capacity c and thermal conductivity k_T^i of pure ice to the temperature T in Kelvin:

$$c = 152.5 + 7.122T \tag{9.1}$$

$$k_T^i = 9.828 \exp\left(-5.7 \times 10^{-3}T\right). \tag{9.2}$$

Thermal conductivity also depends on density but the data are widely scattered, especially for snow with densities less than $500\,\mathrm{kg\,m^{-3}}$ (Sturm et al. 1997). The scatter arises because conductivity depends on snow texture, which influences the area of contact between adjacent ice grains (Adams and Sato 1993). Textures range from small rounded grains to large faceted crystals. The following formulae can be used for thermal conductivity of dry snow, firn, and ice. The Van Dusen (1929) formula,

$$k_T = 2.1 \times 10^{-2} + 4.2 \times 10^{-4}\rho + 2.2 \times 10^{-9}\rho^3, \tag{9.3}$$

Table 9.1: Values for thermal parameters of pure ice.

		Temperature	
		0°C	−50 °C
Specific heat capacity	J kg^{-1}K^{-1}	2097	1741
Latent heat of fusion	kJ kg^{-1}	333.5	
Thermal conductivity	W m^{-1}K^{-1}	2.10	2.76
Thermal diffusivity	10^{-6}m^{2}s^{-1}	1.09	1.73

gives a lower limit in most cases. Low-density hoar layers, however, can be even less conductive (Sturm and Johnson 1992). The Schwerdtfeger (1963) formula,

$$k_T = \frac{2k_T^i \rho}{3\rho_i - \rho},$$ (9.4)

gives an upper limit. The specific (per unit mass) heat capacity of dry snow and ice does not vary with density because the heat needed to warm the air and vapor between the grains is negligible. Thermal diffusivity, α_T, can be calculated for any density and temperature using $\alpha_T = k_T/\rho c$.

9.3 Temperature of Surface Layers

The penetration of seasonal and long-period changes in surface temperature can be analyzed by heat conduction theory. Fourier's law of heat conduction states that the heat flux q (the amount of heat energy flowing across unit area in unit time) at a point in a medium is proportional to the temperature gradient $\partial T/\partial z$, with z measured in the direction of the temperature variation. Thus

$$q = -k_T \frac{\partial T}{\partial z},$$ (9.5)

where k_T denotes the thermal conductivity. The minus sign indicates that heat flows in the direction of lower temperatures.

Consider an element of unit cross-section and thickness δz in a material at rest. Call the heat flowing in one side q; out of the other side flows $q + [\partial q/\partial z]\delta z$. If $\partial q/\partial z$ is positive, more heat flows out than in, so the element cools. By the definition of specific heat capacity c, the change in heat in unit time equals $-\rho c[\partial T/\partial t]\delta z$ where ρ is density and t the time. It follows that, for constant k_T,

$$-\rho c \frac{\partial T}{\partial t} = \frac{\partial q}{\partial z} = -k_T \frac{\partial^2 T}{\partial z^2}, \quad \text{or}$$

$$\frac{\partial T}{\partial t} = \alpha_T \frac{\partial^2 T}{\partial z^2}.$$ (9.6)

The proportionality $\alpha_T = k_T/\rho c$ defines the thermal diffusivity.

Consider a cyclic variation of temperature at the surface ($z = 0$), with fluctuations about the mean value following

$$T(0, t) = A_T \sin(2\pi \omega t).$$ (9.7)

A_T denotes the amplitude and ω the frequency (cycles per time) of the variation. The solution of Eq. 9.6 gives the temperature variation in the subsurface at depth z as

$$T(z, t) = A_T \exp\left(-z\sqrt{\pi\omega/\alpha_T}\right) \sin\left(2\pi\omega t - z\sqrt{\pi\omega/\alpha_T}\right).$$ (9.8)

This solution shows that:

1. The amplitude of the wave decreases as $\exp\left(-z\sqrt{\pi\omega/\alpha_T}\right)$. Thus, the higher the frequency, the more rapid the attenuation with depth. In reality, surface temperature variations are a complicated function of time. They can, however, be expressed as harmonic series. The higher harmonics attenuate most rapidly, and the temperature perturbation at depth approximates a wave of the fundamental frequency. For example, at one-meter depth in cold firn the temperature variations mostly reflect the seasonal cycle, even though large weekly temperature changes occur at the surface. This pattern is seen in field measurements (Figure 9.1).

2. Temperature maxima and minima propagate at a velocity $2\sqrt{\pi\omega\alpha_T}$.

Table 9.2 lists some numerical values. Figure 9.2 shows the seasonal variations in near-surface temperatures. Field observations confirm that seasonal variations are undetectable below a depth of about 20 m. In contrast, Table 9.2 suggests that the 100 kyr ice-age temperature cycle penetrates to the base of an ice sheet, a process we discuss later.

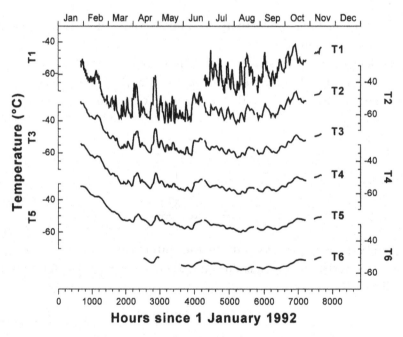

Figure 9.1: Variation of temperature over 10 months at South Pole, at six depths in the near-surface firn. At the end of the experiment, the depths of the six thermistors (called T1 through T6) were about 0.2, 0.4, 0.6, 0.8, 1.0, and 1.2 meters. Due to snowfall, the depths increased by about 0.4 m over the year, beginning around hour 3500. Temperature variability decreases with depth. Adapted from Brandt and Warren (1997).

Table 9.2: Propagation by conduction of a cyclical variation in surface temperature.

$P_\omega(yr)$	$z_5(m)$	$v(m\,yr^{-1})$	$\Delta t(yr)$
1	10.2	21.3	0.48
2500	509	0.427	1192
10^5	3220	0.067	4.8×10^4

$P_\omega = \text{period} = \omega^{-1}$.
$z_5 = \text{depth at which amplitude is 5\% of surface value.}$
$v = \text{velocity of propagation of maxima and minima.}$
$\Delta t = \text{time lag} = z_5/v.$

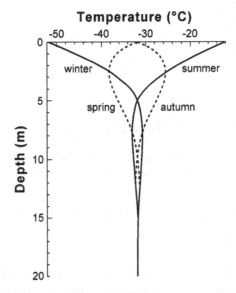

Figure 9.2: Theoretical seasonal cycle of firn temperatures in central Greenland.

In summer, in most places, heat conduction plays only a minor part in heat transfer through the surface layers. Except in the interiors of Greenland and Antarctica or on very high mountains the surface melts in summer and might receive rain. Surface water percolates into the snow and refreezes when it reaches a depth where the temperature remains below melting point. Refreezing of 1 g of water produces enough heat to raise the temperature of 160 g of snow or firn by 1 °C. (the ratio of the latent heat to the specific heat capacity is 160). This process significantly warms the layers near the surface. Figure 9.3 illustrates changes in temperature of the firn at a location on the Greenland Ice Sheet, at about 1600 m elevation on the EGIG survey line (data of N. Humphrey). The firn at 3 to 4 m below the surface warms rapidly, by 8 °C in just 6 days, because meltwater percolates down and refreezes. This process eliminates the winter's "cold wave" much more quickly than would have been possible by heat conduction alone. Cai et al. (1986) and Pfeffer and Humphrey (1996) have made mathematical analyses of this process. The latter study

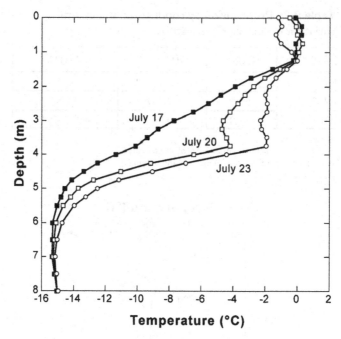

Figure 9.3: Warming of firn by latent heat of refreezing meltwater, at 1600 m elevation on the Greenland Ice Sheet. Data courtesy of N. Humphrey and T. Pfeffer.

also compared measured temperatures to strata in the firn and showed that warming initially concentrates in layers with growing ice lenses.

It is often stated that the temperature measured at a depth of 10 or 15 m in a glacier, beneath the zone of seasonal variation, equals the mean annual air temperature. Though not true in general, this is a good approximation at many cold, dry sites where the maximum air temperature never rises to 0 °C. A comparison of air and 10 m firn temperatures at polar dry-snow locations indicates that, at most sites, the two differ by less than 2 °C (Table 9.3). Firn temperatures are colder on average by only 0.7 °C.

In other cases, the air and firn temperatures differ substantially. Refreezing of surface water warms the firn and raises its temperature above the mean annual value for the air. Measurements at two glaciers in the Alps, for example, found a 0 °C temperature at 30 m depth, even though the mean annual air temperature at both sites was −7 to −8 °C (Hughes and Seligman 1939; Lliboutry 1963). Such warming is not effective in the ablation zone, where most surface water escapes from the glacier. This explains why, in polar glaciers, near-surface temperatures can be lower in ablation zones than accumulation zones, despite their lower altitude (Table 9.4).

On the other hand, warming of the firn in areas of melt is restricted because the glacier surface temperature cannot rise above 0 °C, even if the air temperature does. In contrast, winter snow acts as a blanket that reduces heat loss. Winter snow thus warms the ice, especially in the ablation area; in the accumulation area the net warming is weaker because some of the snow

Table 9.3: Mean annual air temperature and temperature at 10 m depth in dry-snow areas.

Location	Latitude	Longitude	Temperature (°C)	
			Air	Firn
Camp Century	77.2°N	61.1°W	−23.5	−24.5
Site 2	77.1°N	56.1°W	−24.1	−24.1
Station Centrale	70.9°N	40.6°W	−28.3	−27.6
Northice	78.1°N	38.5°W	−30.0	−28.0
Byrd	80.0°S	120.0°W	−28.2	−28.3
Pionerskaya	69.7°S	95.5°E	−38.0	−39.4
South Pole	90.0°S		−49.3	−50.8
Vostok	78.5°S	106.8°E	−56.0	−57.3
Plateau	79.3°S	40.5°E	−56.6	−60.2

Table 9.4: Temperatures at different points in same glacier.

Glacier	Zone	T(°C)	Depth (m)
Jackson Ice Cap[♭]	firn accumulation	−3	20
	ice accumulation	−9	20
	ablation	−8 to −10.5[†]	20
Vestfonna Ice Cap[‡]	accumulation	−3	14
	ablation	−7 to −10[†]	10

[†] Measured at various elevations.
[♭] Franz Josef Land.
[‡] Spitsbergen.
Data from Krenke (1963) and Schytt (1964).

becomes part of the glacier. Other features that produce local temperature variations include water channels within the ice and crevasses. The latter collect water in summer and cold air in winter. In the heavily crevassed shear margin of Whillans Ice Stream, West Antarctica, the ponding of cold winter air depresses mean annual temperatures by about 12 °C at a depth of 30 m (Harrison et al. 1998).

9.4 Temperate Glaciers

9.4.1 Ice Temperature

The definition of a temperate glacier in the introduction is not precise and the often-used term *pressure melting point* misleads; because the ice contains impurities it does not have a distinct melting point, determined solely by pressure. Temperate ice is a complex material consisting of ice, water, air, salts, and carbon dioxide. The water content usually ranges between 0.1% and 2% (Lliboutry 1976). Most of the water occupies the intersections of three or four grains, but it also occurs in lenses on grain boundaries, around air bubbles, and at salt inclusions. Air bubbles typically comprise a few percent of the ice volume. The impurities are distributed between the liquid in the veins along three-grain intersections, the grain-boundary area excluding the veins, and the bulk of the ice (Harrison and Raymond 1976). Although the bulk impurity content is only of the order 10^{-7} by weight, the impurities concentrate enough in the liquid in the veins to affect the equilibrium temperature.

If temperate ice were a mixture of pure ice and pure water, its temperature T at absolute pressure P would be

$$T = T_o - \mathcal{B}P, \tag{9.9}$$

where $\mathcal{B} = 7.42 \times 10^{-8} \text{K Pa}^{-1}$ specifies the rate of change of melting point with pressure, and $T_o = 273.16 \text{ K} = 0.01\,°\text{C}$ denotes the triple-point temperature of water. Strictly speaking, P should be measured relative to its value at the triple point, but this pressure is small enough (600 Pa) to be neglected. Because glacier ice contains air bubbles, the water in it should probably be regarded as air-saturated rather than pure. Dissolved air at atmospheric pressure lowers the equilibrium temperature by $2.4 \times 10^{-3}\,°\text{C}$. Moreover, solubility increases in proportion to pressure. Equation (9.9) should therefore be replaced by

$$T = T_o - \mathcal{B}'P, \tag{9.10}$$

where $\mathcal{B}' = 9.8 \times 10^{-8} \text{K Pa}^{-1}$ or $8.7 \times 10^{-4} \text{K m}^{-1}$ of ice. Stress in a glacier deviates from hydrostatic and this introduces a further complication; P is not necessarily the hydrostatic pressure but rather the stress normal to the liquid-solid interface (LaChapelle 1968; Kamb 1961).

Impurities depress the melting point in proportion to the solute concentration in the liquid inclusions. For small concentrations, this shifts the temperature by $-\mathcal{B}_s C_s / W$, for a fractional water content W by weight, and salt concentration C_s in mol kg^{-1}, and $\mathcal{B}_s = 1.86 \text{ K kg mol}^{-1}$ (Lliboutry 1976). Thus the equilibrium temperature of the ice is

$$T = T_o - \mathcal{B}'P - \frac{\mathcal{B}_s C_s}{W}. \tag{9.11}$$

Curvature of the ice-water interface also influences equilibrium temperature but by a small amount compared to impurities (Lliboutry 1976).

Impurities greatly increase the effective specific heat capacity of ice near the melting point (Harrison 1972). The concentration of salt solution in equilibrium with ice depends on the temperature. Only part of the heat added to the ice raises the temperature – some of it must melt ice to dilute the liquid. The effective specific heat capacity c' thus exceeds the value for pure ice, c:

$$c' = c + L\frac{dW}{dT}, \tag{9.12}$$

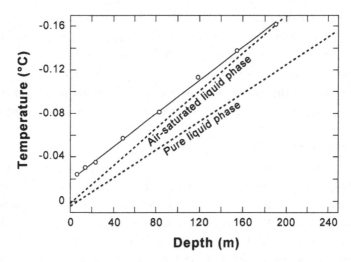

Figure 9.4: Variation of temperature with depth in the temperate Blue Glacier (solid line) and variation expected for pure ice in equilibrium with (a) pure water and (b) air-saturated water. Impurities depress the melting point in the glacier. Redrawn from Harrison (1975).

where L denotes the specific latent heat of fusion, and W is related to T by Eq. 9.11. For typical salt concentrations and $T = -0.01\,°C$, c' can be about one hundred times c. Air bubbles have the same effect as salts (Raymond 1976).

Harrison (1972) proposed a precise definition of temperate ice, based on the concept of effective specific heat capacity: "Temperate ice is ice whose effective bulk heat capacity is significantly greater than that of a pure ice monocrystal." A single crystal is specified to eliminate grain-boundary effects.

Figure 9.4 shows temperatures measured in two boreholes in Blue Glacier, Washington. The glacier is colder than predicted by Eqs. 9.9 and 9.10, due to effects of impurities and air bubbles. In a temperate glacier borehole, the temperature of the water obeys Eq. 9.10, since the solute bulk concentration is negligible; in contrast, the surrounding ice has the colder values given by Eq. 9.11. Thus a water-filled borehole in temperate ice loses heat to the glacier and will refreeze at all depths. This has been observed.

9.4.2 Origin and Effect of Water

For a glacier to be temperate, it must contain heat sources and sinks. These are provided by the freezing of small quantities of water and the melting of small quantities of ice. In the accumulation zone, for example, a parcel of ice experiences an ever-increasing pressure as subsequent snowfalls bury it; its melting temperature therefore decreases and it must cool by giving up heat, which melts a small amount of ice. As ice flows toward the surface in the ablation zone, in contrast, the pressure on a given parcel decreases and so its melting temperature increases. If the ice is to remain temperate, it must be heated.

In a temperate glacier, conduction downward from the surface carries an insignificant amount of heat because the vertical temperature gradient is so small. The direction of this temperature gradient prevents heat conduction from the bed; all the basal heat melts ice there (typically about $1\,cm\,yr^{-1}$). Deformation produces some heat within the glacier but only in about the lower half of the ice thickness is it sufficient to maintain the ice at melting point (Paterson 1971). The necessary heat must therefore be supplied by the freezing of the small quantities of water in the ice.

The water dispersed throughout a temperate glacier, which typically comprises about 1% of the volume, has various sources: percolation and conduit flow from the surface, ice melted by deformational heating, melting induced by pressure changes, and pockets of water trapped when the ice formed. What is their relative importance? Near the base of temperate glaciers, where the shear stress is about $100\,kPa$ and the shear strain rate may be about $0.2\,yr^{-1}$, the heat of deformation could melt about 1% of the ice in $100\,yr$. This mechanism cannot, however, produce this amount of water throughout the bulk of the ice. Paterson (1971) estimated the amounts of water produced by the other sources in Athabasca Glacier and concluded that most of the water was trapped when firn became ice. It does not drain away because the intergranular permeability of temperate ice is extremely low (Chapter 6). Temperature measurements such as those in Figure 9.4 confirm that permeabilities are low; they demonstrate that impurities are not flushed out of the ice. Large quantities of water move through temperate glaciers in fractures, tubes, and other large passageways (Chapter 6). This water carries some heat into the glacier, and a portion of it might freeze. Because such passageways occupy only a small volume of the glacier, their effect on temperatures is often assumed to be negligible, but this topic needs more investigation.

9.4.3 Distribution of Temperate Glaciers

Temperate glaciers are widespread in the literature; how widespread they are in reality is uncertain. It is often assumed that all glaciers in temperate regions are temperate. This is certainly untrue.

For a glacier to be temperate, the previous winter's cold wave must be eliminated by the end of summer. Refreezing of percolating meltwater can accomplish this rapidly in the accumulation zone (Figure 9.3). In the ablation zone, however, the ice is almost impermeable to water. Paterson (1972b) discussed other processes. Heat conduction is slow and the amount of heat is limited because the surface temperature cannot rise above $0\,°C$. Solar radiation does not penetrate deeply enough. Most of the heat at the surface warms the surface ice to $0\,°C$ and then melts it. Whether all the ice attains $0\,°C$ by the end of the summer therefore depends largely on the amount of ablation relative to the depth of penetration of the cold wave, a function of winter temperatures and snowfall. In many glaciers the ice likely remains below melting point in the region of slow ablation immediately below the equilibrium line.

Haeberli (1976) reviewed temperature data from glaciers in the Alps, where accumulation zones at high elevations are not temperate. On Mont Blanc, for example, 15 m temperatures are about $-17\,°C$ at 4785 m elevation and about $-7\,°C$ at 3960 m. Rapid transport of such cold ice may bring it into ablation zones while still subfreezing. Nonetheless, the ice is probably temperate in the lower parts of major glaciers with high ablation rates.

The most likely place to find temperate glaciers is a region with a maritime-temperate climate where intense summer melting follows heavy winter snowfalls. This describes the setting of Blue Glacier, located 75 km from the Pacific Ocean at latitude 48°N in Washington. Here, LaChapelle (1961) found that subfreezing temperatures never penetrated below 4 m in the accumulation zone in winter, and the firn temperature never fell below $-2\,°C$. Summer melt and rain soon remove this small cold wave. Measurements just below the equilibrium line revealed temperate ice there too (Harrison 1975). On the other hand, Mathews (1964) measured temperatures of about $-2\,°C$ in a tunnel under South Leduc Glacier, in western Canada, located 100 km from the coast at latitude 56°N.

In polar regions, some glaciers may have temperate accumulation zones as a result of percolating meltwater, while the ablation zones are cold. Schytt (1969) described this situation in Svalbard: "In many glaciers, previously considered as temperate ... the cooling starts just above the equilibrium line and the ice temperature in the upper several tens of metres stays below 0 °C all the way to the ice edge."

9.5 Steady-state Temperature Distributions

The temperature distribution in a glacier is never in a steady state, but heat flow rapidly reduces large deviations from a steady pattern. A large part of measured temperature profiles can therefore be understood by examining steady-state distributions.

9.5.1 Steady-state Vertical Temperature Profile

Within a glacier, conduction transfers heat both vertically and horizontally, but small temperature gradients usually make the latter negligible. Heat transfer by ice flow, however – the "advection" – is important both horizontally and vertically. Ice moving vertically (z-direction) with velocity w carries a heat flux $\rho c w T$ across a plane of unit area, oriented perpendicular to z. This term must be added to q in Eq. 9.6. A similar term must be included for advection due to ice flow at rate u in the horizontal direction x (choosing x to follow a flow line). Equation 9.6 thus becomes

$$\frac{\partial T}{\partial t} = \alpha_{\mathrm{T}} \frac{\partial^2 T}{\partial z^2} - w \frac{\partial T}{\partial z} - u \frac{\partial T}{\partial x}. \tag{9.13}$$

In thermal steady state, $\partial T/\partial t = 0$. Because the temperature would not remain constant if the ice thickness or velocity changed, the ice sheet is also implicitly assumed to be in a steady state of flow and geometry. In steady state, the temperature profile along a given vertical line, fixed in space, remains unchanged as the ice flows by.

9.5.1.1 *Steady State with No Horizontal Advection*

We first consider a case with negligible horizontal advection, with parameters appropriate for an ice sheet. Robin (1955) obtained the steady-state solution as follows. Position the origin at the base of the ice, assumed flat, and point the x-axis horizontally in the direction of flow. The z-axis is vertical and positive upward. This coordinate system is fixed in space. No ice or heat flows in the transverse direction. The following simplifying assumptions are made:

1. Horizontal conduction can be neglected because the horizontal temperature gradient is small compared with the vertical one. The term $u\,\partial T/\partial x$ cannot be neglected, in general, because, although $\partial T/\partial x$ is small compared with $\partial T/\partial z$, u is normally much greater than w. In Robin's analysis it is neglected, however; the solution should therefore apply near an ice sheet divide, where u is small.

2. The firn layer is replaced by an equivalent thickness of ice.

3. The heat generated by ice deformation is treated as a flux, additional to the geothermal flux, at the base of the ice. This is a reasonable approximation because, in the slow-flowing parts of ice sheets, most of the shearing occurs near the base.

4. The base is colder than melting point.

The assumptions reduce the steady-state heat-transfer equation to

$$\alpha_T \frac{d^2 T}{dz^2} - w \frac{dT}{dz} = 0. \tag{9.14}$$

Let H signify the ice thickness. The boundary conditions are, at $z = H$, $T = T_s$, a constant; at $z = 0$, heat flux $= -k_T [dT/dz]_B = $ constant. One integration gives

$$\frac{dT}{dz} = \left[\frac{dT}{dz} \right]_B \exp \left(\frac{1}{\alpha_T} \int_0^z w\, dz \right). \tag{9.15}$$

To proceed further, w needs to be expressed as a function of z. At the surface w must balance the accumulation (or ablation) to keep the ice sheet in a steady state. At the base w must be zero in the absence of melting. So we take $w = -\dot{b}_i z/H$, with the surface specific mass balance rate $\dot{b}_i$ expressed as thickness of ice per unit time. Integration of Eq. 9.15 using this relation for w gives

$$T - T_B = \left[\frac{dT}{dz} \right]_B \int_0^z \exp\left(-z^2 / z_*^2 \right) dz, \tag{9.16}$$

with $z_*^2 = 2\alpha_T H/\dot{b}_i$. When $\dot{b}_i > 0$, this can also be written

$$T - T_s = z_* \frac{\sqrt{\pi}}{2} \left[\frac{dT}{dz}\right]_B [\text{erf}(z/z_*) - \text{erf}(H/z_*)], \qquad (9.17)$$

in which erf stands for the error function,

$$\text{erf}(z) = \frac{2}{\sqrt{\pi}} \int_0^z \exp\left(-y^2\right) dy. \qquad (9.18)$$

Figure 9.5 illustrates such temperature distributions in terms of the dimensionless variables:

Distance above the bed:

$$\xi = \frac{z}{H} \qquad (9.19)$$

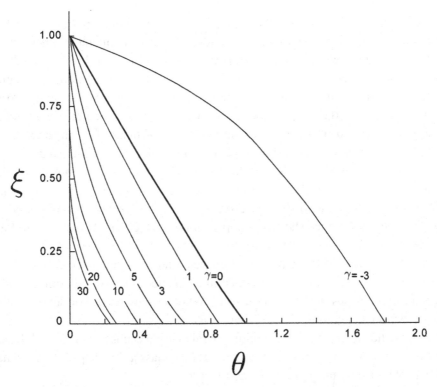

Figure 9.5: Dimensionless steady temperature profiles for various values of the advection parameter (γ). Negative value for γ indicates upward velocity; positive values indicate downward velocity. Equations 9.19–9.21 define the variables; θ refers to scaled temperature and ξ to scaled height above the bed. Adapted from Clarke et al. (1977) and used with permission of the American Geophysical Union, *Reviews of Geophysics and Space Physics*.

Temperature:

$$\theta = \frac{k_T [T - T_s]}{GH} \tag{9.20}$$

Advection parameter:

$$\gamma = \frac{\dot{b}_i H}{\alpha_T} \tag{9.21}$$

The advection parameter γ is, apart from a minus sign in cases of ablation, the same as the *Péclet Number*, a general indicator of the relative importance of advection and conduction. The dimensionless temperature depends on the geothermal flux G, a positive number. In a region of accumulation $\dot{b}_i > 0$. The temperature difference between surface and bed is given by the value of θ at the bed:

$$\theta_B = \left[\frac{\pi}{2\gamma} \right]^{1/2} \mathrm{erf}\,(\gamma/2)^{1/2}. \tag{9.22}$$

For $\gamma > 6.5$, corresponding to $\theta_B < 0.49$, $\mathrm{erf}\,(\gamma/2)^{1/2} \approx 1$, so that $\theta_B = (\pi/2\gamma)^{1/2}$, Figure 9.5 demonstrates that the vertical flow of ice profoundly affects the temperature distribution. In the absence of accumulation or ablation, the temperature-depth profile follows a straight line with slope equal to the geothermal gradient. Downward ice flow, corresponding to accumulation, carries cold ice from the surface and thus decreases the temperature. Ablation carries warm ice upward and increases the temperature. Note that, by Eq. 9.15, the maximum temperature gradient occurs at the bed in the accumulation zone but at the surface in the ablation zone. For $\dot{b}_i = 0.025\,\mathrm{m\,yr^{-1}}$ and $H = 3.5\,\mathrm{km}$, typical values for central East Antarctica, $\gamma = 2.5$ and $\theta_B = 0.7$. For an average geothermal heat flux $(50\,\mathrm{mW\,m^{-2}})$ the temperature difference between surface and bed is $59\,°\mathrm{C}$. For central Greenland, $\dot{b}_i = 0.25\,\mathrm{m\,yr^{-1}}$, $H = 3\,\mathrm{km}$, $\gamma = 21$, $\theta_B = 0.27$, and the temperature difference drops to $20\,°\mathrm{C}$. In Greenland, geothermal heat only affects the lower half of the ice sheet; the upper half is isothermal, as depicted in Figure 9.5. Now consider an ablation zone example: for $\dot{b}_i = -0.5\,\mathrm{m\,yr^{-1}}$, $H = 400\,\mathrm{m}$, and $\gamma = -5.5$, then $T_B - T_s = 36\,°\mathrm{C}$. Because few, if any, glaciers have surface temperatures as low as $-36\,°\mathrm{C}$ in the ablation zone, this calculation suggests that most glaciers, even in the Arctic, tend to reach melting point at their bases in the ablation zone. Whether they do or not depends also on the horizontal advection, not considered in this analysis.

In ice shelves and glaciers that slide rapidly, basal melting cannot be neglected. If $\dot{b}_b$ denotes the ice thickness removed in unit time at the bed (not the mass as in Chapter 4), the temperature distribution is obtained by integrating Eq. 9.15 using

$$w = -\dot{b}_b - \left[\dot{b}_i - \dot{b}_b \right] \frac{z}{H}. \tag{9.23}$$

At an ice divide, the downward flow of ice is slower, for the same depth, than at locations away from the divide (Raymond 1983; see Chapter 8). This reduces the cooling influence of vertical advection and increases the basal temperature; a "warm spot" should be found under the divide. Using the relation for w for an ice divide suggested by Raymond, $w = -\dot{b}_i z^2 / H^2$, rather than the linear one, gives a modified form of Eq. 9.16:

$$T - T_B = \left[\frac{dT}{dz} \right]_B \int_0^z \exp\left(-z^3 / \zeta^3 \right) dz, \tag{9.24}$$

with $\zeta^3 = 3\alpha_T H^2 / \dot{b}_i$. For $\dot{b}_i = 0.25 \, \text{m yr}^{-1}$, $H = 3 \, \text{km}$, and $G = 50 \, \text{mW m}^{-2}$, $T_B - T_s = 27\,°\text{C}$, compared with 20°C derived from Eq. 9.16. This reveals a sensitivity of the temperature distribution to the form of $w(z)$, an important point because very few measurements have been obtained of vertical velocity profiles in ice sheets.

9.5.1.2 Effect of Horizontal Advection

Except near an ice divide, horizontal advection usually exerts a strong influence on temperature profiles; the term $-u \partial T / \partial x$ in Eq. 9.13 needs to be included. Temperatures tend to increase in the direction of glacier flow, because the surface elevation declines. Along the surface, temperature typically increases by 0.4 to 1 °C per 100 m drop in altitude. Glacier flow thus transports colder ice, originating at higher altitudes, into warmer regions – horizontal advection usually reduces temperatures. Figure 9.6 illustrates the effects of horizontal advection, using numerical solutions to Eq. 9.13.

Such cooling does not alter the surface temperature, which is fixed by climate, but depresses the temperature at depth. If sufficiently strong, horizontal advection causes the temperature to decrease with increasing depth, a situation usually referred to as a "negative temperature gradient." Yet geothermal and frictional heat sources still warm the bed, so the net imprint of horizontal advection is a cold spot in the central to upper part of the ice column (Figure 9.6). The "cold spot" is cold *relative* to the profile expected in the absence of horizontal advection (Figure 9.5); a negative temperature gradient does not necessarily develop.

Horizontal advective cooling is strongest where flow is fast and the glacier surface steep. But faster flow also usually increases heat sources from basal friction and ice deformation. Deep in the glacier, such increased heating counteracts the cooling effect of advection (Figure 9.6). The combination of heating near the bed and cooling in mid to upper layers increases the temperature gradient $\partial T / \partial z$ in the lower half of the ice thickness, sometimes by a large amount.

9.6 Measured Temperature Profiles

Figure 9.7 shows some measured temperature profiles in ice sheets and ice caps. Data sources are Vostok, Antarctica (Salamatin et al. 2008); Bindschadler Ice Stream, Antarctica (Engelhardt 2004b); Siple Dome, Antarctica (Engelhardt 2004a); Jakobshavn margin, west Greenland (Lüthi et al. 2002); GISP2, central Greenland (Cuffey et al. 1995); NGRIP, north-central Greenland (Dahl-Jensen et al. 2002); Byrd Station, Antarctica (Gow et al. 1968); Devon Island ice cap,

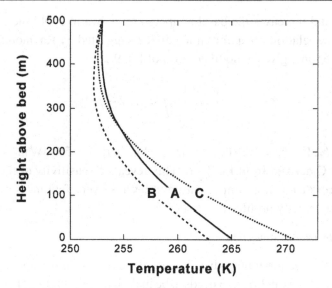

Figure 9.6: Theoretical influence of horizontal advection on the profile of temperature with depth. This example corresponds to the accumulation zone of a 500-m-thick glacier, with 0.4 m yr^{-1} of accumulation and surface velocity of 20 m yr^{-1}. (At depth, the velocity decreases according to the theoretical profile for flow by internal deformation.) Case A: Horizontal flow cools the ice because of an along-glacier temperature gradient of 0.12 °C km^{-1}, but also warms basal layers by frictional heating. Increasing the along-glacier temperature gradient to twice the value in A gives Case B. Temperatures are reduced at all depths, and a "negative temperature gradient" forms in the upper layers. Increasing the ice velocity to 40 m yr^{-1}, but leaving the temperature gradient as in A, gives Case C. Advective cooling and frictional heating both strengthen in this case; the basal zone warms, upper layers cool.

Arctic Canada (Paterson and Clarke 1978). The basal ice has reached melting point at Byrd Station, Bindschadler Ice Stream, Vostok, NGRIP, and Jakobshavn.

Profiles from sites with slow horizontal flow (Vostok, GISP2, NGRIP, Byrd, Siple Dome, and Devon) all show the features of the theoretical curves in Figure 9.5 for the appropriate value of the advection parameter. The Vostok and Siple Dome profiles are close to linear, as expected at dry locations with small vertical velocities. Vertical advection is much greater at the central Greenland sites (GISP2 and NGRIP) and Byrd station; both display nearly isothermal upper layers. Refreezing of percolating meltwater near the surface explains isothermal conditions in the top 80 m of the Devon Island profile.

The Bindschadler Ice Stream (formerly, Ice Stream D) and Siple Dome profiles belong to the same region in Antarctica. Yet they differ greatly because Bindschadler Ice Stream flows rapidly, about 350 m yr^{-1}. It shows the effects of horizontal advection, as expected from the theoretical curves in Figure 9.6: cooler ice at mid-depths and a steeper temperature gradient near the bed. Here, however, the steep basal gradient reflects not only increased heat generation at the

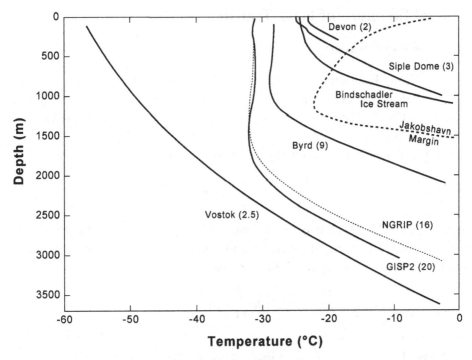

Figure 9.7: Measured temperature profiles in accumulation zones of polar ice sheets and ice caps. For sites with negligible horizontal advection, the number in parentheses gives the advection parameter defined by Eq. 9.21. Data sources are given in the text.

bed, but also vertical thinning of the ice as it passes from the central ice sheet to the thinner, faster-flowing ice stream. In other words, the profile reflects a larger vertical advection than expected from accumulation alone.

The Jakobshavn profile was taken from the side of a rapidly flowing ice stream. It illustrates an extreme case of the influence of horizontal advection. As ice flows rapidly toward the edge of the ice sheet, the surface steepens and the surface temperature increases rapidly. Heat dissipation in deforming ice warms the bed too (Section 9.7.1.1). The core of cold ice in between does not have time to warm, however.

Temperature gradients at the bed are about $0.022\,°C\,m^{-1}$ at Vostok and $0.053\,°C\,m^{-1}$ at Bindschadler Ice Stream. A geothermal flux of $50\,mW\,m^{-2}$ produces a temperature gradient of $0.024\,°C\,m^{-1}$ in ice. However, measurements of basal temperature gradients do not give reliable estimates of geothermal flux without extensive analysis. Frictional heat must be accounted for. Even in the deepest boreholes, the history of climate change affects the basal temperature gradient. Analyzing the temperature profiles in detail requires time-dependent models. The variations of temperature in the upper parts of the GISP2 and Byrd profiles largely reflect past climate changes, though some cooling from horizontal advection is also apparent at Byrd.

9.7 General Equation of Heat Transfer

9.7.1 Derivation of Equation

This analysis follows Paterson and Clarke (1978). The base of the glacier is assumed to be horizontal. As before, the origin sits at the base, with x-axis horizontal and z-axis vertical, positive upward. Choose the y-axis so as to make the system right-handed. The coordinate system is fixed in space.

The equation of conservation of energy in a deforming medium can be written

$$\rho \frac{DE}{Dt} = \dot{S}_E - \frac{\partial q_x}{\partial x} - \frac{\partial q_y}{\partial y} - \frac{\partial q_z}{\partial z} \tag{9.25}$$

(Malvern 1969, pp. 226–231). Here E is internal energy per unit mass, $\dot{S}_E$ the source rate of heat per unit volume, $\vec{q}$ the heat flux vector due to conduction, ρ the density, and t time. D/Dt denotes differentiation following the motion, so that

$$\frac{DE}{Dt} = \frac{\partial E}{\partial t} + u \frac{\partial E}{\partial x} + v \frac{\partial E}{\partial y} + w \frac{\partial E}{\partial z} \tag{9.26}$$

for ice velocity components u, v, w.

In ice sheets, the only component of internal energy that changes significantly is thermal; thus $DE/Dt = c[DT/Dt]$, where T denotes temperature and c is specific heat capacity. Heat flux by conduction, in direction x, amounts to $q_x = -k_T \partial T / \partial x$ for thermal conductivity k_T. Thus

$$-\frac{\partial q_x}{\partial x} = k_T \frac{\partial^2 T}{\partial x^2} + \frac{\partial k_T}{\partial x} \frac{\partial T}{\partial x}. \tag{9.27}$$

Similar relations give the y and z components. Thermal conductivity varies with position because it depends on density and temperature. Variation in the z-direction predominates; we neglect variation in other directions. Thus

$$\frac{\partial k_T}{\partial x} = \frac{\partial k_T}{\partial y} = 0$$

$$\frac{dk_T}{dz} = \frac{\partial k_T}{\partial \rho} \frac{\partial \rho}{\partial z} + \frac{\partial k_T}{\partial T} \frac{\partial T}{\partial z}. \tag{9.28}$$

Equation 9.25 then becomes

$$\rho c \frac{\partial T}{\partial t} = k_T \nabla^2 T - \rho c \left[u \frac{\partial T}{\partial x} + v \frac{\partial T}{\partial y} \right] + \left[\frac{dk_T}{dz} - \rho c w \right] \frac{\partial T}{\partial z} + \dot{S}_E \tag{9.29}$$

with dk_T/dz given by Eq. 9.28. This is the general equation for heat transfer in ice sheets. Note that the dependence of k_T on T makes the equation nonlinear. The terms $\partial k_T / \partial T$ and $\partial k_T / \partial \rho$ can be evaluated from Eqs. 9.2 through 9.4. The temperature dependence matters most near the base of an ice sheet, the location of steepest temperature gradients and smallest w. The term dk_T/dz reinforces the downward advection, reducing basal temperatures slightly below those calculated with a constant value of k_T. The density dependence of k_T is significant in the firn, but nowhere else.

9.7.1.1 Heat Sources

The term $\dot{S}_E$ in Eq. 9.29 consists of heat produced by (1) ice deformation, (2) firn compaction, and (3) freezing of water. The latter two occur primarily near the surface, although (3) sometimes occurs at depth.

Ice Deformation Consider an element with sides of length Δx, Δy, Δz. Applying a stress σ_{zz} causes the sides Δz to lengthen by an amount $\delta z = \epsilon_{zz}\Delta z$, given a strain ϵ_{zz}. The work done to accomplish the strain is $[\sigma_{zz}\Delta x\Delta y]\delta z$. All of this work converts to heat. Thus the rate of heat production, per unit volume per unit time, is $\dot{\epsilon}_{zz}\sigma_{zz}$, where $\dot{\epsilon}_{zz}$ is the strain rate in the z direction. The rate of heat production totals

$$\dot{S}_E^{(1)} = \dot{\epsilon}_{xx}\sigma_{xx} + \dot{\epsilon}_{yy}\sigma_{yy} + \dot{\epsilon}_{zz}\sigma_{zz} + 2\left[\dot{\epsilon}_{xy}\sigma_{xy} + \dot{\epsilon}_{xz}\sigma_{xz} + \dot{\epsilon}_{yz}\sigma_{yz}\right]. \tag{9.30}$$

The factor 2 arises because both of the symmetrical components of each shear stress do work (for example, σ_{xy} and σ_{yx}). In solid ice, but not firn, incompressibility means that the hydrostatic pressure does no work; the stresses can be replaced by their deviatoric counterparts. In the more compact form of indicial notation, an alternative expression is thus $\dot{S}_E^{(1)} = \dot{\epsilon}_{jk}\tau_{jk}$.

Deformational heat production concentrates where both deviatoric stresses and strain rates are highest – usually in basal layers but also in lateral shear margins. For example, measured temperature profiles part way through the rapidly shearing margins of Whillans Ice Stream show significant warming (Harrison et al. 1998).

Firn Compaction Changes in firn density at a fixed depth below the surface are usually small compared to the changes experienced by a given layer of firn during burial. In a steady-state ice sheet, a fixed depth corresponds to a point fixed in space. Mass conservation then requires that $\nabla \cdot [\rho \vec{u}] = 0$, given a velocity vector $\vec{u}$. Since density variations in the x- and y-directions are trivial, this reduces to

$$-w\frac{d\rho}{dz} = \rho\nabla \cdot \vec{u} = \rho\left[\dot{\epsilon}_{xx} + \dot{\epsilon}_{yy} + \dot{\epsilon}_{zz}\right]. \tag{9.31}$$

The hydrostatic component dominates stresses in firn, and so $\sigma_{xx} = \sigma_{yy} = \sigma_{zz} = -P$, with

$$P = g\int_z^{z_s} \rho\left(z'\right)dz'. \tag{9.32}$$

Therefore, by Eq. 9.30, compaction produces heat at a rate

$$\dot{S}_E^{(2)} = \frac{wP}{\rho}\frac{d\rho}{dz}. \tag{9.33}$$

Refreezing of Surface Water in Firn   Refreezing liberates latent heat at a rate, per unit volume and time, of

$$\dot{S}_E^{(3)} = Lw_s\rho_s m_f/z_m. \tag{9.34}$$

Here suffix s indicates surface values, m_f indicates the *melt fraction* – the fraction of the annual firn layer, by weight, formed by refreezing of water – and z_m is the maximum depth to which water penetrates. The refrozen water is assumed to be distributed uniformly over the layer. One small additional term is the heat released as the refrozen water cools to the temperature of the surrounding firn (Paterson and Clarke 1978; Eq. 9.16). The value of m_f, as a function of depth, can be measured in an ice core.

Heat of Sliding Friction   Because this enters the glacier as a heat flux at the base, not by internal heat production, it is not included in the term $\dot{S}_E$. The flux equals the product of displacement and resistive force at the glacier bed, per unit time and area:

$$q_b = u_b\tau_b, \tag{9.35}$$

with u_b and τ_b the magnitudes of basal slip and shear stress, respectively. If $\tau_b = 100\,\text{kPa}$, this flux matches a typical geothermal flux for u_b in the range 15 to $20\,\text{m yr}^{-1}$. Slip rates of large glaciers commonly exceed this range.

Geothermal Heat   Average fluxes on the continents range from $46\,\text{mW m}^{-2}$ for rocks more than 1700 Myr old to $77\,\text{mW m}^{-2}$ for rocks younger than 250 Myr (Sclater et al. 1980). Typical values depend on the tectonic setting. Extensional provinces, like the Basin and Range of western North America, have elevated geothermal heat fluxes (70 to $80\,\text{mW m}^{-2}$). Parts of the West Antarctic Ice Sheet overlie a similar province. Indeed Engelhardt (2004a) estimated a geothermal flux of $69\,\text{mW m}^{-2}$ from analysis of the Siple Dome borehole temperature profile. Similar analyses in central Greenland found a geothermal flux of $53\,\text{mW m}^{-2}$ (Dahl-Jensen et al. 1998). Direct estimates of the geothermal flux beneath ice sheets have only been obtained in a few locations worldwide.

Very much higher geothermal heat fluxes occur locally, on or near volcanic centers. Analyses of melt rates indicate fluxes as high as $7\,\text{W m}^{-2}$ on Mount Wrangell, Alaska, and as high as $50\,\text{W m}^{-2}$ on Vatnajökull, Iceland (Clarke et al. 1989; Björnsson 1988). In north-central Greenland, the North-GRIP ice coring project discovered a previously unknown region of enhanced melt and heat flux beneath the ice sheet; the heat flux ranges from about 100 to $150\,\text{mW m}^{-2}$ at the core site to more than $1\,\text{W m}^{-2}$ nearby (NGRIP 2004; Fahnestock et al. 2001). The former value was obtained from analysis of the measured temperature profile; the latter, from melt rates inferred by down-warping of internal layers seen in radar soundings.

9.7.2 Boundary and Basal Conditions

At the surface, temperature is prescribed as a function of time. In contrast, neither the temperature nor the heat flux at the base of the glacier can be prescribed except in special cases. In the general case, the domain for temperature calculations must extend well into the substrate, to a depth that increases with the timescale considered. At the bottom of the domain, temperature or heat flux is held constant. To prescribe the latter, the temperature gradient is usually set equal to the heat flux divided by the thermal conductivity of the substrate material. This assumes a negligible transport of heat by any circulating fluids.

Three possibilities for conditions at the base of the ice are:

1. For a temperature below melting point, the gradient in the ice often approximates the value expected for a steady state:

$$\left[\frac{\partial T}{\partial z}\right]_B \approx -\frac{G}{k_T} \tag{9.36}$$

for a geothermal flux G. In general, however, the basal gradient changes as temperature transients pass through the bed or the glacier advances onto new terrain. Note that, with z pointing upward, the steady gradient is a negative quantity.

2. If the ice reaches melting point (T_m) only at the glacier base:

$$T_B = T_m \quad \text{and} \quad \left[\frac{\partial T}{\partial z}\right]_B < \frac{\partial T_m}{\partial z}, \tag{9.37}$$

where, given that $\mathcal{B} = 7.42 \times 10^{-8}\,°\text{C Pa}^{-1}$, the maximum (positive) gradient is $\partial T_m/\partial z = \mathcal{B}\rho g \approx 7 \times 10^{-4}\,°\text{C m}^{-1}$. For comparison, note that some of the measurements discussed in Section 9.6 obtain values for $[\partial T/\partial z]_B$ as low as $-0.057\,°\text{C m}^{-1}$. The rate of basal melt or freeze-on, corresponding to the basal specific mass balance $\dot{b}_b$ (dimensions of length per unit time), can be calculated from (see Chapter 4):

$$-\rho_i L \dot{b}_b = G + \tau_b u_b + k_T \left[\frac{\partial T}{\partial z}\right]_B. \tag{9.38}$$

In other words, a net heat flow to the bed – a positive value for the sum on the right-hand side – causes melt, a negative specific balance.

3. Within a basal layer of temperate ice,

$$T = T_m \quad \text{and} \quad \frac{\partial T}{\partial z} = \frac{\partial T_m}{\partial z}. \tag{9.39}$$

Its upper boundary corresponds to the lowest level at which $\partial T/\partial z < \partial T_m/\partial z$. The reversed temperature gradient in the temperate layer prevents the inflow of frictional and geothermal heat; all such heat melts ice at the base, according to Eq. 9.39. The existence of the layer, and all melting within it, results from the heat of ice deformation (Eq. 4.13). All the meltwater is assumed to drain away; otherwise the temperate ice has to be treated as a mixture of ice and water, which increases the complexity of the analysis (Hutter 1982).

9.8 Temperatures Along a Flow Line

Dahl-Jensen (1989) calculated how the steady temperature distribution varies along a flow line in an ice sheet. Her method simultaneously solved the equations for heat flow and ice flow. Only the temperature results are presented here; the corresponding stress and velocity distributions are discussed in Section 8.7.

The assumptions are steady-state, two-dimensional flow, no firn, constant thermal conductivity, internal heating only from shear. Equation 9.29 then reduces to

$$u\frac{\partial T}{\partial x} + w\frac{\partial T}{\partial z} = \alpha_T\left[\frac{\partial^2 T}{\partial x^2} + \frac{\partial^2 T}{\partial z^2}\right] + \frac{2}{\rho c}\dot{\epsilon}_{xz}\tau_{xz}. \tag{9.40}$$

The bed is assumed to be horizontal. At the surface, as elevation drops along the flow line, the specific balance decreases and the temperature increases at prescribed rates. Values of ice thickness, surface temperature, geothermal flux, accumulation, and ablation resemble those for southern Greenland (Figure 8.23). The velocity components are determined, as functions of position, from the solution of the equations for ice flow. All equations are solved numerically.

Figure 9.8 shows computed temperatures along the flow line. The main features are:

1. Basal temperature increases with distance from the ice divide because both surface temperature and heat of deformation increase.

2. Basal melting starts at $X = 0.625$. (Here X denotes the distance along the flow line expressed as a fraction of total length.) The basal temperature gradient, hitherto increasing with X, starts to decrease because some heat goes to melting.

3. A temperate basal layer starts to form at $X = 0.75$, still in the accumulation zone. It first thickens with increasing X but then thins to zero as deformational heating declines near the terminus. (Along a fast-flowing outlet glacier, in contrast, heating would presumably remain important all the way to the front.)

4. Horizontal advection produces a minimum in the vertical profile of temperature. This cold spot strengthens and persists along the flow line, as far as the outer ablation zone.

5. The temperature profile near the terminus resembles that predicted for an ablation zone by Robin's simple analysis (the curve for $\gamma = -3$ in Figure 9.5).

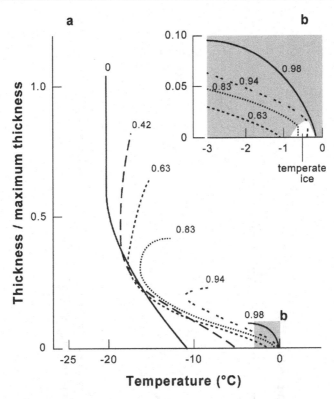

Figure 9.8: (a) Theoretical temperature profiles along an ice sheet flow line. The number on each curve is the distance as a fraction of the flow-line length. The equilibrium line is at 0.91. **(b)** Closer view of profiles near the margin. A temperate layer develops at the bottoms of profiles 0.83 through 0.98. Adapted from Dahl-Jensen (1989) and used with permission of the American Geophysical Union, *Journal of Geophysical Research.*

9.8.1 Observations

A series of temperature profiles along a flow line has never been measured in an ice sheet. However, Figure 9.9 shows temperature profiles measured along the centerline of White Glacier, a polar glacier on Axel Heiberg Island, Canada. Conditions in this valley glacier, which is 15 km long, differ from those assumed in the theoretical analysis in several ways:

1. The glacier is not in a steady state.

2. The bed has an average slope of 6°.

3. The ice thickness varies with distance in an irregular way.

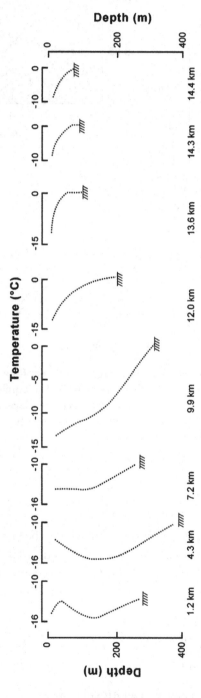

Figure 9.9: Temperature profiles in White Glacier, at various distances along the flow line. Depth to bed is indicated in each case. Note change in temperature scale between 9.9 and 12 km. Data from Blatter (1985).

4. The 10 m ice temperature does not increase steadily with X.

5. The ablation zone is proportionately much larger.

Nevertheless the data show most of the predicted features:

1. An increase in basal temperature with distance down-glacier.

2. A temperate basal layer, in this case restricted to the ablation zone, that does not extend to the terminus.

3. Profiles of the predicted shape near the terminus.

4. A temperature minimum extending into the ablation zone. However, the minimum appears in the first profile, only 1.2 km from the head of the glacier. Thus the cool spot probably represents, in part, a remnant of low temperatures during the Little Ice Age and not just the effects of horizontal advection (Blatter 1987).

Temperate basal layers have also been observed in Laika Glacier and Barnes Ice Cap, Arctic Canada (Blatter and Kappenberger 1988; Classen 1977), and near the ice edge in west Greenland (Stauffer and Oeschger 1979), including the lateral margins of the Jakobshavn Ice Stream (Lüthi et al. 2002).

9.9 Time-varying Temperatures

Steady-state models explain many of the prominent features of temperature profiles, but no real ice sheet is ever in a steady state. Changes of temperature at depth originate with changes of surface temperature, water infiltration, vertical ice flow, horizontal ice flow, and ice thickness. Other factors, such as variations of geothermal flux, are less important in general but locally important.

As Table 9.2 shows, the temperature variations during the 100 kyr ice-age cycle could penetrate through the deepest ice sheets by conduction alone. Advection increases the penetration. The basal temperature gradient therefore changes with time, and calculations must include heat transfer in a bedrock layer of comparable thickness to the ice. In addition, computations have to span a period long enough for the system to "forget" the assumed initial temperature distribution. For ice sheets, this takes several glacial cycles. Most analyses of time-varying temperature distributions require numerical methods.

The response at depth to a time-varying surface temperature can be illustrated with the simple case of an abrupt climate warming. Consider a step increase of surface temperature; that is, the surface temperature maintains a constant, cold value for a long time and then abruptly increases to a warmer value that persists. A wave of warming then propagates downward into the ice sheet by conduction and advection. Figure 9.10 illustrates a calculation of this scenario, a solution to Eq. 9.29 for an ice divide (no horizontal advection) with a constant thickness (of 3 km) and

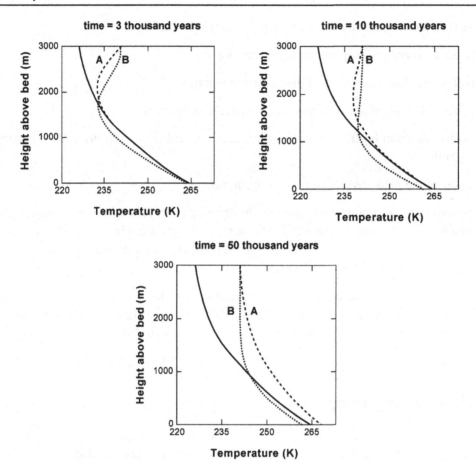

Figure 9.10: Propagation of a temperature wave following a 15 °C step increase of surface temperature at time zero. The initial curve (solid line in all panels) is a steady-state profile for an accumulation rate of 0.07 m yr^{-1}. In all three panels, Case A assumes no change in the vertical ice velocity from the initial state. In Case B, the vertical velocity increases to 0.3 m yr^{-1} at time zero.

density, overlying a 3-km-thick bedrock slab. The temperature profile begins in a steady state, adjusted to parameters appropriate for central Greenland in an ice-age climate. The downward vertical velocity is assumed to equal 0.07 m yr^{-1} at the surface, decreasing linearly to zero at the bed. The curves labelled "A" show the response if the vertical velocity remains constant. If, however, the same temperature change is accompanied by an increase of vertical velocity, to 0.3 m yr^{-1} (surface value), the warming propagates more rapidly (curves "B" in the figure). But the increased advection also causes cooling of the deeper ice, despite the warmer surface. This second scenario is the more realistic one for polar ice sheets, because snowfall increases at transitions to warmer climates, leading to increased vertical velocities. Ice thickness increases for the same reason, a process that, acting alone, leads to warming of deep ice. This opposes the advective cooling (Ritz 1987). But climate warming also results in retreat of the ice sheet's margins, a process that ultimately thins the entire ice sheet (Section 11.4.2.2).

Inspection of Figure 9.10 shows that even ten millennia after the warming event a cold spot persists at depth – a legacy of the earlier, colder climate. Ice sheet temperature profiles taken now, roughly ten millennia after the end of the last ice age, should contain such a signature. Indeed, such cold spots are observed, most clearly in borehole temperature measurements from the Greenland Ice Sheet (Dahl-Jensen and Johnsen 1986; Cuffey et al. 1995).

Climate change entails irregular variations of temperature and accumulation rate over time. Ice cores provide an opportunity to reconstruct the histories of both (see Chapter 15). Such information allows comparisons of measured borehole temperature profiles with predictions from the heat-transfer equation (Johnsen 1977b; Paterson and Clarke 1978; Cuffey and Clow 1997; Lhomme et al. 2005). Ice thickness changes need to be modelled as part of the analysis.

Paterson and Clarke (1978) studied temperatures in a 299-m borehole through the Devon Island ice cap, using the oxygen-isotope record back to 11 kyr B.P. as a measure of surface temperature. They showed that no steady-state model could fit their data well, whereas the model driven by the isotope record matched the data to within 0.04 °C. Their analysis did not include ice thickness changes, but more recent ones have. Cuffey et al. (1995) analyzed the temperature profile from a 3.05-km-deep borehole in central Greenland, using a 100 kyr oxygen-isotope record (δ) to constrain surface temperature and other ice core information to constrain accumulation rate ($\dot{b}_i$). Ice thickness variations were estimated from $dH/dt = \dot{b}_i - w_s$, using a simple ice flow model to calculate w_s. Their calculated temperatures matched the borehole temperature data to within 0.1 °C, provided that the value for isotopic sensitivity (Chapter 15), $\alpha_\star = d\delta/dT$, was adjusted. The value needed for a good match, $\alpha_\star = 0.33$ per mil °C^{-1}, was significantly smaller than the one normally assumed. Using the "normal" value of $\alpha_\star = 0.67$ gave a difference between modelled and measured temperatures of as much as 1 °C. The results were not sensitive to the ice-flow model used to calculate w_s because the depth-age relation for the ice core provided an additional constraint on the vertical displacement of layers.

Lhomme et al. (2005) performed a similar analysis but using a three-dimensional model for the entire Greenland Ice Sheet to constrain flow and thickness changes. The lower resolution of this model reduced the match to the measured temperature profile, but the fit was still good. Moreover, the method permitted a simultaneous comparison between predictions and measurements at multiple sites; specifically, the depth-age relation was calculated for the whole ice sheet and compared to the measured relations at three ice core sites. Good matches were achieved. Salamatin et al. (1998, 2008) analyzed temperatures in the Vostok borehole, East Antarctica, using an approach similar to Cuffey et al. (1995), and achieved a close match to the measured temperatures. The good performance of all these models gives some confidence that the heat-transfer process is well understood and confirms the accuracy of the general pattern of climate changes inferred from ice cores.

Temperature-wave propagation, such as that illustrated by Figure 9.10, occurs throughout an ice sheet after a climate change. In addition, temperatures adjust any time ice flow changes (which alters heat generation and horizontal advection) and whenever the ice sheet thickens or thins. The evolving three-dimensional temperature field is simulated routinely as a component of whole-ice-sheet model calculations; for example, Huybrechts (1996) discussed results for temperatures at the bed of the Greenland Ice Sheet throughout a glacial climate cycle. In such models, temperatures must be calculated because they influence both the ice viscosity and the spatial pattern of basal slip. The simulated temperatures are not well constrained by data, however. More borehole measurements of temperature are keenly needed, not just at ice divides but also on ice sheet flanks.

The only region of an ice sheet flank with temperature measurements in multiple boreholes to the bed is the Siple Coast sector of West Antarctica, a region of fast-flowing ice streams separated by slow-flowing ice ridges (Section 8.9.2.4). The temperatures reveal a complicated thermal legacy of major recent changes of the ice flow. The temperature profile through one of the ridges, Siple Dome, is close to a steady state, with a minor influence from warming at the end of the last ice age (Engelhardt 2004a). Profiles through two ice streams (Kamb and Bindschadler) show the signature of fast glacier flow: strong cooling from horizontal advection at mid-depths and frictional warming near the bed (Figure 9.6; Joughin et al. 2004c; Engelhardt 2004b). But observations show the same type of profile in a large region of stagnant ice (called "the Unicorn"). Other evidence indicates that this region stagnated only two centuries ago (Clarke et al. 2000); the temperature profile is clearly a relict from an extended period of fast flow prior to stagnation. So too with the profile from Kamb Ice Stream, a feature that nearly stagnated a few centuries ago (Retzlaff and Bentley 1993). In contrast, the temperature profile in part of one actively flowing ice stream (B2) adjacent to the Unicorn resembles the profile in the ridge Siple Dome; this indicates that only recently was the ice of B2 entrained in the fast-flowing stream. The Siple Coast region thus contains a startling array of examples of major thermal disequilibrium.

The interaction of ice flow and temperature and the long legacy effects possible in ice due to slow heat conduction suggest that a lot of new information can be gained from analyzing borehole temperature measurements specifically acquired for ice dynamic analyses. Nereson and Waddington (2002) theorized one example. If an ice divide moves, the warm spot beneath it (Section 9.5.1.1) should dissipate only slowly. Migration of an ice divide should therefore leave a thermal trace that might be measured.

9.10 Temperatures in Ice Shelves

How do temperatures vary in a floating ice shelf? At the base of a shelf, the temperature is fixed by the freezing point of seawater, and the heat flux depends on the temperature, salinity, and circulation of the seawater underneath. The flux determines whether ice melts from the shelf or seawater freezes to it. (We have discussed the basal mass balance of ice shelves in Section 4.5.) For a steady state, the vertical velocity component at the base must compensate for the melting or accretion. Ice shelves are fed by glaciers emanating from inland ice sheets; for some distance into the shelf the temperatures reflect the values inherited from the inland ice. Thinning of the shelf as the ice flows out to sea also affects the temperature distribution. Shumskiy and Krass (1976) made a detailed general analysis, treating heat and ice flow simultaneously. They calculated steady-state temperature profiles. Humbert et al. (2004) used a three-dimensional numerical calculation to simulate the coupled heat and ice flow in the Ross Ice Shelf, the world's largest.

Figure 9.11 shows measured temperatures in ice shelves. Data sources are Maudheim (Schytt 1954), Little America (Bender and Gow 1961), Amery (Weeks and Mellor 1978), J9 (Clough and Hansen 1979), and Filchner (Dyurgerov et al. 1988). The measurements have been extrapolated to −2 °C, an estimate for freezing temperature of seawater at the base. The different shape of the Amery curve results from basal freezing (Morgan 1972) in contrast to basal melting at Maudheim, Filchner, and Little America. Because the lowest 6 m of the core from J9 was salty (Jacobs et al. 1979), seawater must be freezing to the shelf upstream, but the rate of accretion is not known. Where melting prevails, the basal temperature gradients are large; at Little America, for example, the gradient amounts to about 15 times a normal geothermal gradient. Melting erodes the basal layers of the shelf, bringing warm seawater in contact with cold ice in the core of the shelf. The resulting steep gradient draws heat from the underlying seawater. For melt to continue, the heat must be continually supplied by circulation of the water beneath the shelf.

Detailed interpretation of the individual curves is difficult because the temperature distribution depends on conditions upstream as well as at the borehole. The report by MacAyeal and Thomas (1979) exemplifies an analysis that tries to account for all the relevant factors. They

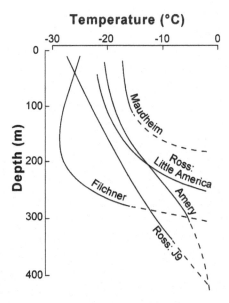

Figure 9.11: Temperature profiles in ice shelves. Broken lines indicate extrapolations. Data sources are given in the text.

obtained a good fit to the measurements for J9 by using a nonsteady-state numerical model that accounts for conditions along the flow line from the ice-shelf grounding line to the borehole.

Temperature variations along-flow and across-flow on ice shelves are not known well from the limited number of borehole studies. Because temperature affects ice viscosity, however, internal temperature distributions leave a mark on the pattern of ice flow throughout a shelf. A surface velocity map of a shelf therefore contains information about temperature. Larour et al. (2005) analyzed extensive flow data for the Ronne Ice Shelf to determine patterns of ice viscosity variation. They showed that the side margins are softer and hence warmer than elsewhere, an expected consequence of concentrated frictional heating where the ice shears rapidly. They also inferred colder and stiffer ice where major glaciers leave the continent and enter the shelf.

CHAPTER 10

Large-Scale Structures

"... what haunted one now was the picture of one of those gulfs opening in the glacier itself: all the horrors of the bottomless pit and the meandering shaft and gallery, beneath the entombing ice from which one would never escape."

Karakoram,[1] **Fosco Maraini**

10.1 Introduction

Deformation in glaciers produces a wide variety of large-scale structures: crevasses, faults, layering of various kinds, folds, and the alternating arcs of white and dark ice known as ogives. Here we describe these features and discuss ideas on how they are formed. An examination of structures sometimes reveals valuable information about the dynamics of a glacier and its history. Folds and the type of layering called foliation reflect the cumulative strain of ice as it passes through a glacier over centuries or millennia. Crevasses reflect the pattern of active deformation. Large strains and the superimposition of different types of deformations can lead to extremely complex structures in glaciers.

Ideas about the formation of these structures have an application broader than the understanding of glaciers because folds, foliation, and related features occur in deformed rocks as well as in ice. Compared to rocks, structures develop more rapidly in glaciers, and the deformations can be measured at different stages in the structures' development. Glaciers thus serve as natural laboratories for the study of geological processes such as deformation – deep in the earth's crust – of rocks that now form mountain belts. Glaciers, for example, demonstrate how different types of structures develop simultaneously in a complex deforming body, rather than in sequential phases as sometimes assumed.

In the accumulation zones of ice sheets, and in glaciers with smooth and regular cross-sections, depositional layers rather than deformational structures predominate. However, features such as folds or thrust planes may form, especially near the bed. By disrupting the layering, these structures greatly complicate the interpretation of the oldest part of a climate record from an ice core. The layering itself is initially parallel to the surface but subsequently develops patterns reflecting the cumulative movement of the ice. Observations of layering thus

---

[1] Translation by J. Cadell.

Copyright © 2010, Elsevier Inc. All rights reserved.
DOI: 10.1016/B978-0-12-369461-4.00010-7

reveal complexities in the flow field related to differences of accumulation rate, basal melt, and other factors.

Crevasses pose a great hazard to travellers on glaciers. This is a particular concern given the increasing use of ground transport in Antarctica. In 1991, a bulldozer pulling a cargo of explosives fell into a crevasse on the Ross Ice Shelf. Superficially, however, the region appeared to be safe (Whillans and Merry 2001). To avoid crevasses, it is never a good idea to rely on theoretical expectations of their occurrence, but an understanding of crevasse patterns assists route-finding. Crevasse formation also plays a role in two processes of great importance to glacier dynamics: the calving of icebergs and the penetration of surface water to the bed.

Deformation also changes small-scale features of the ice such as the size, shape, and orientation of the crystals. These are discussed in Chapter 3.

10.2 Sedimentary Layers

Depositional or *sedimentary layers* are initially parallel to the glacier surface. The sequence of sedimentary layers defines the *stratification*. Boundaries between adjacent layers represent *isochrones*, surfaces of uniform age. In temperate and subpolar settings, annual sedimentary layers consist of alternating thick horizons of bubbly ice originating as winter snow and thin horizons of clear ice originating primarily as refrozen meltwater. Debris horizons, formed when summer melting concentrates rockfall and wind-blown dust, are often the most conspicuous feature of stratification. In cold polar regions, annual layering forms not by melting but by seasonal variation of snow metamorphism and wind deposition.

The pattern of sedimentary layers at depth in a polar glacier can be mapped with radio-echo sounding ("radar" studies); Figure 8.24 shows an example. Radio reflections arise from differences in the acidity, density, or crystal fabric of adjacent layers (Fujita et al. 1999); these factors influence the electrical permittivity or conductivity of the ice or firn. The layers in this case are usually not annual strata but volcanic horizons or multiyear sequences spanning climate transitions. Such layers also correspond to isochrones and therefore parallel the annual strata. The thickness of ice between two isochrones increases with the accumulation during the interval; the patterns seen in radar images therefore reveal differences of accumulation rates from place to place and over time (e.g., Morse et al. 1998). The warping of layers also reveals differences in ice flow. Melting at the bed increases the downward movement of ice and warps layers downward over warm spots (Fahnestock et al. 2001). Flow over and around hills on the bed warps the layering in a pattern that usually conforms to the topography (Robin et al. 1977). The amplitude of such warping decreases with increasing distance above the bed.

10.3 Foliation

Foliation is a planar or layered structure that develops in glacier ice during flow. The layers are characterized by variations in crystal size, and in the size and number of air bubbles. Individual

Figure 10.1: Longitudinal foliation, Comfortlessbreen, Spitsbergen. Photo by M.J. Hambrey.

layers are usually between 10 and 100 mm thick and are seldom continuous for more than a few meters. Figure 10.1 shows typical foliation. Temperate glaciers contain three types of ice: coarse-bubbly, coarse-clear or "blue," and fine-bubbly or "white" (Allen et al. 1960). Crystals of coarse ice usually have diameters in the range 10 to 150 mm: the crystals tend to be larger in clear ice than in bubbly. The average crystal diameter of fine ice is less than 5 mm. Coarse-bubbly ice is the most abundant. Ice below about $-10\,°C$ has fine crystals (as small as 1 mm) and the foliation consists of alternating layers of clear and bubbly ice of similar crystal size. Chapter 3 includes a detailed discussion of processes determining crystal sizes in ice sheets.

Ice near the valley walls and glacier bed displays the strongest foliation, which usually trends parallel to the rock boundaries. Away from the walls, foliation often intersects the glacier surface in arcs, convex in the direction of flow. The three-dimensional pattern of the layers in the lower part of the ablation area has been described as "nested spoons"; the layers dip steeply near the valley walls but have a moderate up-glacier dip near the center line. (The *dip* refers to the largest acute angle between the plane of a geological feature and a horizontal plane.) Glaciers formed by the merging of two or more tributaries often display a separate system of arcs corresponding to each tributary, with longitudinal foliation, parallel to the valley walls, near the boundaries between them (Hambrey and Glasser 2003).

Foliation often forms by the deformation of preexisting inhomogeneities in the ice. The major inhomogeneities are the previously mentioned sedimentary layers, initially parallel to the ice surface, *crevasse traces* initially vertical, *ice glands and lenses* (see Chapter 2),

and *layers of debris* incorporated into the ice at the bed or along the margins. Stratification, being a depositional feature, should be distinguished from foliation. In some glaciers,
however, most of the foliation originates by compression and shearing of sedimentary layers
(Hambrey 1975). Crevasse traces are the scars of old crevasses that have closed or been filled
with snow or refrozen meltwater, or narrow tensional cracks into which water has percolated
and frozen (Hambrey et al. 1980). Much of the "transverse" foliation – oriented across-glacier –
originates as traces of transverse crevasses. Such features are initially vertical, but tilt up-glacier
at progressively lower angles because of shearing between the glacier surface and the ice below;
this history explains the "nested spoon" configuration.

Figure 10.2 illustrates how features that produce foliation deform in two basic strain regimes.
Corresponding deformations of a square and a circle are also shown. (Figure 3.10 illustrates
the relation of these regimes to patterns of ice flow.) Deformation rarely occurs in only two
dimensions; nevertheless the main features of foliation patterns can be understood in terms of
these regimes. In Figure 10.2, row B shows the effect of vertical compression accompanied by
an equal expansion in the direction of flow. This type of deformation, called *pure shear*, occurs
in the upper layers in the accumulation zone, at the onset of an icefall, and at all depths at an
ice divide. A similar system with vertical expansion and horizontal compression applies in the
upper layers in the ablation zone and at the foot of an icefall. Row C illustrates *simple shear*, the
prevailing regime near the glacier bed and sides. Glacier ice normally undergoes a combination
of pure and simple shear with simple shear becoming progressively more important as the ice
moves downward and outward toward the margins.

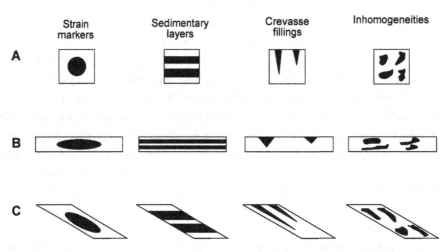

Figure 10.2: Deformation of a square, a circle, and different sources of foliation under
homogeneous strain. (A) Initial state. (B) After pure shear. (C) After simple shear. From Hooke
and Hudleston (1978).

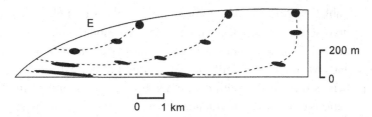

Figure 10.3: Schematic illustration of total strain along particle paths in an ice cap. Flow deforms circles into ellipses. "E" indicates the equilibrium line. From Hooke and Hudleston (1978).

The development of foliation depends on the cumulative strain experienced by the ice; foliation does not reflect strain rates measured at the point of observation (Milnes and Hambrey 1976; Hooke and Hudleston 1978; Hudleston and Hooke 1980). Because glaciers do not contain any features (such as initially spherical inclusions) that indicate the total strain directly, its value has to be estimated from measured or modelled velocity fields by defining particle paths and evaluating the strain rate at every point along them.

Hudleston and Hooke (1980) used measurements of ice thickness and velocity at the surface and at depth along a 10 km flow line from the crest to the edge of the Barnes Ice Cap, Baffin Island. Flow transverse to the line was negligible. They computed total strains in the ice throughout the section. A steady state over the past few thousand years was assumed. Figure 10.3 illustrates the results. Flow deforms circles into ellipses. Basal ice, especially near the terminus, has undergone a large amount of simple shear; this explains the strong elongation of the ellipses. Their major axes roughly parallel the flow lines.

Hudleston and Hooke discussed how foliation of different origins develops. Sedimentary layers persist as near-horizontal sheets throughout the ice cap. Crevasse traces form planes that rotate from a vertical position to one that dips up-glacier. Inhomogeneities such as ice lenses are stretched in the direction of maximum total strain. Numerical calculations predicted that, in ice near the terminus, the total strain was great enough to make the sedimentary layers, crevasse traces, and stretched inhomogeneities parallel to each other and aligned with the particle paths. In other words, there should be a single foliation parallel to the flow lines (which, in the steady state, match the particle paths). This was observed and suggests that foliation is indeed determined by the total strain. Near the bed, simple shear predominates and the direction of maximum total strain coincides with the plane of greatest shear.

The longitudinal section of the ice cap studied by Hudleston and Hooke offers a particularly simple case. Though more complicated, the arc-shaped foliation seen on the surface of the ablation zone in many glaciers can also be understood in terms of total strain. Simple shear predominates near the valley walls and so, after large strains, the foliation approximately parallels the valley sides. On the other hand, ice on the centerline mostly undergoes longitudinal compression in the ablation zone. Thus the major axis of the strain ellipse is oriented across the glacier. After large strains, the surface trace of the foliation acquires this orientation, too (Hambrey and Milnes 1977). As another example, glaciers in which a wide accumulation basin feeds a narrow tongue generally undergo longitudinal extension, and longitudinal foliation develops across the whole width of the ablation zone (Hambrey and Glasser 2003). Two good demonstrations of the correspondence between cumulative strain and foliation orientations were made by Hambrey et al. (1980, 2005). They computed strain ellipses for two valley glaciers, viewed in the map-plane, and compared the results to observations of foliation on the glacier surfaces.

Although foliations of different origin rotate toward the same plane during progressive deformation, in some cases the total strain may not be great enough to complete the process. For example, arc-shaped foliation originating from transverse crevasses may intersect sedimentary layers, or ice that passes through a series of icefalls may have intersecting sets of arc-shaped foliation but no sedimentary layers.

10.3.1 Elongate Bubble Forms

Bubbles in deforming ice should tend toward a shape reflecting the strain of the surrounding ice (Hudleston 1977). They do not usually behave as passive markers, however, because the H_2O saturation vapor pressure depends on the curvature of the adjacent surface. This implies that, in a nonspherical bubble, a pressure gradient drives a diffusive flow of molecules from flatter to more tightly curved regions. Over time, this process, acting alone, would restore a spherical geometry. If the ice continues to deform, the tendency to become spherical competes with the tendency to elongate. Bubble shape should therefore reflect the strain rate instead of the net strain (Alley and Fitzpatrick 1999). The orientation of a highly elongated bubble will nonetheless indicate the direction of net extensional strain. Observations of ice cores reveal elongated bubbles in rapidly shearing layers (Hudleston 1977; Russell-Head and Budd 1979). In the simple shear regime prevailing near the glacier bed, large elongate bubbles lie in the plane of shear, or nearly so. Small bubbles, in contrast, rapidly tend toward spherical form; large surface curvature and a short path for diffusion increase the vapor pressure gradients and speed the process of shape change (Hudleston 1977).

10.3.2 Finite Strain

A small deformation of a region of ice, produced in a time Δt by a strain rate $\dot{\epsilon}_{jk}$, yields a small strain of magnitude $\dot{\epsilon}_{jk}\Delta t$. But many of the structures in glaciers, such as foliation, reflect large cumulative deformations that must be measured by the *finite strain*. Finite strain accounts for the compounding effects of cumulative deformation on further strain.

Consider first a pure shear regime, with extension in direction x (at rate $\dot{\epsilon}_{xx}$) and compression along z (at rate $\dot{\epsilon}_{zz} = -\dot{\epsilon}_{xx}$). A feature in the ice, with initial x-direction length L_o, stretches over time. At any instant, its length increases at rate $dL/dt = \dot{\epsilon}_{xx}L$. Integrating from L_o to

L and $t = 0$ to $t = \Delta t$ shows that

$$L = L_o \exp(\dot{\epsilon}_{xx} \Delta t) \tag{10.1}$$

Thus the length increases exponentially with time, not simply as $\dot{\epsilon}_{xx} \Delta t$. (One measure of the finite strain is $L/L_o - 1$.) The same formulae apply for the compressive strain in the z-direction, but with a minus sign in the exponential. Note that the directions of maximum extension and compression – the *axes of finite strain* – remain aligned with the directions of maximum and minimum strain rate. This describes *coaxial strain*.

Simple shear behaves differently. Consider a layer undergoing shear, with the velocity u (in direction x) increasing with height above the bed (direction z) at a rate $\partial u/\partial z$. Assuming no velocity in direction z, then $\partial u/\partial z = 2\dot{\epsilon}_{xz}$. Extensive and compressive strain rates act along axes inclined at $45°$ to x and z, not parallel and perpendicular to them. A feature in the ice with initial z-direction length L_o rotates and stretches in the shearing flow. All the stretching occurs in direction x. In time Δt, the displacement of one side of the feature, with respect to the other side, amounts to $\Delta x = 2\dot{\epsilon}_{xz} L_o \Delta t$. A line initially parallel to z stretches to a length given by $L^2 = L_o^2 + (\Delta x)^2$, or

$$L = L_o \sqrt{1 + 4\dot{\epsilon}_{xz}^2 (\Delta t)^2}. \tag{10.2}$$

At large strains, L simply increases in direction x as a linear function of $\dot{\epsilon}_{xz} \Delta t$. The axes of finite extension and compression therefore run parallel and perpendicular to the flow, respectively. Thus, in contrast to the case of pure shear, the finite strain axes are rotated $45°$ from the axes of instantaneous strain defined by the strain rates.

In general, the effect of a deformation on a region of ice is described as follows. Before deformation, points in the ice have coordinates X_j (indices j, k, and i will indicate any one of x, y, or z). In time Δt, deformation and flow displace each point along a vector d_j. This displacement varies from point to point. The variations $\partial d_j/\partial X_k$ define the *displacement gradients*, a set of nine numbers. Now consider a pair of nearby points in the ice, initially separated by a vector $\vec{\ell}$. Over the course of deformation, their separation changes to a vector $\vec{L}$, with components

$$L_j = \ell_j + \frac{\partial d_j}{\partial X_k} \ell_k \tag{10.3}$$

(repetition of an index indicates summation). The change of length of the vector can be calculated from $|L|^2 - |\ell|^2 = 2 F_{jk} \ell_j \ell_k$, where F denotes the Lagrangian *finite strain tensor*, with components

$$F_{jk} = \frac{1}{2} \left[\frac{\partial d_j}{\partial X_k} + \frac{\partial d_k}{\partial X_j} + \frac{\partial d_i}{\partial X_j} \frac{\partial d_i}{\partial X_k} \right]. \tag{10.4}$$

(In pure shear, for example, with horizontal extension along x and $\partial d_z/\partial x = 0$, then $F_{xx} = \partial d_x/\partial x + [\partial d_x/\partial x]^2/2$; for a horizontal line segment, $\vec{\ell} = [\ell, 0]$, the formulae reduce to $|L| =$

$\ell [1 + \partial d_x / \partial x]$.) For a small deformation, with displacement gradients much less than one, the final term on the right side of Eq. 10.4 can always be neglected and the equation matches the one expected for a small strain.

10.4 Folds

Glacier ice is often folded (Hudleston 1976; Hambrey 1977; Lawson et al. 1994). Folding involves both original stratification and subsequent foliation. Sections through folds may be seen at the surface or in marginal ice cliffs. Most observed folds are no longer active, having been formed elsewhere and then transported passively. These folds, which cover a wide range of sizes, orientations, and shapes, resemble those in layered crustal rocks (Ramsay and Huber 1987).

An unusual and conspicuous example of folding occurs in the medial moraines on surging glaciers. The moraines are deformed into folds or bulb-like loops a few kilometers apart (Figure 10.4). A loop is formed by flow of ice from a tributary glacier while the main glacier is quiescent. The surge of the main glacier then carries the loop several kilometers down the valley.

Figure 10.4: **Folded moraines in Susitna Glacier, Alaska. Photo by Austin Post, U.S. Geological Survey.**

The tributary forms a new loop after the surge. Additional loops may be formed by surges of tributaries. The moraine patterns can be complex if the glacier has several tributaries.

The terminal lobes of certain Alaskan glaciers, formed where ice emerging from the mountains spreads out over the coastal plain, have spectacular folded medial moraines. Malaspina Glacier is the best-known example (Post and LaChapelle 2000, p. 51). The structure consists of a series of major folds with similar zig zag patterns in each band of moraine. The folds progressively tighten as they move toward the ice margin. Post (1972) studied such folds in Bering Glacier. He concluded that the folds originated farther up the glacier as moraine loops formed during surges. The pattern of flow in the terminal lobe, with compression along each flow line and extension in the transverse direction, transforms even slightly sinuous moraines into large, tight folds.

Folding at a much smaller scale – with dimensions of perhaps 1 or 10 m – is common in valley glaciers. It often causes *transposition* of the foliation, in which progressive tightening of folds eventually produces a layered structure with a new orientation parallel to the axial surfaces of the folds; Figure 10.5 illustrates the process. The small folds develop primarily in simple shear regimes near the glacier bed and sides. Shearing readily accentuates irregularities of certain orientations in foliation or strata. Several processes generate folds, such as: (1) flow over the top of a bedrock bump overriding ice that has taken longer to travel around the sides; (2) overturning of layers at the downstream end of a basal cavity (Boulton 1979); (3) buckling of ice layers with different micro-structures and hence different effective viscosities; (4) irregular warping of layers around ice with different effective viscosities associated with local variations in crystal fabrics (Thorsteinsson and Waddington 2002); (5) warping of layers due to creep-closure of voids such as moulins and tunnels; and (6) changes of the ice velocity field and hence particle paths due to large-scale changes of glacier dynamics (Hudleston 1976).

Folds also develop in association with narrow bands of intense shearing along planes inclined to the foliation or strata (Alley et al. 1997b). Concentrated shearing must warp the layering that

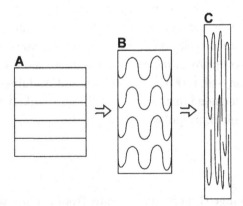

Figure 10.5: How folding transposes the orientation of layers.

it cuts across. Likewise, thrust faulting occurs in glaciers (Section 10.6) and some folding must happen as these structures propagate. Thrusts sometimes propagate up from the glacier bed but do not reach the ice surface (Clarke and Blake 1991); this style of faulting produces many of the folds in crustal rocks.

10.4.1 Folding in Central Regions of Ice Sheets

Recent observations show that folding also occurs in the central regions of ice sheets. A zone of high-viscosity ice occurs near the bed beneath an ice divide. Layers moving toward the bed drape over this zone, forming a large open fold known as a Raymond Bump (Section 8.9.1). If, following a period of stability, a divide moves, the Raymond Bump would deform in simple shear and might form a large overturned fold (Jacobson and Waddington 2005). At a much smaller scale, observations of deep ice cores from central Greenland have revealed folding (Alley et al. 1995b, 1997b). Folding occurs as far as 600 m above the bed, out of a 3 km total ice thickness. Observed features include small z-shaped folds with overturned limbs, and layers dipping more than 20°. The latter indicate folding at scales larger than the 10 cm core diameter. Within 300 m of the bed, structures of unknown shape completely disrupt the stratigraphy (see Section 10.7).

Alley et al. (1997b) measured the crystal fabrics around the small z-shaped folds. They found that the folding occurs on bands where crystal c-axis orientations differ from the prevailing vertical fabric. These bands, referred to as "stripes" because of their appearance in thin sections, tilt at roughly 45° from horizontal (Figure 10.6). They appear to be small shear bands that are oriented to help accommodate the overall state of deformation; beneath a divide, ice deforms in pure shear, with compression and extension in the vertical and horizontal directions, respectively (Section 8.9.1).

The regime changes from pure shear to simple shear away from the divide, within a distance of a few multiples of the ice thickness. Consequently, ice moving along a particle path starting near the divide first experiences pure shear and then simple shear. A tilted layer will flatten in the first regime but will steepen and possibly overturn in the second. The competition between these influences determines whether locally tilted layers develop into persistent folds. The initial tilt could be due to deformation of layers around inhomogeneities of crystal fabric. Folds are more likely to persist, and overturn, if the initial tilt is large, and if the particle path brings ice close to the bed. Waddington et al. (2001) and Jacobson and Waddington (2004) discussed these processes in detail. They concluded that small-scale folding is likely near an ice divide, even in the absence of nonsteady flow or irregular bed topography.

10.5 Boudinage

The structure called *boudinage* (Figure 10.7) results from the stretching and pulling apart of a layer with a viscosity greater than that of the surrounding material. (The term *viscosity* is

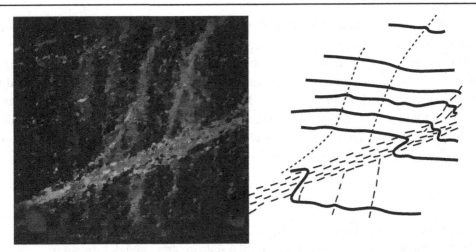

Figure 10.6: Crystal structure seen in a thin section of ice, of vertical orientation, from 2546 m depth in the ice sheet, central Greenland. Width of each panel is about 10 cm. Right panel: Sketch showing the pattern of folded layering (solid lines) and the "stripe" of crystals with anomalous orientations (dashed lines). Left panel: Pattern of crystal orientations revealed by polarized light. Crystals are oriented favorably for shear deformation along the stripe. Adapted from Alley et al. (1997b) and used with permission from the American Geophysical Union, *Journal of Geophysical Research*.

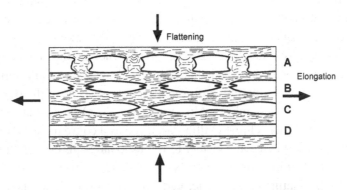

Figure 10.7: Boudinage structures for different viscosity contrasts between the layer and the surrounding matrix. Layer A has the highest viscosity and layer D the same viscosity as the matrix. Note that the "wavelength" increases as the viscosity contrast decreases. Adapted from Hambrey and Milnes (1975).

here used loosely in the sense of resistance to creep deformation.) The cross-section of such a layer resembles – to some observers – a string of sausages, from which the name is derived. Hambrey and Milnes (1975) observed apparent boudinage structures in longitudinal foliation in several Swiss glaciers. They ascribed the implied differences in viscosity to differences in grain size; however, they had no data on crystal fabric or impurity content, which can also affect the mechanical properties.

Smith (1977) made a mathematical analysis of the formation of boudinage. He found that if the creep relation is nonlinear, as for ice, the layer of high viscosity behaves anisotropically; it strains more readily in the direction perpendicular to the layering than in the tangential direction. Small perturbations in the thickness of the stiff layer can grow to form boudinage. In the limit of a very stiff layer, it completely breaks apart rather than stretching. There is a dominant length scale, a function of the layer thickness and viscosity contrast, for which the growth rate of perturbations is a maximum. (Much of the analysis also applies to the formation of folds by buckling, which is a related type of instability.)

10.6 Faults

Shear fractures and faults have been observed in glaciers (Hambrey and Lawson 2000). As in crustal rocks, the style of faulting depends on the state of stress and deformation.

Strike-slip Faults   Horizontal compression and tension, oriented 90° to each other, produce strike-slip faults of vertical orientation at the glacier surface. Such faults are commonly visible where they offset crevasses or crevasse traces (Hambrey 1976). They presumably occur most abundantly in the simple shear zones along lateral margins.

Thrust Faults   Thrust faulting accommodates horizontal shortening. Small steps in the glacier surface produced by differential movement along thrust planes have been observed near glacier termini, in the ablation zones of cirque glaciers, and where moraines impede the advance of the ice (Nye 1951; Goodsell et al. 2005). In a tunnel in the icefall of Blue Glacier, Kamb and LaChapelle (1968) observed that many shallow thrust faults offset the foliation near the bed, in a direction compatible with the direction of shear in the overall flow. During retreat of the valley glacier Pasterzenkees, Austria, ice near the glacier snout partly stagnated and obstructed flow from upglacier. Thrust faulting occurred as a consequence (Herbst et al. 2006). It is probably common for thin ice at a glacier terminus to have this effect; thrusts that intersect the surface in across-glacier lines or arcs have been reported from other glaciers (Hambrey et al. 2005). Thrust faults originating at the glacier bed can transport rock debris up into the glacier, a process that seems particularly important in polythermal glaciers (Clarke and Blake 1991; Hambrey et al. 1999; Swift et al. 2006).

Normal Faults   Normal faulting accommodates horizontal stretching. On Pasterzenkees, normal faulting has developed in stagnating areas of high surface relief along the margins and near the terminus (Herbst et al. 2006). Such regions near the glacier edge are collapsing under their own weight. In icefalls, the downstream wall of a crevasse is often displaced downward relative to the upstream wall.

Faults and fractures of other types might influence glacier dynamics by limiting the buildup of deviatoric stresses near the surface (Lliboutry 2002). Fracturing occurs readily under conditions of low hydrostatic pressure. This implies that ice near the surface, by fracturing, can deform in concert with the underlying glacier without needing to develop large deviatoric stresses. This kind of limitation on stresses may significantly affect the balance of forces in thin glaciers. Fractures also pervade the interiors of some temperate glaciers (Fountain et al. 2005). Many of these fractures originate as tension cracks (Section 10.9), but some are probably fault planes. They can be kept open, despite the weight of overlying ice, by pressurized water.

10.7 Implications for Ice Core Stratigraphy

To interpret the climate record contained in a polar ice core, it is normally assumed that the age of the ice increases steadily with depth and that there are no gaps. These assumptions might fail if some of the structures discussed earlier form at the core site. Folds, thrust faults, and related shear bands could invert and duplicate parts of a record. Even if not well developed, such features might still produce anomalously large gradients in age. Boudinage, normal faults, and associated shear planes could cause discontinuities. In boudinage, furthermore, the thickness of some layers would be anomalously large compared to surrounding layers.

Central Greenland ice cores revealed folds located hundreds of meters above the bed (Section 10.4). More commonly, folding and other strata-disrupting structures should develop near the bed. Johnsen and Robin (1983) have shown how horizontal shear of layers parallel to an undulating bed could produce anomalous "spikes" in an oxygen-isotope profile. Johnsen et al. (1992) suggested that boudinage explains a 4-m-thick layer of constant $\delta^{18}O$ found near the bottom of a 324-m core from the Renland ice cap, east Greenland. The long, continuous reflections seen in radar soundings of ice sheets are usually absent from a basal zone of a thickness comparable with the bedrock relief; irregular flow has probably disrupted the layering (Robin 1983).

Comparisons of neighboring ice cores provide the most precise information about structural disturbances. Such comparisons are rarely possible because deep cores are difficult to extract. Paterson and others (1977) compared the $\delta^{18}O$ profiles in two 300-m cores to bedrock, 27 m apart on the same flow line, from near the ice divide on Devon Island ice cap. The correlation between 50-year mean $\delta^{18}O$ was 0.97 between the surface and 13 m above the bed but only 0.45 between 13 and 5 m. Comparison of the lowest 5-m sections showed gaps of up to 5 cm; construction of a composite record suggested that 25% to 30% of each was missing. A very steep gradient of $\delta^{18}O$ in ice about 60 kyr old was attributed to recent shearing because such a gradient in the original snow would have been smoothed by molecular diffusion. One core had a possible inverted section, and horizontal layers of bubbles were observed to pinch out.

These observations from Devon Island foreshadowed the now infamous case of central Greenland, where two deep ice cores about 3 km in length were obtained from sites 30 km

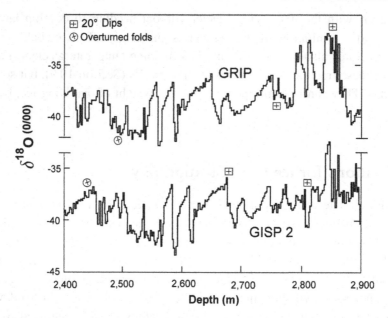

Figure 10.8: Variations of oxygen isotope composition with depth in the Greenland ice sheet, at two locations (GRIP, on the ice divide, and GISP2, 30 km to the west). The correspondence between the two records breaks down at depths greater than about 2700 m. Symbols show where folds and dipping layers were observed in the 10 cm-diameter cores. Adapted from Alley et al. (1995b).

apart. Tilted layers and centimeter-size folds were visible in both cores at hundreds of meters above the bed (Section 10.4). This observation did not prove that the layering had been disrupted at a large scale, but was a warning sign. Indeed, the oxygen isotope records from the two cores were later found to disagree completely in their bottom 300 m (Figure 10.8), despite close agreement throughout the overlying ice (Grootes et al. 1993). Additional evidence subsequently proved that the chronology of the bottom 300 m of both cores was seriously disrupted, invalidating initial findings about climate of the last interglacial period (GRIP 1993). The cores contain globally mixed atmospheric gases, trapped at the time of ice formation. Such gases can be used as age indicators, if their variations over time are known from a core with undisturbed chronology. In the present case, the concentration of methane and the isotopic composition of O_2 were both measured, and compared to Antarctic ice cores (Section 15.2.2.3). Results showed that the bottom 300 m of the ice sheet in central Greenland is a mixture of ices from different time periods, containing repetitions of some layers (Chappellaz et al. 1997). The ice here evidently contains overturned folds, tens of meters in thickness. An undisturbed climate record was later obtained from north-central Greenland. It differed from both of the central Greenland cores (NGRIP 2004). Analyses of gases confirmed that this new core was undisturbed.

10.8 Ogives and Longitudinal Corrugations

Band ogives or *Forbes bands* are alternating bands of light and dark ice that extend across the surface of some glaciers below icefalls. They are arc-shaped, convex in the direction of flow, a pattern that reflects the variation in velocity across the glacier and matches the common arcuate pattern of transverse foliation. The combined width of one light and one dark band corresponds to the distance the ice moves in a year. All ogives form in icefalls. Some ogive systems are more conspicuous than others and some icefalls do not produce ogives at all. Ogives first appear at the foot of the icefall as a series of undulations in the surface, called *wave ogives* (Figure 10.9). The amplitude of these waves, which may be about 5 m initially, decreases as they travel down the glacier until only the bands remain. In some glaciers ogives extend to the terminus; in others they die out.

Figure 10.9: Wave ogives produced by an icefall, Trimble Glacier, Alaska. The light arcs are snow-filled troughs. Photograph by Austin Post, U.S. Geological Survey.

Haefeli (1951a, b) believed that the undulations represent "pressure waves"; each summer, the increase in sliding velocity in the icefall compresses the ice below and a wave crest forms. This mechanism might account for some ogives. However, Nye (1958a) measured the surface strain-rate components in the region where ogives form on Austerdalsbre in Norway. The pattern of strain rates was found to be inconsistent with wave formation by this compression mechanism.

Nye (1958a) proposed an explanation that does not depend on seasonal variations in velocity. Because ice moves much faster in an icefall than elsewhere, it stretches longitudinally as it enters the top of the icefall. Thus any volume element has a larger surface area when it is in the icefall than when upstream or downstream. Elements that pass through the icefall in summer, therefore, lose more ice by ablation than elements in other parts of the glacier. They also collect more wind-blown dust. They become the troughs of the waves that form where the ice is compressed longitudinally at the foot of the icefall. Elements that spend winter in the icefall collect excess snow and become the wave crests. The dark bands should therefore correspond to the troughs. Nye found this relationship on Austerdalsbre. He also observed that the crests became free of snow before the troughs, and ablation on the crests exceeded that in the troughs by 30%. This explains why the waves die out, even though longitudinal compression in the ice below the icefall tends to increase their amplitude. A calculation based on measured velocities and ablation rates correctly predicted the positions and amplitudes of the waves.

Waddington (1986) made a sophisticated mathematical analysis of wave ogive formation to generalize Nye's model. It deals with the interaction between a spatial term, which defines the "forcing," and a temporal term, the specific mass balance, which varies during the year with a pattern that repeats annually. Any section of glacier with a longitudinal gradient of velocity, width, or specific balance acts as a forcing and generates waves by the mechanism described by Nye. But waves generated in different places normally interfere with each other. Only places such as an icefall where the gradient is large and localized can produce observable waves. Moreover, the ice must travel through the critical area – in this case, the zone of rapid stretching at the top of the icefall – in about six months or less; no waves form if the ice takes a whole year to move through the onset of the icefall. This explains why not all icefalls generate ogives.

The preceding theory doesn't explain all the observations of band ogives (Goodsell et al. 2002). At some locations, the dark bands correspond to the down-glacier slope of the wave ogives rather than to the troughs. The rock debris in the dark bands is sometimes too abundant and too coarse to be carried by wind and has characteristics typical of subglacial debris. Goodsell et al. pointed out that the across-glacier foliation in the ogives originates by transposition of crevasse traces. In the longitudinally compressive regime at the foot of an icefall, these foliation planes are oriented favorably for thrusting. Shearing along them might explain how material is carried up from the bed. Such a deep origin for the dark ice would explain why banding sometimes persists far down-glacier despite rapid ablation.

Wave ogives are not the only type of flow structure that involves patterned variations of ice thickness and surface topography. The surfaces of valley glaciers, ice streams, and ice shelves are commonly marked by longitudinal grooves and ridges, sometimes called *lineations* or *flow stripes*. (These terms also refer to bands with distinct ice color or surface roughness; they do not always indicate a topographic feature.) Such corrugations can persist for considerable distances down-glacier – for hundreds of kilometers in the case of large ice shelves (Fahnestock et al. 2000). Individual grooves and ridges are typically tens of meters to a few kilometers wide, and a few meters to tens of meters in amplitude.

How longitudinal corrugations develop is often obscure. They most plausibly originate where the ice is thickened or thinned along a flow line by enhanced extension or compression oriented along-glacier or transversely. The magnitude of the thickness change must differ between adjacent flow lines. The region of origin may have an irregular and rough bed or may have gradients in the rate of basal slip. The latter occur at the heads of ice streams and where weak-bed ice streams flow over sticky spots. Some corrugations begin in heavily crevassed shear margins (Merry and Whillans 1993). A longitudinal groove sometimes begins where two valley glaciers join together, a situation that also produces medial moraines. For a longitudinal corrugation to persist, it must be narrow, smaller than a few multiples of ice thickness (Section 8.7.2) – otherwise, the surface slopes would induce transverse flow and the feature would flatten by diffusion. In ablation zones, interactions between ablation rate and surface aspect or debris concentration might preserve corrugations.

In a steady state, longitudinal corrugations parallel the flow lines. Lack of such a correspondence indicates that the ice flow has recently changed. The orientations of longitudinal ridges on the Ross Ice Shelf, for example, deviate significantly from the present flow vectors, a pattern that reflects changes over the last several centuries in the flow regime of the ice streams feeding the shelf (Fahnestock et al. 2000).

10.9 Crevasses

10.9.1 Patterns and Conditions for Occurrence

Crevasses are tensional fractures in the ice. The patterns they form in simple situations can be deduced from the directions of the principal stresses (Nye 1952a). The glacier surface is assumed to be a plane, the x-axis is taken along the surface in the direction of flow, the y-axis points across the glacier, and the z-axis is perpendicular to the surface. Call the normal stress components σ_{xx}, σ_{yy}, σ_{zz} and the shear stress components τ_{xy}, τ_{yz}, τ_{xz}. Because τ_{xz} and τ_{yz} must be zero at the surface, σ_{zz} is one principal stress. The others are oriented horizontally (surface-parallel, to be precise) and correspond to the roots σ_1, σ_2 of

$$\sigma^2 - (\sigma_{xx} + \sigma_{yy})\sigma + (\sigma_{xx}\sigma_{yy} - \tau_{xy}^2) = 0. \tag{10.5}$$

Define σ_1 as the maximum – the most tensile – of σ_1, σ_2. Commonly, one of these principal stresses is tensile, and the other compressive. Crevasses form where σ_1 defined by the bulk stress exceeds the effective tensile strength of the ice.

10.9.1.1 Patterns

To explain crevasse patterns, Nye (1952a) assumed that crevasses open up in the direction of maximum tension; new crevasses should thus trend perpendicular to σ_1. Observations generally confirm this assumption. Meier (1960) found that crevasses on Saskatchewan Glacier, Canada, are approximately perpendicular to the direction of the extensional principal strain rate. The same configuration occurs on Worthington Glacier, Alaska (Harper et al. 1998). Hambrey and Müller (1978), on the other hand, found this configuration only in certain parts of White Glacier in the Canadian Arctic. They suggested that the value of the second principal stress might also be important, which is certainly true in some cases for reasons given later (Section 10.9.1.2).

Consider a valley glacier of constant width, and assume for simplicity that σ_{xx} takes the same value all the way across the glacier. The shear stress τ_{xy} due to drag of the valley walls is zero on the centerline and increases toward each side.

First suppose that τ_{xy} is the only nonzero stress component at the surface (Figure 10.10a). Then the principal stresses are a tensile stress of magnitude τ_{xy} and an equal compressive stress, each inclined at 45° to the coordinate axes, as shown in the diagram. The crevasses trend at right angles to the tensile principal stress. They do not extend to the center of the glacier because τ_{xy} decreases to zero. The stress state and crevasse pattern on one side of the glacier mirror those on the other side.

Figure 10.10b illustrates extending flow; the velocity increases downstream along the glacier. On the centerline, τ_{xy} again equals zero, but now there is a tensile principal stress σ_{xx}. The crevasses trend at right angles to σ_{xx} (hence in the y-direction). Near the edge, the combination of σ_{xx} and τ_{xy} gives a large tensile principal stress inclined at less than 45° to the x-axis, and a small compressive principal stress as diagrammed. The crevasses trend at right angles to the tensile stress and so make an angle of more than 45° with the side of the glacier. They trend in intermediate directions between the edge and the centerline.

Figure 10.10c illustrates compressing flow; the velocity decreases downstream. Near the edge, the combination of σ_{xx} and τ_{xy} corresponds to a small tensile principal stress inclined at more than 45° to the x-axis, and a large compressive principal stress. The crevasses meet the sides of the glacier at less than 45°. They die out toward the centerline as τ_{xy} goes to zero but σ_{xx} remains compressive. (When σ_{xx} is compressive, the glacier tends to stretch in the y-direction. Steep valley walls prevent this deformation; a transverse compressive stress, as shown in the diagram, must develop.)

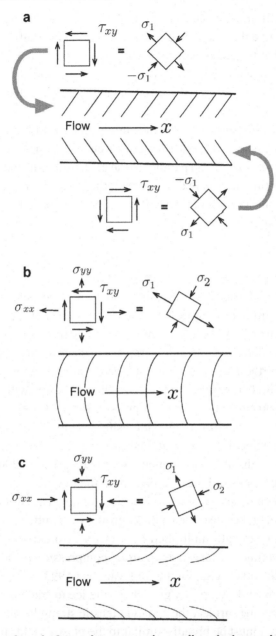

Figure 10.10: Patterns of newly formed crevasses in a valley glacier, and associated stress regimes near the top lateral margin in each case. (a) Effect of shear stress exerted by valley walls only. (b) Shear stress and extending flow. (c) Shear stress and compressing flow. Adapted from Nye (1952a).

Once crevasses form, they are rotated and bent by gradients in the ice velocity. For example, shear-margin crevasses, like those depicted in Figure 10.10a, typically rotate so they no longer point at 45° up-glacier, but instead point across the glacier or even down-glacier. Old deformed crevasses may close, but remain visible as bands of distinct ice or snow – the traces discussed previously. The orientations of deformed crevasses can be used to infer patterns of ice velocity (Vornberger and Whillans 1990).

Allowing for such bending and rotations, the simple patterns in Figure 10.10 do correspond to many observations on glaciers. But other patterns occur. At a glacier snout, for example, flow lines often diverge. This sets up tension parallel to the glacier margin and hence opens crevasses perpendicular to the margin – a so-called *radial* pattern. Irregular crevasse patterns often form in patches on mountain glaciers where the ice flows over a subglacial hill or past a promontory jutting from the valley wall.

10.9.1.2 Occurrence

The stress prevailing locally around a discontinuity in the ice differs, sometimes greatly, from the average stress acting in the larger surrounding region (the "far-field stress" or "bulk stress") (Rist et al. 1999). Such differences reflect the elastic properties of ice. Consider a fracture in the ice, long compared to its width, and terminating in sharp tips at either end. Because the force arising from the far-field stress cannot be transmitted across the fracture, the stresses near the tips are intensified. Furthermore, for an orientation not aligned with σ_1 or σ_2, the sides of the fracture shear past one another, generating tension at the crack tips with an orientation oblique to the fracture. As a second case, consider a compressive far-field stress, σ_2, but $\sigma_1 = 0$. Around a hole in the ice – pictured as an ellipse, say – intensified compression prevails on either side of the hole on surfaces perpendicular to σ_2. But, because of elastic deformation, tension develops on the surfaces intersecting the hole and aligned with σ_2 (Jaeger et al. 2007, pp. 231+). Thus, as long as small preexisting cracks or holes are present in the ice, crevasses could form anywhere that the difference between σ_1 and σ_2 rises to a critical value. (The difference must exceed the tensile strength of the ice, or, because some cracks grow by a combination of shear and tension, it must exceed the tensile strength multiplied by a factor of order one.) Indeed, observations show that crevasses sometimes occur where a strong compression σ_2 acts in one direction but very little stress acts in the other (e.g., Figure 5 of Vaughan 1993).

What level of stress must develop in a glacier for the ice to fracture? For the conditions of low hydrostatic pressure at the surface, the apparent tensile strength – that is, the far-field stress needed to form crevasses – must be broadly similar to the plastic yield strength for creep of ice. Otherwise, either no crevasses would form on glaciers (if the tensile strength greatly exceeded the plastic strength), or crevasses would form even in regions of trivial strain rates (if tensile strength were much smaller than plastic strength). The apparent tensile strength can be inferred from comparisons of crevassed regions with uncrevassed regions; the stress in both cases can be estimated from measured strain rates by using the creep relation for ice. With such an analysis, Vaughan (1993) found tensile strength values of 90 to 150 kPa on two temperate glaciers (Blue and Saskatchewan), but unrealistically high values on the polythermal White Glacier. (The other sites he examined were covered with firn, so the stresses near the surface could not be estimated from the creep relation.)

There is no simple relation between measured creep strain rates and the presence or absence of crevasses. The minimum extensional strain rate where crevasses occur varies considerably from place to place. For example, it varied by a factor of about 500 in Vaughan's compilation of measurements from 17 glaciers and ice shelves in diverse settings. The variation mostly reflected temperature differences. Cold polar glaciers fracture at lower strain rates than do temperate ones, because the creep viscosity increases in colder ice; in a cold glacier a smaller strain rate is needed to exceed a given tensile strength. Because creep strain rates vary as the third power of stress, large differences of strain rate imply much smaller differences of apparent tensile strength. Hambrey and Müller (1978) observed crevasses on White Glacier in one place with strain rates as low as 0.004 yr^{-1}, whereas no crevasses were present at another place with a strain rate of 0.16 yr^{-1}. Such a difference could be explained by a factor of 3.4 difference in the apparent tensile strength. Apparent strength is probably influenced not only by the stress regime but also by inhomogeneities in the ice. Furthermore, the strain rate at the point of origin of a crevasse may differ from that measured some time later at the point of observation.

10.9.2 Crevasse Depth and Propagation

The maximum depth of a crevasse can be predicted by assuming that ice behaves as a perfectly plastic material (Section 3.4.4.6). At the base of a vertical crevasse wall 22 m high, the vertical compressive stress amounts to 200 kPa plus atmospheric pressure. But in the horizontal direction, only atmospheric pressure acts against the wall. This stress system is equivalent to a shear stress of 100 kPa on planes at 45° to the horizontal. Since 100 kPa equals the yield stress in shear, the ice at this depth will flow rapidly and close the crevasse.

Nye (1957) derived a formula for the maximum depth, using the creep relation rather than an approximation of perfect plasticity. He assumed that a crevasse penetrates to the depth d where the maximum horizontal stress changes from tensile to compressive. No accounting was made for the focussing of tensile stress at the bottom of crevasses. This assumption is justified because most crevasses occur in groups, not in isolation; the upper layers of the glacier, once fractured, cannot effectively transmit tension. (See Sassolas et al. 1996 for an analysis of how neighboring crevasses shield one another from the far-field stress.) For a stretching rate $\dot{\epsilon}$, the horizontal deviatoric stress is $\tau_{xx} = [\dot{\epsilon}/A]^{1/n}$, where A and n denote the creep parameters for ice (Chapter 3). The vertical normal stress is $\sigma_{zz} = -\rho g d$, for an ice density ρ and gravitational acceleration g. At the depth with no horizontal tension, $\sigma_{xx} = 0$. But, by definition, $\tau_{xx} = [\sigma_{xx} - \sigma_{zz}]/2$. Equating expressions for τ_{xx} gives Nye's formula:

$$d = \frac{2}{\rho g} \left[\frac{\dot{\epsilon}}{A} \right]^{1/n}. \tag{10.6}$$

This formula indicates that the deepest crevasses should occur in regions of rapid extension and, because the parameter A increases with temperature, in cold glaciers.

Measured crevasse depths in temperate glaciers seldom exceed 25 or 30 m. Few accurate measurements have been made, however, and mountaineering literature contains many descriptions of crevasses deeper than this. Again, crevasse traces often persist on the glacier surface, in spite of ablation, for distances that imply crevasse depths of several tens of meters (Hambrey 1976). Crevasses deeper than 30 m have been encountered in polar regions. Accurate measurements of crevasse depths need to be obtained for further analysis.

The preceding analyses do not apply to a water-filled crevasse. With water of density ρ' filling a crevasse completely, the pressure rises from the atmospheric value P_a at the surface to $P_a + \rho' g z$ at depth z. The static pressure in the surrounding ice (density ρ) is smaller by an amount $[\rho' - \rho] g z$. Because this quantity increases with z, the crevasse can penetrate not only into layers with compressive horizontal stresses, but perhaps to the bottom of the glacier (Weertman 1973a; Robin 1974).

This is an important problem; fractures that span the entire thickness of a glacier contribute to iceberg formation and provide a path for surface water to reach the bed, where it affects glacier motion. To analyze this process, van der Veen (1998, 2007) and Alley et al. (2005) used concepts from fracture mechanics (extending the work of Weertman (1973a), Smith (1976), and Rist (1996), among others). We are concerned here with the rapid initial penetration of fractures into the glacier; whether the resulting crevasse can remain open despite closure by ice creep is no longer the issue. Consider an isolated vertical fracture at the surface of a glacier undergoing extension. At the bottom edge of the fracture, the tensional stress is amplified relative to its far-field value. The effective magnitude of tension is measured by the *stress intensity*, with a value $K_\tau \approx 2\tau_{xx}\sqrt{\pi d}$. Here $2\tau_{xx}$ equals the far-field horizontal tension and d again signifies depth of the fracture. The fracture continues to propagate if K_τ exceeds a material property called the *critical stress intensity*, denoted K_{1c} (also called the *fracture toughness*). Laboratory experiments on ice give values for K_{1c} in the range 100 to 170 kPa m$^{1/2}$, and smaller for firn (Fischer et al. 1995; Rist et al. 1999). In a glacier, the horizontal compressive stress arising from the weight of overburden is also intensified at the bottom of a fracture; the formula $K_o \approx 0.39\rho g d\sqrt{\pi d}$ gives an approximation for the corresponding effective intensity of compression. K_o increases more rapidly with depth than K_τ because, unlike the far-field tension, the overburden stress increases directly with depth. Thus the fracture will not propagate beyond the depth where $K_\tau - K_o = K_{1c}$. After rearranging, this implies a maximum depth, in dry conditions, of

$$d \approx \frac{2.6}{\rho g}\left[2\tau_{xx} - \frac{K_{1c}}{\sqrt{\pi d}}\right].\tag{10.7}$$

Given a τ_{xx} of about 100 kPa, the term with K_{1c} can be neglected; isolated fractures thus should reach a depth of about $5\tau_{xx}/\rho g$, or 2.5 times the depth of open crevasses according to Nye's formula. (These relations are only approximate because, in detail, the stress intensities change with depth according to more complicated formulae; see van der Veen 1998.)

If, in addition, water fills the fracture to a depth d_w, it wedges the crack open and so contributes a tension with effective intensity $K_w \approx \rho' g d_w \sqrt{\pi d_w}$. Now the fracture continues to propagate down into the glacier if $K_\tau + K_w - K_o > K_{1c}$.

With such an analysis, van der Veen (1998) concluded that a water-filled crevasse can reach the glacier bed provided that the tension $2\tau_{xx}$ exceeds about 100 to 150 kPa and the water surface remains within 10 to 20 m of the glacier surface. These results should be interpreted qualitatively. Because a high water level must be maintained in the crevasse as it grows downward, abundant surface melt is an essential component (Alley et al. 2005; van der Veen 2007); calving is therefore connected with the summer climate. Furthermore, when stress conditions allow their growth, fractures propagate rapidly. Thus the rate of water supply, not the kinetics of fracture growth, should control the rate of penetration of a water-filled crevasse. A special case is a crevasse propagating through subfreezing ice. For high but realistic rates of water inflow, simple calculations show that crevasses propagate rapidly compared to the rate of freezing. Thus, water-filled crevasses fed by abundant surface water should penetrate deep into polar glaciers (this is observed in Section 6.2.2).

10.9.3 Related Tensional Features

A *bergschrund* is the crevasse that separates the moving ice at the head of a mountain glacier from the adjacent rock. A large crevasse similar to a bergschrund can form if the highest part of a glacier is frozen to its bed while the remainder can slide (Lliboutry et al. 1976).

A *rift* is a tensional fracture through the whole thickness of an ice shelf. Rifts originate at irregularities on the side margins, at ice rises, or at the front margin, and then propagate into the shelf. Over time, the intersection of rifts forms new tabular icebergs, a process that may take decades on a large ice shelf; Fricker et al. (2002) discussed an example on the Amery Ice Shelf, Antarctica. Rifts reduce the viscous forces arising from flow of an ice shelf past its side margins and ice rises, and so reduce the buttressing force that opposes spreading of the shelf (MacAyeal et al. 1998; see Section 8.9.3). Formation of new rifts may therefore significantly change the dynamics of an ice shelf.

Thermal contraction also forms tension cracks. When the temperature of ice near the glacier surface rapidly decreases due to the onset of cold weather, the ice tries to contract in horizontal directions but is initially prevented from doing so by its cohesion. This sets up tensional stress in horizontal directions, which causes fracture. The process is analogous to the formation of thermal contraction cracks in permafrost (Lachenbruch 1962). Thermal contraction cracks occur in the firn at some low-accumulation sites in East Antarctica (Courville et al. 2007). This may be important for ice core interpretation because such cracks can enhance snow metamorphism and vertical movement of atmospheric gases through the firn.

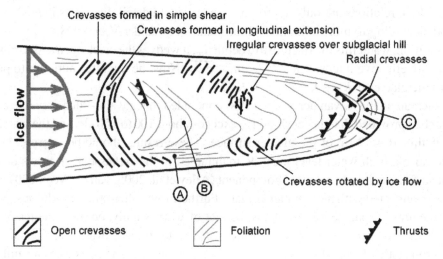

Figure 10.11: Characteristic pattern and assemblages of structures in the ablation zone of a valley glacier.

10.10 Structural Assemblages

Except for stratification in mostly undeformed ice, glacier structures seldom occur in isolation from one another. Different parts of a glacier display different assemblages of structures, reflecting the deformation regimes along the trajectories of ice movement (Hambrey and Lawson 2000, pp. 73+). To illustrate, Figure 10.11 gives a schematic view of characteristic structures at the surface of a valley glacier's ablation zone:

- At location A, depositional strata have been tilted by compressions and extensions upstream, and transformed into longitudinal foliation by shearing against the margins. The foliation is cut at a high angle by crevasse traces.

- Longitudinal compression at the foot of an icefall produces the transverse foliation at location B. The compressed layers consist of crevasse traces and transposed stratification.

- At location C, the ice was foliated by shearing against the bed but the foliation planes are now tilting up-glacier because of longitudinal compression. Thrusts cut through the foliation at a low angle, and radial crevasses cut through it at a high angle. Layers originate, in part, from diverse features such as frozen water channels, basal fractures, and layers of rock debris picked up from the bed.

Further Reading

Hambrey and Lawson (2000) reviewed structures in mountain glaciers. The book by Post and LaChapelle (2000) illustrates many structures with spectacular photographs.

CHAPTER 11

Reaction of Glaciers to Environmental Changes

"The glaciers creep like snakes that watch their prey, from their far fountains...."

Mont Blanc, Percy Shelley

11.1 Introduction

The world's climate[1] is continually changing, with apparently random fluctuations from year to year superimposed on longer term trends: such changes are reflected in variations of glacier extent. The repeated growth and decay of continental ice sheets during the Pleistocene are one example. Another is the shrinking of most glaciers outside the polar regions since the onset of climate warming in the late nineteenth century.

Although advances and retreats of different glaciers appear to be broadly synchronous, the picture becomes more confused on closer examination of the records. Some glaciers in an area may be advancing while others are retreating. Differences in the local climates of individual glaciers may account for some of the differences in behavior. Even with the same climate, however, different glaciers experience different changes of total snowfall and melt, because each glacier has its own distribution of surface area with altitude (Chapter 4). Moreover, glaciers also differ one from another in characteristics such as size, steepness, and substrate, all of which determine how a glacier flows. The reaction of a glacier to a change of climate depends on the flow regime, which itself evolves over time as part of the response.

A climate change influences a glacier most directly by changing the net inflow or outflow of mass determined by snowfall, melt, and related processes – the mass balance. But mass balance is not the only environmental factor that affects the size and flow of glaciers. Melt on warm summer days and intense rainfall increase the volume of water reaching the glacier bed, a process that often changes the ice flow. A climatic temperature change, if it persists over time, modifies

---

[1] We use the term *climate* in the broad sense that includes annual means and interannual fluctuations as well as long-term trends. A formal "climatological mean" usually refers to an average of 30 years.

Copyright © 2010, Elsevier Inc. All rights reserved.
DOI: 10.1016/B978-0-12-369461-4.00011-9

the temperature deep in a polar glacier, altering the ice flow and causing the glacier to thin or thicken. This is an important process for the ice sheets.

The ice sheets also react to changes of sea level. A rise of sea level leads to retreat of marine margins, initiating long-term adjustments of flow and topography that ultimately thin the entire ice sheet. Marine margins also react to changes of the floating ice shelves to which they are attached – ice shelves, in turn, depend on oceanic conditions and climate. Observations over the last two decades have revealed major changes in the flow of polar glaciers, including some of the world's largest ice streams, in response to events at their marine margins. The influence of some of these changes extends a great distance inland. Understanding this process is critical for understanding the future evolution of the ice sheets.

In this chapter we first discuss how a glacier responds to changes of mass balance; we give precise form to some of the steps in the flow chart of Figure 4.2. Specifically, Section 11.2 discusses the magnitude and timescale of the response to small changes and Section 11.3 reviews the physical processes governing changes of flow and ice thickness and describes the response to specific types of mass balance forcings. Section 11.4 then examines response to variables other than mass balance, with an emphasis on evolution of the ice sheets at the end of an ice age. The last part of the chapter, Section 11.5, considers processes at marine margins, including the tidewater glacier cycle and the reaction of ice sheets to changes on fringing ice shelves.

11.2 Reaction to Changes of Mass Balance: Scales

When the mass balance of a glacier changes, its terminus eventually advances or retreats. How far the terminus moves, and how long the glacier takes to adjust to the new mass balance, can be understood from simple arguments.

If the mass balance remained constant for many years the glacier would eventually reach a steady state in which its dimensions remain the same. Consider a glacier that has attained such a condition; we refer to this as the *datum state*. Let $\dot{b}_0(x)$ be the specific mass balance in the datum state, a function of position x along the glacier. Now suppose the specific balance changes to $\dot{b}_0(x) + \dot{b}_1(x)$. If this new pattern is maintained for a long time and there are no instabilities, the glacier will reach a new steady state with a different terminus position and a different surface profile. The terminus position and the surface profile generally change in an interdependent way; at a steady state, the surface profile has a characteristic shape that is largely independent of the specific balances along the glacier (Section 8.10). After an advance, the vertical distance between the original and new profiles increases steadily from the head of the glacier and attains a maximum at the datum terminus. This is because the change in ice flux produced by the change in specific balances accumulates down-glacier. Consequently, lateral moraines and trimlines are often close to the present glacier surface near the equilibrium line but far above it near the terminus (Section 11.3.2.2).

11.2.1 Net Change of Glacier Length

The x-axis points down-glacier. In the datum state, the glacier spans $x = 0$ to $x = L_o$. The terminus is assumed to be wedge-shaped (Figure 11.1) and the glacier width is assumed uniform. Initially no ice flux passes through $x = L_o$; the terminus position is fully adjusted to the specific balance profile $\dot{b}_0(x)$. But after a long period with a specific balance of $\dot{b}_0(x) + \dot{b}_1(x)$, the flux per unit width at $x = L_o$ attains $\int_0^{L_o} \dot{b}_1(x)dx = \langle \dot{b}_1 \rangle L_o$. (We will use $\langle \dot{b}_1 \rangle$ to indicate the spatial average of $\dot{b}_1$.) The terminus will have advanced the distance L_1 needed to remove this amount of ice by ablation. If we call the ablation rate near the terminus $\dot{a}_0$ (equal to $-\dot{b}_0(L_o)$), then $\langle \dot{b}_1 \rangle L_o = \dot{a}_0 L_1$ or

$$L_1 = L_o \langle \dot{b}_1 \rangle / \dot{a}_0. \tag{11.1}$$

For example, if $\dot{a}_0 = 5 \, \text{m yr}^{-1}$ and the specific balance increases uniformly by $0.5 \, \text{m yr}^{-1}$, the length of the glacier will ultimately increase by 10%. If the average width of the glacier is $\overline{Y}$, but near the terminus the width is Y_t, then

$$L_1 = L_o \frac{\langle \dot{b}_1 \rangle}{\dot{a}_0} \frac{\overline{Y}}{Y_t}. \tag{11.2}$$

Thus, especially large terminus fluctuations occur if the glacier drains a wide basin but terminates in a narrow tongue (a large ratio $\overline{Y}/Y_t$). In contrast, for an ice sheet with a circular plan, $\overline{Y}/Y_t = 1/2$; terminus fluctuations are suppressed because the length and the width of the ablation zone increase or decrease simultaneously. (For an ice sheet, Y refers to the width between adjacent flow lines.)

Equation 11.2 explains, in part, why different glaciers in the same region can respond differently to the same change of mass balance. The effect of basin geometry, $\overline{Y}/Y_t$, is one reason. Different values for L_o and $\dot{a}_0$ are another. Long glaciers tend to fluctuate more than short ones; they have a larger L_o. On the other hand, long glaciers also convey a larger flux and so tend to extend to lower altitudes where ablation rates are higher, which has the opposite effect.

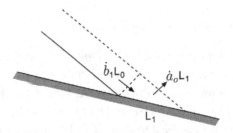

Figure 11.1: The relation between increased flux and advance of a glacier terminus. Adapted from Nye (1960).

11.2.2 Simple Models for Response

We next discuss simple time-dependent models of glacier response and show how characteristic time scales appear in them. *Response time* refers to the time a glacier takes to complete most of its adjustment to a change in mass balance. The qualifier "most" is important because, in principle, a glacier continues to adjust at an ever-decreasing rate for a very long time after the change. For reasons discussed herein, response time typically means the time for a glacier to complete all but a factor 1/e (or 37%) of its net change. With this definition, the time scale depends, though not strongly, on the measure of glacier size; for example, the length probably adjusts more quickly than the volume (Oerlemans 2001, p. 72). Similar presentations, with different emphases and more details, are given by Jóhannesson et al. (1989), Oerlemans (2001, pp. 97+), and Harrison et al. (2003).

Consider a glacier, or a flow band of an ice sheet, of total length L. The specific balance departs from a datum state $b_0(x)$ and becomes $b_0(x) + b_1(x, t)$, causing the glacier to advance or retreat and thicken or thin. To model the response simply, only two relations are needed. First, the total volume of the glacier $V(t)$ changes at a rate dependent on the specific balance perturbation:

$$\frac{dV}{dt} = \int_0^{L_o} Y(x, t)\dot{b}_1(x, t)\, dx + \int_{L_o}^L Y(x, t)\dot{b}(x, t)\, dx, \tag{11.3}$$

with x the distance, and $Y(x, t)$ the width of the glacier or flow band. (The specific balance is measured as ice-equivalent thickness per unit time, equivalent to $\dot{b}_i$ in Chapter 4. Density is assumed uniform and constant). Second, glacier volume varies with ice thickness $H(x, t)$ according to a geometric relation:

$$V(t) = \int_L Y(x, t) H(x, t)\, dx. \tag{11.4}$$

$H(x, t)$ is the mean value of thickness over the cross-section at x and at time t.

11.2.2.1 Response of a Slab with Constant Thickness

As a first model, consider a glacier on a long sloping surface (Figure 11.2a). Except at its terminus, the glacier approximates a parallel-sided slab. The thickness H_o is taken, for now, as constant. In the initial datum state, $dV/dt = 0$. At time $t = 0$ the mass balance increases uniformly from datum value $\dot{b}_0(x)$ to $\dot{b}_0(x) + \dot{b}_1$. Over time t the length varies from datum value L_o to $L_o + L_1(t)$. Equation 11.3 can be approximated, for small changes, as

$$\frac{dV}{dt} = \overline{Y} L_o \dot{b}_1 - Y_t \dot{a}_0 L_1(t). \tag{11.5}$$

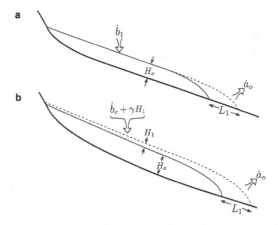

Figure 11.2: Schematic models for advance of a glacier following an increase of mass balance: (a) if thickness changes are negligible; (b) allowing for thickness changes.

As before, $\dot{a}_0$ is the ablation rate near the terminus, and Y_t is the width there. For positive $\dot{b}_1$, this relation simply states that the glacier grows because the perturbation $\dot{b}_1$ adds mass but that the growth is counteracted by increased ablation as the terminus advances.

For an advance $L_1(t)$, the volume of the glacier increases by $Y_t H_0 L_1(t)$. Assuming negligible changes of width, then

$$dV/dt = Y_t H_o\, dL_1/dt. \tag{11.6}$$

Equating this expression to Eq. 11.5, dividing through by $Y_t H_o$, and rearranging gives an equation for $L_1(t)$,

$$\frac{dL_1(t)}{dt} + \frac{L_1(t)}{t_r} = \dot{b}_1 \frac{\overline{Y}}{Y_t} \frac{L_o}{H_o} \quad \text{with} \quad t_r = \frac{H_o}{\dot{a}_0}. \tag{11.7}$$

This is a linear equation with a time scale t_r, a first estimate for the glacier's response time. Equation 11.7 implies that:

1. If the mass balance perturbation remains constant for $t > 0$, then the terminus advances and over time asymptotically approaches a new steady value. The solution is

$$L_1(t) = L_o \frac{\dot{b}_1}{\dot{a}_0} \frac{\overline{Y}}{Y_t} \left[1 - \exp\left(-t/t_r\right)\right]. \tag{11.8}$$

   The distance between the terminus and its final position decreases exponentially with time. After a time t_r, the advance is about 63% complete; after time $2t_r$, about 86%.

2. The response time equals the ratio of ice thickness to the ablation rate at the terminus.

3. For sinusoidal perturbations of mass balance $\dot{b}(t)$, the terminus fluctuations depend on the period of the sinusoid. The terminus fluctuates very little if the period is small compared

to t_r. With a period large compared to t_r, the terminus fluctuates with an amplitude L_1 given by Eq. 11.2, and the advances are in phase with the fluctuations of mass balance. For intermediate periods, advances lag the maxima of mass balance, and their amplitude is smaller than given by Eq. 11.2.

11.2.2.2 How Thickness Changes Affect the Response

Now allow the ice thickness to increase as the glacier advances, as sketched in Figure 11.2b. The increase of ice thickness is $H_1(x, t)$ in general, but $\overline{H}_1(t)$ as a spatial average. The glacier's geometry relates $\overline{H}_1(t)$ to the advance $L_1(t)$. Define a positive number η such that $\overline{H}_1(t)/L_1(t) = \eta \overline{H}_o/L_o$, with $\overline{H}_o$ the average ice thickness in the datum state. (For a long slab shape, $\eta \approx 0$, whereas for a parabolic profile, $\eta \approx 0.5$.)

An increase of thickness alters the situation in two important ways. First, it contributes to the increase of glacier volume, so a term needs to be added to the approximation for volume change:

$$\frac{dV(t)}{dt} \approx Y_t \overline{H}_o \frac{dL_1(t)}{dt} + \overline{Y} L_o \frac{d\overline{H}_1(t)}{dt}. \tag{11.9}$$

Second, as the glacier thickens, its surface elevation increases; the surface therefore cools and the specific balance increases, too. Conversely, when a glacier thins the surface altitude declines; this warms the surface and increases ablation. Thus, in general, a perturbation of specific balance $\dot{b}_1(x, t)$ should be regarded as the sum of two components. Regional climate change determines an initial *climatic* perturbation (magnitude $\dot{b}_c$). The response of the glacier's surface profile, combined with the variation of specific balances with altitude (Section 4.2.5), then contributes a second term called the *altitude-mass-balance feedback*. In sum, $\dot{b}_1(x, t) = \dot{b}_c + \gamma H_1(x, t)$, where γ denotes the gradient of specific balance with altitude, $\partial \dot{b}/\partial S$. Thus Eq. 11.3 is

$$\frac{dV(t)}{dt} \approx \dot{b}_c \overline{Y} L_o + \gamma \overline{Y} L_o \overline{H}_1(t) - Y_t \dot{a}_0 L_1(t). \tag{11.10}$$

(A more detailed treatment would account for variations of H_1 with x in the altitude-feedback term and for effects of the glacier profile shape on the relation for V.)

Equating the two relations for dV/dt (Eqs. 11.9 and 11.10) and rearranging gives another linear equation for $L_1(t)$:

$$\frac{dL_1(t)}{dt} + \frac{L_1(t)}{t_r} = \dot{b}_c \frac{L_o}{\overline{H}_o} \frac{1}{[\eta + \omega]} \quad \text{with} \quad t_r = \frac{\overline{H}_o[\omega + \eta]}{\dot{a}_0 \omega - \gamma \eta \overline{H}_o}. \tag{11.11}$$

For convenience, we have written $\omega = Y_t/\overline{Y}$. The estimated response time again depends on the ratio of ice thickness to ablation rate at the terminus, but now with additional factors, encapsulated in this simple model by three parameters reflecting the

$$\text{balance gradient}\quad \gamma = \partial \dot{b}/\partial S,$$

$$\text{profile shape}\quad \eta = \frac{\overline{H}_1}{L_1} \bigg/ \frac{\overline{H}_o}{L_o},$$

$$\text{plan-view shape}\quad \omega = Y_t/\overline{Y}.$$

Effect of Glacier Profile and Plan   Suppose first that no altitude-mass-balance feedback operates ($\gamma = 0$). Then the response time differs from $\overline{H}_o/\dot{a}_0$ by a multiple $[1 + \eta/\omega]$, a function only of glacier shape. Since η is about 0.5 for a parabolic profile, but zero for a slab, the response time is greater for the parabolic case, by a multiple of 1.5. For an ice cap with a parabolic profile but a circular plan, $\eta \approx 0.5$ and $\omega \approx 2$, so the multiple equals 1.25. These differences are trivial compared to the range of values for $\overline{H}_o/\dot{a}_0$ and in view of the rough approximations leading to the formulae.

A simple approximation can be made in terms of the maximum ice thickness, H_{MAX}. In a slab-profile glacier, H_{MAX} equals the mean thickness. For a parabolic-profile ice cap, H_{MAX} is greater than the mean thickness, by a factor of about 1.3 to 1.5 (Table 1 in Paterson 1972a). Thus, regardless of geometry, an estimate of the response time can be obtained from the simple formula

$$t_r = H_{MAX}/\dot{a}_0. \tag{11.12}$$

Values for Response Time   Table 11.1 lists some response times estimated from Eq. 11.12. These values indicate that temperate mountain glaciers adjust to mass balance changes within a few decades. Observations of glacier changes during the twentieth century confirm this estimate. Many glaciers retreated during the first half of the century as climate warmed, but then readvanced or maintained a constant length in the 1960s and 1970s. In these decades the climate cooled or was constant, depending on the location, but nonetheless remained warm compared

Table 11.1: Glacier response times estimated from Eq. 11.12.

	Thickness (m)	Terminus ablation ($m\,yr^{-1}$)	Response time (yr)
Glaciers in temperate maritime climate	150–300	5–10	15–60
Glaciers in high-polar climate	150–300	0.5–1	150–600
Ice caps in Arctic Canada	500–1000	1–2	250–1000
Greenland Ice Sheet	3000	1–2	1500–3000

to the early twentieth century. If response times were longer than a few decades, retreat would have continued during this period. (Terminus observations indicating response times of several years to a few decades are discussed by Lliboutry 1971a, Röthlisberger 1980, and Oerlemans 2007, for glaciers in the Alps and Norway; and by Harper 1993, for glaciers on Mount Baker in Washington.)

By examining Eq. 11.12 and Table 11.1 we might conclude that, in general, large glaciers always respond more slowly than small glaciers. This is not always the case; large glaciers in temperate climates often extend into warm regions with very high ablation rates. Observations show a weak but positive correlation between glacier length and terminus ablation in temperate regions. Some large glaciers can therefore respond more rapidly than small glaciers nearby (Bahr and Dyurgerov 1999; Bahr et al. 1998).

Geometrical Interpretation: The Filling Time   Jóhannesson et al. (1989) showed that the time scale $H_{MAX}/\dot{a}_0$ is equivalent to the *filling time*: how long it would take a mass balance perturbation to accumulate or remove the difference between the steady-state volumes of the glacier before and after a change in mass balance.

Let t_f be filling time, V the glacier volume, and $A = L\overline{Y}$ the surface area. Suffix 0 will denote the value in the datum state and suffix 1 the perturbation after the adjustment is complete. Thus, for a mass balance perturbation, averaged over the area of the datum glacier, of $\langle \dot{b}_1 \rangle$,

$$t_f = V_1/[\langle \dot{b}_1 \rangle \mathcal{A}_0]. \tag{11.13}$$

For an ice cap of circular plan of radius L and maximum thickness H_{MAX}, $V \propto L^2 H_{MAX}$. With a parabolic surface, H_{MAX}^2/L is a constant and so $V \propto L^{2.5} \propto \mathcal{A}^{1.25}$; this relation provides a reasonable fit to data (Section 8.10.3). It follows that

$$V_1/V_0 = 1.25 \mathcal{A}_1/\mathcal{A}_0 = 1.25\langle \dot{b}_1 \rangle/\dot{a}_0 \tag{11.14}$$

by Eq. 11.2 and $\mathcal{A}_1 = L_1 Y_t$. Therefore from Eq. 11.13,

$$t_f = 1.25\overline{H}_0/\dot{a}_0, \tag{11.15}$$

where $\overline{H}_0 = V_0/\mathcal{A}_0$ is the original average thickness of the ice cap. This time scale matches t_r determined from Eq. 11.11 for the same geometry, if the specific balance does not vary with altitude ($\gamma = 0$).

A data compilation for 144 mountain glaciers gave the relationship $V \propto \mathcal{A}^{1.36}$ (Bahr et al. 1997). In this case

$$t_f = 1.36\overline{H}_0/\dot{a}_0, \tag{11.16}$$

which again resembles $t_r = H_{MAX}/\dot{a}_0$ at this low level of precision.

Importance of the Altitude-mass-balance Feedback   This feedback – corresponding to values $\gamma > 0$ in Eq. 11.11 – increases the accumulation on a glacier that is growing and increases the ablation on a glacier that is shrinking. It reduces the denominator in the formula for t_r (Eq. 11.11), since $\gamma > 0$, and so increases the response time (Oerlemans 2001, p. 97; Harrison et al. 2001). The increase may be significant; numerical simulations of several typical mountain glaciers suggest it increases t_r by a multiple of 2 to 5 (Oerlemans 2001, pp. 89 and 98). The effect is larger for glaciers with small slopes than for steep ones (with reference to Eq. 11.11, a smaller slope implies a larger factor η).

Inspection of Eq. 11.11 shows that with a sufficiently large altitude-mass-balance feedback the response time switches to a negative value. This means that a small increase of mass balance triggers unlimited growth of the glacier rather than an asymptotic approach to a new steady configuration (Bodvarsson 1955). The glacier advance sustains itself because the increase of total accumulation as the surface rises and cools exceeds the increase of total ablation as the terminus advances. Such a process may lead to the initiation of new ice sheets. This unstable response requires that $\overline{H}_o / \dot{a}_0 > \omega / \gamma \eta$, a plausible condition for a polar ice cap (large $\overline{H}_o$, small $\dot{a}_0$) but not for a steep mountain glacier (small η).

11.2.3 Simple Models for Different Zones

Our discussion of response times so far has not considered explicitly the fact that glaciers move; flow was assumed sufficiently rapid to maintain the profile in a characteristic shape. Nor have we considered the possibility that different parts of a glacier can be perturbed in different ways by the same climate change. When climate warms, both melt and snowfall often increase. The mass balance perturbation can vary from increased net ablation near the terminus to increased net accumulation high on the glacier. We should expect such differences on ice sheets and large glaciers. Even moderate-sized mountain glaciers sometimes change this way (Sapiano et al. 1998; see Section 14.2).

Simple models again illustrate important aspects of the response.

11.2.3.1 Response of Accumulation Zone to Small Changes: A Simple Model

Consider a section of accumulation zone, of unit width, extending from the glacier head at $x = 0$ to an arbitrary $x = X$ (Figure 11.3a). For simplicity, regard the specific balance in the section as uniform. At steady state, the total accumulation, $\dot{b}_0 X$, equals the ice flux $q_0(x) = u_0(x)H_0(x)$ at $x = X$. By flow mechanics, flux also depends on ice thickness H and surface slope α. For example, with flow by internal deformation, $q = K H^{n+2} \alpha^n$ (Section 8.3.1). Suppose that, from an initial steady state, the accumulation increases uniformly and permanently by a small amount, from $\dot{b}_0$ to $\dot{b}_0 + \dot{b}_1$ (in this case, not a function of x). The glacier thickens, so $H(x, t) = H_0(x) + H_1(x, t)$, which in turn increases the flux to $q(x, t) = q_0(x) + q_1(x, t)$. Ice volume (per unit width) in the section changes according to the imbalance of accumulation and flux, at rate $dv_1/dt = \dot{b}_1 X - q_1(X)$. Here v_1 denotes the perturbation of ice volume in the section.

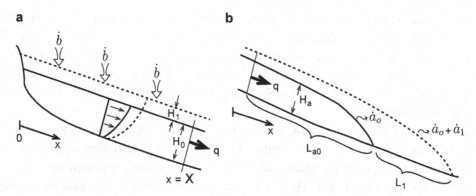

Figure 11.3: Schematic models for: (a) the response of a glacier's accumulation zone to an increase of mass balance; (b) the response of a glacier's lower ablation zone to increased flux or decreased ablation.

Suppose first that the increase of thickness is uniform, $H_1(t)$, so the slope does not change. Thus take $dv_1/dt = X dH_1/dt$ and $q(X, t) = K\alpha_0^n(X)[H_0(X) + H_1(t)]^{n+2}$. For H_1 small compared to H_0, then $q_1(X, t) \approx [n + 2]K\alpha_0^n(X)H_0^{n+1}(X)H_1(t)$. But $K\alpha_0^n(X)H_0^{n+1}(X)$ is simply $q_0(X)/H_0(X)$, which is $\dot{b}_0 X/H_0(X)$. Combining relations and dividing through by X gives an equation for the change of ice thickness (Whillans 1981):

$$\frac{dH_1(t)}{dt} + \frac{1}{t_b}H_1(t) = \dot{b}_1 \quad \text{with} \quad t_b = \frac{1}{n+2}\frac{H_0}{\dot{b}_0}, \tag{11.17}$$

a relation of the same form as Eq. 11.7. Following an abrupt increase of accumulation rate, the glacier initially thickens at a rate equal to $\dot{b}_1$. According to Eq. 11.17, the thickening subsequently slows (the rate decays exponentially with time) as the glacier flux increases in proportion to $[n + 2]H_1(t)$. The ice thickness approaches a new steady value. The time scale t_b is how long the glacier takes to accomplish all but a factor $1/e$ of this change. With $n = 3$, $t_b = H_0/5\dot{b}_0$. Table 11.2 gives some characteristic values, which are similar to the response times t_r for the same glaciers. This similarity should not be misinterpreted; t_b and t_r are not different ways to assess the same response time (for example, t_b can be defined for the East Antarctic Ice Sheet, which has no ablation zone, whereas t_r cannot). The similarity of t_b and t_r instead expresses a climatologically determined relationship between typical rates of ablation and accumulation.

The time t_b provides a rough estimate for how quickly the upper region of a glacier, or the interior of an ice sheet, thickens in response to a snowfall increase (ignoring any effect of ice flow on the temperature). It also describes how quickly the extra snowfall leads to increased transport of ice to the ablation zone.

What if the glacier's surface slope also changes? An increased slope – which must occur if the location of the terminus is fixed by a calving margin – makes the response faster. With a fixed terminus and glacier length L_o, the slope perturbation would scale as $\alpha_1 \sim H_1/L_o$, decreasing the time scale to $[H_0/\dot{b}_0]/[2n + 2] \approx H_0/8\dot{b}_0$. If the terminus advances a great distance, on the

Table 11.2: Accumulation zone response times estimated from
Eq. 11.17.

	Thickness (m)	Accumulation ($m\,yr^{-1}$)	Response time t_b (yr)
Glaciers in temperate maritime climate	150–300	2–4	7–30
Ice caps in Arctic Canada	500–1000	0.5–1	100–400
Greenland Ice Sheet	3000	0.3	2000
East Antarctic Ice Sheet	3500	0.05	14,000

other hand, slope decreases and the response is slower. Because the slope decreases as the terminus advances, the accumulation zone cannot reach a new equilibrium more rapidly than the time for the terminus to find its new position (as implied by the analysis of Section 11.2.2.2).

11.2.3.2 Response of Ablation Zone to Small Changes: A Simple Model

Now consider the lower ablation zone of a glacier terminating on land (Figure 11.3b). For simplicity, again suppose a uniform width. An ice flux $q(t)$ enters the region from upstream, initially with a datum value q_0, but $q_0 + q_1(t)$ thereafter. Call the initial length of the region L_{ao}, and the initial ablation rate near the terminus $\dot{a}_0$. Suppose that the ablation changes, uniformly, so that $\dot{a} = \dot{a}_0 + \dot{a}_1(t)$. The changes $\dot{a}_1(t)$ and $q_1(t)$ determine a change in the ice volume per unit width in the region, of rate dv_1/dt. For small changes, dv_1/dt equals the sum of three terms: gains due to the extra flux from upstream ($q_1(t)$), losses due to lengthening of the ablation zone ($-\dot{a}_0 L_1(t)$), and losses due to increased ablation rate ($-\dot{a}_1(t)L_{ao}$). The volume change $v_1(t)$ covaries with changes of terminus position, defining a thickness scale H_a according to $H_a = dv_1/dL_1$. The value for H_a depends on exact geometry but must be similar to the mean glacier thickness in the ablation zone. Combining expressions gives a relation for terminus advance $L_1(t)$, another linear equation:

$$\frac{dL_1(t)}{dt} + \frac{1}{t_a}L_1(t) = \frac{q_1(t)}{H_a} - \frac{L_{ao}\dot{a}_1(t)}{H_a} \quad \text{with} \quad t_a = \frac{H_a}{\dot{a}_0}. \tag{11.18}$$

As the right-hand side of this equation indicates, a glacier advances or retreats either because the ablation rate on the lower glacier changes ($\dot{a}_1$) or because more or less ice is carried to the lower glacier by flow (q_1). (The models leading to Eqs. 11.7 and 11.11 are special cases in which the specific balance changes by the same amount everywhere on the glacier and the flow changes over time in just the right way to maintain the steady-state profile.)

An abrupt increase of flux, from q_0 to a new constant $q_0 + q_1$, causes the terminus to advance, initially at rate q_1/H_a. The rate of advance then declines exponentially with a timescale t_a, similar to the response time t_r (Eq. 11.12). An abrupt increase of flux is an unusual scenario, however.

If the accumulation increases suddenly over the entire glacier, it takes time for the mass added to the upper glacier to arrive at the front by ice flow. Thus q_1 initially increases over time. It does not fully reflect the new accumulation ($q_1 = \dot{b}_1 L_o$) until after a time equal to a few multiples of t_b, the time scale for thickening of the accumulation zone (Eq. 11.17). Moreover, q_1 continues to increase with time if the altitude-mass-balance feedback operates; this is why the feedback increases the time scale t_r in Eq. 11.11.

If warming causes a sudden increase of ablation near the terminus, with magnitude $\dot{a}_1$, the glacier front immediately begins to retreat at rate $\dot{a}_1 L_{ao}/H_a$ (assuming the glacier was initially in steady state). Likewise, cooling or increased snowfall causes an immediate advance. In general, the time when the effect of a mass balance change *first* appears at the glacier terminus is therefore, as Eqs. 11.7 and 11.11 also imply, much less than the response time t_r.

11.3 Reaction to Changes of Mass Balance: Dynamics

11.3.1 Theoretical Framework

An increase of annual snowfall or a decrease of melt adds extra mass to a glacier. Flow carries the extra mass downstream. But a glacier does not simply act like a conveyor belt, passively transporting material on it; instead, the added mass increases stresses within the ice and drives faster flow. Such flow perturbations, together with the preexisting pattern of velocities, nonuniformly redistribute the mass as it moves down the glacier.

Modern analyses of the general response of a glacier to a change in its mass balance began when Weertman (1958) and Nye (1958a) rediscovered effects pointed out some 50 years earlier by de Marchi, Finsterwalder, and Reid. Nye (1959, 1960, 1963b, 1963c, 1965b) greatly developed the theory; we outline the most important parts of his work in Sections 11.3.1 and 11.3.2. Numerical analyses have largely supplanted use of the solutions worked out by Nye – a development begun by Nye himself – but the older analyses still serve as an essential conceptual framework for understanding the dynamics of glacier response.

Let x denote distance along a glacier, and t denote time. Let $\mathcal{A}_c$ be the cross-sectional area measured perpendicular to the x-axis, Y the width at the surface, and H and u the thickness and velocity on the centerline. Let $\overline{b}$ denote the specific mass balance averaged across the glacier at x and $\dot{b}$ its value at the centerline. (Dimensions of $\dot{b}$ and $\overline{b}$ are length per unit time.) All of these quantities are functions of x and t. Density of the glacier is assumed to be the same everywhere.

The equation of mass conservation for a glacier cross-section takes the form

$$\frac{\partial \mathcal{A}_c}{\partial t} = \overline{b} Y - \frac{\partial Q_*}{\partial x} \tag{11.19}$$

$$\text{with} \quad Q_* = \int_Y q \, dy = \overline{u} \mathcal{A}_c. \tag{11.20}$$

Here Q_* denotes the volumetric flux, the volume of ice passing, in unit time, through the cross-section at x. To analyze glacier response, a relation for Q_* is inserted into Eq. 11.19. Notice that Q_* equals the product of cross-sectional area and mean velocity.

For clarity and simplicity we will consider the case of a glacier or flow band of uniform width. In this case, the flux through the entire cross-section, Q_*, can be replaced by the flux per unit width on the glacier centerline, $q = uH$. (We will not distinguish here between velocity at the surface and its average over depth.) Equation 11.19 reduces to (Eq. 8.79):

$$\frac{\partial H}{\partial t} = \dot{b} - \frac{\partial q}{\partial x}. \tag{11.21}$$

The velocity depends on the mechanism of flow. In terms of the basal shear stress at the centerline, τ_b:

$$u = K'_d \tau_b^n H + K'_b \tau_b^m \tag{11.22}$$

where n and m are constants. The first term represents differential movement within the ice: the second, slip of the glacier on its bed. The parameter $n \approx 3$ but the effective value for m is not well defined. For slow sliding on a hard bed, theory gives $m \approx 2$ to 3. The coefficients K'_d and K'_b may vary along the glacier; the former depends on ice viscosity, the latter on slipperiness of the bed. Taking the basal shear stress as proportional to the driving stress ($\tau_b = f'\tau_d = f'\rho g H \alpha$; see Chapter 8), the velocity is

$$u = K_d H^{n+1} \alpha^n + K_b H^m \alpha^m, \tag{11.23}$$

where the coefficients K_d and K_b (which vary with both x and t in general) now include f', ρ, and g. Because of longitudinal forces, variations of f' over short distances depend on H and α (see Section 8.7). Nye did not include this effect in his analyses. We will ignore it, too; recent studies using numerical methods indicate that longitudinal-force effects have little influence on glacier response in typical situations (e.g., Leysinger-Vieli and Gudmundsson 2004). Note that for fast sliding on a slippery bed, a relation $u = K_b \alpha^m$ is in some cases more appropriate than one with a dependence on H (Section 8.6.1).

Finally, the surface slope α, the surface elevation S, and the bed slope β are related by

$$\alpha = -\frac{\partial S}{\partial x} = \beta - \frac{\partial H}{\partial x}. \tag{11.24}$$

Together with the relation $q = uH$, Eqs. 11.21, 11.23, and 11.24 provide a framework for analyzing the reaction of glaciers to changes in mass balance and flow parameters. All variables in the following discussion should be regarded as varying in both x and t, unless otherwise specified.

11.3.1.1 The Relation for Discharge Variations

For now we suppose that K_d and K_b are constant at a given x; thus q at a given x varies only if H or α varies. Then

$$\frac{\partial q}{\partial t} = \frac{\partial q}{\partial H}\frac{\partial H}{\partial t} + \frac{\partial q}{\partial \alpha}\frac{\partial \alpha}{\partial t} \tag{11.25}$$

Differentiating Eq. 11.21 with respect to x and substituting the result and Eq. 11.21 itself into Eq. 11.25 give an equation for q:

$$\frac{\partial q}{\partial t} = C\dot{b} - D\frac{\partial \dot{b}}{\partial x} - C\frac{\partial q}{\partial x} + D\frac{\partial^2 q}{\partial x^2} \tag{11.26}$$

$$\text{with}\quad C = \frac{\partial q}{\partial H}\quad\text{and}\quad D = \frac{\partial q}{\partial \alpha}, \tag{11.27}$$

where the derivatives defining C and D are calculated at a fixed value of x. This is an advection-diffusion equation; variations of ice flux shift along-glacier at a rate C, and they spread upstream and downstream at a rate determined by an effective diffusivity D. Accumulation is a "source" that increases the flux, either by increasing the ice thickness, if $\dot{b} > 0$, or by increasing the slope, if $\partial \dot{b}/\partial x < 0$. The coefficients C and D vary along the glacier because they depend on K_b, K_d, H, and α.

The next several sections explore the implications of this relation. First we discuss the advective and diffusive movement of flux variations along the glacier. Then we recast the problem to examine thickness variations and consider the response to small changes of mass balance. Finally, we discuss the use of numerical methods, a necessary approach for applications to specific glaciers.

To account for variations of glacier width, the derivation can be repeated to find a relation for Q_*, the total flux (volume per unit time) at x. In the simplest case of a flow band or valley with vertical walls, $A_c = HY$ and Y varies with x but not t. The formula is then

$$\frac{\partial Q_*}{\partial t} = C_*\overline{b} - D_*\frac{\partial \overline{b}}{\partial x} - \left[\frac{C_*}{Y} + \frac{D_*}{Y^2}\frac{\partial Y}{\partial x}\right]\frac{\partial Q_*}{\partial x} + \frac{D_*}{Y}\frac{\partial^2 Q_*}{\partial x^2}, \tag{11.28}$$

with $C_* = \partial Q_*/\partial H$ and $D_* = \partial Q_*/\partial \alpha$. We will not consider the role of variable width further.

11.3.1.2 Kinematic Waves

To illustrate one important aspect of the glacier response, we start by making an unrealistic assumption; we disregard the dependence of q on α in Eq. 11.26. Hence $D = 0$. The coefficient C is $\partial q/\partial H$ evaluated at a location x (Eq. 11.27). But $q = uH$. Thus

$$C = u + H\frac{\partial u}{\partial H} \tag{11.29}$$

and so, by Eq. 11.23,

$$C = [n+2]u_d + [m+1]u_b,$$ (11.30)

where u_d, u_b denote the velocities from ice deformation and slip. As n is about 3, and m is supposed to be about 2 or 3 for hard-bed sliding, C nominally ranges from about 3 to 5 times u according to the relative sizes of u_d and u_b. This statement applies to a hard-bed glacier. For a soft-bed glacier moving only by slip, the term $\partial u / \partial H$ may be as small as zero, in which case C equals the ice velocity.

From Eq. 11.26

$$\frac{\partial q}{\partial t} = C\dot{b} - C\frac{\partial q}{\partial x}.$$ (11.31)

If $\dot{b} = 0$, Eq. 11.31 represents a "wave" on which q is constant, travelling down the glacier with velocity C. (If C is constant then $q = F(x - Ct)$, with F an arbitrary function, would be a solution.) This is an example of a *kinematic wave*, a type of motion studied by Lighthill and Whitham (1955). Their theory had already been applied to problems of flood waves on rivers and traffic flow on roads before Nye applied it to ice flow in glaciers. In the present sense, the word *wave* does not imply a travelling wave train, but merely a set of points moving with velocity C, different from the ice velocity, and carrying with them a particular property; in this case each point has a value for q that remains constant (if $\dot{b} = 0$) or that changes down-glacier at a rate $\dot{b}$ following a point moving with velocity C. In a glacier, the wave velocity C is generally greater than the ice velocity because velocity normally increases with ice thickness. In traffic flow, by contrast, an increase in the concentration of vehicles reduces their speed, and so the wave velocity is less than the vehicle velocity.

Ice velocity varies along a glacier, and therefore C also varies with x. For example, in mountain glaciers C normally attains a maximum near the equilibrium line. In the accumulation zone, where $\partial u / \partial x$ tends to be positive, the set of points defining a wave spreads because C increases with x.

Kinematic waves are one way that a glacier transfers fluctuations of mass toward its terminus. Have such waves been observed in glaciers? There is a component of kinematic wave behavior any time fluctuations of mass balance change the ice thickness. This component is difficult to separate from other aspects of the glacier response. There are, however, observations showing glaciers speeding up as they thicken, and slowing down as they thin, from decade to decade. This is the essence of kinematic wave behavior. Figure 8.3 shows one such case.

11.3.1.3 Diffusive Response

Changes of the surface slope α give rise to the following part of Eq. 11.26:

$$\frac{\partial q}{\partial t} = D\frac{\partial^2 q}{\partial x^2} - D\frac{\partial \dot{b}}{\partial x},$$ (11.32)

a diffusion equation like the one governing temperature change near a glacier's surface (Eq. 9.6). The diffusivity coefficient D is $\partial q/\partial \alpha$, or, given $q = uH$, $D = H\partial u/\partial \alpha$. From Eq. 11.23:

$$D = \frac{H}{\alpha}[nu_d + mu_b]. \tag{11.33}$$

As with the wave velocity, D should be maximum near the equilibrium line on a mountain glacier, where flow is fastest and ice thickest. In an ice sheet, however, D is high throughout the broad central accumulation region, where surface slopes are always small. Equation 11.32 describes a tendency for variations of flux to spread both upstream and downstream.

Diffusion and kinematic wave behaviors act together to determine the glacier response (see also Section 11.3.3). Consider a brief period, perhaps a few years, of unusually large snowfall that blankets a glacier with a uniform layer of extra mass. Figure 11.4 depicts the reaction of a hypothetical hard-bed mountain glacier; the figure plots the discharge perturbation – the excess of discharge over its original steady-state value – calculated from Eqs. 11.21 through 11.24 with a numerical method. The numbers by each curve indicate the years elapsed since the large snowfall event. (The profile of the model glacier is shown later, in Figure 11.7.) With a uniform added thickness, the discharge q immediately increases all along the glacier (curve for time zero), but with a maximum near the equilibrium line where the ice flows fastest. This maximum of "perturbed discharge" then moves down-glacier as a wave, faster than the ice velocity. The transport of mass to the terminus not only increases, but does so more quickly and with greater peak magnitude than if the extra snowfall simply rode along the glacier without changing the flow. The terminus advances temporarily (the wave provides a flux forcing q_1 in

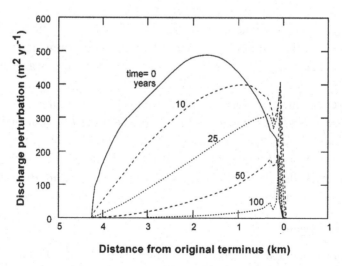

Figure 11.4: Computed evolution of a discharge perturbation on a hypothetical mountain glacier (depicted in Figure 11.7). At time zero, a thin layer of uniform thickness was added to the entire glacier surface.

Eq. 11.18). The peak of discharge moves to the terminus at a rate of about six times the ice velocity; this is faster than the pure kinematic wave speed because of diffusion.

What role does diffusion have in the propagating "wave" behavior seen in the figure? Initially, flux increases most near the equilibrium line. Upstream, the glacier stretches and thins; downstream, there is compression and thickening (Eq. 11.21). With this pattern, surface slopes change such that the flux is reduced in the middle of the glacier but increased near the glacier head and terminus. The flux cannot increase very much on the upper part of the glacier, however, because it is fixed at zero at the head. In contrast, flux can increase significantly near the terminus, where extra mass arriving from up-glacier thickens and steepens the ice. The net effect of diffusive spreading is therefore an increased rate of transfer of mass to the terminus, just as with a kinematic wave.

Note that diffusion permits an increase of flux originating near the terminus to propagate *upstream*. This occurs, for example, when basal slip increases near the terminus or when the terminus steepens as a calving front retreats. Kinematic waves of flux, in contrast, cannot propagate in the direction opposite to the ice flow. The distance over which a flux variation spreads by diffusion in a time interval Δt scales with $\sqrt{D\Delta t}$; this is a general property of diffusive systems.

Furthermore, recall that some glaciers flow entirely by slip on a weak bed. In the limit of negligible basal drag, the flow increases with surface slope but is independent of thickness (Eq. 8.86). In such a case, perturbations of flux are still redistributed by diffusion, but the kinematic wave component simply represents downstream movement of thickness variations, at the same rate as the ice flow.

11.3.2 Ice Thickness Changes

If we assume, again, that the coefficients K_d and K_b in the velocity relation do not vary in space or time, then $\partial q/\partial x$ in Eq. 11.21 can be expanded as $[\partial q/\partial H][\partial H/\partial x]+[\partial q/\partial \alpha][\partial \alpha/\partial x]$. In addition, we write the bed slope in terms of basal elevation B, so that $\beta = -\partial B/\partial x$, and use Eq. 11.24. The result shows that changes of ice thickness H over time obey an advection-diffusion equation similar to Eq. 11.26,

$$\frac{\partial H}{\partial t} = \dot{b} - C\frac{\partial H}{\partial x} + D\frac{\partial^2 H}{\partial x^2} + D\frac{\partial^2 B}{\partial x^2}. \tag{11.34}$$

Specific balance $\dot{b}$ is the "source" that drives ice thickness variations (here, again, dimensions of $\dot{b}$ are length per unit time). Topography of the bed influences ice thickness; ice will fill a depression, where the curvature $\partial^2 B/\partial x^2$ is positive. C and D again represent kinematic wave speed and diffusivity, equal to the derivatives $\partial q/\partial H$ and $\partial q/\partial \alpha$, respectively. Because C and D are functions of H and α, Eq. 11.34 expresses a complicated relationship between glacier geometry and its evolution. The qualitative behavior can most readily be understood by examining the response to changes of small magnitude. Nye elaborated this approach in detail.

11.3.2.1 Response of Ice Thickness to Small Changes of Mass Balance

Consider a glacier at steady state – the datum state defined in Section 11.2. If the specific balance changes by a small amount, $\dot{b}_1$, the thickness and flux also change relative to their datum values. For small deviations, the differential equations can be linearized. Again, express quantities as perturbations (suffix 1) from their values in the datum state (suffix 0):

$$q = q_0 + q_1 \quad b = b_0 + b_1 \quad \alpha = \alpha_0 + \alpha_1 \quad H = H_0 + H_1. \tag{11.35}$$

All of these terms vary with distance along the glacier, and the perturbation terms also vary with time. The conservation relation for the perturbations is (recalling that we are analyzing the case of uniform width)

$$\frac{\partial H_1}{\partial t} = \dot{b}_1 - \frac{\partial q_1}{\partial x}. \tag{11.36}$$

At any given x, expand q_1 in terms of H_1 and α_1:

$$q_1 = C_0 H_1 + D_0 \alpha_1 \tag{11.37}$$

$$\text{with} \quad C_0 = [\partial q / \partial H]_0 \quad \text{and} \quad D_0 = [\partial q / \partial \alpha]_0. \tag{11.38}$$

Further, taking bed slope as constant in time, $\alpha_1 = -\partial H_1 / \partial x$. Thus

$$q_1 = C_0 H_1 - D_0 \frac{\partial H_1}{\partial x}. \tag{11.39}$$

Equations 11.36 and 11.39 are two simultaneous partial differential equations for determining the increases in flux $q_1(x, t)$ and thickness $H_1(x, t)$ due to a given increase $\dot{b}_1(x, t)$ of specific balance.

A relation is needed for the ice flux. For simplicity we will here consider a glacier flowing only by internal deformation; this reduces the number of terms in the following relations without changing the essential behaviors. Then in the datum state, $q_0 = K_d H_0^{n+2} \alpha_0^n$, which is differentiated to find C_0 and D_0. Substitution into Eq. 11.36 gives the relation for thickness change

$$\frac{\partial H_1}{\partial t} = \dot{b}_1 - \frac{\partial}{\partial x} [[n+2] u_0 H_1] + \frac{\partial}{\partial x} \left[\frac{n q_0}{\alpha_0} \frac{\partial H_1}{\partial x} \right]. \tag{11.40}$$

Expanding the derivatives and rearranging yields

$$\frac{\partial H_1}{\partial t} + [n+2]\dot{\epsilon}_0 H_1 = \dot{b}_1 - \left[C_0 - D_0'\right]\frac{\partial H_1}{\partial x} + D_0\frac{\partial^2 H_1}{\partial x^2} \tag{11.41}$$

$$C_0 = [n+2]u_0 \qquad D_0 = \frac{nq_0}{\alpha_0} \qquad \dot{\epsilon}_0 = \frac{\partial u_o}{\partial x}.$$

Here $\dot{\epsilon}_0$ signifies the longitudinal strain rate and D_0' the derivative of D_0 with respect to x. (For a glacier flowing entirely by sliding on a hard bed, $[n+2]$ would be replaced by $[m+1]$, and nq_0 by mq_0.) Equation 11.41 shows the different aspects of a glacier's thickness response to a mass balance perturbation:

1. *General Thickness Change.* Increased mass balance ($\dot{b}_1 > 0$) thickens a glacier and increases the flow. The relation is $\partial H_1/\partial t + [n+2]\dot{\epsilon}_0 H_1 = \dot{b}_1$. How a section of glacier responds to a change in thickness therefore depends on whether flow is extending or compressing; the sign of the coefficient of H_1 determines the style of response. Where flow is extending, a typical state in the accumulation zone, $\dot{\epsilon}_0 > 0$ and the thickness perturbation H_1 approaches a steady final value. This case was analyzed in Section 11.2.3.1, where the same relation was derived from a simple argument.

2. *Instability of Region of Compressing Flow.* Where flow is compressing, the typical state in the ablation zone, $\dot{\epsilon}_0 < 0$ and the thickness perturbation grows exponentially. This unstable response can be understood by considering a small length δx of the ablation zone (Figure 11.5). In a steady state, the flux through the upstream side AA' must exceed the flux through the downstream side BB' by an amount equal to the ablation. If the flux through AA' is $u_0 H_0$, that through BB' is $u_0 H_0 + \dot{b}_0\delta x$ (and $\dot{b}_0 < 0$).

    Now suppose that a layer of thickness H_1 is suddenly placed on the surface. The velocity increases to $u_0 + u_1$. The flux increases by $u_0 H_1$ because of the increase in thickness and by $u_1 H_0$ because of the increase in velocity. But u is proportional to H^{n+1} and so $u_1/u_0 = [n+1]H_1/H_0$. It follows that the flux through AA' has increased by $[n+2]u_0 H_1$; thus

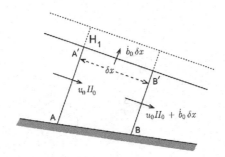

Figure 11.5: Instability of region of compressing flow. Adapted from Nye (1960).

the flux through a section increases in proportion to the steady-state velocity there. With compressing flow, the steady-state velocity at BB' is less than at AA'. Thus the increase in flux through BB', as a result of adding the layer H_1 to the surface, is less than the increase in flux through AA'. Ice therefore accumulates between the two sections and the surface level must rise further. This shows that a region of compression is unstable. However, wave propagation and diffusion will prevent the growth of a pronounced bulge.

3. *Kinematic Wave Propagation.* Because $\partial H_1/\partial t$ depends on $\partial H_1/\partial x$, a thickness perturbation moves along-glacier as a kinematic wave (Section 11.3.1.2). The propagation speed, $[n+2]u_0 - D_0'$, equals the speed for the kinematic wave of discharge if D_0' is small; thus kinematic waves travel faster than ice flow by a factor of about 3 to 5. The term D_0' tends to increase the rate of wave propagation in the ablation zone and decrease it in the accumulation zone.

4. *Diffusion.* The dependence of q on surface slope results in the diffusion of thickness anomalies (as for discharge anomalies; Section 11.3.1.3). Over time, a bulge spreads up- and down-glacier and its amplitude decreases. Thickness changes near a glacier margin propagate up-glacier as a diffusive wave; in an ice sheet these may propagate all the way to the center (Section 11.4.2.2).

    The diffusive behavior of the glacier thickness is most clearly illustrated by considering a bulge; in Figure 11.6, ice thicknesses at A and B are equal, but the surface rises between the two points. The bulge moves down-glacier as a kinematic wave. The slope at B is greater than at A. Because velocity increases with slope, q at B is greater than q at A. The surface elevation between A and B will therefore decrease. The bulge of elevated H and q between A and B is thus diminished; the kinematic wave is diffused. Diffusion moves the leading edge of the wave down-glacier at a velocity higher than the kinematic wave velocity. It therefore shortens the time for thickness perturbations on the glacier to affect the terminus. But, by lengthening the wave, it delays the arrival of the trailing edge at the terminus and thus increases the response time.

    In Figure 11.6, let $\alpha_0 - \alpha_1$ be the surface slope and $u_0 - u_1$ the velocity at A, with $\alpha_0 + \alpha_1$ and $u_0 + u_1$ the corresponding values at B. The ice thickness at both points is H_0 and the

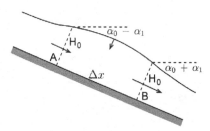

Figure 11.6: Diffusion of a bulge.

distance between them is Δx. Because $u \propto \alpha^n$, then $u_1/u_0 = n\alpha_1/\alpha_0$. But the longitudinal strain rate approximately equals $2u_1/\Delta x$ and so, in plane strain, the vertical strain rate equals $-2u_1/\Delta x$. The vertical velocity at the surface is therefore

$$\frac{-2u_1 H_0}{\Delta x} = \frac{-2n H_0 \alpha_1 u_0}{\alpha_0 \Delta x}. \tag{11.42}$$

But the surface curvature $\partial^2 H_1/\partial x^2 = 2\alpha_1/\Delta x$, and so

$$\frac{\partial H_1}{\partial t} = \frac{n H_0 u_0}{\alpha_0} \frac{\partial^2 H_1}{\partial x^2}, \tag{11.43}$$

another diffusion equation like Eq. 11.32. The preceding argument gives a mechanistic interpretation for why the diffusion term appears in Eq. 11.41, with the diffusion coefficient $D_0 = nq_0/\alpha_0$. The expression for D_0 shows that diffusion operates more strongly on thick, gently sloping, and fast-flowing glaciers than on thin, steep, and slow ones.

11.3.2.2 Combined Response to a Step Increase of Mass Balance

How a glacier thickens and advances when its mass balance increases depends on the combination of the behaviors discussed in the previous section. The heavy lines in Figure 11.7 show the bed and surface profiles of an idealized mountain glacier at steady state. The specific balance varies from $+1\,\mathrm{m\,yr^{-1}}$ in the upper part to $-2.5\,\mathrm{m\,yr^{-1}}$ at the terminus. We first calculated the steady-state profile by holding the specific balance constant and using a numerical method to solve Eqs. 11.21 through 11.24. Next we increased the specific balance uniformly over the glacier, and held it constant at these higher values. (A similar scenario was modelled by van de Wal and Oerlemans 1995, Figure 4). Figure 11.7 shows the glacier profile after 20, 50, 100, and 300 years of increased specific balance. The glacier initially thickens everywhere at the same rate, but its speed and flux also increase. The smallest thickening occurs at the head of the glacier, where longitudinal stretching is most rapid; the largest thickening occurs in the ablation zone. This pattern forms a "bulge" of new thickness, greatest near the terminus and tapering toward the head of the glacier. The glacier front advances, driven by increasing thickness immediately upstream. At all times, thickness changes are spread smoothly along the glacier.

As with the analytical theory described previously, this simulation simplified the flow dynamics by assuming that $\tau_b \propto H\alpha$. Would the results change significantly if we incorporated the complete relations with all stress components? Leysinger-Vieli and Gudmundsson (2004) addressed this question with a full-stress model. Their results showed that application of the more sophisticated model does not lead to any substantial changes in the pattern of terminus advance. In particular, they found that details of flow at the snout do not have much influence on the advance – a conclusion that contradicted some earlier, simpler analyses.

Figure 11.8 summarizes observations of ice thickness changes over several decades along three mountain glaciers that retreated in the twentieth century (Schwitter and Raymond 1993).

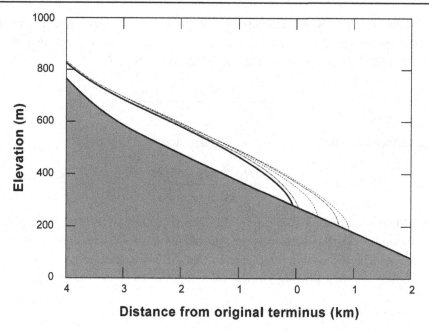

Figure 11.7: Computed advance of a hypothetical glacier following a permanent increase of mass balance. Solid lines show the original, steady-state longitudinal profile. Broken lines show the profile after 20, 50, 100, and 300 years.

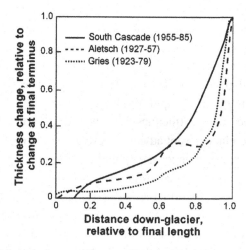

Figure 11.8: Observed thickness changes versus distance along three retreating glaciers, from multidecade measurements. Distance has been normalized to the glacier length at the end of the observation period. Thickness changes have been normalized to the value at the terminus at the end of the observation period. From Schwitter and Raymond (1993).

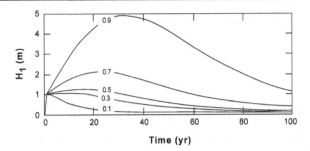

Figure 11.9: Computed response of Berendon Glacier to a unit impulse. The specific balance perturbation was taken as $1\,\mathrm{m\,yr^{-1}}$ from $t = 0$ to $t = 1$ year and zero at other times. The increase in thickness is H_1. The number on each curve is the fractional distance along-glacier, x/L, measured from the head. Adapted from Untersteiner and Nye (1968).

The shape of the curves, tapering toward the head of the glacier, is broadly consistent with the model.

11.3.2.3 Combined Response to an Impulse of Mass Balance

Suppose that a thin, uniform layer of ice is added instantaneously to the entire glacier surface; in other words, the specific balance abruptly rises by an amount b_1 everywhere on the glacier but quickly returns to its original value. The discharge response to such a change was shown in Figure 11.4. The thickness perturbation $H_1(x, t)$ produced by such an event, if applied at time zero and of unit amplitude, defines the *impulse response function* (Nye 1965c). Impulse response functions can be estimated for a glacier using numerical solutions to Eq. 11.41. Figure 11.9 depicts the calculated impulse response functions of Berendon Glacier in western Canada. At points near the head of the glacier the initial increase in thickness begins to decay at once; lower on the glacier a flood wave first builds up and then subsides.

In reality, specific balance on a glacier varies in an irregular fashion over time, though there may be trends. The glacier's response can be understood, conceptually, from its impulse response function; the thickness depends on the sum of individual impulse responses to the year-by-year variations of specific balances. If $\dot{b}_1(t)$ denotes the specific balance perturbations and $e(x, t)$ the impulse response function, the thickness perturbation varies approximately as

$$H_1(x, t) = \int_0^\infty e(x, \tau)\dot{b}_1(t - \tau)d\tau; \tag{11.44}$$

in other words, thickness changes depend on the convolution of mass balance variations with the impulse response. This method has been applied to predict glacier advance (Untersteiner and Nye 1968) and to deduce mass balance histories from observed changes in terminus position

(Nye 1965c). Though a useful guide to understanding glacier behavior, Eq. 11.44 is quantitatively accurate only if changes are small enough to justify the linearization. For most calculations, numerical flow models must be used (Section 11.3.4).

11.3.3 Relative Importance of Diffusion and Kinematic Waves

The relative importance of advective (kinematic wave) and diffusive behaviors in glacier response can be assessed, in a rough fashion, from the Peclet number (Pe). A Peclet number is a dimensionless parameter comparing the rate of advective movement to the rate of diffusion. Specifically, Pe equals the ratio of advective velocity to the diffusivity, multiplied by a characteristic length; for a glacier, $Pe = C_0 L / D_0$, or, for motion by internal deformation,

$$Pe = \frac{[n+2]qH^{-1}L}{nq\alpha^{-1}} \approx \frac{\alpha}{H}L. \tag{11.45}$$

The approximation is also valid for motion by hard-bed sliding, as the ratio $[m+1]/m$ is also about one.

First consider a perturbation affecting a large fraction of the length of a glacier or ice sheet flow line. Taking L to be the entire length, the mean slope is $\alpha = H/L$ for a flat bed and hence $Pe \sim 1$. Such a value indicates that both diffusion and wave behaviors are important. However, α is smaller than its mean over the large interior areas of ice sheets and the central sections of large glaciers. Here $Pe < 1$ and diffusion dominates. In contrast, some mountain glaciers are thin compared to the total elevation drop along their beds. In this case, $H/L \ll \alpha$, $Pe > 1$, and the wave behavior is most important (or, to be precise, changes of thickness play a larger role in the response than changes of slope).

Now consider a local bulge on a glacier. The most appropriate length scale L is only the length of the bulge, so $Pe < 1$ and diffusive smoothing should dominate. Detailed analyses support this conclusion. Gudmundsson (2003) used a numerical glacier model incorporating the full stress equations (but a linear viscous approximation for ice creep) to examine the evolution of bulges originating as local perturbations of specific balance. He found that a bulge flattens rapidly in all realistic scenarios.

There are several records of bulges of increased thickness travelling down a glacier, faster than the ice. Figure 11.10 shows that such a bulge passed down the Mer de Glace at about $800\,\mathrm{m\,yr^{-1}}$ between 1891 and 1899 (Lliboutry 1958a). The ice velocity was $150\,\mathrm{m\,yr^{-1}}$. Similar bulges have been observed on Mer de Glace and the nearby Glacier des Bossons (Finsterwalder 1959) and on Nisqually Glacier in Washington State (Richardson 1973, Figure 4). Such features are sometimes referred to as kinematic waves. The persistence of the bulges, however, suggests that they are not remnants of specific-balance anomalies but instead arise from ongoing dynamical processes. Perhaps they are signatures of locally increased basal slip, originating in anomalous

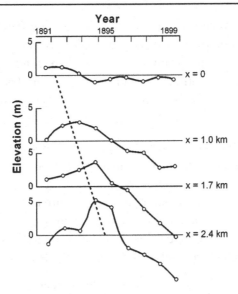

Figure 11.10: Changes of mean surface elevation of Mer de Glace, France, along four across-glacier profiles over a period of nine years. The broken line corresponds to a propagation velocity of 800 m yr⁻¹. Adapted from Lliboutry (1958a).

hydraulic conditions that propagate along the glacier bed (Lliboutry and Reynaud 1981; van de Wal and Oerlemans 1995).

11.3.4 Numerical Models of Glacier Variation

In Sections 11.3.1 and 11.3.2 we discussed the general properties of glacier response to variations of accumulation and ablation. To predict the response of an individual glacier requires numerical methods, in order to treat the complicated geometry, to allow for large perturbations, and to avoid some of the simplifying assumptions. Budd and Jenssen (1975) and Mahaffy (1976) were some of the first to tackle this problem. Bindschadler (1982) gave a concise description of flow-line models, while Oerlemans (2001, pp. 57–81) discussed in detail an approach to flow-line modelling designed for simulating the response of mountain glaciers to climate changes. The inverse problem, to deduce the variations in mass balance that produced an observed time series of terminus advance and retreat, can be treated in the same way.

11.3.4.1 Flow-line Models

Models are founded on the mass conservation relation (Eq. 11.19), which is solved at a set of points, fixed in space, along a transect from the head of the glacier to the snout. The bedrock elevation must be specified, along with initial values for ice thickness or surface elevation. The initial condition might be a steady state, calculated from Eq. 11.19, but the model glacier can be allowed to grow from an arbitrary initial configuration. Setting the flux equal to zero at the head of the glacier gives one boundary condition. Specifying the shape of the terminus leads to another; assuming the terminus is wedge-shaped, for example, fixes the terminus position as a function of the ablation rate and the flux at the last grid point. Six relations, with coefficients unique to each position along the glacier, supplement the mass conservation relation:

1. A relation for specific balance, specified as a function of altitude or distance. The relation should preferably be based on measurements. Altering the values over time then simulates a climate change. The simplest approaches are to shift the specific balance curve up or down in altitude, uniformly, or to add a uniform perturbation at all altitudes. These crude assumptions might be reasonable for small glaciers but are unrealistic for large glaciers or ice caps. An alternative is to derive the specific balance profile, or at least the equilibrium-line elevation, from a climate model (Oerlemans, 1988; Box et al. 2006).

2. A geometric relation describing how valley width – measured across the glacier surface – relates to ice thickness. The width increases as a glacier thickens, which changes the total accumulation or ablation at a given position along the flow line.

3. A similar geometric relation to describe how cross-sectional area relates to ice thickness. For this relation and the previous one, the shape of the glacier cross-section must be known or assumed.

4. A relation for the basal shear stress in terms of the driving stress and hence glacier geometry. Most models use the relation $\tau_b = f \rho g H \alpha$, in which f specifies the "shape factor" that accounts for drag by valley walls (Section 8.6.2). In a few models f is generalized to include longitudinal force coupling.

5. A relation between the flux of ice passing through the whole valley cross-section and the velocity on the flow line of the model (u). For example, $Q_* = r u \mathcal{A}_c$, where r denotes the ratio of velocity averaged over the cross-section to u at that location.

6. A relation between velocity on the flow line (either surface velocity or depth-averaged velocity) and basal shear stress, plus other variables. For example (cf. Eq. 11.22),

$$u = 2A[n+1]^{-1}\tau_b^n H + B\tau_b^m N^{-p}. \tag{11.46}$$

Here A and n are the creep parameters for ice, and parameters B, m, and p determine the rate of basal slip for a given effective pressure N (ice overburden minus basal water pressure).

Given uncertainties in the relation for u, and in the values chosen for specific balance and bed elevation, models must be calibrated by comparing results to measured velocities, ice thicknesses, and surface elevations.

Numerical models simulate glacier response with a level of realism and flexibility that idealized mathematical analyses cannot match. But do numerical models accurately simulate glaciers? Certainly not in some respects. But whether a model's failings prevent it from serving a useful purpose or not depends on the question being asked. Limitations of models include:

1. Basal slip velocities are highly uncertain. The parameterizations do not express rigorous relationships between slip and properties of basal water and sediment, since no such relations have been established and bed properties are usually unknown (Chapter 7). Observations justify defining slip rate as an inverse function of effective pressure, but the calculation of effective pressure is itself very uncertain. In some cases, the volume of basal water, not the pressure, appears to be most important. Furthermore, in yet other cases, conditions at the bed determine not the slip rate but the basal drag; then ice velocities can be calculated only if the model accounts for variations of flow across the glacier (Section 8.6.1).

2. The effective viscosity of ice depends on crystal fabrics and so may be anisotropic. In a temperate glacier, the viscosity depends on water content. Whether such factors change over time in a way that influences glacier response is not known. In addition, viscosities of basal ice layers sometimes differ greatly from typical values, but the presence of such layers cannot be established without borehole studies.

3. Bed elevations – or ice thicknesses – are poorly known for large regions of most glaciers. Ideally, a model uses measured values from radar soundings and boreholes, but in practice often must make do with estimated values based on crude assumptions – for example, that driving stress is a constant.

4. Most models neglect longitudinal strain rates and related force effects. In fact, the basal drag does not vary in direct proportion to the driving stress but rather to a longitudinal average of driving stress (Section 8.7.2). In addition, the velocity component from internal shear increases in the presence of longitudinal stretching or compression, which soften the ice (Section 8.4). To examine whether these effects influence glacier response, Leysinger-Vieli and Gudmundsson (2004) compared results from a standard flow-line model with a two-dimensional one including the full stress terms. They found that the longitudinal effects have little influence on the overall form and timing of the simulated glacier response. They may be important in special cases, however.

Because of these limitations, models have traditionally not said anything convincing about the response of glaciers with slippery beds, but the situation is improving. Moreover, many glaciers develop basal shear stresses close to $100\,\text{kPa}$; the fact that ice deforms readily at this stress level strongly constrains the characteristics of a typical glacier. Inaccuracies in model applications probably arise, for typical glaciers, more from uncertainties in specifying the mass balance than from other factors. In many respects, glaciers are slaves to the mass balance forcings; for example, the change of steady-state length does not depend on how a glacier flows, unless the mass balance changes significantly as a function of glacier thickening or thinning (Section 11.2).

11.3.4.2 Example of Application: Nigardsbreen in Advance and Retreat

For an example of the use of models, we return to the important topic of glacier advance and retreat; a model of the Norwegian glacier, Nigardsbreen, illustrates the dynamical differences between the advance and retreat phases. Nigardsbreen drains a wide accumulation basin and terminates in a narrow tongue. The surface altitude ranges from about 1900 to 350 m, with a corresponding range of specific balances from 2.5 to $-10\,\mathrm{m\,yr^{-1}}$ (see Figure 4.5a). The equilibrium line altitude is roughly 1500 m. A variety of historical sources show how the position of the glacier's terminus has changed over time. It advanced from about 1650 to 1750, and then retreated until the 1990s. The retreat totalled about 4 km. Oerlemans (1997, 2001, p. 91) has modelled these changes. The model specific balances were adjusted as a function of time in order to match the observed history of terminus changes (a technique referred to as *dynamical calibration*).

The model solves the continuity relation using a finite-differences method. The grid follows the centerline of the lower glacier but takes an arbitrary flow line across the broad upper basin. The glacier's cross-section was idealized as a trapezoid. Ice thicknesses are known at points along the lower glacier but needed to be estimated along the upper glacier from an assumed value of driving stress, 140 kPa. Small adjustments of the bed elevations were used to calibrate the model by comparison to the measured surface profile from 1988. Basal slip was calculated from a parameterization suggested by Budd (1979), which specifies that $u_{\mathrm{b}} = k\tau_{\mathrm{d}}^{3}H^{-1}$, with k a constant. Given the usual relation for internal shear, $\Delta u \propto \tau_{\mathrm{d}}^{3}H$, the parameterization implies a predominance of slip where the ice is thin. Slip and internal shear contribute equally if $H = 173\,\mathrm{m}$; thus they contribute about the same amount to the modelled flow of Nigardsbreen, on average, but slip dominates in the ice fall, at the terminus, and near the head. This parameterization for slip is not well justified on theoretical grounds but, Oerlemans found, it allows the model glacier to match observations of the surface profile and rates of advance and retreat at different times. Changing the parameter k in the slip relation leads to a large change in the surface profile unless the specific balance is simultaneously adjusted.

Figure 11.11a compares the late twentieth-century surface profile with the simulated profile at the maximum advance in 1748. The latter compares favorably with the surface profile estimated from trimlines. The length of Nigardsbreen varies strongly with changes in mass balance, because the tongue of the lower ablation zone is narrow compared with the accumulation basin (a large value for $\overline{Y}/Y_t$ in Eq. 11.2).

Figure 11.11b compares the simulated profiles at two times when the glacier was 13 km long: during advance (approximately in year 1690) and during retreat (year 1871). During advance the glacier is thicker and the frontal profile steeper than during retreat. This is consistent with

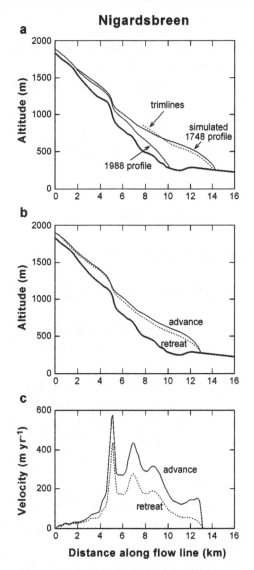

Figure 11.11: Calculated profiles and flow of Nigardsbreen, Norway. (a) Change in profile from 1748 to 1988. Profile of 1748 is compared to observed trimline. (b) Comparison of profiles during advance (year 1690) and retreat (year 1871), showing steeper and thicker front during advance. (c) Model ice velocities along the glacier in these years. Adapted from Oerlemans (1997) and Oerlemans (2001, pp. 90–91).

observations of many glaciers in retreat and advance, and reflects the faster flow and greater flux during advance. Figure 11.11c shows the model estimates for velocities at these times.

Glacier models of this sort have been used to interpret observations of glacier advance and retreat in terms of recent climate change, a topic we discuss in Chapter 14.

11.3.4.3 Plan-view Models

Using both horizontal dimensions rather than a flow line allows a glacier model to capture effects related to complex geometries. The mass conservation relation no longer considers the volumetric flux conveyed by the entire glacier cross-section but the flux per unit width at a point, the vector $\vec{q} = [\bar{u}H, \bar{v}H]$. Thus

$$\frac{\partial H}{\partial t} = \dot{b}_i - \frac{\partial}{\partial x}[\bar{u}H] - \frac{\partial}{\partial y}[\bar{v}H]. \tag{11.47}$$

The driving stress $\vec{\tau}_d$ has components

$$\tau_{dx} = -\rho g H \frac{\partial S}{\partial x} \quad \text{and} \quad \tau_{dy} = -\rho g H \frac{\partial S}{\partial y}, \tag{11.48}$$

and the rates of shearing on a horizontal plane are

$$\frac{\partial u}{\partial z} = 2A(T)\tau_E^{n-1}\tau_{xz} \quad \text{and} \quad \frac{\partial v}{\partial z} = 2A(T)\tau_E^{n-1}\tau_{yz}. \tag{11.49}$$

The preceding relations apply to all situations. To outline the plan-view equivalent of the flow-line model discussed previously, we now consider a special case. First, suppose that velocities vary slowly with distance along and across the glacier; thus, the stresses τ_{xy}, τ_{xx}, and τ_{yy} can be neglected, and τ_{xz} and τ_{yz} increase linearly with depth until, at the bed, they equal the driving stresses given by Eq. 11.48. These assumptions apply only to broad glaciers such as ice caps and large ice fields. The velocity profiles are derived by integrating Eq. 11.49 as in the one-dimensional parallel-flow model (see Section 8.3) but with

$$\tau_E^{n-1} = \left[\tau_{xz}^2 + \tau_{yz}^2\right]^{[n-1]/2}. \tag{11.50}$$

Second, if we restrict attention to the case of temperate ice, then $A(T)$ takes a single value, A_0. Then the depth-averaged velocities, for substitution in the continuity relation (Eq. 11.47), are found by integrating the velocity profiles. This gives

$$\bar{u} = u_b + G\left[-\frac{\partial S}{\partial x}\right] \quad \text{and} \quad \bar{v} = v_b + G\left[-\frac{\partial S}{\partial y}\right] \tag{11.51}$$

with

$$G = \frac{2A_0}{n+2}[\rho g]\alpha_s^{n-1}H^{n+1} \quad \text{and} \quad \alpha_s = \left[\left[\frac{\partial S}{\partial x}\right]^2 + \left[\frac{\partial S}{\partial y}\right]^2\right]^{1/2}. \tag{11.52}$$

In nontemperate cases, temperature varies with depth and these relations cannot be used. Instead the model must resolve the vertical dimension and integrate through it at each time step to find velocities. This is required for application to ice sheets, a subject we take up in Chapter 13.

The ice caps of Iceland, however, are believed to be temperate. Marshall et al. (2005) and Flowers et al. (2005) together made an analysis of Vatnajökull, the largest of the Icelandic ice caps; these reports give an interesting example of the use of an isothermal, plan-view model. Climate parameters governed the model specific balances; from spatial distributions of precipitation, mean annual air temperature, and July air temperature, the degree-days method was used to specify annual snowfall and melt at all locations on the ice cap (see Chapter 5). The model attempted to calculate basal slip rates as a function of water pressures determined with a hydrological model. The model ice cap was forced with some plausible estimates of future climatic warming. Decreased surface balance was found to be the overwhelming control on the evolution of the model ice sheet. With a warming of 2 °C per century, the model ice cap disappeared entirely within 300 years.

11.4 Reactions to Additional Forcings

Mass balance is not the only factor that connects glacier evolution to the changing environment. Glaciers move faster when increased melt at the surface leads to increased water pressures and volumes at the bed. The margins of tidewater glaciers and ice sheets migrate when sea level rises or falls. Except in temperate glaciers, a sustained change of temperature at the surface eventually changes temperatures deep in the ice; this modifies the ice deformation, and might activate basal slip by raising the temperature to melting point. Each of these processes operates with its own characteristic timescale and so, when they act together, the glacier responds in a complex fashion, even to a single shift of climate.

The response of a glacier to any of these variables is fundamentally similar to the response following a change in mass balance; variations of discharge and ice thickness propagate by diffusion and kinematic waves. Thus a change in one region of a glacier eventually affects the entire system. To analyze the response in realistic scenarios generally requires numerical methods, but here we first consider how these processes affect ice thickness using an extension of the linearized theory of Section 11.3.2.1.

11.4.1 Response of Glaciers to Ice and Bed Changes

With modifications, Eqs. 11.34 and 11.41 show how a glacier's thickness responds to perturbations of the flow properties represented by the coefficients K_d and K_b in Eq. 11.23. K_d varies with temperature and, in a temperate glacier, water content (Chapter 3). The basal-motion coefficient K_b depends on how slippery the bed is; weak sediments, abundant water, and high water pressures all increase K_b (Chapter 7).

Consider the gradient of flux along a glacier, $\partial q/\partial x$, as in Eq. 11.21. In general, the expansion of $\partial q/\partial x$, previously written in terms of H and α (Section 11.3.2), should also include terms $[\partial q/\partial K_b][\partial K_b/\partial x] + [\partial q/\partial K_d][\partial K_d/\partial x]$. Thus we write the equation for ice thickness changes as (compare to Eq. 11.34)

$$\frac{\partial H}{\partial t} = \dot{b} - C\frac{\partial H}{\partial x} + D\frac{\partial^2 H}{\partial x^2} + D\frac{\partial^2 B}{\partial x^2} - I\frac{\partial K_b}{\partial x} - J\frac{\partial K_d}{\partial x}, \tag{11.53}$$

where $I = [\partial q/\partial K_b]$ and $J = [\partial q/\partial K_d]$. The terms with I and J act, along with mass balance and bedrock curvature, as forcings on the ice thickness. A down-glacier increase of either K_b or K_d causes the ice to thin over time or, at steady state, to thin down-glacier.

11.4.1.1 Linearized Theory

Suppose that the specific mass balance does not change with time ($\dot{b}_1 = 0$), but other forcings perturb the glacier's flow. For a flow line of uniform width, mass conservation requires that

$$\frac{\partial H_1}{\partial t} = -\frac{\partial q_1}{\partial x}. \tag{11.54}$$

In contrast to the case of mass balance forcing, here thickness perturbations will tend to return the flux to its original value. Let K stand for either K_b or K_d, and assume for illustration that either basal slip or internal deformation determines the flow. As in Eq. 11.37 we consider small perturbations from a steady state; K changes from K_0 to $K_0 + K_1$, and the expansion of q_1 becomes

$$q_1 = C_0 H_1 + D_0 \alpha_1 + I_0 K_1 \quad \text{with} \quad I_0 = \frac{\partial q}{\partial K}. \tag{11.55}$$

Because q varies in proportion to K, then $I_0 = q_0/K_0$. Substituting into Eq. 11.54 and expanding the derivative gives an expression for evolution of the perturbed ice thickness. This expression is identical to Eq. 11.41 except that, in place of $\dot{b}_1$, we now have a forcing function with K_1 and $\partial K_1/\partial x$:

$$\frac{\partial H_1}{\partial t} + [n+2]\dot{\epsilon}_0 H_1 = -K_1 \frac{\dot{b}_0}{K_0} - I_0 \frac{\partial K_1}{\partial x}$$

$$- [C_0 - D_0'] \frac{\partial H_1}{\partial x} + D_0 \frac{\partial^2 H_1}{\partial x^2} \tag{11.56}$$

(The derivation uses the datum state relation $\partial q_0/\partial x = \dot{b}_0$.) Thickness perturbations originating from Eq. 11.56 then propagate by diffusion and kinematic waves, as Eq. 11.41 shows.

Consider first a sudden increase of K (hence a $K_1 > 0$), uniform along a glacier. This could represent, for example, an increased sliding rate due to input of meltwater. The forcing on ice thickness is $K_1 \dot{b}_0/K_0$, so its sign depends on whether accumulation or ablation predominates in the datum state. The accumulation zone thins, because faster flow enhances the down-glacier increase of flux already present in the datum state (arising from $\dot{b}_0 > 0$). In this zone, thickness approaches a new steady value. The ablation zone (where $\dot{b}_0 < 0$ and flux decreases down-glacier) initially thickens and the terminus advances. Over time, however, thinning of the upper glacier propagates all the way to the terminus, making the glacier longer and thinner than its initial configuration. Excess ablation subsequently leads to retreat of the terminus back to near its original position. Finally, in its new steady state, the entire glacier is thin and fast compared to its initial state.

Fountain et al. (2004) invoked a similar process to explain the simultaneous advance and thinning of glaciers in the Antarctic Dry Valleys, observed in the late twentieth century. A combination of thinning and advance is not expected from a uniform mass balance perturbation (Figure 11.7), but would occur if climatic warming softens the ice, increasing K_d. In contrast to our "sudden increase" scenario, however, warming of ice deep in a glacier occurs gradually. K_d would likely increase slowly compared to the time needed for thinning of the accumulation zone to propagate to the terminus. Thus observations could show thinning throughout the response, even in the ablation zone.

Consider next a localized increase of K; for example, suppose that one section of the glacier bed becomes slippery while the rest remains unchanged. In the slippery region, flow increases and the ice either thins or thickens, depending on the sign of $\dot{b}_0$. In addition, the downstream edge of the slippery section rises (where $\partial K_1/\partial x < 0$), while the upstream edge sinks. If the factors responsible for the slippery patch cease, such thickness perturbations would move down-glacier but rapidly diminish by diffusion. If, however, the underlying cause of the slippery section moves down-glacier, then the thickness perturbations might propagate as a distinct wave. This is one possible explanation for observations of travelling bulges on glaciers (Figure 11.10).

11.4.2 Factors Influencing the Reaction of an Ice Sheet to the End of an Ice Age

The last major climate transition – from the cold late Pleistocene to the warm Holocene – did not destroy the ice sheets of Antarctica and Greenland, but they were perturbed in several ways. Moreover, their response is not yet complete. We now discuss the four processes thought to be most important in the response.

11.4.2.1 Reaction to Increase of Accumulation Rate

Annual snowfall on the ice sheets increases by a factor of two to three at the end of an ice age. The interior regions of the ice sheets react by thickening; Sections 11.2.3.1 and 11.3.2.1 described this process. Steady-state ice thickness varies as approximately the eighth root of the accumulation rate (see Section 8.10); if accumulation were the only variable that changed, the thickness would thus increase by 10% to 15%, or a few hundred meters.

11.4.2.2 Reaction to Retreat of the Margin

When an ice sheet margin retreats, because of warming or sea-level rise, thinning propagates inland as a diffusive wave (Section 11.3.1.3). Alley and Whillans (1984) studied this process for the East Antarctic Ice Sheet, where rising sea level at the end of the last ice age caused the margins to retreat by as much as a few hundred kilometers. The basic equations of their model, in terms of deviatoric stress components τ_{jk}, are

$$\frac{\partial}{\partial x}[H\,\overline{u}] = \dot{b}_i - \frac{\partial H}{\partial t} \tag{11.57}$$

$$\frac{\partial u}{\partial x} = A\left[\tau_{xx}^2 + \tau_{xz}^2\right]\tau_{xx} \tag{11.58}$$

$$\frac{\partial u}{\partial z} = 2A\left[\tau_{xx}^2 + \tau_{xz}^2\right]\tau_{xz} \tag{11.59}$$

$$\tau_{xz} = -\rho g\,[H - z]\frac{\partial H}{\partial x} \tag{11.60}$$

The x-axis is horizontal, following a flow line, and the z-axis is vertical, positive upward. The model idealizes the problem by assuming plane strain and a flat bed. Note that, although the deviatoric stress τ_{xx} appears in the expression for the effective shear stress, Eq. 11.60 omits the longitudinal force term $\partial[H\overline{\tau}_{xx}]/\partial x$; in other words, the shear stress on the horizontal plane balances all the driving stress. In Eqs. 11.58 and 11.59, A and τ_{xx} are replaced by their averages over the ice thickness. Numerical solutions to Eqs. 11.57 through 11.59, with τ_{xz} from Eq. 11.60, gave H, $\overline{u}$, and $\overline{\tau}_{xx}$ as functions of x and t.

The model was applied to the flow line originating at Dome C. The present specific balance $\dot{b}(x)$ was used, and A was varied with x in accordance with the present variation of 10-m temperature. The initial surface took the form of the Vialov profile (Eq. 8.131). The initial ice-age margin was placed at the edge of the continental shelf, about 90 km beyond the present ice margin. Sea-level rise at the end of the ice age was represented by a sudden decrease in ice thickness at the present coastline, 15 kyr ago. The amount of the decrease, about 1800 m, equalled the thickness of the Vialov profile 90 km from its edge.

Figure 11.12 illustrates the ice sheet response, assuming no simultaneous change in accumulation rate or temperature. A wave of thinning propagates upstream; it spreads and diminishes as it moves inland. Net thinning decreases steadily from 1800 m at the coast to 160 m at the ice divide. Most of the change at the coast takes place in the first hundred years whereas it takes about 8 kyr to accomplish 63% (or $1 - e^{-1}$) of the change at the divide.

This timescale for divide thinning can be approximated by using the properties of diffusion in a domain with a forcing at one boundary and zero gradient at the other (Cuffey and Clow 1997). For an effective diffusivity D, and for a distance L from margin to divide (where thickness is H_*), the timescale is approximately $L^2/2D$ for the appropriate boundary conditions. The diffusivity is $D = nq/\alpha$ (Eq. 11.41). Representative values for q and α can be taken halfway between the divide and the margin, giving $q = \dot{b}L/2$ and $\alpha = 0.44H_*/L$, the slope for a Vialov profile (Eq. 8.138) with $n = 3$. The $\dot{b}$ here denotes a mean value of specific balance for a broad

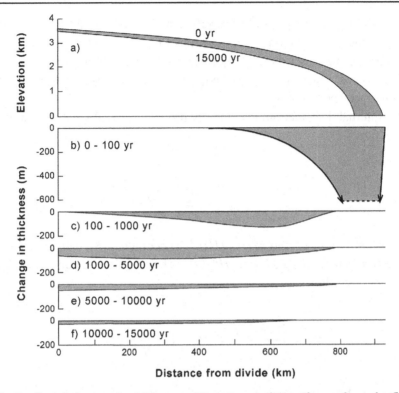

Figure 11.12: Predicted changes in thickness of East Antarctic Ice Sheet, along the flow line starting at Dome C, after a sudden rise in sea level 15 kyr ago. (a) Initial and present profiles. (b) – (f) Thickness changes over different intervals since the sea-level rise, showing how thinning propagates upstream. From Alley and Whillans (1984). Used with permission from the American Geophysical Union, *Journal of Geophysical Research.*

central region of the ice sheet. The timescale $L^2/2D$ is then $0.15 H_*/\dot{b}$. This "margin response" timescale is thus similar to the "accumulation response" timescale given in Table 11.2: a few centuries for small ice caps, about two millennia for the Greenland Ice Sheet, and about ten millennia for East Antarctica.

The thickness of an ice sheet flowing by creep scales with $\sqrt{L}$, regardless of the exponent in the creep relation (see Eq. 8.141). Thus the response time depends primarily on $\sqrt{L}/\dot{b}$ (other factors matter, but the sensitivity to them is low). Nereson and Hindmarsh (1998) gave a more sophisticated analysis of this problem, with essentially similar results.

11.4.2.3 Reaction to Temperature Change

A sustained climatic warming eventually propagates to the bottom of an ice sheet (Section 9.9). Warming softens the ice; the creep parameter A increases with temperature. The ice sheet therefore thins, because a smaller stress and ice thickness are needed to convey the same flux.

(An increase of A implies an increase of the coefficients K_d in Eq. 11.23 and K in Eq. 11.56, and hence thinning in the accumulation zone, as outlined in Section 11.4.1.1.) On the other hand, accumulation rate also increases during major warmings, and eventually leads to faster ice flow and faster advection of cold ice toward the bed; this process lowers temperatures, decreases A, and partly counteracts the thinning.

For flow by internal deformation, the velocity profile at typical locations on the flank of an ice sheet is (Section 8.3)

$$u(z) = 2\tau_d^n \int_0^z A(z) \left[1 - \frac{z}{H} \right]^n dz',$$
(11.61)

with z the height above the bed and $\tau_d = \rho g H \alpha$. The coefficient K_d can be defined in terms of the depth-averaged velocity $(\bar{u})$ as $\bar{u} = K_d H^{n+1} \alpha^n$. Thus K_d depends on the vertical profile of ice softness, $A(z)$, according to

$$K_d = \frac{2[\rho g]^n}{H^2} \int_0^H \int_0^z A(z) \left[1 - \frac{z}{H} \right]^n dz'' dz'.$$
(11.62)

The flow of the ice sheet thus depends much more on the value of A near the bed than on its values in the middle or upper layers; the influence of A is weighted by $[1 - z/H]^n$, which reflects the increase of shear stress with depth and the increase of deformation rates with stress raised to the power n. With $n = 3$, the weighting on A at a distance $0.1H$ above the bed is nearly six times as large as the weighting on A at $0.5H$. Thus most of the effect of a temperature change does not occur until the change has propagated to near the base of the ice sheet. This means that temperature changes considerably lengthen the time for ice sheets to respond at the end of an ice age (Whillans 1981). It takes roughly 10 to 30 kyr for the bed of the Greenland and West Antarctic ice sheets to warm significantly after an ice age ends (see Chapter 9). The delay is even longer in central East Antarctica.

Calculations suggest that the temperature beneath 3 km of ice may vary by about 5 °C during an ice-age cycle. An increase of this amount would increase the ice-softness parameter A by about a factor of 1.7 for a warming from -15 and -10 °C and a factor of 2.6 for warming from -10 to -5 °C (Table 3.4). Because steady-state ice thickness varies inversely with the eighth root of A (Eq. 8.141), these parameter changes correspond to thickness variations of about 200 to 400 m for a 3 km ice thickness. Such variations are comparable with the changes resulting from accumulation increases and sea-level rise. The preceding discussion assumes that the ice remains frozen to its bed. If the basal temperature reaches melting point during part of the ice age cycle, sliding or bed deformation could begin. If this happens over a wide area, substantial thinning might take place.

11.4.2.4 Combined Response

Figure 11.13 shows how changes in accumulation rate and temperature contribute to the response of the Dome C ice divide in the model of Alley and Whillans (compare to Figure 11.12). First the ice thickens because snowfall has increased. Then the rise in sea level starts to have an effect so that, after a few thousand years, the ice thickness begins to decrease. The increase in velocity from the increase in temperature causes further thinning but this is delayed by at least 20 kyr.

Figure 11.13 illustrates the complex interaction between three processes with different time constants. It should not be regarded, however, as a quantitatively accurate simulation of East Antarctica; the accumulation rate doubled at the end of the ice age (Jouzel et al. 1989), whereas Alley and Whillans used an increase of only 10%.

The effects discussed here make it clear that although a steady state ice sheet is a useful theoretical concept it is never attained in reality.

11.4.2.5 Stratigraphic Effect

There is yet another reason that ice sheets are not fully adjusted to the current climate; in Greenland, Arctic Canada, and perhaps West Antarctica, ice deposited during the last ice age deforms in shear about three times as fast as Holocene ice at the same stress and temperature (Section 8.3.2). At present, the layers of Pleistocene ice are being thinned by flow and replaced by Holocene ice. The mean viscosity of the ice column is therefore increasing. To carry off the same accumulation rate, the ice thickness must increase continuously – the opposite effect from a steady warming of the basal ice. An ice cap consisting only of Holocene ice would be about 15% $(= 3^{1/8} - 1)$ thicker than one consisting solely of ice-age ice at the same temperature and for the same accumulation rate and span. Reeh (1985) calculated how much of this change has been accomplished since the end of the ice age roughly 10 kyr ago, on a thin ice cap in Arctic Canada

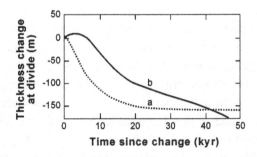

Figure 11.13: Predicted changes in thickness at Dome C as a function of time after sudden rise in sea level. (a) For sea-level rise only. (b) For sea-level rise combined with a 10% increase in accumulation rate and a surface warming of 7 °C. Adapted from Alley and Whillans (1984) by permission of the American Geophysical Union, *Journal of Geophysical Research*.

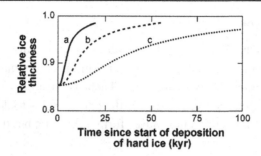

Figure 11.14: Response of ice sheets to gradual thinning of "soft" basal layer of ice-age ice. Thickness is that relative to an ice sheet consisting entirely of "hard" Holocene ice. Curve (a) represents Devon Island ice cap ($H = 300$ m, $\dot{b} = 0.23$ m yr^{-1}), curve (b) Dye 3 ($H = 2037$ m, $\dot{b} = 0.55$ m yr^{-1}), curve (c) Crête, Greenland ($H = 3000$ m, $\dot{b} = 0.28$ m yr^{-1}). Adapted from Reeh (1985).

and at two locations on the Greenland Ice Sheet. Figure 11.14 illustrates the results. In central Greenland, the adjustment takes a time of order 10^5 years. East Antarctica would not undergo a similar process because it contains many alternating layers of ice-age and interglacial ices.

These calculations assume that the factor-of-three viscosity contrast does not change with time, a perhaps questionable assumption. Given enough time, might the Holocene ice develop fabrics similar to those presently found in the ice-age ice? To assess whether the long-timescale behavior depicted in Figure 11.14 is realistic or not requires better models for the structural processes controlling fabrics, including recrystallization and disruption by folding (Chapters 3 and 10).

11.4.3 Ice Flow Increased by Water Input

Numerous observations show that glaciers speed up when water from the surface penetrates to the bed and enhances basal slip (Sections 7.2.6 and 7.3.3). The water may originate as melt or as rain; warm weather and storms both sometimes result in speed-up, provided the basal temperature is at melting point and paths exist for surface water to reach the bed. Observed variations of glacier speed include daily cycles during the melt season, events lasting a few days, and seasonal fluctuations with fast flow usually occurring in early summer. The speedups occur because water accumulates on the bed and within the glacier, increasing the pressure and volume of water in voids at the basal interface and within basal sediments. This lubricates hard-bed sliding or reduces the strength of deformable sediments, processes discussed in Chapter 7. Here we revisit the topic, not to elaborate the mechanisms responsible for speed-up, but to review information about how such events might contribute to the reaction of glaciers to changing weather and climate at a variety of timescales.

The relation between water input and glacier speed lacks consistency, largely because the pressurization and storage of water on the bed depends not only on the rate of input but also on drainage (Chapter 6). In some cases, extra water at the bed enhances drainage, by melting larger and better-connected conduits or by increasing sliding and thus expanding linked basal cavities. Increased drainage, in turn, reduces water pressure and volume and hence reduces glacier motion. This is typical of the seasonal pattern; the early-summer fast-flow period usually ends when an effective drainage system develops. Stabilization also occurs in events of short duration. For example, Unteraargletscher speeds up during heavy rains. After the water has time to drain along the bed, however, the glacier flows more slowly than before the rain started (Gudmundsson et al. 2000).

In other cases, increased glacier flow reduces the drainage by destroying or disrupting existing passageways for water, whether tunnel systems in ice or canals in sediment. This forms a positive feedback between ice flow and water accumulation that leads to large and sustained increases of glacier speed. Surges are the most extreme example of this process (Chapter 12). Such a positive feedback seems most likely for glaciers with deformable beds; on a rigid bed, in contrast, rapid sliding produces large and interconnected cavities that conduct water.

By causing a glacier to speed up, water inputs also change the ice thickness; in general, the flux gradient ($\partial q_1 / \partial x$ in Eq. 11.54) changes with the speed. Faster basal slip is represented by an increase of the coefficient K_b in Eq. 11.23. Equation 11.56 shows how this acts as a forcing on ice thickness, in the case of a small deviation from steady state.

In the simplest scenario, water input increases the glacier speed in proportion to the prevailing rate of basal slip, a situation described by a uniform increase of the coefficient K_b. Figure 11.15 shows such a case, observed at Columbia Glacier, Alaska (Meier et al. 1994). This glacier's speed increases diurnally, due to surface melt, and increases in longer periods of warm and rainy weather. The figure compares flow at two locations along the center line, separated by 7 km, in a region of longitudinal extension. At one site the glacier flows about twice as fast, and variations are nearly twice as large, as at the other site. In this situation, because the velocity increases (Δu) are proportional to time-averaged velocity (u_o), the increases of flux gradient ($\partial \Delta q / \partial x$) are proportional to the prevailing flux gradient ($\partial q / \partial x$). Thus the rate of thickening and thinning is simply proportional to the prevailing flux gradient. A speed-up along the whole glacier, for example, tends to thin parts of the glacier where the flux increases downstream.

The spatial variation of speed-up along a glacier need not follow a simple pattern, however. Thickening and thinning are favored wherever the speed-up decreases or increases down-glacier, respectively. Sugiyama and Gudmundsson (2004) observed several such changes on Lauteraar Glacier, Switzerland. Furthermore, the effect of water inputs on glacier flow extends beyond the region of changing basal water pressures, because longitudinal stretching and compression push or pull the ice. Hanson and Hooke (1994) inferred this process at Storglaciären. (This is the longitudinal stress-gradient coupling discussed in Section 8.7.2.)

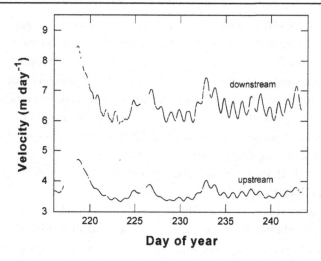

Figure 11.15: Time variations of measured surface velocity at two locations on the centerline of Columbia Glacier. The "downstream" location is 7 km downstream of the other location. Adapted from Meier et al. (1994) by permission of the American Geophysical Union, *Journal of Geophysical Research.*

Changes of glacier speed resulting from variable water inputs are sometimes quite large. This is expected; when water pressures at the bed rise to near the ice-overburden pressure, deforming sediment loses its strength, and the sliding rate responds significantly to small changes in water pressure. Factor-of-two diurnal speed variations occur, for example, on both Lauteraar Glacier and Storglaciären (Hooke et al. 1997). At Columbia Glacier, diurnal fluctuations are as large as $1\,\mathrm{m\,day^{-1}}$, or 10% to 30% of the normally fast flow.

The major question has not yet been answered: how do long-term climatic changes of rain and surface melt affect a glacier's flow? Although water input events appear to increase a glacier's annual average speed, the size of this effect, and how it changes over time, must depend on the feedbacks between ice flow and water drainage, discussed earlier. Because of these feedbacks, the speed-ups depend not only on the quantity of water input but also on its variability and timing, and on the type of substrate.

11.4.3.1 Water Input Effect on Flow of Greenland Ice Sheet

Observations show that meltwater reaching the bed increases the flow of the Greenland Ice Sheet in summer throughout a large region of the ablation zone on the western flank (Joughin et al. 2008; van de Wal et al. 2008). Might this be an important process in the reaction of the ice sheet to climate change?

The region of summertime speed-up extends inland from the margin by more than 50 km. Such increases should be expected on thin, heavily crevassed regions close to the margin. It is less obvious how speed-ups could occur in regions of thick ice; although ice in the ablation zone attains melting point at the bed, temperatures through most of the ice thickness remain subfreezing. Surface water must therefore penetrate through a great thickness of cold ice to reach the bed. This most likely occurs by downward propagation of water-filled fractures, a process we discussed in Section 10.9.2 of Chapter 10. That such a process operates, allowing water to reach the bed, has been proven by observations of rapid drainage of supraglacial lakes together with simultaneous uplift and speed-up of the ice sheet surface (Das et al. 2008). Once connections to the bed form, the conduits probably remain open as long as abundant meltwater drains through them.

Zwally et al. (2002) reported the first extensive measurements of seasonal flow variations, at a site near the equilibrium line with an ice thickness of 1200 m. Over four years of measurement (1996–1999), ice velocity increased during each summer melt season, from a mean rate of about 0.32 m day^{-1} to as much as 0.39 m day^{-1}. The increased flow ceased abruptly at the end of each melt season. Moreover, the warmest two summers had larger speed-ups than the others.

These observations were surprising in part because measurements on Jakobshavn Ice Stream in the early 1990s found no seasonal variations of ice velocity (Echelmeyer and Harrison 1990). Subsequent observations made it clear that fast-flowing outlet glaciers must be distinguished from slower-flowing flank regions of the ice sheet. Joughin et al. (2008) used satellite observations to map velocities in a large "flank" region of the western ablation zone. They found that summertime velocities exceeded annual mean velocities throughout the region. On average, compared to the annual mean velocity of 76 m yr^{-1}, velocities in August 2006 were higher by a multiple of 1.47; in July, by a multiple of 1.93. Ground-based velocity measurements were acquired at two locations. They revealed a clear correspondence between positive degree days – a proxy for melt rate – and ice velocity (Figure 11.16). In addition, the ice velocity peaked briefly when a nearby supraglacial lake drained. Van de Wal et al. (2008) also observed seasonal speed-ups along a transect of stations aligned perpendicular to the ice margin. Ice velocities varied from week to week in correlation with ablation and attained values as high as seven times the long-term average.

Increased ice flow in warmer summers may be an important process in the response of the ice sheet to climate warming, one that leads to faster loss of ice (Chapter 14). But this is not known; as for the case of mountain glaciers, the drainage system can change, too. Van de Wal et al. (2008) acquired annual velocity measurements for 17 years at eight locations along their transect, starting in 1991. Despite increased ablation over this period, annual velocities either showed no change or slightly declined. Thus the only data pertaining to this problem suggest no increase of flow from climate warming, in regions where abundant melt already occurs. On the other hand, inland expansion of the ablation zone would almost certainly lead to increased annual ice flow in regions where little melt occurs at present.

Seasonal variations of ice velocity occur, too, on fast-flowing outlet glaciers (Joughin et al. 2008). These variations, however, are driven not by surface melt penetrating to the bed but by

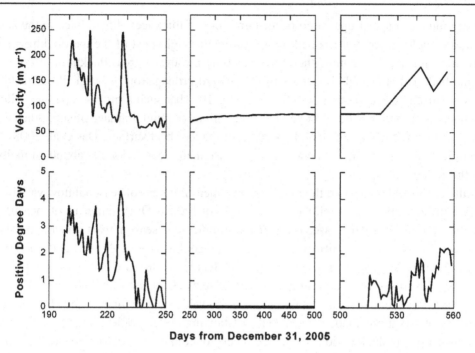

Figure 11.16: Correspondence of Greenland Ice Sheet velocity with surface melt and lake drainage events. Top panel shows the variation of ice velocity at a site about 40 km from the margin of the Greenland Ice Sheet, measured by GPS (the measurement was continuous in the first period, weekly in the later two periods). Bottom panels show corresponding measured positive degree days, a proxy for meltwater generation, at a nearby site. The peak in velocity on Day 210 occurred when a nearby lake drained. Adapted from Joughin et al. (2008).

changes in the position of the glacier terminus (see Section 11.5.1). Fast flow in this case does not cease at the end of the melt season but instead slowly declines as the terminus advances during the onset of winter.

11.5 Changes at a Marine Margin

The analyses in Sections 11.3 and 11.4 assume that the flow of a glacier depends strongly on the local driving stress and hence the local ice thickness and surface slope. In some places this assumption fails, or at best provides an incomplete picture of the factors controlling flow. Such places include not only ice shelves but also valley glaciers and polar ice streams with slippery beds. Ice streams and large glaciers flowing to marine margins are a particularly important case; their beds are often slippery throughout an extensive region of the "inland" ice upstream of the tidewater terminus or grounding line. Many of these regions grade into floating ice shelves. For all of these situations, an analysis of changes in the glacier's force balance provides a necessary supplement to the concepts outlined previously.

   The balance of forces, in the direction of flow, acting on a vertical column through the glacier can be written (Section 8.5.3)

$$\tau_d - \tau_b + \underbrace{\frac{\partial}{\partial x} H\left[2\overline{\tau}_{xx} + \overline{\tau}_{yy}\right]}_{-\tau_L} + \underbrace{\frac{\partial}{\partial y}\left[H\overline{\tau}_{xy}\right]}_{-\tau_W} = 0. \tag{11.63}$$

As usual, τ_d denotes the driving stress, and τ_b, τ_L, and τ_W are the resisting stresses from drag on the bed and from viscous forces due to along-glacier and across-glacier variations of flow. The stresses τ_{xy} and $2\tau_{xx} + \tau_{yy}$ increase with the effective ice viscosity and with the corresponding strain rates $\dot{\epsilon}_{xy}$, $\dot{\epsilon}_{xx}$, and $\dot{\epsilon}_{yy}$.

11.5.1 Conceptual Framework

The essential idea for the following discussion originates with the analysis of Hughes (2003). Van der Veen (2001, pp. 34054+) discussed a partial version. Reports by Thomas (2004) and Howat et al. (2005) contained examples of applications of the theory. Our goal here is to illustrate concepts. For computations, the full relations describing the glacier flow and force balance, reviewed in Chapter 8, need to be inserted into the mass conservation relation (Sections 8.5.2 and 8.5.3). These expressions usually need to be two- or three-dimensional and solved numerically. In applications to specific glaciers, moreover, measurements have so far always been inadequate to constrain well some of the essential terms; observations of basal conditions are particularly difficult to obtain. This explains why we separate theory from observations to an unfortunate degree in the following discussion.

   We will make several assumptions for our discussion: (1) Depth-variations of flow can be ignored, a good assumption because in most cases basal slip far exceeds internal shear. (2) Ice deformation can be written in terms of a viscous creep relation, using an effective viscosity η to connect the rate of stretching ($\dot{\epsilon}_{xx}$) with its corresponding deviatoric stress. In particular we write $\int \tau_{xx}\, dz$ as $2H\eta\partial u/\partial x$; the effects of the nonlinear creep properties of ice, and of any depth variations in the stress, are subsumed in this definition of η. (3) Transverse stretching or compression is negligible ($\partial v/\partial y \approx 0$). Thus in Eq. 11.63 the factor $2\tau_{xx}$ replaces $2\tau_{xx} + \tau_{yy}$.

   Following from our discussion of tidewater glaciers in Section 8.8, we consider a strip of ice that extends along a flow line from a horizontal position $x = x_o$ on the glacier to the margin at $x = X_M$ (Figure 11.17). The domain should be regarded as a zone "near" the margin. To obtain the balance of forces (per unit width across-glacier) on this strip, integrate Eq. 11.63 over x. The net gravitational driving force acting on the strip horizontally in direction x equals $\phi_M + \int \tau_d\, dx$, where ϕ_M is the effective force due to the terminal cliff. (Section 8.2.4 explains ϕ_M; it equals the value of $2\overline{\tau}_{xx} H$ at the margin.) The driving force is resisted by drag on the bed and sides, which contribute a force $\int [\tau_b + \tau_W]\, dx$ in the opposite direction. An excess of the driving force over the drag stretches the ice at x_o and generates a tensional resisting force. Equilibrium of the

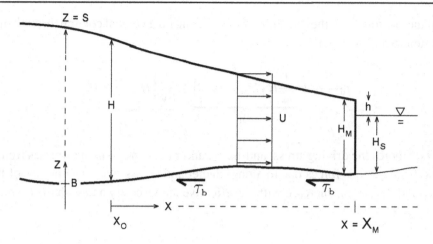

Figure 11.17: Schematic of a longitudinal section through the terminus region of a tidewater glacier.

entire section requires that

$$4\eta H \frac{\partial u}{\partial x} = \phi_M + \int_{x_o}^{X_M} \tau_d \, dx - \int_{x_o}^{X_M} [\tau_b + \tau_w] \, dx. \tag{11.64}$$

In turn, the rate of thinning, $-\partial H/\partial t$, depends on the stretching rate by the continuity relation: $\partial H/\partial t = \dot{b} - \partial q/\partial x$, in which $q = uH$. Furthermore, in terms of glacier geometry, the driving stress is $\tau_d = -\rho g H \partial S/\partial x$ and, from Eq. 8.31,

$$\phi_M = \frac{1}{2} \rho g H_M^2 - F_M. \tag{11.65}$$

Here H_M denotes the ice thickness at the terminus, and F_M the back-force of water and sediment against the terminal cliff. Thus the driving force in Eq. 11.64 takes the form

$$\phi_M + \int_{x_o}^{X_M} \tau_d dx = \frac{1}{2} \rho g H_M^2 - F_M + \rho g \int_{x_o}^{X_M} H \frac{-\partial S}{\partial x} dx. \tag{11.66}$$

Next write the surface elevation as $S = H + B$ in Eq. 11.66, and substitute into Eq. 11.64. The result shows that the ice at x_o thins at a rate given by the two relations

$$-\frac{\partial H}{\partial t} = H \frac{\partial u}{\partial x} + u \frac{\partial H}{\partial x} - \dot{b} \tag{11.67}$$

$$H \frac{\partial u}{\partial x} = \frac{1}{4\eta} \left[\frac{1}{2} \rho g H^2 - F_M + \rho g \int_{x_o}^{X_M} H \frac{-\partial B}{\partial x} dx - \int_{x_o}^{X_M} [\tau_b + \tau_w] dx \right], \tag{11.68}$$

where H, $\dot{b}$ and $u \partial H / \partial x$ are evaluated at x_o. Finally, note that the ice velocity increases with the stretching rate; far inland of the margin, velocity has some value u_o that changes only slowly with time, while along the flow line $u = u_o + \int \partial u / \partial x \, dx$. Thus stretching, thinning, and speed-up of the ice all occur together near the margin, as functions of both space and time. We will use the term "response" to mean a combined change of stretching, thinning, and speed-up in reaction to changes at a glacier's terminus. The zone that reacts immediately to such changes extends inland by only a certain distance – about ten ice thicknesses in the case of a strong bed and temperate ice (Section 8.7.2) but longer with a slippery bed. Over time, however, the response propagates inland as a diffusive wave.

The preceding relations indicate how different factors contribute to a response:

1. Thick ice at the terminus increases the driving force. A thick floating tongue should accelerate and thin substantially if the resisting forces acting on it are neutralized. For a grounded glacier, an increase of driving force near the terminus occurs when the terminus retreats or advances into deeper water. This follows because the plastic strength of ice limits the height of the terminal cliff above water level (the freeboard h); the height is usually not more than about 50 m, and h can be assumed to maintain this value as the frontal ice collapses and flows. The ice thickness at the margin is $H_M = h + H_S$, with H_S the submerged thickness of ice, equivalent to the water depth. The driving force at the margin is, from Eq. 11.65,

$$\phi_M = \frac{1}{2} g \left[\rho [h + H_S]^2 - \rho' H_S^2 \right], \tag{11.69}$$

where ρ' denotes the density of the water. With h a constant, ϕ_M always increases with H_S up to the depth where the ice floats.

Note, however, that the situation differs when water level fluctuates rapidly, as in tidal cycles. In such cases, the ice thickness remains approximately constant but the freeboard fluctuates. A rapidly rising water level thus increases the back-force and so decreases the driving force.

2. Inland from the terminus, thick ice contributes to the driving force (by the term $\frac{1}{2} \rho g H^2$) but only if the surface slopes. (With a flat surface, the term containing $\partial B / \partial x$ balances the increase of $\frac{1}{2} \rho g H^2$; in other words, a glacier sitting in a bowl has no driving stress in the absence of a sloping surface.) Thinning at the terminus increases the surface slope up-glacier, which accelerates flow and extends the zone of thinning inland. This is diffusive propagation, discussed previously.

3. A slippery bed – which means that a given ice velocity produces only a small basal drag τ_b – is an essential factor that permits large responses. A slippery bed, in turn, arises when effective pressures on the bed decline to near zero. Recall that the effective pressure, N, equals $P_i - P_w$, the ice-overburden stress minus the water pressure. Thinning of the ice reduces P_i and hence tends to make the bed more slippery – an important feedback that

makes acceleration and thinning self-sustaining. Conditions that favor high water pressures and hence a slippery bed include deep water at the terminus, low bed elevations, and abundant water sources such as drainage of subglacial melt from a large catchment upstream. A phenomenological relation for basal drag is

$$\tau_b = cNu^{1/m}, \tag{11.70}$$

where c and m denote poorly constrained parameters. In general, the smaller the increase of τ_b with u, the larger the acceleration and thinning of the glacier will be. If an extensive layer of sediment mantles the bed, τ_b may be fixed by the sediment's yield strength and not depend on u at all ($m \rightarrow \infty$).

How τ_b depends on u strongly affects the magnitude of a response. It is unfortunate that relations for τ_b remain poorly constrained; ignorance of them severely limits the accuracy of quantitative predictions.

4. Reduced side drag τ_w enhances a response; increased side drag opposes it. From Eq. 11.63, $\tau_w = -\partial[H\tau_{xy}]/\partial y$. This quantity scales with the velocity u on the glacier centerline, because $\tau_{xy} \propto A_{inv} \dot{\epsilon}_{xy}^{1/n}$, and $\dot{\epsilon}_{xy} \propto \partial u/\partial y$. Thus

$$\tau_w \propto A_{inv} \frac{H}{Y} \left[\frac{u}{Y}\right]^{1/n}, \tag{11.71}$$

where Y indicates one-half the glacier width, n the creep exponent for ice, and A_{inv} an ice viscosity parameter. Processes that reduce side drag therefore include thinning and widening of the glacier and softening of ice in the side margins. These factors all contribute to the response, but faster flow itself increases the side drag and stabilizes the system.

5. Rapid retreat of the terminus – a decrease of X_M – generates a response of the inland ice because the resisting force, $\int[\tau_b + \tau_w]dx$, decreases immediately whereas the driving force, which depends primarily on H, does not. Indeed, observations demonstrate this behavior. Figure 11.18 shows the seasonal advance and retreat of the terminus of Jakobshavn Ice Stream, together with changes of ice velocity. The ice stream speeds up when the terminus retreats. The magnitude of the speed-up decreases inland.

6. Increased ablation ($\dot{b} < 0$), which includes basal melt of floating sections, thins the ice.

7. Advection of thick ice from up-glacier – the term $u\partial H/\partial x$ – opposes thinning of the ice near the margin and might stabilize the system.

8. The nonlinear creep relation for ice implies that the effective viscosity η decreases as the stretching rate increases. The stretching rate, $\partial u/\partial x$, effectively increases as the power $n \approx 3$ of the right-hand side of Eq. 11.68.

A thickness change near the margin originating with these factors propagates inland as a diffusive wave, as discussed in Section 11.4.2.2; Eq. 11.67 gives an ice thickness change that can be

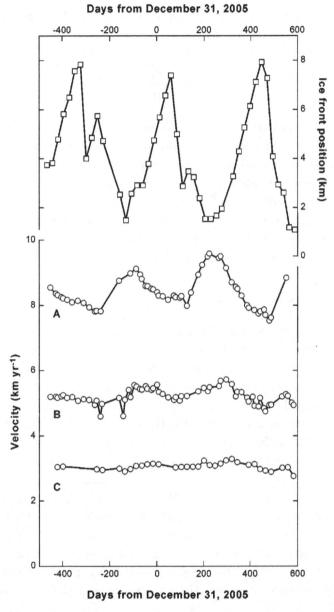

Figure 11.18: Covariation of ice stream terminus position and ice velocity. Top curve shows the seasonal variations of the terminus position of Jakobshavn Ice Stream; larger values indicate advance of the floating terminus. Bottom curves show variations of ice velocity (measured with InSAR) at three locations; A is closest to the terminus (4 to 10 km from the terminus), C is farthest inland (about 25 km). Multiyear trends in velocity have been removed. Data courtesy of I. Joughin. Locations are close to those shown in Figure 3B of Joughin et al. (2008).

regarded as a boundary condition for Eq. 11.34 or 11.41. After a time Δt, thinning propagates inland to a distance Δx given roughly by

$$\Delta x \approx \sqrt{\frac{n'q}{\alpha} \Delta t}, \qquad (11.72)$$

where $n'q/\alpha$ is the effective diffusivity and α the surface slope (Section 11.3.1.3). If the glacier slopes steeply toward the coast, however, the diffusive wave travelling inland may be effectively opposed by a kinematic wave travelling in the opposite direction.

Equation 11.72 applies for either of the two end-member cases relevant to ice stream flow; one with resistance arising only from basal shear, the other only from shearing in side margins. In both cases, the ice velocity depends on α raised to a power n' (see Eq. 8.86). In the latter case, n' refers to the creep exponent for ice, n. In the former case n' is whatever stress-dependence characterizes the basal motion (e.g., m in Eq. 11.23).

11.5.2 The Tidewater Glacier Cycle

A small change in the mass balance of a tidewater glacier can lead to a very large and self-sustaining retreat because of feedbacks involving the balance of forces, the flow, and calving. This contrasts with the processes discussed in Sections 11.2 and 11.3, for which the magnitude of the glacier response scales with the original change of mass balance.

Since the first recorded observations about two centuries ago, many tidewater glaciers in fiords in Alaska have retreated catastrophically. The most spectacular changes have occurred in Glacier Bay. Ice extended to the mouth of the bay in 1750; retreat had begun when Vancouver mapped the area in 1794 and has continued since. Some glaciers have retreated as much as 100 km (Mercer 1961). Columbia Glacier, the only large glacier that recently extended to the mouth of its fiord, began a rapid retreat in 1982. The retreat of these glaciers has contributed significantly to sea-level rise in the twentieth century (Arendt et al. 2002). On the other hand, some tidewater glaciers have been advancing for most of the past one or two hundred years. The rates of advance, typically 20 to 40 m yr^{-1}, are much less than the measured retreat rates of 0.2 to 1.7 km yr^{-1} (Meier and Post 1987). Tidewater glaciers in fiords in Chile appear to behave in a similar way. Tidewater glaciers in Alaska and Chile are temperate and their termini are grounded.

Tidewater glaciers can move very rapidly during retreat, at speeds comparable to surging glaciers and the largest polar ice streams. The velocity near the terminus of Columbia Glacier was about 30 m day^{-1} by the end of the 1990s, for example (Pfeffer 2007). We have discussed the flow of Columbia Glacier in Section 8.8.2. Along its lower 25 km or so, this glacier slips rapidly over its bed, which is below sea level. Before retreat began, the calving front was positioned on a shallow shoal. The front subsequently retreated into deeper water behind the shoal and the glacier began a period of dramatic speed-up, longitudinal extension, and thinning. As of 2008, these processes are continuing, along with further retreat. The net retreat from 1980 to 2008 was about 18 km.

Meier and Post (1987) showed that the advance and retreat behavior of Alaskan tidewater glaciers could be explained if the calving rate (the volume of icebergs discharged per unit time per unit vertical area of the terminus) is proportional to the water depth at the terminus. Observations show that glaciers with stable termini end in shallow water, usually at the head of the fiord, but sometimes on a shoal. In contrast, termini grounded in deep water are usually retreating rapidly because of rapid calving. Greater water depth should favor rapid longitudinal extension near the terminus – for reasons explained in items 1 and 3 of Section 11.5.1 – a process that helps the ice to break apart and calve. Retreat rate, however, is the difference between ice velocity and calving rate, and both are affected by the variables in Eq. 11.68; to explain the observations of rapid retreat, calving rate must increase more than ice velocity.

Tidewater glaciers appear to go through a cycle: stable position with terminus at the head of the fiord, slow advance, stable extended position, rapid retreat. This cycle is not directly related to climatic changes, though such changes can influence the timing, especially by initiating retreat. A glacier ending at the head of a fiord gradually builds a terminal moraine shoal. Because this reduces calving, the mass balance becomes positive and so the terminus starts to advance. The advance can continue into deeper water only if the moraine moves forward to restrict the calving rate. This is accomplished by erosion of material from its upstream side and deposition on the downstream face (Alley 1991b). Because the terminal moraine can advance only slowly, the glacier advances slowly, too. Advance can, however, continue to the mouth of the fiord or at least to a place where the fiord widens. At these points, the increase in the calving rate may be sufficient to stop the advance. In addition, the increase in the ablation area resulting from the advance makes the glacier more sensitive to any small climatically induced increase in the ablation rate. Moreover, a small retreat from such an extended position moves the terminus into deeper water, increasing the calving rate and accelerating retreat.

Meier and Post (1987) estimated that the advance may continue for perhaps 1000 years whereas retreats may be completed in a century or less. Times for individual glaciers depend on the length and depth of the fiord, and on the rate at which erosion provides material to build the terminal moraine. Nearby glaciers may therefore behave asynchronously.

Though incompletely understood, the mechanics of flow in the retreat phase are a subject of great importance. Might a similar process decimate the large ice streams in Antarctica and Greenland whose beds are far below sea level? Fast flow and rapid calving occur together during retreat. Section 11.5.1 enumerates the factors contributing to rapid flow. Particularly important is a low effective pressure at the bed – a condition in which the ice nears flotation – because otherwise rapid flow would generate a large basal drag that neutralizes the force driving stretching (Eq. 11.68). No quantitative calving law has been established; this is an obstacle to modelling the retreat phase. Fracturing due to longitudinal extension probably sustains large calving rates (Section 4.6). Rapid extension, in turn, also accompanies fast flow in conditions of near flotation.

The sustained character of retreats, once begun, indicates that strong positive feedbacks operate in the tidewater glacier system. The preceding discussion and Eqs. 11.67 and 11.68 suggest the following summary, diagrammed in Figure 11.19:

> Some initial retreat leads to a small thinning, as would occur on any glacier. This reduces the effective pressure at the bed, $N = P_i - P_w$, because the water pressure at the terminus is fixed by the water depth. (Up-glacier, water pressures on the bed must normally be higher still to accommodate drainage.) This process, labelled "A" in the diagram, distinguishes tidewater margins from glaciers ending on land. A decreased N reduces basal drag, causing stretching and faster flow. More of the driving stress is balanced by τ_L and τ_w. Stretching thins the ice, further reducing N and completing a positive feedback. The feedback should operate most strongly in places with N already close to zero, because a small decrease of N causes a large increase of stretching. However, in principle, reduction of driving stress due to the thinning (labelled "B") partly counteracts the unstable feedback.
>
> Stretching and perhaps thinning increase calving and hence lead to further retreat of the terminus (labelled "C"). How the response proceeds then depends on the bed topography. With an adverse bed slope (B decreasing inland), the positive feedbacks are reinforced; ice thickness at the terminus increases with retreat, increasing the driving force and, because the terminus moves into deeper water, decreasing N further (assuming that the freeboard h cannot change much). These events also contribute to faster flow and stretching.

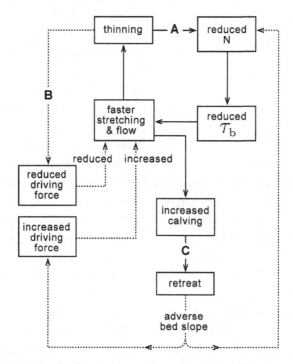

Figure 11.19: Schematic of the feedback processes operating in tidewater glaciers following an initial small retreat and thinning.

The relative importance of all these processes has not been evaluated completely in any specific case. The major limitations are: (1) lack of established relations for calving and basal drag and (2) absence of measurements of basal conditions. An active tidewater glacier is a particularly challenging environment for obtaining measurements of basal conditions.

11.5.3 Interactions of Ice Shelves and Inland Ice

At many of the marine margins in polar regions, the grounded "inland" ice is attached to floating tongues and shelves. In this case Eqs. 11.67 and 11.68 and the concepts outlined in Section 11.5.1 still apply, but it is instructive to replace Eq. 11.68 with the force balance for a section of flow line extending outward from inland coordinate x_o only to the grounding line (at $x = X_G$), rather than to the floating margin. (We have discussed the forces and flow on the seaward floating portion in Section 8.9.3. Readers may wish to review that section for information about buttressing forces.) The stretching rate at x_o, related to thinning by Eq. 11.67, is given by

$$4\eta H \frac{\partial u}{\partial x} = \phi_G [1 - f_b] + \rho g \int_{x_o}^{X_G} H \frac{-\partial S}{\partial x} dx - \int_{x_o}^{X_G} [\tau_b + \tau_w] dx, \tag{11.73}$$

where ϕ_G signifies the contribution of the floating shelf to the driving force; the force ϕ_G pulls the inland ice toward the sea. As before, "force" refers to the force per unit width across-glacier. In terms of ice thickness at the grounding line, H_G,

$$\phi_G = \frac{1}{2}\rho g \left[1 - \frac{\rho}{\rho'} \right] H_G^2 \tag{11.74}$$

(Eq. 8.110). But flow of the ice shelf over grounded spots and past lateral margins provides a resisting force – the back-force or buttressing force – that partly compensates ϕ_G. The ratio of back-force to ϕ_G defines the *degree of buttressing*, given by the parameter f_b (Dupont and Alley 2006). A value $f_b = 0$ means there is no buttressing effect and the shelf gives no resistance to flow. If $f_b = 1$, the shelf is fully buttressed and does not pull on the inland ice. In this case, no stretching occurs at the grounding line, but buttressing by the shelf allows the ice to maintain the thickness H_G.

Perturbations to any of the terms in Eq. 11.73 lead to longitudinal stretching or compression of the inland ice, and hence to a coupled response involving ice flow and thickness, as outlined previously. Important mechanisms for producing a response include:

1. Loss of buttressing when an ice shelf thins, disintegrates, or retreats. Loss of buttressing is represented by a reduction in the value of f_b in Eq. 11.73. The buttressing force is reduced in two ways: the ice loses contact with local grounded regions ("pinning points") that provide basal drag, and the ice loses contact with shearing margins that provide side drag. The latter is particularly important because side drag accounts for most of the buttressing on typical

shelves. Thinning, disintegration, and retreat of the floating ice can occur if, for example, warm ocean waters melt the base, warm air melts the surface and drives fracture growth and calving, or ice-dynamical factors increase calving.

2. Retreat of the grounding line – represented by a reduction in the value of X_G in Eq. 11.73 – when an ice shelf thins or disintegrates. Such a change reduces the resisting force from basal and side drags provided by formerly grounded ice. At locations inland, the total driving force decreases comparatively slowly, at a rate dependent on thinning of the glacier. Thus the loss of resisting force must produce a response of the inland ice. Moreover, this process should impact the inland ice most profoundly if the grounding line retreats into deeper water, because the grounding-line thickness and hence driving force increase.

3. Reduction of basal drag because of hydrological or mechanical changes at the bed: for example, a decrease of effective pressure, or a smoothing of basal topography. As with the case of tidewater glacier retreat, discussed in the previous section, thinning of the grounded ice can be expected to decrease the effective pressure, because the depth of water at the grounding line constrains the water pressure. This mechanism, too, should operate most effectively where the bed elevation declines up-glacier.

Pattyn et al. (2006) and Dupont and Alley (2005, 2006) conducted numerical simulations of the reaction of inland ice to changes on an ice shelf. These simulations used a realistic creep relation for ice. The Pattyn et al. study used a complete force balance relation. These studies concluded that changes of either ice shelf buttressing or grounding line position can force a large response of the inland ice. Such simulations are not predictive – they require ad-hoc assumptions about the magnitude of basal drag as a function of distance inland – but can be regarded as a rigorous demonstration of the ability of floating and grounded ice to interact through force perturbations.

The progress of speed-up, stretching, and thinning of inland ice depends, in principle, on all of the processes discussed in Sections 11.5.1 and 11.5.2, and thus is contingent on factors such as subglacial topography that are unique to each glacier system. One broad generalization, however, concerns how far inland the *initial* response is felt when a loss of buttressing occurs; it depends on the size of the force perturbation ($\phi_G \Delta f_b$) relative to the size of perturbations in the forces opposing the stretching, $\Delta \int [\tau_b + \tau_w] dx$. The latter increases with distance inland, because it is integrated over x; thus the initial response is greatest near the grounding line and decreases inland. It is reasonable to expect that, in general, if $\int [\tau_b + \tau_w] dx$ and $\int \tau_d dx$ are small for a great distance inland, then the initial response occurs to a great distance inland, too. This means that the length of the *transition zone* between shelf flow and the region of normal ice sheet flow (where $\tau_d \sim 100 \, \text{kPa}$) determines how far inland the force perturbation matters (van der Veen 1985; Pattyn et al. 2006). Also important is the driving force magnitude ϕ_G, and hence ice thickness at the grounding line. Consider an ice stream with $\tau_d = 20 \, \text{kPa}$, a grounding line thickness of 1 km, and complete buttressing ($f_b = 1$); these values typify the Siple Coast in

West Antarctica. From Eq. 11.74, the driving force of the shelf is about $\phi_G = 5 \times 10^8$ Pa m. If the shelf disintegrated, leading to complete loss of buttressing, the driving force would increase by 10% far inland – about 250 km from the coast. As a contrasting case, consider a thinner glacier, 500 m thick at the grounding line, with an abrupt transition from "normal" inland ice to the shelf. Given $\phi_G \approx 1 \times 10^8$ Pa m, the driving force perturbation from loss of buttressing decreases to 10% at a distance of only about 10 km inland, compared to 250 km for the Siple Coast example. An example of an abrupt transition occurs at Ekstromisen, East Antarctica, a slow-flowing outlet glacier (Mayer and Huybrechts 1999). Loss of ice shelves has less of an influence on the inland ice in regions like Ekstromisen than in regions like the Siple Coast.

Again, regardless of the spatial extent of the initial response, a thickness change at the original grounding line propagates inland as a diffusive wave. This may, over time, influence the ice sheet all the way to its central divide, as described previously in Section 11.4.2.2.

11.5.3.1 Recent Changes of Glaciers in the Antarctic Peninsula

In 1995, the Larsen A ice shelf on the Antarctic Peninsula rapidly disintegrated, a consequence of warm summer temperatures leading to increased melt and water filling of crevasses (see Chapter 4). Subsequently, its tributary glaciers accelerated, thinned, and retreated (Rott et al. 2002; De Angelis and Skvarca 2003). The major fast-flowing tributaries thinned dramatically and sped up considerably. In contrast, the smaller, slow-flowing tributaries did not change much.

In early 2002, a large section of the Larsen B ice shelf, a larger shelf south of the former Larsen A, also disintegrated by the same mechanism. Within months of this event, the lower parts of the four major glaciers flowing into the collapsed section accelerated and began to thin (Scambos et al. 2004; Rignot et al. 2004). Centerline velocities on these glaciers were as much as eight times larger in 2003 than in 2000, before the collapse. Thinning rates were tens of meters per year. Significant rates of thinning and speed-up occurred more than ten kilometers inland from the new terminus, and their magnitudes increased toward the coast. Moreover, the two glaciers flowing into the surviving part of the ice shelf experienced no similar speed-ups or elevation changes. It is clear from this pattern, from the timing, and from the large magnitude of the dynamical changes, that these events were a response to loss of the ice shelf. Figure 11.20 presents some of the observed changes of flow in this region (Rignot et al. 2004).

The observations are broadly consistent with expectations from the theory outlined in the previous section, if ice shelf disintegration resulted in loss of buttressing at the grounding line, as seems likely. This force perturbation caused stretching, speed-up, and thinning (Eqs. 11.67 and 11.73). The role of feedbacks – for example, that between thinning and effective pressure changes – is not known explicitly.

Rignot et al. (2005) reported speed-up and thinning of glaciers flowing into another shelf that disintegrated, the Wordie Ice Shelf. This shelf was located on the opposite side of the Antarctic Peninsula from the Larsen shelves. The glaciers examined by Rignot et al. were, in 2005, discharging some 80% more ice than required for balance with accumulation in their basins.

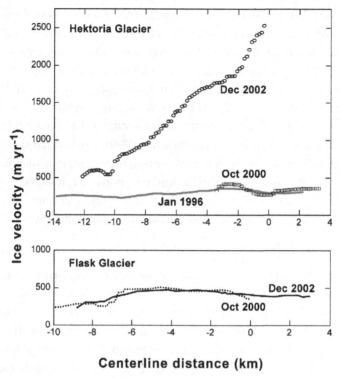

Figure 11.20: Longitudinal profiles of ice velocity along two glaciers before and after the collapse of the Larsen B Ice Shelf in early 2002. Ice flow is from left to right. Hektoria Glacier flowed into the part of the ice shelf that collapsed. Flask Glacier flowed into the surviving part of the ice shelf. Data courtesy of E. Rignot; analysis from Rignot et al. (2004). Data density is similar in all curves, but, for clarity, points are shown explicitly for only two cases.

11.5.3.2 Recent Changes of Pine Island Glacier, Antarctica

Pine Island Glacier (PIG) is the largest ice stream of West Antarctica (see Section 8.9.2.5). Over the last 15 years the ice shelf at the terminus of PIG, and neighboring shelves, thinned at rates of meters per year and their grounding lines retreated (Rignot 1998, 2002; Rignot et al. 2002; Shepherd et al. 2004). Retreat was rapid, as fast as 5 km in 2 years. Comparison of recent observations to photographs taken in the mid twentieth century shows that new cracks and rifts have formed. All of these changes are suggested to be a consequence of warming ocean waters eroding the shelves from below.

As with the Antarctic Peninsula glaciers, thinning of ice shelves and grounding-line retreat should perturb the force balance near the margin and increase the stretching, flow, and thinning of the adjacent grounded ice (Shepherd et al. 2004). Such processes are indeed observed. PIG and its major neighbors, Thwaites and Smith Glaciers, thinned from 1995 to 2004 at rates up to $5\,\mathrm{m\,yr^{-1}}$ near the grounding line and more slowly inland (Shepherd et al. 2001;

Rignot et al. 2002; Thomas et al. 2004). Ice flow measurements show that the thinning is caused by increased stretching and flow rather than increased ablation. Observed thinning rates near the grounding line were much larger in 2002–2003 than in the mid 1990s. The excess of discharge over accumulation in this region contributes about 0.2 mm yr^{-1} to global sea-level rise, according to velocity measurements obtained in 2002–2003.

Observations reveal dynamical thinning occurring at very large distances inland from the coast – more than 200 km (Figure 11.21). PIG is thinning at about 1 m yr^{-1} at 150 km inland. Payne et al. (2004) and Schmeltz et al. (2002) have modelled the dynamical response of the Pine Island Glacier system to changes at the coast. They concluded that the observations, including thinning far inland, can be explained by perturbations of the force balance related either to ice shelf thinning (which reduces the drag on the shelf sides, and hence the buttressing force), or to thinning of the adjacent ice plain. ("Ice plain" refers to a nearly flat zone of the glacier immediately inland of the grounding line.) Partial ungrounding and reduced effective pressures beneath the ice plain would reduce the resistance $\int \tau_b \, dx$ in Eq. 11.73 (Rignot et al. 2002). Thinning of ice at the coast is probably a consequence of melt driven by warming ocean waters (Gille 2002).

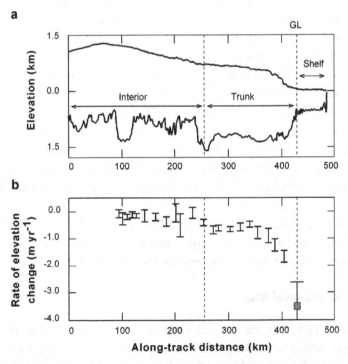

Figure 11.21: Longitudinal profiles of surface and bed elevations on Pine Island Glacier (top), and the corresponding measured rates of thinning (bottom). The dashed line labelled "GL" indicates the grounding line position. Adapted from Shepherd et al. (2001).

Some authors have stated that the observed rapid response of the ice sheet – extending more than 100 km inland – is unexpected from the view of classical glacier response theory. However, the discharge from this system is large and, excepting one region at 30 to 70 km from the grounding line, the surface slope is low. With values typical of the main trunk of PIG ($u = 500$ m yr^{-1}; $H = 1.5$ km; $\alpha = 2 \times 10^{-3}$), a diffusional wave of thinning should propagate a characteristic distance of about 100 km in only one decade (Eq. 11.72).

11.5.3.3 Recent Changes of Glaciers in Southeast Greenland

Helheim and Kangerdlugssuaq are the two largest outlet glaciers in southeastern Greenland (Section 8.9.2.2). They descend steeply through a mountain range, in a state of extending flow, and terminate in tidewater.

From the mid to late 1990s, Kangerdlugssuaq sped up by as much as 1 km yr^{-1} and thinned throughout a 70 km-long section from near the crest of the mountain range to the coast (Thomas et al. 2000). There were no major changes in the calving terminus in this time. The speed-up may have resulted from increased surface melt lubricating the bed, as summers were unusually warm in the late 1990s.

Between 2000 and 2005 the calving termini of both glaciers retreated, likely a consequence of the thinning over the previous several years. With retreat, ice velocities increased strongly, accompanied by further thinning (Luckman et al. 2006; Howat et al. 2005). The correspondence of frontal retreat and speed-up on Helheim Glacier is depicted in Figure 11.22. The most plausible explanation for this response is that, as discussed in Section 11.5.3, retreat of the calving front led to a decreased resisting force ($\int [\tau_b + \tau_w]\, dx$ in Eq. 11.68) without a corresponding decrease of driving force. The rate at which driving force decreases is limited by how quickly the ice can thin, and thinning occurs more slowly than retreat. The speed near the front of Kangerdlugssuaq increased particularly strongly between 2004 and 2006, by about 15 m day^{-1}. Luckman et al. (2006) suggest this was caused by sudden ungrounding of a large section of the glacier near the terminus, which would have abruptly reduced basal drag over that region. In 2006 the speeds decreased back to rates prevailing in 2004, after thinning and retreat of the calving front brought the glacier to a more stable geometry (Howat et al. 2007).

Again, thinning propagates up-glacier as a diffusive wave. The surface slope of Kangerdlugssuaq is about 0.025, and discharge per unit width was about 3 km^2 yr^{-1} in 1999. The nominal propagation distance over one decade thus equals about 60 km (Eq. 11.72).

11.5.4 Forcing by Sea-level Rise

When a large and sustained increase of sea level occurs, grounding lines shift inland and the ice sheet thins over millennia as shown in Figure 11.12. Sea-level rise forced the margins of the Antarctic, Greenland, Laurentide, and Scandinavian Ice Sheets to retreat during the end of the last ice age. Hughes (2002) envisioned how such a retreat proceeds. Calving rates should be greatest where water is deepest; thus the margins should retreat most rapidly along trajectories

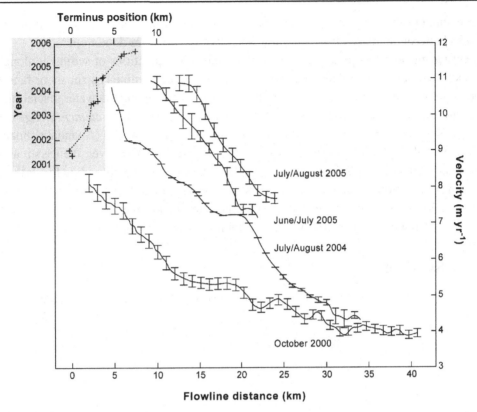

Figure 11.22: Longitudinal profiles of velocity of Helheim Glacier, Greenland. Distances increase inland; terminus positions are shown at the upper left. Adapted from Howat et al. (2005). Used with permission from the American Geophysical Union, Howat, I.M., Joughin, I., Tulaczyk, S., and Gogineni, S., Rapid retreat and acceleration of Helheim Glacier, east Greenland (2005), *Geophysical Research Letters*, Fig. 3/pp. 3703–3712.

of maximum water depth, such as those through the northern entrance to Hudson Bay. Slower retreat occurs in the shallow waters on either side. The ice front would form a giant calving embayment, as observed at a much smaller scale on retreating tidewater glaciers. The role of increased water temperature in these retreats needs more investigation; subaqueous melt might be a driving factor.

At shorter timescales, small changes of sea level shift the position of the grounding line by floating or grounding ice. They also affect the force balance at the ice sheet margin. How so depends on the topography of ice and bed. A small sea-level rise might unground ice shelves from pinning points, decreasing f_b in Eq. 11.73. Grounding lines also shift inland. If the bed slopes downward, inland, then acceleration of flow and thinning will occur as described in Sections 11.5.1 through 11.5.3, and feedbacks can drive unstable retreat as outlined for tidewater glaciers in Figure 11.19. If, instead, the bed rises upstream, the inland ice will temporarily thin and speed up, but the grounding line will presumably soon find a new stable position because feedbacks between increased flow, thinning, and effective pressure do not operate.

Tidewater glaciers and polar ice streams transport large quantities of sediment along their beds. Such sediment accumulates near the grounding line or terminus as shoals or fans. Here the topography of the bed can change rapidly, as when an advancing tidewater glacier moves a shoal along a fiord. Recent observations have revealed a wide shoal or *sediment wedge* (or "till wedge") at the grounding line of Whillans Ice Stream in West Antarctica (Anandakrishnan et al. 2007; Alley et al. 2007). This feature thickens the ice upstream. Moreover, the ice stream does not go afloat until it descends the downstream side of the wedge. The grounding line is therefore positioned at a location where the bed locally slopes upward inland, irrespective of the regional bed slope. Alley et al. pointed out that this configuration makes the ice stream grounding line stable against small rises of sea level, of at least a few meters in amplitude.

Further Reading

The book by Oerlemans (2001) provides an extensive analysis of how mountain glaciers respond to climate change.

CHAPTER 12

Glacier Surges

"The glacier has the appearance of rocky mountains . . . or rather as people might imagine an ocean, frozen in a storm."

Jökull 12 (1962),[1] **Thorvardur Kjerulf**

12.1 Introduction

A few examples of glaciers suddenly and rapidly advancing have been known for about a hundred years. A glacier, after showing little sign of unusual activity for many years, starts to move rapidly and its surface is transformed to a chaotic array of crevasses and ice pinnacles. Typically, the ice in the lower reaches of the glacier moves several kilometers in a period of a few months to a few years. The rapid movement then stops. This is quite different from a normal advance in which the terminus may move forward a few meters or a few tens of meters per year. Figure 12.1 shows the same glacier just before and after a surge.

Such behavior is sometimes called a *catastrophic advance*, while the hallucinatory term *galloping glacier* is popular with journalists. The term *surge* should be used. Before a surge, the lower part of a glacier – typically spanning several kilometers upstream from the terminus – often consists of stagnant ice covered with a layer of rubble. This ice becomes reactivated and thickens greatly during the surge, and the boundary between active and stagnant ice advances as the surge proceeds. But this active *surge front* may not move beyond the limits of the previously stagnant ice. To speak of an advance is not always correct. And in some surges, particularly those of Arctic tidewater glaciers, the entire lower glacier starts to move rapidly all at the same time, without propagation of a surge front. Surges of glaciers in cold regions such as the high Arctic typically are slower and last longer than those elsewhere.

During a surge, a large volume of ice shifts from a reservoir area to the terminal part of the glacier. Many surges are confined to the lower parts of glaciers; others affect the whole glacier. During the quiescent period, converging ice flow thickens the upper reservoir area and the lower part stagnates and thins by ablation. The surge appears to start when the glacier attains some critical profile. Ice deformation cannot produce the high velocities observed in most surges; the

---

[1] Translation by Throstur Thorsteinsson.

Copyright © 2010, Elsevier Inc. All rights reserved.
DOI: 10.1016/B978-0-12-369461-4.00012-0

Figure 12.1: Variegated Glacier, Alaska. (a) Shortly before a surge (29 August 1964). (b) At the end of the surge (22 August 1965). The glacier width is about 1 km in the narrow valley. Photographs by Austin Post, United States Geological Survey. Reproductions are now available through the "Glacier Photograph Collection" of the National Snow and Ice Data Center (University of Colorado, U.S.A.).

flow must be due to basal slip, some combination of sliding and substrate deformation. Although many surges have now been observed, they are nonetheless rare events.

The pattern of medial moraines on surging glaciers is distinctive. Instead of smooth lines, more or less parallel to the valley walls, the moraines can be deformed into large bulb-like loops or folds (see Figure 10.4). A loop forms by flow of ice from a tributary glacier while the main glacier is quiescent. The surge of the main glacier then carries the loop several kilometers down the main valley. The tributary forms a new loop after the surge. Some loops may also form when tributary glaciers surge. These distinctive moraine patterns, combined with a long stretch of stagnant, dirt-covered ice and high trim lines at the terminus, are signs that a glacier surges.

12.2 Characteristics of Surging Glaciers

12.2.1 Spatial Distribution and Relation to Geological Setting

Surging glaciers have been identified only in certain regions. The regions span a wide range of climates and geological settings, but also appear to have one geological factor in common.

Post (1969) identified 204 surging glaciers in western North America, most of them near the Alaska-Yukon border. Surging glaciers are confined to the St. Elias Mountains, the Alaska Range, and parts of the Wrangell and Chugach Mountains. None exist in the United States outside Alaska, nor in the Coast Range, Selkirk Mountains, and Rocky Mountains in Canada. The distribution of surging glaciers within any region is not random. Most of the surging glaciers in the Alaska Range lie along parts of the Denali Fault. In the St. Elias Mountains, most of the glaciers in the valley of Steele Creek surge. Yet in this region no obvious environmental factors correlate with the concentrations of surging glaciers (Clarke et al. 1986). Even in ranges with surging glaciers, only a few glaciers may surge. Glaciers in the same area do not necessarily surge at the same time. The Black Rapids and Susitna glaciers originate in the same ice field, but one surged in 1936–1937 and the other in 1953. Moreover, some glaciers that used to surge no longer do so.

In Iceland, all the main outlets from the Vatnajökull ice cap surge (Björnsson et al. 2003). They are broad, gently sloping lobes, underlain by volcanic material that is probably highly permeable and highly erodible. In contrast, none of the four eastern outlets surge; these are valley glaciers, narrow and much steeper than the main outlets. Gently sloping outlets of the other Icelandic ice caps also commonly surge. The lower parts of many of the surging glaciers rest in overdeepened valleys. Some overlie volcanic bedrock, others on fragmentary volcanic deposits and sediment.

Some 50 glaciers and ice cap outlets in Svalbard have surged, including one that advanced 21 km, the greatest movement yet recorded in a surge (Dowdeswell et al. 1991). Within Svalbard, surging glaciers mostly occur on sedimentary bedrock, in particular on shales and mudstones (Hamilton and Dowdeswell 1996; Jiskoot et al. 1998).

In Asia, at least 40 surging glaciers have been identified in the Pamirs, 21 in the Tien Shan, seven in the Caucasus, and one in Kamchatka (Dolgushin and Osipova 1975). Surges have also been reported from the Karakoram, where Kutiah Glacier advanced 12 km in two months in 1953 (Hewitt 1969). No glaciers in the European Alps surge at present, although there may have been at least one in the past (Hoinkes 1969). There are surging glaciers in the Chilean Andes (Lliboutry 1958c) and Arctic Canada (Hattersley-Smith 1964; Copland et al. 2003a), and a small number in Greenland (Weidick 1988). The latter include an outlet of the Greenland Ice Sheet (Reeh et al. 1994) and a large tidewater valley glacier (Murray et al. 2002).

This summary shows that most surging glaciers occur in tectonically active mountain ranges or geological provinces with weak bedrock – volcanic or sedimentary. Such regions are undergoing, or susceptible to, rapid erosion. This almost certainly means that sediment on a glacier bed significantly assists the surging process. Perhaps a deformable bed is a necessary condition for typical surges. Sediment underlies each of the few surging glaciers whose beds have been reached by boreholes (Black Rapids, Trapridge, Bakaninbreen, and Variegated). On the other hand, a few surging glaciers – one in Alaska and several in northwest Iceland (Björnsson et al. 2003) – have been observed to advance onto bedrock surfaces at their margins.

12.2.2 Distribution in Time

The repeating pattern of looped moraines on many surging glaciers suggests that surges recur at regular intervals. Indeed, direct observations indicate that many, but not all, surges recur at semiregular intervals (Table 12.1). This fact, and the asynchronous surging of adjacent glaciers, indicate that a surge originates from an instability within the glacier rather than an external event such as an earthquake or a change in climate. The data sources for Table 12.1 are Thorarinsson (1964, 1969), Björnsson et al. (2003), Field (1969 and pers. comm.), Post (1969, 1972, and pers. comm.), Krenke and Rototayev (1973), Dolgushin and Osipova (1975), Fatland and Lingle (1998), Eisen et al. (2001), and Haeberli et al. (2004).

Some of the irregularities in period may result from lack of observations: for example, a surge of Variegated Glacier about 1926 could well have passed unnoticed. On the other hand, the 1965–1966 surge of Bering Glacier is well documented. The total ice movement was, however, less than half of that in the 1958–1960 surge (Post 1972). Some deviations from a regular period are expected. How long it takes for the glacier to build up a critical profile must depend to some extent on accumulation and ablation rates, a fact we return to later (Eisen et al. 2001). Other deviations defy understanding. On some Icelandic glaciers, the intervals between surges range from about five years to more than 30 years. Variations of mass balance alone probably cannot account for such a large range.

The duration of surges varies. The 1963 and 1973 surges of Medvezhiy Glacier lasted only three months. The 1982–1983 surge of Variegated Glacier was in two phases, lasting six and

Table 12.1: Dates of observed repeated surges in the same glacier.

Glacier	Location	Dates
Brúarjökull	Iceland	1625, 1730, 1775, 1810, 1890, 1963
Dyngjujökull	Iceland	1900, 1934, 1951, 1977, 1999
Sidujökull	Iceland	1893, 1934, 1963, 1994
Mulajökull	Iceland	1924, 1954, 1966, 1971, 1979, 1986, 1992, 2002
Skeitharárjökull	Iceland	1787, 1812, 1857, 1873, 1929, 1985, 1991
Carroll	Alaska	1919, 1943, 1966, 1987
Tyeen	Alaska	1948, 1966, 1989
Rendu	Alaska	1950, 1966, 1987
Bering	Alaska	1920, 1938–40, 1958–60, 1965–66, 1993–95
Variegated	Alaska	1906, 1947, 1964–65, 1982–83, 1995
Kolka	Caucasus	1834, 1902,[†] 1969, 2002[†]
Medvezhiy	Pamirs	1937, 1951, 1963, 1973

[†] Ice avalanche.

nine months, with an interval of three months in between (Kamb et al. 1985). The most active phase of surges in Iceland typically lasts less than one year, but some enhanced motion occurs over two to three years (Björnsson et al. 2003). In contrast, eight surges in Svalbard lasted between three and 10 years. The quiescent phase in Svalbard is also long, from 50 to 500 years (Dowdeswell et al. 1991). The longest observed surge, a slowly propagating feature in Trapridge Glacier, Yukon, lasted about 25 years (Frappé and Clarke 2007).

The time of onset also varies, but many surges start in fall or winter. The surges of Medvezhiy Glacier start in late winter and the two phases of the surge of Variegated Glacier began in January and October. The surge of West Fork Glacier, Alaska, began in late summer, shortly after the end of the melt season (Harrison et al. 1994). The surge of Sortebrae, a large valley glacier in Greenland, began in mid-winter (Pritchard et al. 2005). Although many surges in Iceland begin in winter, not all of them do (H. Björnsson, pers. comm.).

12.2.3 Temperature Characteristics

Table 12.2 lists temperatures measured during quiescent periods. The first two glaciers are in Alaska, the next three in the Yukon, Vestfonna in Svalbard, and Medvezhiy in the Pamirs. Because significant basal motion occurs at the Variegated, Black Rapids, and Medvezhiy glaciers during quiescent periods, their beds must always be at melting point. On any reasonable extrapolation, the basal ice of the accumulation area of Vestfonna must also be melting. In the Rusty and Trapridge glaciers, deep temperatures were measured. Here only parts of the bed in the center

Table 12.2: Measured 10-m temperatures in surging glaciers.

Glacier	Temperature (°C)	Zone	Reference
Variegated	0	Acc. + Abl.	Bindschadler et al. 1976
Black Rapids	0 −0.6 to −1.7	Acc. Abl.	Harrison et al. 1975
Steele	−2 to −6	Abl.	Clarke and Jarvis 1976
Rusty	−5 to −8	Acc. + Abl.	Clarke and Goodman 1975
Trapridge	−3 to −8	Acc. + Abl.	Jarvis and Clarke 1975
Vestfonna	−0.6 to −0.9 −8 to −10	Acc. Abl.	Schytt 1969
Medvezhiy	−0.9	Abl.	Dolgushin and Osipova 1975

of the glaciers are at melting point; the sides and front are frozen (Jarvis and Clarke 1975). The coldest known surging glacier is Otto Glacier in northwest Ellesmere Island (Hattersley-Smith 1964). Although temperatures have not been measured, values in ice caps in the same region suggest that the 10-m temperature probably lies between −10 and −20 °C. Because the ice is several hundred meters thick, however, the basal ice may be at melting point.

The temperature data thus indicate that both temperate and polythermal glaciers can surge. For rapid slip to occur during the surge, as observed, the basal ice must be at melting point. Between surges, parts of some but not all surging glaciers appear to be frozen to their beds.

12.2.4 Characteristics of Form and Velocity

Glaciers of a wide variety of sizes and types can surge. In North America they range from the 200-km-long Bering Glacier to an unnamed glacier with a length of only 1.7 km. Many surging glaciers are large valley glaciers with low gradients. On the other hand, many small surging glaciers drain hanging valleys or flow down steep slopes (Post 1969). Speeds and displacements also vary widely. High velocities are about $100 \, \mathrm{m \, day^{-1}}$ for short periods, and $5 \, \mathrm{km \, yr^{-1}}$ maintained for one or two years. Low velocities are only several tens to a few hundred meters per year, values typical of many nonsurging glaciers. As a further complication, some large glaciers show small pulses of activity repeated at intervals of 10 years or less (Meier and Post 1969) and other glaciers have both surges and small-scale pulses (Post 1969). In the Saint Elias Mountains, the longer glaciers are more likely to surge than shorter ones (Clarke et al. 1986; Clarke 1991); so too in Svalbard (Hamilton and Dowdeswell 1996). In Iceland, ice cap outlets are most likely to surge if they slope gently. Apparently, the steep ones flow fast enough without surging that they never build up to a critical threshold (Björnsson et al. 2003). On the other hand, long glaciers in Svalbard are more likely to surge if they are steep compared to other glaciers of similar length (Jiskoot et al. 2000).

12.3 Detailed Observations of Surges

12.3.1 Surges of Temperate Glaciers

The first surging glacier studied throughout a full cycle was Medvezhiy Glacier (Dolgushin et al. 1963; Dolgushin and Osipova 1973, 1974, 1975, 1978). Its 1963 and 1973 surges were observed in detail. Each lasted about three months, ending in late May. The glacier area is $25\,km^2$ and its elevation ranges from 2850 to 5500 m. An icefall separates the accumulation area from the tongue, which is about 8 km long, 100 to 200 m thick, and confined in a narrow valley. Its average surface slope is about 5°; the driving stress thus equals about 100 kPa, the typical value for most mountain glaciers, whether they surge or not. Only the tongue, which lies entirely in the ablation zone, surges.

Observations during the surge showed the following features. The stagnant ice in the terminal area was reactivated or overridden by active ice. The new active terminus advanced 1.5 km in less than two months at a maximum velocity of $105\,m\,day^{-1}$. In Figure 12.2, the 1962 and 1964 lines show the surface profile before and after the surge. During the surge, the surface of the upper part of the tongue dropped by up to 100 m while the lower part thickened by as much as 150 m. The increase in volume of the lower part appeared to equal the decrease in

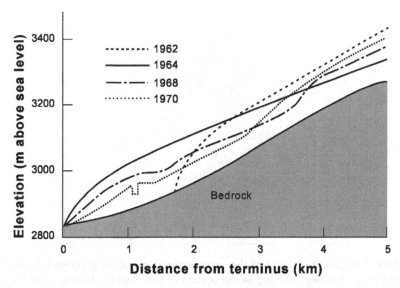

Figure 12.2: Longitudinal profiles of Medvezhiy Glacier in different years. The glacier surged in 1963 and 1973. Adapted from Dolgushin and Osipova (1975).

the upper part. At the foot of the icefall – the upper limit of the area visibly affected by the surge – a large transverse fracture formed, with the ice on the downstream side displaced about 80 m vertically downward. A continuous fracture, 10 m wide and possibly reaching the glacier bed, extended along both sides of the tongue. Blocks of ice were left attached to the valley walls.

The rapid transfer of a large volume of ice from a *reservoir area* to a *receiving area* is a characteristic feature of surges. A plan-view map of the difference in surface elevations before and after a surge makes possible a precise accounting of the transfer; Figure 12.3 depicts one example, the elevation changes resulting from the 1998–2000 surge of Dyngjujökull, an ice cap outlet in Iceland. In general, reservoir and receiving areas are not the same as the accumulation and ablation zones. Indeed, as at Medvezhiy Glacier, surges are sometimes confined to the ablation zone. A break in slope between a steep upper portion and a gently sloping tongue often seems to limit the area affected by surges of valley glaciers.

The total down-glacier displacement of ice during a surge is often a significant fraction of the glacier length. Figure 12.4 shows the total displacements accomplished by a surge of Muldrow Glacier in Alaska. Ice in the vicinity of cross-section AB, the boundary between the reservoir

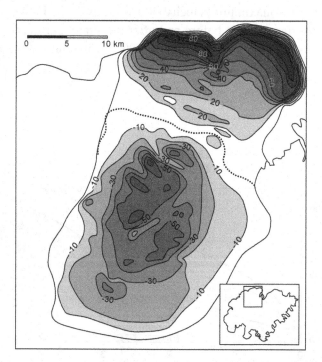

Figure 12.3: Net changes (in meters) of the surface elevation of Dyngjujökull, Iceland, during its 1998–2000 surge, showing thinning of the reservoir area and thickening near the terminus. Inset shows its location on the north flank of Vatnajökull ice cap. Adapted from Björnsson et al. (2003).

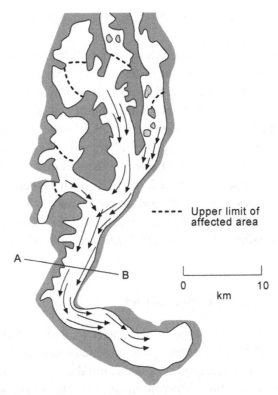

Upper limit of
affected area

A

B

0   10
km

Figure 12.4: **The observed displacement of ice during the surge of Muldrow Glacier, Alaska.**
Adapted from Post (1960) and used with permission from the American Geophysical Union,
Journal of Geophysical Research.

and receiving areas, moved the farthest. However, the amount of movement varied little both
along the glacier and across it. Stanley (1969) observed similar features during the surge of
Steele Glacier in the Yukon.

Ice velocities during most surges are 10 to 100 times those in normal glacier flow, and must be
due to fast basal slip; such high velocities cannot be produced by deformation of a few hundred
meters of ice. Slip must also occur in the quiescent period on some surging glaciers, such as
Medvezhiy and Black Rapids, because they move faster in early summer than in winter.

Waves of increased thickness and velocity have been observed during surges. Four waves,
from 50 to 70 m high and about 2 km long, travelled the length of the tongue of Medvezhiy
Glacier during its surge. This distance is about four times the total ice displacement (Dolgushin
et al. 1963). During the surge of Muldrow Glacier, the region of maximum ice velocity moved
down the glacier at a speed of several times that of the ice (Harrison 1964). (Above this region, the
ice was in tension and heavily crevassed; below, the ice was in compression and thickening.) In
the four-year-long surge of Tikke Glacier in British Columbia, rapid movement was confined to
the upper half of the glacier in the first year, and moved further down-glacier in each subsequent

year (Meier and Post 1969, Figure 2). By the last year, only the lowest quarter of the glacier moved rapidly.

At the end of a surge, the profile of the glacier surface differs markedly from a steady-state profile; the slope is too low. The lower part of the receiving area stagnates but the upper part of the reservoir area soon starts to recover. Velocities in the lowest 5 km of Medvezhiy Glacier dropped to about $0.5 \, \text{m yr}^{-1}$ while ablation thinned the ice at about $7 \, \text{m yr}^{-1}$ and smoothed the broken surface. Continuing flow through the icefall, however, caused the reservoir area to thicken as it compressed longitudinally. Two years after the 1963 surge, the glacier immediately below the icefall was thickening at 15 to $20 \, \text{m yr}^{-1}$. When allowance is made for ablation, this represents an upward ice velocity of about $25 \, \text{m yr}^{-1}$, much higher than in normal glaciers. Upstream thickening produces a step in the surface profile and, as Figure 12.2 shows, the step moves down the glacier. A surface profile in equilibrium with the mass balance is never attained, however, because at some stage of the buildup a critical profile is reached and the next surge begins.

12.3.2 The Role of Water: Variegated Glacier

Variegated Glacier (Figure 12.1) is a temperate glacier in southeastern Alaska, about 20 km long, 1 km wide, and a few hundred meters thick. Detailed observations began in 1973, about halfway through its quiescent period, and continued until the end of a surge in 1983 (Raymond and Harrison 1988; Kamb et al. 1985). Only the uppermost 2 km of the accumulation zone, and a small stretch of stagnant ice at the terminus, were unaffected by the surge. As at Medvezhiy Glacier, the quiescent period saw thickening of a reservoir area and a progressive increase in velocity, more marked in summer than in winter. In the summers immediately preceding the surge, flow occasionally accelerated for periods of perhaps 10 to 20 hours. Similar events, accompanied by increases in basal water pressure, have been observed in nonsurging glaciers.

The surge was in two phases (Figure 12.5). Phase 1, from January to July 1982, was confined to the reservoir area. The increased flux of ice during this phase thickened the lower part of the reservoir area by about 30 m. In Phase 2, from October 1982 to July 1983, this bulge propagated down-glacier as a surge front or kinematic wave with a velocity of a few times the ice velocity. The compressive strain rate at the surge front, where the velocity decreased from the surging value upstream to near stagnation downstream, attained values as high as 0.2 per day. The propagation of the surge front into stagnant ice near the terminus was observed in detail; Raymond et al. (1987) modelled the associated ice deformation. Oscillations in ice velocity, with periods of a few days, occurred during the surge, especially during the final stages (Figure 12.5).

During the surge, the glacier moved as a block with little variation in velocity across a transverse line; deep faults separated the ice from the valley walls and greatly reduced their drag.

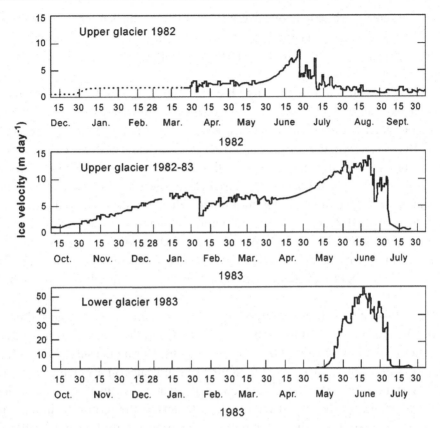

Figure 12.5: Measured velocities of Variegated Glacier during its surge. The curves for the upper glacier are a compilation of data from the zone between 2.8 and 8.3 km from the head of the glacier; lower glacier data are from markers near 15 km. Redrawn from Kamb et al. (1985).

Inclinometer measurements in a borehole, repeated two days later, showed that only 5% of the surface movement of 27.3 m resulted from ice deformation. The remainder was attributed to basal slip. The bed of Variegated Glacier has been described as "hard" (rigid and impermeable) in some reports, but the original investigators believe this is probably a mistake (C. Raymond and W. Harrison, pers. comm.; Harrison and Post 2003). Photography in a borehole revealed dirty basal ice resting on loose rock debris (see Figure 1 of Harrison and Post 2003). The thickness of this layer is not known, however. The foreland of the glacier, exposed by retreat, is mantled with sediment. Thus it is likely that sediment mantles the bed beneath the glacier too, although not necessarily as a continuous layer.

Important features of the work on Variegated Glacier were borehole measurements of basal water pressure, dye-tracing experiments, and monitoring of subglacial streams emerging at the terminus (Kamb et al. 1985; Humphrey and Raymond 1994). During the surge, water pressures remained within 0.5 MPa of overburden, with frequent fluctuations to within 0.15 MPa thereof, and occasional increases slightly above overburden. (These values refer to measurements at a few boreholes, not averages over the bed.) During the surge the greatest velocities appeared to coincide with peaks in water pressure. Before and after the surge, water pressures were considerably lower.

Are the water pressures during the surge high enough to deform basal sediments? Though not discussed by Kamb et al. (1985), a quick calculation shows that the sediments were probably deformable. Sediments deform if a shear stress applied to them exceeds a value $f[P_i - P_w]$. Here $[P_i - P_w]$, the *effective pressure*, is the difference between the ice overburden and the water pressures, and f denotes the friction factor or *internal friction* for the sediment (Section 7.3.3). At Variegated Glacier, the basal shear stress was 100 to 200 kPa during the surge (Raymond 1987, Figure 8). Using mid-range values for shear stress and effective pressure at this time of 150 kPa and 0.3 MPa, to deform sediments would require an f of about 0.5. This is a typical value. Before and after the surge, values for $[P_i - P_w]$ were typically 0.8 to 1.6 MPa, at times of minimum water pressure. Basal sediments would not have deformed at these times. They might have deformed, however, during transient periods of high water pressure.

Figure 12.6 illustrates the results of dye-tracing experiments. Dye was injected into the bottom of a borehole connected to the basal water system, and then measured in water samples collected from the outflow streams at the glacier terminus. During the surge, transit from borehole to terminus took, on average, about 90 hours; this indicates an average velocity for water in the basal system of only $0.02 \, \text{m s}^{-1}$. Moreover, the water was dispersed across the glacier bed because the dye appeared in all three outlet streams at the terminus. In an experiment after the surge, in contrast, the dye appeared as a sharp peak only four hours after injection and was detected only in the main outlet stream. The corresponding speed of water flow, $0.7 \, \text{m s}^{-1}$, is comparable with values measured in surface streams and in normal mountain glaciers in late summer. The solute concentration in the water emerging at the glacier terminus was measured, indirectly, from its electrical resistivity (Humphrey and Raymond 1994). During the surge, daily variations of resistivity were small. After the surge, the variations greatly increased, showing that the daily pulses of surface meltwater now moved swiftly through the glacier.

Floods in the outflow streams accompanied rapid decreases in ice velocity, such as those in the late stages of the surge (Figure 12.5). A particularly large flood coincided with the end of the surge. The outflow streams carried much more silt during the surge than before or afterward, an observation previously made in Iceland (Thorarinson 1964). This suggests that surges are times of rapid erosion of subglacial sediment, and perhaps bedrock too. Humphrey and Raymond (1994) showed that the apparent erosion rate – the volume of rock emerging at the glacier front in unit time, divided by the glacier area – varied approximately $10^{-4}u_b$, where u_b denotes the rate of basal slip. For u_b of $10 \, \text{m day}^{-1}$, as observed, this amounts to an annual rate of more than $0.3 \, \text{m yr}^{-1}$, a very large value compared to typical long-term erosion rates in mountain ranges of 0.1 to $5 \, \text{mm yr}^{-1}$.

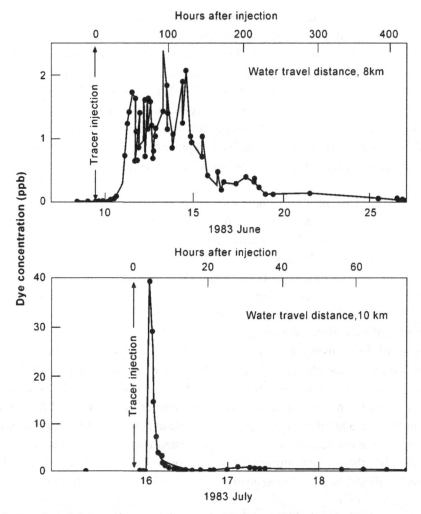

Figure 12.6: Results of dye-tracing experiments in Variegated Glacier, during its surge (top panel) and after the surge (bottom panel). Note the difference in scale between the diagrams. From Kamb et al. (1985).

Observations of water flow through other surging glaciers have confirmed many of these findings from Variegated Glacier (Section 12.4.1).

12.3.3 Surges Where the Bed Is Partly Frozen

Trapridge Glacier is a small surging glacier in the Yukon. Its thermal structure differs from that of the Medvezhiy and Variegated glaciers, where the whole bed is always at melting point. At the base of Trapridge Glacier, ice frozen to the bed surrounds a central area of melting ice. Although 10 m temperatures are between -4 and $-9\,°C$, large quantities of surface meltwater reach the bed in summer (Clarke et al. 1984a). Because the ice is less than 100 m thick, its deformation contributes little to the surface velocity. Sediment covers the bed. Most of the basal slip results from a combination of deformation within the subglacial sediment and sliding on the interface. These processes have been measured – during a slow surge – by instruments inserted in boreholes (Blake et al. 1994; Fischer and Clarke 2001; Kavanaugh and Clarke 2006). The slip rate varied strongly over time as the water pressure fluctuated (Figure 7.11). Typically, during periods of low water pressure the motion was slow and largely occurred by shear within the sediment layer. When water pressures were high, rapid motion occurred, mostly by sliding. The following description of the surge history summarizes information given by Frappé and Clarke (2007).

The glacier surged in the 1930s and 1940s. Between 1941 and 1951, when observations of the glacier front were made, the terminus advanced by about 1 km. The average rate of advance during the most active phase therefore must have been at least 100 m yr^{-1}. The following quiescent phase lasted about 25 years.

Starting around 1976, a new surge began when ice velocities on the lower glacier increased, over one to two years, to four to five times their quiescent values. The increased flow persisted until about 1999, when velocities began a pronounced decline. By 2005, the flow had returned to values typical of the earlier quiescent period.

This surge differed considerably from the earlier surge of this same glacier, and from the surges of the Variegated and Medvezhiy glaciers. It lasted much longer and was much less intense (with velocities of 40 m yr^{-1} and an advance rate of the surging zone of 25 m yr^{-1}). For many years, observers did not regard it as a surge at all, because of the low rates of flow and advance. It never culminated in a highly active phase comparable to that in the 1940s; apparently, climate warming reduced the net accumulation and so prevented the glacier from achieving sufficient mass. Nonetheless, these events differed in no essential way from many surges observed in Svalbard, and so should be regarded as an extreme example of a slow surge (Fowler et al. 2001; Frappé and Clarke 2007).

The front of the surging zone propagated at a rate of about 25 m yr^{-1} into stagnant ice downstream. A large bulge formed at this boundary (Figure 12.7). The boundary between surging and stagnant ice corresponded to the lower boundary between a melting and a frozen bed. Subfreezing temperatures inhibit basal slip, so the bulge propagates downstream only by ice deformation. This was a slow process because, although the front slope of the bulge was steep, its thickness was less than 50 m, and therefore the driving stress was low. Although the boundary between surging and stagnant areas was fixed by the pattern of basal temperatures, the ice flow, the advance of the bulge, and the basal temperature must have been tightly coupled; the arrival of the bulge allowed the bed to warm to melting point, a process driven by the increased ice thickness and the upward ice flow associated with longitudinal compression (Clarke 1976).

Heat to warm the stagnant ice also apparently derives from meltwater flowing beneath the frozen sediments underlying the lower glacier (Clarke et al. 1984a). This is the most plausible

a

b

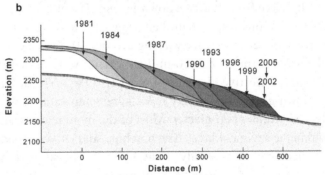

c

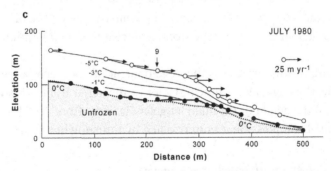

Figure 12.7: continued on next page.

Figure 12.7: The slow surge of Trapridge Glacier. (a) Aerial view of the bulge, 24 June 1980. Photo courtesy of G. Clarke. (b) Year-to-year progress of advance of the bulge. Adapted from Frappé and Clarke (2007) and used with permission from the American Geophysical Union, *Journal of Geophysical Research*. (c) Surface velocity vectors and isotherms (°C) in 1980. The bed is at melting point up-glacier of marker 9; dashed line shows approximate location of 0°C isotherm. Adapted from Clarke et al. (1984a).

explanation for temperature measurements showing a heat flow into the basal ice of roughly ten times an average geothermal flux. This indicates that the frozen bed beneath the terminus did not trap meltwater under the ice upstream.

The slow surge of Trapridge Glacier resembles the surges of Svalbard glaciers that terminate on land. Of these, Bakaninbreen has been studied in greatest detail (Murray et al. 1998, 2000; Porter et al. 1997; Porter and Murray 2001). Bakaninbreen surged from 1985 to 1994, during which time a surge front with a distinct bulge propagated about 6 km down-glacier. The bedrock is mudstone and sandstone, and the glacier rests on a sediment layer, at least 1 to 3 meters thick. Temperatures were measured in two boreholes that reached this layer, one at a location up-glacier of the surge front, and one down-glacier. The effective viscosity of the basal sediment was inferred by inserting rods into the bed. Downstream of the surge front, the bed was frozen and the effective sediment viscosity was high (order 10^{13} Pa s). Upstream of the surge front, the bed and the lower 40 m of the 110-m-thick glacier were at melting point. Here the sediment was much weaker (effective viscosity of order 10^{10} Pa s). The difference in viscosities at the up-glacier and down-glacier sites likely reflects the temperatures, with frozen sediment down-glacier and thawed sediment up-glacier. Most of the motion at the up-glacier site arises from deformation within the sediment layer. The borehole studies were conducted near the end of the surge, when the ice velocity was only about 3 m yr^{-1}. The surge front propagated most rapidly (1 to 1.8 km yr^{-1}) in the first four years of the surge (1985–1989), and subsequently slowed to negligible rates by 1995. Propagation apparently involved fracturing of the stagnant ice and injection of water ahead of the surge front, a process that might limit the rate of advance (Stuart et al. 2005).

An unusual feature of this surge was the development of a surface bulge (referred to as a *fore-bulge*) in the nearly stagnant ice down-glacier of the surge front. This feature was 20 to 30 m high, almost half the height of the main bulge at the surge front. The fore-bulge is thought to form by longitudinal compression accommodated by thrusting in the glacier.

12.3.4 Surges of Polythermal Tidewater Glaciers

Monacobreen is a 40-km-long glacier in Svalbard (Murray et al. 2003). It drains an ice cap and terminates in a marine fiord. Though not known precisely, the ice thickness is probably about 200 m. Radar soundings did not image much of the bed because of strong reflections from within the ice, at 120 to 200 m depth. This reflecting horizon corresponds to a boundary between an upper layer of subfreezing ice and a bottom layer of temperate ice.

The recent surge of Monacobreen lasted more than six years, peaking in 1994, and advanced the terminus by 2 km. Typical ice velocities were about $500 \, \text{m yr}^{-1}$, compared to an expected velocity from internal deformation of only 2 to $3 \, \text{m yr}^{-1}$. In the surge, the ice flow increased for a period of about three years, with half of the increase occurring in five months, and then decreased for more than three years. This long deceleration is very different from the abrupt terminations observed at temperate glaciers such as Variegated, but is similar to the end of the last surge of Trapridge. Moreover, at Monacobreen the surge began simultaneously over the entire lower glacier (Figure 12.8). Thus, in contrast to the temperate glaciers, to Trapridge, and to the land-terminating glaciers in Svalbard, the surge of Monacobreen did not propagate down-glacier as a front. This appears to be characteristic of surges in Svalbard glaciers that terminate at calving margins.

Sortebrae is a 65-km-long valley glacier in East Greenland that terminates in tidewater. Its thermal regime has not been measured, but given the location it is almost certainly polythermal. The lower 53 km of the glacier surged in 1993–1995 (Murray et al. 2002; Pritchard et al. 2005). Surge velocities were typically 10 to $20 \, \text{m day}^{-1}$, values more typical of temperate glaciers than of the polythermal glaciers in Svalbard. The terminus advanced by 10 km. The surge began in the middle of the glacier, in winter, and propagated both up- and down-valley. Termination occurred rapidly, in a few months, and coincided with the onset of the melt season. A freshwater plume appeared in front of the glacier at this time. During the surge, large lakes of sediment-rich water appeared on the glacier margins. These observations suggest that a large volume of basal water was involved in the surge and released at its end, as occurs in temperate glaciers. The bedrock here is strongly fractured basalt, suggesting again a substrate of sediment.

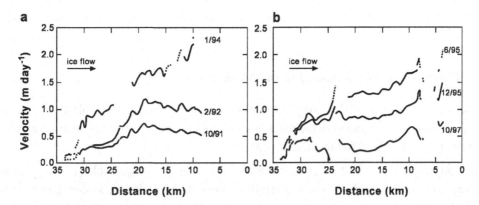

Figure 12.8: Longitudinal profile of velocities on Monacobreen during its surge: (a) At three times during the onset, and (b) three times during the waning phase. The entire lower glacier was affected throughout the duration of the surge. Adapted from Murray et al. (2003).

12.4 Surge Mechanisms

A complete model of the processes responsible for surges should be able to explain why some glaciers surge but others do not, why surges are initiated, interrupted, or terminated at certain times, and why some glaciers have brief *speed-up events* that do not develop into surges. Such a model has not been established, and none is in view. Even if plausible quantitative models are formulated, it may be impossible to validate them, given the difficulty of learning conditions over broad regions at the glacier bed.

Our discussion here focuses on hypotheses for the basic mechanisms, rather than quantitative predictions. To argue for one basic mechanism that can explain all surges may be unrealistic. Surging glaciers have different characteristics. At one extreme are temperate glaciers such as Variegated Glacier; large volumes of surface meltwater flow through and under it every summer. Various thermal mechanisms of surging have been proposed (Schytt 1969; Clarke 1976; Clarke et al. 1977; Fowler et al. 2001); none of these can apply to surges in temperate glaciers. At the other extreme are glaciers with a central area of basal ice at melting point but a frozen bed around the periphery, under relatively thin ice. Surface water still reaches the bed, but its quantity is limited by cool surface conditions and partial refreezing within the glacier. Glaciers in Svalbard with ice thicknesses of several hundred meters may be in this category. Trapridge Glacier, less than 100 m thick, exemplifies an intermediate case with large fluxes of surface meltwater reaching a partly frozen bed. (Many nonsurging glaciers have the same type of temperature distributions as these surging glaciers.)

We must also distinguish between glaciers with hard and soft beds. The relation between glacier flow and basal drag differs for these situations. No direct evidence demonstrates that surges can initiate and develop on glaciers with hard beds. Observations of surging glacier beds are few, however. The fronts of surging glaciers do sometimes advance onto hard beds (Tyeen Glacier in Alaska, and a few glaciers in Iceland). Björnsson et al. (2003) believe that some surges in Iceland originate on hard beds, but this has not been established by direct subglacial observations.

12.4.1 General Evidence Relevant to the Mechanism

The evidence presently available suggests that most surges arise from the plastic properties of subglacial sediments together with feedbacks that disrupt the normal subglacial drainage system. Four lines of evidence lead to this conclusion:

1. The subglacial water system changes drastically over the surge cycle. During the surge, drainage along the bed is slow and spatially dispersed; the hydraulic system conducts water with difficulty. After the surge terminates, on the other hand, much of the basal water moves rapidly in conduits. At Variegated Glacier, dye-tracing experiments showed that, during the surge, water moved along the bed at a speed of only a few centimeters per second. After the surge, the speed was 35 times greater, and typical of flow in open channels. These findings

were replicated in a remarkable natural experiment during a surge of Skeidarárjökull, Iceland (Björnsson 1998). Here, water periodically drains rapidly from the subglacial lake Grimsvötn, producing subglacial floods. One such event occurred during a surge. The water flowed along the bed at about $0.15\,\mathrm{m\,s^{-1}}$. This rate, and the total flux of water, could be explained by flow through roughly 10^4 conduits of $0.1\,\mathrm{m}$ diameter. A long delay separated the release of water from the lake and its emergence at the front of the glacier; during this interval, the stored quantity of water on the bed amounted to a layer of $0.3\,\mathrm{m}$ thickness, averaged over the entire glacier area. Another flood occurred after the surge, and this time the water moved at about $6\,\mathrm{m\,s^{-1}}$, consistent with a single tunnel tens of meters in diameter. These observations confirmed that a tunnel drainage system was absent during the surge but present afterward. Again, observations in Iceland show that when the surge front reaches the front of the glacier, water emerges in numerous small outlets (Björnsson et al. 2003). At other times, in contrast, water emerges in a few large tunnels.

Temperate glacier surges terminate when an efficient drainage system develops. This usually happens in early to mid-summer, when surface melt reaching the bed contributes to the formation of conduits. The newly formed conduits rapidly drain water from the glacier bed, producing floods in the glacier outlet streams during surge termination. Such "terminal floods" were documented in detail for the 1983 surge of Variegated Glacier, discussed earlier and also the 1988 surge of West Fork Glacier, Alaska (Harrison et al. 1994). A later surge of Variegated Glacier, in 1995, terminated after two days of record-high temperatures, which presumably generated a large quantity of surface melt (Eisen et al. 2005). More generally, a surge might terminate when its front reaches thin ice near the terminus, allowing the subglacial water to escape (Sharp 1988). The 1985–1994 surge of Bakaninbreen terminated when water leaked through the frozen zone beneath the glacier and through fractures in the ice leading to the glacier surface (Murray et al. 2000).

Results from theoretical treatments of the subglacial water system provide useful insights (see Chapter 6). For water flow in distributed systems at a glacier bed, theory indicates that the water pressure increases with the water flux. Thus, a focussing of water flux elevates pressures; water tends to leak away, maintaining the distributed configuration of the system. In a tunnel system, in contrast, the water pressure is lowest in passageways conveying the greatest flux. Thus, if a tunnel system forms, conduits should be stable and might coalesce into a few large tunnels. When this happens, some water remains widely distributed on the bed, but some leaks into tunnels and quickly drains away, reducing the overall volume of basal water and its pressure. The formation of a tunnel system thus reduces the lubrication of the bed and reduces the rate of basal slip.

The essential difference between the tunnel system and the distributed system is that, in the tunnel system, large localized fluxes of water concentrate the dissipation of heat and hence the melt. This will not happen in a distributed system, regardless of the details of its geometry. The distributed system might be a system of linked cavities on a hard bed (as envisioned, for example, in the theory of Kamb 1987), or it might be a *macroporous horizon,* or an irregular system of films and blisters on a deformable bed (Flowers and Clarke 2002a). The larger the volume of basal water, the more extensive such distributed systems must be.

2. The presence of deformable sediments on the glacier bed strongly favors surging (cf. Harrison and Post 2003). No other factor explains, simply and plausibly, why surges occur mostly in certain regions – either tectonically active, rapidly eroding mountain ranges, or geological provinces with soft or fragmentary bedrock (including basaltic volcanic rocks, and sedimentary rocks with abundant bedding planes and fractures). Again, within Svalbard the surging glaciers occur preferentially on the weakest bedrock types. Furthermore, sediment underlies each of the surging glaciers where boreholes have reached the bed: Trapridge, Black Rapids, Bakaninbreen, and Variegated. At the first three of these, the sediment layers were probed and found to be at least a few decimeters in thickness.

3. Subglacial sediment deforms, to a good approximation, as a perfectly plastic material (Section 7.3.3). That is, no significant deformation occurs until the applied stress attains a particular value, the yield strength. Once deformation begins, the resistance increases little, or not at all, as the rate of deformation increases. Thus, a glacier flowing over sediment can speed up significantly without producing a counteracting resisting force from its bed – provided that the sediment layer is continuous and the bed smooth. Furthermore, the yield strength, τ_*, obeys the Mohr-Coulomb relation,

$$\tau_* = C + f [P_i - P_w] . \tag{12.1}$$

Here C is the cohesion (a negligible quantity while deformation occurs), f the friction factor (a number usually between 0.2 and 0.8), P_i the normal stress on horizontal shear planes (assumed to equal the overburden of ice), and P_w the water pressure. If P_w rises to equal P_i, the sediment loses all strength ($\tau_* \rightarrow 0$).

A glacier flows on deformable sediment either by sliding along the interface or by riding atop a layer undergoing internal shear; Eq. 12.1 applies to either mechanism.

4. Surges typically occur only after the mass of the glacier in the reservoir area builds to a critical level. At Variegated Glacier, for example, faster mass accumulation reduces the interval between surges (Figure 12.9; Eisen et al. 2001). Likewise, the long interval between surges in Svalbard reflects the low rate of mass accumulation in the island's high-polar climate.

The gravitational driving stress acting on a glacier, $\tau_d = \rho g H \alpha$, is proportional to the mass per unit area. During quiescence, a glacier flows slowly and viscous deformations at its sides provide little resistance to the driving force. Most of the resistance must come from the bed, and so the basal shear stress τ_b approximately equals τ_d. Thus, the observations indicate that the basal shear stress must attain a critical value for surges to occur.

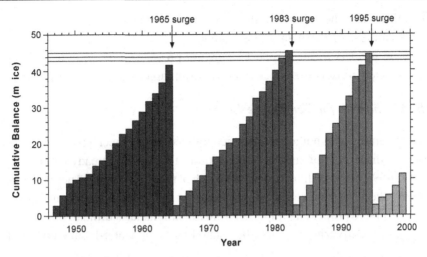

Figure 12.9: Estimated cumulative specific mass balance of Variegated Glacier between consecutive surges. The three horizontal lines indicate the average of the cumulative balance just before each surge (43.5 m) and plus/minus one standard deviation. Adapted from Eisen et al. (2001).

Consider the implications of items 3 and 4 together, with respect to a glacier resting on a bed of sediment. When $\tau_b < \tau_*$, no sediment deformation occurs: the bed maintains rigidity and behaves like porous bedrock. The τ_* fluctuates, however, as water pressure varies; the sediment deforms at some times but not others, and in some patches of the bed but not others. As glacier mass increases, more of the bed reaches a condition $\tau_b = \tau_*$ and for longer periods of time. Perhaps surge initiation occurs when the prevalence of regions with $\tau_b = \tau_*$ ("critical regions") attains a threshold (see later). This explains why surges do not occur until glacier mass builds to a critical level. Furthermore, it explains why some glaciers stop surging when climate warming reduces their size. For example, Svalbard warmed considerably in the early to mid-1900s, and the number of actively surging glaciers then decreased, from 18 in the 1930s to five in 1990 (Dowdeswell et al. 1995). Again, the recent slow surge of Trapridge Glacier never culminated in a rapid surge like the one in the 1940s, because the glacier mass grew too slowly.

In the Yukon and in Svalbard, longer glaciers are more likely to surge than short ones. Factors that inhibit drainage of water along the glacier bed should favor surging. Long glaciers imply that much of the basal water must follow a long path to the terminus, and the water system must deal with comparatively large fluxes. Put simply, long glaciers should tend to trap water. This might explain the observed tendency for surging of long glaciers. On the other hand, a similar argument should apply to glacier surface slope; other factors being equal, gentle surface slopes imply low hydraulic potential gradients and inefficient drainage. Indeed, in Iceland, large ice cap outlets with gentle surface slopes are more likely to surge than steep ones. In contrast, however, slope does not correlate with surge behavior in the Yukon (Clarke 1991), and moreover, steep glaciers (for their length) in Svalbard are more likely to surge (Jiskoot et al. 2000). Because slope and length covary, the validity of these conclusions depends on assumptions underlying the statistical treatment of the data; more analysis is needed. Another possibility is that the tendency for long glaciers to surge reflects a greater prevalence of continuous basal sediment layers; longer glaciers must convey rock material from larger catchment basins.

12.4.2 The Mechanism for Temperate Glaciers

Figure 12.10 diagrams a hypothetical and generalized surge mechanism. It originates from a long history of discussion of surges, in addition to the studies already cited in this chapter. Röthlisberger (1969) proposed long ago that a switch of the basal drainage network from conduits to a distributed "sheet" system could be necessary for a surge. Dolgushin and Osipova (1975) were perhaps the first to propose an important role for subglacial sediment. In summarizing their model for surges of Medvezhiy Glacier, they stated that "the water-loam emulsion lubricating

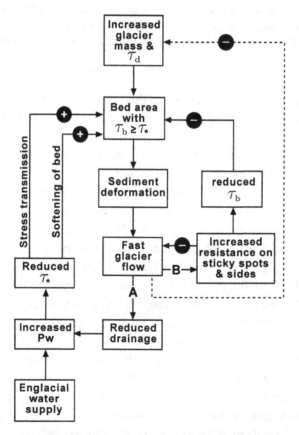

Figure 12.10: Schematic chart of feedback loops in the hypothesized surge mechanism. Dominance of the positive loops (indicated by plus symbols) allows surging. Negative loops (indicated by minus symbols) can prevent surges from developing. The dashed line indicates that a surge ends eventually because of depleted glacier mass.

the glacier bed enables a sharp increase of the speed of flow." Clarke et al. (1984a) first proposed that deformation of sediments would disrupt the drainage system on the bed; hence a drainage switch, as envisioned by Röthlisberger, would occur as an important feedback for surging but not necessarily as the initial trigger. Jones (1979) suggested that surges could be explained by softening of basal sediments by high water content.

In the quiescent phase, glacier mass increases and consequently the basal drag τ_b increases. On some patches of the bed, at some times, τ_b attains the yield stress of the sediment layer ($\tau_b = \tau_*$). The sizes of these patches increase over time, and they appear more frequently. Eventually, during a period of high P_w and hence low τ_*, these patches are large enough and connected enough to permit a large section of the glacier to move by deformation of the sediment, with enough displacement to disrupt the drainage conduits (labelled "A" in the figure). Because high P_w is an essential factor, surge will initiate preferentially at the times of the year when drainage along the glacier bed is inefficient.

Disruption of drainage conduits traps water and increases its pressure upstream. This further reduces the sediment strength τ_* and increases the sizes of the slippery patches. Less of the gravitational driving stress is taken up by drag at the bed in the slippery region, and therefore more of the glacier's weight shifts to ice up-glacier and down-glacier, and to the valley walls. Such stress transmission further expands the region where $\tau_b = \tau_*$. (McMeeking and Johnson 1986 gave a mathematical analysis showing how longitudinal stresses can expand an initially small area of fast slip. Raymond 1987 and Raymond et al. 1987 also analyzed the coupled redistributions of stress and mass that occur during surge propagation.) These processes constitute a positive feedback, expressed as a surge. Once operating, the feedback loop typically breaks only when the glacier mass is depleted (τ_b becomes too small) or an efficient drainage system forms (reducing P_w and increasing τ_*). The latter involves migration of conduits up-valley from the glacier toe, or growth of conduits from large, localized inflows of surface melt.

The arrow labelled "B" in the figure indicates processes that oppose rapid motion of the glacier; they provide resisting forces that increase as the glacier flows faster. Sticky spots on the bed provide regions of high resistance; they might be knobs of bedrock sticking up into the ice, or patches of sediment with low water pressures. (These correspond to the *vestigial roughness* in the model of Fowler et al. 2001.) Viscous drag on the glacier's sides also resists fast flow.

In this view, effective stabilizing processes (loop B in the chart) can also prevent surges from ever developing. As a glacier begins to move rapidly, increased basal drag could abort a surge. A glacier with a discontinuous basal sediment layer offers the most obvious example; bedrock knobs sticking into the ice will provide increasing resistance as the glacier flows faster. Even with a continuous sediment cover, basal resistance might increase with the flow (in other words, the bed might imitate a viscous medium) if, for example, irregular bed topography interferes with shear planes in the sediment, preventing them from expanding. Alternatively, a glacier fails to surge if basal water drainage continues to be effective even while the glacier accelerates. This might be possible, for example, if sediment quantities are insufficient to fill drainage conduits, or if the glacier slopes steeply, implying a large hydraulic gradient. All of these factors might explain why not all glaciers surge, even if they rest on a deformable bed.

12.4.2.1 The Role of Englacial Water

The volume of water released from Variegated Glacier at the termination of its 1983 surge was equivalent to a one-meter-thick layer over the entire glacier area (Humphrey and Raymond 1994). Such a large volume suggests that most of the water was not on the bed when the surge began. Much of it was probably trapped during the subsequent melt season. Some of it was probably stored in the ice prior to surge onset; stored water likely helps surges to begin in winter (Raymond 1987). At the end of the melt season before a surge begins, the discharge in subglacial conduits declines and the conduits close, trapping water on the bed, in fractures and pipes within the glacier, and in the firn. The stored water will tend to leak downward and fill new cavities and pore spaces at the bed when the surge begins. With a large quantity of available englacial water – replenished during the melt season – such filling occurs with little drop in water pressure, allowing the surge feedbacks to develop.

Lingle and Fatland (2003) suggested that a transfer of englacial water to the bed might be the initial trigger for surges. In their view, surge initiation occurs at a critical glacier mass because englacial water volume is proportional to ice thickness. This idea is less plausible than a direct dependence of surge initiation on the magnitude of gravitational force; in contrast to the amount of stored water, the force must scale precisely with the ice thickness.

12.4.2.2 Hydraulic Triggers

Most authors have considered a purely hydraulic phenomenon – one independent of sediment – to be critical for the surge cycle (Kamb 1987 and references in Raymond 1987). Such an explanation does not account for the association of surges with sediment beds. On the other hand, if some surges do occur in glaciers with hard beds, as Björnsson believes to be the case in Iceland, then they must be triggered by a hydraulic switch rather than by initiation of widespread sediment deformation.

In this view, the bed must be rigid so that cavities form in the lee of bumps. As mass accumulates and τ_b climbs during the quiescent phase, the stress difference across bumps increases. Cavity formation thus increases too. (There is a decrease in the *separation pressure*, the minimum water pressure needed to form a cavity in the lee of a bump; see Eq. 7.16.) Over time, the cavities begin to connect and to drain water away from tunnels. The tunnels thus partially close, and the water pressure in them rises. This drives more water out of the tunnels, and the tunnel network might collapse entirely, triggering a surge. The surge terminates, in turn, when the small passageways (the *orifices* of Section 6.3.3) between cavities expand by melt to form tunnels again, a process driven by the large mid-summer fluxes of meltwater from the surface. How randomly oriented orifices would join to form tunnels that follow the direction of the surface slope is not clear.

To investigate these mechanisms, Kamb (1987) made a mathematical analysis of flow in a linked-cavity system. He derived a stability parameter that expresses the condition in which orifices are unstable, meaning that they grow rapidly after a small initial increase in their diameter or in water pressure. Such instability allows drainage and terminates or precludes surging. The stability parameter is a function of hydraulic gradient, sliding velocity, and effective pressure. The analysis predicts that glaciers with gentle hydraulic gradients are most likely to surge, a conclusion consistent with some data. This conclusion is not, however, useful for discriminating between alternative surge mechanisms; all surge mechanisms requiring high water pressures are favored by inefficient drainage and hence gentle hydraulic gradients.

Theories such as Kamb's that analyze linked cavity systems apply only to rigid beds. A sediment bed can behave rigidly; with water pressures low enough that $\tau_b \ll \tau_*$, stable cavities should form in the lee of large clasts. This cannot apply at times of rapid motion, however, when a distributed water system predominates. Even in the quiescent phase, water pressures locally and temporarily rise enough to produce deformation. Thus, it is not clear that rigid-bed theories of cavity formation have any applicability to surging glaciers.

Eisen et al. (2005), elaborating on an idea given by Raymond (1987), discussed why the collapse of the conduit network might be extremely sensitive to the magnitude of τ_b, and hence a plausible mechanism for surge initiation. Consider, once again, that the separation pressure decreases during the quiescent phase as the glacier grows and τ_b rises. For cavities to form on the glacier bed adjacent to a tunnel, τ_b must exceed a value proportional to $P_i - P_w$. According to the theory for a steady-state tunnel, $P_i - P_w$ varies as water flux is raised to a power $\frac{1}{12}$. This extremely weak dependence suggests that the magnitude of τ_b, rather than hydrological factors, determines when conditions allow cavity formation along the edge of a tunnel. A larger τ_b favors the transfer of water from tunnels to cavities. Each year, tunnels collapse when discharges decline late in the melt season. As τ_b rises year after year, the collapse occurs earlier in the season when the glacier holds more water. (This process occurs whether the bed is hard rock or friction-locked sediment; it does not contradict the requirement that a yield stress for sediment be exceeded.) Eisen et al.'s discussion assumes that the hydraulic gradient is fixed by the bed slope and hence independent of the water pressure. It also relies on the relations for steady-state tunnels. The first assumption is inappropriate, while the second might be appropriate depending on whether rapid fluctuations of water pressure play a role; the validity of the argument is unclear.

Fowler (1987b) made a mathematical analysis of surging, based on a phenomenological description of hydraulic effects. He specified that two possible subglacial drainage systems can exist. Each is stable, but one or the other is preferred according to the value of the sliding velocity. The system preferred at high velocities, corresponding to a linked-cavity system, has the higher water pressure. At a given value of basal shear stress, within a critical range, the sliding may be either fast or slow depending on which drainage system exists. The sliding switches from slow to fast when the basal shear stress, increasing in the quiescent phase, exceeds a critical value corresponding to the opening of cavities. Fowler showed that a model with this feature would undergo periodic oscillations resembling a surge. This conclusion may seem obvious from the assumption of two drainage types with different sliding relations. The interesting aspect of the model, however, is that it produces surge-type behavior only if model parameters take values in particular ranges; the bed must be sufficiently rough, the effective pressure in the linked-cavity system must be much smaller than in the tunnel system, and the sliding relations must have appropriate coefficients.

12.4.3 Polythermal Glaciers

The same mechanisms should operate in polythermal glaciers as in temperate glaciers, but with additional constraints on the motion. First, as illustrated by the examples of Trapridge Glacier and Bakaninbreen, the surge cannot propagate rapidly into regions of frozen bed. The surge front advances only as fast as the boundary between melting and frozen beds migrates. The rate of migration is limited by how fast ice deformation thickens the glacier and changes the vertical distribution of temperature, and perhaps by the rate of fracturing and water injection into the stagnant ice. Such limits on migration explain why polythermal surges develop slowly and last a long time compared to temperate ones. A factor possibly contributing to slow surge initiation is the absence of englacial water that can flood the bed when a surge begins. Terminations are also slow. This suggests that basal water cannot drain rapidly, perhaps because of remnants of frozen bed near the terminus, or because surface melt lacks the volume to drive tunnel formation.

In temperate glaciers, and probably in some large polythermal glaciers like Sortebrae, enough basal water is present during the surge to not only saturate basal sediments but also to fill thick films and macroporous horizons at the ice-sediment interface. It is not clear that such volumes of water can accumulate at the beds of typical polythermal glaciers. The volume of water might be small enough to inhibit sliding, prevent rapid motion, and slow down the initiation and termination of a surge.

An important unresolved question concerns *thermal triggers*: to what extent do thawing and refreezing of the bed control the initiations and terminations of polythermal surges? Warming of the bed to melting point, in a large region of the reservoir zone, might trigger a surge (Robin 1955; Clarke 1976). In this case the timing of surges reflects not the accumulation of mass but the timescale for the frozen bed to warm to melting point. Likewise, termination might occur because the bed freezes; during the surge, longitudinal extension thins the ice, increasing the vertical temperature gradient in the glacier and increasing the loss of heat from the bed by conduction.

These processes were simulated by Clarke (1976) in a numerical model of a surge cycle. Though it remains unclear whether a thermal transition initiates surges, Clarke's calculations show how the temperature field evolves as a surge progresses, results that certainly apply to glaciers like Trapridge and Bakaninbreen. Clarke's model has also been adopted, in essence, by MacAyeal (1993a,b) to explain the Heinrich events, episodes of massive iceberg discharge from the Laurentide Ice Sheet into the Atlantic Ocean during the last ice age. Thermally triggered surging is the most plausible explanation for them. (See Chapter 13.)

Fowler et al. (2001) formulated heuristic mathematical model of surge cycles resulting from thawing and freezing of the bed. This model included the conservation of water in subglacial sediment as an important constraint. Further, the water content was assumed to relate to water pressure and hence to the rate of basal slip. The authors concluded that thermally controlled surges should be characterized by slowly propagating boundaries between thawed and frozen beds, as observed, and that such surges are most likely if the sediment layer conducts water poorly. Murray et al. (2003), reviewing the differences between surges in Svalbard and those of temperate glaciers, concluded that the characteristics of surges in Svalbard could be explained by Fowler's model.

Little is known about the mechanisms important for polar tidewater glaciers such as Mona-cobreen. Compared to glaciers terminating on land, basal water pressures must be high near the terminus. This predisposes the lower part of a tidewater glacier to have a slippery bed, and might explain why the surges involve the entire lower glacier from the onset. Further, the high water pressure at the terminus reduces the ability of the bed to drain rapidly, favoring gradual termination of the surge. These subjects need more investigation.

12.5 Surging of Ice Sheets?

Can ice sheets surge? There is no evidence that broad regions of the modern ice sheets surge. (By "broad regions" we mean sectors of an ice sheet much larger than individual outlet glaciers.) In particular, measurements to date have not revealed any broad regions where the ice is nearly stagnant but steadily thickening relative to the region downstream – the signature of surging in mountain glaciers. On the other hand, surging of the Pleistocene Laurentide Ice Sheet provides the most likely explanation for Heinrich events. Unfortunately, no firm information about the glacial dynamics of these events has been obtained.

Most of the outflow from the ice sheets is through outlet glaciers and ice streams moving continuously at surging speeds (Chapter 8). In some of these features, the fast flow occurs by basal slip on deformable sediments. The main difference between these glaciers and surging ones may be that ice sheets provide enough ice to maintain fast flow, whereas surging glaciers exhaust the supply (Weertman 1969b). The boundary conditions also differ; most of the ice sheet outlets terminate at marine margins with water depths of hundreds of meters or more. Water pressures along their beds are thus permanently high.

One major outlet glacier in northeast Greenland is known to surge (Reeh et al. 1994, 2002a, 2003; Mohr et al. 1998). This glacier, Storstrømmen, retreated from 1913 to 1978 and then rapidly advanced for six years, at rates as high as $4 \mathrm{km\,yr^{-1}}$. The surge transferred a large mass of ice from the upper ablation zone, which thinned by up to 80 m, to the lower ablation zone, which thickened by a similar amount. The lower region then stagnated and now thins at the ablation rate of 1.5 to $2 \mathrm{m\,yr^{-1}}$. The upper end is thickening at 2 to $3 \mathrm{m\,yr^{-1}}$ because of ice inflow. Storstrømmen is 200 to 600 m thick, terminates in a floating ice tongue, and is grounded more than 200 m below sea level. Because it drains a catchment basin of some 600 km length, Storstrømmen probably conveys a large quantity of sediment, but the nature of its bed is unknown. It flows faster in summer than in winter, showing that basal motion remains active in the current quiescent phase. Irregular flow, though not surging, has been observed at two other outlet glaciers in northern Greenland (Mock 1966; Joughin et al. 1996).

Distinct surge behavior, as seen at Storstrømmen, is not documented for any ice streams in Antarctica. However, the flow of the Siple Coast ice streams (West Antarctica) changes significantly over centuries (Section 8.9.2.4), and this could be regarded as a type of surging behavior. Most recently, about 200 years ago, Kamb Ice Stream nearly stagnated. Water freezes to the bottoms of these ice streams, driven by steep vertical temperature gradients related to longitudinal stretching in the transition from normal ice sheet flow to streaming flow. The ice stream beds remain at melting point only because of the latent heat supplied by subglacial water originating upstream. Diversion of this water to a neighboring ice stream might explain the stagnation of Kamb Ice Stream (Alley et al. 1994). The widths of the ice streams change, which strongly influences their flow. Neither of these processes occurs in surging valley glaciers.

A question more immediate than the possibility of surges is whether thinning or removal of the ice shelves, causing the ice streams to speed up, could trigger the disintegration of the West Antarctic Ice Sheet (Mercer 1978). Observations in the Antarctic Peninsula, West Antarctica, and Greenland all show that the flow of an ice stream increases dramatically when its terminal ice shelf disintegrates, or when the grounding line retreats (see Chapter 11). This response is analogous to the acceleration and rapid retreat of tidewater glaciers in Alaska. Such a response can probably occur in West Antarctica. It arises primarily from a feedback between increased flow, thinning, and reduced basal drag, dependent on bed elevations well below sea level. This process involves large increases of flow – sometimes referred to as a "surge" – but the mechanism clearly differs from that of surging due to internal instability.

12.6 Ice Avalanches

An ice avalanche is a sudden release of a large mass of ice from a glacier. The ice falls or slides rapidly downslope, while fragmenting into smaller pieces. Such events pose a considerable hazard in steep mountains such as the Alps. Papers by Iken (1977a) and Alean (1985) offer examples of the scientific study of ice avalanches and the glaciers that release them. Margreth and Funk (1999) reported on a problem of hazard mitigation.

We are concerned here only with exceptionally large ice avalanches that involve failure of a large portion of a glacier. Consider one example: Kolka Glacier in North Ossetia, Russia, occupies a deep cirque basin in the high Caucasus Mountains (Krenke and Rototayev 1973; Haeberli et al. 2004). It is about 3 km long, slopes at 5° to 10°, and is covered with rock debris.

In 1902 and 1969 the glacier advanced catastrophically; in 1902, the advance spawned an ice avalanche that killed 32 people. It is not clear whether these events were surges lasting for months, as is typical for temperate glaciers. In summer 2002, avalanches down the steep cirque walls added ice and rock debris to the glacier surface, increasing the glacier mass by about 10%. Abundant water was observed on the glacier surface, possibly indicating elevated basal water pressures. On September 20, a landslide onto the glacier triggered a massive displacement of ice, involving 60% to 75% of Kolka Glacier's entire mass. About 10^8 m^3 of ice and rock debris (equivalent to a cubic block 450 m on a side) slid down-valley, filling the valley bottom over an 18-km-long swath. A mudflow released from this mass traveled further down-valley. More than 100 lives were lost in these events.

For such a large failure to occur, triggered by the addition of a comparatively small mass to the surface, Kolka Glacier must already have been near a critical limit for stability.

Ice avalanches are very different from surges and normal glacier flow because, in avalanches, accelerations and momentum are significant. Consider an ice body – large enough to behave as a fluid – of mass M, uniform density ρ, and area $\mathcal{A}$, resting on a frictional plane with slope β. Call the surface slope α. Assuming that shearing occurs on a basal surface, the ice accelerates at rate du/dt given by

$$M \frac{du}{dt} = \rho g \sin \beta \int_{\mathcal{A}} H [1 + \xi] d\mathcal{A} - \underbrace{\left[\int_{\mathcal{A}} \tau_\mathrm{b} \, d\mathcal{A} + F_W + F_L \right]}_{R}, \qquad (12.2)$$

where H denotes the thickness, τ_b the basal shear stress, F_W the resisting force on the sides (the wall force), and F_L the net resisting force at the head and toe. The factor ξ in the gravitational force accounts for pressure gradients, if the ice surface does not parallel the bed; it is proportional to $\tan(\alpha - \beta) \cot \beta$ (see Section 8.5.3). For an avalanche to occur, the gravitational driving force must exceed all the resisting forces (R) even while the ice begins to flow rapidly. Thus, dR/du must be small or negative.

Dynamically, ice avalanching is essentially the same as landsliding on hillslopes, which involves bedrock or colluvium rather than ice. In all such *mass movements*, catastrophic slip usually occurs because high water pressures reduce or limit the shear strength τ_* of the sliding interface. The basal stress τ_b in Eq. 12.2 cannot exceed τ_*. For a substrate of sediment, τ_* usually obeys the Mohr-Coulomb relation (Eq. 12.1); the same relation applies for sediment beneath ordinary glaciers. For a glacier on hard bedrock, a simple balance of forces shows that unstable avalanching can occur if water pressures exceed a critical value (Eq. 7.22). The criterion also matches Eq. 12.1, but with the effective friction factor (f) related to the slope and roughness of the rigid interface rather than to the internal strength of the substrate.

In contrast to many mass movements on hillslopes, glaciers do not usually move by unstable avalanching. This is partly because the slopes of glaciers tend to be small; most glaciers could not avalanche unless, over a broad region, the water pressures rose simultaneously to nearly equal the ice overburden. Moreover, the resisting forces generated by viscous deformations of the ice increase strongly with the rate of glacier flow (dR/du can be regarded as large). Such forces arise from shears on a glacier's sides, from flow around irregularities on its bed, and from flow through constrictions and bends in the valley. The effective viscosity of ice exceeds that for saturated sediments by several orders of magnitude. Yet ice does not fragment as easily as bedrock, which often contains pervasive preexisting fractures or bedding planes. This too contributes to the stability of glaciers.

Not all mass movements on hillslopes are unstable, however. There are many examples of earthflows that spend long periods in stagnation and then creep during periods of persistently wet weather. Surging glaciers can be regarded as the same sort of phenomenon.

Perhaps it is useful to view glacier surging as an intermediate behavior between ordinary glacier flow and ice avalanches. In all three cases, the stress over a broad region of the glacier bed can be close to, or at, the critical value given by the strength τ_* of the basal interface. In ordinary flow, which includes the quiescent phase for surging glaciers, the various resisting forces operate effectively while the glacier flows slowly. In surges, in contrast, the resisting forces are not effective until the glacier flows rapidly. The rapid flow exhausts the supply of mass, however, so the surge is temporary. (But with an essentially unlimited supply of mass, as in ice sheet outlets and large tidewater glaciers, fast flow continues indefinitely.) Finally, in ice avalanches such as the one of Kolka Glacier, the resisting forces do not increase effectively as the glacier speeds up, causing the whole mass to accelerate catastrophically.

CHAPTER 13

Ice Sheets and the Earth System

"Charpentier was distressed by the depth of ignorance about glaciers that the audience members continued to show." (at Neuchâtel, 1837)

The Ice Finders, **Edmund Blair Bolles**

13.1 Introduction

The major components of Earth's environment – the cryosphere, the atmosphere, the ocean, the surfaces of continents, the biosphere, and others – are inseparably linked by flows of mass and energy, and a common geographic stage. The whole system, now usually referred to as "the Earth system," operates together as a vast machine that converts solar energy and radiogenic heat into a remarkable array of environmental phenomena. The Earth system cannot be placed on a laboratory bench and manipulated in controlled experiments, but its history of natural changes provides a rich source of information. The most relevant and remarkable such "natural experiment" is the Quaternary ice age cycling, the last 2.5 Myr of alternation between cold and warm climates, accompanied by growth and decay of large northern hemisphere ice sheets: the Laurentide and Cordilleran Ice Sheets on North America, and the Fennoscandian Ice Sheet in northern Europe.

Glaciers and ice sheets interact with the rest of the Earth system in numerous ways. Most obviously, they grow and shrink when climate changes and, in turn, affect the sea level worldwide. Ice sheets also influence the global climate, by reflecting solar energy, perturbing atmospheric circulation, and releasing freshwater to the oceans. Glaciers and ice sheets erode their substrates and transport rock material to sites of deposition. The weight of large ice sheets drives viscous flows of rock in the underlying mantle, changing the land surface elevation and the gravitational field – a process known as isostasy.

There are many other examples; a whole book could be written about them, but here we can only introduce a few. We do so largely in the context of the Quaternary ice age cycles and, in the chapter that follows, the current global warming. We begin this chapter with a summary of the interactions between ice sheets and climate, relying on earlier chapters for detailed expositions of the relevant glaciological processes. From this foundation, we turn attention to Quaternary history and to numerical simulations of long-term ice sheet evolution.

Copyright © 2010, Elsevier Inc. All rights reserved.
DOI: 10.1016/B978-0-12-369461-4.00013-2

13.2 Interaction of Ice Sheets with the Earth System

Ice sheets not only react to climate changes but also modify the climate itself through feedbacks. To understand both aspects of the interaction, it is useful to keep in mind a summary of the planetary energy balance. Section 5.1.2 summarizes the mean flow of energy through Earth's atmosphere and the exchange of energy between the atmosphere and the surface. The Earth's surface is heated by two streams of radiant energy: sunlight (or shortwave radiation) and atmospheric "greenhouse" emissions (longwave radiation). As an average over the planet and an annual cycle, about 30% of the total incoming shortwave radiation reflects back to space from clouds, atmospheric aerosols, and the surface. A smaller fraction, about 20%, gets absorbed in the atmosphere by ozone, particles, and clouds. The longwave flux increases with the temperature of the atmosphere and the abundance of clouds and greenhouse gases, primarily water vapor, CO_2, and CH_4. Note that the globally averaged longwave flux absorbed by the surface – about $324\,\mathrm{W\,m^{-2}}$ – is nearly twice as large as the absorbed sunlight; greenhouse radiation must be regarded as a key determinant of Earth's climate.

Figure 13.1 shows how sea level changed with atmospheric CO_2 concentration in the last 50 Myr (Alley et al. 2005). It is clear from a variety of geological indicators that the periods with high CO_2 concentrations were also warm. The indicators include fossil assemblages and isotopic compositions of carbonates and ice. The absorption and emission of longwave radiation by CO_2, and hence the warming, increase roughly as the logarithm of the CO_2 concentration

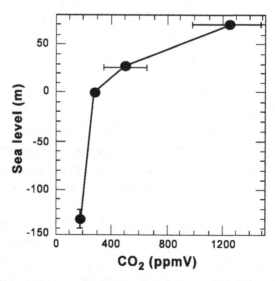

Figure 13.1: Relation between estimated atmospheric CO_2 concentration and the ice contribution to sea level, indicated by geological archives. From highest to lowest sea levels, the four points are: older than 35 Myr ago (a time of no permanent ice); about 32 Myr ago (after an ice sheet formed on Antarctica); the preindustrial modern world; and the Last Glacial Maximum (21 kyr ago). Adapted from Alley et al. (2005).

(Kiehl and Dickinson 1987). Most of the reduction of sea level at times of low CO_2 was due to ice sheet growth: first the establishment of permanent ice on Antarctica about 35 Myr ago; then the expansion of Antarctic ice combined with growth of an ice sheet on Greenland (the modern configuration); and finally the development of large ice sheets on North America and Scandinavia during the cold phases of the Quaternary. The pronounced increase of ice sheet size – and sea level lowering – in the latter periods arises from the large land surface area favorable to ice accumulation in climates slightly colder than the present one. The earlier establishment of extensive ice cover on Greenland, about 3 Myr ago, resulted from climate cooling driven by declining CO_2 levels (Lunt et al. 2008).

Global climate varies not only with CO_2 concentrations, but with the other factors that determine the cascade of energy through the atmosphere and its interaction with the surface. High-latitude climate, in addition, depends on factors that redistribute energy from place to place, such as the circulation of the ocean and atmosphere, and the exchange of heat between the surface ocean and the deep ocean. We next summarize the pathways by which climate and related changes work their influence on ice sheets.

13.2.1 Processes Driving Ice Sheet Change

Climate and sea level drive ice sheet evolution through the action of numerous mechanisms. Chapters 4 and 5 discuss the mass balance processes. Chapter 11 discusses the ice dynamical ones. The most significant are:

1. Increased snowfall directly increases the surface mass balance. In cold polar regions, snowfall tends to increase when the climate warms, but shifts of atmospheric circulation also alter the pattern and magnitude of precipitation (Section 4.2).

2. Increased summer air temperatures enhance surface melt and decrease the fraction of precipitation reaching the surface as snow rather than rain. Both mechanisms decrease the surface mass balance. On ice sheets, they operate primarily in ablation zones and near equilibrium lines; they make little difference in polar accumulation zones, where most of the water – if any is produced – refreezes in the firn (Sections 4.2 and 5.4).

3. Melt, thinning, and breakup of the marine termini of ice streams reduce the forces restraining their flow; this causes faster flow and thinning of the inland ice (Section 11.5).

4. In particular, warming of ocean waters increases bottom melt of ice shelves; as a consequence, grounding lines retreat and ice shelves thin. On the surface of ice shelves, the persistence of summer air temperatures above the freezing point generates meltwater and drives fractures downward through the ice. This process leads to complete disintegration of ice shelves. Shrinkage or disappearance of ice shelves by either process results in faster flow and thinning of the adjacent grounded ice (Sections 4.6, 10.9, and 11.5).

5. Surface melt penetrates to the ice sheet bed, increasing the flow of ice to calving margins and low-elevation ablation zones (Sections 6.2.2, 10.9, and 11.4.3.1).

6. Any factors that increase ice flow by reducing the ice viscosity or making the bed more slippery thin an ice sheet. The most important examples are: rising basal water pressures, increased abundance of deformable basal sediment, warming ice, and strengthening of c-axis fabrics favorable to shear (Section 11.4).

7. Rising sea levels drive retreat of calving margins and grounding lines. How much retreat occurs depends strongly on the topography of the bed (Section 11.5.4).

8. Subglacial sediment accumulates in shoals at the termini and grounding lines of marine margins. These features permit such margins to advance and might stabilize them against small increases of sea level (Section 11.5.4).

9. Ice sheets can retreat rapidly through deep subglacial basins because of feedbacks between ice sheet geometry, force balance, basal slip, and calving (Section 11.5.2). Such basins are formed by isostatic depression and, over geological time, by subglacial erosion.

Other factors control ice sheet response indirectly and must operate through mechanisms already listed. Most importantly:

1. Isostatic depression reduces the elevation of the whole ice sheet. Near the margins, this increases the rate of surface melt and the losses by calving into lakes and the sea (Section 13.2.2.2).

2. High geothermal heat fluxes, related to tectonic setting or volcanic activity, increase subglacial melt and so increase basal slip. The Northeast Greenland Ice Stream offers the clearest modern example of this connection (Section 8.9.2.3).

3. Expansion of the sea-ice cover on nearby oceans leads to cooling and perhaps reduced snowfall on an ice sheet (e.g., Li et al. 2005; Otto-Bliesner et al. 2006).

4. Variations of solar energy flux influence air temperatures and ablation rates at ice sheet margins. Especially important on millennial timescales are increases in summertime solar flux originating with variations of the orientation and tilt of Earth's rotation axis (Section 13.3.1).

5. Numerous factors acting together determine variations of Earth's high-latitude climate, and so influence melt, snowfall, sublimation, and calving. The most important are the clouds and gases that control the greenhouse effect; the concentration of atmospheric particulates and aerosols that reflect sunlight; the overall reflectivity of the planet related to snow, sea ice, clouds, and vegetation; atmospheric heat transport by mixing of polar and subtropical air masses; and heat transport by ocean currents. See Hartmann (1994) and Wallace and Hobbs (2006) for an overview of these climatic processes.

13.2.1.1 Timescales for Ice Sheet Growth and Decay

Surface Mass Balance Simple arguments provide an estimate for how rapidly a large ice sheet could grow or shrink when the surface mass balance changes by itself. As a first approximation, the volume of an ice sheet on a flat horizontal base of circular plan and parabolic surface profile can be shown to be

$$V = 0.53 \pi L^2 H_* \quad \text{with} \quad H_* = K L^{1/2}. \tag{13.1}$$

Here L denotes the radius, H_* the thickness at the center, and K a constant with a value of about $4\,\text{m}^{1/2}$ (Eq. 8.133). It follows that

$$\frac{dV}{dt} = 1.325 \pi K L^{3/2} \frac{dL}{dt}. \tag{13.2}$$

But this must equal $\pi L^2 \overline{b} - \dot{B}_c$, where $\overline{b}$ represents the specific mass balance averaged over the whole ice sheet and $\dot{B}_c$ the volumetric calving flux. Equating these expressions for dV/dt gives a differential equation for the ice sheet growth. For $\dot{B}_c = 0$, the solution shows that a change from size L_o (and center thickness H_o) to size L requires a time

$$t = 2.65 \frac{K}{\overline{b}} \big[L^{1/2} - L_o^{1/2} \big] = \frac{2.65}{\overline{b}} [H_* - H_o]. \tag{13.3}$$

For $\overline{b}$ of $1\,\text{m}\,\text{yr}^{-1}$ it would take 8 kyr to grow an ice sheet with a center thickness of 3 km. Because this is a rather high value for $\overline{b}$, it suggests that growing a large ice sheet takes a few tens of millennia. The timescale would be shorter if, rather than growing outward from a single origin, the ice sheet builds up from a broad region of initial accumulation. Such a pattern of growth seems necessary to explain the rapid growth of ice sheets, amounting to about 50 m of sea-level equivalent in 10 kyr, at the start of the last glacial period.

How rapidly could decreased surface mass balance destroy an ice sheet? Ablation rates in the Arctic are often three or four times a typical accumulation rate. This does not, by itself, imply that ice sheets shrink more rapidly than they grow, because ablation zones cover a much smaller area than accumulation zones. More pertinent is the sensitivity of ablation and accumulation rates to climate shifts. Suppose the ablation zone covers a fraction f_a of an ice sheet's area. Call the mean value of $\dot{b}$ in the ablation zone $\dot{b}_a$ and that in the accumulation zone $\dot{b}_c$. For an ice sheet in equilibrium with climate, $-f_a \dot{b}_a = [1 - f_a] \dot{b}_c$, and $\overline{b} = 0$. Taking for example $\dot{b}_c = 0.5\,\text{m}\,\text{yr}^{-1}$ and $f_a = 0.15$, then $\dot{b}_a = -2.8\,\text{m}\,\text{yr}^{-1}$. (These represent characteristic values for Greenland.) Two simultaneous effects lead to shrinkage of the ice sheet when the climate warms: intensification of ablation and broadening of the ablation zone. The average surface balance changes by $\Delta \overline{b}$, or

$$\Delta \overline{b} \approx f_a \Delta \dot{b}_a + \Delta f_a \big[\dot{b}_a - \dot{b}_c + \Delta \dot{b}_a \big], \tag{13.4}$$

assuming no change in $\dot{b}_c$. A large climatic warming (take $5\,^\circ C$) would increase ablation at the terminus by about $5\,\mathrm{m\,yr^{-1}}$ (Section 5.4.3) and increase the mean ablation rate by about half as much. For our example ice sheet, using $\Delta \dot{b}_a = -2.5\,\mathrm{m\,yr^{-1}}$, and simultaneously doubling the extent of the ablation zone ($\Delta f_a = 0.15$) gives $\overline{b} = -1.24\,\mathrm{m\,yr^{-1}}$. From Eq. 13.3, an ice sheet comparable to that on Greenland ($H_* = 3\,\mathrm{km}$; $\dot{b}_c = 0.5\,\mathrm{m\,yr^{-1}}$) would disappear in about 6000 years under such a warming (this corresponds to a sea-level rise of order $1\,\mathrm{mm\,yr^{-1}}$). This crude result is similar to estimates for how quickly a $5\,^\circ C$ sustained warming would destroy the Greenland Ice Sheet, according to whole-ice-sheet models that are sophisticated but do not simulate dynamical changes of outlet glaciers (Figure 5 of Alley et al. 2005).

Dynamical Changes   At the other end of the spectrum is an ice sheet that decays, not because of any change in surface balance, but because of increased flow of outlet glaciers to calving margins. Marine ice sheets can, in principle, rapidly disintegrate due to feedbacks between retreat of calving margins, increased ice flow, and increased calving. This process has been observed during retreat of large tidewater glaciers in Alaska and Greenland, with retreat rates of up to $1\,\mathrm{km\,yr^{-1}}$ and increases of calving velocity by as much as $7\,\mathrm{km\,yr^{-1}}$. Because of this behavior, and other processes that increase ice flow, the rate of ice sheet decay can be considerably greater than indicated by Eq. 13.3. For example, the geological record reveals that rapid deglaciations occurred in the past; during the termination of the last ice age, sea level rose by about $20\,\mathrm{m}$ in $500\,\mathrm{yr}$ around $14.5\,\mathrm{kyr\,b.p.}$, an event referred to as "Meltwater Pulse 1A" (Weaver et al. 2003).

Consider, again, an ice sheet of circular plan and radius L, with variables defined as before. If increased calving alone drives ice loss then

$$\frac{dV}{dt} = \underbrace{\pi b L^2 - \dot{B}_c^*}_{\text{"normal flow"}} - \Delta \dot{B}_c \tag{13.5}$$

where $\dot{B}_c^*$ represents a component of calving due to normal flow and $\Delta \dot{B}_c$ the enhancement of calving related to external influences. We assume that the terms labelled "normal flow" always sum to zero; there is always a component of ice flow, adjusted to the size of the ice sheet, that just carries away all the ice gained from the surface balance. The enhanced calving depends on the cross-sectional area and increased velocity of outlet glaciers. If the outlet glaciers, with thickness H_g, account for a fraction f_g of the total ice sheet perimeter, then

$$\Delta \dot{B}_c = 2\,\pi L f_g H_g \Delta u, \tag{13.6}$$

where Δu denotes the increase of velocity. By combining Eqs. 13.5 and 13.6 with Eq. 13.2 and $H_* = K L^{1/2}$, integration gives an expression for the time needed to reduce the ice sheet size from L_o to L. For complete disappearance ($L = 0$), the time is

$$t = 0.038\,\frac{H_o^3}{f_g H_g}\,\frac{1}{\Delta u}, \tag{13.7}$$

where H_o again stands for the initial thickness in the ice sheet center. Further, if we call u_o the outlet glacier velocity prior to retreat, when the ice sheet existed at a steady state, then $\pi \bar{b} L_o^2 = 2\pi L_o f_g H_g u_o$; rearranging gives an expression for $f_g H_g$ to substitute into Eq. 13.7. Finally, assuming that $f_g H_g$ does not change with time, complete destruction of the ice sheet occurs over an interval

$$t = 0.88 \frac{H_o}{\bar{b}} \frac{u_o}{\Delta u}. \tag{13.8}$$

Using $H_o = 3\,\text{km}$ and $\bar{b} = 0.1\,\text{m yr}^{-1}$, values appropriate for the modern Greenland Ice Sheet, gives $t \approx 3 \times 10^4 u_o / \Delta u$. By how much can ice streams speed up? In the last decade, flow of the outlet glaciers of the Greenland Ice Sheet has doubled (Rignot and Kanagaratnam 2006). Taking $\Delta u = u_o$ gives $t = 30\,\text{kyr}$. The retreat of Columbia Glacier, Alaska, into a deep subglacial basin – a topography analogous to that beneath the West Antarctic Ice Sheet – led to a six-fold increase of calving rate. Taking $\Delta u = 5u_o$ gives $t \approx 6\,\text{kyr}$. Thus a dramatic increase of ice flow acts about as quickly as surface melt after a large climate warming.

The operation of both mechanisms together would significantly hasten an ice sheet's demise. Even the combined rate, however, does not seem to explain the most rapid sea-level increases, such as Meltwater Pulse 1A. If, however, a calving margin retreated from a narrow outlet into a broad embayment, the total rate of mass loss to calving could increase greatly; rather than being constant with time, $f_g H_g$ would increase significantly. Such an episode of retreat into a "calving embayment" offers a likely explanation for the extraordinary episodes of past sea-level rise (Hughes 1992). In addition, a polar marine ice sheet such as West Antarctica would be more responsive than one like Greenland. In the former, the absence of surface melt leads to greater calving fluxes. For West Antarctica, specifically, $\bar{b} \approx 0.3\,\text{m yr}^{-1}$, as opposed to 0.1 for Greenland. Thus, from Eq. 13.8 with $H_o = 3\,\text{km}$, a six-fold velocity increase destroys the ice sheet in only about 1.8 kyr. Broadening of calving bays would make it even faster.

In general, the rate of decay of an ice sheet depends on the climate, the relation between climate and surface mass balance, the distribution of surface area with altitude, and the calving rate. The rate of mass loss from a sector of the ice sheet, extending from margin to divide, can be written in terms of the specific surface mass balance $\dot{b}_s$ and area-altitude distribution $\mathcal{A}(\text{S})$ (Section 4.2.2.4)

$$\dot{M} = \int \dot{b}_s(\text{S}) \, \mathcal{A}(\text{S}) \, d\text{S} - \dot{B}_c. \tag{13.9}$$

The rate of mass addition to the surface, $\dot{b}_s$, has units of mass per unit time per unit area. The integral is taken over all elevations S of the surface. How surface balance varies with elevation depends on latitude, direction of prevailing winds, and other factors (Section 4.2). A typical grounded ice sheet is steep near its margins, reducing the area $\mathcal{A}(\text{S})$ at low altitudes where ice rapidly ablates. When climate warms, the ablation zone (where $\dot{b}_s < 0$) extends to higher elevations and, because of the convex surface profile, its area expands more rapidly than a linear function. If the ablation zone expands into the vast, gently sloping interior region, the ice sheet will rapidly expire.

In the absence of calving margins, changes of ice flow may either counteract or enhance the decay of an ice sheet. Early in the process, a retreat of the margin (driven by ablation at low elevations) leads to an increased surface slope and hence increased flow into the ablation zone. Such enhanced flow maintains the steep slope near the margin, preventing the ablation zone from expanding inland. On the other hand, increased ice flow to the margins thins the middle of the ice sheet and draws down the elevation there. Eventually, thinning of the interior propagates to the margin, helping the ablation zone to expand inward. In both of these processes, ice flow acts by changing $\mathcal{A}(S)$ in Eq. 13.9. Moreover, as discussed previously, ice flow transfers mass to calving margins for disposal, a dominating process.

In some situations, the bed is very slippery near the ice sheet margin, leading to a flat and low surface profile (Section 8.10.2). This implies an unusually large surface area at low elevations and allows the ablation zone to expand rapidly when the climate warms. This was probably an important factor in the rapid retreat of the southwestern margin of the Laurentide Ice Sheet at the end of the last ice age.

13.2.2 Feedback Processes

The ice sheets themselves modify the rest of the planet through several important processes.

13.2.2.1 Sea Level

Currently, every $1 \, \text{km}^3$ of ice (water-equivalent volume) transferred from land to the ocean raises mean sea level by about $1/360 \, \text{mm}$ (see Section 14.1.1 for more information). This number is inversely proportional to the surface area of the global ocean, and so changes by about 10% as ice sheets wax and wane on North America and Eurasia over the ice-age cycles (Marsiat and Berger 1990). Changes of sea level are not globally uniform – Earth's rotation and gravitational field both vary as water moves between ice sheets and ocean – but for the large changes that occurred during the Quaternary the absolute sea level change can be regarded as uniform for glaciological analyses. In contrast, the local *relative* sea level – the water level measured relative to a reference point on the adjacent land – varies from place to place by tens to hundreds of meters, primarily because of isostasy.

13.2.2.2 Isostasy

Isostatic depression occurs because the weight of an ice sheet induces pressure gradients in the underlying mantle. Such gradients cause viscous creep of mantle rock away from the high pressure center, which lowers the surface above. Following removal of an ice sheet, the mantle rock flows back in, but this process takes millennia; rebound of the surface is delayed. Because isostatic movements change the altitude and slope of an ice sheet's surface, this process feeds back on specific balances, ice flow, and calving into marginal lakes and oceans.

The simplest model for isostatic compensation of ice-covered terrain treats it as a local process; that is, the compensation at a given location depends only on the loading at that point. Call Z_o the bedrock elevation that would be observed in the absence of any isostatic displacements and B the actual bedrock elevation. The magnitude of isostatic depression is $Z_o - B$. Compensation is complete when this depression equals the loading $\rho_i H / \rho_m$ for ice thickness H and densities ρ_i, ρ_m of ice and mantle rock. (The ratio $\rho_i / \rho_m \approx 1/3$.) If the depression is too small given the current loading, the bed subsides further; if too large, it rebounds. Thus

$$-\frac{\partial B}{\partial t} = \frac{1}{t_i} \left[\frac{\rho_i}{\rho_m} H(t) - [Z_o - B] \right], \tag{13.10}$$

where parameter t_i defines a lag time, usually assigned a value of 3 to 5 kyr. Many models use this local relation.

In general, however, the isostatic compensation at a point depends not only on the local loading, but also on the loading in a broad surrounding region, a few hundred kilometers in radius. The lithosphere behaves as a strong elastic plate, so any loading at a point deflects the lithosphere into a broad bowl-shaped depression. The subsidence or uplift given by Eq. 13.10 thus should depend on a spatial convolution of $H(t)$ with a filter describing the depression due to a unit loading at a point (e.g., Huybrechts and DeWolde 1999, p. 2184).

Yet more-sophisticated versions account for factors such as variations in viscosity of the upper mantle (e.g., Tarasov and Peltier 2004). The added complexity is only well motivated if the model results are being compared to rebound or gravity data. In that case it remains unclear whether the "viscous" response of the mantle – which involves nonlinear and anisotropic creep – is understood well enough to take full advantage of the information.

Isostatic adjustments occur worldwide throughout the ice-age climate cycles (Peltier 1998; Lambeck 2004). The growth and expansion of ice sheets depress the land underneath and around the edges, but also bend the surrounding lithospheric plate to form an upwarp called a forebulge that parallels the ice edges. In addition, as ice sheets grow and shrink, the partial emptying and filling of ocean basins change the load forces over much of the planet's surface. Loadings from changing ice thickness and ocean depth are inextricably linked in both space and time (Peltier 2004). The linkage occurs not only through mass directly (water to grow ice sheets is extracted from the ocean) but also through the effects of mass redistribution on Earth's rotational inertia. The rate of rotation and the position of the planetary axis both change.

13.2.2.3 *Albedo and Cooling Effect of Ice Sheets*

Reflection of sunlight from the snow and ice surfaces of ice sheets reduces the global mean climatic temperature; such surfaces are more reflective – they have a higher albedo – than rock, vegetation, or water. Ice sheet albedos are typically very high (0.7 to 0.85) because dry snow covers their vast interior regions. Nonetheless, the present ice sheets contribute only a few percent to the total planetary albedo, because they sit at high latitude, where sunlight is a minimum, and cover a small area compared to other reflective surfaces such as seasonal snow and sea ice, and clouds.

On millennial timescales, however, changes of ice sheet area significantly increase reflection in periods of cold climate. This is a positive feedback, critically important for understanding the Quaternary ice ages (Hansen et al. 1984). During the Last Glacial Maximum, the expansion of the Greenland and Antarctic Ice Sheets, and especially the growth of the Laurentide and Scandinavian Ice Sheets, decreased the global average absorption of solar energy by about $3\,\mathrm{W\,m^{-2}}$, as an average over the planet (see Section 13.3.2).

The "ice sheet albedo effect" can be defined as the change in the shortwave radiation reflected from the regions covered by ice sheet. The magnitude (in $\mathrm{W\,m^{-2}}$) can be defined as

$$\Delta E = \frac{1}{\mathcal{A}_E} \int_{\mathcal{A}} E_S^\downarrow [\alpha_o - \alpha_i]\, d\mathcal{A}, \tag{13.11}$$

with $\mathcal{A}_E$ the surface area of Earth, $\mathcal{A}$ the area covered by ice sheets, and $E_S^\downarrow$ the solar energy flux at the top of the atmosphere, per unit area of ground surface. The albedos α_i and α_o include all reflections from the surface as well as clouds and atmospheric aerosols. Here α_i and α_o symbolize the albedos with and without the ice sheet present. (The latter usually refers to the twentieth century, but in general α_o depends on the climate period chosen for comparison.) For example, the modern value for summertime albedo in North America at 40°N, the southern limit of the former Laurentide Ice Sheet, is about 0.35, including clouds. On the ice sheet it would probably have been 0.6 to 0.8. Thus, in this region, $\alpha_o - \alpha_i \approx 0.35 - 0.7 = -0.35$.

In general, more solar radiation reaches the surface at low latitudes than high latitudes. It is currently about 40% higher at 40°N than in northern Canada at the ice sheet's center of origin. The albedo effect, per unit area of ice cover, thus gains strength as ice sheets expand toward the equator. The decrease of seasonal snow cover at lower latitudes magnifies this property (α_o in Eq. 13.11 decreases toward the equator).

Albedo is the most important but not the only factor accounting for cooling by ice sheets. Because of the high altitude of an ice sheet surface, the overlying atmosphere is thin and dry. This facilitates the transmission of longwave radiation to space, and so provides a cooling influence that adds to Eq. 13.11. Climate model calculations suggest that this effect only matters in summer and makes little impact beyond the ice sheet itself; a simultaneous decrease of cloud cover (Section 13.2.2.4) counteracts the cooling by increasing solar absorption (Felzer et al. 1996). Ice sheets also cool the summer climate because the temperature of their surface never rises above the melting point; during ablation, energy from the atmosphere melts ice rather than warms the ground. This effect has no global-scale importance, because it is limited by the small size of ablation zones, but it strongly influences summer climate near the ice sheet margin.

Global climate model simulations with idealized ice sheets placed at high latitudes on the northern hemisphere continents suggest that the reduction in annual mean temperature around the globe scales with the area of the ice sheet, and depends little on the ice sheet height

(Felzer et al. 1996). In other words, the albedo effect is the important cooling agent for regions remote from the ice sheet. The cooling extends throughout the entire planet but concentrates in the northern hemisphere (Broccoli and Manabe 1987; Felzer et al. 1996). Expansion of sea-ice cover magnifies the cooling.

On the ice sheet surface, the atmospheric lapse rate reduces temperatures as a function of altitude. This has no direct influence on climate elsewhere on the planet.

13.2.2.4 Forcing of Atmospheric Circulation

Because of the albedo and altitude effects discussed in the previous section, the atmosphere above a large ice sheet tends to be cold for the latitude and the altitude. In other words, its potential temperature (Θ) is low. Θ is defined as the temperature of the air if it were to descend to sea level without exchanging energy with its surroundings:

$$\Theta = T\left[\frac{P_o}{P}\right]^{\omega}.$$

(13.12)

(Here T is the *in situ* temperature, P and P_o denote the *in situ* and sea-level pressures, and the constant ω equals the universal gas constant divided by the constant-pressure specific heat capacity of air.) Low potential temperature, in turn, implies a denser and thinner air column. Such compression of the air column, localized over the ice sheet, sets up horizontal pressure gradients in the upper troposphere that drive convergent winds above the ice sheet. This inflow increases the total mass of the air column above the ice sheet and so generates high pressure at the surface. High pressure, in turn, drives divergent air flow along the surface toward the ice sheet's edges. The inflow aloft and outflow at the surface are connected by subsidence, which reduces cloud cover. The interiors of large ice sheets thus tend to be regions of sinking air, clear skies, and high surface-level pressure, with diverging flow on the surface. In the northern hemisphere, diverging flow also rotates clockwise (anticyclonic flow) because of Earth's rotation. On a small ice sheet, however, transient cyclonic storms frequently disrupt these patterns. Regardless of size, the margins of an ice sheet are often cloudy because of orographic lifting associated with the passage of cyclones.

Ice sheets also influence the atmospheric circulation over a large surrounding region by topographic blocking of winds. This is analogous to the blocking effect of large mountain ranges (Cook and Held, 1992). For example, the North American Cordillera blocks westerly winds, which turn northward and so warm the Pacific Northwest coast. The return flow to the south, over eastern North America, significantly cools New England and the Canadian maritime provinces. Further east, over the Atlantic, the flow again turns northward, warming western Europe (Seager et al. 2002). This sequence of oscillations of wind direction is referred to as a *stationary wave* pattern, to distinguish it from the highly variable wind patterns of the eddies that mix polar and subtropical air and generate cyclonic storms.

An ice sheet at mid to high latitudes in the northern hemisphere should likewise block westerly winds. The air piles up on the western side of the ice sheet, and the winds turn northward.

Such deflection is a consequence of Earth's rotation. A column of air stretches vertically as it enters the "pile-up" on the western side of the ice sheet. The spin of the air column, derived from planetary rotation, must increase, resulting in the northward acceleration. The air column then thins as it crosses the high-elevation middle of the ice sheet, and the winds deflect back to the south. (Formally, these deflections of winds express the conservation of absolute potential vorticity; see Holton 2004, Chapter 4.)

The resulting stationary wave pattern affects climate far downwind (Figure 13.2; Roe and Lindzen 2001). Over the ice sheet, the wind pattern increases temperatures and ablation on the western margin, where warm air moves northward; on the eastern margin, in contrast, cold air moves southward, decreases temperatures, and suppresses ablation. At the same time, however, the orographic effect enhances precipitation on the western margin, increasing snowfall. Roe and Lindzen pointed out that, because of the stationary wave pattern, two ice sheets at the same latitude might affect each other's mass balances. The Laurentide and Fennoscandian Ice Sheets could have interacted this way.

Cooling of the air above a northern hemisphere ice sheet increases the thermal contrast with warm air to the south. This steepens the south-to-north pressure gradients in the upper

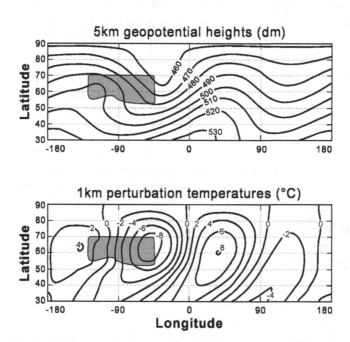

Figure 13.2: Theoretical calculation of atmospheric stationary waves and temperature anomalies generated by a northern hemisphere ice sheet (location shown in gray). Top panel: Variations of pressure at about 5 km altitude (given as equivalent height). Wind directions are approximately parallel to these contours. Bottom panel: Temperature anomaly at elevation of 1 km; southward flow causes cooling, northward flow causes warming. Adapted from Roe and Lindzen (2001).

troposphere and increases westerly air flow along the southern flank of the ice sheet. Thus two zones of concentrated air flow might form; one north of the ice sheet, because of the stationary wave pattern, and one south of the ice sheet. This situation is referred to as a "split jet" (Shinn and Barron 1989; Manabe and Broccoli 1985). Cyclonic storms, which mix cold polar air with warmer subtropical air, will be concentrated along the southern jet. At a larger scale, the effects on albedo and winds combine to shift the thermal equator southward in the Atlantic; thus the northern hemisphere ice sheets influence rainfall, winds, and sea surface temperatures in the tropics (Chiang et al. 2003).

13.2.2.5 Meltwater Impacts on Climate

The meltwater from ice sheets often accumulates in *proglacial lakes* along the ice margin. The largest lakes develop as retreat exposes subglacial basins formed by erosion and isostatic depression. The quantity of water can be very large; for example, Lake Agassiz – which formed at the southern margin of the Laurentide Ice Sheet at the end of the last ice age – was far larger than any modern freshwater lake and probably larger than the Caspian Sea.

The release of water from such lakes varies in space and time. The ice margin migrates, exposing new outlets and blocking old ones. Isostatic rebound continuously changes the bedrock topography. In a landscape of low relief, this process can significantly alter the sizes of lake basins and the elevations of outlets on timescales of only centuries. Lakes can also drain subglacially to a lower-elevation water body on the opposite side of the ice sheet, or across an ice lobe. This glacier-flooding process, called *jökulhlaup*, occurs most commonly now in Iceland (Section 6.4.1). Lake drainage typically occurs in a catastrophic flood, whether the water flows subglacially, over the ice surface, or along the margin. Flooding arises from a positive feedback that operates once drainage begins; the heat generated by frictional dissipation in the flowing water melts surrounding ice, enlarging subglacial tunnels or deepening surface spillways, and further increasing the rate of drainage.

Episodic lake drainage from the ice sheets on North America and Eurasia delivered an irregular supply of freshwater to the North Atlantic ocean during the ice ages. It is now generally accepted that this process triggered climate variations that, although strongest in the North Atlantic region, extended throughout the hemisphere and even into the tropics. The freshwater floods influence climate by changing ocean circulation and sea-ice formation.

Ocean circulation is partly driven by winds and tides, and partly by density contrasts related to salinity and temperature differences (the *thermohaline circulation*; see Rahmstorf 2006 for a summary). The thermohaline circulation involves a planetary-scale overturning of the ocean, with subsidence of water in concentrated plumes and more widely distributed turbulent mixing that allows upwelling. Geographical variations of density in the near-surface ocean arise from variable heating and cooling, freshening by precipitation and continental runoff, and concentration of salts by evaporation and sea-ice formation. The density variations, in turn, induce pressure gradients that drive currents in the underlying ocean. The deep currents must be fed by sinking surface waters. The sinking plumes occur at only a few sites of *deep water formation*, located around Antarctica and in the polar North Atlantic, where densities at the surface are unusually large because the water is both cold and salty.

A key point is that this system has two stable modes and the potential to switch between them (Stommel 1961; Weaver et al. 1993). The water at the sites of deep water formation must be salty. At the North Atlantic sites, the saltiness derives from evaporation in the subtropics, followed by northward transport in surface currents. The dense surface waters not only feed descending plumes but also help to drive the return currents in the deep ocean. This describes one stable mode. If the currents were to stop, the North Atlantic surface waters would freshen because of precipitation and runoff. Freshening, in turn, shuts off the descending plumes and partly neutralizes the pressure gradients driving the deep circulation. This is a second stable state, in which the surface currents into the North Atlantic are much reduced. A switch from the first stable state to the second reduces the northward flow of warm subtropical waters and therefore cools the ocean surface.

An unusually large influx of fresh water to the North Atlantic would cap the ocean surface with a low-density layer; this is one way to switch from the first stable state to the second (Broecker 1997; Manabe and Stouffer 1988). The consequent cooling of the ocean surface, and also perhaps the freshening itself, would expand the sea ice, reducing air temperatures further; sea ice both reflects sunlight and, especially in winter, insulates the air from underlying "warm" ocean waters. Though having the greatest impact in the North Atlantic and regions down-wind in Eurasia, the cooling would also change the patterns of wind and precipitation in the tropics, by changing the position of the intertropical convergence zone (Chiang et al. 2003). The climatic effects of meltwater releases are thus widespread. Ice core measurements show that atmospheric methane concentrations and the isotopic composition of atmospheric O_2 both varied in correlation with the climate at these times (Sections 15.2.2.3 and 15.8.1). Because methane and oxygen are globally well-mixed gases with widespread surface sources, these measurements demonstrate the global, or at least hemispheric, scale of such events.

Releases of fresh water from the Laurentide and Fennoscandian Ice Sheets triggered climate changes. Some climate events during the termination of the last ice age coincided with the formation of new outlets from Lake Agassiz (Teller and Leverington 2004). These events consisted of, among other changes, abrupt coolings in Greenland and Europe, increased winds and aridity in the tropical Atlantic, and reductions of global atmospheric methane concentrations. The largest such event was the onset at about 12.7 kyr b.p. of the *Younger Dryas*, a major cold interval that disrupted the overall warming trend of deglaciation.

Perhaps the most interesting event, because it occurred late in the deglaciation when warm Holocene climate was already established, occurred at around 8.2 kyr – the so-called "8k event" (Alley and Agustsdottir 2005). It immediately followed the final drainage of Lake Agassiz, an event whose timing is known from geological evidence. The details of drainage remain unknown, but topographic constraints indicate the water took a path northward into Hudson Bay along the bed of the disintegrating remnant ice sheet. Clarke et al. (2004) modelled this event using the theory for subglacial tunnel formation originally developed for Icelandic floods, with modifications to account for the input of lake waters warmer than melting point (see Chapter 6, and Clarke 2003). They concluded that the large size of the lake and the feedbacks between drainage and melt acted together to produce an enormous flood when drainage began. Specifically, the model suggested an initial flood lasting less than 1 year and attaining a peak discharge of approximately $5 \times 10^6 \, \mathrm{m}^3 \, \mathrm{s}^{-1}$ – about 25 times the discharge of the Amazon. The model also suggested that drainage would have continued through stable tunnels after the initial flood lowered the water level in the source lake.

Meissner and Clark (2006) simulated the climate effects of such drainage events using an "Earth-system model" – an ocean circulation model linked to separate models for the atmosphere, sea ice, and land surfaces. Results indicated that the large climate effects were driven not by the initial large flood, but by the less dramatic but persistently high drainage in the centuries that followed, which they called *routing events*. Large ice-dammed lakes that formed in Eurasia are another likely source of drainage floods (Mangerud et al. 2004). According to one hypothesis, occasional releases of water from the ice sheets on both continents interacted with other periodic climate forcings, possibly solar, to generate large millennial-scale variations of climate throughout the last glacial period (Alley et al. 2001).

13.3 Growth and Decay of Quaternary Ice Sheets

A characteristic feature of the past 2.5 Myr was the growth and decay of ice sheets in the northern hemisphere. Figure 13.3 illustrates this with a geological proxy for changes of ice volume. Specifically, the figure plots the oxygen-isotope ratio measured in fossil foraminifera ("forams") in cores from the ocean floor. More specifically

$$\delta^{18}\mathrm{O} = \left[\frac{R}{R_s} - 1 \right] \tag{13.13}$$

where R denotes the ratio of the concentrations of $\delta^{18}\mathrm{O}$ and $\delta^{16}\mathrm{O}$ in the sample and R_s the ratio in "standard mean ocean water." Values for $\delta^{18}\mathrm{O}$ are reported as parts per thousand or "per mil." The measured $\delta^{18}\mathrm{O}$ reflects the isotopic composition of the ocean water at the time the forams were alive. For reasons explained in Chapter 15, the ice in the polar ice sheets is depleted in $\delta^{18}\mathrm{O}$; thus, $\delta^{18}\mathrm{O}$ of the ice is negative. It follows that the $\delta^{18}\mathrm{O}$ of ocean water varies as water transfers back and forth from the oceans to the ice sheets as they grow and shrink. Relative to its long-term mean, the ocean's $\delta^{18}\mathrm{O}$ is positive in a glacial age and negative in an interglacial. Most of the variability of the curve in Figure 13.3a reflects changes in the volume of the ice sheets. But water temperature also affects the oxygen-isotope ratio in the foram records; roughly 30% of the variation records cooling of the ocean during glacial ages rather than increases of ice volume (Adkins et al. 2002). For the past 1 Myr, quasi-periodic oscillations with a mean wavelength of about 100 kyr dominate the curve. (This is often referred to as the "100 kyr cycle.") Prior to this time, in contrast, the dominant feature is an oscillation with a period of about 40 kyr. The change

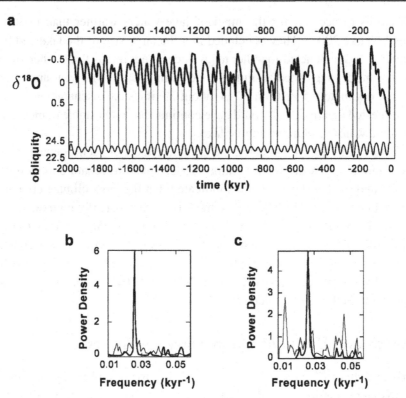

Figure 13.3: Variations of ice volume over the last two million years. (a) The marine $\delta^{18}O$ record (heavy line); low values correspond to greater ice volumes. For each major deglaciation, a thin vertical line shows the timing relative to the obliquity variations, shown along the bottom. Note the larger and longer-period fluctuations of $\delta^{18}O$ in the second million years compared to the first. Adapted from Huybers (2007). (b) Power spectra of variability from 2 Myr to 1 Myr ago. Thin line is for rate of change of $\delta^{18}O$, a proxy for rate of ice volume change. Thick line is the spectrum for insolation integrated over the summer. Both show strong peaks at a 41-kyr period (frequency $0.024\,\mathrm{kyr^{-1}}$). Adapted from Huybers (2006). (c) Same as (b), but for the most recent 1 Myr. The 41-kyr period remains clear, but the ice volume changes strongly at a 100-kyr period with no corresponding insolation change. Higher-frequency variations also appear in both spectra. Adapted from Huybers (2006).

to 100 kyr oscillations defines the *mid-Pleistocene transition*. Note that in the later era the ice volume varied with periods of about 20 and 40 kyr, in addition to the larger 100 kyr oscillation.

Figure 13.4 shows the last glacial cycle in detail. It also includes another indicator of ice sheet volume: variations in sea level as measured by the elevations of dated coral reefs on tectonically uplifted coasts. The two curves show the same overall trends: a slow buildup of ice followed by rapid decay. At the end of the ice age, a volume equivalent to a rise in world sea level of about 130 m was returned to the ocean in some 8 kyr. This rapid decay of ice sheets is a major feature of the 100-kyr cycle and it makes them markedly asymmetric. At the termination of a cycle, the

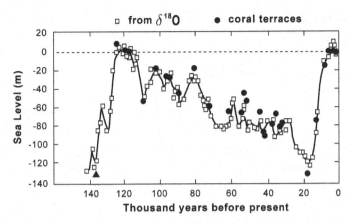

Figure 13.4: Variation of sea level over the last 140 kyr, estimated from two independent proxies. Adapted from Chappell et al. (1996).

Laurentide and Fennoscandian ice sheets did not simply melt; they rapidly disintegrated as the ocean reclaimed areas that are now Hudson Bay and the Baltic Sea. Rapid retreat also occurred in the extensive ice-dammed lakes around the southern margins.

Although the buildup occurred slowly on average, Figure 13.4 shows periods of rapid growth. For example, around 115 kyr b.p. sea level dropped by about 50 m in 10 kyr. This requires a large flux of moisture to the growing ice sheets on North America and Scandinavia. One contributing factor was warm ocean surface waters, which would heat the overlying atmosphere and recharge it with moisture (Ruddiman and McIntyre 1979). Ocean cores provide evidence for such conditions. Another likely contributing factor was a steepened equator-to-pole temperature gradient related to the variation of insolation with latitude (Khodri et al. 2001). A steep temperature gradient would strengthen the atmospheric circulations responsible for moisture transport.

13.3.1 Relation to Milankovitch Forcings

The Milankovitch hypothesis attributes the alternation between glacial and interglacial periods to variations in the amount of solar radiation received at northern latitudes in summer. These variations result from the precession of the Earth's axis, from variations in the tilt of the axis relative to the plane of Earth's orbit around the sun, and from variations in the eccentricity of the orbit. These variations have periods of 19 and 23 kyr, 41 kyr, and about 100 kyr. Oscillations at all these periods appear in the ice-volume record (Ruddiman and Wright 1987; Huybers 2007). The dominance of the 100 kyr period in the last 1 Myr (Figure 13.3) is unexpected because, compared to the other variations, the changes in eccentricity have little effect on the receipt of solar radiation in summer (Imbrie et al. 1993).

The solar radiation probably influences the volume of northern hemisphere ice sheets by controlling summer temperatures and, hence, ablation rates (Weertman 1976b). Thus it is most meaningful to compare insolation to the rate of change of ice volume, which can be estimated from the marine sediments as $d\delta^{18}O/dt$. As Figure 13.3b shows, $d\delta^{18}O/dt$ varies with a period of about 41 kyr through the entire record and, for the last 1 Myr, the other frequencies also appear. Moreover, $d\delta^{18}O/dt$ varies inversely with insolation; in particular, the major deglaciations occur at times of high insolation (Roe 2006). Milankovitch variations clearly are one important factor governing the ice-age cycles.

Nonetheless, controversy surrounds the role of the Milankovitch forcings. The 100 kyr eccentricity cycle has no direct role, because its influence on summer radiation is too small. It does modulate the magnitude of the precession cycles (19 and 23 kyr); their amplitude increases at times of high eccentricity. This probably influences the timing of some deglaciations and glacial inceptions. The obliquity cycle (41 kyr) matters, a point elaborated most clearly by Huybers (2006, 2007; also Huybers and Wunsch 2005). Most of the deglaciations (or *terminations*) throughout the last 2 million years occur in times of high obliquity (Figure 13.3a). The most prominent exception is the third-to-last termination, about 240 kyr b.p., which occurred at a time of low obliquity. It was, however, followed only 20 kyr later by another termination when the obliquity was high. This is the only example of an "interglacial" period split in two by a brief return to nearly full glacial conditions.

The obliquity and precession act in different ways on the summer insolation (Berger and Pestiaux 1984; Huybers 2006). A high obliquity increases the insolation integrated over the summer. This is plausibly the most important factor for ablation. Present mean summer temperatures are closely related to the integrated insolation. In comparatively warm regions – such as the ice sheet margins when they advanced into the mid-latitudes – the integrated insolation is the most plausible proxy for the positive-degree-day index, which correlates strongly with ablation on modern glaciers. This index accounts, in a simple way, for both the duration and intensity of melt (Section 5.4.3).

In contrast to obliquity, the precession modulates the peak value of insolation, which occurs in early summer – but does not change the total energy input over the summer. This is a consequence of Kepler's second law; Earth moves most rapidly along its orbit when closest to the sun. Thus, the peak insolation and the duration of summer are anticorrelated in precession cycles. Because the precession determines the peak values of insolation, it might be the best proxy for ablation in extremely cold regions, where days with melting are comparatively rare.

The regular 40-kyr cycle of ice volumes in the early Quaternary (before 1 Myr) can be regarded as a direct consequence of ice sheets expanding when the obliquity declines, and then rapidly ablating in the next period of high obliquity. In this view, the longer cycles of the late Quaternary reflect "skipping" of deglaciations that should have happened during obliquity highs but did not. If so, the late Quaternary cycles would be approximately 80 kyr or 120 kyr in duration, modified in some cases by the timing of precession peaks. Based on the marine isotopic record, the last several full glacial-interglacial climate cycles – going back in time – lasted approximately 120, 80, 20, 90, 90, 120, 80, 80, 80, 80, and 80 kyr. (To get these numbers we defined full climate cycles as extending from the onset of clearly interglacial conditions, through a clearly glacial minimum, to the next transition back to an interglacial. These cycles are delimited by terminations during the obliquity cycles numbered 1, 4, 6, 7, 9, 11, 14, 16, 18, 20, 22, and 24 in Figure 5 of Huybers 2007.) Thus the cycles are best described, not as a single 100 kyr oscillation, but as a series of fluctuations with durations from 80 to 120 kyr. Though not perfect, the data support Huybers' view that obliquity tends to set the timing of terminations. But to complicate matters, the precession evidently cannot be ignored.

The apparently important role of obliquity leads to two major questions, neither of which has been answered satisfactorily:

Why Is There Skipping Only in the Late Quaternary? Ice sheets almost never survived periods of high obliquity in the early Quaternary, but regularly did so in the most recent 1 Myr. This might simply reflect a continuation of the general trend of the last 40 Myr toward cooler conditions, most likely a consequence of declining atmospheric CO_2 concentrations (Raymo et al. 1988). A different explanation invokes cumulative changes in the substrate for the Laurentide and Fennoscandian Ice Sheets (Clark and Pollard 1998). If, at the start of the Quaternary the continents were mantled with regolith (unconsolidated weathered bedrock and soil), the beds of the early ice sheets would be deformable, and the buildup of the ice limited by the shear strength of this material. The ice sheets could be spatially extensive, but their surface profiles would be low. This, in turn, implies they were vulnerable to ablation when climate warmed. Over many glacial cycles, the regolith mantle would be eroded and transported to the margins of the ice sheets, leaving a core region of hard, undeformable bedrock beneath the ice sheet centers (the present configuration). In the central regions of ice cover, the surface profiles would thus be steeper in the late Quaternary than in the early Quaternary. The later ice sheets would be more likely to survive partial retreat in warm periods.

Why Are Glacial Cycles Asymmetric? Essentially, this is equivalent to the question of why deglaciation occurs rapidly and completely from conditions of maximum ice extent. The rapid deglaciations are the most striking and enigmatic feature of the late Quaternary ice ages. They imply a peculiar vulnerability of large or long-lived ice sheets to climate warming.

There are several possible mechanisms, which probably act in coordination:

1. Delayed isostatic rebound of bedrock. As an ice sheet starts to retreat into the bowl-shaped depression it created, the surface elevation at its terminus decreases and so ablation intensifies. Increased ablation also steepens the surface and therefore increases the ice flux, which accelerates thinning and retreat. Calving may also increase if the terminus rests in the ocean or large lakes. Conversely, rebound of the bedrock after the ice sheet has disappeared favors snow accumulation and the growth of a new ice sheet. Weertman (1961b) proposed this plausible mechanism. Oerlemans (1980) examined the behavior of a simple ice sheet model with delayed isostatic adjustment and a specific mass balance dependent on surface elevation. The model ice sheet waxed and waned spontaneously even in the absence of climate variations, and retreat occurred rapidly compared to growth.

2. Onset of basal melting. When a growing ice sheet reaches a critical thickness, the base of the ice in the central region may warm to melting point. Basal slip then contributes to ice flow, and the ice sheet interior thins. In addition, ice streams can form where the frictional heating is large enough to warm the bed to melting point. This heating, in turn, increases with the ice flux and hence the size of the ice sheet. Ice streams make the ice sheet more vulnerable to ablation by lowering the surface profile and decreasing the time for ice to be drawn from the interior (see next item). Marshall and Clark (2002) modelled the thermal evolution of the Laurentide Ice Sheet throughout a glacial cycle. They found that the time-scale for the development of large areas of basal melting was about 100 kyr. This is therefore a plausible contributing factor to deglaciation.

3. *Downdraw*. This term means the thinning of the central part of an ice sheet by fast flow of ice streams. Basal melting is a prerequisite. Denton and Hughes (1983) and Hughes et al. (1985) assert that this was the main cause of the decay of the Laurentide Ice Sheet. They believe that this ice sheet, at its maximum, was fringed by ice shelves in the Labrador Sea, Baffin Bay, and the Arctic Ocean, and that it was drained by ice streams principally in the Hudson Strait and the Gulf of St. Lawrence. Thinning of these ice shelves by increased basal melting, or reduction of grounded areas by rising sea level, would reduce the back pressure on the ice streams. This would increase their velocity and thus also the rate at which the ice sheet thins (Section 11.5.2). The feedbacks between thinning, faster flow, and grounding line retreat could rapidly destroy the marine part of the ice sheet. (See Hughes 1992 for a description of how such destruction might proceed.) This is a likely mechanism of rapid deglaciation.

4. Albedo-precipitation feedback. The amount of solar radiation that is reflected rather than absorbed at the surface increases as ice sheets grow and the sea-ice cover expands (Section 13.2.2.3). The resultant cooling favors further expansion of the ice cover and yet more cooling. But cooling reduces precipitation on the ice sheets, which leads to thinning of the ice and, depending on the magnitudes of temperature and precipitation changes, perhaps to warming of the bed. Both factors increase the ice sheets' susceptibility to subsequent warming. Although this feedback probably exists, its strength and importance are uncertain. How precipitation changes near the ice sheet margin must be a critical factor.

5. Variations in atmospheric CO_2 concentration. The concentration of CO_2 in the atmosphere during the coldest parts of the ice age was about 80 p.p.m. less than during the interglacials (Section 15.8.1). The rise back to interglacial values occurred rapidly and so reinforced the climate warming and deglaciation by strengthening the greenhouse effect. (Reduced albedo from ice retreat contributed likewise, to an even greater extent: Section 13.3.2.) The rise of CO_2, however, appears to have lagged the increase of temperature by centuries to a few millennia (Caillon et al. 2003). Thus rising CO_2 did not initiate deglaciations, but would have contributed to the warming over millennia. At present, it remains unknown in detail

why CO_2 is removed from the atmosphere when the climate cools and returned during deglaciations. The large quantity involved can only be stored in the ocean through a change in its carbonate solution chemistry; several proposed hypotheses can account for the change (Sigman and Boyle 2000; Archer et al. 2000).

Several investigators have simulated the variations of ice volume in recent ice-age cycles and were able to match some aspects of the marine isotope data (Birchfield et al. 1981; Pollard 1983; DeBlonde and Peltier 1993). In these studies, Milankovitch insolation variations are essential for determining the timing and sign of ice volume changes. But the overall form of the ice-age cycle – the shape of the curve of ice volume against time – reflects not the insolation variations but the interaction of ice sheets with climatic and geophysical processes. First, specific balance increases with surface elevation, so a thickening ice sheet can sustain its own expansion (Oerlemans 1981b). Second, isostatic depression and delayed rebound contribute to rapid decay at the end of glacial cycles, and to regrowth after rebound. Third, there must be additional processes, such as those enumerated here, that increase the rate of deglaciation.

13.3.2 Climate Forcings at the LGM

How cold was Earth at the Last Glacial Maximum (LGM), 20 kyr ago, compared to present? At the latitude of the large northern hemisphere ice sheets, it was very much colder; about 20 °C in central Greenland, according to ice core indicators (Cuffey et al. 1995). Temperatures in the Antarctic are less well constrained, but ice core analyses suggest LGM temperatures about 10 °C colder than present (Blunier et al. 2004). But the low latitudes account for a much larger fraction of Earth's surface area, so tropical temperatures contribute greatly to the global mean. Isotopic and element-ratio indicators in groundwater, coral skeletons, and forams indicate a low-latitude LGM cooling of 3 to 5 °C (Lea et al. 2000; Bush and Philander 1998; and references therein). Snowline depression of tropical glaciers and altitudinal shifts of vegetation revealed by pollen suggest a similar amount of cooling (Rind and Peteet 1985). Thus a reasonable estimate for the global mean cooling at LGM is 5 or 6 °C.

What accounts for such a reduction in global temperature? Any factor that changes the annual global mean energy flux to the surface (the "net energy flux") ultimately contributes to a change in the mean temperature (Hansen et al. 1984). However, some of the contributing factors are effectively fixed by the climate itself on timescales of only days to a few years. Examples include clouds, atmospheric water vapor, and seasonally varying sea ice and snow. These are called *fast feedbacks*, a term used to distinguish them from some of the underlying climatic influences that – although independent of the climate from year-to-year – evolve as a function of climate on long timescales. Examples of the latter – the *slow feedbacks* – are changes in atmospheric CO_2 concentration and shortwave reflection from ice sheets.

Call the change in the net energy flux ΔF_{net}. It sums the contribution from the fast feedbacks with the contribution from the underlying *climate forcings* (size ΔF) which include slow feedbacks and independent variables such as the brightness of the sun. The relationship can be written as $\Delta F_{\text{net}} = [1 + f]\,\Delta F$, where the parameter f, the *feedback factor*, accounts for the amplification of climate changes by the fast feedbacks. (Note that in some treatments the term "feedback factor" refers to $1 + f$.) A positive ΔF_{net} causes the planet to warm, which increases the loss of longwave radiation to space and reestablishes an equilibrium. The mean temperature change, $\Delta \overline{T}$, then equals $\Delta F_{\text{net}}/\lambda$ where λ is a constant known as the black-body radiative damping coefficient (about $3.8\,\text{W}\,\text{m}^{-2}\,{}^{\circ}\text{C}^{-1}$). However, because the fast feedbacks do not vary independently of the temperature, it is most useful to ask how much warming occurs for a given magnitude of the underlying climate forcing. This defines the *climate sensitivity*, ξ:

$$\Delta \overline{T} = \xi \cdot \Delta F = \frac{[1 + f]}{\lambda}\,\Delta F \tag{13.14}$$

(Hansen et al. 1984, 1993; Hoffert and Covey 1992).

Table 13.1 lists estimates for the known climate forcings at the LGM, compared to the late Holocene before the industrial revolution. Each of these terms is the estimated difference (LGM minus Holocene) in an annual global mean radiant energy flux to the surface.

Greenhouse gas concentration refers to CO_2 and methane. The seasonal distribution of insolation at the LGM differed from today's. The annual total, however, was nearly the same; the slight difference shown in the table is due to the interaction of the seasonal distribution with the latitudinal variation of albedo. The albedo forcings are estimated from GCM simulations, which include ice sheets delimited by geological reconstructions. The calculated reflection by atmospheric aerosols is constrained by measurements of impurities in ice cores, but is still poorly known.

Table 13.1: Climate forcings at the LGM.

Variable	Radiative forcing $\text{W}\,\text{m}^{-2}$	Reference
Greenhouse gas concentration	−2.8	Hoffert and Covey 1992
	−2.2	Hewitt and Mitchell 1997
Insolation	0.1	Hewitt and Mitchell 1997
Surface albedo:		
total	−3.0	Hoffert and Covey 1992
ice sheets only	−2.9	Hewitt and Mitchell 1997
land	−0.6	Hewitt and Mitchell 1997
total	−3.5	Hewitt and Mitchell 1997
Atmospheric aerosol and dust	−0.9	Hoffert and Covey 1992
	−1 to −2.3	Claquin et al. 2003
Sum of Forcings:		
low value	−6.0	
high value	−8.7	

The comparison of LGM with late Holocene conditions provides an estimate for Earth's climate sensitivity, ξ, and the feedback factor, f (Hansen et al. 1984, 1993; Lorius et al. 1990; Hoffert and Covey 1992). In one such comparison, Hansen et al. (1993) derived a value $\xi \approx 0.75 \pm 0.25\,°C/(W\,m^{-2})$. Likewise, our tabulated values suggest that $\xi \approx 0.6$ to $1\,°C/(W\,m^{-2})$ and therefore $[1 + f]$ lies in the range 2.2 to 3.8. Thus the fast feedbacks double or quadruple the temperature change caused by an initial radiative forcing. These values are in the range implied by climate models (Solomon et al. 2007).

13.3.3 Onset of Quaternary Cycles

The first large fluxes of sediment carried by icebergs into the northern oceans occurred 2.4 to 2.6 Myr ago (Raymo 1994). This followed a general increase of marine $\delta^{18}O$ composition (about 1 per mil) from 3.0 to 2.6 Myr, suggesting ice growth and cooling climate. Regular oscillations of the $\delta^{18}O$ began around 2.7 Myr. The first of these cycles to reach a $\delta^{18}O$ value equivalent to the *average* of the most recent cycle (0 to 0.12 Myr) was about 2.4 Myr ago. By this time, apparently, large ice sheets were periodically forming on the northern continents.

Little is known about the extent of these early ice sheets. Old glacial deposits are difficult to date. Recently, Balco et al. (2005) used a cosmogenic isotope technique to determine an age for one of the oldest deposits in North America: the "Atlantic Till," found in central Missouri State at a latitude of about 39°N. The Laurentide Ice Sheet margin rarely advanced so far; the Atlantic Till is nearly the southernmost deposit. Remarkably, its age proved to be 2.41 ± 0.14 Myr. Thus the ice sheet attained its full extent very early in the history of glacial cycling. If, as suggested by Clark and Pollard (1998), the entirety of this early ice sheet were underlain by deformable sediments its surface profile would have been low. Hence, its volume would not have matched that of the later ice sheets. Such a geometry would explain why the marine $\delta^{18}O$ at 2.4 Myr was still 0.5 per mil lower than at the LGM, even if the Laurentide Ice Sheet reached Missouri.

13.3.4 Heinrich Events

The North Atlantic marine sediment record includes six conspicuous and widespread layers of ice rafted debris, containing carbonates, deposited between 65 kyr and 14 kyr b.p. The time between each of these *Heinrich events* averages about 7 kyr, but the interval varies. Heinrich events are attributed to brief but exceptionally large discharges of icebergs from the Laurentide Ice Sheet (Heinrich 1988; Hemming 2004). The most likely source is a surge and/or catastrophic retreat of the outlet glacier in Hudson Strait. It is assumed, as seems necessary, that between each such event the ice slowly advanced and regrew.

This behavior resembles that of the large tidewater glaciers in Alaska. In Alaska the glacier termini rest on sediment shoals as they advance. When they retreat into deeper water behind the shoal, feedbacks between flow and thinning cause extensive, rapid, and inexorable further retreat (Section 11.5.2). Although they cannot advance, the bedrock sills in Hudson Strait might play an analogous role to the sedimentary shoals at the onset of retreat.

A remarkable aspect of the Heinrich events is their timing; they occurred in periods of exceptionally cold climate, after long intervals of decreasing air and sea-surface temperatures. Thus the most "obvious" mechanism for triggering these events – calving induced by surface melt and climate warming – can be discounted. Another remarkable feature is the large quantity of rock material carried in the icebergs. A convincing Heinrich-event model needs to explain how this debris was originally entrained in the basal ice.

MacAyeal and Alley proposed the *binge-purge model* for Heinrich events (MacAyeal 1993a, 1993b; Alley and MacAyeal 1994). In this model, Heinrich events are essentially thermally triggered surges as outlined by Clarke (1976) (Chapter 12). During the several millennia between surges, the ice sheet in Hudson Strait would regrow. Its bed is frozen during this time. The bed temperature increases steadily in response to geothermal heat. More precisely, the ice warms because it receives an "excess heat flow," a difference between the geothermal flux from the underlying crust (G) and the conduction upward to the surface. During this phase, the surface elevation increases and so the surface cools at the atmospheric lapse rate (ℓ). The excess heat flow amounts to $G - k_T \ell$ for a thermal conductivity of ice k_T. MacAyeal showed that, assuming heat transport by ice flow can be neglected, the time needed for the bed to warm to melting point from an initial value T_o is

$$t_{\mathrm{L}} = \frac{\pi}{4\alpha_{\mathrm{T}}} \left[\frac{k_T T_o}{G - k_T \ell} \right]^2 , \qquad (13.15)$$

where T_o denotes the sea-level temperature (in °C) when the ice sheet begins to reform, and α_{T} is the thermal diffusivity of ice. When the bed reaches melting point, water saturates the underlying sediment and makes it very slippery, triggering a surge. For the time between surges to be about 7 kyr, the T_o in Eq. 13.15 needs to be about $-10\,°C$, a reasonable value. Thinning of ice during the surge increases the temperature gradient and causes the bed to refreeze when the surge ends. Basal freeze-on entrains sediment that is subsequently released in icebergs at the start of the next Heinrich event. To support this model, MacAyeal (1993b) evaluated its crude assumptions, while Alley and MacAyeal (1994) analyzed the sediment entrainment problem.

The binge-purge model is plausible in the broad behavior it describes; a thawed bed is a prerequisite for rapid ice flow, and rapid thinning should drive refreezing. Beyond this broad picture, however, the glaciological structure of Heinrich event cycles is not known. One recent hypothesis holds that the bed first reaches melting point over a broad region, but a frozen fringe prevents surging at first. The frozen zone is where ice shelves that formed during the cold climate preceding the surge are grounded on sills (Alley et al. 2006). Only when the ice over these sills thickens enough to thaw the underlying bed do surges and Heinrich events occur. This scenario was motivated by observations of sediments deposited near Hudson Strait during the Heinrich events; these sediments are not coarse ice-rafted debris but fine-grained sorted sediment deposited by flowing water, probably in a flood. The grounding and freezing of ice shelves on sills would trap water upstream, which would flood out during the surge.

13.4 Ice Sheet Evolution Models

The flow and evolution of an entire ice sheet can be simulated with a numerical model. Such a *whole-ice-sheet model* attempts to be comprehensive; to include all important processes, a realistic representation of geography, and an initial state that reflects the enduring legacy of prior conditions. Though far from achieving this ideal, current ice sheet models are essential tools for glaciology. They reveal the precise implications of a defined set of assumptions about how ice sheets work. They produce quantitative hypotheses about how ice sheets respond over millennia to climate changes and other events. They make predictions that can be compared to measurements; disagreements then show what is unknown and what additional data are needed.

Whole ice sheet models are often likened to the general circulation models (GCMs) used in climate forecasts – a valid analogy. However, it must be appreciated that the development of GCMs was guided by comparisons to meteorologic data sets of vastly greater scope and completeness than the data available for ice sheet models. Furthermore, the development of GCMs involved collaborations between many scientists with diverse expertise, and with computer scientists and programmers with deep knowledge of the technology. In contrast, the development of ice sheet models was until recently the work of only a few pioneering individuals.

Ice sheet modelling began with Mahaffy (1976) and Jenssen (1977). The modern phase, using the remarkably increased computational power of recent decades to simulate Antarctica and Greenland, began with Huybrechts (1992, 1993, 1994). Some recent modelling efforts using innovative approaches to tackle diverse questions include Marshall and Clarke (1997a, 1997b), Greve (1997b), Payne (1999), Huybrechts and de Wolde (1999), Hulbe and MacAyeal (1999), Cuffey and Marshall (2000), Ritz et al. (2001), Pattyn (2003), Tarasov and Peltier (2004), Clarke et al. (2005), Otto-Bliesner et al. (2006), and Pollard and DeConto (2009).

13.4.1 Model Components

Most ice sheet models use the finite difference method to solve governing equations on a grid (a few models use finite elements). The horizontal grid size is typically 20 to 50 km, and the vertical grid, with a few dozen layers, is adjusted to match the ice thickness. It also extends far into the bed for temperature calculations.

We now summarize the basic relations used in most models. (This topic follows from Section 11.3.4.3, a discussion of numerical models of valley glaciers and temperate ice caps.) Define x and y as horizontal coordinates, z as vertical and positive upward. The bed and ice-equivalent surface elevations are B and S, and $H = S - B$ the ice-equivalent thickness (density is treated as uniform). The index j means either horizontal coordinate, x or y. The index k means x, y, or z. Repetition of j or k in the same term implies summation.

Conservation of Ice Mass (Section 8.5.2) Ice thickness changes if the specific mass balance at a point, $\dot{b}_i$, differs from the divergence of the ice flux. Call the ice velocity components $\vec{u} = [u_x, u_y, u_z]$. Then

$$\frac{\partial H}{\partial t} = \dot{b}_i - \frac{\partial}{\partial x_j} \int_B^S u_j \, dz \quad \text{and} \quad u_z(z) = -\frac{\partial}{\partial x_j} \int_B^z u_j \, dz + u_z(\text{B}) \qquad (13.16)$$

The second relation states that the ice moves downward, at velocity u_z, if the ice flux in the layers beneath is divergent. Here $\dot{b}_i$ is the rate of ice addition to the glacier at a point, with dimensions of thickness per unit time.

Ice Flow (Section 8.3)   Horizontal velocity is the sum of basal slip ($\vec{u}_b$) and the net internal deformation. The latter corresponds to the vertical integral of the strain rate, $\dot{\epsilon}_{jz} = \frac{1}{2} \partial u_j / \partial z$, so that:

$$u_j(z) = u_{bj} + 2 \int_B^S \dot{\epsilon}_{jz} \, dz. \qquad (13.17)$$

The strain rate depends on shear stresses by the creep relation for ice (Chapter 3):

$$\dot{\epsilon}_{jz} = E \, A(T) \, \tau_E^{n-1} \, \tau_{jz} \quad \text{with} \quad \tau_E = \left[\tau_{xz}^2 + \tau_{yz}^2 \right]^{1/2}. \qquad (13.18)$$

A is a temperature-dependent coefficient, E a number to correct the ice viscosity for influences other than temperature, τ_E the effective stress magnitude, and n a number in the range 2 to 4, usually taken as 3. A varies with depth because temperature (T) does. In a general case, longitudinal deviatoric stresses would be included in τ_E, but they are assumed to be negligible in this approximation.

Conservation of Energy (Chapter 9)   Temperature must be calculated in order to evaluate A, and to delimit regions of thawed bed. Energy conservation requires that:

$$\rho c \frac{\partial T}{\partial t} = \frac{\partial}{\partial z} \left[k \frac{\partial T}{\partial z} \right] - \rho c u_k \frac{\partial T}{\partial x_k} + \dot{S}_E \qquad (13.19)$$

where the first term on right represents vertical diffusion, the second term advection, and the final term the heat generated by frictional dissipation. Horizontal heat diffusion is neglected because horizontal temperature gradients are usually trivial.

Conservation of Momentum (Section 8.5)   Most models assume that, for grounded ice, the "shallow-ice approximation" can be used (Section 8.5.2). In this approximation, basal drag ($\vec{\tau}_b$) balances all the gravitational driving stress ($\vec{\tau}_d$), and

$$\tau_{jz} = \tau_{bj} \left[1 - z/H \right] \quad \text{with} \quad \tau_{bj} = \tau_{dj} = -\rho \, g \, H \frac{\partial S}{\partial x_j}. \qquad (13.20)$$

The driving stress in the horizontal direction, τ_{dj}, results from pressure-gradient forces arising from gradients of surface elevation.

Basal Boundary Condition (Chapter 7) Basal slip occurs where the basal temperature reaches melting point. Most models use a Weertman-type relation, in which basal motion increases as the second or third power of basal shear stress. These relations should also express, in principle, the effect of basal water and sediment on slipperiness of the bed. In the simplest approach consistent with some data, slip rate depends inversely on the basal effective pressure (the ice overburden pressure minus the water pressure). Most models do not calculate basal water pressure, however, and such a calculation would be uncertain in any case. The widely cited models of Huybrechts (see Huybrechts and De Wolde 1999, p. 2183) use a relation

$$u_{bj} = c\,\tau_{bj}^n/Z \tag{13.21}$$

with c a constant. The parameter Z equals the ice thickness if the bed elevation is above sea level. With a bed below sea level, Z equals the ice thickness minus a factor $\rho_w\,\Delta Z/\rho_i$, where ΔZ is the elevation below sea level. This allows for increased basal motion in deep subglacial basins and near grounding lines. (Z is also assigned a minimum value to prevent division by zero.) It is essential for models to simulate fast basal motion near grounding lines, so the authors cannot be faulted for using a relation such as Eq. 13.21. It is, however, not mechanistic. The basal motion in real ice streams depends on their width, the properties of their substrates (sediment strength and coverage, or bedrock roughness), and the basal effective pressure. Other models use relations more consistent with mechanics (e.g., Marshall and Clarke 1997b; Tarasov and Peltier 2004), but all such relations are poorly constrained.

Isostasy Whole-ice-sheet models must calculate isostatic adjustments of the land surface, as described in Section 13.2.2.2.

Specific Mass Balance (Chapters 4 and 5) Different models use a great diversity of approaches for calculating the specific mass balance. Except on ice shelves, they all treat the specific balance as equivalent to the difference between snow accumulation (rate $\dot{a}_s$) and surface melt (rate $\dot{m}_s$). Most approaches adopt some form of the relation

$$\dot{b}_i = \dot{a}_s - \dot{m}_s = f_s\,P - f_m\,D \tag{13.22}$$

in which f_s signifies the fraction of precipitation falling as snow, and P the precipitation. D is the annual positive degree day index (Section 5.4.2), and f_m a coefficient expressing how much melt occurs in 1 day for every degree in excess of $0\,°C$. The fraction f_s depends on air temperature. This, the precipitation, and the positive degree days all need to be calculated for each location in the model domain. The most elaborate models use output from climate models to do so. Most adopt a simpler approach, such as prescribing D as a function of mean annual temperature and calculating P from the observed modern precipitation, P_o; for example, from $P = P_o[1 + p]^{\Delta T}$, with p a small positive number (of order 0.05) and ΔT the temperature change relative to modern.

Additional Components   Some models apply the following components:

1. Ice shelves and low-stress zones. To model Antarctica it is essential to include ice shelves and, for West Antarctica, the adjacent regions of low driving stress and slippery beds. For these regions, a more complete vertically integrated force balance relation replaces Eq. 13.20 (see Chapter 8). For the x-direction:

$$-2\frac{\partial}{\partial x}[H\,\tau_{xx}] - \frac{\partial}{\partial y}[H\,\tau_{xy}] + \tau_{bx} = -\rho g H\frac{\partial S}{\partial x}. \qquad (13.23)$$

A similar relation applies for the y-direction. By using the creep relation for ice (Chapter 3), the stresses in these formulae are replaced by velocity gradients. This gives two differential equations for the velocity components u_x and u_y, which are taken to be uniform in the vertical direction (MacAyeal 1989). For an ice shelf, $\tau_{bx} = \tau_{by} = 0$. For the "transition zone" in West Antarctica – the region of grounded ice near the coast – Ritz et al. (2001) used these relations with a small but nonzero value for $\vec{\tau}_b$. Huybrechts instead modelled the transition zone using Eqs. 13.17 and 13.18, but including the stresses τ_{xx}, τ_{yy}, and τ_{xy} in τ_E. This gives faster flow in the transition zone, as desired. That approach probably generates a reasonable simulation for parts of the East Antarctic coast, but does not resemble the real situation in West Antarctica or along low-stress ice streams in East Antarctica where basal shearing is negligible.

2. Pattyn (2003) developed a technique for using the complete force balance, Eq. 13.23, resolved at all vertical levels (Sections 8.5.1 and 8.9.5). This is the best approach for modelling places such as onsets of ice streams and transitions to ice shelves where all the stresses can be important. This approach, however, is computationally expensive compared to using the standard approximations of "shallow-ice" models.

3. Marshall and Clarke (1997a, 1997b) developed a technique for including ice streams implicitly as subgrid-scale features. The approach regards each grid cell as consisting partly of fast ice stream flow and partly of regular ice sheet flank flow, with rules to link the two. The partitioning of each cell into these two types is a free variable. Marshall and colleagues applied this method to simulate the Laurentide Ice Sheet.

4. Modifications of ice viscosity (Chapter 3). Most Greenland Ice Sheet models reduce the viscosity of ice deposited in the last glacial period, relative to that for Holocene ice, by a factor of 2.5 or 3. (In other words, the factor E in Eq. 13.18 is increased by a multiple of 2.5 or 3 compared to its value for Holocene ice.) The boundary of the two is tracked through time. In order to account for layers of temperate ice – features known to exist in Greenland – Greve (1997a) used a more general energy conservation relation than Eq. 13.19, accounting for melting and refreezing within temperate ice. A temperate layer shears more readily than a layer of subfreezing ice.

5. Clarke et al. (2005) developed a technique for tracking layers of a given age (isochrones) as they are swept through the ice sheet. They applied this to a simulation of Greenland through the last glacial cycle and produced a three-dimensional map of isochrones. They compared this result to known values at ice core sites (Lhomme et al. 2005).

In the near future, model development will focus on improving the representation of ice streams, their interaction with ice shelves, and ice shelf mass balance processes including calving and basal melt. Model calibration techniques must also advance.

13.4.2 Model Calibration

Models of Greenland and Antarctica are calibrated by comparing the calculated surface elevations and ice thicknesses with measured ones. Even without calibration, the models should reproduce the topography reasonably well, since the gross ice sheet form is determined largely by the creep properties of ice and the outline of the land mass, except in coastal regions with slippery beds (Section 8.10). With simple calibrations, the models very closely match the measured topography averaged over a few grid cells. Calibration is usually achieved by adjusting the fluidity of the ice, the enhancement factor E in Eq. 13.18. For Greenland, calibrated E values are typically in the range 3 to 6, probably due to softening of the ice by fabrics favorable to simple shear, the predominant deformation in basal layers of ice sheets. Furthermore, adjusting the specific mass balance relation, in particular how $\dot{b}_i$ depends on temperature, improves the match between modelled and observed ice margin positions in Greenland models.

By using the tracer-transport method (Lhomme et al. 2005), it is possible to construct model predictions of the internal stratigraphy of an ice sheet, including the layering of stable isotope values and the impurity horizons that reflect radio waves. Comparing these predictions to stratigraphic data from ice cores and radio-echo soundings might be a useful basis for choosing among climate and mass balance histories. Likewise, measured borehole temperatures, though available at only a few sites, can be compared with model temperatures to adjust geothermal fluxes and the imposed climate history. Ice surface velocity data are now being acquired over broad regions of the ice sheets using satellite-based methods. Huybrechts has suggested that these data will be useful for constraining whole-ice-sheet models. Velocities, temperatures, and stratigraphy all offer a more rigorous test of model predictions than does ice sheet topography. Such constraints have been applied widely in studies of small regions such as individual ice streams (Section 8.9.2).

13.4.3 Simulations of Quaternary Ice Sheets

Whole-ice-sheet models can be used to simulate the evolution of ice sheets through the late Quaternary. Although the uncertainties are large, such models offer a way to bring together a wide range of paleo-environmental information and glaciological concepts to produce broadly ranging and internally consistent hypotheses about ice sheet histories. Here we briefly examine examples of models of the Antarctic, Greenland, and Laurentide ice sheets.

Antarctic   Huybrechts (2002) modelled the Antarctic ice sheet over the last 400 kyr using a climate history based on the Vostok ice core and a global sea-level history from the marine isotope record. In Antarctica, the primary direct effect of climate change is assumed to be a reduction of snowfall during cold periods. Cold periods also lead, over time, to decreased temperatures at depth and hence increased ice viscosity. These two processes have counteracting effects on ice thickness in the model (as explained in Section 11.4.2). Grounding lines are located where the ice thickness decreases to the flotation point, and are thus determined in the model by a combination of sea level, ice thickness, and bedrock elevation. The latter varies over time because of isostasy. Ice thickness at the grounding line depends, through the mass conservation relation, on flow of both the inland ice and the ice shelves. Ablation in Antarctica occurs almost entirely by iceberg calving and bottom melting of ice shelves. The model calculates neither process, but instead allows ice shelves to extend to the limits of the numerical grid.

Figure 13.5a shows the model ice sheet topography at three times: the last interglacial ("Eem"), the LGM, and now. The main results of the model analysis are:

1.   The ice sheet expands when sea level falls. The volume of the ice sheet is largely controlled by sea level, which itself is determined by the extent of ice in the northern hemisphere. This is a long-held belief (Penck 1928; Hollin 1962).

2.   Variations of Antarctic ice volume account for about 20 m of sea-level change on glacial-interglacial timescales (out of a total of 130 m).

3.   The surface elevation changes very little in the interior of East Antarctica, the site of the longest ice-core paleoclimate records.

4.   Grounded ice in West Antarctica and the Antarctic Peninsula expands considerably in cold periods. Geological evidence shows that such expansion did occur (Denton and Hughes 2002; Lowe and Anderson 2002; Anderson et al. 2002).

5.   The retreat of the grounding lines in West Antarctica continued through the Holocene and is still ongoing; West Antarctica is expected to be thinning at present. Many indicators, including sedimentary deposits, exposure ages of bedrock, and the internal layering of ice domes, show that grounded ice in West Antarctica indeed retreated and thinned throughout the Holocene (Conway et al. 1999; Stone et al. 2003; Waddington et al. 2005).

6.   During the last interglacial period, an era warmer than present, the model ice sheet thins and retreats in West Antarctica – but its central region remains intact. This is broadly consistent with evidence for higher sea levels, 4 to 6 m above late Holocene ones (Overpeck et al. 2006). Retreat of both Greenland and West Antarctica could have contributed. Whether the center of West Antarctica survived this period is a major unanswered question. Pleistocene diatom fossils and high concentrations of cosmogenic beryllium isotopes in sediments beneath the

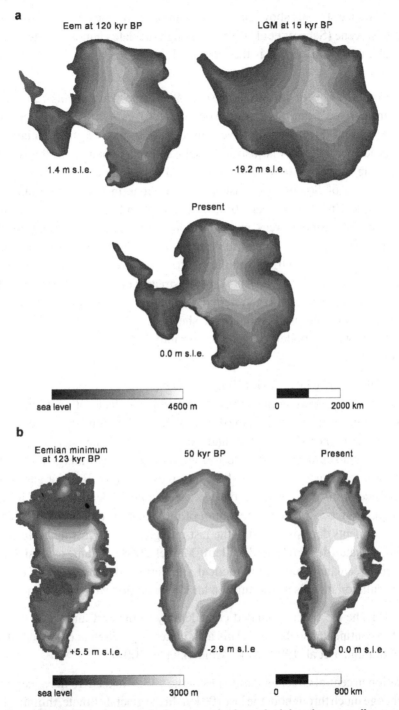

Figure 13.5: Snapshots of ice sheet evolution over the last glacial cycle, according to the model of Huybrechts (2002). (a) The Antarctic Ice Sheet. (b) The Greenland Ice Sheet. Numbers for each panel indicate the contribution to global sea-level change. Adapted from Huybrechts (2002).

West Antarctic ice streams show that this was an open marine environment at some time in the late Pleistocene (Scherer et al. 1998), although the date cannot be determined. The long interglacial around 400 kyr b.p. is the most likely time.

Greenland   To model Greenland for the last ice-age cycle, Huybrechts (2002) again used a sea-level history derived from the marine isotope record. A hybrid of ice core data gave the climate history; central Greenland data were used back to the last interglacial, and Antarctic data for the earlier period. (This is justified because climate changes in the two hemispheres appear to correlate well over time scales of several millennia and longer.) The model does not permit formation of floating ice. Instead, the ice margin is never permitted to advance beyond the coastline determined at each time by sea-level and bedrock elevation (which adjusts isostatically). Implicitly, this enforces an assumption that calving acts efficiently and removes all ice that flows to the margin. In contrast to Antarctica, much of the ablation occurs by surface melt. Melt decreases when the climate cools, and this allows the ice sheet to expand. Since the model cannot simulate the narrow, fast-flowing ice streams or the floating tongues fed by them, it does not resemble the real ice sheet system dynamically. The model nonetheless captures the overall form and dimensions of the modern ice sheet.

Figure 13.5b shows the model ice sheet topography at three times. The main results of the model analysis are:

1. During the last interglacial period, 120 to 130 kyr b.p., the ice sheet was substantially diminished by temperatures warmer than present, although a high elevation dome remained in the center. The model cannot constrain the details well, but these findings are consistent with ice core evidence (Koerner 1989; Cuffey and Marshall 2000). The deepest layers in ice cores from western and southern Greenland formed during the last interglacial, at low altitude, and suggest a readvance of the ice sheet into these regions. But the gas content of last-interglacial ices from the center of the ice sheet indicate persistence of a high elevation dome. In addition, ocean cores from south of Greenland reveal a large increase of spruce pollen in sediments deposited during the last interglacial, suggesting both ice sheet retreat and warm climate (de Vernal and Hillaire-Marcel 2008). Again, extensive retreat of ice in southern Greenland occurs in numerical simulations that use climate parameters from general circulation models as forcings on an ice sheet model (Otto-Bliesner et al. 2006).

2. At the LGM, the ice sheet expanded considerably to the east and west. Its margins were close to the continental shelf edge. This also matches geological evidence (Andrews et al. 1997, 1998; Funder et al. 1998, 2004; Weidick et al. 2004).

3. The elevation in central Greenland, the sites of the GISP2 and GRIP deep ice coring projects, did not change much throughout the last 100 kyr. In the glacial climate, thinning from reduced snowfall was opposed by thickening due to colder ice and, most importantly, to expansion of the ice sheet margins.

Laurentide Tarasov and Peltier (2004) modelled the Laurentide Ice Sheet in the last glacial cycle. Although the simulation ran through the entire glacial period, most of this time was used for initialization; the report focused on the LGM and subsequent deglaciation. The analysis was rather different from those of Huybrechts, because Tarasov and Peltier constrained the model by adjusting it to match several constraints: (1) the history of ice margin positions during

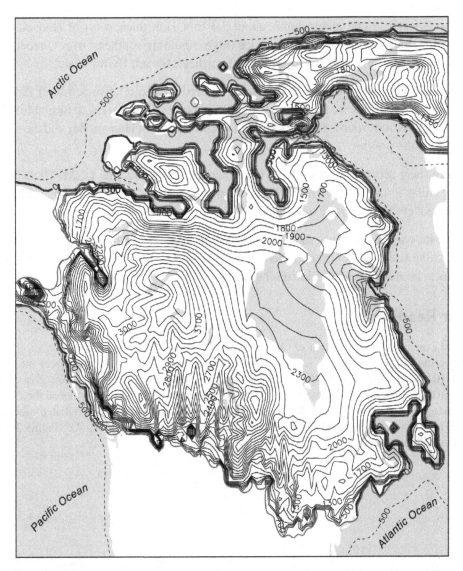

Figure 13.6: A model hypothesis for the configuration of the Laurentide Ice Sheet at the Last Glacial Maximum. Contour labels give elevations in meters above sea level. Adapted from Tarasov and Peltier (2004). (Refer to the insert for a color version of this figure)

deglaciation known from geomorphologic evidence; (2) the relative sea-level histories at more than 40 locations in northeastern Canada; (3) the measured rate of crustal uplift at one location in the interior of western Canada; and (4) measurements of absolute gravity along a transect from Hudson Bay south to Iowa. Isostatic adjustment was calculated with a variable viscosity spherical-earth model.

Figure 13.6 illustrates the modelled ice sheet topography at the LGM. The main results from the analysis are:

1. The ice sheet had one major dome, about 3.3 to 4.3 km thick, west of Hudson Bay (the Keewatin Dome), and a broad, thinner ice ridge extending southeast over Quebec. The ice surface sloped gently over Hudson Bay and northeast toward Baffin Island.

2. To achieve this configuration, which was required by the relative sea-level data, it was necessary to model the ice flow in some regions as basal slip on a weak deformable substrate. These regions corresponded to the sediment-mantled areas of Hudson Bay and the southwest margin on the prairies.

3. The ice sheet accounted for 60 to 75 m of reduced sea level at the LGM. These limits are very similar to the 56 to 76 m deduced much earlier from standard ice sheet profiles and geological evidence of ice extent (Paterson 1972a).

4. The relative sea-level curves were best matched if the ice over Hudson Bay thinned dramatically at the times of Heinrich events. Uncertainties are still too large, however, for strong claims to be made about this intriguing result.

Further Reading

For a good example of the forefront of "Earth system modelling" that includes an ice sheet model as a subroutine, see Vizcaino et al. (2008); they discussed how the Greenland Ice Sheet interacts with North Atlantic circulation and local albedo. A paper by Pollard and DeConto (2009) concerns a new forefront of ice sheet modelling applied to West Antarctica. A classic set of works on the Pleistocene ice sheets is the compendium by Denton and Hughes (1981). A widely used detailed model for the worldwide ice cover at the LGM and subsequent isostatic adjustments is "ICE-5G" (Peltier 2004, Ann. Rev. Earth Planet. Sci. 32, 111–149).

CHAPTER 14

Ice, Sea Level, and Contemporary Climate Change

"With the aid of Langley's figures it is possible to calculate the change in temperature which would result from a known definite increase in the carbonic acid content of the air."

Nature's Heat Usage,[1] **Svante Arrhenius (1896)**

14.1 Introduction

The global mean sea level rises when glaciers and ice sheets shrink. This superficially simple fact links the central themes of this book – the physical properties and mechanisms governing glacier behavior – with many topics of broad scientific and societal interest. Through variations of sea level, coastal landscapes worldwide are influenced by high-latitude climate and oceanographic conditions, and by the internal dynamics of glaciers and ice sheets. A rise or fall of sea level alters patterns of coastal erosion and sediment accumulation at river mouths, the interaction of estuaries and marshes with tides, and the ecology of barrier islands, coral reefs, and other coastal systems. Human populations and infrastructure are concentrated in coastal regions; this fuels interest in the topic of ongoing and future sea-level rise. In conjunction with storms and waves, sea-level rise has the potential to cause great harm or, at the very least, to extract a severe price for mitigation. Sea-level rise of only a few meters would inundate large portions of low-lying coastal regions, such as the Gulf Coast of North America, the Ganges and Mekong deltas, and many oceanic islands.

Concerns now focus on *global warming* – the ongoing planetary-scale uptake of heat by the oceans, atmosphere, and land surface. Global warming is a (mostly) unnatural experiment of great importance. Changes observed in the last century such as retreat of glaciers and thawing of permafrost are only the beginnings of a large sustained and unplanned transformation of Earth's environment with diverse and problematic consequences. One result will be continued shrinking of glaciers and polar ice sheets, a process that accelerated in recent years. The extent of future changes is highly uncertain, but recent observations show that glaciers and ice sheets are reacting rapidly.

---

[1] Translation by H. Rohde et al., *Ambio* 26, p. 4.

Copyright © 2010, Elsevier Inc. All rights reserved.
DOI: 10.1016/B978-0-12-369461-4.00014-4

After setting the stage with basic information about the relationship between ice and sea level, the present chapter reviews how mountain glaciers and small ice caps have responded to warming, and how they might respond in the century ahead. We then discuss the polar ice sheets: their current state, their mechanisms of response, and their possible future evolution. Forecasts of sea-level rise remain speculative; our approach is to summarize the governing mechanisms and emphasize extrapolations of observations. Throughout, we cite earlier chapters for thorough discussions of processes and, in some cases, observations. In particular, Chapters 4 and 5 review mass balance processes, Chapter 11 analyzes how glaciers respond to climate changes, and Chapter 13 summarizes mechanisms of interaction between ice sheets and global climate.

Information about this subject is rapidly accumulating. Much of what we write about the future of sea level will no doubt need revision in short order.

14.1.1 Equivalent Sea Level

Currently, oceans cover about 71% of Earth's surface. The area of ocean thus totals about $\mathcal{A}_o = 0.71[4\pi R_E^2]$, where $R_E \approx 6370\,\mathrm{km}$ is the planet's radius. Every gigaton of ice ($1\,\mathrm{Gt} = 10^{12}\,\mathrm{kg} = 1\,\mathrm{km}^3$ water-equivalent volume) transferred from land to the ocean raises the sea level by about $1/\mathcal{A}_o$ kilometers, or $\frac{1}{362}$ mm. For a marine ice sheet or tidewater glacier, however, only the mass of ice above the flotation level contributes. Consider an ice body of thickness H and depth-averaged density $\overline{\rho}$, resting on a bed at depth $|B|$ below sea level. Per unit area, the mass potentially contributing to sea-level rise is then $\overline{\rho}H - \rho'|B|$ for a sea-water density of ρ', nominally $1028\,\mathrm{kg\,m}^{-3}$.

Melt or disintegration of a floating ice shelf has no *direct* effect on sea level, but can trigger increased flow of the glaciers feeding it (Section 11.5). This draws ice from inland and thereby contributes to sea-level rise.

In what follows, we always use the global mean sea-level rise as a convenient point of reference. It should be kept in mind, however, that sea level does not rise uniformly; the ocean surface is considerably more complex than the rising water level in a bathtub (Meehl et al. 2007). As climate changes, so do the distributions of temperature and salinity in the ocean, and the currents driven by winds and density gradients. All of these factors influence the worldwide pattern of ocean surface elevations. From place to place on the globe, differences in net sea-level rise of a few decimeters are expected by the end of this century. In addition, shrinkage of an ice sheet involves a shift of mass from one local region to the ocean basins worldwide. Such a redistribution of mass changes Earth's gravitational field; the pattern of sea-level change therefore depends on where ice loss occurs (Mitrovica et al. 2001). And all of these effects are in addition to spatial patterns of longer-term sea-level change related to sediment compaction and to deformations of the solid earth by isostatic and tectonic processes.

14.1.1.1 Contemporary Ice Volumes

Table 14.1 gives current estimates for the sea-level-equivalent volumes of major ice reservoirs. Values for the ice sheets were calculated from spatial interpolation of measured ice thicknesses,

Table 14.1: Equivalent sea-level volumes of glacial reservoirs.

Reservoir	Sea-level equivalent (m)	Reference
East Antarctic Ice Sheet	52	Lythe et al. (2001)
West Antarctic Ice Sheet	5	Lythe et al. (2001)
Greenland Ice Sheet	7.3	Bamber et al. (2001)
Glaciers and small ice caps: total excluding peripheral bodies°	$0.7^\dagger \pm 0.2$ 0.37 ± 0.06	Meier et al. (2007) Dyurgerov and Meier (2005)

† Given as volume of $250 \times 10^3 \, km^3$. Excludes Antarctic Peninsula. Error estimate from Dyurgerov and Meier (2005).
° Peripheral bodies are glaciers and ice caps adjacent to the main ice sheets on Antarctica and Greenland.

mostly obtained in radar soundings. For glaciers and small ice caps, surface areas – the observed quantities – were converted to ice volumes using statistical relationships. Relationships between volume and area generally conform to simple power laws (one for glaciers, one for ice caps), a consequence of the plastic properties of ice and constraints on dynamics related to mass and momentum conservation (Bahr et al. 1997; Section 8.10.3). These statistical relations were calibrated using data from more than 100 ice masses.

14.1.2 Recent Climate and Sea-level Change

Earth's mean surface temperature began to increase in the late nineteenth century. From 1900 to 2007, the net warming was about 0.8 °C. Temperatures on land and at the ocean surface, measured independently, both increased. The gain of heat energy occurred mostly in the ocean, because of its large area and the rapid mixing of waters between the surface and the layers beneath. Natural causes, especially a small increase in solar radiation, contributed to the early part of this *global warming*. By the end of the twentieth century, however, the majority of the cumulative warming – including essentially all the warming after 1970 – was due to increased absorption of longwave radiation by greenhouse gases. The concentrations of the important greenhouse gases (including carbon dioxide, methane, nitrous oxide, tropospheric ozone, and various industrial compounds) have all increased from human activities. Increased concentration of water vapor and reduced snow and ice cover amplified the warming as feedbacks. The evidence for these statements is reviewed extensively in Solomon et al. (2007). Recent analyses of critical factors include Domingues et al. (2008) on oceanic heat uptake, Santer et al. (2007) on the water vapor feedback, and Stroeve et al. (2007) on diminishing sea ice.

As Figure 14.1 illustrates, sea level rose over the same period, at an average rate of $1.7 \pm 0.5 \, mm \, yr^{-1}$. A review by Bindoff et al. (2007) summarized the evidence concerning sea-level rise. Thermal expansion from increased water temperatures accounts for part of the rise, called the *steric* component, which is 15% to 30% of the total. Much of the remaining rise must be due to increased water mass (the *eustatic* component), primarily from melting glacial ice. Processes such as isostasy, tectonics, and sedimentation also influence sea level by changing the sizes of the ocean basins; this may amount to a few tenths of a millimeter per year. The long-term record

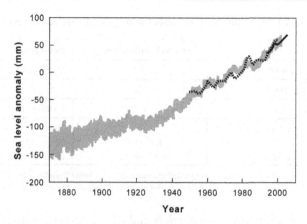

Figure 14.1: Observed rise in global mean sea level since the mid-nineteenth century. Light gray band shows range of reconstructed sea levels since 1870, based on sparse tide gauge measurements assimilated into patterns of spatial variability observed in recent studies. Dashed line shows reconstructions directly from coastal tide gauges, starting in 1950. The black line was derived from satellite altimetry, and so includes open oceans. Adapted from Bindoff et al. (2007).

of sea-level changes was constructed by synthesizing records from tide gauges. Since 1992, satellite altimetry has provided a second metric.

Recent sea-level rise has been faster than the century-long trend. In the period from 1993 to 2003, the level rose at 3.1 ± 0.7 mm yr^{-1}, of which 1.6 ± 0.5 can be attributed to thermal expansion. Thus, the magnitude of the eustatic component may have been as little as 0.3 or as high as 2.7 mm yr^{-1} (mid-range value of 1.5). The rate of sea-level rise varies from decade to decade for reasons that are not fully understood. Thus, at this timescale, sea-level measurements can place only rough bounds on the contribution from evolving glaciers and ice sheets.

14.2 Global Warming and Mountain Glaciers

Outside of the Antarctic, the majority of mountain glaciers have reacted to the warming by shrinking (Lemke et al. 2007; Oerlemans 2005). Glacier retreat is apparent in all the great mountain ranges, including the Alps, Himalayas, Andes, Canadian Rockies, and ranges of southern Alaska. In general, retreat occurred most rapidly from about 1920 to 1940 and again from 1980 to the present. These intervals correspond to the episodes of most rapidly rising global temperatures. Little warming occurred in the middle period, from 1940 to 1975. Many small mountain glaciers stabilized or readvanced in these decades.

Increased snowfall or local coolings have allowed some glaciers to advance, but two processes driving retreat have dominated glacier behavior:

1. Each 1 °C of summer season warming increases annual ablation rates at the termini of typical mid-latitude and subpolar glaciers by about 1 to 1.5 m yr^{-1} (Section 5.4.3). (A smaller increase occurs in the severely cold regions at high altitudes and latitudes.) To counteract such an increased rate of mass loss, snowfall rates would typically need to increase by 20% to 50% for the same warming, an unlikely change (Oerlemans et al. 1998; de Woul and Hock 2005). Moreover, precipitation changes related to warming lack coherence over the globe; some regions become drier while others become wetter. To a first approximation, climate warming intensifies existing patterns of precipitation.

2. Some glaciers flow faster, which conveys ice more rapidly to low-altitude ablation zones and, in particular, to the calving termini of tidewater glaciers. This is an important process driving mass loss from the large tidewater glaciers of southern Alaska and Patagonia (Section 14.2.2.1). Summertime warming increases surface melting, which then increases movement by basal slip as the water makes its way to the bed (Section 11.4.3). Moreover, increased surface melting and warmer ocean waters both erode the termini of tidewater glaciers, through increased calving and frontal melt. By thinning the terminus or forcing retreat into deeper water, such erosion reduces the forces restraining the glacier, triggering increased ice flow (Section 11.5.2). Together these processes constitute the *dynamic effect*.

Glaciers thin and retreat in response to increased ablation (Chapter 11). The thinning is greatest near the terminus, but extends far up-glacier. If snowfall also increases, the upper part of the glacier may thicken even while the lower part thins. Figure 14.2 shows this pattern in the measured net thickness changes of Bear Creek Glacier, Alaska, from 1957 to 1994 (Sapiano et al. 1998). The glacier as a whole lost mass. These observations appear to be typical for subpolar and mid-latitude mountain glaciers. However, such precise data over multiple decades are available for only a small number of glaciers, and generally do not extend back prior to the middle of the twentieth century.

14.2.1 History of Glacier Lengths

More information is available about changes of glacier lengths, especially in the European Alps. The longest record, of Untere Grindelwaldgletscher, began in 1534. Oerlemans (2005) has compiled and analyzed the glacier length records from around the world. From 1700 to 1860, the average glacier length first increased slightly and then decreased, yielding no cumulative change. After 1860, most glaciers retreated; from 1860 to 1900, 35 of the 36 monitored glaciers retreated, while from 1900 to 1980, 142 of the 144 monitored glaciers retreated. Retreat was not steady, however, and many glaciers partly readvanced in the middle of the century. The average cumulative decline of glacier length, which amounted to about 1.5 km, is shown in Figure 14.3a.

Oerlemans recognized that the glacier lengths can be used as a crude thermometer to learn how the climatic temperature varied over the last few centuries. He used the following simple

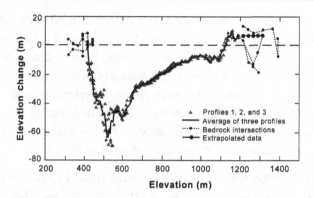

Figure 14.2: Observed changes of the surface elevation of Bear Creek Glacier, Alaska, from 1957 to 1994. The lower glacier thinned while the highest part of the glacier thickened slightly. Adapted from Sapiano et al. (1998).

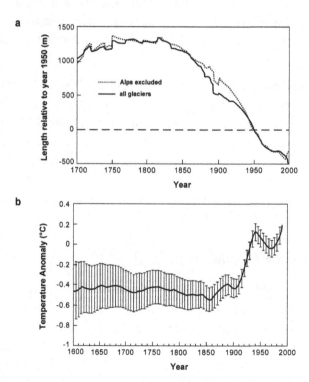

Figure 14.3: (a) Average observed change in the length of mountain glaciers, for all available locations and for all locations outside of the European Alps. (b) The global mean temperature change estimated from the glacier retreat observations. Adapted from Oerlemans (2005).

approach. Suppose that a glacier, initially at equilibrium with the climate, retreats or advances in response to a temperature change T'. Call the change in glacier length L'. When the glacier attains a new equilibrium, $L' = cT'$, where the parameter c defines a *climate sensitivity*. In terms of equilibrium values for L',

$$c = \frac{dL'}{dZ_E} \frac{dZ_E}{dT'}, \tag{14.1}$$

for an equilibrium line altitude Z_E. The factor dL'/dZ_E depends inversely on the glacier slope; a given change of Z_E yields a larger change of total annual mass input on a gently sloping glacier than on a steep one. (If the glacier is nearly flat, a small rise in Z_E can transform the entire surface into an ablation zone.) The second factor, dZ_E/dT', depends on the climate and generally is largest in warm, wet environments with strong melt and precipitation. Oerlemans used detailed flow-line models (Section 11.3.4) for seven well-studied glaciers to obtain a calibration for c, which was $c = 2.3 P^{0.6} \alpha^{-1}$. Here P denotes the annual precipitation ($\mathrm{m\,yr^{-1}}$) and α the mean surface slope. Values of c for different glaciers range from about 1 to $10\,\mathrm{km\,°C^{-1}}$.

The change of glacier length is not achieved immediately. For mountain glaciers the adjustment typically takes a few decades. Further, the more L' deviates from its equilibrium value, the faster the change will be. Thus, Oerlemans used the relation, for time t,

$$\frac{dL'}{dt} = -\frac{1}{t_\ell} \left[cT' + L' \right], \tag{14.2}$$

with parameter t_ℓ representing a lag time. (Note that a warming, $T' > 0$, drives a retreat, $L' < 0$.) The lag time is closely related to the response time, $H/\dot{a}$, for a glacier of characteristic thickness H and terminus ablation rate $\dot{a}$. (See Section 11.2.) As H varies inversely with slope α, and $\dot{a}$ is proportional to the elevation drop along the glacier (αL) times the mass balance gradient with altitude ($\gamma = d\dot{b}/dS$), the lag time should depend inversely on all three of these variables: α, γ, and L. Indeed, such a dependence appears in Oerlemans' relation for t_ℓ. He derived it, however, not from $H/\dot{a}$ but from the ratio of a glacier's length to its characteristic velocity: the timescale for ice flow to carry mass from the head of the glacier to the snout. As with c, values for t_ℓ were calibrated against detailed flow-line models for a few glaciers. Most of Oerlemans' values for t_ℓ fall in the range of 40 to 100 years.

Using Eq. 14.2 and observations of retreat ($L'(t)$), Oerlemans derived temperature histories for all 169 glaciers with length records. Results were combined to obtain temperature histories for each of several major regions: the Alps, Asia, Northwestern America, the North Atlantic, and the southern hemisphere. Further combination gave an estimate for global mean temperature (see Figure 14.3b). It matched, reasonably well, the global mean temperature curve based on meteorologic records – an important result because the two methods are entirely independent. Unfortunately, the temperature reconstruction from glacier lengths cannot yet include the most recent 15 years, due to the glacier lag time and the delay in reports of observations. Mass balance studies, however, reveal that glacier volumes have diminished at an increasing rate in this most recent period (Section 14.2.2).

How the warming varies with altitude has been a subject of much controversy. Oerlemans' temperature reconstructions showed no significant difference for low-altitude sites compared to high-altitude ones.

14.2.2 Worldwide Mass Balance of Mountain Glaciers and Small Ice Caps

Although fluctuations of a glacier's length provide a general indication of changes in its mass, a rigorous assessment of mass balance – the change in total mass of a glacier over an interval of time – requires a much more extensive set of observations. We have summarized the methods of such mass balance studies and discussed examples from specific glaciers in Chapter 4. Here we consider the composite of mass balance measurements from mountain glaciers and small ice caps worldwide. The goal is to quantify their net contribution to sea-level rise. Large uncertainties remain.

Efforts to monitor glacier mass balances at the global scale began in the 1960s. Dyurgerov (2002), Dyurgerov and Meier (2005), and Cogley (2005) made thorough compilations of the data. Annual data are available for at least 60 glaciers since 1964 and for at least 90 glaciers since 1981, peaking at 109 glaciers in 1998 (Dyurgerov 2002). These compilations make use of short records from many different glaciers; only 40 glaciers have at least 30 years of annual mass balance data at the time of writing. Dyurgerov and Meier (2005) used information from 304 different glaciers, and Cogley (2005) listed 330 glaciers with at least one year of data in his inventory. This is a small fraction of the estimated global population of more than 160,000 glaciers. Thus, claims about global mean mass balance rest on the behavior of a small subset of glaciers. This is not necessarily a problem, because glacier variations tend to be regionally coherent (Section 4.3.3), but neither is it desirable.

Extrapolating from the global set of available mass balance data, and the areal distribution of glaciers known from remote-sensing, Dyurgerov and Meier (2005) made regional, continental, and global estimates of glacier changes in the period from 1961 to 2003. Kaser et al. (2006) produced a consensus update, merging several global datasets and interpretations. Figure 14.4 shows Dyurgerov and Meier's estimate for the cumulative ice volume change, averaged globally, and the corresponding annual mean balances. The persistently negative values show that Earth's mountain glaciers are losing mass, consistent with the observations of glacier retreat discussed previously. The mean global balance was, according to this analysis, -0.24 m yr^{-1} in this four-decade period, equivalent to an average rate of sea-level rise of 0.50 mm yr^{-1}. The rate of loss increased toward the end of the period. Furthermore, the losses occurred despite a moderate increase in annual snowfall. Glaciers are conveying more ice, but melt is increasing more rapidly than snowfall (Dyurgerov 2003).

Compared to the mass balance records for individual glaciers (e.g., Figures 4.9 and 4.10), interannual variability in this global series is muted, largely because of the averaging of different

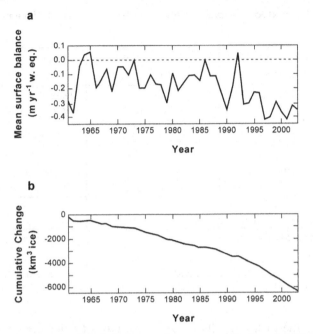

Figure 14.4: Estimate for recent global changes of mountain glaciers. (a) The yearly average of specific surface mass balance. (b) The cumulative change in total ice volume. From data given by Dyurgerov (2002) and Dyurgerov and Meier (2005). Data compilation by S.J. Marshall.

regions with asynchronous climate variations. Nonetheless, effects of some global climate events are discernible, in addition to the persistent negative values associated with global warming. Most prominent is the brief cooling that followed the 1991 eruption of the Phillipine volcano Mount Pinatubo. Sulfur gases released by the volcano oxidized to form acid droplets in the upper atmosphere; such aerosols reflect sunlight. This brief cooling produced the only year with a positive mass balance since 1965.

Most of the data in the Dyurgerov and Kaser compilations were acquired by the *glaciologic method*, in which surface mass balances are measured at representative points on a glacier and then extrapolated to find the glacier-wide total (Section 4.3). Such measurements do not include losses by increased calving, which must be discerned separately. Increased calving – mostly a consequence of faster ice flow – accounts for much of the recent ablation of large marine-terminating glaciers, especially in southern Alaska. To assess such *dynamic losses*, the compilations also include data obtained with the *altimetric method*: the direct measurement of glacier volume changes by repeat mappings of surface elevation. Though generally too sparse to show year-to-year variations of the sort depicted in Figure 14.4, such information must be included to evaluate overall trends.

The current rate of mass loss from glaciers significantly exceeds the mean rate in the 1990s. Thus, Meier et al. (2007) updated the global synthesis, focussing on the period from 1996 to 2006. They also improved the accounting for dynamic losses.

Table 14.2: Estimated sea-level contribution from mountain glaciers and small ice caps.

Region	Time interval (Approximate[†])	Equivalent sea-level rise (mm yr^{-1})	Reference
Globe	1961–2003	0.5	Dyurgerov and Meier (2005)
	1961–1990	0.38 ± 0.19	Kaser et al. (2006)
	1991–2004	0.77 ± 0.26	
	2001–2004	0.98 ± 0.19	
	2006	1.1 ± 0.24	Meier et al. (2007)
Alaska	1955–1995	0.14 ± 0.04	Arendt et al. (2002)
	1995–2001	0.27 ± 0.10	
Patagonia	1970–2000	0.042 ± 0.002	Rignot et al. (2003)
	1995–2000	0.11 ± 0.011	
Observed global sea-level rise:			Bindoff et al. (2007)
20th century		1.7 ± 0.5	
1993–2003		3.1 ± 0.7	

[†] For Patagonia and Alaska, data are not available for all places in any single year, so starting years for the interval vary.

Table 14.2 lists some of the results from these studies (under the category "globe"). The retreat of mountain glaciers and small ice caps contributed moderately to sea-level rise in the twentieth century: apparently about 2 to 3 cm total in the second half of the century. The global rate of mass loss has increased significantly since 1970, a trend clearly related to warming (Kaser et al. 2006).

These estimates for the global contribution have large uncertainties. They have been revised a number of times over the last several years, generally upward, due in part to new data from Alaska and Patagonia, where changes have been rapid and widespread.

14.2.2.1 Balance of Glaciers in Alaska and Patagonia

The mountains of southern Alaska and Patagonia contain much of the glacier mass outside the polar regions. This mass has diminished significantly over the last half-century. (The global numbers in Table 14.2 include this component.) Arendt et al. (2002) used the altimetric method to assess balances of glaciers in Alaska. For 67 glaciers, they compared data from airborne laser surveys in the mid-1990s to topographic data from air photogrammetry from earlier decades. For 28 glaciers, the laser surveys were then repeated in 2000 and 2001. On average, the glaciers thinned in both periods, more rapidly in the later one. The results were extrapolated to all glaciers in southern Alaska, using the observed relationships between altitude and rate of thinning – a procedure that should be regarded skeptically. The estimated net contribution to sea-level rise is one of the largest inferred for a region (Table 14.2).

A similar analysis for Patagonia also used the altimetric method, with data from photogrammetric surveys in the years 1968 to 1975 and 1995 compared to radar topography measurements from the Space Shuttle in 2000 (Rignot et al. 2003). The radar topography data are spatially complete, so it was not necessary to use an extrapolation procedure as Arendt et al. (2002) did for Alaska. The total contribution of Patagonian ice to sea-level rise (Table 14.2) was smaller than the contribution from Alaska. But, per area of glacier, the contribution from Patagonia was 1.5 times larger.

In both Patagonia and Alaska, the thinning of ice reflects climate warming. In both places, however, much of the thinning results, not from increased melt rates, but from faster glacier flow – the dynamic effect defined at the start of Section 14.2. The portion of the mass lost from tidewater glaciers reflects the instability discussed in Section 11.5.2; although initially triggered by warming, the large magnitude of the response is a manifestation of dynamical feedbacks.

14.2.2.2 Retreat of Tropical Glaciers

Over the last century, significant retreat occurred in all the redoubts of glaciers in the tropics: the Equatorial Andes, the Ruwenzori Mountains and high volcanoes of East Africa, and the highest summits of Irian Jaya (Kaser 1999; Kaser and Osmaston 2002; Lemke et al. 2007). Many of these glaciers lost half of their entire area, or more. Tropical glaciers have little influence on sea level, but their fate captivates attention because of their extraordinary contrast with most elements of tropical landscapes. They also contribute significantly to local water supplies in some cases and, in a complicated fashion, measure changes of tropical climate.

Tropical glaciers are confined to high-altitude mountains, mostly 5 km and higher. Many of these sites are cold and dry enough that sublimation contributes significantly to ablation, in addition to melt (Section 5.3.6.8). Where melt is important, warming air should increase ablation by enhancing sensible heat transfer and incoming longwave radiation. Such an increase may play a significant role in this century, and probably was important for the lower-altitude tropical glaciers in the last (including many of the glaciers in the Andes). In East Africa, however, at the high altitudes where glaciers occur, temperatures have not increased appreciably but atmospheric humidity has decreased (Mölg et al. 2006). In the Equatorial Andes, furthermore, climate depends strongly on fluctuations in Pacific Ocean sea surface temperatures (the ENSO events), which influence many atmospheric variables in addition to temperature.

Detailed studies have identified several mechanisms controlling glacier wastage in the Tropics (Kaser 1999; Francou et al. 2003, 2004; Favier et al. 2004; Mölg et al. 2003b, 2006). Reduced snowfall not only decreases accumulation directly, but also significantly enhances melt by reducing the surface albedo. Reduced atmospheric humidity inhibits cloud formation and therefore increases the penetration of solar radiation to the surface, enhancing melt. Reduced humidity also directly increases ablation by sublimation. The largest effect of increased sublimation, however, is to absorb energy that would otherwise generate melt; even a small rate of sublimation can prevent a significant amount of melting (Section 5.4.5.3). Increased air temperatures change precipitation from snow to rain, again driving increased melt by reducing surface albedos.

How this complex web of interrelated processes drove retreat of tropical glaciers is not entirely clear, in part because dominant processes have varied from place to place. Reduced snowfall and increased temperatures, both related to warming of the Pacific sea surface, were important in the Andes (Favier et al. 2004; Francou et al. 2003, 2004). On Kilimanjaro, the likely causes were drying of the atmosphere and associated increases in solar radiation, sublimation, and melt on vertical ice cliffs (Mölg et al. 2003b; Cullen et al. 2006). Rapid retreat of the glaciers on Kilimanjaro began early in the twentieth century and has continued steadily since; the glaciers of Kilimanjaro are probably remnants of an earlier, moister climate that prevailed until the late nineteenth century (Hastenrath 2001; Cullen et al. 2006). For this reason, the famous glaciers of Kilimanjaro are not a good icon for the effects of global warming.

14.2.2.3 Mountain Glacier Changes in the First Half of the Twentieth Century

Direct measurements of mass balance are not available for the first half of the century, but the observations of glacier retreat show that significant mass losses occurred in this period too. Furthermore, the increase in temperature from 1900 to 1940, about 0.3 °C, was comparable to the increase from 1950 to 1990. (A 0.1 °C cooling occurred between 1940 and 1950.)

Rough estimates for the total sea-level contribution have been obtained by two methods. The first relies on model calculations to combine observed temperature histories and glacier areas with simple descriptions of glacier response (similar to Eq. 14.2). The calculation is done separately for different regions of the globe (Zuo and Oerlemans 1997). In the second method, mass balance records are extended back in time by using regression equations relating accumulation and ablation to precipitation and temperature at weather stations (Meier 1984). For this analysis, the results must be extrapolated to include regions with no long observational records, using assumptions about how net balances scale with the climatic regime.

Meier concluded that mountain glaciers and small ice caps contributed about 0.2 to 0.7 mm yr^{-1} to sea-level rise from 1900 to 1961 (1 to 4 cm total). Zuo and Oerlemans estimated a total of 1 to 2 cm for the same period. Adding these numbers to the estimates for the latter part of the century suggests a sea-level increment of 3 to 5 cm in the twentieth century (roughly 15% to 30% of the observed total) – and possibly as much as 7 cm.

14.2.3 Sea-level Forecasts: Mountain Glaciers and Small Ice Caps

The principal concern about global warming is not the impact of the twentieth century's moderate warming, but the impact of continued warming expected in the centuries ahead. The magnitude of future warming cannot be predicted well – it depends on socioeconomic factors as well as uncertain climate parameters – but most likely the global mean temperature will increase by another 1 to 5 °C by year 2100, and more beyond. The world's mountain glaciers will shrink dramatically as a result.

In principle, we can calculate the future sea-level rise by combining regionally specific climate predictions and mass balance models with maps of glacier distribution, thickness, and topography (e.g., Gregory and Oerlemans 1998; van de Wal and Wild 2001; Raper and Braithwaite 2006). In practice, the uncertainties are too large for this method to provide a convincing quantitative forecast, though such studies illuminate data gaps and provide a rigorous analysis of likely surface mass balance changes. Even if the glaciological parts of this problem – especially changes of calving – were tightly constrained, which they are not, the large uncertainties in regional climate forecasts would remain a major obstacle. Furthermore, because water temperatures sometimes determine the retreat of calving margins, the forecast would need to be accurate for conditions in the ocean as well as in the atmosphere.

Alternatively a rough idea can be obtained from simple extrapolations of observations. Meier et al. (2007) adopted this approach. Their synthesis of mass balance data (Section 14.2.2) gave the recent (year 2006) rate of mass loss from all mountain glaciers and small ice caps as $\dot{M} = -402 \pm 95\,\mathrm{Gt\,yr^{-1}}$, or $1.1 \pm 0.24\,\mathrm{mm\,yr^{-1}}$ of sea-level rise. They also compared mass balance data from recent years with data from the 1950s through 1990s to calculate the "acceleration" of mass loss; that is, the rate of increase of $\dot{M}$ with time. They found $d\dot{M}/dt = -11.9 \pm 5.6\,\mathrm{Gt\,yr^{-2}}$. These numbers gave two projections of sea-level rise, one assuming that $\dot{M}$ remains constant, the other assuming that $d\dot{M}/dt$ remains constant (Table 14.3).

Future mass loss will depend on the magnitude of warming; more warming implies faster melt (at a fixed location) and, if temperatures steadily rise, a larger area of ice exposed to melt (at higher altitudes and latitudes). Meier et al's compilation suggests an increase of mass loss rate, from $\dot{M} \approx 100 - 200\,\mathrm{Gt\,yr^{-1}}$ around 1980 to $\dot{M} \approx 300 - 450\,\mathrm{Gt\,yr^{-1}}$ in 2002–2006. From the global mean temperature increase over this interval of $\Delta T \approx 0.4\,°\mathrm{C}$, this implies an increase of mass loss per degree warming in the range $d\dot{M}/d\Delta T \approx 250 - 900\,\mathrm{Gt\,yr^{-1}°C^{-1}}$.

To make a simple extrapolation that depends on the magnitude of warming, we use the following approach. Call the year-by-year change of global temperature $\Delta T(t)$. A permanent warming of size ΔT immediately generates increased melt from glaciers, but the rate of melt decreases over time as the glaciers retreat toward a new equilibrium. (For a single glacier, the timescale of adjustment is the response time discussed in Section 11.2.2, or the lag time in Eq. 14.2.) By analogy with Eq. 14.2, the global rate of melt decreases roughly in proportion to $\exp(-\tau/t_g)$, with τ the time elapsed since the warming occurred and t_g a characteristic

Table 14.3: Projected sea-level rise from mountain glaciers and small ice caps by year 2100.

Assumption	Sea-level rise by 2100 (cm)	Reference
Constant $\dot{M}$	10 ± 3	Meier et al. 2007
Constant $d\dot{M}/dt$	24 ± 13	Meier et al. 2007

Linear response of $d\dot{M}/d\Delta T$
  2 °C warming
  4 °C warming
	
	20
	35
		This section

$d\dot{M}/d\Delta T$ from models and observations	7–17	Meehl et al. 2007 (IPCC)
Model forced by GCM[†]	8	Van de Wal and Wild 2001

[†] The authors report the rise over 70 years, which we multiplied by 1.4 to extend to year 2100.

timescale. This parameter now refers not to a single glacier but to the global response of all glaciers (following Greene 2005). Thus, starting from a state of equilibrium at year t_o, the rate of melt in year t depends on the history of temperature changes by:

$$\dot{M}(t) = C_g \int_0^{t-t_o} \Delta T(t - \tau) \cdot \exp(-\tau/t_g)\, d\tau, \qquad (14.3)$$

where C_g specifies the instantaneous increase of mass loss per degree of warming (units $\mathrm{Gt\, yr^{-1} {}^\circ C^{-1}}$). We have found that, using values for ΔT given by the global temperature series beginning in 1890, the values $t_g = 100\,\mathrm{yr}$ and $C_g = 700\,\mathrm{Gt\, yr^{-1} {}^\circ C^{-1}}$ yield a good match to the global rates of mass loss in the different periods shown in Table 14.2. Greene (2005) found the same value for t_g.

Cumulative melt can be calculated by integrating Eq. 14.3 itself over time. Using the preferred values for t_g and C_g gives a net sea-level rise of 3.5 cm from the end of the nineteenth century to present. Suppose that temperature continues to increase, linearly, until year 2100. Table 14.3 lists the net additional sea-level rise calculated with Eq. 14.3 for this scenario (labelled "Linear response ..."), using a total warming over the coming century of either 2 or 4 °C. Notice that using a constant C_g for projections implicitly assumes there is no reduction of ablation due to loss of glacier surface area; the areas contributing most of the melt are assumed to expand to higher altitudes and latitudes as termini retreat and mid-latitude glaciers disappear.

The final value in the table was derived in a detailed analysis by Van de Wal and Wild (2001), which accounted only for changes in surface mass balance. They used a general circulation model to estimate climate changes due to a doubling of CO_2, a mass balance model that was previously calibrated against observations, and a scaling relationship between ice volume and surface area. These were applied to every glaciated region on the planet, except Antarctica and the Greenland Ice Sheet, assuming a size distribution of glaciers within each region. Most, about 70%, of the predicted mass loss occurs in central Asia, northwestern North America, and the Canadian Arctic. The rate of sea-level rise predicted in this analysis for the coming century is lower than the rate already observed in year 2006; this prediction is almost certainly too small. The discrepancy most likely reflects the important contribution of glacier dynamical effects to ongoing mass losses.

The first value in the table can also be dismissed. The small ice caps and largest mountain glaciers contain most of the ice volume outside of the ice sheets. They are relatively thick and therefore cannot retreat rapidly by melt (although some retreat rapidly by calving). A steady increase of temperature will therefore increase the disequilibrium between these glaciers and the climate. Thus, although some small glaciers will disappear entirely and no longer contribute melt, the global rate of mass loss will almost certainly increase, not remain the same. At the other extreme, a 35 cm rise implies a significant reduction of glacier area, so extrapolations would overestimate the rate of loss. The most likely sea-level rise by year 2100 from wastage of mountain glaciers and small ice caps is therefore between 10 and 35 cm. The total sea-level equivalent volume of mountain glaciers and small ice caps is about 70 cm (Table 14.1). The numbers in Table 14.3 suggest that a significant portion of these glaciers will be lost in the coming century.

In general, the future worldwide mass loss from mountain glaciers will be a complicated function of how much warming occurs in different places, how much ice exists in particular climatic regimes, and how strongly dynamical effects contribute. Consider further the value for C_g in Eq. 14.3. Studies of individual glaciers suggest that, typically, every one degree of warming reduces the annual surface balance, averaged over the glacier, by an amount $C_T = 0.2-0.6\,\mathrm{m\,yr^{-1}\,{}^{\circ}C^{-1}}$ for subpolar glaciers and $C_T = 0.8-1.2\,\mathrm{m\,yr^{-1}\,{}^{\circ}C^{-1}}$ for maritime temperate glaciers (Section 4.2.6). The parameter C_g is the effective globally averaged value of C_T (but should also include dynamical losses). The total area of glaciers and small ice caps is currently about $7.63 \times 10^{11}\,\mathrm{m^2}$ (Meier et al. 2007, p. 1067). For C_g of 0.4 or 1.0 (mid-range values of C_T for subpolar and temperate glaciers), reduced surface balances would contribute a global total of about 300 or $800\,\mathrm{Gt\,yr^{-1}\,{}^{\circ}C^{-1}}$.

Thus an important factor determining global mass loss is the relative areas of glaciers in different regimes; the larger the fraction of glacier area in warm, wet environments, the larger the total mass loss for a *uniform* global warming. For this reason, the preponderance of glacier area in subpolar and polar environments holds down the total mass loss. Indeed, rigorous assessments of the global-averaged value of C_g, accounting for the geographical distribution of glacier area but not including dynamical losses, find $C_g \approx 0.3-0.4$, characteristic of subpolar glaciers (Oerlemans and Fortuin 1992; Dyurgerov and Meier 2000; Raper and Braithwaite 2005). An opposing factor, however, is the spatial variation of warming; models and observations both indicate that the high latitudes will warm more than the global average. (Such *polar amplification* arises from feedbacks between warming, the extent of snow and sea ice, and the abundance of dark vegetation on the land surface.) A final factor is the seasonal pattern of warming; warming concentrated in winter should cause much smaller mass losses than warming concentrated in summer, the season of melt.

Estimates for C_g using regional and seasonal patterns of warming forecast with global climate models give values of $C_g \approx 0.2-0.3\,\mathrm{m\,yr^{-1}\,{}^{\circ}C^{-1}}$ (Meehl et al. 2007, p. 814; given as $0.4-0.6\,\mathrm{mm\,yr^{-1}\,{}^{\circ}C^{-1}}$ sea-level equivalent). This low value presumably reflects a preponderance of wintertime warming.

The mass balance measurements, however, indicate a much higher current value. The optimal value $C_g = 700\,\mathrm{Gt\,yr^{-1}\,{}^{\circ}C^{-1}}$ we identified for Eq. 14.3 corresponds to $C_g \approx 0.9\,\mathrm{m\,yr^{-1}\,{}^{\circ}C^{-1}}$. Faster flow of tidewater glaciers explains some of the additional mass loss; none of the model estimates of C_g include dynamical losses. Errors in the surface balance calculation might also contribute.

The most recent IPCC projections (Table 14.3) assumed that the rate of mass loss from glaciers and ice caps is proportional to the global mean temperature rise (with a proportionality equivalent to $C_g \approx 0.3-0.5\,\mathrm{m\,yr^{-1}\,{}^{\circ}C^{-1}}$), times a factor to account for reduced glacier area as glaciers retreat (Meehl et al. 2007, p. 844). Their value for the first proportionality was based on a simple regression of mass loss rate versus temperature, rather than a formula like Eq. 14.3, and could not incorporate the most recent observed rates (Meier et al. 2007). Furthermore, to account for reduction of glacier extent they assumed a decrease of melt-contributing area in simple proportion to glacier volume – a dubious assumption given that ablation zones of large glaciers should transiently expand in a steady warming. This increases the sensitivity of each such glacier to further warming. For both reasons, their projections are probably too low, but a longer time series of observations is needed to judge. The higher values in Table 14.3 arise from the observations of increasing mass loss since year 2000. If this trend reflects a temporary anomaly, then the highest values in the table over-predict the future changes.

14.3 The Ice Sheets and Global Warming

14.3.1 Greenland

The Greenland Ice Sheet will shrink as the climate warms, a process already under way. Air temperature measurements, available for only a few sites, indicate that Greenland warmed early in the twentieth century, returned to cooler conditions in the 1960s to 1980s, and recently warmed again (Box et al. 2006). Little is known about the response to the earlier warming. The floating terminus of Jakobshavn Ice Stream, the largest outlet glacier on the west coast, retreated by about 30 km from 1860 to 1960, but about half of the retreat was accomplished before 1900 (Weidick 1992). The mass balance of the ice sheet before 1960 cannot be assessed well because information about calving rates is too sparse. However, air temperature, surface balance, and calving have all correlated since 1960; if these correlations applied in the past, even approximately, the ice sheet would have lost mass during the warming of the early twentieth century (Rignot et al. 2008b).

14.3.1.1 Contemporary State of the Ice Sheet

From 1980 to 2006, coastal stations warmed about 2 °C. On the ice sheet, the area of melt expanded and the duration of the melt season increased, but the snowfall increased too (Box et al. 2006). The largest observed change was thinning, retreat, and increased flow of the outlet glaciers (Rignot and Kanagaratnam 2006). Increased flow thins the outlet glaciers by increasing the rate of longitudinal extension; the glaciers stretch horizontally, and so must compress vertically (Section 8.4.1). If calving rates increase together with the flow to the margin, the ice sheet loses mass. In fact, calving not only kept pace with the outflow but exceeded it; observations show that many termini retreated.

The combination of increased flow and melt near the coast with increased snowfall in the interior implies that the ice sheet surface simultaneously rose in the interior and deflated around the edges. Altimetry measurements from both satellite and aircraft show this pattern clearly (Krabill et al. 2004; Zwally et al. 2005).

The net balance of the whole ice sheet was near zero but slightly negative from 1960 to 1990 (Table 4.3); most of the mass loss has occurred in subsequent years. The rate of loss

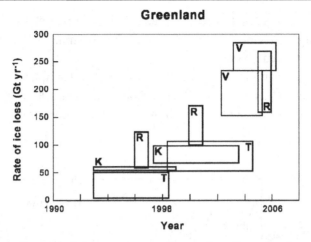

Figure 14.5: **Various estimates of the recent rate of mass loss from the Greenland Ice Sheet. The vertical dimensions of boxes indicate the published uncertainties; the horizontal dimension, the period of observation. Sources are: K, Krabill et al. (2004); R, Rignot and Kanagaratnam (2006); T, Thomas et al. (2006); and V, Velicogna and Wahr (2006a). A less selective version of this plot was presented by Cazenave (2006).**

generally increased up through year 2006, although year-to-year fluctuations, possibly large, are superimposed on the general trend. Figure 14.5 shows several estimates for mass balance of the whole ice sheet for different time periods, made with different techniques: specifically, the flux-gate method, applied to all the major drainage basins of the ice sheet (Rignot and Kanagaratnam 2006); the gravitational method (Velicogna and Wahr 2006a); the altimetric method using an airborne laser (Krabill et al. 2004); and the altimetric method using a satellite-based laser (Thomas et al. 2006). (See Section 4.7 for descriptions of the techniques.) We have excluded several additional published estimates: the gravitational estimate of Luthcke et al. (2006), in which the interannual variability is too large for reliable trend detection; the satellite-based radar altimetry estimate of Zwally et al. (2005), in which elevation changes of coastal regions are not well resolved; and meteorological estimates such as that of Box et al. (2006), which provide information only about surface mass balance and not calving.

In order to apply the flux-gate method, Rignot and Kanagaratnam (2006) used InSAR to map ice velocities on all the major outlet glaciers of the ice sheet, an achievement without precedent. This information was combined with earlier surface balance estimates to assess the overall balance of drainage basins. In this study, the surface mass input was the most uncertain term. The Velicogna and Wahr gravitational method probably overestimates the mass loss from the ice sheet because it also senses changes of the glaciers and small ice caps peripheral to the main ice sheet. On the other hand, the altimetric studies underestimated the mass loss because they did not acquire measurements on all the glaciers; losses from the unmeasured glaciers were estimated from surface melt alone. True uncertainties on all the estimates are probably larger than shown in the figure. Nonetheless, it is clear that Greenland has been losing mass, and the rate of loss has increased in recent years. The mean rate of loss over the last decade corresponded to, roughly, a global sea-level rise at 0.2 to 0.4 mm yr^{-1}, and probably increased to 0.5 to 0.6 mm yr^{-1} by year 2006.

Figure 14.6a plots the variations of mass of the ice sheet and peripheral regions inferred from the gravitational method (Velicogna and Wahr 2006a; Velicogna 2009). In addition to showing the multiyear trend of loss, these data also show how the ice sheet grows and shrinks over the annual cycle.

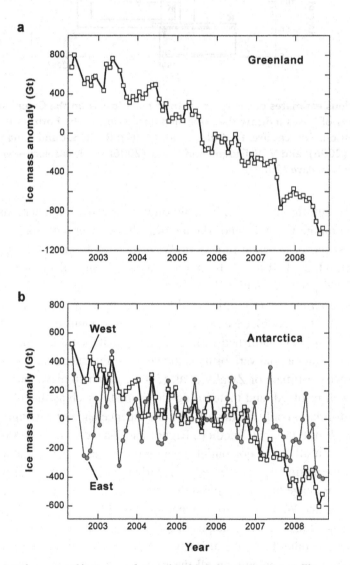

Figure 14.6: Recent changes of ice mass determined by the GRACE satellite experiment. (a) Greenland. (b) East and West Antarctica. Updated from Velicogna and Wahr (2006a, 2006b) and Velicogna (2009). Data courtesy of I. Velicogna.

14.3.1.2 Mechanisms of Response: Greenland

Figure 14.7 summarizes the most important processes governing the rate of ice loss from Greenland in a warming climate. Details of the mechanisms are discussed in earlier chapters, as cited in the following text. The critical surface mass balance processes are:

1. Melt ablation increases with increasing summer temperatures (Chapter 5). The energy flux to the surface that is available for melt, integrated over the summer, increases in rough proportion to the positive degree days. Reduced surface albedo, related to earlier exposure of ice each summer and increased ponding of water, enhances the absorption of solar radiation.

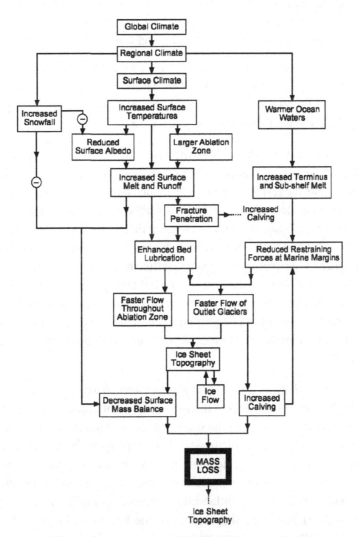

Figure 14.7: Schematic summary of processes determining rate of mass loss from the Greenland Ice Sheet in a warming climate.

A key question is how the positive degree days at low elevations around the ice sheet will vary as a correlate of global climate indicators.

2. As warming raises the equilibrium line altitude, the ablation zone expands inland, increasing the total runoff. The pace of inland expansion depends not only on the rate of warming but also on the ice sheet surface profile.

3. Increased snowfall counteracts increased melt to some extent. By how much is not clear. Greenland ice core records show no consistent increase of snowfall with temperature in the Holocene, at timescales from decades to millennia (Kapsner et al. 1995; Cuffey and Clow 1997). But snowfall has increased slightly in the interior of Greenland in recent decades (McConnell et al. 2000; Box et al. 2006). Cyclonic storms generate most of the snowfall on Greenland; a key question is how the intensity and trajectories of such storms will change.

Increased surface melting accounts for approximately one third of the mass loss from Greenland in the last decade (Rignot and Kanagaratnam 2006). The rest was lost by increased calving related to glacier dynamics, and such processes will probably continue to play an important – perhaps dominant – role. The critical ice dynamical processes are:

1. *Meltwater lubrication effect.* Throughout the ablation zone, ice flow speeds up when surface melt penetrates to the bed and lubricates basal motion (Section 11.4.3.1). The few observations so far available suggest that this process, once begun, does not continue to strengthen as melt increases (van de Wal et al. 2008), probably because the subglacial hydraulic system adjusts to accommodate the increased water fluxes. At the same time, however, inland expansion of the ablation zone should allow this process to operate in an increasingly large area of the ice sheet. Observations also suggest that the size of the effect on the slow-flowing flank regions of the ice sheet is modest; the annual average of ice velocity increases by less than 10%. Further investigation is needed to answer the important question of how this effect contributes to the speed-up of the fast-flowing outlet glaciers. But evidence so far implicates changes at glacier termini (see next item) rather than melt lubrication as the important factor (Joughin et al. 2008).

2. *Destabilization of marine margins.* Outlet glaciers with marine margins accelerate, thin, and retreat rapidly (Section 11.5). This is the most important process accounting for mass loss from the ice sheet in the last decade; the speeds of many of the outlet glaciers doubled (Rignot and Kanagaratnam 2006). Most of the speed-up appears to be a response to reduction of the restraining forces acting near tidewater margins. Reasons for reduced restraining forces probably include: (1) retreat of floating termini when warming air and ocean waters increase calving (Holland et al. 2008); (2) inland migration of grounding lines ("ungrounding") as glaciers thin; and (3) increased slipperiness of the bed as a glacier thins, or retreats into deeper water, or melts more rapidly at the surface. Outlet glaciers respond dramatically to such changes because of feedbacks between increased flow, stretching, thinning, and

reduced restraining forces. The response can be highly nonlinear. With the exception of bed lubrication by surface melt, all of these processes operate only if the outlet glacier ends in a deep water body. Thus, this process shuts down when the margin retreats onto bedrock above sea level, or above the level of deep proglacial lakes. The western and northern margins of the ice sheet are therefore most susceptible to this process, in the long term. In contrast, the southeastern margin is guarded by steep mountains.

3. *Effect on ablation zone extent.* The size of the ablation zone changes as ice flow redistributes mass and so changes the surface profile. In the long term, the outflow of ice will draw down the surface in the ice sheet interior, allowing the ablation zone to expand greatly. Early in the response, however, the ablation zone might shrink as melt steepens the margin. The model of Huybrechts and DeWolde (1999) showed this behavior. The model did not include outlet glaciers, however, and so the result only applies to flank regions between outlets.

14.3.2 Antarctica

The Antarctic Peninsula warmed about 1 to 2.5 °C in the second half of the twentieth century (Vaughan et al. 2001). For the mainland of Antarctica, ice core data suggest a small warming trend overall (about 0.6 °C over the same period), concentrated in West Antarctica, and strong decade-to-decade variability (Schneider and Steig 2008; Steig et al. 2009). The Southern Ocean entirely surrounds Antarctica and partly buffers it from climate changes. Because the oceans warm slowly compared to the large northern hemisphere continents, warming in mainland Antarctica is expected to lag the warming on other land masses, although sea-ice changes should amplify the warming that does occur. More importantly, mainland Antarctica is cold enough to render surface melt negligible, even at sea level. Mid-summer temperatures will have to increase by roughly 5 °C for surface melt to produce much ablation, and such a large change is not likely in the coming century. Thus – for the mainland – the most important climate variable in the near future is not air temperature but the temperature of ocean waters; warm water melts the bottoms of ice shelves, causing thinning and grounding line retreat. On the Antarctic Peninsula, on the other hand, conditions are warm enough that surface melt already matters; air and water temperatures both influence the mass balance.

14.3.2.1 Contemporary State of the Ice Sheet

Figure 14.6b shows, together with the curve for Greenland, the recent evolution of mass in East and West Antarctica, determined by the gravitational method (Velicogna and Wahr 2006b; Velicogna 2009). West Antarctica as a whole is losing mass, but East Antarctica has no detectable trend. Table 14.4 lists values for the equivalent sea-level rise from 2002 to 2008.

The flux-gate method (Section 4.7) offers an entirely independent view – one that, unlike the gravitational method, cannot be affected by isostatic adjustments. Rignot et al. (2008a) made the first comprehensive mapping of velocities on all the major ice streams draining Antarctica.

Table 14.4: Regional changes in Antarctica.

Region	Time interval	Sea-level rise (mm yr^{-1})[†]	Reference
West Antarctica	2000	0.29 ± 0.17	Rignot et al. (2008a)
	2002–2005	0.4 ± 0.2	Velicogna and Wahr (2006b)
	2002–2008	0.37 ± 0.06	Velicogna (2009)
	2006	0.43 ± 0.22	Rignot et al. (2008a)
East Antarctica	2000	0.01 ± 0.17	Rignot et al. (2008a)
	2002–2008	0.01 ± 0.15	Velicogna (2009)
Amundsen Sea sector (West Antarctica)	1996	0.11 ± 0.08	Rignot et al. (2008a)
	2000	0.18 ± 0.08	Rignot et al. (2008a)
	2002/2003	0.24	Thomas et al. (2004)
	2006	0.25 ± 0.08	Rignot et al. (2008a)
Interior East Antarctica (north of 81.6°S)	1992–2003	−0.12 ± 0.02	Davis et al. (2005)
Antarctic Peninsula	2000	0.08 ± 0.12	Rignot et al. (2008a)

[†] Converted, in some cases, from Gt yr^{-1} using 360 Gt per mm.

(As for Greenland, velocities were obtained with InSAR. For Antarctica, the comprehensive map was made for year 2000, but velocities for 1992 and 2006 were obtained in important locations.) For all basins, Rignot et al. calculated the mass outflow where the grounded ice sheet transitions to ice shelves and compared this to the area-integrated surface balance, estimated with a regional climate model. The latter was verified against an extensive set of field measurements (van den Broeke et al. 2006; van de Berg et al. 2006). Figure 14.8 depicts Rignot et al.'s results for the six regions of the most active change. Table 14.4 gives combined totals. Results match those from the gravitational method; to within uncertainty, the mass of East Antarctica remained steady, but West Antarctica and the Antarctic Peninsula lost mass. Most losses occurred in the Peninsula and Amundsen Sea regions of West Antarctica. The Ross Sea and Weddell Sea regions both gained mass. These patterns are broadly consistent with results of satellite altimetry measurements (Zwally et al. 2005; Helsen et al. 2008).

We now summarize information about such regional changes in Antarctica.

The Antarctic Peninsula Several major ice shelves fringing the northern peninsula recently disintegrated (Doake and Vaughan 1991; Vaughan and Doake 1996; Rott et al. 1996, 2002; Scambos et al. 2004; Rignot et al. 2005). Mercer (1978) was the first to suggest that this process would occur as a consequence of anthropogenic global warming. He observed that ice shelves on the Antarctic Peninsula exist only south of the 0 °C isotherm for January, the warmest month. Mercer concluded that,

> "One of the warning signs that a dangerous warming trend is under way in Antarctica will be the breakup of ice shelves on both coasts of the Antarctic Peninsula, starting with the northernmost and extending gradually southward."

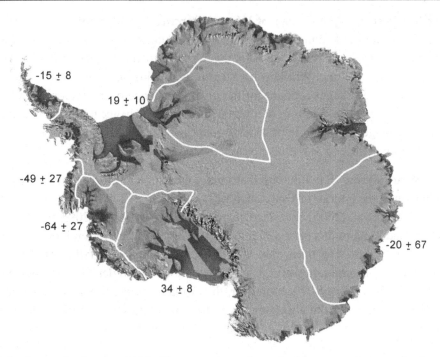

Figure 14.8: Antarctica in year 2000. The map shows ice velocities around the entire Antarctic continent (faster flow indicated by red color), and the six regions of largest inferred mass changes. Numbers give the mass changes in $Gt\,yr^{-1}$. Adapted from Rignot et al. (2008). Velocity map courtesy of E. Rignot. (Refer to the insert for a color version of this figure)

As Mercer predicted, these events were related to melt driven by warm conditions (Scambos et al. 2000; van den Broeke 2005). Moreover, breakup of the ice shelves increased the outflow of the adjacent grounded ice, and hence contributed to sea-level rise. Table 14.4 and Figure 14.8 give the estimated mass loss for this region according to Rignot et al. (2008a). For a discussion of the mechanisms of ice shelf break-up, see Section 4.6.3; for the response of inland ice on the Peninsula, see Section 11.5.3.1.

The Amundsen Sea Region of West Antarctica The Amundsen Sea region of West Antarctica faces the central Pacific. Several major fast-flowing ice streams drain it; Pine Island Glacier and Thwaites Glacier are the largest (Section 8.9.2.5). This region of the ice sheet is thinning, a process discussed in Section 11.5.3.2. The thinning concentrates along the axes of the major ice streams, which are speeding up. The Rignot et al. (2008a) analysis derived mass loss rates in this region, for three years: 1996, 2000, and 2006. Earlier, Thomas et al. (2004), also using the flux-gate method, derived a loss rate for 2002/2003. The results show an increasing rate of mass loss over time (Table 14.4). Altimetry measurements confirm the loss of mass from this region (Helsen et al. 2008).

The Ross Sea Sector (Siple Coast) of West Antarctica   Five prominent ice streams drain this region. Their flow and configuration change over centuries because of internal instabilities (Section 8.9.2.4). One of these ice streams, Kamb Ice Stream, stagnated 200 years ago and now gains mass. Another one, Whillans Ice Stream, is slowing down and hence starting to gain mass. Together, these changes account for the positive balance of the region, shown in Figure 14.8. These events are manifestations of internal instability and hence unrelated to contemporary climate and its changes. At the millennial timescale, the region is still undergoing a slow, general retreat originating at the end of the last ice age (Section 13.4.3).

Interior East Antarctica   Satellite-based altimetry measurements reveal increasing elevations in a large region of the interior of East Antarctica (Zwally et al. 2005; Davis et al. 2005; Helsen et al. 2008). Davis et al. (2005) examined changes from year 1992 to 2003 and concluded that the region gained mass in this period (Table 14.4). The rate of elevation increase is small, but the affected area is vast. The mass gain must reflect increased snowfall rates. The magnitude of the mass gain is uncertain, however, because variations of firn density complicate the signal (Helsen et al. 2008; see Eq. 8.77). Moreover, the available evidence suggests that the average snowfall on East Antarctica did not change over the longer interval of 1955 to 2004 (Monaghan et al. 2006).

14.3.2.2   Mechanisms of Response: Antarctica

Two processes compete in the overall evolution of the Antarctic Ice Sheet in a sustained climate warming: mass loss related to retreat of marine margins, and mass gain due to increased snowfall. Perhaps the most likely scenario – over many centuries – is an initial phase of mass loss as marine margins retreat followed by slower mass gain from accumulation of excess snow. The retreat process will stop once ice is removed from the deep subglacial basins. Simple constraints suggest that it will be difficult for increased snowfall to overcome the mass deficit at this stage (see the following). Furthermore, the geological record shows that Antarctic ice volume decreased in warmer climates (see Chapter 13). For both reasons, a significant warming of Antarctica will most likely diminish the ice sheet in the long term, with a potentially major impact on global sea level.

   Figure 14.9 summarizes the most important processes governing how fast the mass of Antarctic ice changes in a warming climate. In particular:

1.   The surface mass balance increases throughout the continent because of increased snowfall. The increase will be small in the dry interior, only a few centimeters per year, but will affect a large region. If snowfall on Antarctica everywhere increased by 5%, the immediate effect, before ice flow adjusts, would be a sea-level fall at about $0.3 \, \mathrm{mm \, yr^{-1}}$. Climate models suggest that snowfall increases by 5% to 9% per °C of warming (Gregory and Huybrechts 2006; Meehl et al. 2007, p. 817). This occurs primarily because the water content of the air rises with the saturation vapor pressure (Section 4.2.1). In addition, strengthening cyclonic storms and increased evaporation from ice-free oceans enhance the atmospheric transport of moisture from the Southern Ocean onto the continent (van Lipzig et al. 2002). The increase

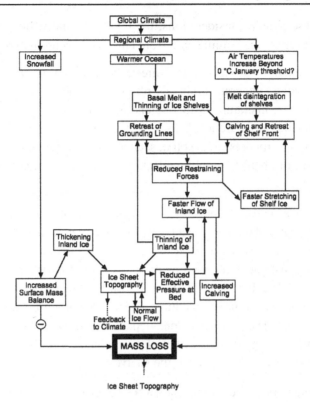

Figure 14.9: Schematic summary of processes determining rate of mass loss from the Antarctic Ice Sheet in a warming climate.

of snowfall is a powerful effect; with a 5 °C warming, snowfall rises, in theory, by roughly 30% to 50% relative to present values. Such a warming, if it occurs instantaneously, initially draws down the sea level at about 2 to 3 mm yr^{-1}, a much larger rate than the current sea-level rise related to ice dynamical changes.

On the other hand, consider the ultimate effect on the ice sheet's volume of a 30% to 50% snowfall increase. (Take 40% as a mid-range value.) Because of the plastic properties of ice, the thickness at the ice sheet center varies roughly as the one-eighth power of the accumulation rate (Eq. 8.141). Hence, the volume of the ice sheet would, if the margins remain fixed, increase by only about 4% for a 40% increase of accumulation. Given the current sea-level equivalent volume for East Antarctica of 52 m (Table 14.1), this amounts to a sea-level drop of 2.2 m. In comparison, West Antarctica contains about a 4 m sea-level equivalent of ice sitting in basins with floors below sea level. East Antarctica contains at least a few meters in analogous regions of Wilkes Land. Furthermore, retreat of East Antarctic margins will thin the ice throughout the interior of the continent; the thickness in the center of an ice sheet varies as approximately the square root of the ice sheet's radius (Eq. 8.141).

According to these simple considerations, the total potential volume loss from retreat of marine margins exceeds the volume gain from enhanced accumulation.

2. If summer air temperatures rise past the threshold necessary to cause significant surface melt, ice shelves will disintegrate as already observed on the Antarctic Peninsula. A critical question is whether temperatures will rise enough for this to happen on the ice shelves fringing the mainland of Antarctica. This would require mid-summer temperatures to rise by about 5 °C. Climate models suggest that such a change would, in turn, require a globally averaged warming of about 5 °C, although the uncertainty on this figure is large (Meehl et al. 2007, p. 819).

3. The basal melt of ice shelves increases dramatically as ocean waters warm (Section 4.5.2). Whereas surface melt will not become a factor for the Antarctic mainland until the warming amounts to several degrees, basal melt can increase immediately as oceans warm. This process apparently instigated the ongoing mass loss from the Amundsen Sea region (Gille 2002; Shepherd et al. 2004). It is likely to be the dominant process driving dynamical changes of the ice sheet over the next century. The relation of this process to global climate is particularly complicated because basal melt depends not only on the magnitude of warming in the oceans surrounding Antarctica, but also on changes in the ocean's circulation and density stratification, especially beneath the shelves.

4. Thinning or disintegration of ice shelves and retreat of grounding lines reduce the *buttressing* forces that restrain the inland ice (Section 8.9.3.7). Loss of buttressing, in turn, increases the flow of grounded ice to the sea (Section 11.5.3). This, in conjunction with erosion by warming ocean waters, explains the observations of ongoing grounding line retreat and mass loss in both the Antarctic Peninsula and the Amundsen Sea region. As with Greenland's outlet glaciers, such events involve feedbacks between thinning, faster flow, and retreat. Where the bed remains below sea level for a considerable distance inland, as in most of West Antarctica and a few regions of East Antarctica, this process might lead to rapid and inexorable retreat. (The retreat of tidewater glaciers in coastal Alaska over the last century is a possible analogue: Section 11.5.2.) Whether it does or not depends on the dynamics of the ice streams conveying ice to the grounding line – in particular, on changes of the streams' widths, longitudinal extensions, and basal drags. Recent observations show that the ice streams in the Amundsen Sea region are thinning as they flow faster. Thinning propagates far inland as a diffusive wave (Section 11.4.2.2 and Fig. 11.21). Thinning is a necessary component of sustained retreat, because it not only causes further retreat of grounding lines but also increases the discharge of ice streams by bringing them closer to flotation, reducing basal drag.

Before any observations showed the process operating in Greenland or Antarctica, Hughes (1977, 1992) envisioned how sustained retreat of marine margins might hollow out an ice sheet. Mercer (1978) connected the possibility to anthropogenic global warming; loss of buttressing following ice shelf disintegration would instigate retreat and ultimately destroy the West Antarctic Ice Sheet. (This is why he referred to the warming trend as potentially "dangerous" in the previous quotation.) Warner and Budd (1998) modelled such a process, instigated by bottom melt of ice shelves. All of these early discussions adopted a rather crude picture of the ice dynamics at marine margins. It is still not possible to model them without dubious assumptions (Section 14.3.3.1).

Increased snowfall counteracts sustained retreat by thickening and steepening the inland ice. Flow of thick ice toward a marine margin could short-circuit the feedbacks responsible for ongoing retreat. Thus, interactions between increased snowfall and retreating marine margins must be considered. The different timescales of these processes are relevant. Increased flow at a marine margin propagates inland within decades, if the bed is slippery. Increased ice thickness, on the other hand, is rate-limited by the magnitude of snowfall changes and is unlikely to matter until centuries have passed.

14.3.3 Model Forecasts of Ice Sheet Contributions to Sea-level Change

Ideally, future sea-level rise would be forecast using whole-ice-sheet models accounting realistically for all relevant processes of ice flow and mass exchange. Section 13.4 summarizes the architecture of whole-ice-sheet models.

This Century   First consider the contribution from changes of surface balance alone. The most comprehensive estimates use climate predictions from general circulation models as inputs for mass balance relations. These are evaluated on a high-resolution topographic model of the ice sheet (see Gregory and Huybrechts 2006, for a discussion of the technique). Meehl et al. (2007, p. 820) have summarized results from such calculations. By the end of this century, increased snowfall in Antarctica is predicted to lower sea level by between 2 and 14 cm. In Greenland, increased melt dominates the surface balance, and the predicted net effect is a rise of sea level between 1 and 12 cm. Thus, surface balances of Greenland and Antarctica partly or perhaps entirely – cancel one another.

The large range of predicted values arises from the large number of uncertain variables needed for the calculation, including the climate change itself, which must be calculated for a range of plausible forcing factors. Nonetheless, the order of magnitude of these numbers reflects simple characteristics of the dominant underlying processes, as the following calculations show:

- In Antarctica, the dominant process is the increase of snowfall with the concentration of moisture in saturated air. Assuming that accumulation rate increases as a linear function of saturation vapor pressure, the net surface balance of Antarctica increases by approximately 18 mm yr^{-1} °C^{-1} (van Lipzig et al. 2002). This corresponds to an increasing mass of about 250 Gt yr^{-1} °C^{-1}, and hence a global sea-level decline of about 0.7 mm yr^{-1} °C^{-1}. For a 3 °C warming of Antarctica by year 2100 (mean of 1.5 °C over the century), this implies a sea-level drop totalling about 10 cm.

- In Greenland, the dominant process is increased melt ablation. Field measurements in Greenland's ablation zone show that melt increases by about 0.75 m for every 100 positive degree days (Figure 5.8). Given a three-month melt season, a 1 °C warming increases the positive degree days at sea level by about 100. If the melt rate, averaged over the ablation zone, is only half of the sea-level value, this implies an average increase of 0.37 m yr^{-1}°C^{-1}. The ablation zone spans about 15% of the total ice sheet area, or 2.6×10^5 km^2, implying a total mass loss of about 96 Gt yr^{-1}°C^{-1}, or roughly 0.3 mm yr^{-1}°C^{-1} of sea-level rise. For a 5 °C warming of Greenland by year 2100 (larger than the value for Antarctica, because temperature changes in the Arctic are amplified), this implies a net sea-level rise of 7 cm.

The total sea-level contribution from the ice sheets depends also on ice dynamical processes (Sections 14.3.1.2 and 14.3.2.2). The most advanced whole-ice-sheet models used in forecasts have predicted negligible ice dynamical changes in the first century of warming (e.g., Huybrechts and DeWolde 1999; Gregory and Huybrechts 2006). Observations have already falsified this prediction for Greenland, the Antarctic Peninsula, and West Antarctica (Sections 14.3.1.1 and 14.3.2.1). Thus, the models are inadequate for predicting the total sea-level contributions for the next century and beyond. They nonetheless illuminate some of the interactions between the climate and the evolving ice sheet form, and provide a plausible minimum scenario for the ice sheet response. It is therefore worthwhile to examine their long-range projections.

Long-Range Forecasts   Currently, the surface balance of the Greenland Ice Sheet is positive and calving removes the surplus (Tables 4.2 and 4.3). If the edges of the ice sheet retreat to a configuration with no calving margins, the average surface balance determines whether the ice sheet survives or not. This average depends not only on the climate but also on the surface profile (Eq. 13.9); specific balances decrease with time as the surface subsides. The models predict that a warming in the range 2.7 to 4.5 °C would change the average surface balance from positive to negative and the ice sheet would disappear entirely (Huybrechts et al. 1991; Janssens and Huybrechts 2000; Gregory and Huybrechts 2006).

A sustained warming of 5 to 10 °C over Greenland is plausible if CO_2 concentrations rise to three or four times the preindustrial value of 280 ppm. Table 14.5 shows a few model predictions of the net sea-level rise from Greenland over two millennia of sustained warmth. The

Table 14.5: Projected sea-level rise from Greenland by A.D. 3000.

Assumption	Net sea-level rise (m)	Reference
5.5 °C warming	3	Huybrechts and de Wolde 1999
8 °C warming	6	Huybrechts and de Wolde 1999
CO_2 concentration of 4× preindustrial	6	Ridley et al. 2005

Ridley et al. (2005) calculation specified a constant atmospheric CO_2 concentration and calculated the climate from a general circulation model. As time progressed, the representation of Greenland was updated using results from the ice sheet model. Note that 6 m of sea-level rise corresponds to disappearance of all the ice on Greenland, except for remnants on the high mountains.

Similar calculations can be made for Antarctica, but here it is of primary importance to model ocean temperatures, their effect on melt rates at marine margins, and the consequences for ice dynamics. Warner and Budd (1998) conducted one such analysis. Rather than modelling ice stream dynamics explicitly they assumed that – in the absence of buttressing forces from ice shelves – the longitudinal strain rate at a marine margin increases as the third power of the ice thickness (as expected for an unconfined ice shelf: Section 8.9.3.2). Thus, the grounded ice is rapidly drawn to the sea as the margin retreats into deep subglacial basins. But first the ice shelves must be destroyed by warming ocean waters. Warner and Budd found that increasing the bottom melt of ice shelves by 5 m yr^{-1} would lead to destruction of the marine portions of West Antarctic in about one millennium. The available data on how bottom melt varies with temperature suggests this would require only a 0.5 °C warming of ocean waters (Section 4.4.2), although the effects of such warming on circulation must also be evaluated. In this scenario, after the major ice shelves disappear West Antarctica would contribute a sea-level rise of a few millimeters per year – roughly ten times greater than its current contribution.

14.3.3.1 Challenges for Model Forecasts

For ice sheet models to forecast sea-level contributions, the following limitations will need to be addressed. The same limitations apply to simulations of the past.

1. The models are not well validated. They adequately reproduce the modern configuration of the ice sheets, after judicious calibration, but their transient behavior has never been rigorously tested. Simulations of the last ice-age cycle demonstrate broad consistency with geological evidence (Section 13.4.3), largely because a few variables such as the plastic strength of ice determine the gross morphology of ice sheets. This does not prove that the models provide useful information about the response at the timescale of a few centuries.

2. The models reproduce well the modern configuration of drainage basins, which depend strongly on bedrock topography. However, the resolution of the models is too coarse to represent realistically the outlet glaciers and ice streams that convey most of the ice to the margins.

3. There is no reason to believe that the models meaningfully predict changes of basal slip in fast-flowing ice streams. The processes controlling important variables such as subglacial effective pressure and sediment strength are only partly understood, and not at all represented in models. The motion of ice streams appears to depend critically on the distribution and nature of regions of high drag (*sticky spots*). It is not known what controls the present configuration of these features, let alone how they might change over time. The motion also

depends critically on an ice stream's width, which can change over time if not controlled topographically. (See Sections 8.2, 8.5.1, 8.6, and 8.9.2 for information about ice stream flow.)

4.   Standard models are not capable of calculating changes of force balance and flow at marine margins. Yet these processes govern the rapid response of outlet glaciers to changes of ice shelves and calving termini (Section 11.5.2).

5.   The bed topography at a marine margin continually evolves because of sediment deposition and scour. Models do not represent these processes, which are difficult to predict in any case. Yet they probably influence the ice sheet's response to sea-level rise (Section 11.5.4).

6.   Although quantitative relations for calving flux have been proposed, none has been established (Section 4.6). Thus, the extent of ice shelves – and hence their force-effect on the ice sheet – is not predictable, and neither are advances or retreats of grounded tidewater termini.

7.   The geothermal heat flux strongly influences meltwater production beneath ice sheets, and hence their flow. Yet the heat flux has never been measured in most regions of Antarctica and Greenland. Likewise, the bed material – sediment or bedrock – remains unknown throughout vast regions.

8.   How a global climate change influences the specific mass balance at each location on an ice sheet and its fringing shelves depends on a complex chain of causal relations. The precipitation rates and the air and ocean temperatures that most influence the mass exchange processes depend on regional climate and on ocean stratification and circulation. Predicting regional climate changes is more difficult than predicting global trends. Even with a well-known regional climate, the calculation of annual specific mass balance remains uncertain; for example, large adjustments based on field measurements are required to calculate Greenland surface balances from even the best validated, highest-resolution weather models (Box et al. 2006). Climatological forecasts are considerably more tenuous, and predictions of melt beneath ice shelves are highly speculative. The rate of frontal and basal melt at marine margins can be large compared to surface melt, not only on ice shelves but also on tidewater glaciers (Section 4.5.2; Motyka et al. 2003). The interaction of ice and ocean waters must therefore be modelled accurately.

This list of challenges is formidable. At present, quantitative forecasts of ice sheet changes should not be accepted as anything more than qualitative scenarios of uncertain validity.

14.3.4 Simple Approaches to Forecasts for the Century Ahead

Given the inability of current models to make a rigorous calculation of dynamically driven losses, it is worthwhile to frame the problem by examining the implications of several simple assumptions.

14.3.4.1 Loss Rates that Scale with Temperature

As a point of reference, first suppose that the dynamical losses from both Greenland and Antarctica simply continue at a constant rate. The IPCC authors adopted this approach (Meehl et al. 2007). They took a constant ice sheet contribution of $0.32 \pm 0.35 \, \text{mm} \, \text{yr}^{-1}$, the supposed observed value for the decade beginning in 1993, which gave a net sea-level increment of 3.5 ± 3.9 cm for the twenty-first century.[2] This is additional to the contribution from surface balances. These authors also asked what would happen if dynamical losses simply scale with the rise of global temperature. For example, regarding the recent (1993–2003) loss rate of $0.32 \, \text{mm} \, \text{yr}^{-1}$ as a consequence of the warming since the late nineteenth century (0.63 °C to the mid-1990s) implies that dynamical losses increase at about $0.5 \, \text{mm} \, \text{yr}^{-1} \, {}^{\circ}\text{C}^{-1}$. Thus, for an additional 3 °C warming by year 2100, dynamical losses would contribute a sea-level increment of about 10 cm. (In this example, sea level rises at a mean rate of $0.32 + 0.5(1.5) \, \text{mm} \, \text{yr}^{-1}$, which assumes 1.5 °C of warming by mid-century.) For a range of temperature scenarios and ice loss sensitivities, the IPCC authors used such arguments to suggest a range of cumulative dynamical losses over the twenty-first century of about 3 to 20 cm.

If we use, instead, the year 2000–2006 observed loss rates from combined surface balance and dynamical changes – about $0.4 \, \text{mm} \, \text{yr}^{-1}$ from Greenland and a similar amount from Antarctica – the sea-level rise over the twenty-first century would total about 8 cm, assuming a constant rate. Assuming a linear scaling with global temperature implies a roughly 15 to 30 cm rise for a net warming of 2 to 4 °C by century's end.

14.3.4.2 Kinematic Scales

An alternative approach to calculating dynamically driven losses is to consider constraints on the rate of ice stream flow. This approach was suggested by Pfeffer et al. (2008); here we explore a simplified but more robust version of their analysis that uses known surface balances to constrain steady-state outflow. Consider the effect of increased flow to calving margins. If we assume that rates of retreat are small compared to ice velocities – a conservative assumption – then the ice sheet's rate of mass loss equals the increase of total discharge, $\rho \Delta Q$ (with ΔQ measured as volume of ice per unit time). Given an average increase of ice velocity of Δu, then $\Delta Q \approx \mathcal{A}_g \Delta u$, where $\mathcal{A}_g$ denotes the cross-sectional area of all ice streams at their grounding lines or calving fronts. This area can be approximated readily, because the net balances of the ice sheets were close to zero in the late twentieth century; we can equate the outflow of mass to the input at the surface. Thus, $\mathcal{A}_g u_o = \mathcal{A}_i \bar{b}$, with u_o the average velocity when net balance was zero, $\mathcal{A}_i$ the surface area of the ice sheet, and $\bar{b}$ the mean specific surface balance in the late twentieth century. The $\bar{b}$ has dimensions of length per unit time; it is the spatial average of $\dot{b}_i$ as defined

---

[2] Meehl et al. gave projections for the decade 2090–2099 rather than for year 2100, a distinction that can be ignored throughout this discussion, given the crude assumptions.

in Chapter 4). Thus, the rate of mass change is simply (in volume equivalent)

$$-\Delta Q = -\mathcal{A}_i \bar{b} \frac{\Delta u}{u_o}. \tag{14.4}$$

Observations show that velocities can increase greatly at marine margins: the flow of Greenland's outlet glaciers recently doubled; the flow of glaciers on the Antarctic Peninsula increased by as much as eight times when ice shelves collapsed; and the flow of Columbia Glacier, Alaska, increased by at least six times during retreat. Mechanisms allowing such large increases include reduction of basal drag as the ice approaches flotation, increases of glacier width, and loss of restraining forces when ice shelves shrink (Section 14.3.2.2).

Let us insert values into Eq. 14.4. For Greenland, $\mathcal{A}_i \approx 1.7 \times 10^6\,\mathrm{km}^2$ and $\bar{b} \approx 0.1\,\mathrm{m\,yr}^{-1}$ (Table 4.2), giving $\mathcal{A}_i \bar{b} \approx 170\,\mathrm{Gt\,yr}^{-1}$. Thus, a doubling of velocity (hence $\Delta u = u_o$) starting from a state of zero balance implies mass loss at about $170\,\mathrm{Gt\,yr}^{-1}$ (consistent with the detailed calculations by Rignot and Kanagaratnam, 2006, which used measurements of ice thickness and velocity for all of Greenland's outlet glaciers). This corresponds to $0.5\,\mathrm{mm\,yr}^{-1}$ of sea-level rise. A six-fold increase in flow rates ($\Delta u = 5u_o$) would generate $2.4\,\mathrm{mm\,yr}^{-1}$ of sea-level rise from Greenland.

For Antarctica, $\mathcal{A}_i \approx 1.4 \times 10^7\,\mathrm{km}^2$ and $\bar{b} \approx 0.16\,\mathrm{m\,yr}^{-1}$ (Table 4.2), giving $\mathcal{A}_i \bar{b} \approx 2240\,\mathrm{Gt\,yr}^{-1}$. A doubling of all Antarctic outflows would thus contribute about $6\,\mathrm{mm\,yr}^{-1}$ to sea-level rise. This illustrates Antarctica's enormous potential capacity to generate sea-level rise – *if* warming of the surrounding oceans and ice shelf surfaces reaches a level that triggers dynamical changes all along the coast of the mainland. In fact, for the next century, it is unlikely that the ice streams flowing into the large ice shelves would be so affected. Vulnerable areas are the Amundsen Sea sector and the Peninsula in West Antarctica, and the eastern region of East Antarctica, stretching from Cook Glacier, past the Totten Glacier, to the Phillipi Glacier. (We consider these as vulnerable because they are not protected by large ice shelves or mountain ranges.) A doubling of velocity on all outlets from these regions, relative to a steady state, would generate about $1.5\,\mathrm{mm\,yr}^{-1}$ of sea-level rise. A six-fold velocity increase would generate roughly $7\,\mathrm{mm\,yr}^{-1}$. (For West Antarctica and the Cook-Phillippi portion of East Antarctica we used $\bar{b}$ of 0.35 and $0.16\,\mathrm{m\,yr}^{-1}$, respectively, and areas of 0.564 and $2.05 \times 10^6\,\mathrm{km}^2$, respectively.)

Pfeffer et al. (2008) developed a kinematic analysis of this sort, with greater detail. Rather than using surface mass balances to constrain initial ice stream discharges, they used measured velocities and cross-sections for all marine-terminating outlets in Greenland. For Antarctica, they considered only the Pine Island and Thwaites glaciers (West Antarctica), and, in some scenarios, the Lambert Basin ice streams in East Antarctica. One scenario was a *low-range* estimate, in which velocities were doubled in Greenland and on the two West Antarctic ice streams. In another scenario (*high-range*), Greenland and Lambert velocities were increased by ten times, and West Antarctic velocities were held at a constant value of $14.6\,\mathrm{km\,yr}^{-1}$, the greatest speed so far observed on any ice stream. (A ten-fold increase of velocity on Pine Island Glacier would imply a speed greater than $25\,\mathrm{km\,yr}^{-1}$. Such a rate seems excessive but is not precluded by physical constraints, if the widths of ice streams can increase.) The calculated rates of sea-level rise in these scenarios, due only to the dynamical changes, totalled 0.9 to 4.7 mm yr^{-1} from Greenland and 1.1 to 3.9 mm yr^{-1} from West Antarctica. Combining these numbers with various crude estimates for surface mass balance, and for dynamical contributions from the Antarctic Peninsula, gave total sea-level contributions from Greenland and Antarctica of 16 to 54 cm and 15 to 62 cm, respectively, by year 2100. This calculation assumed that the dynamical effects maintain their full strength throughout the century, after a one-decade period of onset.

The preceding kinematic analyses make the assumption that the outflow of ice increases only because ice streams speed up. In fact, some ice streams – those not confined to bedrock valleys – can also widen. Widening of ice streams occurs when calving margins retreat into basins. The formation of such *calving embayments* is a plausible reason for the rapid disintegration of ice sheets at the end of ice ages (Section 13.2.1.1). Widening has already been observed on at least one West Antarctic ice stream (Thwaites Glacier; Rignot 2008), and might play a significant role in the coming century.

14.3.4.3 Meltwater Lubrication Effect

Parizek and Alley (2004) included the meltwater lubrication effect (Section 14.3.1.2) in a model of the Greenland Ice Sheet responding to global warming. This process operates throughout the ablation zone and so should be distinguished from acceleration of outlet glaciers with calving margins, although it might contribute to such accelerations. The lubrication effect would act primarily by drawing ice toward the margin and hence reducing elevations around the equilibrium line – thereby increasing melt and reducing the surface mass balance.

Parizek and Alley assumed that basal slip increases in proportion to the quantity of surface melt. They calibrated a relationship by using observations reported in Zwally et al. (2002), with the assumption that melt depends on positive degree days. Results suggested that, for a warming by year 2100 of 2 or 4 °C, melt-lubrication increases the net sea-level rise by 6 or 14 cm, respectively, over the century. This is a useful illustrative result, but speculative; the long-term effect of melt production on annual mean ice velocities is still unknown.

14.4 Summary

14.4.1 Recent Sea-level Rise

Table 14.6 summarizes contributions to sea-level rise for the decade beginning in 1993 and for the year 2006. The observed rise over the decadal period matches the known sources, to within uncertainties. Thermal expansion and diminishing ice contributed roughly equal quantities. The ice component increased with time. The contribution from polar ice sheets appears to be roughly similar to that from mountain glaciers and small ice caps.

Table 14.6: Recent rate of sea-level rise (mm yr^{-1}).

Source	Years 1993–2003	Year 2006	Refer to
Glaciers and ice caps	0.5–1.0[b]	0.8–1.4[ä]	Section 14.2.2
Greenland Ice Sheet	0.1–0.4	0.3–0.6	Section 14.3.1.1
Antarctica	0.1–0.5[†]	0.2–0.7[‡]	Section 14.3.2.1
Total (ice sheets)	0.2–0.9	0.5–1.3	
Total (all Ice)	0.7–1.9	1.3–2.1	
Thermal expansion	1.1–2.1		Bindoff et al. (2007)
Total ice + thermal	1.8–4.0		
Observed	2.4–3.8		Bindoff et al. (2007)

[b] Kaser et al. (2006) value for 1991-2004.
[ä] Meier et al. (2007).
[†] Rignot et al. (2008a).
[‡] Rignot et al. (2008a) value for West Antarctica.

14.4.2 The Twentieth Century

For the whole twentieth century, thermal expansion and melting of glaciers and small ice caps do not account for all the observed sea-level rise (Table 14.7). The missing component includes the contribution from the polar ice sheets, which is poorly known. Ice core evidence suggests that West Antarctica warmed over both the first and second halves of the century (Schneider and Steig 2008; Steig et al. 2009). It is possible that warming triggered retreat of marine margins throughout the century. Data from the Amundsen Sea ice streams indicates mass loss at a low rate in the 1970s, and greatly accelerated mass loss since the early 1990s (Rignot 2008). In Greenland, the limited available data show that a significant warming occurred on the coast between 1920 and 1940 (Box et al. 2006). This probably increased runoff and led to retreat of marine margins (Rignot et al. 2008b). The missing component shown in the table is not constrained tightly enough to infer the net ice sheet losses. Other factors, including the worldwide filling of reservoirs and withdrawal of groundwater, also need to be included for an accurate budget.

14.4.3 This Century

As the numbers in Table 14.6 show, if glacial ice continues to diminish at the rate observed in year 2006, sea level will rise another 12 to 20 cm by year 2100, from this source alone. This number is almost certainly too small, because continued warming can be expected to increase the rate of loss. The contribution of thermal expansion must be included too.

Together, the arguments arrayed in Sections 14.2 and 14.3 illustrate why plausible sea-level forecasts span a wide range, and what the possible values are. Table 14.8 shows the components of four scenarios for net sea-level rise during the twenty-first century. The "Low" and "High" scenarios illustrate the range that we consider most likely; these should be viewed as equally plausible.

Table 14.7: Estimated mean sea-level rise (mm yr^{-1}) from 1900–2000.

Source		Refer to
Glaciers and small ice caps	0.3–0.5	Section 14.2.2.3
Thermal expansion[†]	0.2–0.6	Section 14.1.2
Observed total	1.2–2.2	Section 14.1.2
Missing component	0.1–1.7	

[†] Low value is lower estimate of DeWolde et al. (1995) for 1891–1991.
High value is upper estimate for 1961–2003 from Bindoff et al. (2007).

Table 14.8: Illustrative scenarios for sea-level rise (cm) by 2100.

Source	Very Low Meehl et al. (2007)	Low	High	Very High Pfeffer et al. (2008)
Greenland:				
margin dynamics	0	5	10	47
lubrication	0	0	10	
surface balance	1	5	10	7
Antarctica:				
margin dynamics	0	5	30	61
surface balance	−5	−10	−5	1
Glaciers and ice caps	7	15	30	55
Total from ice	3	20	85	171
Thermal expansion[†]	10	10	40	30
Total	13	30	125	201

[†] Meehl et al. 2007, pp. 812 and 820.

For our low scenario we suppose that both Greenland and Antarctic outlet glaciers continue to flow at their current rate (their fronts continue to retreat but the force imbalance remains constant), the melt-lubrication effect is negligible, and surface balances are in the lower half of the range predicted by models (Section 14.3.3). The mountain glacier/small ice cap contribution is at the lower end of the range we consider likely (Section 14.2.3). The resulting contribution from ice lies at the upper end of the range for a constant mass balance equal to the observed year 2006 value.

For our high scenario we suppose that the contribution from Greenland outlet glaciers doubles; it does not increase more because ice velocities are limited by the narrow topographic troughs along which the glaciers flow. The wide Antarctic glaciers, on the other hand, speed up to six times their steady value by the end of the century (Section 14.3.4.2). This means the fastest ice streams in West Antarctica attain speeds comparable to the highest velocities ever observed (the limit used by Pfeffer et al. 2008). In this scenario, in addition, the melt-lubrication effect in Greenland contributes an additional mass comparable to that modelled by Parizek and Alley (Section 14.3.4.3). The surface balances, the mountain glacier/small ice cap contribution, and the thermal expansion are all near the upper ends of predicted ranges. The largest differences between the high and low scenarios are the Antarctic ice stream and thermal expansion terms.

In general, the ice stream dynamics in both hemispheres constitute the largest – and least well constrained – plausible source of accelerated sea-level rise (Pfeffer et al. 2008). The "Very High" scenario represents the upper end of the range considered plausible by Pfeffer and colleagues, which assumed a ten-fold increase in ice stream velocities. Their low-end scenario (about 80 cm total) falls within our most-likely range. Even the "Very High" scenario cannot be discounted entirely, however, because Antarctic ice streams can widen as they retreat, forming calving embayments. Such a process is likely when margins retreat into deep and broad subglacial basins – but may take centuries not decades to develop.

The "Very Low" scenario uses the lower limits for all the terms, except for the Antarctic surface balance, as assessed by the IPCC (Meehl et al. 2007). We used a moderate value for Antarctic surface balance because, for this term only, a smaller warming corresponds to a higher (less negative) number. The "Very Low" scenario illustrates a plausible situation if the outlet glaciers quickly stabilize and the net warming over the century is strictly limited by socioeconomic changes or unanticipated natural events (such as a marked increase in volcanism).

We conclude – in agreement with Pfeffer et al. (2008) – that a sea-level rise of order 1 m by year 2100 is plausible. This does not mean, however, that such a large increase is more likely than one closer to 30 cm. The IPCC suggested a maximum of about 60 cm if no rapid changes of ice flow occur, but also pointed out that such dynamical changes must be included to establish a true upper bound.

The ice sheets of Greenland and West Antarctica both shrank significantly during some interglacial periods of the late Quaternary (Section 13.4.3). That they will shrink again as global climate warms seems nearly certain. The preceding analyses illustrate the possibility of certain sea-level rise scenarios. Such arguments must eventually be replaced with predictive models.

CHAPTER 15

Ice Core Studies

"The present conditions of life have not yet erased the traces of the past."

Eugenio Montale

15.1 Introduction

The year-after-year accumulation of ice on a glacier creates a record showing how environmental conditions have changed over time. Records from the cold interior regions of polar ice sheets provide an exceptionally rich view. Here melt does not corrupt the record of trapped gases or blur the record of other impurities. Moreover, vast intervals of time can be examined, at a high resolution.

Polar ice reflects the environment at the time of deposition in numerous ways. Bubbles preserve the atmospheric compositions of gases such as CO_2 and CH_4, but retain no memory of other gases such as ozone and water vapor. The isotopic composition of the ice depends on atmospheric temperatures in a broad region encompassing not only the ice sheet but also the oceanic vapor sources. Some of the physical properties of the ice – the temperature, the abundance of bubbles, the thickness of annual layers – are sensitive to climatic temperature and snowfall rate at the site. A wide range of environmental processes, as diverse as forest fires, wind erosion, and microbial activity, modulate the small concentrations of particulate and soluble impurities in the ice. In most cases, the relationship between the ice properties and the environmental variables is complex. The ice properties are *proxies*; they vary in response to environmental conditions but do not preserve them directly. On the other hand, a few parameters such as the concentrations of CO_2 and CH_4 in bubbles directly record the past environment.

Time series of environmental changes are extracted by analyzing ice cores and boreholes. These studies illuminate major questions. How variable is the climate, and how rapidly can it change? What, in detail, is the progression of climate change through the ice-age cycles? How have the atmospheric compositions of the greenhouse gases CO_2, CH_4, and N_2O changed with time? What was the atmosphere like before the industrial revolution?

Useful information bearing on many of these topics is preserved not only in the ice sheets, but also in ice caps in Arctic Canada, Alaska, and Patagonia, and in high-altitude glaciers in Tibet, the Andes, and other mountainous regions.

Copyright © 2010, Elsevier Inc. All rights reserved.
DOI: 10.1016/B978-0-12-369461-4.00015-6

In this chapter we can only introduce the mechanisms by which glaciers create their archives, and illustrate the scope and potential of ice core analyses. We do not attempt a comprehensive review of the results, or discuss every one of the methods. Figure 15.1a and b show the locations of important polar ice core sites.

15.1.1 Some Essential Terms and Concepts

There is no perfect way to organize this chapter because of interrelationships between many of the topics. The following brief summary introduces essential terminology and concepts for readers unfamiliar with the subject.

In an ice sheet, the age of the ice increases steadily with depth (see Section 15.2). In accumulation zones, layers are usually stretched horizontally by longitudinally extending flow (Sections 8.4 and 8.7.1). The layers are thus compressed vertically and progressively thin toward the bed. The ice-isotopic composition – the relative abundance of isotopes in the water molecules composing the ice – varies with the temperature and other climate parameters at the time of snowfall (Section 15.5). Ice-isotopic composition shows the timing and amplitude of climate changes, as a high-resolution and continuous function; it serves as a climatic template for all other analyses of the core. Likewise, electrical conductivity can be measured continuously along a core, providing a template that reflects aspects of impurity content rather than climate (although the two are often closely related) (Section 15.10.1). Of the numerous kinds of impurities found in ice cores – which include soluble chemicals, small particles of rock and soot, and microbes – some are picked up by winds from the surfaces of continents and oceans, whereas others are produced in the atmosphere (Section 15.10). They either fall onto the glacier surface directly or arrive attached to snowflakes.

The trapping of gases into bubbles occurs at the base of the firn, some 40 to 120 m below the surface. The age of the gas thus differs from the age of the surrounding ice (Section 15.2.3). This difference is generally small compared to the duration of the whole climate record, but is crucial for determining the relative timing between changes in climate and in gas concentrations. The air in bubbles largely matches the composition of stable gases in the atmosphere at the time of trapping, but fractionation processes in the firn create subtle deviations of elemental and isotopic compositions from atmospheric values (Section 15.3). An understanding of these processes leads to several powerful techniques for ice core interpretation.

The gases and impurities trapped in the ice serve a dual function; they not only record past environmental conditions, but also offer a variety of time markers useful for assigning an age scale to a core. In this chapter we first discuss how the age of ice varies with depth – the processes controlling the relation and the methods for determining it – and then review how various quantities measured in ice cores relate to climate and other environmental factors.

15.1.2 Delta Notation

The isotopic compositions of ice, gases, and impurities are usually discussed in terms of their "delta values" (δ), a measure of the ratio of two quantities. Commonly, the quantities of interest

a

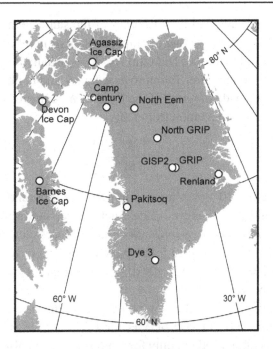

b

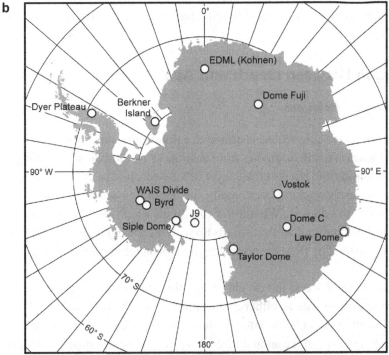

Figure 15.1: Locations of the most important ice-core sites in (a) the Arctic, and (b) the Antarctic.

are the concentrations of heavy and light isotopes of the same element, such as ^{18}O and ^{16}O, D and H, or ^{40}Ar and ^{36}Ar. Suppose that a sample contains a number N_h of the heavy isotope, and a number N_ℓ of the light isotope. The ratio N_h/N_ℓ is usually a very small number, and its absolute value is difficult to measure. To improve accuracy and to obtain convenient numerical values, measurements are made in terms of the δ value; δ compares the ratio $R = N_h/N_\ell$ to a standard ratio R_s and finds its deviation from unity,

$$\delta = \left[\frac{R}{R_s} - 1 \right]. \tag{15.1}$$

Multiplying this number by 10^3 gives a value in parts per thousand (‰) or "per mil," just as multiplying by 10^2 would give a percent value. Such conversion factors are sometimes carried through mathematical formulations, a convention that produces clumsy coefficients in analyses like the one discussed later in Section 15.5. We avoid the use of conversion factors; for example, in our notation a value $\delta = -50‰$ (or $\delta = -0.05$) implies that $1 + \delta = 0.95$, not $-49‰$, unless specified otherwise with "‰."

Delta notation usually specifies the isotope pair being discussed by indicating the rare or heavy isotope: $\delta^{18}O$ for $^{18}O/^{16}O$; δD or δ^2H for D/H; $\delta^{40}Ar$ for $^{40}Ar/^{36}Ar$, and so forth. Finally, delta notation is often used not only for isotope ratios but also, in the same manner, for ratios between the concentrations of different elements (such as Kr/Ar) or different molecules (such as O_2/N_2).

15.2 Relation Between Depth and Age

15.2.1 Theoretical Relations

The best place for an ice core record is near an ice divide; the slow or negligible horizontal flow means that all the ice at depth originates from near the same location on the surface. Figure 15.2a illustrates the situation and shows how a layer of ice deposited at the surface moves downward, thinning vertically and stretching horizontally. Such thinning and stretching also occur on either side of the divide (the flanks). Whether at the divide or on the flanks, horizontal stretching occurs because the horizontal motion speeds up with increasing distance from the divide, a case of longitudinally extending flow, the normal regime for an ice sheet accumulation zone (Sections 8.4 and 8.7.1).

Call the thickness of ice at the divide H and the accumulation rate $\dot{b}$ ($m\,yr^{-1}$ of ice), and define the vertical coordinate as either z (elevation above the bed) or $\hat{z} = H - z$ (depth below the surface). Essential variables to know include the current depth $\hat{z}(a)$ of layers of age a, and the current thickness $\lambda(a)$ of the annual layer of age a. Define t as time, with a value 0 at present and $-a$ when the layer was deposited.

A layer moves downward according to the vertical velocity w of the surrounding ice (a negative quantity if measured with respect to z; positive with respect to $\hat{z}$). Layer thickness

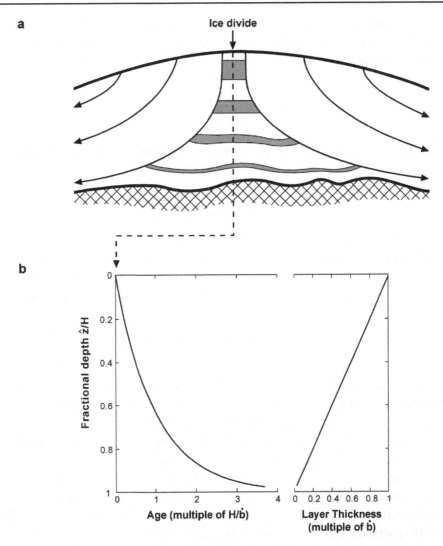

Figure 15.2: (a) Schematic cross-section through an ice divide, showing flow paths and the effect of extending flow on the thickness of a layer (shown in gray). (b) The age of ice and thickness of annual layers as a function of depth, according to Eqs. 15.7 and 15.8. Substituting the second equation into the first shows that layer thicknesses decrease linearly with depth: $\lambda(\hat{z}) = \overline{b}[1 - \hat{z}/H]$. Equation 15.6, with $\lambda_o = \overline{b}$, implies the same.

evolves because the ice deforms. Assuming that layers remain horizontal, the instantaneous rate of thinning is

$$\frac{d\lambda}{dt} = \dot{\epsilon}_{zz}\lambda \quad \text{with} \quad \dot{\epsilon}_{zz} = \frac{\partial w}{\partial z}. \tag{15.2}$$

The strain rate of the layer, $\dot{\epsilon}_{zz}$, varies with time and depth as the layer moves. Integrating from initial thickness $\lambda_o = \dot{b}(t = -a) = \dot{b}(a)$ to present thickness $\lambda(a)$, and from time $t = -a$ to present, gives, after rearrangement,

$$\lambda(a) = \dot{b}(a) \exp\left(\int_{-a}^{0} \dot{\epsilon}_{zz} \, dt\right). \tag{15.3}$$

The current depth to the layer equals the sum of the thicknesses of the overlying, younger layers, or

$$\hat{z}(a) = \int_{0}^{a} \lambda(a') \, da'. \tag{15.4}$$

In combination, Eqs. 15.3 and 15.4 show how the depth of a layer of given age depends on the subsequent accumulation, integrated over time, and the cumulative strain of those layers. Alternatively, the depth can be obtained from the history of vertical velocity of the layer:

$$\hat{z}(a) = \int_{-a}^{0} w(\hat{z}(t)) \, dt + \underbrace{H(0) - H(-a)}_{\Delta H}. \tag{15.5}$$

The correction factor ΔH accounts for changes of ice thickness between $t = -a$ and the present. Here vertical velocity is measured relative to the bed of the ice sheet; subsidence or uplift of the whole ice sheet does not, of course, directly change the depths of layers.

All of these relations are general and, for application to realistic cases, must be solved by tracking layers in a numerical ice-flow model to obtain strain rates or vertical velocities along their paths. Simple hypothetical cases, however, illuminate the essential points. In the firn, compaction enhances strain rates and downward velocities (Eq. 8.75 and Fig. 8.17) and layer thicknesses depend inversely on their density. To circumvent these complications, all vertical distances in the following discussions are ice-equivalent values.

15.2.1.1 Uniform Strain Rate

The simplest model (Nye 1963a; Haefeli 1963) is based on only two assumptions: the vertical strain rate along any vertical line in the ice is uniform at any given instant and there is no melting at the base. It follows that the total vertical strain of any layer equals the total vertical strain of the ice beneath it, and hence:

$$\frac{\lambda}{\lambda_o} = \frac{z}{H_o} = f \frac{H}{H_o}. \tag{15.6}$$

Here λ_o denotes the original thickness of the annual layer now at height z above the base. H_o is the total ice thickness at the time and place at which the layer was deposited. Note that no assumption of steady state is made.

Referring to the height of the layer as a fraction f of the current ice thickness H gives the second equality. This shows that layers at a given position f are relatively thick if the ice sheet has thickened since deposition of the layer. Likewise, thinning of the ice sheet leads to thinner layers; because we are dealing with the case of zero basal melt, thinning can only occur by increased downward flow of ice causing vertical compression.

15.2.1.2 Steady-state, Uniform Strain Rate

Suppose that climate is constant; $\dot{b}$ varies from year to year but its mean value over many years $(\overline{b})$ holds steady. If the ice sheet thickness remains constant, averaged over many years, the vertical velocity at the surface must equal $-\overline{b}$ (the negative sign indicating downward flow). The depth-averaged strain rate is thus $\dot{\epsilon}_{zz} = -\overline{b}/H$ (the negative sign indicating vertical compression).

Assuming again a uniform $\dot{\epsilon}_{zz}$ over depth, Eq. 15.3 gives

$$\lambda(a) = \dot{b}(a) \exp\left(-\frac{\overline{b}}{H}a\right). \tag{15.7}$$

Thus, annual-layer thicknesses decrease exponentially with age but remain proportional to their initial values, the accumulation a years ago. Substituting this result into Eq. 15.4, integrating and rearranging yields the age-depth relation (Nye 1963a):

$$a(\hat{z}) = \frac{H}{\overline{b}} \ln\left(\frac{1}{1 - \hat{z}/H}\right). \tag{15.8}$$

Ages increase rapidly in the deepest layers; Figure 15.2b shows the patterns of both ages and layer thicknesses for this simple case. (Because the flow regime is assumed steady, Eq. 15.8 can also be obtained by integrating the inverse of velocity, $1/w$, from the surface at $\hat{z} = 0$ to depth $\hat{z}$; in the present case of uniform strain rate, $w = \overline{b}[1 - \hat{z}/H]$. Yet another way to obtain Eq. 15.8 is to combine Eq. 15.6 with $d\hat{z} = \lambda da$ and then integrate.)

Layers close to the bed are sometimes too thin or disrupted for paleoclimate interpretation. In addition, layers close to the bed often represent great intervals of time. Although this matches expectations, the simple theory cannot predict the intervals well, as they depend sensitively on local variations of strain rate and long-term changes of the ice sheet. The simple theory does, however, give a useful rough estimate for the age interval represented by most of an ice core. One indicator, for example, is the theoretical age of ice at a depth of $\hat{z} = 0.8H$. As Eq. 15.8 shows, this age scales with the ratio of ice thickness to accumulation rate; records of great duration are therefore obtained from the thickest parts of the ice sheets, in regions of slow accumulation. The logarithmic factor equals 1.6 for $\hat{z} = 0.8H$. Table 15.1 compares theoretical and measured ages at a depth of $0.8H$ in several ice cores.

As the table shows, the timescale $H/\overline{b}$ accounts for much of the variation of ice-core ages from site to site, but the predicted ages are too small by a factor of two or more. In part, this discrepancy reflects reduced accumulation rates in cold ice-age climates; for cores extending

Table 15.1: Ice core ages at 80% of total depth.

Core Site	Ice thickness (m)	Accum. rate (m yr^{-1})	Theoretical age (kyr)	Actual age (kyr)	Reference
Agassiz Ice Cap	111[†]	0.098	1.8	5	(1)
Law Dome	1220	0.7	2.8	4	(2)
Taylor Dome	550	0.06	15	50	(3)
GISP2	3050	0.23	21	50	(4)
EDML[††]	2774	0.07	64	105	(5)
Dome C	3275	0.027	195	350	(6)

[†] Ice-equivalent value.
[††] EDML stands for "EPICA Dronning Maud Land."
References: (1) Fisher et al. 1995; (2) Morgan et al. 1997; (3) Steig et al. 2000; (4) Meese et al. 1997; (5) EPICA Community members 2006; (6) Parrenin et al. (2007).

into the last glacial period, the time-averaged value of $\bar{b}$ should be only half or less of the modern one. More generally, the discrepancy arises from the erroneous assumption of uniform $\dot{\epsilon}_{zz}$ over depth; in fact, $\dot{\epsilon}_{zz}$ must decrease to zero at the bottom if the ice does not slip on its bed. The theory can be modified to take account of nonuniform $\dot{\epsilon}_{zz}$; the simplest approach is the Dansgaard-Johnsen model, discussed in the next section.

15.2.1.3 Depth-variable Strain Rate

The depth-variation of $\dot{\epsilon}_{zz}$, and hence the thinning of layers, depends intimately on the vertical profile of *horizontal* velocity (u) upstream and downstream of the core site (or, at a divide, downstream of the core site on both flanks). In plane strain, the incompressibility of ice requires that $\dot{\epsilon}_{zz} = -\dot{\epsilon}_{xx} = -\partial u/\partial x$, where x denotes the horizontal coordinate. Along a flow line, the shape of the velocity profile $u(z)$ tends to vary much less than the magnitude of u. Thus $\partial u/\partial x$ is approximately proportional to $u(z)$ itself. The velocity profile $u(z)$, in turn, includes a component of basal slip (u_b) and a component of internal deformation, implying that

$$\dot{\epsilon}_{zz} \propto -\left[u_b + 2 \int_0^z \dot{\epsilon}_{xz} dz \right], \tag{15.9}$$

where $\dot{\epsilon}_{xz}$ is the shear strain rate $\frac{1}{2}\partial u/\partial z$. The magnitude of $\dot{\epsilon}_{zz}$ is smallest at the bed (zero on a frozen bed) and increases upward; the typical pattern is a rapid increase in the deepest layers but a nearly constant value in the upper half of the ice thickness (Sections 8.3 and 8.9.1). Based on this concept, Dansgaard and Johnsen (1969) developed a simple model that has been widely used. They assumed that the strain rate $\dot{\epsilon}_{zz}$ takes a constant value from the surface down to some height $z = h$ above the bed, and from there decreases linearly to zero at the base. Thus, for r a positive constant,

$$h \geq z \geq 0 \quad \dot{\epsilon}_{zz} = -rz \tag{15.10}$$

$$H \geq z \geq h \quad \dot{\epsilon}_{zz} = -rh. \tag{15.11}$$

Integration gives w at the surface; equating it to $\bar{b}$ shows that

$$r = 2\frac{\bar{b}}{h}[2H - h]^{-1}. \tag{15.12}$$

The age scale is then

$$H \geq z \geq h \qquad a = \frac{2H - h}{2\bar{b}} \cdot \ln\left(\frac{2H - h}{2z - h}\right) \tag{15.13}$$

$$z = h \qquad a' = \frac{2H - h}{2\bar{b}} \cdot \ln\left(\frac{2H - h}{h}\right) \tag{15.14}$$

$$h \geq z \geq 0 \qquad a = a' + \frac{2H - h}{\bar{b}} \cdot \left[\frac{h}{z} - 1\right]. \tag{15.15}$$

For a velocity profile $u(z)$ of typical shape, the appropriate value of h lies between about $0.2H$ and $0.5H$. Choosing $h = 0.33H$, as an example, gives the age of ice at $\hat{z} = 0.8H$ as $a \approx 2.4H/\bar{b}$, as compared to $1.6H/\bar{b}$ for the uniform strain-rate model (Section 15.2.1.2). The velocity profile very close to an ice divide – within a distance of about $2H$ – does not necessarily resemble a "typical" shape, however, because strain rates are reduced in a low-stress zone near the bed (Section 8.9.1). Both theory and measurements indicate that, in this setting, the strain rate $\dot{\epsilon}_{zz}$ decreases approximately linearly all the way from the surface to the bed. This situation corresponds to the Dansgaard-Johnsen model with $h = H$, and hence the age of ice at $\hat{z} = 0.8H$ increases to $a \approx 4H/\bar{b}$. On the other hand, the location of an ice divide likely shifts during major climate transitions, so few locations experience such a "divide flow" regime in perpetuity.

Measurements of tilting boreholes in ice sheets show that the shearing rate $\dot{\epsilon}_{xz}$ sometimes varies significantly between adjacent layers, due to differences in ice viscosity (Figure 8.9 illustrates an example). As Eq. 15.9 indicates, the rate of layer thinning ($\dot{\epsilon}_{zz}$) depends on the integral of the shear rate, not its local value. Thus the observed variations of viscosity do not imply irregular variations of $\dot{\epsilon}_{zz}$; a plot of $\dot{\epsilon}_{zz}$ versus height would generally show a continuous function, increasing monotonically, but with abrupt changes of slope at the boundaries between layers with contrasting shear rates. Superimposed on this general pattern may be irregular variations at a small scale, associated with structural processes like boudinage and folding (Chapter 10). Conformity of measured depth-age relations with simple theoretical models implies that such

Annual Layer Thickness (m)

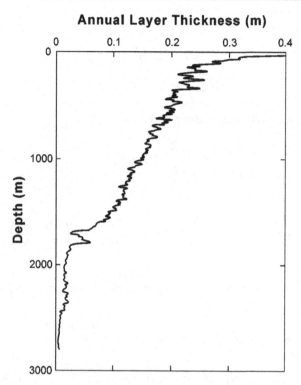

Figure 15.3: Decrease with depth of observed annual layer thicknesses in the GISP2 central Greenland core (calculated from Meese et al. 1997 data). The data have been smoothed for clarity. The approximately linear decrease through the Holocene layer (above 1700 m) conforms to the simple model shown in Figure 15.2b because the accumulation rate, averaged over centuries, was close to constant. Increased thicknesses near the surface correspond to low-density firn. Ice-age layers (below 1700 m) are thinner than expected from extrapolation of the Holocene layers because accumulation rates were reduced. The pronounced variation between 1700 and 1900 m shows the deglacial climate changes discussed later in the chapter (Section 15.11.1).

variations, if they exist, have little effect on the overall pattern of thinning (Figure 15.3). Techniques for measuring depth-age relations will be summarized in Section 15.2.2.

Flow-line Models Many ice core sites, including Vostok in Antarctica and GISP2 in Greenland, are not located on ice divides but rather some distance down the flank (about 300 km for Vostok and 30 km for GISP2). The deformation regime discussed so far still applies – as a first approximation – but it is necessary to account for variations in ice thickness and accumulation rate along the flow path. This requires a model that calculates ice flow as a function of both depth and horizontal position. Particle paths from the surface to different depths at the site, and the time a particle takes to travel along them, must be determined. Annual layer thicknesses can then be calculated directly from the depth-variation of the calculated ages. If isochrones are not parallel to the surface, then not only vertical strain but also shearing changes layer thicknesses. The rate of thinning is (compare to Eq. 15.2)

$$\frac{1}{\lambda}\frac{d\lambda}{dt} = \frac{\partial w}{\partial z} + \theta \frac{\partial u}{\partial z}. \tag{15.16}$$

Here θ denotes the slope of the isochrone relative to the horizontal at the point considered, positive when the isochrone slopes down in the direction of flow. As before, z points upward.

Figure 15.4 illustrates dating by a flow-line model (Paterson and Waddington 1984). The particle paths were computed for a steady state with present values of accumulation rate and ice thickness. In this example, the calculated annual layer thicknesses agreed with those measured in the core down to a depth corresponding to an age of 1300 years. Below this depth, the calculated thicknesses were too small, by an amount that increased downward. This discrepancy could reflect either a decrease in accumulation rate toward the present, or thickening of the ice (Reeh 1989). Model calculations of particle paths need not be restricted to a single flow line or to a steady state. Huybrechts et al. (2007), for example, used a three-dimensional and time-dependent model of the whole Antarctic Ice Sheet. Nested within this model was another model with a high spatial resolution, covering only the region feeding ice to the EDML deep coring site. By calculating particle paths upstream from the core site, the authors estimated how the temperature history recorded in core parameters was influenced by thickness changes over time and by upstream variations in the surface elevation. The model depth-age relation closely matched the measured one in the upper 2.35 km of the 2.7-km-long core.

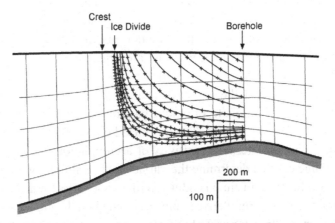

Figure 15.4: Particle paths to a borehole near the crest of Devon Island ice cap computed by a finite element model. Ice takes 200 years to travel between two marks on a path. Vertical exaggeration is two times. Adapted from Paterson and Waddington (1984), and used with permission from the American Geophysical Union, *Reviews of Geophysics and Space Physics.*

Flow-line models have also been made for most of the other important ice core sites. For example, Hvidberg et al. (1997) analyzed flow between the GRIP and GISP2 sites in central Greenland, and Salamatin et al. (2008) analyzed the flow line leading to Vostok in East Antarctica.

15.2.2 Determination of Ages

An age scale is constructed by assimilating information from many different methods. The accuracy of age scales can be assessed by comparing to features of known age, such as volcanic layers or climate events that have been dated radiometrically in other types of geological records. Most age scales use an ice flow model with depth-varying strain rate (Section 15.2.1.3) as a starting point and template. Model parameters related to flow, ice thickness, and accumulation rate are adjusted in order to match points of known age.

The method for determining an age scale depends on whether or not the ice contains distinct annual layers that can be counted. At sites with annual layers, flow models add information about ages only in the lower part of a core, below the level at which layering becomes indistinct. The central Greenland GISP2 core provides a good example of an age scale at a site with distinct and countable annual layering (Alley et al. 1997a; Meese et al. 1997). The uncertainty of ages was estimated by Meese et al. as 2% at the end of the Pleistocene (11.5 kyr ago) and 10% at 60 kyr ago. This can be compared to the chronology for the more recent core from NGRIP in north-central Greenland, another site with countable layers (Svensson et al. 2008); this core displays the same climate changes as the GISP2 core. In the 40–60 kyr interval, the NGRIP chronology differs from the Meese et al. scale by less than 2.5 kyr, while the uncertainties on its absolute ages are estimated as less than ±1.5 kyr.

Parrenin et al. (2007) summarized a good example of an age scale at a site with no annual layers: Dome C in East Antarctica. For this core, the longest record so far obtained from ice, the age scale is constructed from an ice flow model modified to match constraints (shown later, in Figure 15.7). In the model, past accumulation rates are assumed to vary as a function of temperature. The model has been modified in three steps. First, poorly known model parameters are optimized so that the age scale matches one set of independent constraints (which include volcanic layers, horizons of unique gas composition, and other marker horizons discussed in Sections 15.2.2.2–15.2.2.4). The optimized parameters include, for example, the mean Holocene accumulation rate, the sensitivity of accumulation rate to temperature in earlier periods, and parameters that determine the variation of strain rate with depth. Second, the accumulation rate as a function of time is adjusted to give a perfect match with another set of independent constraints. Third, cumulative strains calculated by the model are adjusted *a posteriori* to give a perfect match to a third set of independent constraints in the bottom 500 m of the core. For the Dome C age scale, the absolute ages are estimated to be accurate to within 6 kyr, back to 800 kyr before present. An accuracy of about 6 kyr has apparently also been achieved for the 400-kyr Vostok core. Here different recent chronologies disagree with one another by up to 6 kyr, although typical mismatches are only about 2 kyr (Suwa and Bender 2008). The recent chronologies disagree, however, by up to 15 kyr with the age scale used in publications up to year 2000. For another record, the Dome Fuji core, Kawamura et al. (2007) estimated a typical uncertainty of 1 to 3 kyr. Vostok and Dome Fuji are, like Dome C, East Antarctic cores without annual layers.

15.2.2.1 Annual Layer Counts

Like the rings in a tree, annual layers in an ice core can be counted to determine ages. Properties that vary annually include concentrations of soluble impurities (nitrate, hydrogen peroxide), electrical conductivity (a proxy for acidity), concentrations of microparticles, and the ratios of oxygen and hydrogen isotopes of the ice. In some locations, physical characteristics such as the size and number of bubbles define annual layers visible to the naked eye. Using several indicators together increases the accuracy of layer-counting techniques.

Different parameters vary with the seasons for different reasons. Ice isotopes are sensitive to temperature (Section 15.5). Unique summer layers defined by physical properties originate as depth hoar (Section 2.6) formed when solar heating drives metamorphism (Alley et al. 1990). The advance and retreat of sea ice on nearby oceans lead to changes of marine salt concentrations (Section 15.10.2). Changes of winds and soil moisture modulate the quantity of dust picked up in desert source regions and transported to the ice sheets (Section 15.10.2). Atmospheric nitrate production varies with convective storm activity; lightning is a major source (Section 15.10.3). Hydrogen peroxide production varies directly with the intensity of sunlight (Sigg and Neftel 1988). Figure 15.5 illustrates one example of annual layering.

Annual layer counting does not work in regions with low yearly accumulation, such as central East Antarctica, where most of the snowfall may occur in a few storms, and where any layers that do form are scoured and homogenized by winds. At all sites, identifying the layers becomes increasingly difficult with depth as thinning progresses. Molecular diffusion smoothes the annual signal in ice isotopes and can eliminate it altogether (Johnsen 1977a; Johnsen et al. 2000).

15.2.2.2 Volcanic, Cosmogenic, and Other Marker Horizons

Horizons of known age provide a rigorous constraint on depth-age relations and are used to tune them. Ash or acid layers from volcanic eruptions are the most abundant such "marker horizons" (Hammer et al. 1980). Most volcanic horizons are acidic layers rich in sulfate (Section 15.10.3), usually identified as pronounced increases in measured sulfate concentrations and electrical conductivity. A few volcanic horizons, seen in cores close to the eruptive source, consist of rock microparticles ejected during eruption – such layers are usually referred to as "ash" or "tephra." Dates of recent eruptions, known from historical records, include A.D. 1883 (Krakatoa, Indonesia), A.D. 1815 (Tambora, Indonesia), and A.D. 1600 (Huaynaputina, Peru). The ages of many eruptions in the last 10 kyr are known well from annual layer counting of ice cores from high-accumulation sites (e.g., Zielinski et al. 1994). The Mount Mazama eruption at around

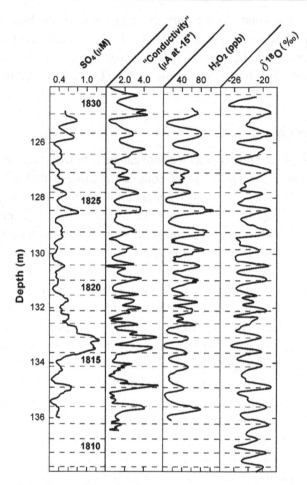

Figure 15.5: Two decades of annual layering in a core from Law Dome, Antarctica. Measured parameters are sulfate concentration, electrical current (a measure of the conductivity), hydrogen peroxide concentration, and $\delta^{18}O$ of ice. Fallout from the A.D. 1815 eruption of the Indonesian volcano Tambora appears around 133 m. Adapted from Morgan et al. (1997).

4400 B.C. appears in northern hemisphere cores. In Greenland, Zielinski et al. (1996) identified a volcanic layer in the GISP2 core as the 71 ± 5 kyr eruption of Toba, while Ram et al. (1996) found ash from an Icelandic volcano with an age of about 57 kyr. Records in West Antarctica contain at least a dozen radiometrically dated events (Dunbar et al. 2008a; Wilch et al. 1999). The ash from an eruption of Mt. Berlin, West Antarctica, has been dated radiometrically to 92.5 ± 2 kyr before present (Narcisi et al. 2006).

Cosmogenic isotopes are another source of marker horizons (Raisbeck et al. 1981, 2002; Beer et al. 1985). Cosmic radiation produces the isotopes ^{10}Be and ^{14}C by nuclear reactions with atmospheric nitrogen and oxygen. ^{10}Be has a half-life of 1.5 million years. It attaches to atmospheric aerosols and is removed by precipitation within a year or two of formation. Its concentration in ice, though minuscule, can be measured with accelerator mass spectrometry, a technique that counts individual atoms. Earth's geomagnetic field shields the atmosphere from cosmic radiation, but the effectiveness of the shield varies with solar activity and with internal fluctuations of the planetary dynamo; thus the production of ^{10}Be varies over time. Its concentration in ice varies with both the cosmic-ray flux and with the accumulation rate (Section 15.7). Measured ^{10}Be concentrations in ice show the sunspot cycle, and hence indicate the thickness of eleven years of accumulation (Steig et al. 1998). In time periods with relatively constant accumulation rates, such as the Holocene, variations of ^{10}Be concentration mostly reflect variations of cosmic-ray flux and so the same patterns should be seen in different cores. Such patterns provide a template for identifying which volcanic layers correspond to which known eruptions (e.g., Udisti et al. 2004). Moreover, because ^{10}Be and ^{14}C are both cosmogenic, the Holocene variations of ^{10}Be have themselves been assigned absolute ages by calibration against the ^{14}C age scale known from tree rings (e.g., Muscheler et al. 2004). Thus, in principle, measurements of ^{10}Be along a core are sufficient by themselves to determine the depth-age relation through the Holocene, with an accuracy of a few centuries or better. To do so, effects of fluctuations in accumulation rate need to be removed from the ^{10}Be concentration record.

Occasional large changes in the geomagnetic field leave a mark not only on ^{10}Be concentrations in ice but also on the magnetic properties of lava flows around the world. The "Laschamp geomagnetic excursion" produced a large increase in ^{10}Be production that peaked around 41 kyr ago. Its age is known from layer counting of Greenland cores (Raisbeck et al. 2002) and from radiometric dating of lava flows (Guillou et al. 2004). The Brunhes-Matuyama Reversal, dated radiometrically to 776 ± 12 kyr, appears in the ^{10}Be record at Dome C, Antarctica (Raisbeck et al. 2006).

Major global-scale climate transitions of known age offer some useful reference points because they appear in ice cores as well as a variety of other records. For example, radiometric dating of carbonate deposits in caves gives an age of 129 to 131 kyr b.p. for the abrupt end of the penultimate glacial-to-interglacial transition (Parrenin et al. 2007 and references therein). Such events, of course, provide a precise age constraint only if they occurred simultaneously at the ice core site and the other locations.

A useful constraint on the age of recent, near-surface firn is the depth to mid-twentieth century radioactive horizons. The explosion of a nuclear bomb releases large amounts of radioactive debris, mainly products of the fission of uranium. These disperse worldwide in the stratosphere and eventually fall out. Radioactive layers in firn cores can be detected by measuring gross β-activity, which results mainly from the decay of strontium-90 and cesium-137 (Picciotto and Wilgain 1963). Peaks in β-activity in Greenland and Arctic Canada occur one to two years after the tests; the delay is three to four years in Antarctica. The most prominent horizons, 1954 and 1963 in Greenland (1955 and 1964 in Antarctica), have been widely used as reference levels for measurements of accumulation rate. Measurements of tritium concentrations also reveal the fallout from nuclear tests. The burn-up of a U.S. satellite (SNAP-9A) containing ^{238}Pu in 1964 and the Chernobyl nuclear accident of 1986 also left a record in Greenland firn (Koide et al. 1977; Pourchet et al. 1988).

15.2.2.3 Markers Related to Globally Mixed Gases

Gases with atmospheric lifetimes of several years or longer vary synchronously everywhere on the planet. This is the basis for a powerful set of methods for cross-dating ice cores and correlating them between hemispheres and with other geological records such as marine sediments and cave deposits.

Methane Methane concentrations fluctuate on timescales of decades to millennia and longer (Section 15.8). The atmospheric residence time of CH_4 molecules is only about ten years, so a change in the source flux – primarily the release of CH_4 by microbial activity in soils – rapidly produces a change in the atmospheric concentration. Because concentrations varied in an irregular fashion throughout the last glacial period (Brook et al. 1996, 2000), and even varied considerably in the Holocene, detailed measurements of methane along an ice core allow detailed correlations to gas records from other cores (Blunier and Brook 2001). The methane variations have also been calibrated in terms of absolute ages, using layer-counting and marker-horizon techniques. Blunier et al. (2007) summarized the technique for synchronizing Greenland and Antarctic cores for the most recent 50 kyr using CH_4. Typical uncertainties are estimated as 200 to 500 yr, but even greater accuracy can be achieved at times of rapid changes.

Isotopes of Diatomic Oxygen The isotopic composition of atmospheric O_2 (symbolized $\delta^{18}O_2$) varies over time due to complex processes related to photosynthesis and respiration. As with CH_4, these variations are global and permit correlation of ice cores. Measurements of $\delta^{18}O_2$ complement those of CH_4; $\delta^{18}O_2$ varies more slowly, and most of the variation occurs over centuries to millennia, not decades. The slow variation is a consequence of the approximately 1000-year atmospheric residence time of O_2, compared to only about 10 years for CH_4. Combining CH_4 and O_2 data in an ice core often provides a unique indicator of time that eliminates ambiguity in correlations to other records with established age scales.

The causes of variations in $\delta^{18}O_2$ are secondary to their use as a correlation tool, but nonetheless deserve attention for their indirect links to climate. When an organism produces organic compounds by photosynthesis, it takes up water and releases O_2 with an isotope ratio matching that of the consumed water. But in respiration organisms preferentially use isotopically light O_2, leaving the atmosphere with an excess of ^{18}O. Currently, the $\delta^{18}O_2$ is about $+23.9‰$ relative to seawater (Barkan and Luz 2005); the offset is known as the *Dole Effect*. As ice sheets grow and shrink, the $\delta^{18}O$ of seawater slowly changes (see Chapter 13), so the composition of oxygen released by the biota varies, too. This is true on land as well as at sea, because precipitation on land originates in the ocean. The magnitude of the Dole Effect also varies with time, largely because of variations in the Asian monsoon (Wang et al. 2008; Severinghaus et al. 2009). About half of the total variation of $\delta^{18}O_2$ arises from changes in the Dole Effect, and about half from changes in seawater. Initial efforts attempted to use $\delta^{18}O_2$ as a proxy for seawater $\delta^{18}O$ (Sowers et al. 1993; Shackleton 2000), but subsequent work has shown that the Dole Effect fluctuates erratically and aperiodically because of the complexity of factors influencing the monsoons. The influences include astronomical precession cycles and abrupt climate changes in the North Atlantic region (Kawamura et al. 2007; Severinghaus et al. 2009).

15.2.2.4 Gas Indicators Used for Astronomical Dating

Astronomical dating is possible if a measured parameter fluctuates over millennia in synchrony with solar radiation. Measures of seasonal radiation, such as the mean June insolation at 65°N, vary with the configuration of Earth's orbit and rotational axis. The timing of such variations in the Quaternary are known well from orbital mechanics calculations, the Milankovitch theory (Section 13.3.1). A general difficulty for establishing reliable astronomical dating techniques is that most ice-core parameters depend not on radiation directly, but on climate and ocean composition – which have their own complex relationship to astronomical radiation cycles. In particular, $\delta^{18}O_2$ varies with the $\sim$20-kyr precession cycle but the relationship is now recognized as too complex to allow precise astronomical dating. Nonetheless, the cycles of $\delta^{18}O_2$ correlate strongly enough with precession to be used for low-precision dating, with quoted uncertainties of 6 kyr (e.g., Parrenin et al. 2007). Even if this underestimates the true uncertainty by some few kyr, the technique provides a valuable constraint on the ages of deep and old ice from East Antarctica, where ages attain several hundred kyr.

Oxygen to Nitrogen Ratio and Total Air Content The amounts of N_2 and O_2 in the atmosphere remain nearly constant over time, and so we expect the ratio of their concentrations to remain essentially constant along an ice core. In theory, changes in biomass and biological productivity across major climate transitions could change the O_2/N_2 ratio by a small amount, about 1‰ (Sowers et al. 1989). In fact, measurements have revealed much larger fluctuations of about 10‰ in the O_2/N_2 ratio along the core from Vostok, with periods of about 20 kyr (Bender 2002). Moreover, the shape and nominal timing of the fluctuations appear to match the timing of variations in the strength of summer sunlight at the core site, related to precession of Earth's orbit and changes of its obliquity (Figure 15.6). Bender concluded that "the Vostok O_2/N_2 curve is a strip chart recording of local solar insolation." The same phenomenon was subsequently found at another East Antarctic site, Dome Fuji (Kawamura et al. 2007).

Measurements of bubbles in a variety of cores show that, in general, O_2 is depleted compared to the atmosphere, with $O_2/N_2 \approx -10‰$ (Sowers et al. 1989). The most likely explanation for the depletion is preferential escape of O_2, because of its small molecular size, through the minuscule passageways or thin ice walls between bubbles just before close-off (see Section 15.3). Bender proposed that fluctuations of O_2/N_2 occur because of changes in the physical dimensions or configurations of ice grains that modulate the escape of O_2 from the close-off zone. Most likely,

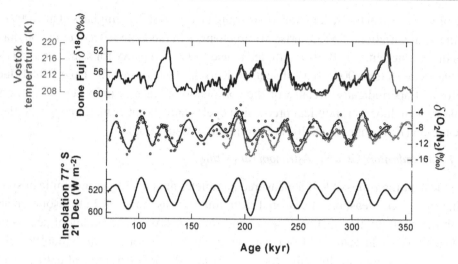

Figure 15.6: Measured long-term variations of O_2/N_2 ratio in East Antarctic ice (middle curves), compared to summer insolation at 77°S (bottom curve). Top curves show variations of ice-isotopic composition for context. In top and middle curves, darker lines are data from Dome Fuji; lighter lines, from Vostok (Bender 2002). Adapted from Kawamura et al. (2007).

the intensity of sunlight at the surface drives metamorphism of the firn, producing variations of grain shape, size, or layering that are partially preserved many years later at the firn-ice transition, where the gas exclusion occurs (Severinghaus and Battle 2006). If this explanation is correct, there should be little or no time lag of the measured O_2/N_2 with respect to the insolation, a considerable strength of the method. Note, too, that according to this explanation other small molecules such as monatomic neon should escape from the close-off zone, producing fluctuations of Ne/N_2 that coincide with those of O_2/N_2 (Severinghaus, pers. comm.); this prediction can be tested if improvements in analytical techniques make possible measurements of ice-core neon concentrations.

Suwa and Bender (2008) used correlation of O_2/N_2 ratios with theoretical December insolation at 78°S to make an age scale for the Vostok core, for ages greater than 112 kyr. Comparisons to the ages of prominent global climate transitions showed that the chronology matched radiometrically dated speleothem records to within a few millennia. Kawamura et al. (2007) used this technique to compare the timing of Antarctic climate changes with summertime insolation variations in both hemispheres.

A related technique for astronomical dating has been proposed by Raynaud et al. (2007). They believe that the total volume of trapped air per unit volume of ice (Section 15.4) varies with the 41-kyr cycle of Earth's obliquity (again, a consequence of changes in metamorphism related to summer insolation). With this assumption, measured fluctuations of total air content have been used as a constraint on the age scale for the Dome C core (Parrenin et al. 2007), at a claimed accuracy of 4 kyr.

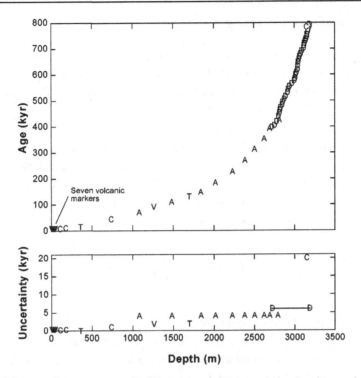

Figure 15.7: Depths and ages of marker horizons used to constrain the timescale (version "EDC3") for the Dome C core; numbers are from Table 1 of Parrenin et al. 2007. Different symbols indicate different types of markers: V = volcanic horizons; C = cosmogenic isotope constraints; T = climate transitions; A = total air content periodicity; D = $\delta^{18}O_2$. Values for uncertainties represent two standard deviations. True uncertainties for "A" and "D" indicators are doubtless larger than quoted by Parrenin et al.

Figure 15.7 summarizes all the marker horizons used to constrain the Dome C age scale, as of year 2007.

15.2.2.5 Argon Gas Indicator of Absolute Ages

Radioactive decay in the Earth's crust and mantle releases ^{40}Ar, which seeps to the surface and accumulates in the atmosphere. The atmospheric ratios $^{40}Ar/^{36}Ar$ and $^{40}Ar/^{38}Ar$ are therefore increasing over geological time. In comparison to the modern atmosphere, the air in old ice should be slightly depleted of the heavy isotope; indeed, measurements reveal this effect (Bender et al. 2008). The older the ice, the larger the depletion. The depletion of ^{40}Ar relative to ^{38}Ar is about 0.05‰ in 700-kyr-old ice. This provides a way to estimate the absolute age of a sample of very old ice, with an uncertainty of about ±180 kyr for ice 1 million years old. The measurements are corrected for gravitational settling (Section 15.3) by also measuring the ratio $^{38}Ar/^{36}Ar$, a constant in the atmosphere.

15.2.3 Difference of Gas and Ice Ages

The trapping of gases occurs at tens of meters depth in polar firn, where the density reaches about $830 \, \mathrm{kg \, m^{-3}}$, the density of pore close-off. Consequently, gases are younger than the enclosing ice, by an amount ranging from a few decades to a few millennia; Table 2.2 lists some measured values. The difference between the gas age and ice age, usually denoted Δ_{age} or D_{age}, must be known to compare the timing of changes in gases, such as CO_2, to the timing of changes of indicators in the ice, such as the stable isotope proxy for temperature.

The Δ_{age} depends on how rapidly snowfall buries layers and how rapidly the firn densifies; we have discussed these processes in Section 2.5. The age of ice (also Δ_{age}) at the depth of pore close-off ($\hat{z}_c$) can be estimated by equating the mass of ice between the surface and $\hat{z}_c$ with the mass of accumulation over the last Δ_{age} years:

$$\rho_i \overline{b} \Delta_{\mathrm{age}} = \int_0^{\hat{z}_c} \rho \, d\hat{z} \tag{15.17}$$

where $\overline{b}$ indicates the mean annual accumulation ($\mathrm{m \, yr^{-1}}$ of ice) in that interval, ρ_i the density of pure ice, and ρ the density of firn at depth $\hat{z}$. Because $\overline{b}$ differs by as much as 50 times between cold sites and warm sites, its variations account for most of the differences in Δ_{age} from place to place. Not all bubbles seal off at the same depth, so there is also a spread of ages. In Greenland, at present, the spread ranges from 20 to 75 yr (and Δ_{age} from 100 to 400 yr), whereas at Vostok the spread is about 600 yr and $\Delta_{\mathrm{age}} \approx 2500$ yr (Schwander and Stauffer 1984).

One method for estimating Δ_{age} through time is to use a firn densification model to calculate the depth-density relationship (Section 2.5.2). The accumulation rate and temperature need to be specified, using information from analyses of the core. To account for changes of the density profile in the years after an abrupt climate change, the densification model must be solved simultaneously with a model for temperatures. The primary weakness of the modelling approach is the absence of modern analogues for the extremely cold conditions in central Antarctica during the ice-age climates. Densification models appear to overestimate Δ_{age} for these conditions (Loulergue et al. 2007). They probably work better in central Greenland (Schwander et al. 1997).

Another method uses the isotopic composition of unreactive gases to estimate Δ_{age} directly from an ice core; the following section explains the techniques. Such information can be used to validate or calibrate densification models.

15.3 Fractionation of Gases in Polar Firn

Gases in the firn move freely by diffusion and, near the surface, by wind-driven convection (Sowers et al. 1992). The mobility of gases decreases with depth as the firn densifies and constricts the passageways that connect air-filled pores (Section 2.5.3). Bubbles close off and trap gases in a meters-thick zone at the base of the firn (the *close-off zone*), where density ranges from about 790 to 830 kg m^{-3} (Schwander et al. 1993; Severinghaus and Battle 2006). Diffusion may cease even before close-off begins, defining a thin *nondiffusive zone*. At the top end of the firn colum, the *convective zone* varies in thickness from about 1 m to more than 10 m (Kawamura et al. 2006). In the middle is the *diffusive zone*, which accounts for most of the firn thickness. Its vertical extent defines the *diffusive column height*.

A change of gas composition in the overlying atmosphere propagates into the firn; complete exchange occurs within about two decades (Schwander et al. 1993). Even with a steady atmospheric composition, a slight difference in gas concentrations develops between the top and bottom of the firn layer, due to gravitational settling of heavy gases and separation driven by temperature gradients. In the latter process, known as *thermal fractionation*, the heavy isotopes of a gas concentrate at the colder end of the firn column (Severinghaus et al. 1998). By thermal fractionation, seasonal fluctuations of the temperature gradient produce measurable gas-isotope anomalies in the upper firn; such features have been used to validate models of the process (Severinghaus et al. 2001).

Consider a gas with light and heavy isotopes – for example, argon with both ^{36}Ar and ^{40}Ar, or diatomic nitrogen consisting of both ^{14}N^{14}N and ^{15}N^{14}N. For a relative concentration δ of heavy and light isotopes, conservation of mass and the properties of diffusive transport (net fluxes proportional to gradients in concentration) together imply that concentrations vary over time in the diffusive zone according to:

$$s\frac{\partial \delta}{\partial t} = -\frac{\partial}{\partial \hat{z}}\left(sD(\hat{z}, T)\left[-\frac{\partial \delta}{\partial \hat{z}} + \frac{g\Delta m}{RT}[1+\delta] - \Omega\frac{dT}{d\hat{z}}\right]\right). \tag{15.18}$$

(This relationship is from Kawamura et al. (2006) but we have included the factor $1 + \delta$ in the gravitational term.) Here $\hat{z}$ indicates depth, T temperature in K, g the gravitational acceleration, Δm the mass difference between isotopes, R the gas constant, and Ω the thermal diffusion sensitivity (in $\%_0$ K^{-1}). The factor s indicates the effective or "open" porosity, the volume of air-filled voids through which gases move (per unit volume of firn). The parameter D is the effective molecular diffusivity of gas in firn, or density (see Eq. 2.6). The three terms on the right-hand side of Eq. 15.18 correspond to three factors driving diffusion: molecules move down concentration gradients; heavy molecules preferentially settle to the bottom of the firn; and heavy molecules preferentially move from warm to cold regions.

A first implication of Eq. 15.18 is that gas isotope ratios record the diffusive column height (Schwander 1989; Sowers et al. 1992; Craig and Wiens 1996). Relative to atmospheric values, a thicker layer produces a larger anomaly of heavy isotopes concentrated by gravitational settling. The anomaly increases with depth according to the *barometric equation*:

$$\delta = \exp\left(\frac{g\hat{z}\Delta m}{RT}\right) - 1 \approx \frac{g\hat{z}\Delta m}{RT}. \tag{15.19}$$

At the base of the firn, the enrichment of heavy argon should therefore be four times larger than the enrichment of heavy nitrogen (δ^{40}Ar $\propto \Delta m = 40 - 36 = 4$ whereas δ^{15}N $\propto \Delta m = 29 - 28 = 1$). Moreover, for each gas pair the enrichment should vary in proportion to the diffusive column height; measurements of gas isotopes along an ice core thus give a record of firn thickness, if the thickness of convective and nondiffusive zones can be assumed constant or negligible. Primarily, firn thickness depends on temperature and accumulation rate, so gas isotope measurements give information about either one if the other is known (Schwander et al. 1997; Huber et al. 2006b). However, a most unusual history of diffusive column height was discovered in the core from Siple Dome, West Antarctica, from gas isotope data (Severinghaus et al. 2003). The inferred height late in the last ice age mostly ranged between 40 and 60 m, typical values. Around 15.3 kyr b.p., however, the δ^{40}Ar briefly dropped to near zero, indicating a temporary loss of the gravitational signal, perhaps due to deep air convection.

A second implication of Eq. 15.18, first recognized by Severinghaus et al. (1998), is that gas isotope ratios preserve a signal of rapid temperature changes (Severinghaus et al. 2003; Lang et al. 1999). An abrupt climate warming, for example, produces a temperature contrast across the firn layer that persists for decades to centuries. During this time, heavy isotopes concentrate in the relatively cold horizons at the base of the firn, where gases are trapped. The concentration anomaly subsides over time as heat flow equalizes temperatures throughout the firn. Gases trapped in the ice preserve this sequence as a spike in the concentration of heavy isotopes; Figure 15.8 illustrates an example (Landais et al. 2004).

The position along the core of the thermal fractionation spike gives another measure of firn thickness at the time of an abrupt climate change. Because gases are trapped at the base of the firn, not the surface, the thermal spike will be located farther down the core – in older ice – than the ice-isotopic shift for the same climate change (Figure 15.8). The distance between the two signals shows the thickness of firn, if corrections are made for the cumulative strain and the shape of the density profile at the time of gas entrapment. The number of annual layers in the ice over the same interval gives a direct measure of the difference between the gas and ice ages. Leuenberger et al. (1999), for example, used this technique to find a Δ_{age} of 209 ± 15 yr in central Greenland in the early Holocene (at 8.2 kyr b.p.). In contrast, Δ_{age} was 809 ± 20 yr at 11.6 kyr b.p., in the cold and dry climate just before the termination of the ice age (Severinghaus et al. 1998).

The magnitude of a thermal fractionation anomaly depends proportionately on the temperature difference, ΔT, between the top and bottom of the diffusive column: $\delta = \Omega \cdot \Delta T$. The *thermal diffusion sensitivity*, Ω, is determined in laboratory experiments and has a unique value for each gas pair. In the decades following a rapid change of climate, ΔT depends closely on the magnitude of the temperature change. Thus, by using models for heat transfer and gas diffusion, we can calculate the magnitude of the temperature change from measurements of the gas isotope spike. Grachev and Severinghaus (2005) summarized ways to do this. Information from the ice-isotope record about the form and timing of the surface temperature change can be incorporated in the analysis.

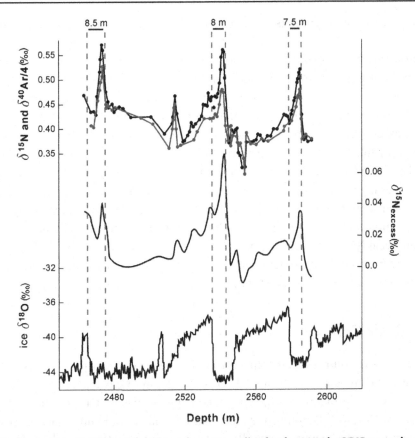

Figure 15.8: Examples of thermal fractionation anomalies in the North-GRIP core, in the depth interval corresponding to ages of 62–76 kyr. Time moves forward from right to left. The bottom curve is the $\delta^{18}O$ proxy for temperatures, which shows the abrupt warmings responsible for the gas-isotope anomalies. Adapted from Landais et al. (2004) and used with permission from the American Geophysical Union, *Geophysical Research Letters.*

Changes in the thickness of firn, which follow changes of either accumulation rate or temperature, complicate the analysis; the gas isotope spike contains a combined signal of perturbed gravitational settling and thermal fractionation. The two contributions can be inferred by measuring the isotope spike for two different gases simultaneously. The quantity depicted in Figure 15.8,

$$\delta^{15}N_{\text{excess}} = \delta^{15}N - \frac{1}{4}\delta^{40}Ar, \tag{15.20}$$

isolates the thermal signal from the gravitational one. (The factor 4 corresponds to the ratio of mass differences, noted previously.) A further correction can be made for small fractionations caused by leakage of argon through fractures during core processing, or through narrow passageways during bubble close-off (Severinghaus et al. 2003). The correction uses the ratio of krypton to argon; krypton is a larger molecule than argon and hence less susceptible to leakage.

Concerning this latter point, measurements of air extracted from firn show not only the expected signals of gravitational enrichment and thermal fractionation, but also a pronounced enrichment of some gases in the bottom few meters, just above the close-off depth (Severinghaus and Battle 2006; Huber et al. 2006a). This *close-off fractionation* must reflect escape of some gases, but not others, through narrow passageways or thin ice walls as bubbles close off. Changes in the strength of this process with time probably account for the correlation of O_2/N_2 with insolation, the dating technique discussed in Section 15.2.2.4. The enrichment does not depend on molecular mass; it has no signal in isotope ratios. It appears in elemental ratios like O_2/N_2 and He/Ar and depends on molecule size. The smallest molecules, He and Ne, strongly fractionate, whereas large molecules like Kr, Xe, N_2, CO_2, and CH_4 do not fractionate at all (O_2 and Ar fractionate weakly). This process explains the observation that O_2/N_2 and Ar/N_2 ratios in bubbles are lower than the atmospheric values, although gas leakage during core handling may contribute (Sowers et al. 1989; Bender et al. 1995).

15.4 Total Air Content

Measurements of the volume of air trapped in polar ice may reveal past changes in ice sheet surface elevations (Raynaud and Lorius 1973). The amount of gas enclosed in a given volume of bubbles depends, in principle, on the atmospheric pressure and hence on the elevation of the site at the time of close-off. Defining V as the total volume of gas in unit mass of ice, measured at standard temperature T_o and pressure P_o, and V_c as pore volume per unit mass at close-off, then

$$\frac{V}{V_c} = \frac{P_c}{T_c}\frac{T_o}{P_o}.$$

$$(15.21)$$

Here T_c and P_c are the temperature and atmospheric pressure at close-off. The value of V_c for present-day ice can be found by comparing the density at the close-off point (about 830 $kg\,m^{-3}$) with that of bubble-free ice. In the first application of the method V_c was assumed to be constant (Raynaud and Lorius 1973). This assumption proved to be wrong, however, because pore volume depends on grain size, a function of variables such as temperature and the process that originally deposited the layer. An empirical approach was therefore adopted.

Martinerie et al. (1992), following Raynaud and Lebel (1979), assembled measurements of V in recently formed ice at sixteen sites spanning a wide range of elevation. Because V_c was unknown, they looked for a correlation between V and site elevation, although the expected correlation is with V/V_c, the air content per unit pore volume, rather than with V, the air content per unit mass. The authors found a good relationship, as Figure 15.9 shows, with an approximate decrease of V by 1.7 $mm^3\,g^{-1}$ for a 100 m increase of elevation. But this relationship cannot

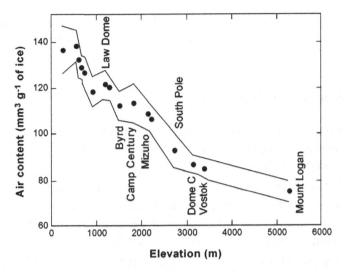

Figure 15.9: Total air content versus site elevation. The lines show the maximum uncertainties. All the sites except Camp Century (Greenland) and Mount Logan (Yukon) are in Antarctica. Adapted from Martinerie et al. (1992).

be used, by itself, to convert measurements of V in old ice to past surface elevations; some of the change in V arises from differences in the temperature T_c. To correct for this, V_c must be known.

For the data used in the figure, Martinerie et al. (1992) calculated V_c from Eq. 15.21 and present values for P_c and T_c. The result showed a good positive correlation between V_c and T_c, implying an increase of 0.76 mm^3 g^{-1} for a 1 K increase of T_c. Thus to calculate past elevations requires a correction not only for the direct temperature dependence in Eq. 15.21 but also for temperature effects on V_c. Temperature variations would occur with changes of both elevation and regional climate. But temperature, unfortunately, proves to be not the only determinant of V_c. Measurements of over 1000 samples from the Vostok core showed large and rapid fluctuations in V that are not plausibly explained by elevation and do not correlate with other indicators of temperature (Martinerie et al. 1994). Air content records from two other long-duration East Antarctic ice cores revealed analogous variations (Kawamura 2000; Raynaud et al. 2007). Such variations most likely arise from differences in the microstructural development of firn near the surface, which affect the size and abundance of bubbles and hence V_c at pore close-off. Possible causative factors include wind speed (Martinerie et al. 1994) and snow metamorphism related to summer insolation (Raynaud et al. 2007). The latter idea gains support from analysis of the periodicity of air content variations in the EPICA-Dome C core (Section 15.2.2.4); there is a signal at about 40 kyr, which corresponds to orbital obliquity and hence to insolation integrated over the summer.

Because no reliable elevation proxy has yet been established, further development of the total air content method to account for these other factors should remain a major priority.

Nonetheless, the technique may already be useful at sites where large elevation changes have occurred – at some of the ice domes in coastal Antarctica, for example (E. Wolff, pers. comm.).

15.5 Stable Isotopes of Ice

Most of the water in the atmosphere originates as evaporation from the ocean. The isotopic composition of seawater is nearly uniform, with relative abundances of $H_2{}^{16}O$, $H_2{}^{18}O$, and $HD{}^{16}O$ of $0.9977 : 0.0020 : 0.0003$. Because the vapor pressures of the heavier waters are slightly lower than that of the common water, the heavier molecules evaporate less rapidly and condense more readily from the vapor. The isotopic compositions of condensate and associated vapor thus differ, a phenomenon known as *fractionation*. As evaporation and precipitation redistribute water between the surface and the atmosphere, repeated fractionations have a cumulative effect on the isotopic compositions of air masses and the precipitation derived from them. Climate changes that alter this system of repeated fractionations also alter the isotopic content of snow falling on the ice sheets. The first long records of ice isotopes were measured on cores from Camp Century in Greenland (Dansgaard et al. 1969) and Byrd Station in Antarctica (Epstein et al. 1970; Johnsen et al. 1972).

The $^{18}O/^{16}O$ and D/H ratios of water or ice are measured as $\delta^{18}O$ and δD, defined by Eq. 15.1. The standard ratios for water are known as "standard mean ocean water." In the following discussion, "δ" stands for either $\delta^{18}O$ or δD unless either one is specified.

15.5.1 Conceptual Model

The global atmospheric circulation acts as a giant filtration system that depletes heavy-isotope bearing waters from high-latitude and high-altitude air masses (Dansgaard 1964; Johnsen et al. 1989). Figure 15.10 depicts the system schematically. Water evaporates from the oceans in the subtropics (some moisture also derives from continents and polar oceans) and moves toward the poles. Cooling of the air mass drives condensation and precipitation. At each stage, the heavy isotopes are preferentially removed from the vapor when it condenses to form liquid droplets or ice particles. Thus the vapor, and the precipitation derived from it, becomes isotopically lighter with distance along the path. Finally, the snow accumulating on the ice sheet is substantially lighter than the original oceanic source. In the following sections we examine how this net distillation depends on temperatures and other climatic variables.

15.5.1.1 Fractionation

Consider a parcel of air in the atmosphere depicted in Figure 15.10. The parcel contains n_o moles of common, light water vapor ($H_2{}^{16}O$) and n_j moles of heavy water vapor (either $H_2{}^{18}O$ or $HD{}^{16}O$). The isotopic ratio of the vapor is $R_v = n_j/n_o$ and, by definition, $1 + \delta_v = R_v/R_s$ (R_s being the standard ratio). As condensate forms, its isotopic ratio, R_p, is offset from R_v by

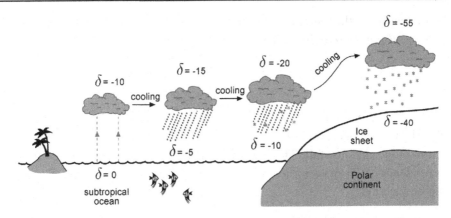

Figure 15.10: Schematic of moisture transport from mid to high latitudes, showing associated change in δ values for the air mass and the precipitation. Adapted from Cuffey and Brook (2000).

a small amount:

$$R_p = \alpha R_v \quad \text{or} \quad 1 + \delta_p = \alpha \left[1 + \delta_v \right]. \tag{15.22}$$

The condensate is assumed to be removed immediately from the air mass as precipitation. The parameter α, a number slightly greater than one, is known as the *fractionation factor*. Fractionation factors increase with decreasing temperature (Merlivat and Nief 1967; Horita and Wesolowski 1994). Across the range of Earth surface temperatures, α increases from about 1.008 to 1.025 for $^{18}O/^{16}O$ fractionation. For D/H fractionation, $\alpha \approx 1.07$ to 1.25. These values assume equilibrium between vapor and condensate. Note that the hydrogen isotopes fractionate much more strongly than the oxygen isotopes. This is not explained by the relative mass differences; the mass ratio of $HD^{16}O$ to $H_2{}^{16}O$ is only 1.056 compared to 1.11 for $H_2{}^{18}O$ and $H_2{}^{16}O$. Instead, the stronger fractionation of hydrogen reflects the comparatively large effect on molecular vibration frequencies when a D atom replaces an H atom in a water molecule.

The value for α depends on whether condensation forms liquid droplets or ice crystals. In both cases α rises as temperature falls, but a larger fractionation occurs for ice (when ice replaces liquid, $\alpha - 1$ increases by about 10% and 30% for $^{18}O/^{16}O$ and D/H, respectively). Another process, *kinetic fractionation*, modifies the value for α when ice forms in cold air (Jouzel and Merlivat 1984). Crystals grow in air that is supersaturated. Excess vapor pressures drive a diffusive flow of water molecules to growing crystals. Because $H_2{}^{16}O$ moves by diffusion more readily than the heavy waters, less fractionation occurs than expected for equilibrium – and the value for α is reduced. Because of this process, models for precipitation isotopes contain parameters describing the degree of supersaturation in clouds. Such parameters must be calibrated against data because supersaturations are unknown and diffusivity parameters are uncertain (Cappa et al. 2003).

During condensation, the amounts of vapor removed from the air mass, dn_o and dn_j, are related by $dn_j = R_p dn_o = \alpha R_v dn_o$. Thus, because $R_v = n_j/n_o$,

$$\frac{dn_j}{n_j} = \alpha \frac{dn_o}{n_o}. \tag{15.23}$$

The increase of α at low temperatures implies that cold air masses lose their heavy isotopes more quickly than warm ones, other factors being equal. This is one reason for the temperature sensitivity of isotopes in snowfall. The most important reason, however, arises from the temperature dependence of air-mass moisture content at mid and high latitudes. The next section explains this connection.

15.5.1.2 Distillation

As condensation extracts water vapor from the air mass, the isotopic composition of the remaining vapor changes (Dansgaard 1964; Merlivat and Jouzel 1979; Johnsen et al. 1989). From the definition $R_v = n_j/n_o$, differentiation gives, after rearrangement,

$$\frac{dR_v}{R_v} = \frac{dn_j}{n_j} - \frac{dn_o}{n_o} = [\alpha - 1]\frac{dn_o}{n_o}. \tag{15.24}$$

(The second equality follows from Eq. 15.23.) Changes of the isotopic compositions of vapor and condensate covary; because $R_p = \alpha R_v$, then $dR_p = R_v d\alpha + \alpha dR_v$. Dividing through by αR_v, and using Eq. 15.22 and the definition of δ, shows that

$$\frac{d\delta_p}{1+\delta_p} = \frac{d\alpha}{\alpha} + \frac{dR_v}{R_v}. \tag{15.25}$$

The second term on the right-hand side can be replaced with Eq. 15.24. But first note that, because the abundance of light water greatly exceeds that of the heavy water ($n_o \gg n_j$), the total quantity of water vapor very nearly equals the quantity of the light species. Call W the total moles of water vapor in the air parcel. Using $W \approx n_o$,

$$\frac{d\delta_p}{1+\delta_p} = \frac{d\alpha}{\alpha} + [\alpha - 1]\frac{dW}{W}. \tag{15.26}$$

This relation describes how the δ value of the condensate – and hence the precipitation – decreases as the air mass loses water ($dW < 0$).

As an instructive approximation, we replace the temperature-dependent α with its mean value, $\bar{\alpha}$. Then integration of Eq. 15.26 between initial and subsequent values of vapor content (W_o and W) and composition (δ_{po} and δ_p) gives the *Rayleigh distillation* relation:

$$\delta_p = [1 + \delta_{po}]\left[\frac{W}{W_o}\right]^{\bar{\alpha}-1} - 1. \tag{15.27}$$

The fractional reduction in water content determines the net distillation of heavy isotopes. Taking, for example, a value $\bar{\alpha} = 1.1$ (appropriate for deuterium) and an initial $\delta D = 0$, reducing the water vapor content to $0.9 W_o$, $0.5 W_o$, and $0.1 W_o$ reduces the precipitation's δD to about $-10‰$, $-70‰$, and $-200‰$. Distillation increases sharply as W/W_o approaches zero.

Atmospheric condensation is driven by cooling of saturated air; thus the reduction of W depends on a change of temperature. According to the gas law, the quantity of water vapor in a given quantity of air (moles of water per total moles of gas) is directly proportional to the saturation vapor pressure e_s and inversely proportional to the total pressure P; in other words, $W \propto e_s/P$. But e_s varies only with temperature, according to the Clausius-Clapeyron relation. The latter approximates an exponential function (Section 4.2.1) such that $e_s \propto \exp(-\Lambda/T)$, for Kelvin temperature T, with $\Lambda \approx 5400\,\text{K}$ (over water) or $\Lambda \approx 6150\,\text{K}$ (over ice). Thus, from Eq. 15.27, a change from initial temperature of condensation T_o to final temperature T, and from pressures P_o to P, results in a final isotopic value of precipitation given by:

$$1 + \delta_p = \left[1 + \delta_{po}\right]\left[\frac{P_o}{P}\right]^{\bar{\alpha}-1} \exp\left(-\Lambda\,[\bar{\alpha}-1]\frac{[T_o - T]}{T T_o}\right). \tag{15.28}$$

The temperature drop from T_o to T depletes the air mass of heavy isotopes. The absolute value of temperature, which varies along the path, also influences the magnitude of depletion, through the factor $1/T T_o$ and the temperature dependence of α.

Figure 15.11 plots two examples of the relation between temperature and δ_p according to Eq. 15.26, using values for α and W that vary with temperature. In one case, α values are those for equilibrium, including a switch from liquid to ice condensate. The second case shows the effect of the kinetic fractionation related to supersaturation in clouds.

In general, a change of climate or weather affects the temperature over oceanic source regions, T_o, less than the temperature over the continental ice sheets – at all timescales from seasonal to multimillennial. In addition, the geographical distribution of sources can change, with a tendency to maintain evaporation from warm regions of the sea surface; this, too, tends to minimize changes in T_o. Thus the temperature drop $T_o - T$ usually fluctuates in correlation with T. Several types of data demonstrate that on the polar ice sheets precipitation isotopes vary with temperature as expected from this theory. Figure 15.12a shows the modern spatial covariation; whereas Figure 15.12b shows the correspondence over the seasonal cycle on the ice sheet in central Greenland. Moreover, comparisons of ice-core isotope records with independent measures of temperature demonstrate that isotopes and temperatures correlate strongly though imperfectly when climate changes; borehole temperatures prove the point for changes over millennia, and gas-phase isotopes do the same for decadal-scale abrupt warmings (Section 15.3 and Section 15.6.1; Cuffey et al. 1995; Grachev and Severinghaus 2005; Landais et al. 2004).

Nonetheless, many factors in addition to temperature modulate the isotopic composition of precipitation and its ice-core archives. The observed isotope-temperature correlations are surprisingly strong and must reflect, in part, the pervasive interconnections between many different

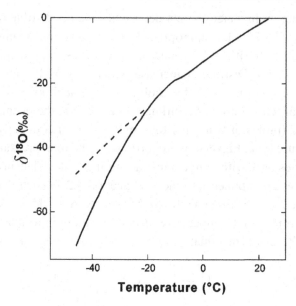

Figure 15.11: Calculated variation of $\delta^{18}O$ due to cooling of an air mass, assuming the water content reflects saturation at each temperature. Condensate changes from liquid to solid around $-10°C$. Solid curve uses equilibrium values for fractionation factors. Dashed line accounts for kinetic fractionation during growth of ice crystals, using the relation proposed by Jouzel and Merlivat (1984), assuming a steady increase of water-vapor supersaturation as temperature decreases.

aspects of global climate; many variables affect the isotopic composition, but, because they all tend to vary together the isotopes nonetheless retain a surprisingly strong correlation with temperature (Cuffey 2000). Thus, for many applications in polar regions a simple relation between δ and temperature T at a site is well justified; most commonly, the assumption is made that $\delta = \alpha_* T + \beta$, with constant coefficients α_* and β. The relationship between δ and T that applies over time at a site – the *temporal relationship* – differs, in general, from the *spatial relationship* observed in samples of modern precipitation. In some regions, the spatial relationship is weak or nonexistent; for example, in the Canadian Arctic excluding the margin of Baffin Bay (Koerner 1979). To appreciate why the temporal and spatial relationships are not the same requires an understanding of the geographical and climatological context for isotopic distillation. The following two sections engage this topic.

15.5.1.3 Geographical View

Averaged over months and years, the composition of precipitation at an ice-core site is determined, not by a single trajectory as shown in Figure 15.10 and described by Eq. 15.28, but by an ensemble of many such trajectories and mixing of vapor along them. The overall effects can be examined using constraints on the spatial variations of δ. Fisher developed a theory for this

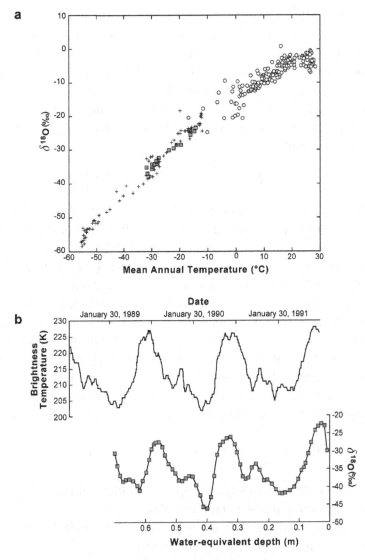

Figure 15.12: Observed relationships between precipitation isotopes and surface temperatures. (a) Worldwide spatial covariation of mean annual temperatures and $\delta^{18}O$ values (data compilation from Alley and Cuffey 2001). Plus symbols represent Antarctic sites; squares, Greenland sites; circles, lower-latitude sites. (For a recent and thorough compilation of Antarctic data, see Masson-Delmotte et al. (2008).) The spatial pattern implies a change of $\delta^{18}O$ with surface temperature of about 0.7‰ °C^{-1} at high latitudes. In the Tropics, excluding cold regions on high mountains, $\delta^{18}O$ varies little with surface temperature. (b) Variation of surface temperature over annual cycles in central Greenland (measured from satellite), compared to variation of measured $\delta^{18}O$ with depth in the firn that accumulated over the same time period (adapted from Shuman et al. 1995). Note that the isotopes vary not only with the seasonal cycle but also with the most prominent of the higher-frequency temperature fluctuations.

purpose (Fisher 1990, 1991, 1992; Fisher and Alt 1985). Hendricks et al. (2000) recast it in the form shown here, and Kavanaugh and Cuffey (2003) contributed corrections.

Consider the balance of atmospheric moisture in a vertical column extending upward from a coordinate on the surface. The column contains a total amount of moisture W, while winds carry through it a vertically integrated moisture flux $\vec{Q}_w$. Water enters the column's base as evaporation, at rate E, and exits as precipitation, rate P. If we consider a time interval long enough for a steady-state to apply, conservation of mass requires that the horizontal divergence of the atmospheric flux balances the net source of water:

$$E - P = \nabla \cdot \vec{Q}_w = \nabla \cdot \left[\vec{u} W - k \nabla W \right]. \tag{15.29}$$

Viewed as a large-scale process, atmospheric moisture fluxes take the form of either advective transport (in which winds of persistent direction carry a flux $\vec{u}W$) or diffusive eddy transport (in which winds of variable direction result in a net flux down the gradient in moisture content: $-k\nabla W$). Together, these fluxes give the second equality in Eq. 15.29. Here $\vec{u}$ denotes wind velocity and k the *eddy diffusivity*. Eddy diffusion dominates atmospheric transport in the subpolar zones of cyclonic storm activity that feed moisture to the ice sheets.

Similar balance relations can be written for each type of heavy water. Referring to the isotopic ratios of evaporating water, precipitating water, and atmospheric water as R_e, R_p, and R_a, respectively, then $R_e E - R_p P = \nabla \cdot \left[\vec{u} R_a W - k\nabla[R_a W] \right]$, recalling that the amount of $H_2^{16}O$ nearly equals the total, W. Expanding and regrouping the right-hand side give $R_a \nabla \cdot \vec{Q}_w + \vec{Q}_w \cdot \nabla R_a - \nabla \cdot [kW\nabla R_a]$. But $\nabla \cdot \vec{Q}_w$ equals $E - P$. Making that substitution, replacing each R with its equivalent $1 + \delta$, and dividing through by W leads to the formula

$$[\delta_e - \delta_a] \frac{E}{W} - [\delta_p - \delta_a] \frac{P}{W} = \vec{u} \cdot \nabla \delta_a - k \nabla^2 \delta_a \tag{15.30}$$

for the case of uniform k. At a global scale, the spatial variation represented by $\nabla \delta$ and $\nabla^2 \delta$ can be regarded as the variation from low to high latitudes.

Equation 15.30 shows how the spatial variations of water isotopes in the atmosphere, averaged on climatological timescales, depend on various factors:

1. Precipitation depletes the atmosphere of heavy isotopes, at a rate $[\delta_p - \delta_a]P/W$. In terms of the fractionation factor, α, this equals $[\alpha - 1][1 + \delta_a]P/W$. The rate of precipitation is governed by the rate of condensation throughout the atmospheric column. In the absence of evaporative recharge, P corresponds to dW in Eq. 15.26 (with $E = 0$ and $k = 0$, the model describes Rayleigh distillation). In turn, the condensation causing the change dW is driven by cooling; P in Eq. 15.30 is a consequence of the temperature drop in Eq. 15.28. Some of the cooling occurs when winds carry moisture from warm to cold regions, or when warm air mixes with cold air. In this case, the amount of isotopic depletion generally correlates with temperatures at the land surface as they change over time and with distance; the data in Figure 15.12 demonstrate such a relationship. But cooling also occurs in localized

upwelling air currents driven by land-surface heating ("thermal convection" that produces cumulus clouds). In this situation, although the air cools greatly as it moves from the surface to high altitude, the temperature of the surface may not change at all. The amount of isotopic depletion then correlates not with the surface temperature but with the rate of precipitation. The isotopic composition is said to be determined by the *amount effect*. The correlation with precipitation rate arises because rain partially equilibrates with vapor in the lower atmosphere, and the vapor is recycled into storms (Lee et al. 2007; Risi et al. 2008).

2. Evaporation recharges the atmosphere with isotopically heavy water. Specifically, the composition of evaporate from the ocean surface depends on the marine-water composition δ_m according to $1 + \delta_e = \alpha^{-1}[1 + \delta_m]$. The marine composition (and thus δ_e) varies little from low to high latitudes whereas δ_a decreases steadily toward the poles; thus $[\delta_e - \delta_a]$ is a positive quantity, larger over polar than subtropical oceans. At low latitudes, abundant evaporation prevents the depletion of isotopes by precipitation, except in individual storms. This accounts for the relative uniformity of precipitation isotopes at warm temperatures (corresponding to low latitudes) seen in Figure 15.12a.

   In evaporation from the sea surface, values for α depend on a kinetic effect and are usually estimated as (Merlivat and Jouzel 1979)

$$\alpha = \frac{1 - c\mathcal{H}}{1 - c}\alpha_{eq}. \qquad (15.31)$$

Here α_{eq} denotes the equilibrium fractionation factor, $\mathcal{H}$ the relative humidity above the sea surface, and c a constant dependent on wind speed regime and the diffusivities of water molecules in air. Low relative humidity increases evaporation rates. Light vapor diffuses more readily than heavy vapor into the overlying air, enhancing the fractionation. Thus the composition of evaporate depends on four factors: the seawater composition (δ_m), the temperature (via α_{eq}), the relative humidity, and the wind regime (via c). Cappa et al. (2003) emphasized that cooling of the water surface by evaporation affects the value for α_{eq}; the temperature of the surface cannot be assumed equal to the temperature of the bulk. They also made corrections to the parameters inferred originally by Merlivat and Jouzel.

3. A given rate of precipitation or evaporation has a larger influence on the atmospheric isotopes where W is small than where large. Temperatures govern W through the relation for water content of saturated air. Although relative humidities vary greatly from place to place because of patterns of atmospheric circulation, the general pattern of W is a decrease from low to high latitudes as a function of temperature.

4. The character of atmospheric moisture transport – whether advective or eddy-diffusive – affects the distribution of isotopes, a point first discussed by Eriksson (1965). Compared to the case of advection, eddy diffusion reduces the depletion of isotopes in polar regions, because it mixes depleted with undepleted air. To explain observed values of δ_p

in Antarctica, eddy diffusion must be the dominant process of moisture transport to the continent (Kavanaugh and Cuffey 2003), as expected from meteorological observations.

For a comprehensive analysis of how water isotopes vary with atmospheric conditions, a general circulation model can be equipped with tracers for each type of water molecule (Joussaume et al. 1984). These models adequately simulate the modern observed spatial distributions and seasonal cycles of isotopes in precipitation, and make informative if somewhat dubious simulations for past climates (Jouzel et al. 1991; Joussaume and Jouzel 1993; Hoffmann et al. 2000; Lee et al. 2007). The models have difficulty simulating the subtle differences observed in the behaviors of deuterium and oxygen isotopes (the deuterium excess; see Section 15.5.2.1), indicating that some aspects of fractionation in the atmosphere remain poorly understood.

15.5.2 Interpretation of Records

Ice-isotopic records constitute the heart of ice core analysis; no other measurement gives such a continuous and high-resolution view of climate history along most of a core. Figure 15.13 gives two examples, one from north-central Greenland, the other from central East Antarctica. Both show the end of the last interglacial period, the cold but variable conditions of the last glacial period, the deglacial warming, and the Holocene. From millennium to millennium, climate fluctuated more in Greenland than in Antarctica, and the timing of fluctuations differed between

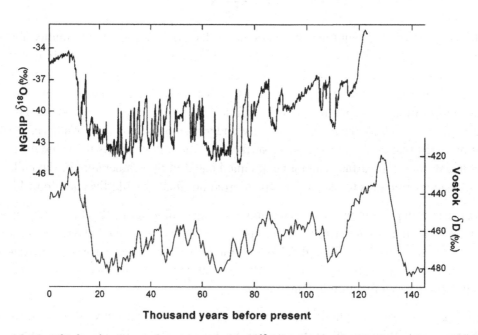

Figure 15.13: The last ice-age cycle as recorded by $\delta^{18}O$ in the North-GRIP core from Greenland (top) and by δD in the Vostok core, central East Antarctica (bottom). Data sources are North-GRIP Project members (2004) and Petit et al. (1999), respectively.

them. Antarctica often warmed gradually while Greenland remained cold, but then cooled after Greenland abruptly warmed (Blunier et al. 1998). The record from Greenland is the oldest one currently available from this ice sheet; at present, no record from the northern hemisphere shows the whole of the last interglacial.

The modern observed correlations of δ with temperature (Figure 15.12) and the predominance of the temperature-distillation effect in cold-region precipitation (Section 15.5.1.2) together justify assuming that, in general, δ varies in proportion to the climatic temperature in the region of an ice-core site (call it T_i). The δ record can first be corrected for known variations in the composition of marine source waters (δ_m) and then regarded as a composite of two terms:

$$\Delta\delta_i = \Delta\delta - \epsilon\Delta\delta_m = \alpha_\star \Delta T_i + \Theta, \qquad (15.32)$$

where Δ signifies a change from the present value, and the equation defines δ_i. The function Θ lumps together all variations of $\Delta\delta_i$ that do not correlate with changes of site temperature T_i. The $\Delta\delta_m$ is assumed to equal the marine composition $\delta^{18}O_m$ recorded by marine sediments (in the case of deuterium, the marine composition can be assumed equal to $8\delta^{18}O_m$; Section 15.5.2.1). The coefficient ϵ accounts for the net removal of heavy isotopes during atmospheric transport, and ranges from $\epsilon \approx 1$ for $\delta^{18}O$ in southern Greenland to $\epsilon \approx 0.6$ for δD in central East Antarctica. (Many older analyses of data from the Vostok core, such as Petit et al. (1999), erroneously used $\epsilon = 1$. Vimeux et al. (2001) first made the correction.)

The *isotopic sensitivity*, $\alpha_\star$ (usually given in units of $‰\,°C^{-1}$), must account not only for temperature changes at the site but also for simultaneous changes in all other climate variables ($v_1, v_2, \ldots$) that influence the value of δ_i preserved in the ice:

$$\alpha_\star = \frac{d\delta_i}{dT_i} = \underbrace{\frac{\partial\delta_i}{\partial T}}_{A} + \frac{\partial\delta_i}{\partial v_1}\frac{dv_1}{dT_i} + \frac{\partial\delta_i}{\partial v_2}\frac{dv_2}{dT_i} + \cdots \qquad (15.33)$$

Term A indicates the sensitivity expected if the temperature over the ice sheet changes by itself. At sites remote from their vapor sources (such as the interiors of ice sheets), this sensitivity reflects the increase of distillation as temperature drops (Section 15.5.1.2) and should be similar to the present spatial covariation of isotopes and cloud temperatures, roughly $0.7‰\,°C^{-1}$ for 18-oxygen. At coastal sites, the composition of atmospheric vapor cannot deviate much from the composition of evaporation from nearby oceans, so a smaller sensitivity is expected. Considering the processes discussed in Sections 15.5.1.2 and 15.5.1.3, variables $v_1, v_2, \ldots$ of likely importance include (see also Dansgaard et al. 1973, p. 33):

1. *Source region temperature* (T_o in Eq. 15.28). Because isotopic distillation depends foremost on the temperature difference between the vapor source region and the ice core site, δ_i varies less if the source and site cool and warm together than if the site temperature changes by itself. Note that the vapor source region can include a broad swath of subtropical and polar oceans and, for Greenland, evaporation from North America.

2. *Strength of evaporative recharge.* An increase of evaporation relative to precipitation in nearby ocean areas (Eq. 15.30) should reduce distillation and raise δ_i values, especially at coastal sites.

3. *Source region location.* The geographical pattern of vapor sources may change along with atmospheric circulation, sea-ice extent, sea surface temperatures, and other factors (e.g., Johnsen et al. 1989; Charles et al. 1994). This alters the effective initial temperature and vapor composition and the pattern of temperature change along transport paths.

4. *Seasonal timing of accumulation.* Isotopes record temperature only when it snows. If more snow falls in summer than winter, mean δ_i values are higher than expected for the mean temperature. A change in the ratio of summer to winter snowfall shifts δ_i. For Greenland, climate models suggest that the ratio of summer to winter snowfall decreases when the climate warms; in very cold climates, almost no snow falls in winter. Thus, if v stands for the fraction of snow in summer, $\partial \delta_i / \partial v > 0$ but $dv/dT_i < 0$, and this factor reduces the value of $\alpha_\star$ (Fawcett et al. 1997; Krinner et al. 1997; Masson-Delmotte et al. 2005). Removal of snow by winds, if concentrated in certain seasons, has the same effect as a change in the seasonal timing of snowfall (Fisher et al. 1983).

5. *Warm-weather bias.* On the polar ice sheets, snowfall typically occurs in warm weather for the time of year. As with the seasonal timing, a change in this bias shifts δ_i. Based on experiments with a climate and isotope model, Sime et al. (2008) predict that such a change will prevent δ_i from increasing very much in Antarctica as climate warms over the next century.

6. *Cloud properties.* Isotopic fractionation in the atmosphere depends on cloud properties such as the degree of supersaturation of water vapor and whether condensation forms liquid droplets or ice crystals. How these properties relate to temperature might change with the climate. For example, the prevalence of ice crystals depends on the abundance of wind-borne particles that act as nucleation sites. Isotopic distillation might therefore proceed differently in a dusty atmosphere than in a clean one (Fisher 1991). Over a certain temperature range the condensate in clouds exists in both solid and liquid forms. Their relative abundances affect the supersaturations and fractionations (Ciais and Jouzel 1994) and might change with atmospheric chemistry or dynamics.

7. *Characteristics of cyclones.* Cyclonic storms are responsible for much of the moisture transport to, and precipitation on, polar ice core sites. An increase in the vigor of these storms might increase the overall diffusive character of moisture transport and hence decrease the mean isotopic distillation. The vertical structure of cyclones might also change (Holdsworth 2001). Cyclones juxtapose different air masses and mix moisture at the boundaries between them. Measurements of snow on high mountains reveal a pronounced change in the trend of isotopic values with altitude as a certain elevation is crossed, a pattern that reflects the layered structure of cyclones (Holdsworth 1991).

8. *Postdepositional alteration.* After snow accumulates on the ground, but before it is deeply buried, water molecules in the ice crystals and pore spaces can exchange with vapor in the overlying atmosphere. At sites with low snowfall rates and hence slow burial, sublimation and condensation can alter the mean isotopic value of the accumulated ice (Neumann and Waddington 2004; Neumann et al. 2005).

9. *Cloud and ground temperatures.* Isotopes of precipitation depend on temperatures in the clouds where condensation occurs. Temperatures at the ground surface can differ significantly from temperatures at cloud height. The ground is sometimes warmer, because of the normal atmospheric lapse rate, and sometimes colder, because inversion layers develop by radiative cooling of the surface in calm weather. Because borehole temperatures, firn densification rates, and other processes depend on surface temperatures, the difference between cloud and ground temperatures might be significant in some analyses.

Experiments with climate models can be used to explore how each of these factors impacts the value of $\alpha_\star$ and also generates an uncorrelated signal $\Theta(t)$ in Eq. 15.32. The most desirable approach, however, is to reconstruct variations directly from measurements of an ice core. Unfortunately, such an approach is usually not possible because most of the factors leave no discernible signature. The exception is source region temperature, which can be estimated by analyzing both $\delta^{18}O$ and δD together, as discussed in the next section. But first, Table 15.2 lists some empirical values for $\alpha_\star$ at ice core sites, determined by comparing isotopic changes to temperature changes constrained by independent indicators. All of these pertain to climate changes, not to seasonal cycles or year-by-year variations.

Table 15.2: Empirical values of $\alpha_\star$ for $\delta^{18}O$.

Location	Time period	$\alpha_\star$ (‰ °C^{-1})	Reference
Antarctica:			
Peninsula	Twentieth century	0.56[‡]	(1)
coastal mainland (Law Dome)	Last 1000 years	0.6	(2)
central mainland (Vostok)	LGM-Holocene	0.9	(3)
Central Greenland			
(abrupt warmings)	38–64 kyr b.p.	0.36–0.46	(4)
(abrupt warming)	11.5 kyr b.p.	0.2–0.5	(5)
	LGM-Holocene	0.40[†]	(6)
	8.2 kyr event	0.41–0.82	(7)
	Mid to late Holocene	0.25	(6)
	Last 200 yr	0.46	(6)

[†] Given as 0.33, but here the ocean water change has been removed.
[‡] Value originally given for δD has been divided by eight to find $\delta^{18}O$ equivalent.
References: (1) Aristarain et al. (1986); (2) Dahl-Jensen et al. (1999); (3) Calculated from isotope models (Vimeux et al. 2002) but supported by analysis of firn thickness (Blunier et al. 2004); (4) Huber et al. (2006b); (5) Grachev and Severinghaus (2005); (6) Cuffey et al. (1995); (7) Kobashi et al. (2007).

15.5.2.1 Deuterium Excess and Source-region Temperature

Measurements of precipitation worldwide show that $\delta^{18}O$ and δD values for the same sample lie close to a line,

$$\delta D = 8\delta^{18}O + 10\%o, \tag{15.34}$$

referred to as the *meteoric water line* (Craig 1961). Equation 15.26 indicates that the slope is, approximately, the ratio of $[1 + \delta D][\alpha_D - 1]$ to $[1 + \delta^{18}O][\alpha_{18} - 1]$. Both α_D and α_{18} (the fractionation factors for the respective isotope pairs) vary with temperature, but they do so proportionately; hence the linear relationship. But the linear relationship starts to break down in the center of Antarctica; the depletion of deuterium is large enough that $[1 + \delta D]$ deviates significantly from one, decreasing the slope and increasing the deuterium excess. *Deuterium excess* refers to deviations from the line of Eq. 15.34: by definition,

$$d = \delta D - 8\delta^{18}O. \tag{15.35}$$

Deuterium excess is not a fundamental property of the water but rather a convenient way to examine the nonredundant information contained in both oxygen and hydrogen isotope ratios.

Evaporation from the ocean surface – with its rate-dependent fractionation given by Eq. 15.31 – followed by equilibrium fractionation during condensation in the overlying atmosphere together produce a value of $d \approx 10\%o$ in tropical and subtropical precipitation over the ocean (Merlivat and Jouzel 1979). A change of relative humidity or, more strongly, of sea-surface temperature, shifts the value of d in precipitation. A shift occurs not only in subtropical precipitation but also in precipitation all the way to the interior of the polar ice sheets, though the magnitude of the shift varies along the path (Johnsen et al. 1989; Kavanaugh and Cuffey 2003). Evaporation from cold polar oceans leads to a smaller value of d over the ice sheet. In addition, cooling over the ice sheet can increase d; cooling increases distillation and hence reduces the factor $[1 + \delta D]$. This effect is, however, strongly limited by the kinetic fractionation accompanying ice crystal growth in supersaturated air. Only in East Antarctica are depletions strong enough for d to rise above its subtropical value; in central East Antarctica, $d \approx 15\%o$ in modern precipitation.

For ice-core site temperature T_i and vapor source-region temperature T_o, we may write, for small perturbations Δ (Vimeux et al. 2002),

$$\Delta\delta = \gamma_i \Delta T_i - \gamma_o \Delta T_o + \gamma_m \Delta\delta_m \tag{15.36}$$

$$\Delta d = -\beta_i \Delta T_i + \beta_o \Delta T_o - \beta_m \Delta\delta_m. \tag{15.37}$$

The γ and β coefficients, all positive numbers as defined here, have been estimated for central East Antarctica using a wide variety of isotopic models, from simple Rayleigh distillation models to global general circulation models equipped with isotopic tracers. Solving these two relations together, using measurements of both $\delta^{18}O$ and δD from an ice core, gives histories of temperature for both the ice sheet and the source region. Figure 15.14 illustrates the result

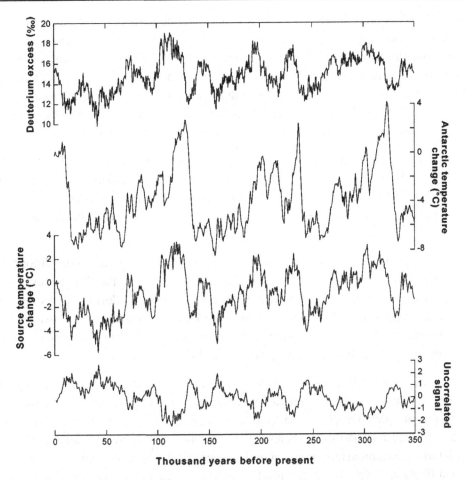

Figure 15.14: A 350-kyr history of deuterium excess from the Vostok ice core (Vimeux et al. 2001), and the temperature changes for Vostok and the vapor-source region inferred from Eqs. 15.36 and 15.37 (Vimeux et al. 2002; Cuffey and Vimeux 2001). Bottom curve shows the contribution of source-temperature variations to the function Θ, the part of the isotopic signal not accounted for by temperature changes at the ice core site or changes in seawater composition.

for the Vostok ice core (Vimeux et al. 2002; Cuffey and Vimeux 2001), and the contribution of source temperatures to the function Θ defined by Eq. 15.32.

15.5.2.2 *Additional Complexities in Interpretations of Records*

All of the preceding discussion concerns mechanistic influences on the relation between ice-core δ values and climate at the time of accumulation. To interpret an ice core record in terms of climate history, additional complexities must be considered:

1. *Elevation of the surface.* The ice thickness and hence the elevation of the ice core site change over time with the accumulation rate, temperature, and positions of the ice sheet

margins (see Chapter 11). Because temperatures decline with increasing altitude, a change of elevation affects the local climate at the site. This effect needs to be removed to discern the regional climate change. Elevation changes can be assumed to be minimal at sites where thin ice sits on bedrock summits, such as small domes near the edges of the main ice sheets. Comparing records from such sites to records from the interior of the ice sheet can therefore reveal the effects of elevation changes (Vinther et al. 2009). Large elevation changes will also be revealed by measurements of total air content (Section 15.4).

2. *Site of accumulation.* With increase of depth in a core, the ice originates progressively further inland and at higher elevations. Thus δ becomes more negative as depth increases, even if the climate doesn't change. This problem can be eliminated by drilling at an ice divide, provided that its position has not changed significantly during the period covered by the core.

3. *Large-scale pattern of ice flow.* The ice-flow pattern may have changed. Although major changes in flow since the end of the last ice age are unlikely, the flow patterns in many existing ice caps may differ significantly from flow patterns during the ice age. In the Canadian Arctic islands, for example, some small ice caps were probably overridden by a larger ice sheet. This would explain the very large increase of $\delta^{18}O$ at the end of the last ice age – about 15‰ – at Barnes Ice Cap (Hooke and Clausen 1982).

4. *Small-scale irregularities of deformation.* Irregular flow can disrupt the age scale in the lowest part of the ice column, generating variations of isotopic composition with depth that can be misinterpreted as climate changes (see Section 10.7). Flow over rough bedrock can produce folds, faults, and shear planes in the ice immediately above. Changes in flow pattern and deformation around inhomogeneities in the ice can also produce folds. Boudinage, which is most likely at an ice divide, can produce gaps in a climatic record but leave the layers in chronological order.

15.6 Additional Techniques of Temperature Reconstruction

15.6.1 Borehole Temperatures

The vertical variations of temperature through an ice sheet, measurable as temperatures in a borehole, preserve a direct record of climatic temperature changes at the surface. The surface signal propagates downward by diffusion and advection, a process discussed in Chapter 9. Because diffusion smoothes variations, only recent or low-frequency climate changes leave any trace (MacAyeal et al. 1993; Cuffey 2006a).

Such changes can be reconstructed using a heat flow model (Section 9.9 provides additional information about such reconstructions). The surface temperature, serving as a boundary condition, can be adjusted to optimize the match between calculated and measured borehole temperatures. A perfect match, however, can only be made with large and unrealistic fluctuations in the reconstruction; mathematically reversing the diffusion process amplifies uncertainties in measured temperatures and in model variables such as ice velocity and thickness.

The simplest approach that avoids this difficulty is to form a reconstruction as a sequence of mean temperatures in a small number of intervals (e.g., Dahl-Jensen and Johnsen 1986). A better approach is to make the reconstruction proportional to variations in the ice-isotopic composition (Paterson and Clarke 1978; Johnsen 1977b; Cuffey et al. 1995). At cold polar sites, especially, the ice-isotope records contain an enormous amount of information on the timing and magnitude of climate changes at all frequencies; this information should be used. The reconstruction in this case is based on a calibration of the isotope record; an optimal value of $\alpha_\star$ in Eq. 15.32 or a series of such values applied in consecutive intervals. A third method uses a Monte Carlo scheme (Dahl-Jensen et al. 1998). A very large number of temperature histories are constructed at random and applied to the heat flow calculation. Those histories generating a good match – within a specified tolerance – between model and measured borehole temperatures are compiled. This defines a range of acceptable temperature values at each time in the past.

Analyses of borehole temperatures have shown that the ice sheet surface in central Greenland warmed by some 20°C between the Last Glacial Maximum and the Holocene (Cuffey et al. 1995; Dahl-Jensen et al. 1998). They have also quantified recent warming at sites in central Greenland and coastal Antarctica (Alley and Koci 1990; Dahl-Jensen et al. 1999). Borehole temperature reconstructions are one way to determine the isotopic sensitivity parameter $\alpha_\star$ (Section 15.5.2); the values given by references (2) and (6) in Table 15.2 come from this method.

Because the downward flow of ice carries heat, all temperature reconstructions using borehole measurements must also account for variations in ice flow over time. Such variations are closely tied to the history of accumulation rate (Chapter 11), so temperatures and accumulation rates are ideally inferred together (e.g., Cuffey and Clow 1997).

15.6.2 Melt Layers

In the percolation zone of a glacier, meltwater penetrates into the snow and refreezes to form a layer of coarse firn and an irregular pattern of ice layers, glands, and lenses. The amount of melting each year probably depends on the duration and magnitude of warm weather in summer, as measured by the positive degree-day index (Section 5.4.2.1). This index typically correlates with the mean, and perhaps the maximum, summer temperature. In the interior of the Greenland and West Antarctic ice sheets, occasional melt events produce single ice layers of a few millimeters' thickness. Experiments and weather station data suggest that such layers form in West Antarctica in warm weather events with a positive-degree-day value exceeding 1°C-day (Das and Alley 2005). Here, too, the likelihood of such events increases with the mean summer temperature.

Down to a certain depth, ice formed by refreezing of meltwater can be distinguished from that formed by firn compaction; it usually contains few or no bubbles. The variation with depth of the percentage of ice originating as meltwater, or of the fraction of annual layers containing a melt layer, thus provides a record of past variations in summer warmth.

In Canadian Arctic ice caps and the Greenland Ice Sheet, abundant melt layers in early Holocene strata show this was a time of warm summers (Koerner and Fisher 1990; Alley and Anandakrishnan 1995). The frequency of melt decreased through the Holocene, coincident with a decrease of maximum summer insolation related to orbital precession. Melting increased again in the most recent century as the climate warmed. The melt-layer record from Devon Island ice cap, for example, shows that the period since 1925 had the warmest summers in the past 700 years (Koerner 1977). Summers were generally cold between about 1600 and 1860, with exceptionally cold periods from 1685 to 1710, and from 1820 to 1860. During this last period various British naval expeditions were trying to traverse the Northwest Passage; the cold summers and consequent severe ice conditions undoubtedly contributed to their failure.

15.6.3 Thermal and Gravitational Fractionation of Gases

As explained in Section 15.3, the size of a thermal fractionation anomaly in gas isotopes reveals the magnitude of the rapid temperature change that caused it. Using this technique, Grachev and Severinghaus (2005) demonstrated that central Greenland warmed by $10 \pm 4°C$ in just a few decades at the end of the Younger Dryas cold interval, 11.6 kyr ago. The same technique showed that an even larger warming, about 16°C, occurred 70 kyr ago (Lang et al. 1999; Landais et al. 2004).

Section 15.3 also discussed gravitational fractionation of the constituents of firn air. Because of this process, measurements of gas isotopes can be used to estimate past firn thicknesses. Through the mechanisms of densification discussed in Chapter 2, firn thickness depends, in turn, on both temperature and accumulation rate. If accumulation rate is known, the temperature can therefore be estimated from reconstructed firn thickness (Schwander et al. 1997; Leuenberger et al. 1999; Huber et al. 2006b). Results from this technique should be regarded as rough estimates, because models of densification are uncertain.

15.7 Estimation of Past Accumulation Rates

Fractional changes in accumulation rate from year to year are revealed directly by differences in measured thicknesses of adjacent annual layers; the strain factor in Eq. 15.3 – the correction factor accounting for cumulative thinning since deposition at the surface – can normally be assumed constant at this scale. As one example, annual layer thicknesses observed in ice cores from central Greenland increase abruptly across the climate transition at the end of the Younger Dryas cold period, 11.6 kyr ago. Layer thicknesses double within about ten annual layers; thus the accumulation rate at this site must have doubled within a decade (Alley et al. 1993).

On the other hand, to infer an absolute value for accumulation rate or changes over millennia requires multiplying layer thicknesses by the strain factor in Eq. 15.3. Factors determining the net vertical strain – the thinning – are discussed in Section 15.2.1. The size of the correction factor generally increases exponentially with age. Simple models of strain can provide a useful rough estimate for thinning, but in general a detailed model of ice flow between the ice divide and the core site, accounting for variations over time, must be used. Because ice flow interacts with both accumulation rate and temperature (Chapter 8), model-based reconstructions of accumulation rate should ideally be linked to reconstructions of climatic temperature (Cuffey and Clow 1997; Huybrechts et al. 2007). Regardless of the level of sophistication of the model, uncertainties in strain estimates increase drastically in the bottom 30% or so of the ice thickness; alternative approaches must be used to reconstruct accumulation rates.

A different approach, independent of ice flow models, relies on dilution of the cosmogenic beryllium isotope, ^{10}Be. ^{10}Be is produced in the atmosphere by cosmic radiation (Section 15.2.2.2). The concentration of ^{10}Be in the ice varies with the accumulation rate; faster accumulation dilutes the isotope. But ^{10}Be concentrations also vary with the rate of production and hence the cosmic-ray flux, which is modulated by solar activity and the strength of the geomagnetic field. Measured ^{10}Be concentrations in ice show the sunspot cycle, a 50% increase during the Maunder minimum of solar activity, and occasional peaks reflecting increased production (Beer et al. 1985; Raisbeck et al. 1981, 1987, 2006). But variations in solar activity cannot explain why the concentration in ice-age ice is two to three times that in Holocene ice. Moreover, in central East Antarctica the concentrations in ice-age ice correlate with variations of ice-isotope ratios, with highest concentrations during the coldest periods – the times when the least precipitation is expected. ^{10}Be concentrations therefore appear to provide valid estimates of long-term trends in accumulation rate, at least in the dry central part of East Antarctica (Raisbeck et al. 1987) and probably in central Greenland, too (Yiou et al. 1997). The ^{10}Be concentration should work better as a proxy for accumulation rate at dry sites than at high-precipitation sites (Alley et al. 1995c). At the dry sites, much of the ^{10}Be falls to the surface directly from the atmosphere. As precipitation increases, more of the ^{10}Be arrives at the surface attached to snowflakes, a process that counteracts the diluting effect of increased precipitation (in terms defined later, in Section 15.10, "wet deposition" increases relative to "dry deposition").

A third method has been used to estimate past accumulation rates in Antarctica. In the interior of the continent, the present precipitation rate appears to be determined by the water vapor content of the overlying atmosphere (Robin 1977). This content, in turn, depends on the saturation vapor pressure and hence the temperature at which precipitation forms. But the same temperature correlates with the δD and $\delta^{18}O$ of precipitation (Section 15.5), suggesting that the ice isotopes and the accumulation rate should covary with time. Thus accumulation rate can be approximated as a correlate of the ice-isotopic composition, assuming that their present relationship applied in the past (Jouzel et al. 1993).

This latter approach is useful for analyses such as ice sheet modelling that must apply a continuous and long-term history of accumulation rate; the method is far superior to assuming a constant rate, for example. Of course this method also defeats one of the major goals of ice core analysis: to determine how different climate variables change in relation to one another. Moreover, precipitation and temperature are unlikely to correlate strongly in places like Greenland and coastal Antarctica where cyclonic storms dominate moisture transport in the atmosphere. Indeed, over the small changes of climate that occurred during the Holocene, accumulation rate and temperature in central Greenland did not correlate on timescales of decades to a few centuries (Kapsner et al. 1995). At longer timescales they even varied inversely during the Holocene (Cuffey and Clow 1997).

A newly proposed method for estimating accumulation rates uses measurements of the abundance of bubbles. The configuration of grains and bubbles at pore close-off appears to be independent of the mean size of the grains (Gow 1969). Specifically, the ratio of the number of bubbles to the number of grains, in a given volume of ice, is approximately constant; there are about two bubbles for each grain (Spencer et al. 2006). Assuming that the number of bubbles per volume of ice remains fixed after close-off – a reasonable assumption for spherical bubbles at depths too shallow for clathrate formation – the grain size at close-off can therefore be inferred by counting bubbles. This grain size, in turn, depends on the temperature in the firn (which controls the rate of grain growth; Section 3.3.3.1) and the age at close-off. The latter depends on the temperature and the accumulation rate (Chapter 2). Thus, if past temperatures are known, measured bubble densities can be used to calculate past accumulation rates. The calculation must be done with a model for firn densification. Using known accumulation rates and ice samples from shallow cores, Spencer et al. showed that the method discriminates present-day large accumulation rates from small ones. The data are noisy, nonetheless, and more work is needed to validate the method.

15.8 Greenhouse Gas Records

15.8.1 Histories of Atmospheric Concentration

At cold polar sites, the concentrations of carbon dioxide (CO_2), methane (CH_4), and nitrous oxide (N_2O) initially trapped in air bubbles closely match their atmospheric concentrations at the time of bubble formation. This is confirmed by comparisons between ice core measurements and samples taken directly from the atmosphere in the twentieth century (Etheridge et al. 1992, 1996). It is also confirmed by extensive analyses of the processes of gas transfer in firn and the associated chemical anomalies (Section 15.3). At sites too cold for melting, such anomalies are far too small to explain most of the variation of gas concentrations measured along ice cores.

Ice core measurements have revealed the timing and magnitude of atmospheric CO_2, CH_4, and N_2O changes caused by the industrial revolution and the recent intensification of agriculture (Figure 15.15). When atmospheric concentrations rise rapidly, as in recent decades, the increase of concentrations in new bubbles lags by a few decades because the gases must diffuse through the firn. The most precise view of recent atmospheric changes is thus achieved by combining measurements of air trapped in bubbles with measurements of air sucked out of pore spaces in the overlying firn, at sites with rapid accumulation (Battle et al. 1996).

The history of CO_2 concentrations is now known back to 800 kyr before present, from measurements on a core from Dome C (Lüthi et al. 2008), combined with measurements for the

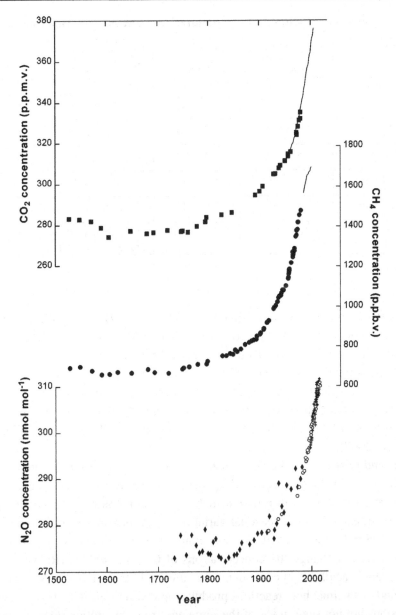

Figure 15.15: Changes in recent centuries of atmospheric greenhouse gas concentrations inferred from ice core samples (filled squares, circles, and diamonds), direct atmospheric samples (continuous curves and, for N_2O, the plus signs), and firn-air samples (open symbols on N_2O plot). Data sources are: CO_2, Etheridge et al. (1996); CH_4, Etheridge et al. (1992) and Blunier et al. (1993); N_2O, Battle et al. (1996).

Table 15.3: Measured CO_2 concentration: LGM vs. early Holocene.

Site	CO_2, 20–25 kyr (ppmv)	CO_2, 9–11 kyr (ppmv)	References
Taylor Dome	185–205	260–270	Smith et al. (1999) Indermühle et al. (1999, 2000)
Dome C	185–195[†]	260–270	Monnin et al. (2001)
Dome F	185–200	260–265	Kawamura et al. (2007)
Vostok	185–195	260–265	Petit et al. (1999)
Siple Dome	185–200	265–290	Ahn et al. (2004)
Byrd	185–205	245–285	Neftel et al. (1988) Staffelbach et al. (1999)

[†] Only for period 20 to 22 kyr.

most recent 350 kyr from Vostok and Dome Fuji (Petit et al. 1999; Kawamura et al. 2007). All three of these cores are from East Antarctica. Six separate Antarctic cores show the rise of CO_2 concentrations from the LGM to the Holocene (Table 15.3). The good correspondence between measurements from multiple sites shows that the overall amplitude and form of CO_2 variations are known well, although the accuracy of measurements needs improvement. Sporadic differences between data sets of up to 20 ppmv arise from problems with techniques for extracting all the gas from the ice or from dissolution in small amounts of meltwater (e.g., Ahn et al. 2004; Kawamura et al. 2007).

At warmer and dustier sites, however, ice-core CO_2 concentrations do not accurately preserve the atmospheric values. Because CO_2 dissolves readily in water, dissolution from the air enhances the concentration in refrozen melt layers. For this reason, the Holocene record from Dye 3 in Greenland shows seasonal variations with maxima in the summer melt layers (Stauffer et al. 1985). Furthermore, if the ice contains enough carbonate dust to make it alkaline, chemical reactions can change the concentrations of CO_2 in bubbles (Raynaud et al. 1993). Oxidation of organic acids might contribute (Tschumi and Stauffer 2000). Much of the ice-age ice in Greenland is alkaline; here reactions produce rapid variations in CO_2 in some ice-age ice. Similar variations are not seen in ice of the same age from Antarctica (Oeschger et al. 1988). The atmospheric concentration of CO_2 is of course almost the same, worldwide. All the ice in Antarctica is acidic. For this reason, and also because the surface never melts, the mainland of Antarctica is the best place for ice core CO_2 measurements.

Over the 800 kyr preceding the industrial revolution, the CO_2 concentration never exceeded 300 ppmv; in comparison the current concentration (year 2008) has attained 385 ppmv. The most striking feature of the long CO_2 history is the pronounced variation through the ice-age climate cycles (Figure 15.16); for early reports of this discovery, see Neftel et al. (1982) and

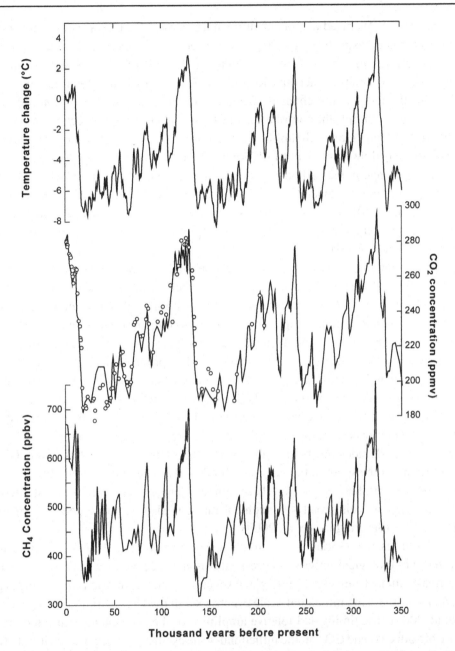

Figure 15.16: A 350-kyr history of ice-age cycling from the Vostok ice core (Petit et al. 1999; Vimeux et al. 2002). Points on the CO_2 plot show measurements from the Dome Fuji core for comparison (Kawamura et al. 2007).

Barnola et al. (1987). Averaged over a few millennia or longer, the CO_2 generally covaries with temperature indicators, rising by 60 to 100 ppmv from cold glacial periods to warm interglacials. The covariation is strong, with r^2 of 0.84 to 0.89 for the last 350 kyr (Cuffey and Vimeux 2001). In contrast, the CO_2 variations account for only 30% to 40% of the global radiative forcing responsible for the temperature changes (Section 13.3.2). Thus the covariation reflects, to a significant degree, not simply the contribution of CO_2 to the greenhouse effect but also processes by which reduced temperatures themselves drive removal of CO_2 from the atmosphere. The mechanism for CO_2 drawdown is not fully known, but must involve increased storage in the ocean through a change in the carbonate solution chemistry (Sigman and Boyle 2000; Archer et al. 2000). During the last glacial period, CO_2 concentrations fluctuated by about 20 ppmv over timescales of 5 to 10 kyr. These variations correlate with Antarctic temperatures rather than northern hemisphere temperatures, indicating an important role for processes in the Southern Ocean (Ahn and Brook 2008).

Figure 15.16 shows too the long-term variation of methane concentrations from the Vostok core (Petit et al. 1999); it has also been measured back to 800 kyr at Dome C (Loulergue et al. 2008). As with CO_2, CH_4 varies in synchrony with glacial-interglacial climate cycles, but its correlation with long-term temperature changes is comparatively weak. The atmospheric residence time of CH_4 is only about one decade, so concentration variations over decades and longer closely reflect changes in the source flux. The primary methane sources are biotic emissions from wetlands in the tropics and northern-hemisphere continents. Reduced outgassing of methane during the glacial climates resulted from some combination of drying, freezing, and cooling of the source regions, and coverage of boreal wetlands by ice. During the last glacial period, CH_4 fluctuated over centuries and millennia in correlation with temperature in Greenland, although with a few decades' lag behind abrupt warmings (Brook et al. 1996, 2000). Given the widespread nature of the sources, the large-amplitude CH_4 fluctuations imply that a significant fraction of the northern hemisphere felt these climate changes; they were not restricted to the North Atlantic region. This broad geographical range is also demonstrated by variations of atmospheric $\delta^{18}O_2$ (Severinghaus et al. 2009).

Like CH_4, nitrous oxide variations record changes in the strength of sources. N_2O emissions are a byproduct of biological activity in terrestrial soils and, secondarily, in the ocean. The N_2O concentration fluctuated with glacial-interglacial cycles (low concentrations prevailing in glacial periods) and with the same millennial-scale events seen in the CH_4 record (Sowers et al. 2003; Spahni et al. 2005). The timing and relative amplitude of the N_2O changes differed, however, from those of both CH_4 and CO_2. In the millennial-scale events, N_2O began to change before the onset of spikes in CH_4 and continued to change after their terminations, presumably because of contributions from ocean sources. Understanding how the flux of N_2O emitted by soils varies is now a task of considerable importance. One much-discussed strategy for reducing anthropogenic CO_2 emissions is to replace fossil fuels with biofuels derived from crops. But increased crop production would likely lead to significant increases of N_2O emissions (Crutzen et al. 2008). Per molecule, N_2O contributes significantly more to the greenhouse effect than does CO_2.

In some periods, N_2O measurements in Antarctica appear to be corrupted by artefacts related to high concentrations of dust in the ice. Localized spikes of high N_2O concentrations are also observed in Greenland cores (Flückinger et al. 2004). Such anomalies might arise from microbial activity (Rohde et al. 2008), which is known to perturb the gas contents of some tropical ice cores (Campen et al. 2003).

15.8.2 Isotopic Compositions of Greenhouse Gases

The isotopic compositions of gases contain additional information about their sources; in some cases, different sources emit compounds with distinct isotopic signatures. Measuring isotopes of trace gases is difficult because of the high precision required, but advances in analytical and sampling techniques have made them possible. The $^{13}C/^{12}C$, $^{14}C/^{12}C$, and D/H compositions of methane have all been measured across the climate warming at the end of the last ice age (Schaefer et al. 2006; Petrenko et al. 2009; Sowers et al. 2006; Fischer et al. 2008). The D/H ratio decreases, suggesting enhanced wetland sources and a relative decrease of fossil hydrocarbon sources. The $^{14}C/^{12}C$ decreased, but only slightly. Both of these measurements negate a hypothesis proposed by Kennett et al. (2003) that marine sediments released large quantities of fossil methane; such a change would increase D/H and significantly decrease $^{14}C/^{12}C$. The $^{13}C/^{12}C$ decreased significantly, indicating that boreal sources turned on when the climate warmed.

For N_2O, neither the $^{15}N/^{14}N$ nor the $^{18}O/^{16}O$ composition changed significantly over the last 30 kyr (Sowers et al. 2003). This probably means that the end of the ice age did not alter the proportion of N_2O emitted by the marine and terrestrial biospheres, a surprising result. The $^{13}C/^{12}C$ composition of CO_2 did not change during deglaciation either, showing that the rise of CO_2 concentration was due to the release of carbon dissolved in the ocean, rather than a change in the terrestrial biosphere (Smith et al. 1999) – a conclusion already implied by the large quantity of CO_2 involved.

15.9 Gas Indicators of Global Parameters

15.9.1 Global Mean Ocean Temperature

Gases dissolve more readily in cold water than in warm water – the solubility varies inversely with temperature. Some amount of gas should therefore shift from the atmosphere to the ocean when the global climate changes from interglacial warmth to ice-age chill. This effect will be more pronounced for heavier, more soluble gases than for lighter, less soluble ones. Thus the relative abundance of different gases in the atmosphere will change by an amount dependent on the global mean temperature change of the ocean. In principle, measurements of gas ratios in an ice core could reveal the history of mean ocean temperature, provided that the atmospheric concentrations of the gases are not affected by other factors. Headly and Severinghaus (2007) developed such a method using the unreactive gases Kr and N_2, the former being more soluble.

They measured differences of the Kr to N_2 ratio between the Last Glacial Maximum (20 kyr b.p.) and the present in air bubbles in ice from central Greenland. The LGM ratio was $1.34 \pm 0.37 \permil$ lower than the modern one. With a correction for gravitational fractionation, this indicated a mean temperature change of the whole ocean of $2.7 \pm 0.6°C$. This new method promises to yield extraordinary information about changes in the global-scale state of the ocean.

15.9.2 Global Biological Productivity

As discussed in Section 15.2.2.3, the uptake of oxygen by the biosphere in respiration preferentially removes light oxygen isotopes, leaving the atmosphere with a positive $\delta^{18}O_2$ anomaly. An increase of the global rate of biological productivity would increase the anomaly. Because $\delta^{18}O_2$ can be measured in ice cores, it might be possible to estimate biological productivity in the past. But a second process modifies the isotopic composition of O_2. Photochemical reactions profoundly change the composition of stratospheric O_2, which mixes down into the troposphere. To make use of the $\delta^{18}O_2$ measurements, this effect must be taken into account.

The key is to acquire measurements of the other heavy oxygen isotope, ^{17}O (Luz et al. 1999). It behaves in similar fashion to ^{18}O, producing a positive anomaly in $\delta^{17}O_2$ in the troposphere. The isotopic fractionations in the biosphere depend on mass, so the ^{17}O anomaly is smaller than the ^{18}O anomaly. But the stratospheric processes affect both isotope pairs about equally. Thus by measuring both $\delta^{17}O_2$ and $\delta^{18}O_2$ (together giving the "triple isotopic composition") the two effects can be separated.

Interpreting such measurements in terms of global biological productivity is not a straightforward task. It must be done using a model accounting for spatial and seasonal variations in type of vegetation, climate, and isotopic composition of rainwater – all of which influence the isotopic effects of photosynthesis and respiration. Blunier et al. (2002) and Landais et al. (2007) made analyses of this type, with a focus on asking what change in global biological productivity can explain the measured difference in atmospheric oxygen isotopes between the Last Glacial Maximum and the present. These two analyses found that productivity was lower at the LGM than at present by 76–83% or 60–75%, respectively.

15.10 Particulate and Soluble Impurities

Glacial ice contains a record of past variations in atmospheric fallout, although the concentrations are usually extremely small (Delmas 1992; Legrand and Mayewski 1997). The major sources are terrestrial dusts, the ocean surface, and gases emitted by biological activity. Anthropogenic pollution has contributed significantly since the beginning of the industrial era. Extraterrestrial matter has also been found, consisting of both interplanetary dust and the products of ablation of meteorites as they fall through the Earth's atmosphere.

In theory, measurements of impurity concentrations in ice can be deciphered to reveal the histories of atmospheric composition and circulation: specifically, the atmospheric burdens of a great variety of aerosols and the wind-borne fluxes of materials from source regions to the ice sheets. In practice, the extraordinary diversity of factors contributing to the records – encompassing the geography and strength of sources, atmospheric dynamics, processes of nucleation and scavenging in the atmosphere, incorporation in the ice, and postdepositional alteration – often preclude strong interpretations. Unlike the gases discussed in Sections 15.2.2.3 and 15.8, most of the impurities remain in the atmosphere for only days to weeks and are therefore not globally mixed. The records from Antarctica and Greenland thus give unique information about portions of each hemisphere.

Impurities travel from atmosphere to ice sheet surface either attached to snowflakes or as independent aerosols (Junge 1977). These two modes are called *wet deposition* and *dry deposition*, respectively. The latter mode dominates only in very dry regions, such as present-day central East Antarctica. The net flux of impurity into the ice can be calculated from core measurements as the product of concentration in the ice, C_i, and ice accumulation rate, $\dot{b}$. The simplest plausible model describes this net rate as the sum of separate dry and wet deposition fluxes, both of which are assumed to increase in proportion to the atmospheric concentration of the impurity, C_a:

$$\dot{b}C_i = k_d C_a + k_w \dot{b} C_a \tag{15.38}$$

where the coefficients k_d and k_w express the efficiencies of dry and wet deposition, respectively (Legrand 1987). If an impurity reaches the surface mostly by dry deposition, snowfall acts as a dilutant and the ice-core concentration C_i varies inversely with the accumulation rate over time. If wet deposition predominates, C_i varies independently of accumulation rate but in direct proportion to the atmospheric concentration C_a. In the general case, a plot of $\dot{b}C_i$ versus $\dot{b}$ from ice core data will, according to this simple model, define a line whose slope and intercept are both proportional to C_a during the period of accumulation, and whose intercept indicates the importance of dry deposition. By how much C_a changed across a climate transition can be determined from the ratios of slopes and intercepts *if* the values k_d and k_w do not themselves change with the climate. For many species, this is a dubious assumption (Davidson 1989). On the other hand, ice core records show that the fluxes of some impurities changed by factors of ten or more. Such large changes are unlikely to be explained by atmospheric washout processes; instead, they must primarily reflect variations of source fluxes. Thus valuable information can be obtained about the history of source fluxes even if atmospheric concentrations cannot be discerned.

Most of the impurities in ice were once atmospheric aerosols. Aerosols raining onto a glacier surface are of two types: *primary aerosols*, substances like continental dust and sea spray that are picked up by winds and become aerosols directly; and *secondary aerosols* which form in the atmosphere from gases. In addition to both of these aerosol categories, some soluble gases in the atmosphere (such as HCl, NH_3, HNO_3, and H_2O_2) adsorb directly onto ice. They are measured in a core as soluble ions rather than as constituents of trapped air in bubbles. The chemistry of primary aerosols sometimes changes in the atmosphere because of reactions. In particular, acids react with salts in water droplets, separating the salts into metal ions and HCl gas. In modern snow on Greenland and Antarctica, the secondary aerosols – especially acids – dominate the chemistry, except close to the coasts where sea salts are abundant.

Detailed measurements of impurities have been made through the last glacial climate cycle in Greenland (Mayewski et al. 1997) and through eight cycles in East Antarctica (Wolff et al. 2006). Steffensen et al. (2008) achieved subannual resolution in measurements across rapid climate transitions that occurred between 11 and 15 kyr ago in north-central Greenland.

15.10.1 Electrical Conductivity Measurement (ECM)

The impurities in polar ice determine its electrical properties. This observation is the basis of a quick way of measuring impurity content along a core (Hammer 1980; Neftel et al. 1985; Taylor et al. 1993). Two electrodes about 10 mm apart and with a high potential difference between them (usually 1250 V) are moved along the clean flat surface of a core. The current, at fixed temperature, is measured (a quantity referred to as "ECM"). Because volume conduction predominates over surface conduction, the measured current is closely related to the D.C. electrical conductivity of the ice. The current increases with the concentration of acids (Wolff et al. 1997), but is still measurable even in alkaline ice. The relation between ECM values and acidity, determined by measuring the pH of melted samples, is nonlinear. Primarily, ECM is useful not as a measure of absolute acidity but as a generic, quick, and reproducible measure of the variations in overall impurity content. Similar to the ice-isotope records, but related to impurity deposition rather than to climate, the ECM gives a continuous view of the timing and scale of environmental changes – a useful framework to which all other measurements on the core can be referenced. ECM measurements also identify volcanic layers as spikes clearly emerging above the varying background.

15.10.2 Primary Aerosols

15.10.2.1 Sea Salts

Many primary aerosols, especially those containing Na, Mg, K, Cl, and some of the Ca and SO_4^{2-}, originate as sea salts (Legrand 1987; DeAngelis et al. 1987; Herron and Langway 1985). In both Greenland and Antarctica, concentrations of sea salts are much higher in ice-age ice than in Holocene or other interglacial ice. For example, LGM ice in central Greenland contains about six to ten times more marine salt than does Holocene ice (Mayewski et al. 1997). Because accumulation rates were only about three to four times lower, the flux of sea salt to the ice sheet was also enhanced at the LGM, by a factor of two to four. From an analysis based on Eq. 15.38, Alley et al. (1995c) concluded that atmospheric concentrations were enhanced by a factor of about three. In East Antarctica, the flux of sea salt likewise increased by a factor of about three between glacial maxima and interglacials (Wolff et al. 2006).

In one interpretation of these records, the salts are supposed to be incorporated in the atmosphere as sea spray in windy regions of open water. In this case, their deposition onto ice sheets should increase with the speed of winds on the ocean surface and decrease with the extent of sea ice on polar oceans. Because sea ice expands during cold climates, the higher ice age fluxes would reflect stronger winds and slower washout (e.g., Petit et al. 1981; Herron and Langway 1985). In contrast, Rankin et al. (2002) suggested that enhanced salt fluxes reflect *increased* sea ice cover. Observations of newly formed sea ice show that a brine layer accumulates on its top surface and is redistributed in various ways. For example, surface tension draws the brine up onto hoar-frost crystals, called *frost flowers*. These crystals form in great profusion on the sea-ice surface due to the juxtaposition of liquid water in the brine with cold overlying air. Wind scour of such crystals and other salty surfaces produces sea-salt aerosols. A chemical fractionation during the formation of sea ice depletes the brine of sulfate (mirabilite, a hydrous form of sodium sulfate, precipitates out). Such a depletion of sulfate has been measured in aerosols and ice cores from coastal and interior Antarctica, indicating a sea-ice source for the salt, not an open water source. In addition, the concentrations in aerosols reach a seasonal minimum in summer, the time of minimum sea-ice extent, rather than winter (Wagenbach et al. 1998). Thus, sea-salt fluxes to ice core sites most likely indicate the rate of formation of new sea ice rather than the extent of open water, at least in Antarctica (Wolff et al. 2006). With this interpretation, the 740-kyr record from East Antarctica shows a close and consistent relationship between winter sea-ice production and temperature, if averaged over millennia.

15.10.2.2 Continental Dust

Other primary aerosols are microparticles ("dust") picked up from the continents by winds and carried through the troposphere. About 95% of the microparticles have radii in the range 0.1 to 2 μm. Their composition reflects the types of rocks found on continents; most are insoluble particles of silicate rocks (for which Fe, Al, or Si are taken as an index), but some are soluble $CaCO_3$ and $CaSO_4$. Matching the isotopic composition of trace elements in the particles (especially rubidium, strontium, and neodymium) with compositions of possible source materials shows that much of the dust in Greenland originates in central Asia, while Patagonia supplies most of the dust to Antarctica (Biscaye et al. 1997; Delmonte et al. 2008). Antarctic ice contains less dust than does ice in Greenland or Arctic Canada. The concentrations of numerous trace elements derived from rocks and soils can be measured with high-precision techniques (Marteel et al. 2008), but most of this information awaits interpretation.

In both hemispheres, the concentration and flux of dust increased dramatically during the cold glacial periods. Responsible factors include drying and strengthening of winds in source regions and erosion by glaciers. The increase of dust mass flux, by a factor of more than 20 in central Antarctica (Röthlisberger et al. 2002a; Lambert et al. 2008) and a factor of more than 10 in central Greenland (Mayewski et al. 1997), is much too large to reflect only changes in atmospheric circulation or the rate of washout; it must primarily reflect stronger sources (Reader et al. 1999;

Lambert et al. 2008). Likewise, large increases in the fluxes of continental-dust tracers like Fe and non-sea-salt Ca primarily reflect more prolific sources (Wolff et al. 2006).

The ice-age dust in the northern hemisphere contains a large component of $CaCO_3$, probably derived from the exposed beds of ephemeral desert lakes. The $CaCO_3$ neutralizes acids and makes the ice alkaline. Calcium also increases in the ice-age horizons in Antarctica, but the total concentration remains small enough that all the ice is acidic (Hammer et al. 1985).

15.10.2.3 Black Carbon

Because of incomplete combustion, the burning of fossil fuels, vegetation, and other organic materials releases black carbon or soot to the atmosphere. A small concentration – a few nanograms of carbon per gram of ice – appears in recent Greenland firn. McConnell et al. (2007) reported measurements of black carbon from cores spanning the last two centuries. Concentrations were more than twice as large in the period 1890 to 1950 as before or since, reaching a peak enhancement in year 1910 of about ten times. Adding black carbon to snow increases the absorption of solar energy, especially in early summer, contributing to climate warming. This may be one reason for rapid warming of the Arctic observed in the first half of the twentieth century. McConnell et al. also measured the concentration of vanillic acid, a compound released in forest fires. These measurements showed no significant enhancement; the black carbon therefore derived from fossil fuel combustion rather than forest fires.

15.10.2.4 Microbes

Microbes are carried by winds along with mineral particles. The extraordinary worldwide diversity of microbes suggests that they might be valuable tracers of source region location and climatic conditions, and of atmospheric pathways. Differences of cell numbers or genetic types between different layers of ice have been observed in Antarctica, the Andes, Tibet, Greenland, and elsewhere (e.g., Abyzov 1998; Priscu et al. 1999; Zhang et al. 2006; Miteva et al. 2009). How such observations reveal characteristics of climate or source regions has yet to be worked out. Unlike dust particles, microbes in ice can live and multiply, using the small quantities of liquid water in intergranular veins as a habitat and deriving energy and carbon from ions in solution or from particle surfaces (Price 2000; Tung et al. 2006). With some microbes reproducing and perhaps consuming material from others, the difficulty of using microbes as environmental tracers could attain dizzying heights. On the other hand, in polar ice such activity should be severely limited by the small quantities of water and available nutrients.

15.10.3 Secondary Aerosols

15.10.3.1 Sulfur Compounds

One of the important acids in ice is H_2SO_4 (Legrand 1997). It forms in the atmosphere, primarily by dissolution of SO_2 in water droplets. Another important sulfur compound is the acid MSA (methane-sulfonic acid) and its constituent methanesulphonate (MS^-). Both SO_2 and MSA are produced by oxidation of dimethyl sulfide gas (DMS), emitted by organisms in the near-surface ocean. This system is interesting, in particular, because of its potential connection to climate; marine stratus clouds nucleate on H_2SO_4 aerosols, so an increased biogenic production of sulfur gases might increase cloud cover and hence increase reflection of sunlight (Charlson et al. 1987). Originally, it was believed that MSA measurements on ice cores give a relatively direct measure of DMS production and possible related climate effects. The fraction of DMS exuded by the ocean that becomes MSA instead of H_2SO_4 is not fixed, however, complicating interpretations. Moreover, it was found that MSA is lost at low-accumulation sites by evaporation from snow.

Another reason for interest in MSA is that, at some high-accumulation sites in coastal Antarctica, the concentration of MSA correlates strongly with the extent of sea ice (Curran et al. 2003). Apparently, the most active phytoplankton blooms – and hence the largest production of DMS – occur when seasonal sea ice breaks up and melts. More extensive winter sea ice thus leads to larger MSA concentrations in snowfall on the continent. Such a correlation does not appear at all sites, however (Abram et al. 2007).

Although most of the sulfur compounds in ice originate as biogenic gas, some derive from sea salt, volcanic activity, and pollution (Legrand 1997). The best measure of the contribution of biogenic gases to the total sulfur input probably depends on the setting. At high-accumulation sites, one proposed indicator is the MSA fraction, defined as the ratio $MSA/[MSA + SO_4^{2-}]$ (Whung et al. 1994). In Greenland and West Antarctica, the MSA fraction, concentration, and flux were lower during the LGM than the Holocene (Saltzman et al. 1997, 2006). In Greenland, MSA fraction rose slowly from the LGM through most of the Holocene, reaching a maximum about 3.5 kyr ago. This slow response contrasts with the rapid shifts of most other climate parameters during deglaciation. It may reflect an ecological change, such as a change in relative abundances of different marine organisms that produce sulfur compounds.

To avoid problems with postdepositional changes, a better indicator of biogenic sulfur at low-accumulation sites is the flux of non-sea-salt SO_4^{2-} (Wolff et al. 2006). Remarkably, this indicator varies little – by only about 20% – over the entire 740-kyr record from central East Antarctica, suggesting that DMS emissions vary little over time. Furthermore, the small variations that do occur have no clear correspondence to the glacial-interglacial cycles. These observations suggest that the proposed feedback between biogenic sulfur, clouds, and climate is weak or inactive, at least in the ocean sector sampled by East Antarctic cores.

Compared to biogenic emissions, the volcanic supply of sulfur to the ice sheets is small on average, but can be large for a few years following an eruption. Major volcanic eruptions inject large volumes of gases and silicate microparticles into the stratosphere. The microparticles settle out quickly and are often not detectable in ice cores, but the gases are widely dispersed. Among the gases are hydrogen sulphide and sulfur dioxide, which again oxidize to form H_2SO_4 aerosols. These reenter the troposphere and are washed out by precipitation. Eruptions are therefore recorded in the ice sheets by layers of increased acidity; Figure 15.17 illustrates two

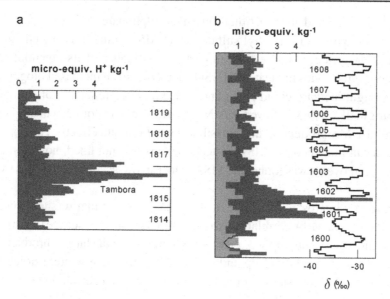

Figure 15.17: Acid fallout from two volcanic eruptions in an ice core from Crête, Greenland: (a) Tambora, Indonesia; (b) source uncertain but possibly Huaynaputina, Peru. Panel (b) also shows seasonal variations in oxygen-isotope ratio (curve on right) and concentration of nitrate ions (lightly shaded curve on left). The Tambora eruption also appears in Antarctic cores (see Figure 15.5). Redrawn from Hammer et al. (1980).

examples. The fallout from Tambora reached a maximum in 1816, which historical records describe as "the year without a summer" in eastern North America; back-scattering of solar radiation by sulfate aerosol caused the cooling. Such volcanic layers help to constrain the age of ice in a core (Section 15.2.2.2).

The recent and substantial increase of SO_2 production by industrial activities, especially coal burning, has produced a clear signature of increased SO_4^{2-} concentrations in Greenland firn, but not at sites in northwestern Canada or the Antarctic (Mayewski et al. 1993).

15.10.3.2 Nitrogen Compounds

Atmospheric nitrogen oxide compounds – often referred to collectively as NO_x – play an important role in the environment. NO_x compounds help the atmosphere to cleanse itself of trace constituents by oxidizing them to other forms: for example, methane is eliminated by oxidation to carbon dioxide. Oxidation of NO_x compounds themselves produces various nitrate (NO_3^-) compounds including nitric acid (HNO_3), another important acid in polar ice (Legrand 1987; Mayewski et al. 1997; Röthlisberger et al. 2000; Wolff et al. 2008).

A wide variety of initial sources contribute to NO_3^- in polar ice, including bacterial emissions, biomass burning, lightning, and photochemical reactions. Some of these sources derive primarily from continents at low to mid latitudes. In addition, a small amount of NO_3^- is present in primary aerosols derived from salts. This very complicated mixed source inhibits clear interpretations of ice-core NO_3^- measurements. So does the complexity of depositional and postdepositional processes. Significant quantities are lost after deposition. The concentrations of NO_3^- in snow sometimes fluctuate substantially from day to day, probably because of trapping of HNO_3 gas by ice surfaces (Wolff et al. 2008; Röthlisberger et al. 2002b). To quote Wolff et al., "it seems difficult to relate the concentrations unequivocally to an important environmental parameter." The case of nitrate illustrates how much work still needs to be done to fully interpret the impurity record in ice cores.

Ammonium (NH_4^+), another important nitrogen compound, also derives from low- and mid-latitude continental sources; primarily emissions from soils, biota, and fires. Whereas most impurities correlate well with major climate parameters like temperature and accumulation rate, NH_4^+ follows its own path. In Greenland, NH_4^+ concentrations in LGM and late Holocene ice are similar, but higher concentrations occur in between, suggesting a correlation with northern-hemisphere summer insolation (Meeker et al. 1997).

15.11 Examples of Multiparameter Records from Ice Sheets

15.11.1 Deglacial Climate Change

Figure 15.18 assembles a few ice-core records from both Greenland and Antarctica that span the end of the last ice age. In Greenland, temperature and accumulation rate changed dramatically, with warmings of about 10°C and doublings of accumulation rate in only a few decades. The corresponding variations of CH_4 concentration – with an amplitude of about 40% of the mean value – demonstrate the hemispheric-scale reach of these changes. Indeed, these events left signatures in other geologic records from sites as diverse as the tropical Atlantic ocean, the Pacific coast of North America, and China. Yet the timing of temperature changes in Antarctica differed greatly; Antarctica warmed progressively while Greenland remained cold, and then held steady or slightly cooled after Greenland's abrupt warming at 14.5 kyr. All of these features can be explained by abrupt reorganizations of circulation and stratification in the Atlantic Ocean, with the climatic consequences amplified by rapid variations of sea ice and shifts of the thermal equator (Ahn and Brook 2008; Chiang et al. 2003; Stocker and Johnsen 2003; Alley et al. 2001). The resemblance of the CO_2 record to Antarctic temperatures suggests that changing conditions in the Southern Ocean governed the return of CO_2 from ocean to atmosphere, a hypothesis supported by analyses of marine sediments (Anderson et al. 2009) and earlier parts of the ice-core record (Ahn and Brook 2008).

15.11.2 A Long Record of Climate Cycling

The longest ice core record so far acquired, from Dome C in East Antarctica, spans some 800 kyr (Figure 15.19). It shows the variations of Antarctic temperature, global greenhouse

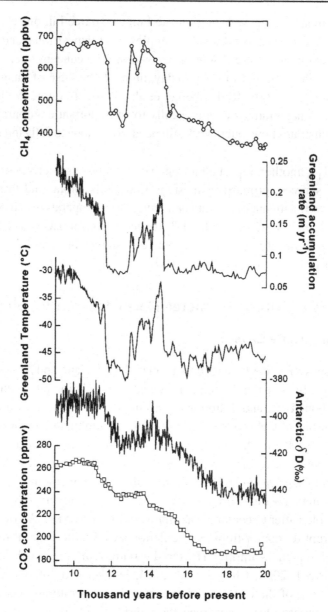

Figure 15.18: Changes at the end of the last ice age: atmospheric methane concentration (Blunier and Brook 2001); accumulation rate and temperature in central Greenland (Cuffey and Clow 1997); δD proxy for temperature in East Antarctica and atmospheric carbon dioxide concentration, both from the Dome C core (Monnin et al. 2001).

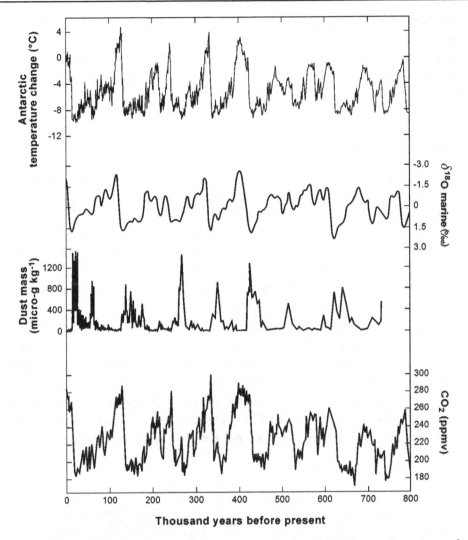

Figure 15.19: 800-kyr history of parameters measured in the Dome C core, East Antarctica, and marine $\delta^{18}O$ from a compilation of sedimentary records (EPICA Community members 2004; Jouzel et al. 2007; Lüthi et al. 2008).

gas concentrations, and other parameters through eight complete glacial-interglacial cycles – about 30% of the entire Pleistocene. The patterns of variation resemble those of marine water isotopes, a proxy for the total volume of the ice sheets (Chapter 13). Variations of ice sheet volume lagged the climate changes by some 5 kyr, however, because of the slow pace of ice sheet growth and decay, which are limited by the rates of snowfall, melt, and outflow to calving margins. A noteworthy feature of Pleistocene history, seen in the figure, is the smaller amplitude of glacial-interglacial changes before 450 kyr, primarily because the interglacials were not as warm or as ice-free as more recent interglacials. Marine isotope records show that before 1 Myr b.p., each glacial-interglacial cycle lasted for only about 40 kyr rather than 80 to 120 kyr (Figure 13.3a). Obtaining ice core records extending into this earlier time period, a major target for future ice core studies, should be possible in some thick-ice regions in East Antarctica.

15.12 Low-latitude Ice Cores

Although some of the parameters measured in polar cores represent global-scale quantities, most represent only the high-latitude climate. Records from ice caps on high mountains in the tropics and subtropics greatly expand the domain of deep-time ice core studies. In recent decades, Thompson et al. (1986, 1989, 1995, 1997, 1998, 2000, 2002, 2006a, b; also Thompson 2000) have taken cores from three sites in the Andes, four in Tibet, and one in East Africa (the summit of Kilimanjaro). Ramirez et al. (2003) took an additional core from the Andes. Figure 15.20 plots their locations. Despite ice thicknesses of only a few hundred meters, at least three and maybe four of these records extend into the last glacial period.

The depth-age relations for these cores are mostly based on annual layer counts in their upper parts (NO_3^-, dust, and $\delta^{18}O$) and, in their lower parts, based on flow models, marker horizons, and correlations to other records. Some of the cores contain enough organic material from remnants of plants and insects to obtain a few ^{14}C dates. The Andean cores seem particularly well dated, with annual layer counts validated against volcanic horizons and the older parts constrained by ^{14}C dates and $\delta^{18}O_2$ measurements. Except for the annually layered upper zones, however, the age scales are not known with the precision achieved in polar cores.

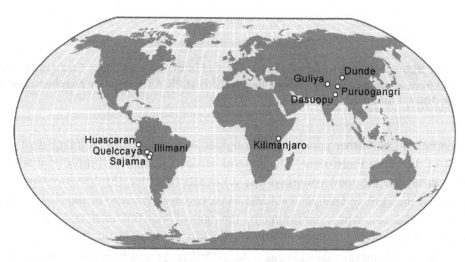

Figure 15.20: Locations of some low-latitude, high-altitude ice core sites.

Interpretations of the low-latitude records are also generally less precise than for their polar counterparts. The ice isotopes, while certainly providing an objective measure of the climate state, may not correlate strongly with temperatures at the core site; at warm sites close to vapor sources the temperature-distillation effect can be swamped by other factors listed in Section 15.5 (Alley and Cuffey 2001; Vimeux et al. 2005, 2008). Although recent variations of accumulation rate can sometimes be reconstructed with clarity from annual layers, their interpretation in terms of precipitation rate is problematic because of significant losses to sublimation and wind erosion (Vimeux et al. 2008). Furthermore, melt at the surface and biological activity at depth can perturb gas compositions (Campen et al. 2003).

Three Andean cores, Sajama and Illimani from Bolivia and Huascaran from northern Peru, show the pattern of climate changes across the end of the last ice age. Together they illustrate that large-amplitude climate changes occurred in the low latitudes; ice $\delta^{18}O$ values shifted by 4‰ to 6‰ from the late glacial to the present at these sites (Vimeux 2009). The patterns of $\delta^{18}O$ variation at Sajama roughly follow the pattern seen in Greenland cores, with two rapid warmings (nominally at 11.5 and 14.5 kyr) and a cold "Younger Dryas" interval from 11.5 to 12.5 kyr. At Illimani and Huascaran on the other hand, the $\delta^{18}O$ variations resemble those from Antarctica. In contrast to the polar regions, Bolivia experienced wet conditions in the late glacial period, recorded in the ice as low dust concentrations and high accumulation rates. Wet glacial-age conditions are also indicated by other geological evidence in the region, such as elevated lake levels. Increased rain-out of atmospheric moisture upstream of the ice core sites can explain the observed shifts of $\delta^{18}O$ (Vimeux et al. 2005, 2008). These shifts probably represent a combination of changes in precipitation, temperature, and moisture source location; all of these factors influence isotopic variability in the present (Vuille et al. 2003).

The Dasuopu core gives a 1-kyr record from a site perched at 7200 m altitude in the Himalayas on the southern edge of Tibet, a location dominated by summer monsoon precipitation. It shows a pronounced recent warming – an increase of ice $\delta^{18}O$ with no comparable change in accumulation rate – and contains peaks in Cl^- and dust concentrations at the times of disastrous historical droughts in A.D. 1790–1796 and 1876–1877. Further north in Tibet, the records from Dunde and Guliya contain pre-Holocene ice. The depth-age relations at both are atypical because of an unusual focusing of deformation into upper layers, as though the deeper ice were trapped in a bowl. Thompson et al. (1997) believe that the Guliya record not only spans the entire last glacial period but also contains basal ice older than 500 kyr. Both of these claims are dubious, as most of the core lacks any rigorous age controls and the features used to date it can be explained in other ways.

A composite, covering the last millennium, of the ice-isotope records from seven sites in Tibet and the Andes gives the curve in Figure 15.21. This was calculated by removing the mean from each record, normalizing the fluctuations to the standard deviation, and averaging with the corresponding curves from the other sites. These data show that late twentieth-century climate was anomalous and different in character from the Medieval Warm Period around A.D. 1000. (Here "climate" should be regarded as a mixture of temperature and hydrological factors that

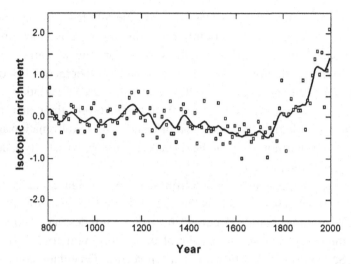

Figure 15.21: Composite of low-latitude ice-isotope records from Thompson's three sites in South America and four sites in Tibet. Data of Thompson et al. (2006b); figure from North et al. (2006).

control the isotope values for the cores; see Vimeux et al. 2008 for a summary of interpretations and their limitations for the Andes.) The pattern is interesting, in part, because it resembles hemisphere-averaged temperature reconstructions for the same period (North et al. 2006). Such reconstructions are difficult to justify on their own because of the sparseness of data in their early parts.

Cores from Kilimanjaro suggest an age for the summit ice cap of about 11.7 kyr, the age of the basal ice (Thompson et al. 2002). This conclusion relies on a simple flow model calibrated by comparing the ice $\delta^{18}O$ record with an isotope record from an eastern Mediterranean cave. One ^{14}C date of 9.4 kyr was also obtained. This ice cap has diminished rapidly over the last century and, if the present rate continues, will disappear in a few decades, a remarkable event considering the suggested antiquity of the ice cap. Stronger constraints on the age of ice are needed, however, before this interpretation is accepted.

15.13 Surface Exposures in Ablation Zones

As an ice sheet transports mass to its margins, ice buried in the accumulation zone is re-exposed at the surface in the ablation zone. (Ablation zones ring the Greenland Ice Sheet and the Canadian Arctic ice caps, but occur in only a few sectors around Antarctica.) One consequence, long recognized, is that the vertical sequence of layers beneath the ice sheet interior will be displayed horizontally at the surface in the ablation zone. The oldest ice outcrops at the margin, and ages progressively decrease inland and upward toward the equilibrium line.

Ancient ice exposed in this situation can be sampled for paleoclimate analyses. Such a surface exposure (sometimes called a "horizontal ice core" or "marginal ice mine") is particularly useful if matched to ice cores in the accumulation zone upstream with continuous and dated layering. Other surface exposures might be dated using their own sequence of globally mixed gases and radiometrically dated volcanic ashes. Marginal ice layers are intensely deformed – a significant drawback – but permit large volume sampling. Small sample sizes strictly limit the measurements made on normal ice cores. Using surface exposures opens the possibility of unprecedented measurements of low concentration variables in ancient ice, including rare trace gases and their isotopic compositions, and rare microparticles and microbes. In addition, some margin sites might expose ice of great antiquity.

Reeh et al. (1991, 2002b) pioneered this approach. At Pakitsoq, a location on the western edge of the Greenland Ice Sheet (Figure 15.1a), they found a semicontinuous 40-kyr record spanning a distance of about 500 m. The record was originally believed to extend further back in time, until comparisons to new deep ice cores from central Greenland showed otherwise. At this site, Petrenko et al. (2006, 2008, 2009) have recently analyzed gases on large samples of ice extracted with chain saws, chisels, and shallow coring devices. Ages were established from sequences of CH_4, $\delta^{18}O_2$, $\delta^{15}N$, and ice-$\delta^{18}O$ compositions. Large-volume samples taken here allowed them to measure, for the first time, the ^{14}C composition of CH_4 across the termination of the last ice age (Section 15.8.2).

Antarctic sites have advantages over Greenland ones, especially for gas studies: no summer melt, lower concentrations of organic and mineral impurities, and lower carbonate content. A few sites have been studied in Antarctica, but not yet for paleoclimate. The sites all lie in "blue-ice zones" where ablation occurs only by sublimation (Section 5.4.5.1). The flank of the summit crater of Mount Moulton, West Antarctica, exposes a 400-m-long blue-ice section containing some 50 ash layers from an active volcano nearby (Dunbar et al. 2008b). Radiometric dating gives ages ranging from 10.5 kyr to 495 kyr, and the layers, though tilted, appear to be in proper order. This easily accessible record thus covers about 60% of the time captured by the Dome C deep core.

Along the Transantarctic Mountains in Victoria Land, the Allan Hills blue ice zones are famous as a source of meteorites (Whillans and Cassidy 1983). The configuration of the ice margin here has changed little throughout the late Quaternary, and perhaps for much longer. Flow continually brings new ice to the surface, where sublimation strips it off. Having nowhere to go, meteorite fragments accumulate on the surface. At another site in this region, Taylor Glacier, Aciego et al. (2007) examined the variation of ice $\delta^{18}O$ and δD along a 28-km-long flow line, using a statistical sampling technique. Comparisons to the nearby Taylor Dome ice core, supplemented with a few measurements of $\delta^{18}O_2$, showed that ages range from 10 kyr b.p. back to perhaps 70 kyr b.p. This site exposes a vast amount of ice from the end of the last ice age; the deglacial transition spans about 10 km on the surface. This is a promising candidate site for gas measurements, although the pattern of disruption of the chronology needs to be worked out in detail.

Further Reading

Interesting topics of ice core research not discussed in this chapter include *in situ* production of cosmogenic isotopes (Lal et al. 1990), records of variations of solar activity (Vonmoos et al. 2006), records of climatic forcing by volcanic activity (Gao et al. 2008), and records of methyl halide gases that play a role in ozone chemistry (Saltzman et al. 2009). A collection of papers on the physics of ice cores, edited by T. Hondoh (2000), discusses many topics in detail.

A book by Alley (2002) introduces ice core paleoclimate studies to a general audience.

A Primer on Stress and Strain

Definition of Stress

The study of stress is concerned with the balance of forces between two parts of a body. For example, at any horizontal plane in a column of rock, the weight of the rock above must be supported by an equal upward force.

Consider a point P within a body, choose some direction $\vec{R}$, and consider a small flat surface of area δA perpendicular to $\vec{R}$ (Figure A.1). The material on one side of the surface exerts a force on the material on the other side. The resultant will be a force $\delta\vec{F}$ acting in some definite direction (often different from $\vec{R}$). The limit of $\delta\vec{F}/\delta A$ as δA tends to zero is called the *stress vector* $\vec{S}_R$ at the point P across the plane perpendicular to $\vec{R}$. The material on the positive side of the surface exerts a force $\vec{S}_R\,\delta A$ on the material on the negative side; conversely, the material on the negative side exerts a force $-\vec{S}_R\,\delta A$ on the material on the positive side. This applies in all directions at every point in the body. The quantity $\vec{S}_R$ is a vector with dimensions of force per unit area. The S.I. unit is 1 Pascal $= 1\,\mathrm{N\,m^{-2}}$ (and $100\,\mathrm{kPa} = 1\,\mathrm{bar}$).

Stress Components

Define a rectangular coordinate system with x-axis in direction $\vec{R}$; δA therefore lies in the yz-plane. The quantity $\vec{S}_R$ can be resolved into components denoted σ_{xx}, τ_{xy}, τ_{xz}. The first suffix indicates the direction perpendicular to the small surface δA, the second the direction in which the component acts. The component σ_{xx}, which points perpendicular to δA, is called a *normal stress* component. By convention a tensile stress is usually taken as positive (because tension causes a material to stretch) and a compressive stress as negative. The components τ_{xy} and τ_{xz}, which are in the plane δA, are called *shear stresses*. Similarly, the stress vector at P across

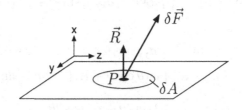

Figure A.1

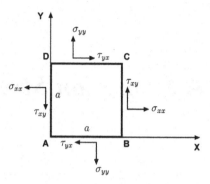

Figure A.2

a plane perpendicular to the y-axis has components τ_{yx}, σ_{yy}, and τ_{yz}. Likewise, components perpendicular to z are τ_{zx}, τ_{zy}, and σ_{zz}. These nine components specify completely the state of stress at the point P. They are the components of a *second-order tensor*. Note that all of the shear stresses τ_{xy}, τ_{yx}, τ_{xz}, ... are also denoted σ_{xy}, σ_{yx}, σ_{xz}, ..., and the full stress tensor called σ.

To specify the state of stress within a body, we need to know the nine components at every point. In fact, only six components are independent. Figure A.2 shows the stresses on the sides of a small square in the xy-plane. Associated with each stress is a force per unit length (the unit length being perpendicular to the page). For example, a force $a\sigma_{xx}$ pulls the square in the x-direction by acting on side BC. An opposing force $a\sigma_{xx}$ pulls the square in the negative x-direction by acting on side AD. Likewise, a force $a\tau_{xy}$ acting on BC pushes the square in the y-direction, while a force $a\tau_{xy}$ acting on AD pushes the square toward negative y. To prevent rotational acceleration of the element, the tangential forces on adjacent faces must balance: thus, $\tau_{xy} = \tau_{yx}$. In three dimensions, furthermore, $\tau_{xz} = \tau_{zx}$ and $\tau_{yz} = \tau_{zy}$. Thus, the value of each shear stress component is independent of the order of its subscripts.

Rotation of Axes

The stress components depend on the definition of axes. A rotation of axes by an angle θ from the original x- and y-axes changes the stress components as follows. Consider a small triangular element as shown in Figure A.3. Balance of forces in the x- and y-directions gives

$$[a\sigma_{x'x'}]\cos\theta - [a\tau_{x'y'}]\sin\theta = \sigma_{xx}[a\cos\theta] + \tau_{xy}[a\sin\theta] \qquad (A.1)$$

$$[a\sigma_{x'x'}]\sin\theta + [a\tau_{x'y'}]\cos\theta = \sigma_{yy}[a\sin\theta] + \tau_{xy}[a\cos\theta], \qquad (A.2)$$

and so

$$\sigma_{x'x'} = \sigma_{xx}\cos^2\theta + \sigma_{yy}\sin^2\theta + 2\tau_{xy}\sin\theta\cos\theta \qquad (A.3)$$

$$\tau_{x'y'} = [\sigma_{yy} - \sigma_{xx}]\sin\theta\cos\theta + \tau_{xy}[\cos^2\theta - \sin^2\theta]$$

$$= \frac{1}{2}[\sigma_{yy} - \sigma_{xx}]\sin 2\theta + \tau_{xy}\cos 2\theta. \qquad (A.4)$$

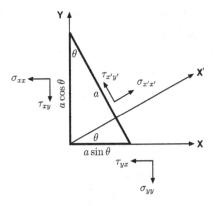

Figure A.3

Replacing θ by $\theta + \pi/2$ in the formula for $\sigma_{x'x'}$ gives the final component

$$\sigma_{y'y'} = \sigma_{xx} \sin^2 \theta + \sigma_{yy} \cos^2 \theta - 2\tau_{xy} \sin \theta \cos \theta. \tag{A.5}$$

Note that $\sigma_{x'x'} + \sigma_{y'y'} = \sigma_{xx} + \sigma_{yy}$. In other words, the quantity $\sigma_{xx} + \sigma_{yy}$, or, in three dimensions, $\sigma_{xx} + \sigma_{yy} + \sigma_{zz}$, is *invariant* when the axes are rotated.

Principal Stresses

Equation A.4 shows that $\tau_{x'y'} = 0$ when $\tan 2\theta = 2\tau_{xy}/[\sigma_{xx} - \sigma_{yy}]$. This equation defines two directions at right angles to each other; planes with normals in these directions have zero shear stress. It may also be shown, by differentiating Eq. A.3 and equating to zero, that these directions maximize the normal stresses. Such maximal values are called the *principal stresses*, denoted by σ_1 and σ_2. The corresponding directions are called the *principal axes*. Using principal axes as reference axes simplifies calculations.

If we take the x-axis in the direction of σ_1, the greater principal stress, the normal and shear stresses across a plane whose normal makes an angle θ with it are

$$\sigma = \sigma_1 \cos^2 \theta + \sigma_2 \sin^2 \theta = \frac{1}{2}[\sigma_1 + \sigma_2] + \frac{1}{2}[\sigma_1 - \sigma_2] \cos 2\theta, \tag{A.6}$$

$$\tau = \frac{1}{2}[\sigma_1 - \sigma_2] \sin 2\theta. \tag{A.7}$$

It follows that the shear τ has its greatest magnitude $[\sigma_1 - \sigma_2]/2$ when $\theta = 45°$ or $135°$, that is, across planes whose normals bisect the angles between the principal axes.

In three dimensions, we can find three mutually perpendicular directions in which the stress is purely normal. Its components, the principal stresses, are the three roots σ of the equation

$$\begin{vmatrix} \sigma_{xx} - \sigma & \tau_{yx} & \tau_{zx} \\ \tau_{xy} & \sigma_{yy} - \sigma & \tau_{zy} \\ \tau_{xz} & \tau_{yz} & \sigma_{zz} - \sigma \end{vmatrix} = 0. \tag{A.8}$$

The state of stress at a point can be specified either by the six independent components or by the orientation of the principal axes and the values of the three principal stresses.

It can be shown that the shear has stationary values across planes whose normals bisect the angles between the principal axes. In particular, the greatest shear stress is across a plane whose normal bisects the angle between the directions of greatest (σ_1) and least (σ_3) principal stress. Its value is $[\sigma_1 - \sigma_3]/2$.

Stress Deviators

The mean value of the normal stresses at a point is

$$\sigma_M = \frac{1}{3}[\sigma_1 + \sigma_2 + \sigma_3] = \frac{1}{3}[\sigma_{xx} + \sigma_{yy} + \sigma_{zz}]. \tag{A.9}$$

The two expressions are equivalent because the sum of the normal stresses is invariant under rotation of axes. (Multiplying σ_M by -1 gives the *pressure*, the same quantity but defined as positive for compression.) Subtracting the mean value from the normal stress components defines the *stress deviators* or *normal deviatoric stresses*:

$$\tau_{xx} = \sigma_{xx} - \sigma_M = \frac{1}{3}\left[2\sigma_{xx} - \sigma_{yy} - \sigma_{zz}\right]$$

$$\tau_{yy} = \sigma_{yy} - \sigma_M = \frac{1}{3}\left[2\sigma_{yy} - \sigma_{xx} - \sigma_{zz}\right] \tag{A.10}$$

$$\tau_{zz} = \sigma_{zz} - \sigma_M = \frac{1}{3}\left[2\sigma_{zz} - \sigma_{xx} - \sigma_{yy}\right]$$

The three stress deviators sum to zero.

Stress-equilibrium Equations

Figure A.4 shows the stresses acting in the x-direction on a small element, assumed to have unit length perpendicular to the page. The net force due to the stresses is

$$\left[\frac{\partial \sigma_{xx}}{\partial x}\delta x\right]\delta y + \left[\frac{\partial \tau_{xy}}{\partial y}\delta y\right]\delta x. \tag{A.11}$$

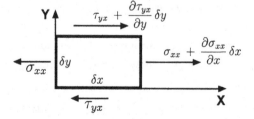

Figure A.4

These forces act on the surface of the element. Additional forces such as gravity act on all parts of the body. Let X be the x-component, per unit mass, of these. The x-component of the body force is therefore $\rho X \delta x \delta y$, for a density ρ. Thus, for equilibrium,

$$\frac{\partial \sigma_{xx}}{\partial x} + \frac{\partial \tau_{xy}}{\partial y} + \rho X = 0. \tag{A.12}$$

Similarly for the y-components,

$$\frac{\partial \tau_{xy}}{\partial x} + \frac{\partial \sigma_{yy}}{\partial y} + \rho Y = 0. \tag{A.13}$$

Note that these two equations, by themselves, cannot determine the three stress components.

Normal Strain

Stresses cause solids to deform. We now outline the different kinds of deformation and how to measure them. The deformations are assumed to be small. Our discussion is restricted to two dimensions, but the results are easily generalized to three.

A rectangular element can be deformed by changing the lengths of its sides but maintaining its rectangular shape (Figure A.5). Suppose that the element, originally in position $P_1 P_2 P_3 P_4$, moves during deformation to position $P_1' P_2' P_3' P_4'$. For a small deformation, the *normal strain* refers to the change in length of a line in the element divided by its original length. Its components are $\epsilon_{xx} = [\delta x' - \delta x]/\delta x$, and $\epsilon_{yy} = [\delta y' - \delta y]/\delta y$. (If the deformation is not small, $\epsilon_{xx} = \ln(x'/x)$, $\epsilon_{yy} = \ln(y'/y)$.) Thus, a positive strain indicates extension, and a negative strain indicates compression. The strain components can be expressed in terms of *displacements*. Let (u, v) be the components of displacement. If P_1 has coordinates (x, y), the coordinates of P_1' are $(x + u, y + v)$. Because the lengths of the sides have changed, u and v are functions of x and y. Thus, P_2 has a displacement $u(x + \delta x, y)$ in the x-direction. For small deformations, as assumed,

$$u(x + \delta x, y) = u(x, y) + \frac{\partial u}{\partial x} \delta x. \tag{A.14}$$

It follows that the length of $P_1' P_2' = \delta x' = u(x + \delta x, y) - u(x, y) = [\partial u/\partial x]\delta x$, and so $\epsilon_{xx} = \partial u/\partial x$. Similarly, $\epsilon_{yy} = \partial v/\partial y$.

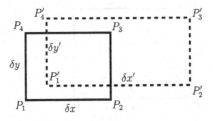

Figure A.5

Shear Strain

Deformation can also change the angular shape of an element, as in Figure A.6. The *shear strain* is defined to be one-half the decrease in a right angle after deformation: $\epsilon_{xy} = [\theta_1 + \theta_2]/2$. In terms of displacements

$$\tan \theta_1 = \frac{v(x + \delta x, y)}{\delta x} = \frac{\partial v}{\partial x}$$

$$\tan \theta_2 = \frac{u(x, y + \delta y)}{\delta y} = \frac{\partial u}{\partial y},$$

because in this case $u(x, y) = v(x, y) = 0$, as shown. For a small deformation, the tangents can be replaced by the angles and

$$\epsilon_{xy} = \frac{1}{2} \left[\frac{\partial u}{\partial y} + \frac{\partial v}{\partial x} \right]. \tag{A.15}$$

In general, shear not only distorts an element but also rotates it. The rotation is defined by

$$\omega = \frac{1}{2}[\theta_1 - \theta_2] = \frac{1}{2} \left[\frac{\partial v}{\partial x} - \frac{\partial u}{\partial y} \right]. \tag{A.16}$$

If $\theta_1 = \theta_2$, then $\omega = 0$ and the strain is *irrotational*. In this case, $\epsilon_{xy} = \theta_1 = \theta_2$ and the shear strain is the angle through which a line element in the x-direction is rotated anti-clockwise. Because rotation by itself does not change the size or shape of a body, no stresses are involved and so it is not of much interest in most analyses of glacier flow.

Components of Strain

Most deformations are a combination of extension, contraction, and shear. If $u(x, y)$ and $v(x, y)$ denote the components of the displacement of a point P_1, the displacement of a neighboring

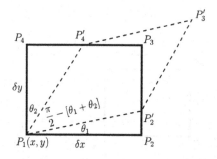

Figure A.6

point P_2 has components

$$u(x + \delta x, y + \delta y) = u(x, y) + \frac{\partial u}{\partial x}\delta x + \frac{\partial u}{\partial y}\delta y,$$

$$v(x + \delta x, y + \delta y) = v(x, y) + \frac{\partial v}{\partial x}\delta x + \frac{\partial v}{\partial y}\delta y.$$

Deformation changes the length of the line $P_1 P_2$ by $[[\Delta u]^2 + [\Delta v]^2]^{1/2}$, in which

$$\Delta u = \frac{\partial u}{\partial x}\delta x + \frac{\partial u}{\partial y}\delta y,$$

$$\Delta v = \frac{\partial v}{\partial x}\delta x + \frac{\partial v}{\partial y}\delta y.$$

In terms of strain components and rotation

$$\Delta u = \epsilon_{xx}\delta x + [\epsilon_{xy} - \omega]\delta y, \qquad (A.17)$$

$$\Delta v = [\epsilon_{xy} + \omega]\delta x + \epsilon_{yy}\delta y. \qquad (A.18)$$

Because the three strain components are derived from two components of displacements, they are not independent. Their derivatives satisfy the *compatibility condition*

$$\frac{\partial^2 \epsilon_{xx}}{\partial y^2} + \frac{\partial^2 \epsilon_{yy}}{\partial x^2} = 2\frac{\partial^2 \epsilon_{xy}}{\partial x \partial y}. \qquad (A.19)$$

In any material there is some relation between stress and strain or strain rate. The compatibility condition for the strains therefore implies some relation between the stress components. This gives another equation, in addition to the two stress-equilibrium equations, to determine the three stress components.

In glacier studies we most commonly deal with strain rates, denoted $\dot{\epsilon}_{xx}$, $\dot{\epsilon}_{yy}$, and $\dot{\epsilon}_{xy}$, rather than strains. In this case u and v are velocities rather than displacements. Strain is dimensionless, but strain rate has dimensions of $[\text{time}]^{-1}$. Equations similar to Eqs. A.3–A.5 relate components relative to two sets of axes inclined at angle θ to each other. As in the case of stresses, we can find a set of *principal axes* for which strains are purely normal. These are the *principal strains*. The quantity $\dot{\epsilon}_{xx} + \dot{\epsilon}_{yy} + \dot{\epsilon}_{zz}$ is invariant under rotation of axes. The quantity $\epsilon_1 + \epsilon_2 + \epsilon_3$ is called the *dilatation* because, for small strains, it is the ratio of the change in volume to the original volume. For an incompressible material the dilatation is zero.

Pure Shear and Simple Shear

Deformation in glaciers can often be approximated as the simple end-member cases known as pure and simple shear, illustrated in Figure A.7. Pure shear often applies near the surface and simple shear near the bed (see also Figure 3.10, which situates these cases in example flow

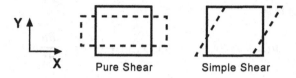

Figure A.7

fields). In pure shear, $\dot{\epsilon}_{xx} = -\dot{\epsilon}_{yy}$ and $\dot{\epsilon}_{xy} = 0$. Simple shear is the opposite case: $\dot{\epsilon}_{xx} = \dot{\epsilon}_{yy} = 0$. Simple shear corresponds to the case $\dot{\theta}_1 = 0$, $\dot{\theta}_2 = 2\dot{\epsilon}_{xy}$ in Figure A.6 ($\dot{\theta}$ indicating the rate of change of an angle θ). If in pure shear the axes are rotated through 45° (using Eqs. A.3–A.5 but for strain rates, not stresses), we obtain $\dot{\epsilon}_{x'x'} = \dot{\epsilon}_{y'y'} = 0$, $\dot{\epsilon}_{x'y'} = -\dot{\epsilon}_{xx}$. In other words, rotating the axes through 45° transforms a pure shear into a simple shear. Conversely, rotation through 45° transforms a simple shear $\dot{\epsilon}_{xy}$ into a pure shear with $\dot{\epsilon}_{x'x'} = \dot{\epsilon}_{y'y'} = \dot{\epsilon}_{xy}$.

Although infinitesimal strain theory, as outlined here, is adequate for many analyses of glacier flow, some topics such as formation of folds and related structures require the more complex theory of finite strain (Section 10.3.2).

Index

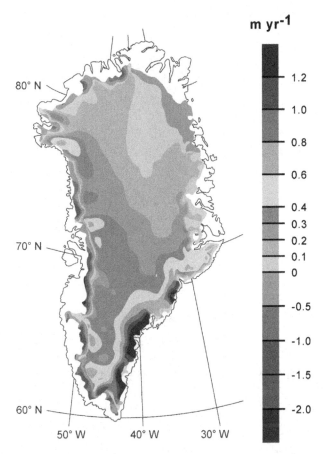

Figure 4.15 Estimated specific surface balance of the Greenland Ice Sheet, averaged for years 1988–2005 (data courtesy of J. Box; map courtesy of N. Schlegel). The balances are a revision of those given by Box et al. (2004 and 2006) using a degree-day method to calculate melt.

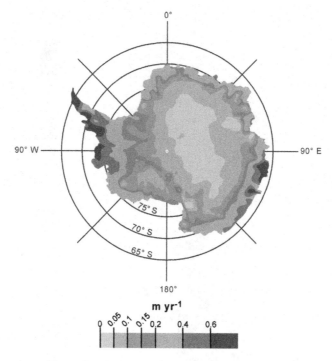

Figure 4.16 Estimated specific surface balance of Antarctica in the late twentieth century (adapted from Vaughan et al. 1999)

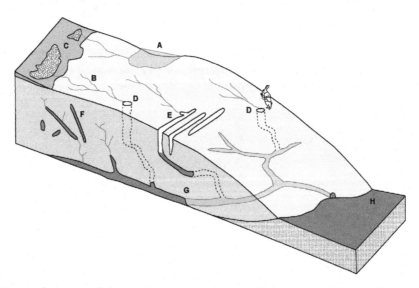

Figure 6.1 Some elements of the glacier water system: **(A)** Supraglacial lake. **(B)** Surface streams. **(C)** Swamp zones near the edge of the firn. **(D)** Moulins, draining into subglacial tunnels (for scale, white rabbit is about 10 m tall). **(E)** Crevasses receiving water. **(F)** Water-filled fractures. **(G)** Subglacial tunnels, which coalesce and emerge at the front. **(H)** Runoff in the glacier foreland, originating from tunnels and also from upwelling groundwater. Though not depicted here, water is also widely distributed on the bed in cavities, films, and sediment layers. Sediment and bedrock beneath the glacier contain groundwater.

Figure 6.2 A meltwater lake on the surface of the Greenland Ice Sheet, 30 km from the western margin, August 2005. The lake's diameter is 1.4 km on its long axis; the volume is about 30×10^6 m^3 (Box and Ski 2007).

Figure 6.3 A tunnel (R-channel) emerging at the terminus of Pastaruri, Peru. It formed during drainage of a lake. Photo courtesy of M. Hambrey.

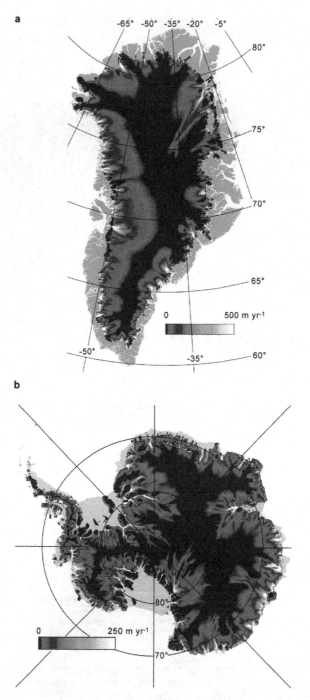

Figure 8.2 Balance velocities for (a) the Greenland Ice Sheet (Bamber et al. 2000a) and (b) the Antarctic Ice Sheet (updated from Bamber et al. 2000b by J. Bamber). Images courtesy of J. Bamber.

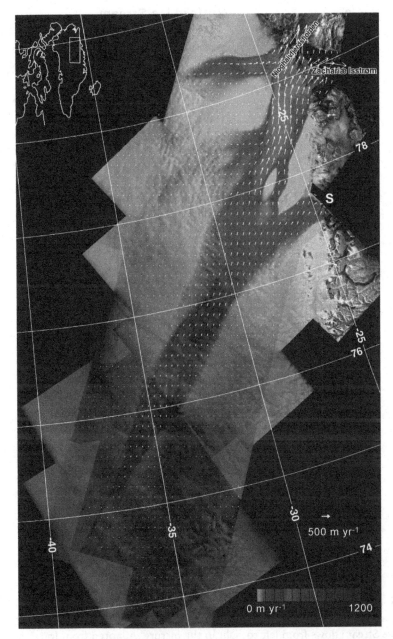

Figure 8.4 Surface velocity map of the Northeast Greenland Ice Stream and surroundings, measured by InSAR (Joughin et al. 2001). Inset panel at upper left shows the location of the figure. Image courtesy of I. Joughin and used with permission of the American Geophysical Union, *Journal of Geophysical Research*.

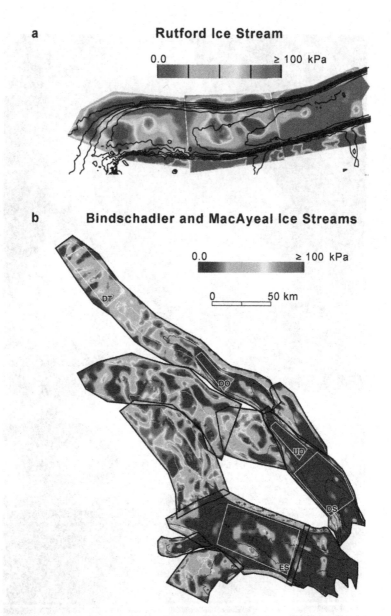

a   Rutford Ice Stream

0.0   ≥ 100 kPa

b   Bindschadler and MacAyeal Ice Streams

0.0   ≥ 100 kPa

0   50 km

Figure 8.15 Maps of basal drag τ_b beneath three major ice streams in Antarctica. Values are only approximate, and are obtained by calibrating a flow model against measured velocities (Section 8.5.3.2). (a) Rutford Ice Stream flows from left to right in the picture. Adapted from Joughin et al. (2006). (b) Ice streams D (Bindschadler) and E (MacAyeal) flow toward the lower right. Adapted from Joughin et al. (2004b). Images courtesy of I. Joughin, and used with permission from the American Geophysical Union, *Journal of Geophysical Research*.

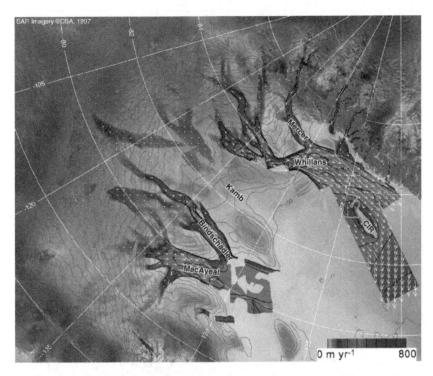

Figure 8.28 Surface velocities of the Siple Coast region, Antarctica, superimposed on radar imagery. From analysis of Joughin et al. (2004b). Image courtesy of I. Joughin.

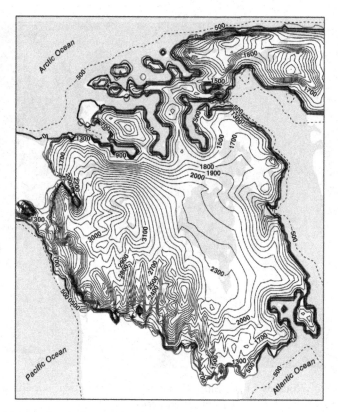

Figure 13.6 A model hypothesis for the configuration of the Laurentide Ice Sheet at the Last Glacial Maximum. Contour labels give elevations in meters above sea level. Adapted from Tarasov and Peltier (2004).

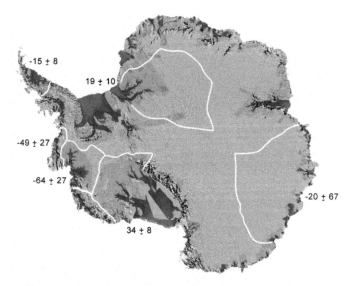

Figure 14.8 Antarctica in year 2000. The map shows ice velocities around the entire Antarctic continent (faster flow indicated by red color), and the six regions of largest inferred mass changes. Numbers give the mass changes in $Gt \, yr^{-1}$. Adapted from Rignot et al. (2008). Velocity map courtesy of E. Rignot.

Printed in the United States
By Bookmasters

inted in the United States
Bookmasters